MATHEMATICS OF BUSINESS

For example,

$$(+5) + (+7) = +12$$
$$(-5) + (-7) = -12$$

Rule 2:   When adding two numbers with different signs, subtract the numbers, then attach the sign of the greater number to the sum.

For example,

$$(+5) + (-7) = -2$$
$$(-5) + (+7) = +2$$

Rule 3:   When subtracting two numbers, change the sign of the number being subtracted, and add the two resulting numbers.
For example:

$$7 - 5 = 7 + (-5) = +2$$
$$7 - 10 = 7 + (-10) = -3$$
$$-8 - 2 = (-8) + (-2) = -10$$
$$-8 - (-2) = (-8) + (+2) = -6$$

Rule 4:   When multiplying signed numbers, multiply the two numbers together and attach the sign as follows:

A positive times a positive equals a positive.

A positive times a negative equals a negative.

A negative times a negative equals a positive.

A negative times a positive equals a negative.

For example,

$$(+5) \times (+7) = +35$$
$$(+5) \times (-7) = -35$$
$$(-5) \times (-7) = +35$$

CALCULATOR   SOLUTION

Keyed entry	Display
AC 5 × 7 =	35
AC 5 × 7 +/− =	−35
AC 5 +/− × 7 +/− =	35

Rule 5:   When dividing signed numbers, divide the two numbers and attach the sign as follows:

A positive divided by a positive equals a positive.

A positive divided by a negative equals a negative.

A negative divided by a negative equals a positive.

For example,

$(+35) \div (+7) = +5$

$(+35) \div (-7) = -5$

$(-35) \div (-7) = +5$

Here are eight more equations to reinforce these concepts:

$(-10) + (+7) \qquad = -3$

$(-10) + (+7) + (-4) = -7$

$(-10) \times (+7) \qquad = -70$

$(-10) \times (-7) \qquad = +70$

$(-10) \div (+2) \qquad = -5$

$(-10) \div (-2) \qquad = +5$

$(-10) - (+7) \qquad = -17$

$(-10) - (-7) \qquad = -3$

Order of Operations

Occasionally, we encounter expressions that contain several mathematical operations $(+, -, \times, \div)$. For example, the expression $12 - 10 + 6 \times 8 \div 2$ has all four operations. To properly evaluate this expression, we perform all multiplication and division first, and then complete any addition and subtraction.

Order of operations rule

With an expression that has some or all of the operations of addition, subtraction, multiplication, and division, but no parentheses, first complete all multiplication and division operations and then complete any additions and subtractions.

Example 2

Recall in algebra we often use a letter to represent a quantity, such as the result of this calculation. Solve for x: $x = 12 - 10 + 6 \times 8 \div 2$.

Solution 1

$x = 12 - 10 + 6 \times \underbrace{8 \div 2}$

$x = 12 - 10 + 6 \times \quad 4$

Step 1: Divide 2 into 8.

$x = 12 - 10 + \underbrace{6 \times 4}$

$x = 12 - 10 + \quad 24$

Step 2: Multiply 6 and 4.

$x = \underbrace{12 - 10} + 24$

$x = \quad 2 \quad + 24$

Step 3: Subtract 10 from 12.

$x = 2 + 24$

$x = 26$

Step 4: Add the 2 and 24.

Note: We can do either division *or* multiplication first; when performing addition and subtraction, we can also do either first.

Solution 2

$x = 12 - 10 + \underbrace{6 \times 8} \div 2$

$x = 12 - 10 + \quad 48 \quad \div 2$

Step 1: Multiply the 6 and the 8.

$x = 12 - 10 + \underbrace{48 \div 2}$

$x = 12 - 10 + \quad 24$

Step 2: Divide 48 by 2.

$x = 12 + \underbrace{(-10) + 24}$

$x = 12 + \quad 14$

Step 3: Add the -10 and 24.

$x = 12 + 14$

$x = 26$

Step 4: Add the 12 and 14.

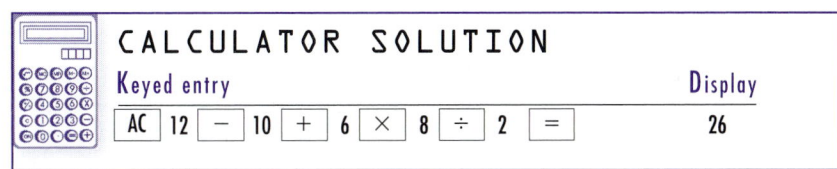

CALCULATOR SOLUTION

Keyed entry	Display
AC 12 − 10 + 6 × 8 ÷ 2 =	26

This order of operations rule always applies unless parentheses are used to override the rule. Always work inside the parentheses first.

Parentheses rule for operations

When performing operations on expressions that contain parentheses, complete the operations in the innermost parentheses first, then the operations inside the next innermost parentheses, and so forth.

Example 3

Let the variable symbol x represent the final result of this calculation. Solve for x: $x = ((12 - (10 + 6)) \times 8) \div 2$.

Solution

$x = ((12 - \underbrace{(10 + 6)}) \times 8) \div 2$ *Step 1:* Evaluate the innermost

$x = ((12 - \quad 16 \quad) \times 8) \div 2$ set of parentheses first.

$x = (\underbrace{(12 - 16)} \times 8) \div 2$ *Step 2:* Evaluate the next

$x = (\quad -4 \quad \times 8) \div 2$ innermost set of parentheses.

$x = \underbrace{(-4 \times 8)} \div 2$ *Step 3:* Evaluate the next

$x = \quad -32 \quad \div 2$ innermost set of parentheses.

$x = -16$

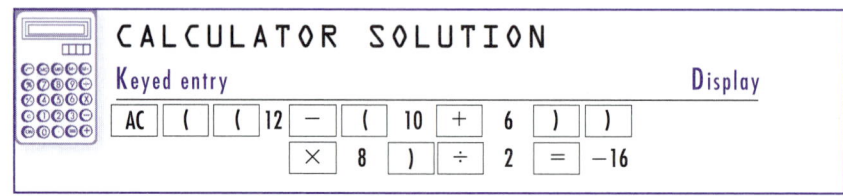

Notice that the answer to Example 3 is different from the answer to Example 2. Both problems are correctly evaluated, but the parentheses used in Example 3 overrode the order of operations rule, which resulted in a different final value.

Exponents

When a quantity is multiplied by itself, we express the multiplication by the use of **exponents.** Exponents are superscripts placed to the right of a quantity to indicate a repeated multiplication of the quantity by itself. The exponent is also referred to as the *power* of the quantity. Each of the following multiplications can be represented by the exponential expression to the right of the equals sign.

$$2 \times 2 = 2^2$$
$$2 \times 2 \times 2 = 2^3$$
$$2 \times 2 \times 2 \times 2 = 2^4$$
$$2 \times 2 \times 2 \times 2 \times 2 = 2^5$$

.

.

.

$$\underbrace{2 \times 2 \times 2 \times 2 \times 2 \times 2 \times \cdots \times 2} = 2^n$$
if there are n 2s multiplied together

In each of these expressions, the number that is raised to the power (in this case, 2) is called the *base* of the exponential expression.

Example 4

Evaluate 3^5.

Solution

3^5 means that there are five 3s multiplied together (3 is the base, and 5 is the power). Let x represent this, as of yet unknown, product.

$$x = 3^5$$
$$x = 3 \times 3 \times 3 \times 3 \times 3$$
$$x = 243$$

CALCULATOR SOLUTION

Keyed entry	Display
AC 3 y^x 5 =	243

Example 5

Let x represent the sum of two exponential expressions.
Solve for x: $x = 2^3 + 3^2$.

Solution

2^3 means $2 \times 2 \times 2$; 3^2 means 3×3, so:

$$x = 2^3 + 3^2$$
$$x = 2 \times 2 \times 2 + 3 \times 3$$
$$x = 8 + 9$$
$$x = 17$$

Recall by our order of operations rule that we multiply first and then add.

Example 6

Solve for x: $x = (4^2 + 2^4)^2$.

Solution

$$x = (4 \times 4 + 2 \times 2 \times 2 \times 2)^2$$
$$x = (16 + 16)^2$$
$$x = 32^2$$
$$x = 32 \times 32 = 1,024$$

Step 1: Evaluate inside the parentheses first.

Step 2: Evaluate this result to get the solution.

General Algebraic Expressions

Recall that we use letters of the alphabet to represent unknown quantities, or quantities that might assume different values.

We used x in the previous examples to represent the results of calculations. For example, we let $x = 2^3 + 3^2$ until we multiplied $2 \times 2 \times 2 + 3 \times 3$ to determine the final result of 17. In algebra we use letters and symbols of various types to represent unknowns, or general concepts and principles. We combine these algebraic symbols with the operations of arithmetic to form algebraic expressions.

Example 7

If x represents a first number and y represents a second number, use these symbols to represent (a) the sum of the two numbers, (b) the first number subtracted from the second number, (c) the product of the two numbers, (d) the first number divided by the second number, (e) the first number raised to the fourth power, and (f) the sum of the first number doubled and the second number tripled.

Solution

(a) the sum of the two numbers	$x + y$
(b) the first number subtracted from the second number	$y - x$
(c) the product of the two numbers	xy
(d) the first number divided by the second number	x/y
(e) the first number raised to the fourth power	x^4
(f) the sum of the first number doubled and the second number tripled	$2x + 3y$

If someone now told us that $x = 5$ and $y = 10$, then we could evaluate these expressions to be $5 + 10 = 15$, $10 - 5 = 5$, $5 \times 10 = 50$, $5/10 = .5$, $5^4 = 625$, and $2 \times 5 + 3 \times 10 = 40$.

We can also express *relationships* algebraically. For example, the expression

$$2x + 6 = 24$$

tells us that the two quantities separated by the equals sign have the same value; both represent the value 24. In the next section we will review how we determine that x must equal 9 in order for this relation to be a true statement.

Three mathematical relationships are very helpful to us in calculations:

- The **commutative principle** for addition and multiplication tells us that we can add or multiply two numbers without concern for order (e.g., $2 + 3 = 3 + 2$, or $2 \times 3 = 3 \times 2$). We express this relationship in general terms as $a + b = b + a$, and $a \times b = b \times a$.

- The **associative principle** for addition and multiplication tells us that if we are adding or multiplying three or more numbers together, we can add any two first, or multiply any two first (e.g., $2 + 3 + 4 = (2 + 3) + 4 = 9$, or $2 + (3 + 4) = 9$. For multiplication, $2 \times 3 \times 4 = (2 \times 3) \times 4 = 24$, or $2 \times (3 \times 4) = 24$). Expressed generally, this relationship is $(a \times b) \times c = a \times (b \times c)$, and $(a + b) + c = a + (b + c)$.

- The **distributive principle** tells us that if we are multiplying a sum by a number, we can calculate the sum first and then multiply, or we can multiply the number by each term of the sum and then add (e.g., $2 \times (3 + 4) = (2 \times 3) + (2 \times 4)$). This relationship expressed in general terms is $a(b + c) = ab + ac$.

This third principle allows us to *combine like terms,* such as $3x + 4x$. If we reverse the distributive principle, we develop an operation called *factoring out of the common factor.* To demonstrate this operation generally, we start with the distributive principle,

$$a(b + c) = ab + ac$$

and rewrite it to be

$$ab + ac = a(b + c),$$

which can be rewritten using the commutative principle of multiplication to be

$$ba + ca = (b + c)a,$$

factoring out the common factor a from each term ba and ca. This factoring out of a common factor helps us to combine like terms in an algebraic expression. Because the expression $3x + 4x$ has two terms ($3x$ and $4x$) that each have a number and an x, we refer to the two terms as *like terms* ($3x$ and $4y$ would not be like terms), factoring out the common factor x allows us to "combine" the two terms into one term,

$$3x + 4x = (3 + 4)x = 7x$$

Our ability to combine like terms will be very useful to us in the solution of business problems.

CHECK YOUR KNOWLEDGE

Algebraic Expressions and Operations

Evaluate the following expressions:

1. $(-12) + (+9)$
2. $(-1) + (+9) + (-11)$
3. $(-22) \times (+2)$
4. $(-22) \times (-2)$
5. $(+22) \div (-2)$
6. $(-22) \div (+2)$
7. $(+4) + (+2)$
8. $(+4) + (+2) + (-5) + (-3)$
9. $(+4) \times (-2)$
10. $(-4) \div (-2)$
11. $(+4) \div (-2)$
12. $(-4) \div (+2)$
13. $(+4) - (+8)$
14. $(+4) - (-8)$
15. $15 - 8 + 7 \times 9 \div 3$
16. $24 + 16 - 4 \times 5 - 9$
17. $(15 - (8 + 7) \times 9) \div 3$
18. $((24 + 16) \div 4) \times (5 - 9)$
19. $4^3 + 7^2$
20. $5^2 - 3^3 + 2^4$
21. Write an algebraic expression for the following: "Two times x plus four times y is equal to eight."
22. Evaluate: $3x + 2y - 4$ if $x = 5$ and $y = 7$.
23. Evaluate: $x^3 + y^2$ if $x = 2$ and $y = 5$.
24. Combine like terms: $5x + 9x$.
25. Combine like terms: $3x + 4y - 2x + 7y$.

1.1   EXERCISES

Complete the following operations:

1. $4 - 9$
2. $-4 + 9$
3. $4 \times (-9)$
4. -4×9
5. $-4 \times (-9)$
6. $20 \div 5$
7. $-20 \div 5$
8. $20 \div -5$
9. $-20 \div -5$
10. $-2 - 7$
11. $3 - 4 + 5 - 12$
12. $44 + 21 - 55 - 14$
13. $-30 + 15 + (-45) + 20$
14. $-44 \div (-11)$
15. $50 \div -20$
16. -2.4×10
17. -5.5×-3
18. $-31 \div 6$
19. $-42 \times .05$
20. $50 \div (-.5)$
21. $6^3 + 8^2$
22. $4^4 - 2^3 + 3^2$
23. $(-3)^4 + (-6)^3$
24. $(-5)^2 + (-3)^4 + (-2)^4$
25. $(-2^3) - 9^2$
26. $(-8)^2 - (-3)^4 - (-2)^5$
27. $2 - 8 + 12 \div 3 + 5 \times 2$

Answers to CYK: **1.** −3   **2.** −3   **3.** −44   **4.** 44   **5.** −11   **6.** −11   **7.** 6   **8.** −2   **9.** −8   **10.** 2   **11.** −2   **12.** −2   **13.** −4   **14.** 12   **15.** 28   **16.** 11   **17.** −40   **18.** −40   **19.** 113   **20.** 14   **21.** $2x + 4y = 8$ **22.** 25   **23.** 33   **24.** $14x$ **25.** $x + 11y$

28. $16 \div 2 - 7 \times 3 + 10 \div 5$
29. $19 - 12 \times 3 + 27 \div (-3)$
30. $38 \div 19 + (-7) \times 5 + (-18) \div (-6)$
31. $4^2 \div 2^3 + (-8) - (-10) \div 2$
32. $(2 - 8 + 12) \div 3 + 5 \times 2$
33. $(16 \div (2 - 6)) \times (3 + 10) + 5$
34. $(19 - 12) \times 3 + 27 \div (-3)$
35. $(36 \div (19 + (-7))) \times 5 + (-18) \div (-6)$
36. $4^2 \div (2^3 + (-6)) - (-10) \div 2$
37. Combine the following into one term: $7x + 12x - 4x$.
38. Evaluate the expression in problem 37 if $x = 7$.
39. Combine the following into two terms: $3x - 5y + 7y - 6x + 4y - 8x$.
40. Evaluate the expression in problem 39 if $x = 5$ and $y = -3$.
41. Combine like terms in the following expression: $3x + 2y + 3(4x + y)$.
42. Evaluate the expression in problem 41 if $x = 2$ and $y = 4$.
43. Combine like terms in the following: $2^5 x + 5^2 y + 3^3 x - 9y$.
44. Evaluate the expression in problem 43 if $x = 1$ and $y = 2$.
45. Combine the following into one term: $3x^3 + 5x^3 - 7x^3$.

In Problems 46 through 50, write an algebraic expression in x and y that represents each statement.

46. Three times x minus five times y
47. Two minus x plus five times y
48. Nine times x minus four times y plus eight
49. Seven times x minus y divided by four equals 10
50. Nine times x plus thirteen times y equals eight times x minus five times y

1.2 EQUATIONS AND FORMULAS

Equations

A numeric sentence that contains an equals sign is called an **equation.** The left side of the equation must have the same numeric value as the right side. Some examples of equations are as follows:

$$3 + 2 = 5$$
$$4 \times 3 = 12$$
$$-9 + 3 = -6$$
$$\frac{10}{-5} = -2$$

These equations consist entirely of constants. A **constant** is a symbol that has a fixed value; this value will not change. In other words, the number 3 will always be equal to 3; 5 will always have the value of 5.

Sometimes equations contain variables. A **variable** is a *symbol* that may represent *any* number. The equation $x = -4$ contains one variable called *x*. In order to make this equation true, we must find the number that *x* represents. We call this "solving for x." In this case, x must be -4 so that $-4 = -4$.

The following rules are very helpful when solving equations.

Rule of opposite operations

An "opposite operation" will *undo* the original operation. The following pairs of operations are "opposite operations" of each other.

addition $\longleftrightarrow$ subtraction

multiplication $\longleftrightarrow$ division

Note: These pairs of "opposite operations" are also called "inverse operations."

For example, if we multiply 5 by 3, we get 15.

$$5 \times 3 = 15$$

To undo this multiplication, we use division, which is the opposite operation of multiplication: 15 divided by 3 brings us back to the original 5.

$$\frac{15}{3} = 5$$

Rule of equals

If equals are added to equals, the sums are equal.

If equals are subtracted from equals, the differences are equal.

If equals are multiplied by equals, the products are equal.

If equals are divided by equals, the quotients are equal.

Always do to the right side of an equation what you do to the left.

An easy way to begin to solve for x is to isolate all the variables on the left side of the equals sign and all the constants on the right. To remove an unwanted variable or constant, perform the opposite operation.

Example 8

Using subtraction, remove the quantity 3 from the equation $3 + x = 5$.

Solution

$$3 + x = 5$$
$$\underline{-3 \quad\quad = -3}$$
$$x = 2$$

Original equation

Subtract 3 from both sides of the equation to eliminate $+3$ from the left side of the equation, which will keep all constants to the right of the equals sign.

Learning objective
Use algebra to determine the value of an unknown variable when given an equation with one unknown variable.

Check: Always check to make certain your solution is correct. This can be done by substituting the answer you got for x in the original equation.

$$3 + x = 5$$
$$3 + 2 = 5$$
$$5 = 5$$

True; therefore $x = 2$ is the correct solution.

Example 9

Using addition, remove the quantity -3 from the equation $x - 3 = 2$.

Solution

$$x - 3 = 2$$
$$\underline{+ 3 = +3}$$
$$x = 5$$

Original equation

Add $+3$ to cancel out the -3, which will keep all constants to the right of the equals sign.

Check:

$$x - 3 = 2$$
$$5 - 3 = 2$$
$$2 = 2$$

True!

Example 10

Using division, remove the quantity 4 from the equation $4 \times x = 12$.

Solution

$$4 \times x = 12$$
$$\frac{4 \times x}{4} = \frac{12}{4}$$
$$x = 3$$

Original equation

Divide both sides by 4 to keep all constants to the right of the equals sign.

Check:

$$4 \times x = 12$$
$$4 \times 3 = 12$$
$$12 = 12$$

True!

Note: Instead of writing $4 \times x$ to represent 4 multiplied by x, it is common practice to write $4x$.

Example 11

Using multiplication, remove the quantity 4 from the equation $\frac{x}{4} = 3$.

Solution

$$\frac{x}{4} = 3$$ Original equation

$$\frac{4}{1} \cdot \frac{x}{4} = \frac{3}{1} \cdot \frac{4}{1}$$ Multiply both sides of the equation by 4 to keep all variables to the right of the equals sign.

$$x = 12$$

Check:

$$\frac{x}{4} = 3$$

$$\frac{12}{4} = 3$$

$$3 = 3$$ True!

Example 12

Solve for x by combining like terms: $2x + 3x = 20$.

Solution

$$2x + 3x = 20$$ Original equation

$$5x = 20$$ $2x$ and $3x$ are *like terms:* they each have a number and the same *letter(s);* they can be added or combined into the single term $5x$.

$$\frac{5x}{5} = \frac{20}{5}$$ Divide both sides of the equation by 5 to isolate the variable.

$$x = 4$$

Check:

$$2x + 3x = 20$$

$$2(4) + 3(4) = 20$$

$$8 + 12 = 20$$

$$20 = 20$$ True!

Example 13

Solve for x: $4 + 4x - 3 = 11 - 1x$.

Solution

$$4 + 4x - 3 = 11 - 1x$$ Original equation

$$1 + 4x = 11 - 1x$$ Combine like terms in the equation.

$$\underline{-1 -1}$$ Subtract 1 from both sides of the equation to begin isolating all constants to the right of the equals sign.

$$4x = 10 - 1x$$

$$\underline{+\, 1x +\, 1x}$$ Add $1x$ to both sides of the equation to isolate all variables to the left of the equals sign.

$$5x = 10$$

$$\frac{5x}{5} = \frac{10}{5}$$ Divide by 5 to finish isolating all constants to the right of the equals sign.

$$x = 2$$

Formulas

Formulas are algebraic expressions that represent general principles that have many applications.

For example, we represent the area of a rectangle by the expression $A = L \times W$ to tell us that we can determine area by multiplying the length of the rectangle by the width of the rectangle. So if we then wanted to calculate the floor area for three different rooms with dimensions 10 feet by 12 feet, 15 feet by 20 feet, and 30 feet by 20 feet, the one formula tells us:

Area of floor = length $\times$ width

Area of floor 1 = 10 feet $\times$ 12 feet = 120 square feet

Learning objective
Solve for any variable in a formula.

Area of floor 2 = 15 feet $\times$ 20 feet = 300 square feet

Area of floor 3 = 30 feet $\times$ 20 feet = 600 square feet

Some other formulas that you most likely are familiar with include:

Area of a square = $(\text{side})^2 = s^2$

Area of a triangle = $1/2(\text{base})(\text{height}) = 1/2bh$

Area of a circle = $\mathrm{p}r^2$

Formulas tell us how to go about calculating quantities, such as area, in many different circumstances.

We will have occasion to want to solve a formula for different variables that are a part of the formula. Since formulas are equations, we use the same rules as we used to solve equations.

Example 14

Solve the formula $A = L \times W$
  a.   for the variable W
  b.   for the variable L

Solution

a. $A = LW$ Original equation

$\dfrac{A}{L} = \dfrac{LW}{L}$ We want to isolate the W, so we must remove the L by dividing both sides of the equation by L.

$\dfrac{A}{L} = W$ Note: Because the two quantities are equal, we can interchange them from side to side.

$W = \dfrac{A}{L}$

b. $A = LW$ Original equation

$\dfrac{A}{W} = \dfrac{LW}{W}$

$L = \dfrac{A}{W}$

Example 15

The perimeter of a rectangle is given by the formula $P = 2L + 2W$. Solve for W.

Solution

$P = 2L + 2W$ Original equation

$P = 2L + 2W$ Isolate the term with the W in it by subtracting $2L$ from both sides of the equation

$\underline{-2L = -2L}$

$P - 2L = 2W$ Next, divide both sides by 2

$\dfrac{P - 2L}{2} = \dfrac{2W}{2}$

$W = \dfrac{P - 2L}{2}$ Final solution

Example 16

In Chapter 2 we will work with an important business formula involving base, rate, and part.

Part = Base × Rate

$P = B \times R$

Solve the formula for R.

Solution

$P = BR$	Original equation
$\dfrac{P}{B} = \dfrac{BR}{B}$	Divide both sides by B
$R = \dfrac{P}{B}$	Final solution

Example 17

We will eventually work with simple interest. The formula for simple interest is

$I = PRT$

Solve the formula for T.

Solution

$I = PRT$	Original equation
$\dfrac{I}{PR} = \dfrac{PRT}{PR}$	Since T is multiplied by both P and R, divide both sides by PR to isolate T.
$T = \dfrac{I}{PR}$	Final solution

CHECK YOUR KNOWLEDGE

Equations and Formulas

Solve the following equations.

1. $9 + x = 13$

2. $x - 7 = 5$

3. $5 \times x = 15$

4. $x/11 = 3$

5. $6x + 2x = 48$

6. $5x - 7x = -10$

7. $8 + 8x - 6 = 22 - 2x$

8. $3.1x + 3.2 = 1.1x - 2.8$

9. $x = 15 + 9 - 6 \times 12 \div 4$

10. $(4x + 8 - 3x) \times 2 = 18$

11. Solve $P = BR$ for B.

12. Solve $I = PRT$ for P.

13. Solve $A = P + I$ for I.

14. Solve $P = R - C$ for C.

15. Solve $P = 2L + 2W$ for L.

1.2   EXERCISES

Solve the following equations.

1. $7 + x = 14$

2. $x - 8 = 11$

3. $5 \times x = 40$

4. $x/11 = 4$

5. $7x + 3 + 2x = 21$

6. $-x - 3 + 6x = 23$

7. $-4 + 6x - 10 = 40$

8. $3x + 7 - 5x + 5 = 22$

9. $\dfrac{4}{5}x = 8$

10. $\dfrac{1}{2}x + 1 = 2$

11. $-42 - \dfrac{1}{2}x + 4 = 1\dfrac{1}{2}x - 3 + 68$

12. $21 + 3 - 2x = 4x + 5x + 32$

13. $3x - 5 = 5x + 45$

14. $6x - 32 + 8 = 20 - 2x + 8$

15. $-6.5 - \dfrac{3}{4}x + 14 = -32 - \dfrac{1}{2}x + 49$

16. $x = .6x + 320$

17. $y - .4y = 7.2$

18. $2z = 1.3z + 66$

19. $b - .5b = 3b + 156$

20. $w + 3.2w = -4.3w + 170$

In problems 21 through 30, solve the formulas for the variable indicated. You will work with several of these formulas later in this text. (The names or meanings of these formulas are indicated in parentheses.)

21. Solve $G = HR$ for R.   (gross earnings equals number of hours $\times$ rate per hour)

22. Solve $G = VR$ for V.   (gross earnings on commission)

23. Solve $N = G - D$ for G.   (net earnings equal gross minus deductions)

24. Solve $A = L - E$ for E.   (assets equal liabilities minus equity)

25. Solve $D = PR$ for R.   (cash discount equals price times discount rate)

26. Solve $S = C + M$ for M.   (selling price equals cost plus markup)

27. Solve $M = P + I$ for I.   (maturity value of investment equals principal plus interest)

28. Solve $M = P + PRT$ for P.   (maturity value of investment equals principal plus interest)

29. Solve $A = P(1 + i)^n$ for P.   (compound interest formula)

30. Solve $R = P - M$ for M.   (reduced price equals price minus markdown)

1.3   SOLVING WORD PROBLEMS

Normally, the problems we encounter in daily life are not presented in the form of equations; problems are usually presented in the form of words. In this section, we will demonstrate how to systematically solve word problems.

Answers to CYK: **1.** $x = 4$ **2.** $x = 12$ **3.** $x = 3$ **4.** $x = 33$ **5.** $x = 6$ **6.** $x = 5$
7. $x = 2$ **8.** $x = -3$ **9.** $x = 6$ **10.** $x = 1$ **11.** $B = P/R$
12. $P = I/RT$ **13.** $I = A - P$ **14.** $C = R - P$ **15.** $L = (P - 2W)/2$

Solving word problems

First:	Read and reread the problem.
Second:	Write down the information given in the problem.
Third:	Write down what the problem is asking for.
Fourth:	Write down any additional information that is needed in order to solve the problem.
Fifth:	Write an equation that states the problem, what information is given, what information is missing, and any additional information.
Sixth:	Solve the equation and check the answer.

Example 18

What is the total cost of a gross of paperback books if the price per book is $5.95?

Solution

Given:	Cost per book = $5.95
Asking for:	Total cost of a gross of books
Additional information:	144 units = one gross

Learning objective
Set up an algebraic equation from a word problem and solve it.

The cost of a gross of books equals the cost per book times 144 books. Let x = total cost of a gross of books.

$$x = \$5.95 \times 144$$
$$x = \$856.80$$

Example 19

A couple bought a house in Colorado for $210,000. One-fifth (20%) of the price of the house was needed for a down payment. What is the dollar amount of the down payment?

Solution

Given:	House cost = $210,000
	Down payment = $\frac{1}{5}$ of house cost
Asking for:	Amount of down payment
Additional information:	None needed

Amount of down payment needed is $\frac{1}{5}$ of the house cost. Let x = amount of down payment needed.

$$x = \tfrac{1}{5} \times \$210,000 \qquad \text{or} \qquad x = .20 \times \$210,000$$
$$x = \$42,000 \qquad\qquad\qquad\qquad x = \$42,000$$

Example 20

The two leading companies in the laundry detergent industry are "Quick Clean" and "Easy Wash." The profits of Easy Wash are four times that of the profits of Quick Clean. If the total profits for these two companies are $120,000, what is the profit for each company?

Solution

Given: Total profits = $120,000

Total profits = Easy Wash + Quick Clean profit

(Easy wash profits = 4 × Quick Clean profit)

Asking for: Profit for Quick Clean and profit for Easy Wash

Additional information: None needed.

Total profits = Easy Wash profit + Quick Clean profit

$$\$120,000 = 4(\text{Quick Clean profit}) + \text{Quick Clean profit}$$
$$x = \text{Quick Clean profit}$$
$$4x = \text{Easy Wash profit}$$
$$\$120,000 = 4x + x$$
$$\$120,000 = 5x$$
$$\frac{120,000}{5} = \frac{5x}{5}$$
$$\$24,000 = x = \text{Quick Clean profits}$$
$$\$24,000(4) = x(4)$$
$$\$96,000 = 4x = \text{Easy Wash profits}$$

Following are two helpful tips for translating word problems into equations. The word *is* is represented by an equal sign (=); for example, $2x$ is 12 can be translated

$$2x = 12$$
$$x = \frac{12}{2}$$
$$x = 6$$

The word *of,* especially used with percent, suggests multiplication; for example, 10 percent of x is 20.

$$10\% \text{ of } x \text{ is } 20$$
$$.10x = 20$$
$$x = \frac{20}{.10}$$
$$x = 200$$

Some other phrases and their representations are shown in the following box.

Phrase	Representation (with x as the unknown)
5 more than a number	$x + 5$
7 times a number	$7x$
Subtract 6 from a number	$x - 6$
Four times the sum of a number and 3	$4(x + 3)$
Eight less than a number	$x - 8$
The difference of a number and 9	$x - 9$
The quotient of a number divided by 4	$\dfrac{x}{4}$
One eighth "of" a number	$\dfrac{1}{8}x$

Direct and Inverse Variation and the Relationship between Two Variables

Applied mathematics often studies the relationships that may exist between two variables. We sometimes refer to two variables *varying directly,* or being in a **direct variation** relationship. For example, the total price that you pay for gasoline for your car varies directly with the number of gallons of gas that you buy: Cost = (Price per gallon) $\times$ (Number of gallons). If the price per gallon is \$2.00, then this relationship is $C = \$2.00\, G$, where C represents the total cost of your purchase and G represents the number of gallons you purchase. We refer to cost as the **dependent variable** and number of gallons as the **independent variable** in the relationship. We also refer to the independent variable as the *input variable* and the dependent variable as the *output variable.* The word *direct* is used to express the fact that as G *increases, C increases* as well. If G *decreases, C* does the same thing; it *decreases* also. In our cost example, the more gallons of gas, the greater the cost of the total purchase.

If two variables *vary inversely,* or are in an **inverse variation** relationship, then they do just the opposite of each other (recall our earlier reference to opposite operations as inverse operations). For example, if $y = -3x + 12$, and we let x values increase from 1 to 3, we can see that y does just the opposite, it decreases from 9 to 3. Similarly, if we let the x value decrease from 4 to 1, we see that y does just the opposite, it increases from 0 to 9. We say x and y in this case *vary inversely.*

The word *inverse* is used to express the fact that as *x increases, y decreases* (i.e., it behaves in just the opposite way that *x* does.) If *x decreases, y* does the opposite, it *increases.*

Graphs of Relationships in Two Variables

Graphs are the pictures of mathematical relationships between two variables. The concepts of direct and inverse variation between two variables are two of the most basic relationships that can occur between two variables, and they can be pictured with graphs.

For example, we could picture the relationship between cost of a gasoline purchase and the number of gallons purchased as $C = 2G$, when gasoline costs $2.00 per gallon, with the following graph:

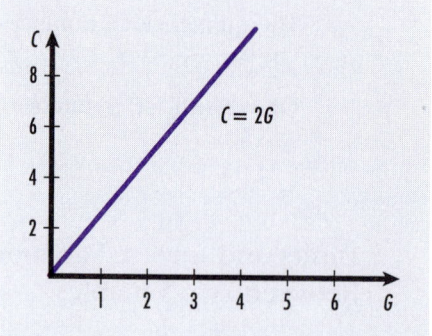

The graph (picture) of this relationship is a straight line. If you imagine yourself moving along the straight line from left to right, you would be mov-

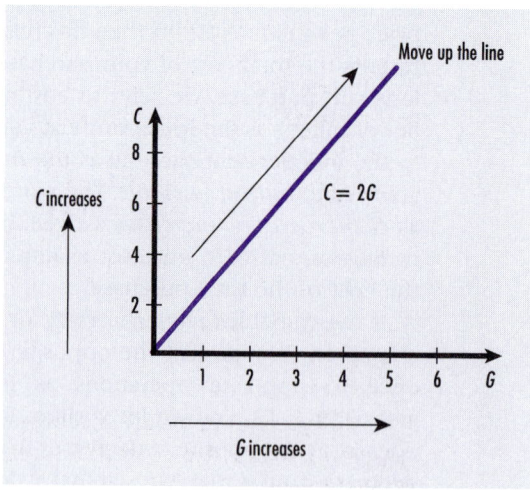

ing up as you move to the right. That is, as G increases in value, C increases in value too. You can see the direct variation relationship. Similarly, if you moved from right to left along the line, the values of G would be decreasing and the values of C would be also decreasing.

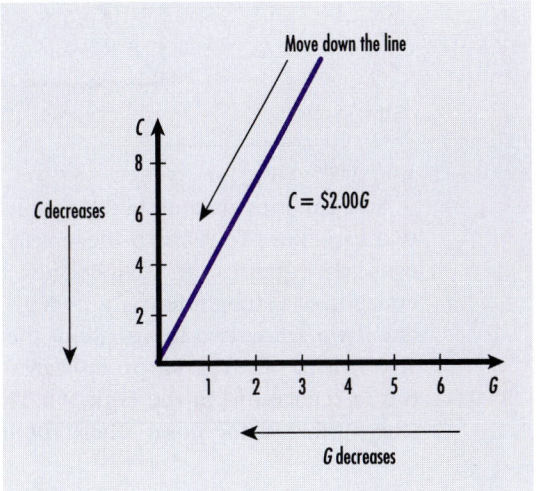

We usually set up the graph with the variable on the horizontal axis as the independent variable and the variable on the vertical axis as the dependent variable. We would say that the cost of the gasoline purchase depends on the number of gallons of gas purchased. *Cost* is the dependent variable and is represented on the vertical axis; the *number of gallons* is the independent variable, which is represented on the horizontal axis.

We can draw the graph of this straight line by building a table of values and then plotting the points on the graph:

Number of gallons	Cost of gas purchase
1	$2
2	$4
3	$6
4	$8

This table produces ordered pairs of numbers (C, G), which we can plot on the graph. We then connect the points with a smooth straight line to complete the graph:

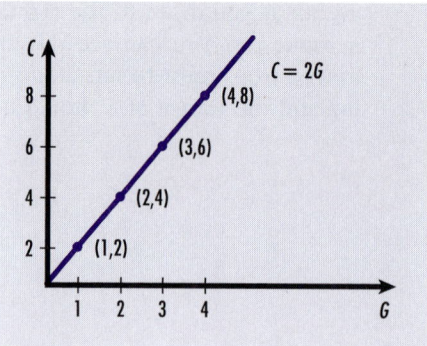

Straight lines are among the simplest relationships that can exist between two variables. We refer to these relationships as **linear relationships** because their graphs are straight lines. The most commonly used form of the equation of a straight line is $y = mx + b$. We can draw the graph of a straight line if we know two things about the line: its direction and a point through which it passes. The **slope** indicates the direction of a straight line and is the m referred to in the equation. The b in the equation is the y-intercept (the y value of the point where the line intersects the vertical axis):

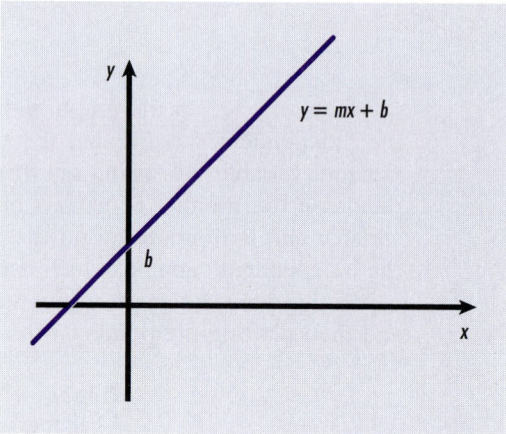

The slope is a comparison of how much the line changes vertically (the change in y) with how much it changes horizontally (the change in x).

Slope of a line

$$\text{Slope} = \frac{\text{Change in } y}{\text{Change in } x}$$

We could calculate the slope of the line in the gasoline example used earlier by using any two points from the table which we formed earlier:

Number of gallons	Cost of gas purchase
1	$2
2	$4
3	$6
4	$8

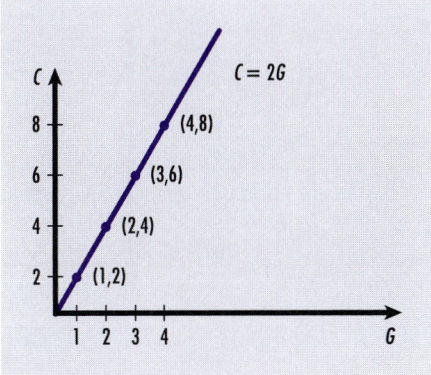

Using the first two points, (1, 2) and (2, 4),

$$\text{Slope} = \frac{\text{Change in cost}}{\text{Change in gallons}} = \frac{4 - 2}{2 - 1} = 2$$

Or, using the points (2, 4) and (4, 8),

$$\text{Slope} = \frac{\text{Change in cost}}{\text{Change in gallons}} = \frac{8 - 4}{4 - 2} = \frac{4}{2} = 2$$

Any pair of points would result in the same slope calculation:

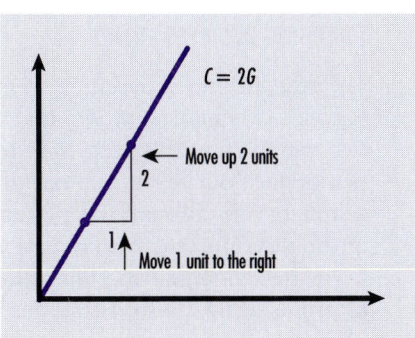

In this case, if we compare the equation $C = 2G$ with the general form $y = mx + b$, we can see that the C-intercept is 0:

$$C = 2G$$

$$y = mx + b$$

$$C = 2G + 0$$

The inverse variation relationship can be seen in the graph of the linear equation $y = -3x + 12$. In this equation, the slope is -3 and the y-intercept is 12:

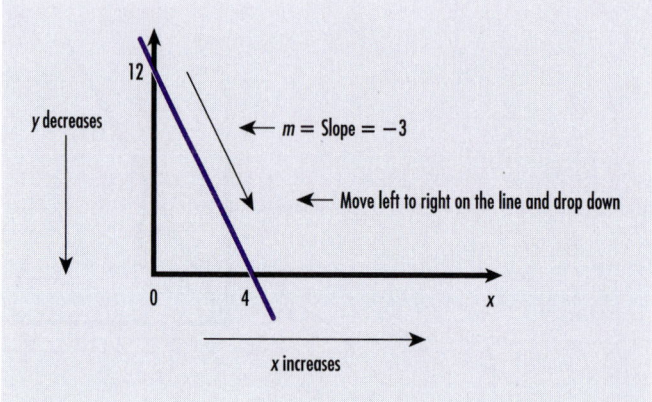

We can envision the inverse variation relationship between x and y here by imagining the movement of a point from left to right on the line—i.e., x is increasing—and we would see that the y values of the points would decrease.

The concepts of direct variation, inverse variation, dependent variable, independent variable, slope, equation of a straight line, and graphs are all important to the study of business and economics.

For Your Information

GRAPHING BUSINESS TRENDS

The straight line equation, $y = mx + b$, is indeed a very common and simple mathematical model that describes the relationship between two variables, in this case, x and y. Even when connecting successive points does not result in a straight line being formed, we often envision a straight line through the points in order to analyze whether the trend of the data indicates an increase or decrease in the dependent variable.

In the accompanying illustration of Hornpiper's Shoppes, the graph of profits against advertising costs with unconnected points would be called a *scatterplot*. The graph with each pair of points connected by a straight line segment is called a *broken line graph*. The straight line superimposed on the broken line graph is called the *trendline*, or *best fitted line*. The trendline shows the general trend that as advertising costs increase, so too do profits increase. The slope of the equation of the trendline, $y = 3.75x + 10,000$, suggests that for each additional dollar spent on advertising the profits increased by \$3.75.

Owen Hornpiper's Celtic Shoppes are located in eight malls located in eight different towns. During the month of October, each shop was given a different amount of money to spend on newspaper advertising in its area. The profits for the month were noted at each store after the advertisements were put in the paper. The results were:

Advertising	Profits
\$200	\$10,000
\$400	\$11,000
\$600	\$13,000
\$800	\$12,000
\$1000	\$17,000
\$1200	\$14,000
\$1400	\$15,000
\$1600	\$15,000

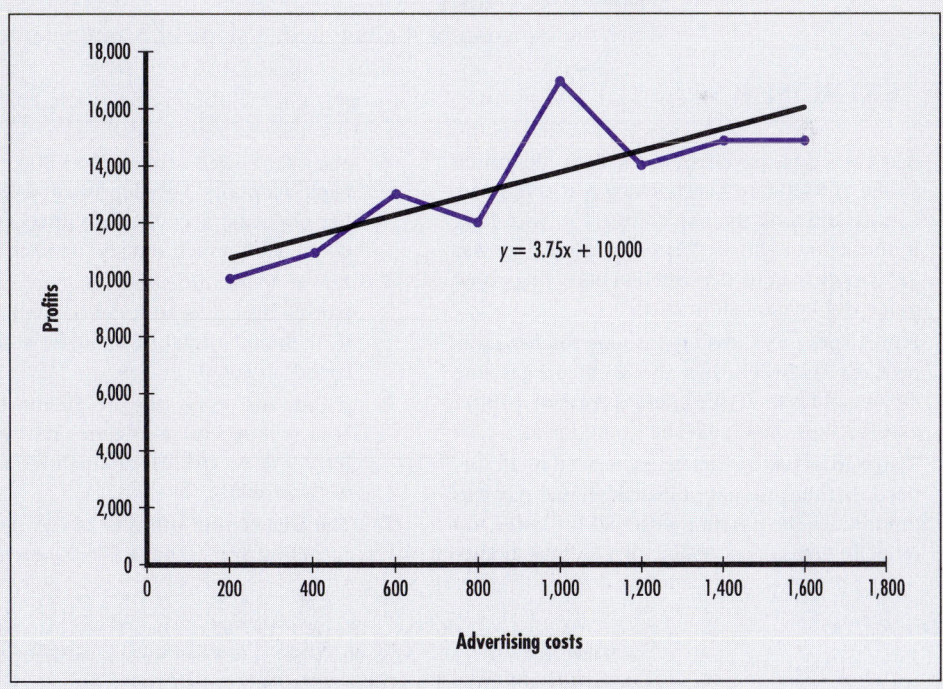

CHECK YOUR KNOWLEDGE

Solving Word Problems

1. Bob earns $3 per hour less than his father at the local woolens mill. How much does Bob earn per hour if their total hourly wage is $15?

2. After paying a percentage of his earnings to his attorney, his manager, and his agent, Johnny Hollywood receives 13% of the money made by his films. If Johnny received $260,000 from his latest film, how much money did the film make?

3. A florist has found that there are $4\frac{3}{4}$ times as many red roses purchased for Valentine's Day as white roses. If the total number of roses purchased is 782, find the number of red roses purchased.

4. The top salesperson for Alvarez Realty had sales of $150,000 more than the next salesperson for the month of June. If the total sales of these two people was one-and-a-half million dollars, how much did each sell?

5. The Meadville athletic center has a 48,000-seat capacity, comprised of reserved seats, box seats, and general admission seats. There are three times the number of reserved seats as box seats, and 13,000 more general admission seats than box seats. How many of each type of seat are in the center?

6. Calculate the slope between the points (7, 3) and (12, 13).

7. Write the equation of the line with a slope of 5 and a y-intercept of 9.

1.3   EXERCISES

1. Bart cashed his paycheck at the bank, deposited a total of $450 into his checking and savings accounts, and took the remaining $150 with him. If the amount he deposited in checking was $250 more than he deposited in his savings, how much did he deposit in each?

2. Bonita spent $30 less on gasoline for her compact car last week than she spent on gasoline for her minivan. If she spent a total of $86 on gasoline, how much did she spend for each car?

3. Brad's gross pay for one week was $640. He noticed that the amount of his take-home pay was three times the amount withheld for taxes and other benefits. How much did he take home?

4. Mary deposited a total of $1,000 into her two bank accounts. She deposited $300 more into her checking account than into her savings account. How much did she deposit in each?

5. Anwar spent three times as much money on repairing his car as he spent on clothes last week. If he spent a total of $360, how much did he spend on each?

6. If Tom and Jerry win the Million Dollar Lotto, Tom will receive four times the winnings that Jerry will receive. How much money will each of them win?

7. The area of your office is 24 feet by 21 feet. The carpeting you want is $36.99 per square yard.

Answers to CYK: **1.** $x = \$6$ **2.** $x = \$2,000,000$ **3.** $x = 136$ white roses; $4\frac{3}{4}x = 646$ red roses
4. $x = 675,000$; $x + 150,000 = 825,000$ **5.** $x = \#$ box seats; $x = 7,000$ box seats;
$3x = 21,000$ reserved seats; $x + 13,000 = 20,000$ general admission seats   **6.** $m = 2$
7. $y = 5x + 9$

How much will it cost to carpet your new office? (1 square yard = 9 square feet)

8. You fill your car with gasoline, then travel 271.4 miles. When refilling your car you notice that you put 11.8 gallons of gasoline in the tank. How many miles per gallon did you get?

9. You are about to cook a roast beef that weighs 5 pounds. If the roast needs to be cooked 45 minutes per pound, what's the total cooking time required?

10. Wilbur and JB have $12.00 to spend on fishing bait. If minnows are $.75 each and worms are $.05 each, how many of each can they purchase if they want to buy the same number of minnows as worms?

11. A business wishes its average sales to be $88,000 per month for a six-month period. If sales during the 5 months were $90,000, $98,000, $79,000, $80,000, and $87,000, what should sales be in the sixth month to achieve their objective?

12. You originally paid $50,000 for your home. Ten years later, you sold it for $80,000. From the money you made selling your house, you must pay the realtor 6% of the selling price and the attorney 1% of the selling price. How much money is left for you?

13. The state symphony orchestra has four-fifths the number of men this season as it does women. If the total number of players is 45, how many men are in the orchestra?

14. A computer salesman needs to sell an average of $2 million worth of equipment per month. In January and February, his sales were $2.4 million and $1.2 million, respectively. What do his sales need to be during the month of March in order for him to maintain his average?

15. A local music store sells compact disks and cassette tapes. If there are a total of 38,000 CDs and tapes in inventory, and there are nine-tenths as many tapes as CDs, how many tapes are in inventory?

1.4 RATIOS AND PROPORTIONS

Ratios

Learning objective
Determine the ratio that occurs between two numbers or between each number and the whole of which it is a part.

A **ratio** is a comparison of one number to another. The comparison can be of a part to the whole, or of one part to another part. A ratio shows how large/small one number is relative to another number. Ratios are expressed as fractions such as 6/3, or they are expressed horizontally using a colon such as 6 : 3; they are read as "a ratio of 6 to 3." We can simplify the fraction and express the ratio in different but equivalent terms. For example,

$$\frac{6}{3} = \frac{2}{1} \times \frac{3}{3} = \frac{2}{1}$$

Example 21

The data processing department of Allied Industries has 50 computers, of which 10 are Compaqs, 25 are IBMs, and the remainder are Macs. What is the ratio of (1) IBMs to total computers, and (2) Macs to Compaqs?

Solution

1. $\text{ratio} = \dfrac{\text{number of IBMs}}{\text{total number computers}} = \dfrac{25}{50} = \dfrac{1}{2}$, or $1 : 2$

2. $\text{ratio} = \dfrac{\text{number of Macs}}{\text{number of Compaqs}} = \dfrac{15}{10} = \dfrac{3}{2}$, or $3 : 2$

Quite often in business applications, we like the second number of the ratio (the denominator of the ratio when written as a fraction) to be a 1. Especially with financial ratios, we like to compare the number of dollars to *one* dollar. We can express a ratio in an equivalent form with the second number a 1 by dividing both the numerator and denominator by the value in the denominator. For example,

$$\frac{3}{2} = \frac{3/2}{2/2} = \frac{1.5}{1}, \quad \text{or} \quad 1.5:1$$

Example 22

Jeb Wilcox made a $7,000 purchase with $2,000 cash and the remainder on credit. What is the ratio of credit to cash used for Jeb's purchase?

Solution

$$\text{ratio} = \frac{\text{amount of credit}}{\text{amount of cash}} = \frac{\$5,000}{\$2,000} = \frac{5}{2} = \frac{2.5}{1}, \quad \text{or} \quad 2.5:1$$

We could then say that Jeb used $2.50 of credit for every $1 of cash to make the purchase.

We must always be careful about the quantities being compared in a ratio. We generally want to compare quantities that are expressed in the same units so that we have a clear understanding of the magnitude of the ratio. For example, if we were comparing $4.50 to 30 cents, we would compare *cents* to *cents;* that is,

$$\frac{450 \text{ cents}}{30 \text{ cents}} = \frac{15}{1}, \quad \text{or} \quad 15:1$$

If we were comparing $3\frac{1}{3}$ hours to 45 minutes, we would use minutes for the comparison:

$$\frac{3\frac{1}{3} \text{ hours}}{45 \text{ minutes}} = \frac{3\frac{1}{3} \text{ hours} \times 60 \text{ minutes/hour}}{45 \text{ minutes}}$$

$$= \frac{200}{45} = \frac{40}{9}, \quad \text{or} \quad 40:9$$

Most of the problems involving ratios require the calculation of the ratio itself. However, you should be prepared to determine variable quantities when the relationship between variables is expressed as a ratio. Example 23 illustrates the solution to this type of problem.

Example 23

If a firm had total sales for the week of $12,000 with a 5:3 ratio of cash sales to credit sales, what is the amount of cash sales and credit sales for the week?

Solution

Since the sum of the parts equals the whole, we can note that total sales are divided into 8 parts (5 + 3), of which 5 parts are cash sales and 3 parts are credit sales. Next, we can express the ratios as fractions, or their decimal equivalents, to determine what amount of the total each part represents. Therefore, total sales are divided into 8 parts, of which 5 parts $\left(\frac{5}{8}\right)$ are cash and 3 parts $\left(\frac{3}{8}\right)$ are credit sales.

$$\frac{5}{8}, \quad \text{or} \quad .625 \times \$12,000 = \$7,500 \text{ cash sales}$$

$$\frac{3}{8}, \quad \text{or} \quad .375 \times \$12,000 = \$4,500 \text{ credit sales}$$

This analysis of parts that make up the whole can be used to solve problems in which the whole is made up of more than two parts.

Example 24

Martin, Browne, Seubert, and Rigal invested a total of $256,000 in a new business venture. The ratio of each investor's share is $7:4:3:2$, respectively. How much did each invest?

Solution

The whole is broken into 16 parts (7 + 4 + 3 + 2)
7 of those 16 $\left(\frac{7}{16}\right)$ represents Martin's share

$$\frac{7}{16} \times \$256,000 = \$112,000$$

4 of those 16 $\left(\frac{4}{16}\right)$ represents Browne's share

$$\frac{4}{16} \times \$256,000 = \$64,000$$

Similarly 3 of the 16 $\left(\frac{3}{16}\right)$ and 2 of the 16 $\left(\frac{2}{16}\right)$ represent Seubert's and Rigal's share, respectively; Seubert invested $48,000 and Rigal invested $32,000.

Proportions

When two ratios a set equal to each other, the resulting expression is called a **proportion.** This can be expressed in either of the following forms:

$$\frac{6}{3} = \frac{2}{1} \quad \text{or} \quad 6:3 = 2:1$$

Learning objective

Solve a proportion for one of the four parts of a propor- tion, given the other three parts of the proportion.

The proportion above would be expressed in words as "6 is to 3 *as* 2 is to 1." In general, the proportion is written:

$$\frac{a}{b} = \frac{c}{d} \qquad \text{or} \qquad a:b = c:d$$

If we know three of the values of a proportion, we can use our equation-solving techniques to solve for the fourth value.

Example 25

Solve the following proportion for x:

$$\frac{x}{12} = \frac{5}{6}$$

Solution

$$\frac{x}{12} = \frac{5}{6}$$

First, let's clear out the denominators of each side of the equation by multiplying by the common denominator of $12 \times 6 = 72$.

$$72 \times \frac{x}{12} = \frac{5 \times 72}{6}$$

$$\overset{6}{\cancel{72}} \times \frac{x}{\cancel{12}} = \frac{5}{\cancel{6}} \times \overset{12}{\cancel{72}}$$

$$6x = 60$$

Then divide both sides by 6 to get

$$x = 10$$

Check:

$$\frac{10}{12} = \frac{5}{6}$$

A shortcut method to the same solution is to cross multiply in the following way:

$$\cancel{\frac{x}{12}} \times \cancel{\frac{5}{6}}$$

$$6x = 60$$

$$x = 10$$

This *cross multiplication* method is just a shortcut for the first method using our rules for equation solving, but it allows us to also solve proportions expressed in the alternate form of

$$x:12 = 5:6$$

We identify the x and the 6 as the *outer part of the proportion* and the 12 and the 5 as the *inner part of the proportion:*

$$x:12 = 5:6$$

inner

outer

Multiply the inner parts and the outer parts, set them equal to each other, and solve.

$$6x = 60$$
$$x = 10$$

The cross multiplication method can make solving proportion problems relatively simple, compared to the equation solving method.

Example 26

Solution

Solve the proportion $\dfrac{8}{x} = \dfrac{9}{27}$ for x.

$$\frac{8}{x} = \frac{9}{27} \qquad \text{or} \qquad 8:x = 9:27$$

Cross multiply either proportion to get:

$$8 \times 27 = x \times 9$$
$$216 = 9x$$
$$x = 24$$

Check:

$$\frac{8}{x} = \frac{9}{27}$$
$$\frac{8}{24} = \frac{9}{27}$$
$$\frac{1}{3} = \frac{1}{3}$$

Example 27

Sylvia Jacobi and Sandy Black opened a Unisex Beauty Salon with $25,000 of Sylvia's money and $15,000 of Sandy's money. They agreed that after all costs for the business were paid, they would distribute the profits between them proportionally to the amount of their investment. If Sandy's profit share last year was $10,000, how much of the profits did Sylvia receive?

Solution

Let x = Syliva's share of the profits.

Since they agreed to split the profits proportionally with the amount of their investment, we can form the proportion:

Sylvia's share is to Sandy's share as $25,000 is to $10,000.

$$x : \$10,000 = \$25,000 : \$15,000$$

or

$$\frac{x}{\$10,000} = \frac{\$25,000}{\$15,000}$$

Cross multiply and solve:

$$\$15,000x = \$250,000,000$$
$$x = \$16,666.67$$

Obviously, we would have made our work easier if we had simplified the right side of the proportion to 5/3 before cross multiplying.

$$\frac{x}{\$10,000} = \frac{\$25,000}{\$15,000} = \frac{5}{3}$$

$$\frac{x}{\$10,000} = \frac{5}{3}$$

$$3x = \$50,000$$

$$x = \$16,666.67$$

Ratios and proportions will be very useful to us in our later study of finance.

CHECK YOUR KNOWLEDGE

Ratio and Proportion

Solve each of the following problems.

Write a ratio for each phrase in problems 1 through 3, and express the ratio as a part to 1.

1. 60 wins to 15 losses

2. 150 rejected units to 2,500 total units produced

3. 45 minutes to 3 hours

4. A university received 5,425 applications for admissions last year. If the admissions office accepted 3,000 of those who applied and 1,800 students enrolled in the freshman class,

   a. What is the ratio of "freshman enrollment" to "acceptances"?

   b. What is the ratio of "acceptances" to number of "applications"?

5. Morris Pump International Inc. employs 2,750 people and has a labor-to-management ratio of 20 to 5. How many workers are in each category?

6. Solve the proportion for s: $\dfrac{s}{5} = \dfrac{6}{15}$

7. The proportion of women employees to men employees at Montauk Import Exporting Agency is 5 to 2. If there are twenty women employed, how many men are employed?

1.4 EXERCISES

Solve each of the following problems.
Write a ratio for each phrase in problems 1 through 10, and express the ratio as a part to 1.

1. Marcie Brookes had 21 wins at tennis last year and 42 losses.

   a. Write the ratio of her wins to losses.

   b. Write the ratio of Marcie's wins to the total number of matches she had last year.

2. Harrison Electronics supplies large quantities of resistors to other manufacturers of electrical appliances. The quality control department of Harrison maintains regular records on the ratio of resistors that do meet buyers' specifications. They are proud to claim that 98% of their resistors meet those specifications. The last sampling off of the production line indicated that 147 of the 150 resistors sampled did meet specifications. Write the ratio of acceptable resistors to the total sampled as a ratio of the part to 1. Did Harrison maintain its claim in this sample?

3. Jackson's Women's Apparel store had 150 credit card sales and 100 cash sales last month.

   a. Write the ratio of credit card sales to cash sales for the month.

   b. Write the ratio of cash sales to credit card sales for the month.

   c. Write the ratio of credit card sales to total sales for the month.

4. Winona College has been concerned about its enrollment for the past few years. In the past, the college administrators have been able to boast that 90% of those students accepted for their freshman class have indeed enrolled at the college. This past fall, 800 of the 950 students who were accepted did enroll. Write the ratio of students who did enroll to the total number accepted as a part to 1 ratio.

Answers to CYK: **1.** 4:1   **2.** .06:1   **3.** .25:1   **4. a.** 3:5 **b.** 120:217 or .55:1   **5.** 2200 labor and 550 management   **6.** $s = 2$ **7.** 8 men

5. Winona College has been encouraged by its board of trustees to have at least 20% of its freshman class be made up of minorities and international students. Last fall's class had 175 students who met this criteria. Write the ratio of minorities and international students to the total size of the freshman class as a ratio of the part to one.

6. Bartlesville Central Schools have 112 women elementary school teachers out of a total of 140 elementary teachers. Write the ratio of men elementary teachers to women elementary teachers at Bartlesville.

7. If the ratio of men to women in the Budtown Kiwanis Club is six to one, how many women are in the 63-member club?

8. Marilou Arnold and Marilyn Jefferson own M&M Cleaning Inc., which cleans homes for people. Marilou invested $10,000 in the company and Marilyn invested $5000 when they first began the company. They agreed to divide the profits from the business each year in the same ratio as their original investments.

    a. What is the ratio of Marilou's investment to Marilyn's investment?

    b. If the business had a profit of $36,000 last year, how much of the profits did each receive?

9. Sid Jackson, Ben Johnson, and Hal Washington formed a partnership in which they purchased a number of townhouses, which they then rented. Sid, Ben, and Hal each invested $50,000, $80,000, and $70,000, respectively, in the partnership. They agreed to divide any profits they made in the same ratio as their investment. Last year they made $50,000 profit. How much did each receive of the profits?

10. Mark Witter, Donna Brown, and Willie B. Wright own small businesses in tax preparation, home improvement, and personal investment, respectively. They decided to jointly rent office space in a large open building that was once a supermarket with total floor space of 3,000 square feet. Mark's office occupies 1,000 square feet, Dean's occupies 1,500 square feet, and Willie's occupies 500 square feet. They agreed to share the rental cost of $2,400 per month in the same ratio as the space their offices occupy. How much does each pay per month?

11. Solve the following proportion for x: $\dfrac{x}{8} = \dfrac{27}{12}$.

12. Solve the following proportion for t: $\dfrac{9}{t} = \dfrac{14}{252}$.

13. Solve the following proportion for p: $\dfrac{4}{120} = \dfrac{15}{p}$.

14. Solve the following proportion for q: $\dfrac{15}{225} = \dfrac{q}{30}$.

15. Solve the following proportion for y: $y:12 = 14:84$.

16. Solve the following proportion for z: $16:z = 9:36$.

17. Solve the following proportion for w: $154:7 = 11:w$.

18. Solve the following proportion for r: $32:288 = r:9$.

19. The proportion of boys to girls enrolled at the Playtime Daycare Center is 4 to 5. If there are twenty girls, how many boys are enrolled at PDC?

    a. Write a proportion expression for this problem.

    b. Solve the proportion to determine the number of boys enrolled.

20. The proportion of young women in the Beeville Police Explorer Post to young men is two to seven. If there are fourteen men in the Post, how many women are there?

    a. Write a proportion expression comparing the number of men to women.

    b. Solve the proportion to determine the number of women.

21. J. R. Noteworthy has recently inherited $30,000, some of which he wants to invest. He has decided to invest the money proportionally between stocks and bonds, so that for every four dollars that he invests in bonds he will invest nine dollars in stocks. If he invests $18,000 in stocks, then (a) how much will he invest in bonds, and (b) how much of his inheritance will be invested in stocks and bonds?

22. Aiesha Lock and Lexie Bagel have begun a small catering business in which Aiesha has invested seven dollars for every two dollars that Lexie invested in the business.

    a. Write a proportion expression that compares the amounts of money each invested.

    b. If Aiesha invested $11,200, how much did Lexie invest in the business?

23. McGuinness's Bakery revenues show that for every three dollars that they sell in pastries they receive about eleven dollars for breads and bagels. If their revenues for pastries sold last week were six hundred sixty dollars, what were the revenues for bread and bagels?

24. Pleasant Rest Nursing Home has five women for every two men residents. If there are a total of 140 residents at the Home, how many are men, and how many are women?

25. Westside Manor Adult Home has three women for every two men. If there are seventy men at the Home, how many women are there?

CHAPTER REVIEW EXERCISES

The following problems review all mathematics and business concepts presented in the chapter. If you experience any difficulty solving a problem, go to the appropriate chapter section and review the examples provided. Express all answers as decimals rounded to the nearest one hundredth, or dollars rounded to the nearest cent, unless otherwise instructed.

Complete the following problems.

1. $6 - (-3) =$

2. $6 \times (-5) =$

3. $18 \div (-9) =$

4. $24 - 18 + 5 \times 6 \div 3 =$

5. $(18 - (4 + 6) \times 3) \div 6 =$

6. $4^2 - 3^3 + 2^4 =$

7. Write an algebraic expression for the following: "Five times x minus 7 times y is equal to 11."

8. Evaluate: $5x - 3y - 4$ if $x = 4$ and $y = 3$.

9. Evalute $x^4 + y^3$ if $x = 2$ and $y = 3$.

10. Combine like terms: $5x + 3y - 9x + 6y$.

11. Solve for x: $8 + 5x = 23$.

12. Solve for y: $\dfrac{3y}{14} = 9$.

13. Solve for z: $12 + 4z - 14 = 33 - 3z$.

14. Solve $I = PRT$ for R.

15. Mary, VanAhn, and Sari, each sold Girl Scout cookies for their troop. Mary sold twice as many boxes as Sari, and VanAhn sold ten more boxes than Sari. Altogether the girls sold 90 boxes of cookies. Let x represent the number of boxes that Sari sold.

   a. Write algebraic expressions in terms of x for the number of boxes sold by Mary and by VanAhn.

   b. Write an equation using the expressions in (a) that could be solved to determine how many boxes each sold.

   c. Solve the equation in (b) to determine how many boxes each sold.

16. Martha and Bart earn the same wage at the local Dog and Burger Works. Last week the total of their pay before any deductions was one hundred twenty dollars, and Martha worked twice as many hours as Bart. How much did each earn before deductions?

17. Sarah earns one dollar per hour more than Myra at Hafners Flower Mart. Both Sarah and Myra worked forty hours last week, and the total of their paychecks before any deductions was three hundred sixty dollars. What is the hourly wage of each of the women?

18. The Sydney Smash Soccer team won 45 of the 120 games that they played last year. Write the ratio of games won to games lost.

19. Alexandria Bay College was interested in the ratio of students who enroll as freshmen at the college that receive a degree four years later. Alexandria had a freshman class of 832 four years ago and 672 are graduating this year. What is the ratio of graduations to freshman enrollment?

20. Solve the proportion for m: $m : 12 = 7 : 2$.

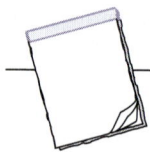

EXPRESS YOUR UNDERSTANDING

Compose one or two well-written sentences to express the requested information in your own words.

1. Explain how you would add two numbers that have different signs: one positive and one negative.

2. Express how you would divide (-12) by (-3).

3. Describe how you would multiply (-3), $(+5)$, and (-8) together.

4. Describe the steps you would take to solve the equation $3x - 9 = 4x - 11$.

5. Explain how you would evaluate the expression: $4 + 9 \times 8 + 5 \times (3 - 5) + 3^2$.

6. Write the rule for order of operations.

7. If $x =$ Dick's age and $x - 5 =$ Mike's age, write the following expressions in words with no mention of x. (a) $x + (x - 5) = 2x - 5$ (b) $2x - 5 = 11$

8. Express in words the following equation in terms of a percentage of the unknown.

   $.35x = 105$

9. If x represents Juan's daily wage and $(x + 15)$ represents Mario's daily wage, state how the two daily wages compare.

10. What does the exponent 4 applied to the number 7 tell you to do?

11. Describe the steps you would follow to evaluate $(5^2 + 4^3)^2$.

12. Describe in your own words what the commutative principle tells you about multiplying or adding two numbers.

13. What is a ratio?

14. What is a proportion?

15. Describe the steps you would follow to solve the proportion $5 : W = 9 : 45$.

SELF-TEST

A. Terminology

Complete the following items using the key terms presented at the beginning of the chapter. Check your responses against the answer key at the end of the test.

1. A numeric sentence in which the values on the left of the equals sign must equal the values on the right is called a(n) _____.

2. Positive and negative numbers are often called _____ numbers.

3. A number whose value is fixed is called a _____.

4. A symbol that represents any number is called a _____.

5. In the expression 7^4, the 4 is called the _____ of the expression.

6. An equation that represents a general principle or rule, such as $A = LW$, is called a _____.

7. When we compare two numbers either as a fraction, 4/5, or with a colon, 4 : 5, we refer to this form of comparison as a(n) _____.

8. A proportion is an expression of the equality of two _____.

B. Calculation

The following concepts and short problems are designed to test your understanding of the objectives identified at the beginning of the chapter. Answers are provided at the end of the test.

9. $-32 \times (-3)$

10. $42 \div (-2)$

11. $30 \div 10 + 7.5$

12. $4x + 10 = 20$

13. $-2 + 6x = 24 - 8x$

14. $4.5x - 12.5 = -44 - 10.5x$

15. $100x + 3 + 4x = -200 - 50x + 303$

16. A basic business formula is that profit equals revenue minus cost. (a) Write this as an algebraic formula using the letters P, R, and C, and (b) solve the formula for C.

17. The sum of two numbers is 102. One of the numbers is five times the other number. Find the two numbers.

18. A car salesman wishes to sell an average of $11,000 per day. If his sales during the first four days of the week are $6900, $10,500, $18,500, and $13,000, what do his sales need to be on the fifth day of the week in order to achieve the desired average?

19. Marlie's Speedy Fill gas and minimart had a total of $30,000 in gasoline sales for the month of August. Regular octane gas sold for $1.10 per gallon and ultra octane sold for $1.40 per gallon. If credit sales accounted for 40% of the sales and the remaining were cash sales, determine the amount of credit sales and the amount of cash sales that occurred during that August.

20. El Hombre Community College noted that its ratio of minority faculty to nonminority faculty is

the same as the ratio of minority students to non-minority students. The faculty ratio of minority to nonminority is 5 to 11.

a. Write a proportion that would represent the proportion of minority to nonminority students.

b. What is the number of minority students if there are 4,400 students who are not regarded as minority students?

Answers to Self-Test:

1. equation   *2.* signed   *3.* constant   *4.* variable   *5.* exponent
6. formula   *7.* ratio   *8.* ratios   *9.* 96   *10.* −21   *11.* −18.5
12. 2.5   *13.* 13/7   *14.* −2.1   *15.* .649   *16.* a $P = R - C$
b $C = R - P$ *17.* first = 17   second = 85   *18.* $6100
19. credit sales = $12,000 cash sales = $18,000   *20.* a $m{:}n = 5{:}11$
b $m{:}4400 = 5{:}11, \; m = 2000$

PERCENTAGE

Learning Objectives

After completing the exercises in this chapter, you will be able to:

1. Convert fractions to decimals and percents interchangeably.

2. Define the elements of the percent formula.

3. Identify the base, rate, and part of a percent problem.

4. Use the percent formula to solve for the part.

5. Use the percent formula to solve for the base.

6. Use the percent formula to solve for the rate.

7. Calculate the rate of increase and the rate of decrease.

8. Develop a price index and use the Consumer Price Index to calculate the rate of inflation.

9. Define the key terms.

INTRODUCTION

This chapter is one of the most important in the text because the concepts we discuss will reappear as part of the quantitative procedures in subsequent chapters. Percentage computations are used by every segment of our society to express the degree of efficiency and/or effectiveness achieved between objectives and performance. As consumers, we are concerned with the percent change in the cost of living when compared to our annual income and the percent of interest on home mortgages, saving accounts, and taxes. Governments report the percents of economic growth and unemployment, and express population demographics as percentages. Businesses summarize their percent increases or decreases in sales, costs, expenses, and profits. Managers of today's businesses are evaluated on their applications of percent calculations when making decisions.

The chapter begins with a discussion of converting fractions and decimals to percents, followed by an in-depth presentation of the percent formula. It concludes with the application of percentage to specific types of business problems.

2.1   CONVERTING FRACTIONS AND DECIMALS TO AND FROM PERCENTS

A fraction is a mathematical expression of a part of the whole. The fractional presentation of a value is very common in business. For example, stock quotations (up $5/8$), interest rates $10\frac{1}{2}$%), retailing ($6\frac{3}{4}$ doz. or $1\frac{1}{8}$ gross), and payroll ($48\frac{1}{2}$ hours) are data reported with fractions.

If we are to use technology as a tool to increase the quality and quantity of our productivity, we must possess the arithmetic skills to convert data to appropriate input formats. For example, microcomputers and calculators analyze and print data in decimals; therefore, we must be able to express fractional parts of a whole as decimals or as percents. As students of business mathematics, this section of the chapter is extremely important. If you are to solve business problems successfully, using state-of-the-art technology, you must be capable of converting fractions to decimals or percents.

Converting a Fraction to a Decimal

A **decimal number** is any number that is written using the proper selection and combination of any digit that is part of the **decimal system** (0, 1, 2, 3, 4, 5, 6, 7, 8, 9). The decimal system is a **place value** system because the value of each digit depends on its position relative to the decimal point. The decimal point in a decimal number separates whole numbers from their fractional parts. For example, 1,325.675 would be read or written as one thousand three hundred twenty-five and six hundred seventy-five thousandths. We will now look at how to convert a fraction to its **decimal equivalent.** The procedure involves dividing the numerator by the denominator and rounding the answer to the required place value.

Example 1

Solution

Convert the fraction ⅝ to a decimal rounded to the nearest thousandth.

$$\begin{array}{r} .6250 \\ 8\overline{)5.0000} \\ \underline{48} \\ 20 \\ \underline{16} \\ 40 \\ \underline{40} \\ 0 \end{array}$$

Add as many 0s as necessary to round your answer to the required place value.

Therefore, ⅝ = .625.

```
        CALCULATOR   SOLUTION
  Keyed entry                              Display

   AC   5   ÷   8   =                        0.625
```

If a fraction can be converted to a decimal, a decimal can be converted to a **decimal fraction.** A decimal fraction has a denominator of 1, 10, 100, 1,000, and so on. It is sometimes necessary to report data in fractional form even though it may be easier to perform the calculations with decimals. To convert a decimal to a decimal fraction, use one of the following methods:

Method 1   Write the decimal in fractional form exactly the way it is read and reduce to lowest terms.

Example 2

Solution

Convert the decimal .875 to a fraction and reduce to lowest terms.

.875 = Eight-hundred seventy-five thousandths, or

$$\frac{875}{1,000} = \frac{875 \div 5}{1,000 \div 5} = \frac{175 \div 5}{200 \div 5} = \frac{35 \div 5}{40 \div 5} = \frac{7}{8}$$

Note: By the rules of divisibility, both 875 and 1,000 are divisible by 5, as are 175, 200, 35, and 40.

Method 2

> *Step 1:* Write the decimal as a whole number in the numerator.
>
> *Step 2:* Write 1 in the denominator and add as many 0s as the number of digits in the original decimal.

Example 3

Write .075 as a fraction and reduce to lowest terms.

Solution

Step 1: $.075 = \dfrac{75}{?}$ (drop the 0, write as a whole number)

Step 2: $.075 = \dfrac{75}{1 + 3\ 0s} = \dfrac{75}{1,000} = \dfrac{3}{40}$

When we convert fractions or mixed numbers to decimal equivalents, the basic arithmetic functions of addition, subtraction, multiplication, and division can be carried out rapidly by calculators and microprocessors, as most systems automatically account for the decimal if the data is entered properly.

Example 4

Find the sum of $3\frac{2}{5} + 6\frac{3}{8} + 12\frac{11}{20}$ Round your answer to thousandths.

Solution

Convert each fraction to a decimal and enter the data as decimal numbers with the instructions to add. (Remember that $3\frac{2}{5}$ is $3 + \frac{2}{5}$, or $3 + .40 = 3.40$; $6\frac{3}{8}$ is $6 + \frac{3}{8}$, or $6 + .375 = 6.375$; $12\frac{11}{20}$ is $12 + \frac{11}{20}$, or $12 + .55 = 12.55$.)

Mixed number	Decimal equivalent
$3\frac{2}{5}$	3.400
$6\frac{3}{8}$	6.375
$12\frac{11}{20}$	12.550
Total	22.325

The total is 22.325. If we were required to report the answer as a mixed number, we would convert the decimal number to $22\frac{13}{40}$ using the procedure explained earlier in this chapter.

CALCULATOR SOLUTION

Keyed entry										Display
AC	3	+	2	÷	5		=			3.4
							Min	*		3.4
	6	+	3	÷	8		=			6.375
							M+			6.375
	12	+	11	÷	20		=			12.55
							M+			12.55
							MR			22.325

*Key clears memory and enters new value in memory.

The same procedure would be followed if we were required to perform calculations involving multiplication or division. Let's look at an example that involves multiplication.

Example 5

Multiply 12 1/8 by 4 3/4. Round to hundredths.

Solution

Mixed number	Decimal equivalent
12 1/8	12.125
4 3/4	4.75
Total	57.59375
	or
	57.59 (rounded to hundredths)

CALCULATOR SOLUTION

Keyed entry								Display
AC	12	+	1	÷	8		=	12.125
							Min	12.125
	4	+	3	÷	4		=	4.75
							×	4.75
							MR	12.125
							=	57.59375

Converting a Fraction to a Percent

Learning objective
Convert fractions to
decimals and per-
cents interchangeably.

A percent also represents a part of the whole. Percents are hundredths, or parts of a 100. A percent is a numeral written with a percent sign (%). For example, 10% is 10 parts of 100, and 50% is 50 parts of 100. Because 100% refers to the whole, 125% indicates that more than one whole unit (each unit divided into 100 parts) is involved; 125 of those parts are under consideration.

To convert a fraction to its **percent equivalent,** first convert the fraction to a decimal, then move the decimal point two places to the right and add the % sign; or multiply the decimal by 100 and add the percent sign (.50 × 100 = 50 or 50%).

Example 6

Convert the fractions ½, ⅞, and 1¼ to percents.

Solution

Fraction	Decimal equivalent	Percent equivalent
½	.50	50%
⅞	.875	87.5%
1¼	1.25	125%

Note: To change a decimal to a percent, move the decimal point 2 places to the right and add the percent sign. If the decimal number extends beyond the hundredths position (⅞ = .875 or 87.5%), the percent equivalent includes *parts of a percent.*

Any fraction can be converted to a percent, which also means any percent can be converted to a fraction. To convert a percent to a fraction, drop the % sign and move the decimal two places to the left. Then convert the decimal number to a fraction and reduce to lowest terms.

Example 7

Convert the percents 2% and 12.5% to fractions and reduce to lowest terms.

Solution

Percent	Decimal equivalent	Fraction equivalent
2%	.02	²/₁₀₀, or ¹/₅₀
12.5%	.125	¹²⁵/₁₀₀₀, or ⅛

Note: To change a percent to a decimal, drop the percent sign and move the decimal 2 places to the left; or divide the number by 100 and drop the percent sign (50 ÷ 100 = .5).

If the percent also contains a fraction, the conversion is a little more difficult. The procedure explained in Example 7 can be used if the fractional ending of the percent is an aliquot part. An **aliquot part** is defined as any number that will divide evenly into another number. For example, the fraction $3/4$ is an aliquot part because the numerator (3) can be divided by the denominator (4) evenly. Therefore, if we wished to convert $5\,3/4\%$ to a fraction, we would follow the procedure that was explained in Example 7, as shown next.

Example 8

Solution

Convert $5\,3/4\%$ to a fraction and reduce to lowest terms.

Fractional percent	Decimal percent	Decimal equivalent	Fractional equivalent
$5\,3/4\%$	5.75%	.0575	$\dfrac{575}{1,000}$, or $\dfrac{23}{400}$

To convert a percent that contains a fractional ending that is not an aliquot part, follow the procedure shown in Example 9.

Example 9

Solution

Convert $4\,2/7\%$ to a fraction and reduce to lowest terms.

$$4\,2/7\% = \frac{30}{7}\%$$
Convert to an improper fractional percent.

$$= \frac{\frac{30}{7}}{100}$$
Divide by 100 to convert to a decimal.

$$= \frac{\frac{30}{7}}{100} \times \frac{\frac{7}{1}}{\frac{7}{1}}$$
Multiply both the numerator denominator by 7 to simplify the fraction.

$$= \frac{30}{700} \div \frac{10}{10}$$
Divide both the numerator and the denominator by the largest common factor (10).

$$= \frac{3}{70}$$
Reduce to lowest terms.

Table 2.1 Fraction, decimal, and percentage equivalents

Denominator	Numerator														
	1	2	3	4	5	6	7	8	9	10	11	12	13	14	15
2	.50 50%														
3	.333 33 1/3%	.666 66 2/3%													
4	.25 25%		.75 75%												
5	.20 20%	.40 40%	.60 60%	.80 80%											
6	.166 16 2/3%	.333 33 1/3%	.50 50%	.666 66 2/3%	.833 83 1/3%										
7	.142 14 2/7%	.285 28 4/7%	.428 42 6/7%	.571 57 1/7%	.714 71 3/7%	.857 85 5/7%									
8	.125 12 1/2%	.25 25%	.375 37 1/2%	.50 50%	.625 62 1/2%	.75 75%	.875 87 1/2%								
9	.111 11 1/9%	.222 22 2/9%	.333 33 1/3%	.444 44 4/9%	.555 55 5/9%	.666 66 2/3%	.777 77 7/9%	.888 88 8/9%							
10	.10 10%	.20 20%	.30 30%	.40 40%	.50 50%	.60 60%	.70 70%	.80 80%	.90 90%						
11	.090 9 1/11%	.181 18 2/11%	.272 27 3/11%	.363 36 4/11%	.454 45 5/11%	.545 54 6/11%	.636 63 7/11%	.727 72 8/11%	.818 81 9/11%	.909 90 10/11%					
12	.083 8 1/3%	.166 16 2/3%	.25 25%	.333 33 1/3%	.416 41 2/3%	.50 50%	.583 58 1/3%	.666 66 2/3%	.75 75%	.833 83 1/3%	.916 91 2/3%				
13	.077 7 9/13%	.153 15 5/13%	.230 23 1/13%	.307 30 10/13%	.385 38 6/13%	.462 46 2/13%	.538 53 11/13%	.615 61 7/13%	.692 69 3/13%	.769 76 12/13%	.846 84 8/13%	.923 92 4/13%			
14	.071 7 1/7%	.142 14 2/7%	.214 21 3/7%	.285 28 4/7%	.357 35 5/7%	.428 42 6/7%	.50 50%	.571 57 1/7%	.642 64 2/7%	.714 71 3/7%	.786 78 4/7%	.857 85 5/7%	.929 92 6/7%		
15	.066 6 2/3%	.133 13 1/3%	.20 20%	.266 26 2/3%	.333 33 1/3%	.40 40%	.466 46 2/3%	.533 53 1/3%	.60 60%	.666 66 2/3%	.733 73 1/3%	.80 80%	.866 86 2/3%	.933 93 1/3%	
16	.062 6 1/4%	.125 12 1/2%	.1875 18 3/4%	.25 25%	.3125 31 1/4%	.375 37 1/2%	.4375 43 3/4%	.50 50%	.5625 56 1/4%	.625 62 1/2%	.6875 68 3/4%	.75 75%	.8125 81 1/4%	.875 87 1/2%	.9375 93 3/4%

Note: Table 2.1 shows the decimal and percentage equivalents for many fractions used in this text. If your instructor encourages you to use a calculator or a computer as part of the instructional resources of the course, you could use either Table 2.1 or your calculator to find decimal equivalents.

CHECK YOUR KNOWLEDGE

Converting Fractions to Decimals and Percents

Convert each fraction or mixed number to a decimal rounded to the nearest thousandth.

1. $\frac{1}{3}$ 2. $6\frac{1}{5}$ 3. $\frac{5}{9}$ 4. $28\frac{13}{15}$

Convert each decimal to a fraction and reduce to lowest terms.

5. .8 6. 3.4166 7. .025 8. .08

Convert each decimal to a percent.

9. .2 10. 5.05 11. .0016 12. 3

Convert each percent to a decimal.

13. 39% 14. $12\frac{7}{8}$% 15. 6.125% 16. 225%

Convert each percent to a fraction or mixed number and reduce to lowest terms.

17. 50% 18. 2.25% 19. .02% 20. $5\frac{5}{6}$%

Convert each of the following fractions or mixed numbers to decimal and percent equivalents. Round decimals to thousandths.

21. $\frac{3}{8}$ 22. $18\frac{11}{12}$ 23. $\frac{2}{50}$ 24. $3\frac{1}{3}$

2.1 EXERCISES

Convert each fraction or mixed number to a decimal, rounded to the nearest thousandth where appropriate.

1. $\frac{2}{3}$ 2. $8\frac{1}{3}$ 3. $148\frac{9}{64}$
4. $\frac{13}{16}$ 5. $\frac{2}{5}$ 6. $15\frac{1}{15}$

Convert each decimal to a fraction reduced to lowest terms.

7. .5 8. 2.4 9. .80
10. .1875 11. .125 12. .008

Answers to CYK:

1. .333 2. 6.200 3. .556 4. 28.867 5. $\frac{4}{5}$ 6. $3\frac{2083}{50000}$ 7. $\frac{1}{40}$
8. $\frac{2}{25}$ 9. 20% 10. 505% 11. .16% 12. 300% 13. .39 14. .12875
15. .06125 16. 2.25 17. $\frac{1}{2}$ 18. $\frac{9}{400}$ 19. $\frac{1}{5,000}$ 20. $\frac{7}{120}$
21. .375; 37.5% 22. 18.917; 1,891.7% 23. .040; 4% 24. 333.333; 333.3%

Convert the following decimals to percents.

13. .3   **14.** .125   **15.** 1.50
16. .0075   **17.** .05   **18.** .032

Convert the following percents to decimals.

19. 6%   **20.** 17.5%   **21.** 1/2%
22. 3,500%   **23.** 20%   **24.** 3.33%

Convert the following fractions or mixed numbers to percents.

25. 1/4   **26.** 8 2/3   **27.** 22/32
28. 8/24   **29.** 7/10

Convert the following percents to fractions or mixed numbers reduced to lowest terms.

30. 12%   **31.** 640%   **32.** 6 2/9 %
33. .5%   **34.** 5.25%

Complete the following conversions as indicated. Round decimal answers to thousandths and reduce fractions to lowest terms.

	Fraction	Decimal	Percent
35.	1/5	_____	_____
36.	_____	.625	_____
37.	_____	_____	4%
38.	_____	1.8	
39.	17 1/2	_____	_____

Convert the following mixed numbers and fractions to decimals and solve as indicated. Round all answers to thousandths.

40. 6 5/8 + 9 3/4   **41.** 11/22 × 3/6
42. 110 17/52 − 87 15/32   **43.** 285 ÷ 4 1/3
44. 3/8 × 6/7 × 2/9
45. 10 3/14 + 75 3/10 + 9 17/22
46. (105 2/8 ÷ 8 1/6) × 5 2/21

Solve the following word problems. Express all answers as decimals rounded to hundredths unless otherwise instructed.

47. Ahmed Al-Hindi owns a small farm consisting of six lots of the following acreage: 5 3/4, 10 1/8, 27 1/2, 18 2/5, 15 1/6, and 42 1/4. What is the total acreage of Mr. Al-Hindi's farm?

48. A truck gets 17 1/8 mpg. How many miles will the driver be able to go before stopping for gas if the gas tank contains 12 2/5 gallons?

49. The enrollment this fall at Ourtown Community College is 1,275 of which 969 are local students, 255 are out-of-state students, and 51 are foreign students. Express the student population as fractional parts of the total student enrollment.

50. The Wildcat youth football program is planning a fall picnic. If 266 attend the picnic and each is estimated to eat 1 1/2 hot dogs, how many pounds of hot dogs must be purchased if there are 8 hot dogs per pound? Express your answer as a mixed number.

51. In 1998, Allied Manufacturing employed 368 people. Today the company employs 920. Express as a fraction the number of employees working today compared to 1998.

52. Buildrite Construction Company purchases 20 3/10 acres of land for a housing development. Two and four-fifths acres of the property will be used for roads. If each building lot is to be 5/8 of an acre, how many houses will be constructed in the tract?

53. Last winter, fuel oil prices averaged $1.03 3/8 per gallon. This winter, fuel oil is expected to average 89 1/2 cents. If a homeowner's average fuel oil consumption is 986 4/5 gallons per heating season, how much money will be saved this year over last year?

54. Market Research Associates conducted a telephone survey that involved the purchase of a new product being test-marketed in New York, Pennsylvania, and New Jersey. Of the 1,600 households contacted, 573 indicated they had purchased the product, and 250 indicated they would purchase the product regularly. What part of those surveyed would use the new product regularly? Express your answer as a fraction, a decimal, and a percent.

55. Jim, Juan, and Giles formed a business partnership to sell sporting goods. Jim owns 1/2 of the company, and Juan 1/6. If the ownership is based on the amount of money invested, and the total investment was $90,000, what fractional part of the company is owned by Giles? Also, how much money did he invest?

56. Ann Nolan plans to attend a conference in Buffalo, New York. She lives in Albany and estimates it will take her 2½ hours to drive from Albany to Syracuse, 1¼ more hours to drive from Syracuse to Rochester, and an additional 1¾ hours from Rochester to Buffalo. If the distance from Albany to Buffalo is 288 miles, what is Ann's estimated speed if she makes a nonstop trip?

2.2   THE PERCENT FORMULA

Learning objective
Define the elements of the percent formula.

All percent problems can be solved by using the **percent formula.** The formula consists of three key variables: the *base,* the *rate,* and the *part.*

The percent formula

part = base × rate

or

$P = B \times R$

Learning objective
Identify the base, rate, and part of a percent problem.

The **base** (*B*) in the formula is the *total,* or *whole,* and always equals 100%. For example, the base in the equation shown in Figure 2.1 is net sales, which equals $15,000. In word problems, you can usually identify which quantity is the base by looking for the quantity preceded by the preposition "of" or expression "part of."

The **rate** (*R*) is the **percent,** decimal, or fractional part of the base to be calculated. It is easy to identify when it is expressed with the symbol %

Figure 2.1

The percent formula

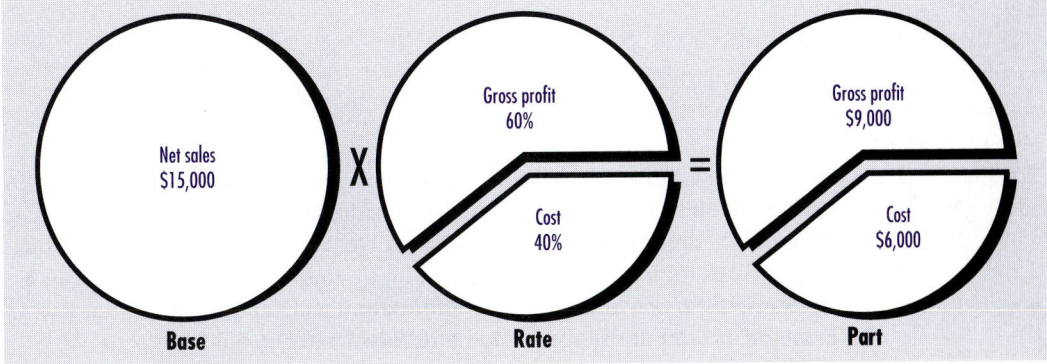

Net sales $15,000

Base

X

Gross profit 60%

Cost 40%

Rate

=

Gross profit $9,000

Cost $6,000

Part

(percent). For example, in Figure 2.1, 40% of net sales is cost and 60% is gross profit.

The **part** (*P*) is the *portion* of the base that is determined by the rate. The base can be divided into as many parts as required by the data. The part is always expressed as a number and is generally a value less than the base. Returning to Figure 2.1, net sales of $15,000 (the base) multiplied by .40 (the rate of 40% converted to a decimal) equals $6,000 (the part, or portion, of net sales that represents cost).

Now that we understand the key variables of the percent formula, we can begin solving percent problems. A number of examples will help to differentiate between the given variables and the unknown variables that must be calculated.

Computing the Part

Learning objective
Use the percent formula to solve for the part.

We find the part (portion) by multiplying the base times the rate as expressed by the basic percent formula.

part = base × rate

Example 10

The student body of a university is 23,000, and 54% of it is female. How many female students attend the university?

Solution

Step 1:   To solve this problem we must first identify the variables given.

base = 23,000   total student body
rate = 54%   percent of female students
part = unknown   number of female students

Step 2:   We insert the variables into the formula and solve for the unknown part.

$P = B \times R$
$P = 23,000 \times .54$ (change percent to decimal)
$P = 12,420$ female students

The base can be divided into more than one part, however, as shown in Figure 2.1: Net sales of $15,000 included a gross profit of $9,000 and a cost of $6,000. Therefore, you must read the problem carefully to determine which part you are required to find.

Example 11 contains the same information as Example 10, except that in Example 11 you are asked to find a different part of the base.

Example 11

Solution

The student body of a university is 23,000, and 54% of it is female. What portion of the student body is male?

Step 1: The first step in solving this problem is to identify the variables provided.

base = 23,000   total student body

rate = 46% (100% − 54%)   percent of male students

part = unknown   number of male students

The rate given in the example refers to the part of the student body that is female. We are asked to find the portion that is male; therefore, we must subtract the rate of those students who are female (54%) from the rate of the base (100%) to find the rate of those students who are male (46%).

Step 2: Now insert the values into the equation and solve for the unknown.

$P = B \times R$

$P = 23,000 \times .46$ (change percent to decimal)

$P = 10,580$ male students

Check:

$P = B \times R$

$10,580 = 23,000 \times .46$

$10,580 = 10,580$

It should be noted that although the part is usually smaller in value than the base, it can also be larger. The numerical value of the part is smaller than the base when the rate is less than 100%, and larger than the base when the rate exceeds 100%. In Examples 12 and 13, notice how the rate is converted to its decimal equivalent to determine the value of the part.

Example 12

Solution

What is 18¼% of 12,000?

$P = B \times R$

$P = 12,000 \times .1825$

$2,190 = 12,000 \times .1825$

The part (2,190) is smaller than the base (12,000).

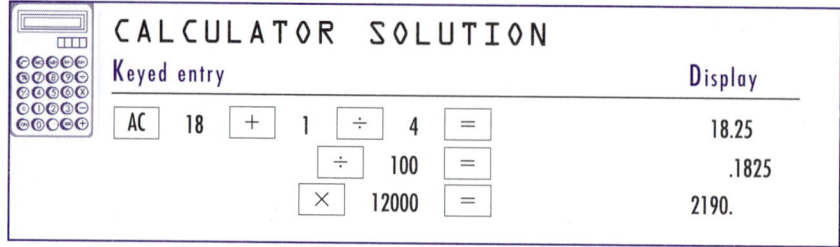

CALCULATOR SOLUTION

Keyed entry   Display

AC	18	+	1	÷	4	=	18.25
				÷	100	=	.1825
				×	12000	=	2190.

Example 13

What number is 115% of 500?

Solution

$$P = B \times R$$
$$P = 500 \times 1.15$$
$$575 = 500 \times 1.15$$

The part (575) is larger than the base (500).

CHECK YOUR KNOWLEDGE

Calculate the Part

1. What is 45% of 300?

2. A worker's gross pay for the week was $500. If 23% was withheld for various payroll deductions, how much did the worker take home?

3. Sales for Rojon Enterprises in September were $250,000. What will be the dollar amount of October's sales if they are estimated to be 120% of September's sales?

4. A local hospital estimates that approximately $5\frac{3}{4}$% of its billed services are uncollectible each year. If the hospital billed patients a total of $3,785,260 for medical services, what amount will be uncollectible?

5. Find the sales tax on an automobile priced at $12,345.00 if the rate of sales tax is $6\frac{1}{4}$% of the sales price.

Computing the Base

Many business transactions provide the given variables rate (R) and part (P). In such problems, the base (B) is the unknown. We can find the base

Answers to CYK: *1.* 135   *2.* $385   *3.* $300,000   *4.* $217,652.45   *5.* $771.56

Learning objective
Use the percent for-
mula to solve for the
base.

(B) by using a *variation* of the basic formula, $P = B \times R$. We can develop the formula by dividing both sides of the equation by R, as explained in Chapter 1.

$$\frac{P}{R} = \frac{B \times \cancel{R}}{\cancel{R}} \quad or \quad \frac{P}{R} = B \quad or \quad P \div R = B$$

Now let's solve a base problem with the formula. We will use the data provided in Example 14.

Example 14

At a local university, 10,580 male students (46% of the student body) are enrolled. What is the total number of students?

Solution 1

Step 1: Identify the given variables.

base = unknown	total number of students
rate = 46% (.46)	percent which are male
part = 10,580	number of male students

Step 2: Insert the variables into the formula and solve. Because the base is the unknown, use the formula variation and divide.

$$B = \frac{P}{R}$$

$$B = \frac{10,580}{.46} \qquad \text{(or } 10,580 \div .46)$$

$$B = 23,000 \text{ students}$$

Check:

$$P = B \times R$$
$$10,580 = 23,000 \times .46$$

In Chapter 1, we learned how to solve for the unknown, which we called x. We also know that we can change the order of numbers when we multiply. To solve percent problems algebraically, we must apply this knowledge to the percent formula. When computing the base, we will substitute the letter B for x as the unknown.

$$P = BR$$

To solve Example 14 with algebra, we will use the same analysis.

Solution 2

Step 1: Identify the given variables.

B = unknown	total student body
R = 46% (.46)	% of students male
P = 10,580	number of male students

Step 2:   Insert variables into equation and solve.

$$P = BR$$
$$10,580 = (B)(.46)$$
$$10,580 = .46B$$
$$\frac{10.580}{.46} = \frac{.46B}{.46}$$
$$23,000 \text{ students} = B$$

We can use algebra to solve all percent problems reducing the number of formulas you must learn to one: the basic equation ($P = BR$).

Example 15

Fred Chang has a new yearly salary that is 8% more than last year's salary. If Fred's new salary is $37,962, what was his salary last year?

Solution

Step 1:   Identify the given variables.

base = unknown   last year's salary
rate = 108% (100% + 8% = 108%)
part = $37,962   current year's salary

Step 2:   Insert the variables into the equation and solve.

$$P = BR$$
$$\$37,962 = (B) \times (1.08)$$
$$\$37,962 = 1.08B$$
$$\frac{37,962}{1.08} = \frac{1.08B}{1.08}$$
$$\$35,150 = B$$

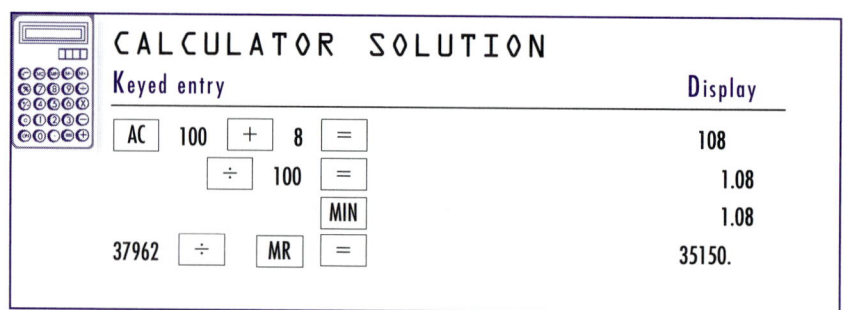

CALCULATOR SOLUTION

Keyed entry	Display
AC 100 + 8 =	108
÷ 100 =	1.08
MIN	1.08
37962 ÷ MR =	35150.

The part ($37,962) is greater than the base ($35,150) because the rate (8%) represents the size of increase to be added to the base amount. In other words, the part (new salary) is 108% of the base (last year's salary).

Check:

$$P = B \times R$$
$$\$37,962 = \$35,150 \times 1.08$$
$$\$37,962 = \$37,962$$

Example 16 illustrates how to calculate the base when the rate given indicates a decrease. In this type of problem, the part (which is given) has a value less than the base (the unknown).

Example 16

Peco Products employed 396 people after implementing a 12% reduction in the work force (a layoff). How many people were employed by Peco before the layoff?

Solution

Step 1:   Identify the given variables.

base = unknown   number of employees before layoff
rate = 88% (100% − 12% = 88%)
part = 396   employees currently employed

Step 2:   Insert the variables into the equation and solve.

$$P = BR$$
$$396 = (B) \times (.88)$$
$$396 = .88B$$
$$\frac{396}{.88} = \frac{.88B}{.88}$$
$$450 = B$$

Check:

$$P = B \times R$$
$$396 = 450 \times .88$$
$$396 = 396$$

The rate 88% (100% − 12%) must be used because the base is unknown and the part (396) represents 88% of the base as shown in the check.

CHECK YOUR KNOWLEDGE

Calculate the Base

1. 640 is 20% of what number?

2. If 12½% of a number is 800, what is the number?

3. 338 is 130% of what number?

4. 80 is 25% more than what number?

5. 266 is 24% less than what number?

6. Stephen Francis earns 10.5% on a certificate of deposit. He received his first annual statement showing interest of $525.00. How much money did he originally invest?

7. An employee has $41.30 withheld for social security. This amount represents 7.51% of total wages for the week. What was the employee's weekly wage?

8. The student tuition of a community college amounted to $2.5 million. If the tuition revenue amounts to 33.3% of the college operating budget for the year, what is the amount of the operating budget? (Round your answer to nearest tenth of a million dollars.)

9. Jill Clark's commissions this year are 20% more than last year's commissions. If her commissions this year are $12,000, what were last year's commissions?

10. The food concession sales at the Fireman's Field Days this year were 4% less than last year. If this year's sales were $15,260, what were last year's sales?

Computing the Rate

Like the base, the rate can be found by using a variation of the basic percent formula. In problems where the rate (R) is the unknown, we will be given the base (B) and part (P). To convert the basic formula $P = B \times R$ into the rate formula, we divide both sides of the equation by B.

Learning objective
Use the percent formula to solve for the rate.

$$\frac{P}{\cancel{B}} = \frac{\cancel{B} \times R}{B} \quad \text{or} \quad \frac{P}{B} = R \quad \text{or} \quad P \div B = R$$

The result of the calculation is a decimal value that must be converted to a percent or decimal percent. For example, .25 is 25% and .075 is 7.5%. (This procedure was explained in Section 2.1.)

Answers to CYK: *1.* 3,200   *2.* 6,400   *3.* 260   *4.* 64   *5.* 350   *6.* $5,000   *7.* 549.93
8. 7.5 million   *9.* $10,000   *10.* $15,895.83

Example 17

A university has a total student body of 23,000, of which 10,580 are male students. What percent of the student body is male?

Solution 1

Step 1:   Identify the given variables.

base = 23,000   total number of students
rate = unknown   % of male students
part = 10,580   number of male students

Step 2:   Insert the variables into the formula and solve.

$$R = \frac{P}{B}$$

$$R = \frac{10,580}{23,000}$$ (or 10,580 ÷ 23,000)

$$R = .46, \text{ or } 46\%$$ Always express the rate as a
                                       percent (%).

Check:

$$P = B \times R$$
$$10,580 = 23,000 \times .46$$
$$10,580 = 10,580$$

Solution 2

Step 1:   Now use algebra to solve Example 17. Identify the given variables.

B = 23,000   total number of students
R = unknown   percent of male students
P = 10,580   number of male students

Step 2:   Insert the variables into the equation.

$$P = BR$$
$$10,580 = (23,000)(R)$$
$$10,580 = 23,000R$$
$$\frac{10,580}{23,000} = \frac{23,000R}{23,000}$$
$$.46, \text{ or } 46\% = R$$

CHECK YOUR KNOWLEDGE

Calculate the Rate

1. 1,500 is what percent of 6,000?

2. What percent is 780 of 600?

3. In a recent basketball game, Matthew Ryan scored 20 points. If he made 8 of 12 field goal attempts and 4 of 9 free throws, what is his rate of accuracy in each category?

4. Rick's Appliance Store reported sales of console TVs at 30 units and portable TVs at 90 units for the month of December. What percentage of total sales is represented by portable TVs?

5. A real estate agent received a sales commission of $5,525 for selling a house to a client for $85,000. What was the rate of commission paid to the real estate agent?

6. Outdoor Sports purchased 40 snowmobiles for the season. If they sold 25 of the snowmobiles, what percent of the snowmobiles remain in inventory?

2.2   EXERCISES

Solve for the part (round answers to the nearest whole cent or hundredth).

1. What is 15% of 500?

2. 35.5% of $1,200 is _____.

3. What is $7\frac{3}{4}$% of $89.45?

4. 125% of 52,785 is _____.

5. 13.2% of 12,856 is _____.

Solve for the base (round answers to the nearest whole cent or hundredth).

6. 30 is 60% of what number?

7. 120% of what number is 19,200?

8. 33.125 is $6\frac{5}{8}$% of what number?

9. 325 is 30% more than what number?

10. 500 is 5% of what number?

Solve for the rate (round answers to the nearest tenth of a percent).

11. 45 is what percent of 150?

12. What percent is 924 of 560?

13. 1,200 is what percent of 1,800?

14. 15.4 is what percent of 48?

15. .33 is what percent of 3?

Solve the following problems and express your answers to the nearest tenth of a percent, nearest whole cent, or nearest whole number.

Answers to CYK: *1.* 25%   *2.* 130%   *3.* $66\frac{2}{3}$%; 44.4%   *4.* 75%   *5.* 6.5%   *6.* 37.5%

Supply the missing information in the table.

Part	Base	Rate
16. _____	1,500	20%
17. 240	_____	30%
18. $3,267.00	$9,900	_____
19. $2.38	_____	.7%
20. _____	3,000	1/5 %
21. $817.50	$6,540	_____
22. $58.50	_____	130%
23. 15	60	_____
24. _____	$640	.8%
25. $15.60	$240	_____

Calculate the base, rate, or part.

26. What percent of 336 is 112?

27. 425 is 12.5% of what number?

28. 7.9% of $6,379 equals what number?

29. 15% of what number is 120?

30. .2 is what percent of .8?

31. What number is 12¾% of $35,000?

32. 6 is ½% of what number?

33. What percent of .25 is 4?

34. 150% of what number is 1,275?

35. 1,500% of .5 is what number?

Solve problems 36–45 using the percentage equation. Round your answers to the nearest tenth of a percent or nearest whole number.

36. Bill's Sub Shop reported profits of $12,000 in 1999. What will the profit be for 2000 if profits are estimated to be 130% of the 1999 profits?

37. A car dealer sold 125 cars last month to people entering his showroom. How many customers visited the dealership if the 125 who purchased cars represent 2½% of those who visited the showroom?

38. If an investment of $12,000 earned $1,260 in annual interest, what was the rate of interest earned?

39. Ann Randall's annual salary was $15,000 before she received a promotion to head teller. The bank has a policy of allowing employees to select a minimum dollar increase of $1,000 or 6% of their annual salary. Which type of raise would Ann have selected? Why?

40. A real estate salesperson receives 3½% commission on gross sales per month. What would be the amount of commission if the salesperson sold three properties during August for the following amounts: $54,900, $89,000, $70,600?

41. Sally Zimmerman sells automobiles for Eastside Imports. Sales records show that she sold 15% more cars in June than she sold in May. If Sally sold 46 cars in June, how many cars did she sell in May?

42. An inspector rejected 15 units due to defective assembly, which represented 1½% of the total day's production. How many units passed inspection?

43. Kyle Bush sold his 4-year-old automobile for $5,550, which is 62.5% of what he paid for it. How much did Kyle pay for the car?

44. Mrs. Adams, Miss Ramìrez, and Ms. Elliott are business partners. Mrs. Adams invested $5,000, which represented 12% of the total investment. How much money was invested by Miss Ramìrez and Ms. Elliott if they invested 38% and 50%, respectively?

45. A house contains 1,800 square feet of living space. If the family room measures 12 feet by 20 feet, what percent of the total living space is devoted to other living areas?

2.3   COMPUTING THE RATE OF INCREASE OR DECREASE

In the preceding section, we discussed how to calculate the rate. The rate (%) is an important measure of performance because it can be used to reflect the rate of increase or decrease in just about every type of business

Learning objective
Calculate the rate of
increase and the rate
of decrease.

activity. To calculate the rate of increase or decrease, we use the rate formula $R = P/B$. However, the variables must be adapted in the following manner:

Calculating the rate of increase or decrease

$$\text{rate (increase or decrease)} = \frac{\text{part (difference)}}{\text{base (original value)}}$$

In this formula, the **difference** is found by subtracting the **original value** from the **current value.** The original value is always the base; when divided into the difference, the base produces the rate of change. If the current value is greater than the original value (positive difference), the rate reflects an *increase.* A *decrease* results when the current value is less than the original value (negative difference). When the rate is converted to a percent, we refer to the change as a **percent increase** or a **percent decrease.**

The following example illustrates the logic in calculating the rate of increase or decrease.

Example 18

Country Pride Foodservice reported profits for the first and second quarters of 1998 as $15,000 and $12,000, respectively. What was the percent of change if profits for the same quarters of 1999 were $18,000 and $10,000, respectively?

Solution

First Quarter

Step 1: Identify the given variables.

base = $15,000	earlier period profits
rate = unknown	percent of change
part = $3,000	the difference in profits between the two periods

$$\$18,000 \quad - \quad \$15,000 \quad = \quad \$3,000$$

current value   original value   difference

Step 2: Insert variables into the rate formula.

$$\text{rate (increase)} = \frac{\text{part (difference)}}{\text{base (original value)}}$$

$$R = \frac{\$3,000}{\$15,000}$$

$$R = .20, \text{ or } 20\% \text{ increase}$$

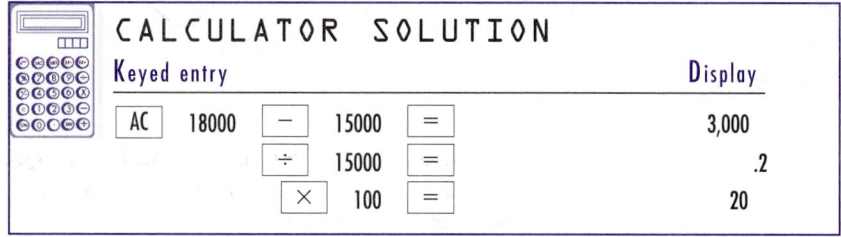

CALCULATOR SOLUTION				
Keyed entry				Display
AC 18000	−	15000	=	3,000
	÷	15000	=	.2
	×	100	=	20

Rate of change problems can also be solved with the percent formula by adapting the variables, as illustrated in the solution for the second quarter of Example 18.

Second Quarter

Step 1:   Identify the given variables.

base = $12,000	earlier period profits
rate = unknown	percent of change
part = −$2,000	the difference in profits between the two periods

$$\begin{array}{ccccc} \$10,000 & - & \$12,000 & = & -\$2,000 \\ \uparrow & & \uparrow & & \uparrow \\ \text{current value} & & \text{original value} & & \text{difference} \end{array}$$

Step 2:   Insert variables into the percent formula.

$$P = BR$$

$$(\$10,000 - \$12,000) = (\$12,000)(R)$$

$$-\$2,000 = \$12,000R$$

$$\frac{-\$2,000}{\$12,000} = \frac{\$12,000R}{\$12,000}$$

$$-.1666 = R \text{ or } 16.7\% \text{ decrease}$$

```
┌──────────────────────────────────────────────────────────────────┐
│ ▭▭▭   CALCULATOR   SOLUTION                                        │
│ ◎◎◎◎◎                                                              │
│ ◎◎◎◎◎  Keyed entry                                      Display    │
│ ◎◎◎◎◎ ─────────────────────────────────────────────────────────── │
│ ◎◎◎◎◎   ┌──┐        ┌─┐         ┌─┐                                │
│ ◎◎◎◎◎   │AC│ 10000  │─│ 12000   │=│              −2000.            │
│         └──┘        └─┘         └─┘                                │
│                     ┌─┐         ┌─┐                                │
│                     │÷│ 12000   │=│                    −.166666    │
│                     └─┘         └─┘                                │
│                     ┌─┐         ┌─┐                                │
│                     │×│   100   │=│                   −16.6666     │
│                     └─┘         └─┘                                │
└──────────────────────────────────────────────────────────────────┘
```

Notice that the answer in this solution was rounded to the nearest tenth of a percent (.1666 becomes 16.66%, rounded to 16.7%). Unless otherwise indicated, apply the following rules when rounding your answers to percentage problems:

1. When an answer is expressed in monetary form, round to the nearest cent.

2. When an answer is expressed in basic number form (decimal), round to the nearest tenth.

If you do not understand the rounding of values explained in this presentation, you should refer to Appendix A before proceeding to the next section.

Example 19

The Grinding Department of Santos Manufacturing Company produced 40,000 units of part no. 1385 during the month of May. If the total units produced in May represent an increase of 5,000 units more than were produced in April, what was the rate of increase for the period?

Solution

Step 1: Identify the given variables.

$$40,000 \quad - \quad 5,000 \quad = \quad 35,000$$

current value   difference   original value

base = 35,000   April's production

rate = unknown   percent of change

part = 5,000 units   the difference in production
                       between the two periods

Step 2: Insert the variables into the rate formula.

$$rate\ (increase) = \frac{part\ (difference)}{base\ (original\ value)}$$

$$R = \frac{5,000}{35,000}$$

$$R = .1428,\ or\ a\ 14.3\%\ increase$$

CHECK YOUR KNOWLEDGE

Calculate the Rate of Increase or Decrease

1. If you reported taxable income of $15,000 in 1998 and $17,000 in 1999, what was the percent change in your taxable earnings?

2. A coat that you have been interested in buying has been advertised for $79.00. The advertised price is $19.95 less than the original price. What percent of savings will you realize if you purchase the coat during the sale?

3. Last month, the average price for unleaded gasoline was $1.25\%/10$. This month the average price dropped to $1.19\%/10$. What was the percent of change during this period?

4. If the Miller Paper Company reported net sales of $142,600 this year, which was $15,800 less than net sales for the same period last year, what was the percent change in sales for the period?

5. A student received a grade of 65 on a recent math exam. On a makeup exam, the student achieved a score of 90. Calculate the percent of change in test scores.

6. Last month the number of unemployed workers in the central region of the state was 18,500. This month the number of unemployed workers was reported to be 20,750. What was the percent of change in unemployment during this period?

Price Indexing—Applying Percent of Change

Learning objective
Develop a price index and use the Consumer Price Index to calculate the rate of inflation.

Price indexes provide businesses and individuals with information necessary to measure the impact of inflation on the purchasing power of the dollars they spend for goods and services. For example, a business may determine the percent increase or decrease in the price it charges for goods or the wages to be paid to employees based on changes in price indexes. You could

Answers to CYK:							
1.	13.3% increase	**2.**	20.2% decrease	**3.**	4.8% decrease	**4.**	10% decrease
5.	38.5% increase	**6.**	12.2% increase					

use a price index to measure the extent your purchasing power improves or diminishes each year by comparing your change in income to the change in the index.

To develop a price index for a single item (often referred to as a **price relative**) we select a price and time as the base and compare any price change to the base year price. An index is a percent and is calculated by dividing the current year price by the base year price and multiplying by 100. Although indexes are actually percents, the percent sign is normally omitted.

$$\frac{\text{price index}}{\text{number}} = \frac{\text{current period price}}{\text{base period price}} \times 100$$

Example 20

Solution

Develop a price index for a gallon of regular gasoline that had an average price of $1.159 in 1990, $1.189 in 1991, $1.229 in 1992, $1.249 in 1993, $1.289 in 1994, $1.239 in 1995, $1.259 in 1996, $1.329 in 1997, $1.219 in 1998, and $1.269 in 1999. Use 1990 as the base period price.

Year	Price	(Calculation)	Index
1990	$1.159	(1.159 ÷ 1.159)100	100.0
1991	1.189	(1.189 ÷ 1.159)100	102.6
1992	1.229	(1.229 ÷ 1.159)100	106.0
1993	1.249	(1.249 ÷ 1.159)100	107.8
1994	1.289	(1.289 ÷ 1.159)100	111.2
1995	1.239	(1.239 ÷ 1.159)100	106.9
1996	1.259	(1.259 ÷ 1.159)100	108.6
1997	1.329	(1.329 ÷ 1.159)100	114.7
1998	1.219	(1.219 ÷ 1.159)100	105.2
1999	1.269	(1.269 ÷ 1.159)100	109.5

There are many different types of indexes; however, the most widely used index is the **Consumer Price Index (CPI).** The U.S. Bureau of Labor Statistics (BLS) began to publish complete indexes in 1919 but used 1967 as the reference base year to report the rate of inflation until 1982. The Bureau selected 1982–84 = 100 as the new base year to reflect the updated expenditure weights developed from consumer expenditure surveys. The current Consumer Price Index is based on 1982–84 prices of food, clothing, shelter, fuels, transportation fares, and medical charges purchased for day-to-day living. To calculate the index, each item is assigned a weight based on its importance in consumer budgets. Price changes for these items are compared monthly and annually to the 1982–84 prices of these same items to update the CPI. Indexes for these goods and services are prepared for the United States as a whole and for selected cities and urban areas as illustrated in Tables 2.2 and 2.3, respectively.

Using the data provided in the Consumer Price Index (see Table 2.2) we can find the rate of inflation between 1982 and 1997 by subtracting the index for the base year (1982) from the 1997 index (166.0 − 100 = 66.0). This means that prices have increased 66.0% between 1982 and 1997. Remember an index number is a percent, but the percent sign is omitted in the index. We can convert the index to monetary terms by moving the decimal point two places to the left. (166.0 ÷ 100) = 1.66) Therefore, we had to pay $1.66 to purchase the same quantity and value of goods in 1997 that would have cost us $1.00 in 1982.

If we want to use the CPI to compare the price changes for any two of the years presented in the index, we use the percent of increase or decrease procedure presented in the preceding section of this chapter. First, we find the difference between the two index values; then we divide the difference by the index value of the earlier year.

Example 21

Verify the rate of inflation between 1994 and 1995 in the Consumer Price Index (Table 2.2).

Solution

Step 1:   Identify the given variables.

base = 153.4	earlier period index
rate = unknown	percent of inflation
part = 4.3	difference in indexes between the two periods (current period − earlier period = difference)

157.7 − 153.4 = 4.3

Step 2:   Insert the variables into the formula and solve.

$$R = \frac{\text{part (difference)}}{\text{base (earlier period value)}}$$

$$R = \frac{4.3}{153.4}$$

$$R = .0280312 \text{ or } 2.8\%$$

Table 2.2

Consumer Price
Index

Year	Index	% Inflation
1982	100.0	6.2
1983	103.2	3.2
1984	107.6	4.3
1985	111.5	3.6
1986	113.6	1.9
1987	117.7	3.6
1988	122.5	4.1
1989	128.4	4.8
1990	135.3	5.4
1991	140.9	4.2
1992	145.3	3.0
1993	149.5	3.0
1994	153.4	2.6
1995	157.7	2.8
1996	162.3	2.9
1997	166.0	2.3
1998	168.7	1.6

*All figures (1982–84)

A 2.8% increase in inflation suggests that a person would need to have had at least a 2.8% wage increase in 1995 to maintain the same purchasing power he or she had in 1994. Any percent less would have resulted in a loss of purchasing power; any percent greater would have indicated a gain in purchasing power.

Example 22 illustrates another useful application of the CPI.

Example 22

If the CPI for college tuition was 119.9 in 1989 compared to a CPI of 264.8 in 1999, how much would it cost in 1999 for college tuition that would have cost $5,000 in 1989?

Solution

Step 1:   Identify the given variables.

base = 119.9   earlier period index

rate = unknown   rate of inflation for the period

part = 144.9   difference in indexes between two periods

Step 2:   Insert the variables into the formula and solve.

$$R = \frac{\text{part (difference)}}{\text{base (earlier period index)}}$$

$$R = \frac{144.9}{119.9}$$

$R = 1.208507$ or 120.9%

Step 3:   Multiply the earlier cost (base) by the rate of increase plus 1 to find the current cost.

$P = B \times R$

$P = \$5,000 \times 2.209$ (1.00 + 1.209)

$P = \$11,045.00$

Because the part is greater than the base, we add the percent of increase (120.9%) to the base percent (100%) to accurately state the rate (220.9%). The part (\$11,045.00) then represents 220.9% of the base.

The price index numbers in Table 2.3 can be used to compare prices for a city. The index numbers however, cannot be used to determine price comparisons between cities. For example, the cost of "all items" in New York in 1995 was 158.3% of what they were in 1982–84. The cost of "all items" in Los Angeles in 1995 was 149.4% of what they were in 1982–84. Therefore, the data does not suggest that "all items" in New York cost more than in Los Angeles. What the data does say is, the items prices have increased at a more rapid rate in New York than in Los Angeles. If we use Table 2.3 to calculate the change in price of any item, we would need to know the base year price of the item, the time period of the comparison, and the corresponding price index number. Example 23 illustrates how to use the Consumer Price Index for selected urban areas to calculate price changes.

Table 2.3

Consumer Price Index—Selected areas: 1995

Item	Boston	Chicago	Los Angeles	New York	Miami
All items	157.4	148.4	149.4	158.3	146.9
Food and beverages	152.1	151.6	153.5	155.3	155.7
Housing	150.0	141.4	146.4	158.0	139.2
Apparel and upkeep	150.7	125.1	126.8	121.8	149.4
Transportation	140.1	133.5	138.4	146.0	138.9
Medical care	259.9	227.0	218.8	227.6	200.0
Entertainment	169.9	160.1	142.1	158.1	141.9

*All figures (1982–84 = 100)

Example 23

In 1982, a movie theater ticket cost \$3.75 in New York City. Determine the price of the same ticket in 1995 (use Table 2.3).

Solution

Step 1: Identify the given variables.

base = \$3.75	1982 ticket price
rate = 158.1	price index number
part = unknown	1995 ticket price

Step 2: Insert variables into formula and solve.

$P = B \times R$

$P = \$3.75(1.581)$

$P = \$5.92875$

$P = \$5.93,$ or $\$6.00$ (rounded to the nearest dollar)

For Your Information

INDEXING: MEASURING THE IMPACT OF CHANGE

Business Week Index

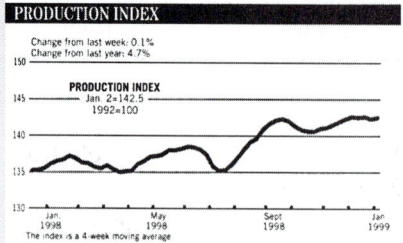

PRODUCTION INDEX

Change from last week: 0.1%
Change from last year: 4.7%

PRODUCTION INDEX
Jan. 2=142.5
1992=100

Jan. 1998   May 1998   Sept 1998   Jan 1999

The index is a 4-week moving average

The production index rose in the week ended Jan. 2. The unaveraged index was also up, to 143.2, from 141.8 in the previous week. The monthly index for December increased 0.1%, to 142.4, from 142.3 in November. After seasonal adjustment, production of oil, coal, lumber, and rail-freight traffic posted healthy gains considering that this was a holiday week. Only electric power output was down. Because of holiday plant closings, there was no output of autos and trucks this week.

BW production index copyright 1999 by The McGraw-Hill Companies.

LEADING INDICATORS

	LATEST WEEK	WEEK AGO	YEARLY % CHG
STOCK PRICES (1/8) S&P 500	1275.09	1229.23	37.4
CORPORATE BOND YIELD, Aaa (1/8)	6.34%	6.23%	-3.2
MONEY SUPPLY, M2 (12/28) billions	\$4,433.9	\$4,424.1r	9.1
INITIAL CLAIMS, UNEMPLOYMENT (1/1) thous.	350	372r	6.4
MORTGAGE APPLICATIONS, PURCHASE (1/8)	254.8	249.0	-4.2
MORTGAGE APPLICATIONS, REFINANCE (1/8)	1,494.2	1,468.5	-18.9

Sources: Standard & Poor's, Moody's, Federal Reserve, Labor Dept., Mortgage Bankers Assn. (Index: March 16, 1990=100)

INTEREST RATES

	LATEST WEEK	WEEK AGO	YEAR AGO
FEDERAL FUNDS (1/12)	4.78%	4.54%	5.42%
COMMERCIAL PAPER (1/12) 3-month	4.79	4.80	5.37
CERTIFICATES OF DEPOSIT (1/13) 3-month	4.91	4.93	5.53
FIXED MORTGAGE (1/8) 30-year	6.89	6.99	7.08
ADJUSTABLE MORTGAGE (1/8) one-year	5.73	5.75	5.68
PRIME (1/8)	7.75	7.75	8.50

Sources: Federal Reserve, HSH Associates, Bloomberg Financial Markets

PRODUCTION INDICATORS

	LATEST WEEK	WEEK AGO	YEARLY % CHG
STEEL (1/9) thous. of net tons	1,843	1,787#	-17.2
AUTOS (1/9) units	110,410	0#	6.9
TRUCKS (1/9) units	138,121	0#	22.3
ELECTRIC POWER (1/9) millions of kilowatt-hrs.	70,366	64,171#	13.8
CRUDE-OIL REFINING (1/9) thous. of bbl./day	15,171	15,350#	2.0
COAL (1/2) thous. of net tons	18,852#	15,104	-6.0
LUMBER (1/2) millions of ft.	263.5#	245.8	-6.8
RAIL FREIGHT (1/2) billions of ton-miles	21.5#	19.2	-6.5

Sources: American Iron & Steel Institute, Ward's Automotive Reports, Edison Electric Institute, American Petroleum Institute, Energy Dept., WWPA1, SFPA2, Association of American Railroads.

PRICES

	LATEST WEEK	WEEK AGO	YEARLY % CHG
GOLD (1/13) \$/troy oz.	285.450	287.650	1.0
STEEL SCRAP (1/12) #1 heavy, \$/ton	84.00	84.00	-41.5
COPPER (1/8) ¢/lb.	68.2	69.6	-12.8
ALUMINUM (1/8) ¢/lb.	59.0	60.0	-18.4
COTTON (1/8) strict low middling 1-1/16 in., ¢/lb.	57.57	58.47	-9.9
OIL (1/12) \$/bbl.	12.72	11.85	-21.7
CRB FOODSTUFFS (1/12) 1967=100	207.98	201.63	-8.1
CRB RAW INDUSTRIALS (1/12) 1967=100	263.91	263.47	-10.7

Sources: London Wednesday final setting, Chicago market, Metals Week, Memphis market, NYMEX, Commodity Research Bureau.

FOREIGN EXCHANGE

	LATEST WEEK	WEEK AGO	YEAR AGO
BRAZILIAN REAL (1/13)	1.3100	1.2089	1.1184
BRITISH POUND (1/13)	1.65	1.65	1.63
CANADIAN DOLLAR (1/13)	1.53	1.51	1.43
EUROPEAN EURO (1/13)	1.1672	1.1602	NA
JAPANESE YEN (1/13)	113.57	113.15	131.19
KOREAN WON (1/13)	1174.0	1158.5	1671.0
MEXICAN PESO (1/13)3	10.510	9.770	8.178
TRADE-WEIGHTED DOLLAR INDEX (1/13)	105.4	104.6	110.9

Sources: Major New York banks. Currencies expressed in units per U.S. dollar, except for British pound in dollars. Trade weighted dollar via J.P. Morgan.

#Raw data in the production indicators are seasonally adjusted in computing the BW index (chart); other components (estimated and not listed) include machinery and defense equipment. 1=Western Wood Products Assn. 2=Southern Forest Products Assn. 3=Free market value NA=Not available r=revised NM=Not meaningful

Source: Reprinted from the January 25, 1999, issue of *Business Week* by special permission, copyright ©1999 by The McGraw-Hill Companies, Inc.

Every week McGraw-Hill publishes this index in its periodical, *Business Week,* to provide important financial information to its readers. Each section of the index reports current prices or rates and past prices or rates to illustrate the degree of change for each item of the index. Some of the sections also report the yearly percent change (rate of change) in prices or rates such as, the Production Index, Production Indicators, Leading Indicators, and Prices. The Production Index graph illustrates changes in industrial production for a one-year period. As shown, the index rose from 137.8 in January 1998 to 142.5 in January 1999, for a cumulative net increase of 4.7% for the period. This index is important because there is a reasonably high correlation between production and consumer demand for goods and services.

Source: Business Week, January 25, 1999.

CHECK YOUR KNOWLEDGE

Price Indexes

1. Calculate the Price Index on a textbook that sold for $57 in 1998 compared to $32 in 1994.

2. What is the rate of inflation of an item with a CPI of 120.8 in 1990 and a CPI of 238.4 in 1998?

3. Compute the yearly index and inflation rate of a half-gallon of milk for each of the following years:

Year	Price	Index	Inflation Rate
1996	$1.10	_____	_____
1997	1.18	_____	_____
1998	1.24	_____	_____
1999	1.39	_____	_____

4. The price index for a 19-inch color television set was 172.6 in 1990 and 365.8 in 1999. How much would a comparable set cost in 1999, if the 1990 price was $129?

Answers to CYK: *1.* 178.1%   *2.* 97.4%   *3.* 1996 = 100.0, 0%; 1997 = 107.3, 7.3%; 1998 = 112.7, 5.1%; 1999 = 126.4, 12.1%   *4.* $273.35   *5.* 26.2% decrease   *6.* $224.10

5. If the price index for a gallon of unleaded gasoline was 162.9 in July 1997 compared to 120.3 in July 1998, what was the rate of inflation for the period?

6. A man's suit cost $150 in 1982 in Miami, Florida. What was the price of the suit in 1995 dollars? (Use Table 2.3.)

2.3 EXERCISES

For problems 1 through 5, calculate the difference and the percent increase or decrease from original to current amounts. Round your answers to the nearest tenth percent.

Item	Original amount	Current amount
1. Gasoline (per gal.)	$1.31⁹/10	$.89⁹/10
2. Automobile	$8,399	$8,785
3. Student enrollment	6,480	7,695
4. Personal computer	$1,499	$1,029
5. Fixed-interest rate	12.95	9.50

For problems 6 through 10, provide the missing information where indicated. Round answers to nearest tenth percent, whole cent, or number.

Item	Original amount	Current amount	Amount of difference	Percent change
6. Sport coat	$79.00	_____	−$20.00	_____
7. Common stock share	_____	67⅞	2¼	_____
8. Net profit	$128,600.00	_____	_____	12.8%
9. Units of production	250,000	233,750	_____	_____
10. Hourly wage	$10.50	_____	_____	−3%

Using the following prices, calculate the Price Index for each year. Round your answers to tenths.

Year	Price	Index
11. 1995	17.00	_____
12. 1996	21.50	_____
13. 1997	18.75	_____
14. 1998	20.45	_____
15. 1999	24.00	_____

For problems 16 through 20, use the Consumer Price Index numbers in Table 2.3 to determine the price of the following items by selected urban area.

	Urban area	Consumer item	1982 price	Current price
16.	Boston	Transportation	$20,000	_____
17.	Chicago	Food and beverage	45	_____
18.	Los Angeles	Entertainment	75	_____
19.	Miami	Medical	500	_____
20.	New York	Housing	95,000	_____

Solve the following percent increase or decrease problems and round your answers to the nearest tenth of a percent.

21. Maria Torres's annual salary was $15,000 before she received a promotion to head teller. She now receives an annual salary of $16,500. What is the rate of change in her salary?

22. Last season, Silver Bluffs Ski Association was open for business a total of 50 days. This season, weather conditions were more favorable, which permitted the association to offer 65 days of skiing. What is the percent change in the number of days Silver Bluffs was open for business?

23. The property tax in Richland County increased this past year to $180.00 per $1,000 of assessed evaluation. What is the percent change if last year's rate was $160.00?

24. Agri-Mart sold 340 lawnmowers this year, which was 60 more than they sold last year during the same period. What is the percent change in sales of mowers for the period?

25. Last year, the Dudick family received utility bills amounting to $1,345.00, which included $960 for natural gas to heat their home. This year the Dudick's heating costs decreased to $840 after their home was insulated. What was the percent change in heating costs?

26. The Allied Manufacturing Company recently reduced its work force from 5,200 to 3,700. Allied intends to recall 500 of the laid-off workers over the next 2 weeks. What is the percent change in the size of the work force as a result of these decisions?

27. Over the past 2 years, the liability insurance premiums for the West End Football Association increased from $1,600 per year to $2,400 per year. What was the percent change in insurance premiums during the 2-year period?

28. Bart's Service Station sold 3,200 fewer gallons of gasoline during the month of December than it sold in November. If gasoline sales for December were 62,845 gallons, what was the rate of change in gasoline sales for the month?

29. A pair of sneakers cost $60.00. This same pair of sneakers sold for $25.00 5 years ago. What was the rate of inflation over the 5-year period?

30. The CPI for 1995 was 157.7, compared with a CPI of 135.3 for 1990. What was (a) the rate of inflation during the 5-year period, (b) the cost of an item in 1995 if that same item cost $45.00 in 1990?

31. The price of a quart of motor oil in 1990 was $1.39. Determine the price index for the oil if the 1998 price was $2.19.

32. Use the CPI (Table 2.2) to determine the price of an item in 1996 that sold for $15.00 in 1982.

33. An airline ticket from Boston to Miami was priced at $349 in 1982. What was the price of the ticket in 1995? (Use Table 2.3.)

34. In 1985 the CPI for all items purchased in Denver Colorado was 107.1. The CPI for all items in 1997 increased to 149.6. What was the 1997 price of a refrigerator, if the 1985 price was $359?

35. The price index for cable television service is currently 268.4. What is the current monthly charge, if the 1982 price for cable service was $12.50?

CHAPTER REVIEW EXERCISES

The following problems review all mathematics and business concepts presented in this chapter. If you experience any difficulty solving a problem, go to the appropriate chapter section and review the examples provided. Express all answers as decimals rounded to tenths, or dollars rounded to cents, unless otherwise instructed.

1. Sally Matthews bought an automobile that was priced at $18,000. If she made a downpayment of 15% and financed the balance at $8\frac{1}{2}\%$, what was the amount of the downpayment?

2. A gourmet blend of coffee contains the following amounts of each blend: $12\frac{7}{8}$ pounds Brazilian; $40\frac{2}{5}$ pounds Colombian; and $10\frac{3}{4}$ pounds Mexican. What percent of the blend was made from Colombian coffee?

3. If Alex Kowalski received a $2,466.75 commission for selling a house, and the commission rate was $3\frac{1}{4}\%$ of the selling price, what was the selling price?

4. A sporting goods store sold a set of golf clubs for 40% more than their cost. How much was the selling price, if the clubs cost the store $125?

5. Suburban Lawn and Garden purchased 130 lawnmowers in March. If they sold 80 of the lawnmowers by July, what percent of the lawnmowers remain unsold?

6.. Susan Steinbaugh owns 150 shares of stock. This year she received a dividend of $1.35 per share, which represents a $5\frac{3}{8}\%$ return on her investment. What was the total value of her shares of stock at the time the dividend was paid?

7. A homeowner pays school taxes based on the assessed value of her home. If her home is assessed for $80,000 and her school taxes are $3,500, what is the school tax rate?

8. A business decreased its operating expenses this year by 10%. If this year's operating expenses are $30,800, what were last year's operating expenses?

9. The current price of a health club membership is 12% more than last year's price. If last year's price was $175, what is the membership price this year?

10. On May 1, International Equipment Inc. stock was listed on the New York Stock Exchange at

$92\frac{3}{8}$. Three months later, the stock was listed at $104\frac{1}{2}$. What was the percent change in the stock's listed price for the period?

11. Carey Transportation Ltd. reduced its work force to 1,675 people, which is 430 people less than it employed last year at the end of the second quarter. Find the percent change in personnel.

12. The spring semester student enrollment at Empire Community College is 6,845. If the fall semester enrollment was 6,460, what was the percent change in student enrollment for the academic year?

13. A pair of walking sneakers is priced at $79.99. The price is $20.99 more than the original cost of the sneakers. Find the percent change in cost to price.

14. Determine the price index for a CD that sold for $14.99 in 1998, compared to $12.99 in 1996.

15. The average starting salary for an elementary school teacher in the northeast in 1995 was $21,500. Calculate the price index if the salary for the same position in 1998 was $27,900.

16. Calculate the rate of inflation of an item with a CPI of 148.9 in 1992 and a CPI of 157.4 in 1998.

17. The Consumer Price Index for a new automobile in 1985 was 106.1 and 141.0 in 1995. How much would a new automobile cost in 1995 if the 1985 price was $8,500?

18. Matthew Geluso's hourly wage in 1982 was $5.25. What would his hourly wage have been in 1995 to have the same purchase power with his 1995 wages that he had with his 1982 wages? (Use Table 2.2.)

19. A specific prescription drug cost $7.25 in Chicago in 1982. Determine the cost of the same prescription drug in Chicago in 1995. (Use Table 2.3.)

20. Carol Moses purchased a house in Boston for $125,000 in 1985. She sold the house in 1992. How much did she sell the house for if she based the selling price on the 1985 cost and the change in the Consumer Price Index? (Use Table 2.2.)

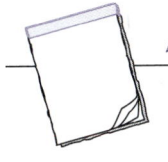

EXPRESS YOUR UNDERSTANDING

Compose one or two well-written sentences to express the requested information.

1. Explain in words the process required to convert a fraction to its percent equivalent.

2. Develop a step-by-step explanation of how to convert 62.5% to a fraction reduced to lowest terms.

3. Identify and briefly define the three key variables of the percent formula.

4. If a percentage problem contained the numbers 20 and 80, and you were asked to find the rate, how would you determine which of those two numbers is the base?

5. Explain how you would solve the following:

15 is 25% of what number?

6. Develop a step-by-step procedure to determine which variables are given and which variable is missing in the following:

What number is 120% of 400?

7. Explain in detail the procedure required to calculate the rate of increase or decrease.

8. Describe how you would use the Consumer Price Index to determine the rate of inflation between 1982 and 1990.

9. Explain how to convert an index of 298.4 to monetary terms.

10. Describe how you would develop a price index for an item that sold for $1.25 in 1996, $1.32 in 1997, and $1.44 in 1998.

SELF-TEST

A. Terminology

Complete the following items using the key terms presented at the beginning of the chapter. Check your responses against the answer key at the end of the test.

1. When the numerator of a fraction is divided by the denominator, the result is called the _____ of the fraction.

2. The _____ is the percent of the base to be calculated.

3. We use the terms *whole* and *total* when we are referring to the _____.

4. The _____ is a portion of the base and is always expressed as a number.

5. The expression $P = B \times R$ is referred to as the _____.

6. The rate is also referred to as a _____ of the base.

7. When the current value is larger than the original value, the rate will be stated as a _____.

8. A _____ results when the current value is less than the original value.

9. We are solving for the _____ when we divide the part by the rate.

10. The _____ is found by dividing the part by the base.

11. The difference in a percent-of-change problem is found by subtracting the _____ value from the _____ value.

B. Calculation

The following concepts and short problems are designed to test your understanding of the objectives identified at the beginning of the chapter. Answers are provided at the end of the test. Solve for each.

12. A community college employs 630 people. If 30% of the employees are faculty, how many faculty members are employed by the college?

13. If you are required to make a down payment of 5% on the purchase of a new home that amounts to $4,250 of the selling price, what is the selling price?

14. The budget of the Reddy Corporation consists of the following items and respective amounts: salaries $250,000, equipment $85,000, supplies $12,000, administrative expenses $25,000, maintenance $10,000, advertising $18,000, and production material inventory $600,000. What percent of the budget is allocated for salaries, supplies, advertising expense, and equipment, collectively?

15. If you earned $30,000 this year, which represents a 6% increase over last year's annual income, what was your annual income last year?

16. If a manufacturer sells an item for $40 per unit, which includes a unit labor cost of $8, what percent of the unit price is the labor cost?

17. Find the percent change in sales for Lakeside Marina if it sold six more Lazer sailboats this year than last year. Last year, the marina sold 24 sailboats.

18. If an employee's absences last year accounted for 4,000 lost work-hours compared to 5,500 lost work-hours this year, what was the percent change?

19. Last month you paid $.89 for a head of lettuce at the grocery store. This month the selling price for a head of lettuce is $.69. What is the rate of inflation?

20. If sales are $1,250,000 after a 15% decrease, what were sales before the decrease?

21. A hamburger at Art's Diner cost 75 cents in 1985. Art adjusts his prices based on the CPI. If the CPI was 161.2 in 1995, how much did Art charge for a hamburger in 1995 with a CPI of 322.2?

22. Find the price index for a pound of grapes that sold for $1.39 in September compared to $1.99 in December.

23. If a pocket calculator cost $8.50 in 1982, what would the same model calculator cost in 1997 based on the change in the CPI? (Use Table 2.2.)

24. The price of a six-pack of soda was $2.49 in 1998. Determine the price index for the soda if the 1995 price was $1.79.

25. A one-way bus ticket from New York City to Buffalo, New York, cost $39 in 1982. What was the cost of the ticket in 1995? (Use Table 2.3.)

Answers to Self-Test: *1.* decimal equivalent   *2.* rate   *3.* base   *4.* part   *5.* percent equation   *6.* percent   *7.* percent increase   *8.* percent decrease   *9.* base   *10.* rate   *11.* original, current   *12.* 189   *13.* house cost = $85,000; salary = $28,333.33   *14.* 36.5%   *15.* $28,301.89   *16.* 20%   *17.* 25%   *18.* 37.5%   *19.* −22.47%   *20.* 1,470,588.20   *21.* $1.50   *22.* 143.2   *23.* 14.11   *24.* 139.1   *25.* $56.94

3

Key Terms

data
statistics
stem and leaf display
frequency distribution
relative frequency
    distribution
percent frequency
    distribution
bar graph
line graph
circle graph
pie chart
numerical summary
    measures
measures of central
    tendency
mean
trimmed mean
median
mode
weighted mean
measures of dispersion
range
variance
standard deviation
inference
probability
outcome
chance

BUSINESS STATISTICS AND GRAPHS

Learning Objectives

After completing the exercises in this chapter you will be able to:

1. Form a relative frequency distribution from a frequency distribution.

2. Form a percent frequency distribution from a relative frequency distribution.

3. Draw a bar graph using the information from a frequency distribution.

4. Draw a percent bar graph using the information from a percent frequency distribution.

5. Draw a line graph using the information from a frequency distribution.

6. Draw a percent line graph using the information from a percent frequency distribution.

7. Draw a pie graph using the information from a percent frequency distribution.

8. Calculate the mean and trimmed mean from a set of data.

9. Calculate the median from a set of data.

10. Determine the mode of a set of data.

11. Calculate the weighted mean from a set of grouped data.

12. Calculate the range, variance, and standard deviation for a given set of data.

13. Calculate the probability that an event will occur.

14. Define the key terms.

INTRODUCTION

In the business world we are often confronted with data about our particular business, and we are expected to use that data to make business decisions. **Data** are information—for our use here, they are numerical information or numerical data. Such numerical data might indicate to us retail cost, wholesale cost, number of units of a product that sold in a week, number of women who bought a particular item on sale, the age of the buyer of a certain product, or the weekly wages of the workers in a certain industry, for example.

Statistics is the mathematics that gives us ways to collect data, organize that data, analyze it, interpret the meaning of the data, and make predictions and decisions about the future based on that data. We will consider several tools that help us to analyze and interpret data: frequency distributions, graphs, and numerical measures of the data. Graphs allow us to picture the data and obtain a visual image of the data that often helps us better understand the situation from which the data was undertaken. Numerical measures allow us to summarize large sets of numbers into a few numbers that describe the data's center of location and its spread.

3.1 FREQUENCY DISTRIBUTIONS AND GRAPHS

We will look at a set of data, and work our way through its organization, analysis, and interpretation. We will also compare it with a second set of data later on.

> Twenty-five applicants interested in working for a national chain of retail outlets took an exam in New York City designed to measure their aptitude for sales and interpersonal relations skills. The following scores were earned by the applicants:
>
> 84, 59, 71, 78, 94, 82, 81, 80, 68, 74, 98, 62, 73,
> 51, 66, 70, 89, 93, 62, 71, 69, 84, 84, 98, 84

If someone were to ask, "How well did these applicants perform on the exam, assuming that 100 is a perfect score?" most of us would find it rather difficult to respond, looking at the data in this format.

We could organize the data into a slightly different format by putting the scores in order from low to high (ascending order) (see Figure 3.1).

This format gives us a little better means of deciding how well these applicants performed on the exam. The data are still considered *raw* data, however, and are difficult to comprehend.

Figure 3.1

Data arranged in ascending order.

51, 59, 62, 62, 66, 68, 69, 70, 71, 71, 73, 74, 78, 80, 81, 82, 84, 84, 84, 84, 89, 93, 94, 98, 98.

Stem and Leaf Display

A very basic picture of our data can be obtained through a **stem and leaf display.** This statistical device uses the tens place of the two-digit number $87 = (8 \times 10 + 7 \times 1)$ as the stem and the ones place as the leaf. tens place   ones place   Figure 3.2 shows this in table form.

Figure 3.2

Stem and leaf display

Stem	Leaf	
5	1 9 ←	51 and 59 are represented here
6	2 2 6 8 9 ←	62, 62, 66, 68, and 69 are here
7	0 1 1 3 4 8	etc.
8	0 1 2 4 4 4 4 9	
9	3 4 8 8	

Now we can see that several applicants scored in the 60s, 70s, and 80s.

Frequency Distribution

We can take this grouping one step further by forming a **frequency distribution,** which arranges our raw data into a *grouped* data format made up of *classes* and corresponding *frequencies* for those classes.

Example 1

Build a frequency distribution from the New York applicants' aptitude test scores, grouping the scores into classes of 50–60, 60–70, 70–80, 80–90, and 90–100. A score will be included in a class if it is greater than or equal to the smaller class limit and less than the larger class limit of the class. For example, scores from 50 to 59 will be included in the 50–60 class.

Solution

First, we form classes into which our scores will fall. Second, we indicate the frequency, f, which is the number of scores that fall into each class.

Class scores	Frequency f	
(lower limit ≤ score < upper limit)*		
50–60	2	— There are 2 scores from 50 through 59
60–70	5	— There are 5 scores from 60 through 69
70–80	6	etc.
80–90	8	
90–100	4	

*In order for a score to fall into a class, its value must be greater than or equal to the lower limit of the class and less than the upper limit of the class.

The frequency distribution allows us to consolidate our large set of data into a smaller, more manageable set. However, when we group the data, we lose sight of the specific values that occur in any one class. We know there are six pieces of data ranging in value from 70 to 79, but we don't know any of the specific values, because we no longer have the original data set of New York applicant scores available to us.

We can also use our knowledge of ratios and percents to form a **relative frequency distribution** and a **percent frequency distribution.**

Learning objective
Form a relative frequency distribution from a frequency distribution.

Example 2

Form a relative frequency distribution from the frequency distribution of New York applicants' scores in Example 1.

Solution

To form a relative frequency distribution we form a ratio between each frequency and the total number of scores in the set.

Relative Frequency Distribution

Class scores	Frequency f	Relative f
(lower limit ≤ score < upper limit)		
50–60	2	2/25 = .08
60–70	5	5/25 = .20
70–80	6	6/25 = .24
80–90	8	8/25 = .32
90–100	4	4/25 = .16
Totals	25	1.00

These can be written as percents to form the **percent frequency distribution.**

Example 3

Form a percent frequency distribution from the relative frequency distribution in Example 2.

Solution

To change from the relative form of a percent to the percent form, we multiply each decimal number in the relative frequency column by 100 and add the percent symbol. That is, $.20 \times 100 = 20\%$ scored in the 60s, $.32 \times 100 = 32\%$ scored in the 80s, and so on.

Learning objective
Form a percent frequency distribution from a relative frequency distribution.

Percent Frequency Distribution

Class scores	Percent %f
(lower limit ≤ score < upper limit)	
50–60	8%
60–70	20%
70–80	24%
80–90	32%
90–100	16%
Totals	100%

We could include all of these distributions on one composite frequency distribution, shown in Figure 3.3.

Figure 3.3

Composite of all three frequency distributions

Class scores	Frequency f	Relative f	Percent %f
(lower limit ≤ score < upper limit)			
50–60	2	2/25 = .08	8%
60–70	5	5/25 = .20	20%
70–80	6	6/25 = .24	24%
80–90	8	8/25 = .32	32%
90–100	4	4/25 = .16	16%
Totals	25	1.00	100%

The relative frequency distribution and the percent frequency distribution are useful if we are comparing the performance of two or more groups of applicants, especially if the sizes of the groups are not the same. We will next consider a second set of scores of 50 people who took the same aptitude test in Seattle.

A group of 50 applicants for positions with the same company as the New York applicants took the aptitude exam in Seattle on the same day. Their results are listed as follows:

51, 55, 55, 59, 64, 65, 66, 67, 69, 69, 70, 72, 75, 77, 78, 78, 79,

79, 80, 81, 82, 83, 84, 85, 86, 87, 88, 88, 88, 88, 90, 93, 94, 95,

95, 95, 97, 98, 98, 99

Example 4

Form a composite frequency distribution for the aptitude test scores of the Seattle applicants, using the same class limits as were used for the New York applicants.

Solution

Using the same techniques that we used in Examples 1 through 3, we form the following composite distribution.

Class scores	Frequency f	Relative f	Percent %f
(lower limit ≤ score < upper limit)			
50–60	5	5/50 = .10	10%
60–70	10	10/50 = .20	20%
70–80	20	20/50 = .40	40%
80–90	12	12/50 = .24	24%
90–100	3	3/50 = .06	6%
Totals	50	1.00	100%

The numbers of applicants taking the exam in the two cities are different, so it is difficult to compare the frequency columns. However, we can easily compare the *percentages* of applicants that scored in each class and thus compare the performances of the two groups. That is, 16% of the New York applicants scored in the 90s, whereas only 6% of the Seattle group scored in the 90s, 32% of the New York group scored in the 80s and 24% of the Seattle group scored in the 80s, and so on.

Graphs

In mathematics, graphs are our pictures of mathematical relationships. We will work with three types of graphs: the bar graph, the line graph (also called a *frequency polygon*), and the circle graph (also called a *pie chart*).

Learning objective
Draw a bar graph using the information from a frequency distribution.

Bar Graphs. We construct a *bar graph* by using the class limits of the frequency distribution on the horizontal axis of the graph and the frequencies in the vertical axis.

Example 5

Draw a bar graph using the frequency distribution of the New York applicants' scores in Example 1.

Solution

First we set up axes perpendicular to each other, then we label the vertical axis with the frequencies and the horizontal axis with the class limits. Finally, we draw in the bars to the appropriate heights for each class frequency. The final graph is shown in Figure 3.4.

Figure 3.4

New York applicants' scores—bar graph

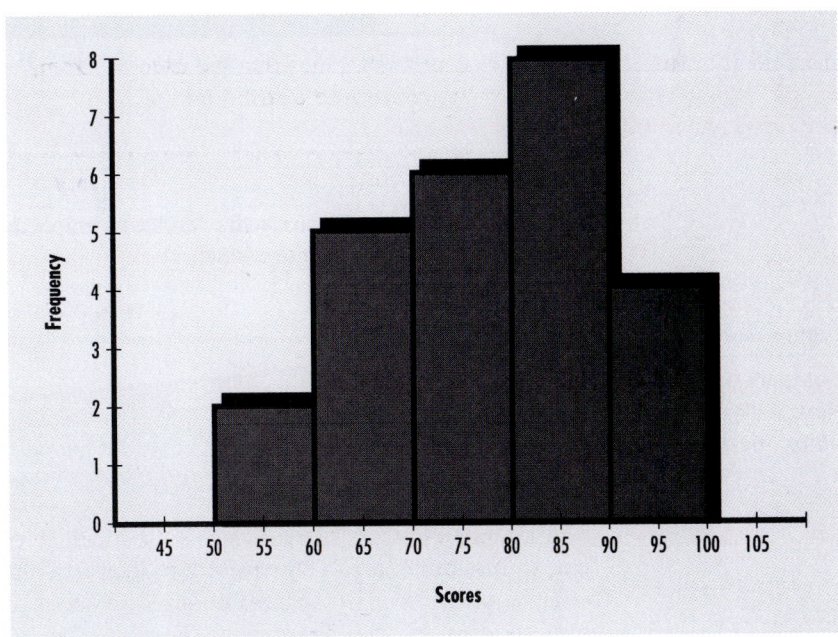

Example 6

Draw a bar graph using the frequency distribution of the Seattle applicants' scores in Example 4.

Solution

Using the same techniques as in Example 5, we can form the bar graph of Seattle applicants' scores, shown in Figure 3.5.

Figure 3.5

Seattle applicants'
scores—bar graph

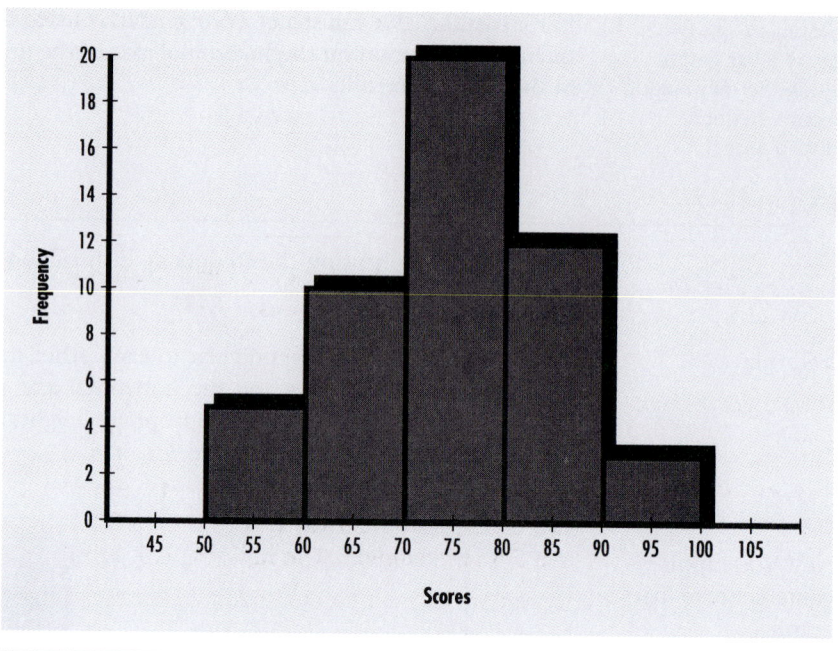

Some computer software will even let us graph these three-dimensionally
on the same axes (see Figure 3.6).

Figure 3.6

Comparison of
New York and Seat-
tle frequencies

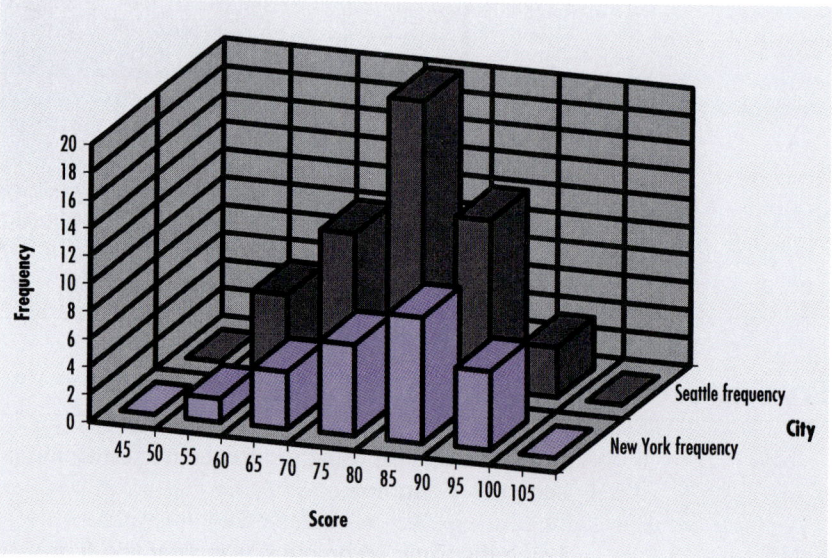

Notice that we can easily make a visual comparison of the performance of the two groups who took the examination.

We can make each of these a *percent bar graph* by simply changing the labeling of the vertical axis to *percents*.

Example 7

Learning objective
Draw a percent bar graph using the information from a percent frequency distribution.

Use the percent frequency distribution to form a percent bar graph of the New York applicants' scores.

Solution

To form a percent bar graph we label the vertical axis of the graph with percents and the horizontal axis with the class limits, and we draw the bars as we did in Example 6. The final graph is shown in Figure 3.7.

Figure 3.7

New York applicants' scores—percent bar graph

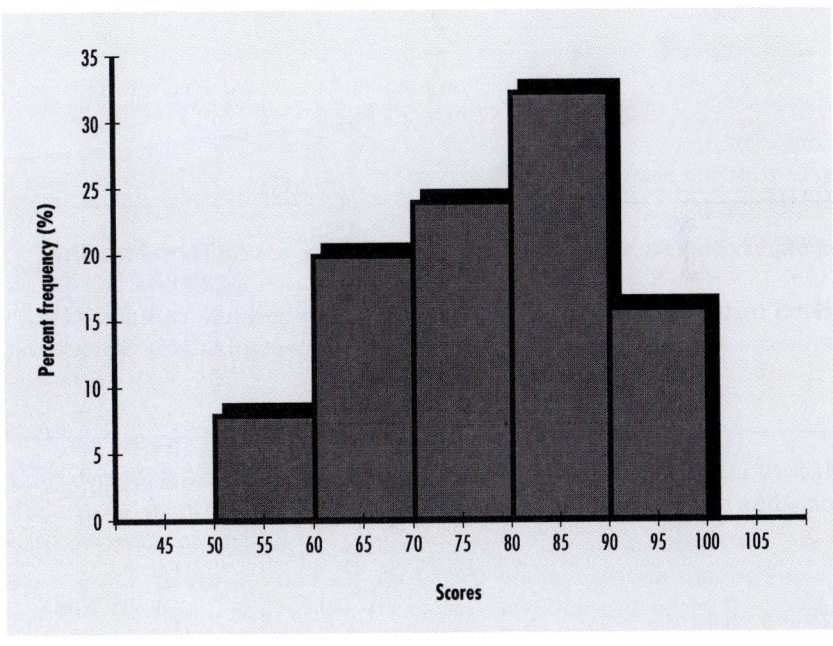

Figures 3.8 and 3.9 show the bar graphs for the Seattle applicants and the combined three-dimensional graph of the scores of both groups.

Figure 3.8

Seattle percent bar graph

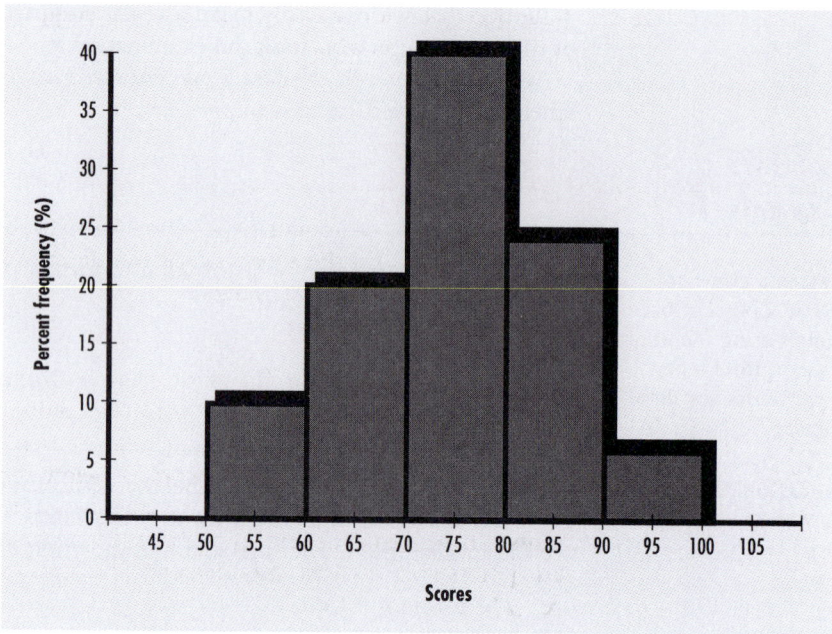

Figure 3.9

Comparison of New York and Seattle percent frequencies

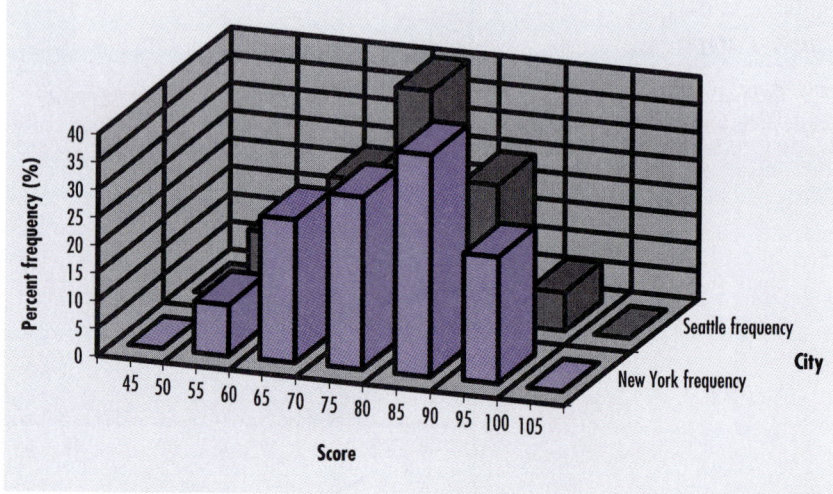

Learning objective
Draw a line graph using the information from a frequency distribution.

Line Graphs (frequency polygons). A **line graph,** also called a *frequency polygon,* can be formed from the bar graph. To form the line graph we find the middle values of each class (also called the *midpoints* of each class), and mark the midpoint at the top of each bar. Then we connect the points with straight line segments.

Example 8

Form line graphs, or frequency polygons, for the scores of the New York applicants, and also for the scores of the Seattle applicants. Form percent line graphs for each data set as well.

Solution

Learning objective
Draw a percent line graph using the information from a percent frequency distribution.

As already noted, we set up the vertical and horizontal axes in the same way we did for the bar graph. We then mark the midpoints of each class at the height that is appropriate for the frequency of each class. We also note what would be the midpoint of the next class below the lowest class (e.g., 45) and the midpoint of the next class above the highest class, and mark the points on the zero line. Finally we connect these points to their adjacent class points with straight line segments. The final graphs are shown in Figures 3.10 through 3.13.

Figure 3.10

New York applicants' scores—line graph

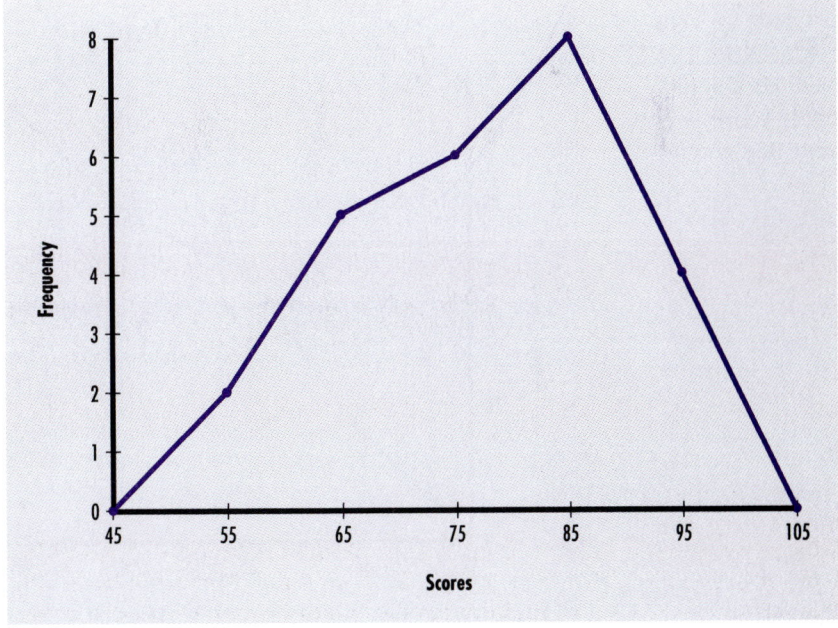

Figure 3.11

Seattle applicants' scores—frequency polygon (line graph)

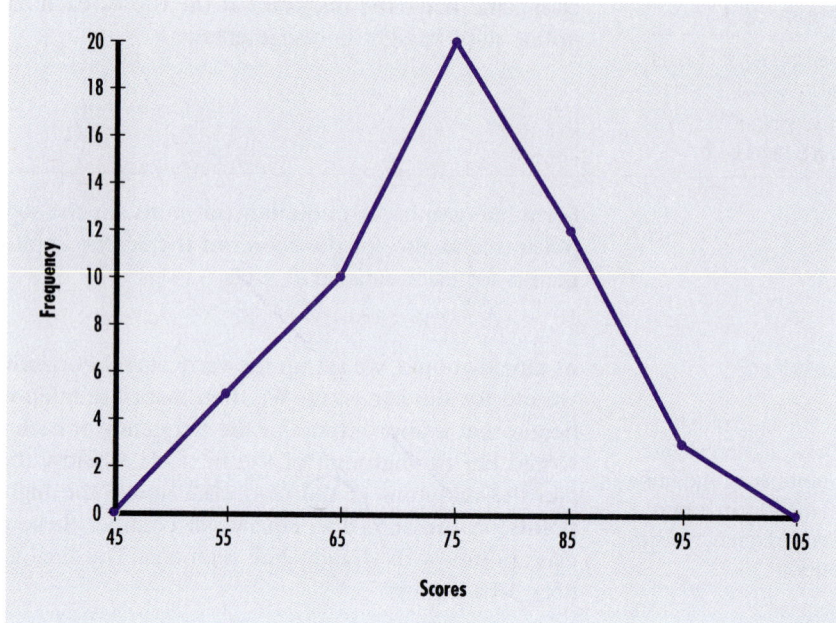

Figure 3.12

New York applicants' scores—percent line graph

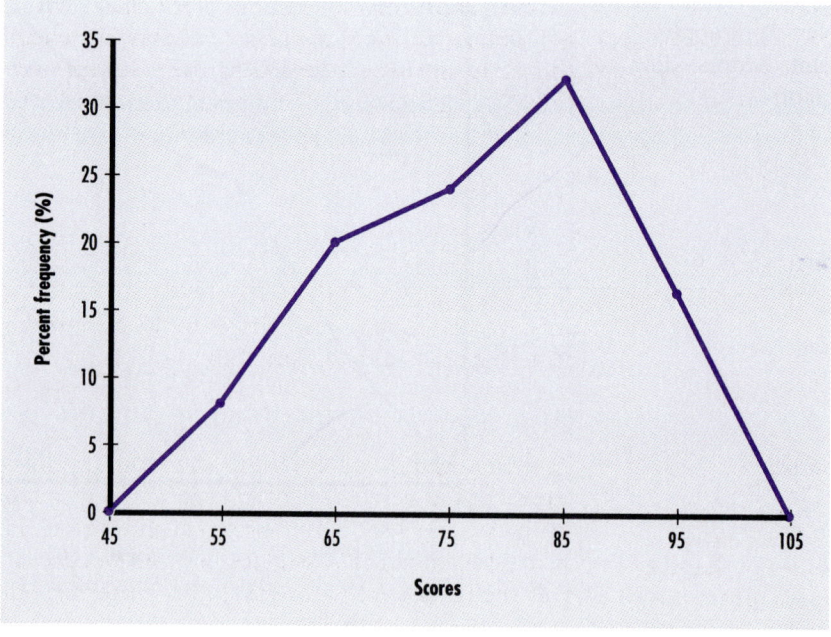

Figure 3.13

Seattle applicants' scores—percent line graph

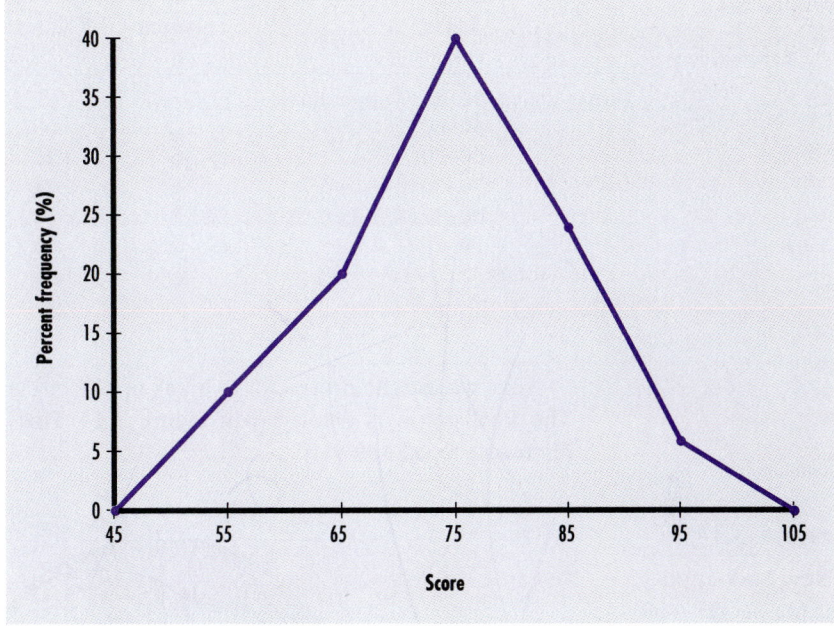

Circle Graphs or Pie Charts. Recall that a circle has 360 degrees (360°). We use our relative frequency column to determine how many degrees should be graphed for each class. Then we form sectors of the circle for each class. The end result, a **circle graph,** looks like a pie cut into pieces, which is why it is often called a **pie chart.**

Example 9

Form a circle graph of the scores of the New York applicants, using the relative frequency distribution of Example 2.

Solution

Learning objective
Draw a pie graph using the information from a percent frequency distribution.

The circle graph will be divided into five sectors that will proportionately represent the percentage of the applicants who scored in each class range. We divide the 360° of the circle proportionately to accomplish this. To determine the number of degrees that will be allotted to each sector, we multiply the relative frequency by 360° to determine the number of degrees in each sector.

Class scores	Frequency f	Relative f	Degrees (rel. f) × 360°
(lower limit ≤ score < upper limit)			
50–60	2	.08	.08 × 360 = 28.8°
60–70	5	.20	.20 × 360 = 72.0°
70–80	6	.24	.24 × 360 = 86.4°
80–90	8	.32	.32 × 360 = 115.2°
90–100	4	.16	.16 × 360 = 57.6°
Totals	25	1.00	360°

Next, we use the protractor to break up the circle and complete the graph. The final graph is displayed in Figure 3.14. This graph was drawn using Microsoft Excel software.

Figure 3.14

New York applicants' scores—pie chart

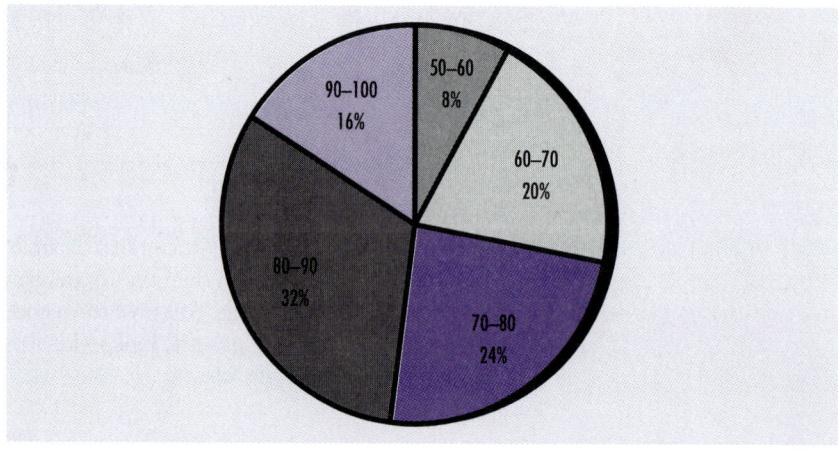

For Your Information

PICTOGRAMS

Pictograms are pictorial representations of numerical data. They are used to provide another way to convey numerical information in a visual format. Just as with any type of graphical presentation, care must be taken in interpreting the scale of the pictogram. The pictogram to the right depicting the *Ratio of Workers to Retirees* seems to accurately depict the ratios 42:1, 3.3:1, and 2.4:1. Now look at the second pictogram depicting the increasing *Cost of Long Term Care*. If the number of individual stacks of money above each year is meant to represent the yearly costs indicated just above the year, how much does each individual stack represent? Always be careful when reading any graphical representation of data to see that the pictorial impression is a reasonable representation of the numbers represented.

Ratio of Workers to Retirees

Year	Workers		Retirees
1945	(worker figures)	42 to 1	(retiree figure)
1984	(worker figures)	3.3 to 1	(retiree figure)
2020	(worker figures)	2.4 to 1	(retiree figure)

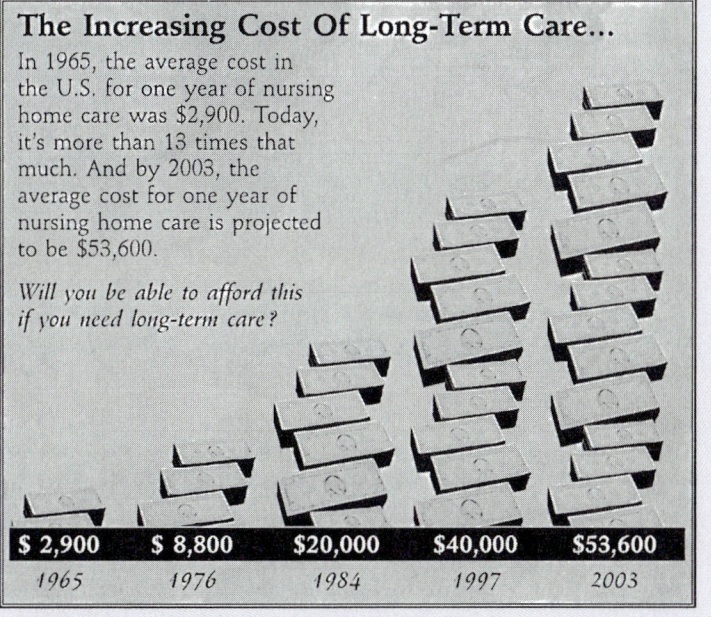

The Increasing Cost Of Long-Term Care...

In 1965, the average cost in the U.S. for one year of nursing home care was $2,900. Today, it's more than 13 times that much. And by 2003, the average cost for one year of nursing home care is projected to be $53,600.

Will you be able to afford this if you need long-term care?

$ 2,900	$ 8,800	$20,000	$40,000	$53,600
1965	*1976*	*1984*	*1997*	*2003*

Source: Courtesy of Successful Money Management Seminars Inc. and TIAA-CREF.

We can form a similar pie chart for the Seattle group, as well.

Class scores	Frequency f	Relative f	Degrees: rel. f × 360
(lower limit ≤ score < upper limit)			
50–60	5	.10	.10 × 360 = 36.0°
60–70	10	.20	.20 × 360 = 72.0°
70–80	20	.40	.40 × 360 = 144.0°
80–90	12	.24	.24 × 360 = 86.4°
90–100	3	.06	.06 × 360 = 21.6°
Totals	50	1.00	360°

Figure 3.15

Seattle applicants' scores—pie chart

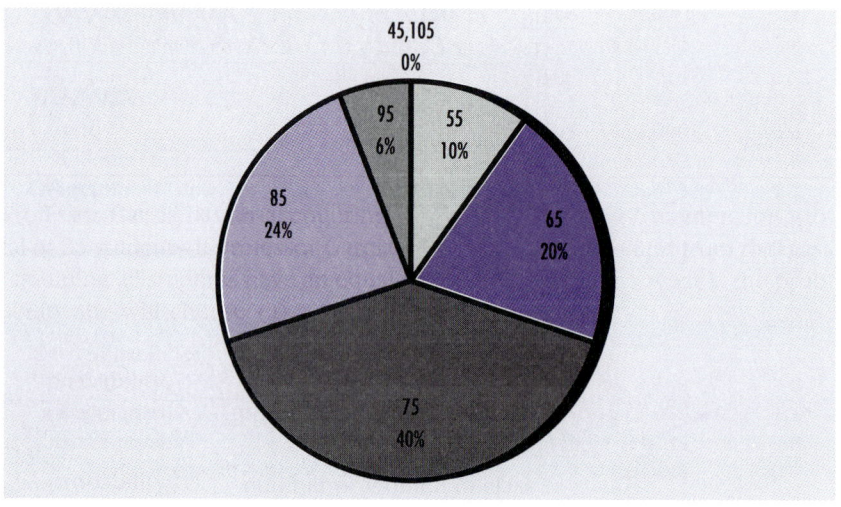

Bar graphs, line graphs, and circle graphs have many uses in business, and they are formed in different ways using different information. Consider the following example.

Example 10

The three types of graphs that we have studied are often used to represent data at a single point in time (e.g., a day, a month, a year), or to represent data taken over a longer period of time (e.g., several years). Consider a home improvement store and three departments in that store—the wallpaper department, the paint department, and the hardware department—and the advertising expenses for those departments from 1996 through 1999.

Year	Wallpaper department	Paint department	Hardware department	Total expenses
1996	$2,000	$10,000	$8,000	$20,000
1997	$3,000	$8,000	$9,000	$20,000
1998	$4,000	$5,000	$4,000	$13,000
1999	$5,000	$3,000	$7,000	$15,000

a.   Draw a bar graph of the expenses for the three departments for 1999.

b.   Draw a line graph of the expenses for the three departments for 1999.

c.   Draw a circle graph (pie chart) of the total expenses of these three departments for 1999.

d.   Draw a bar graph and a line graph of the advertising expenses of the wallpaper department from 1996 through 1999.

e.   Draw a bar graph and a line graph of the advertising expenses of the wallpaper department from 1996 through 1999.

Solution

a.   Draw a bar graph of the expenses for the three departments for 1999. We label the vertical axis with the cost, and horizontal axis with the department, and then draw the bars as before (see Figure 3.16).

Figure 3.16

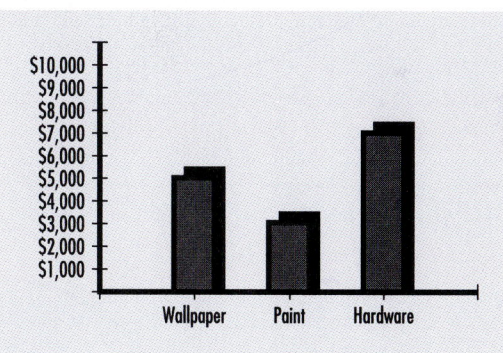

b.   Draw a line graph of the expenses for the three departments for 1999. We label the axes similarly to the bar graph, but draw a line graph (see Figure 3.17).

c.   Draw a circle graph (pie chart) of the total expenses of these three departments in 1996 (see Figure 3.18). We need to form the relative frequency distribution and determine the appropriate number of degrees for each department.

Figure 3.17

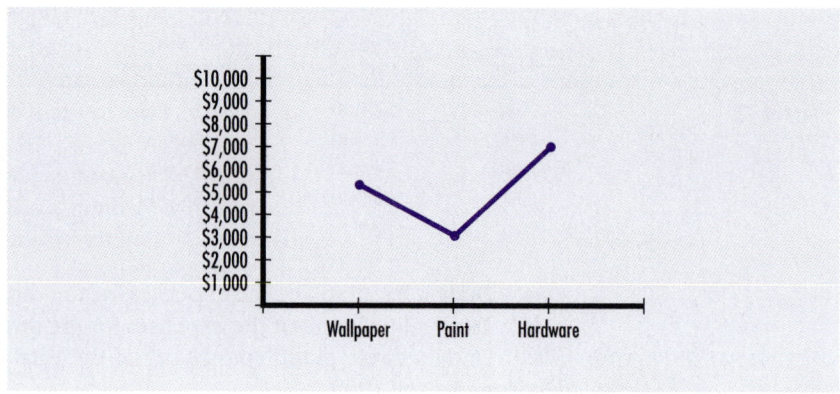

Department	Expenses	% of total	Degrees
Wallpaper	$ 5,000	33%	.33 × 360° = 120°
Paint	$ 3,000	20%	.20 × 360° = 72°
Hardware	$ 7,000	47%	.47 × 360° = 168°
Totals	$15,000		

Figure 3.18

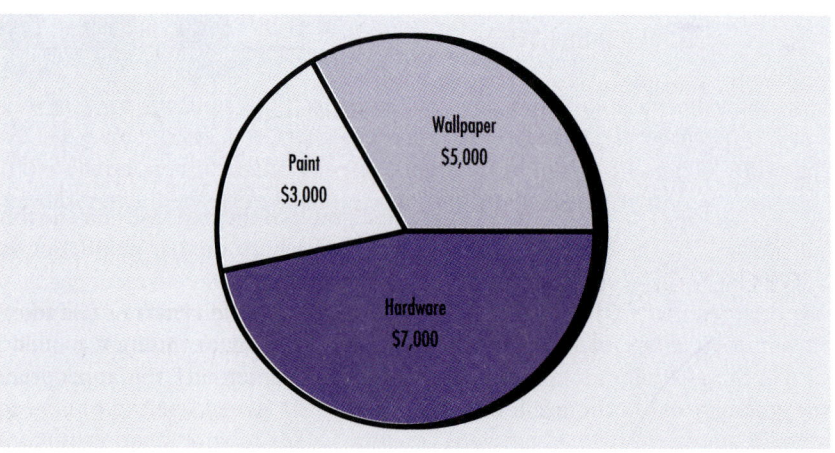

d. Draw a bar graph and a line graph of the advertising expenses of the wallpaper department from 1996 through 1999 (see Figure 3.19).

Figure 3.19

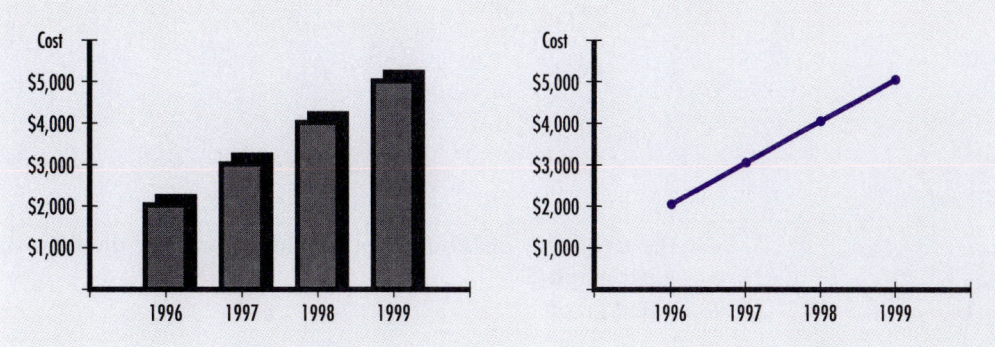

e. Draw line graphs for each department over the 4-year period, and put them on the same set of axes. This is called a composite graph (see Figure 3.20).

Figure 3.20

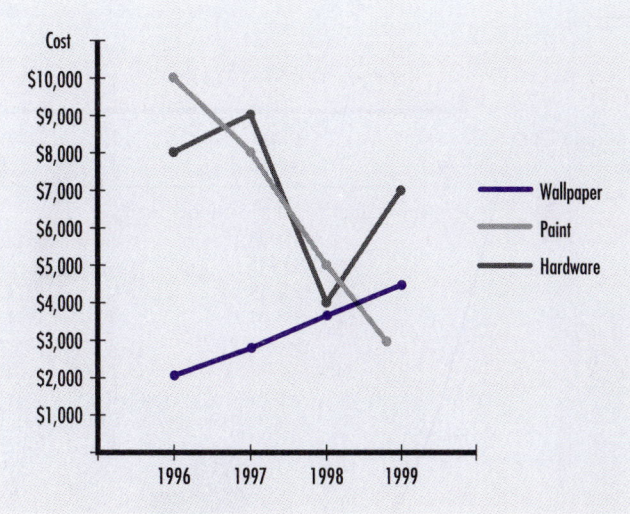

CHECK YOUR KNOWLEDGE

Frequency Distributions and Graphs

1. Given the following frequency distribution, complete the columns for relative frequency and percent frequency.

Class scores	Frequency f	Relative f	Percent %f
(lower limit ≤ value < upper limit)			
25–35	4		
35–45	8		
45–55	12		
55–65	10		
65–75	6		
Totals	40		

2. Use the frequency distribution of problem 1 to draw the following:
    a. a bar graph
    b. a line graph

3. Use the percent frequency distribution in problem 1 to draw the following:
    a. a percent bar graph
    b. a percent line graph
    c. a pie chart

Answers to CYK:

Class scores	Frequency f	Relative f	Percent %f
(lower limit ≤ value < upper limit)			
25–35	4	.10	10%
35–45	8	.20	20%
45–55	12	.30	30%
55–65	10	.25	25%
65–75	6	.15	15%
Totals	40		

2. *a.*

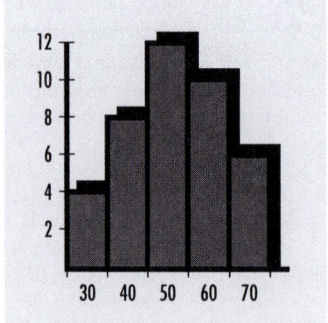

    b.

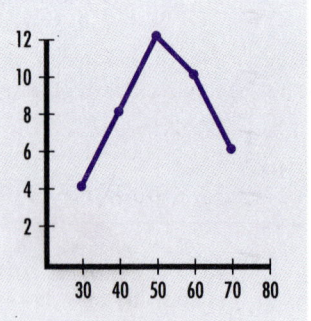

Answers to CYK: *3.* **a.**   **b.**
(continued)

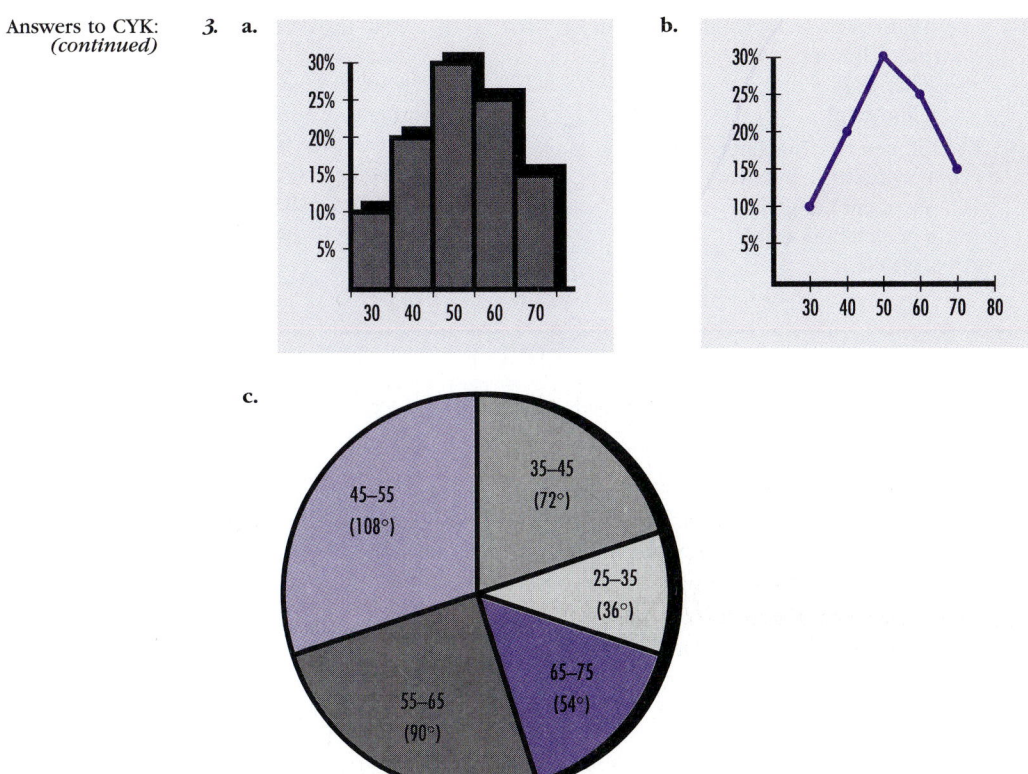

c.

3.1 EXERCISES

1. Given the following frequency distribution of ages of patients in the rehabilitation unit of Community General Hospital, complete the columns for relative frequency and percent frequency.

Class ages	Frequency f	Relative f	Percent %f
(lower limit ≤ age < upper limit)			
5-15	3		
15-25	6		
25-35	15		
35-45	20		
45-55	10		
55-65	3		
65-75	3		
Totals	60		

2. Use the frequency distribution of exercise 1 to draw the following:
   a. a bar graph
   b. a line graph

3. Use the percent frequency distribution in exercise 1 to draw the following:
   a. a percent bar graph
   b. a percent line graph
   c. a pie chart

4. Well Built Construction Company focuses its business on commercial remodeling and repair work. The management developed the following graphs of anticipated revenues from projects for which they have contracts during the next year. Complete the columns for the frequency, relative frequency, and percent frequency distributions in the following table.

Class revenue (× $1,000)	Frequency f(# contracts)	Relative f	Percent %f
(lower limit ≤ value < upper limit)		.24	4
10–30		.06	6
30–50		.10	10
50–70		.13	13
70–90		.15	15
90–110		.24	24
110–130		.15	15
130–150		.13	13
150–170			
Totals			

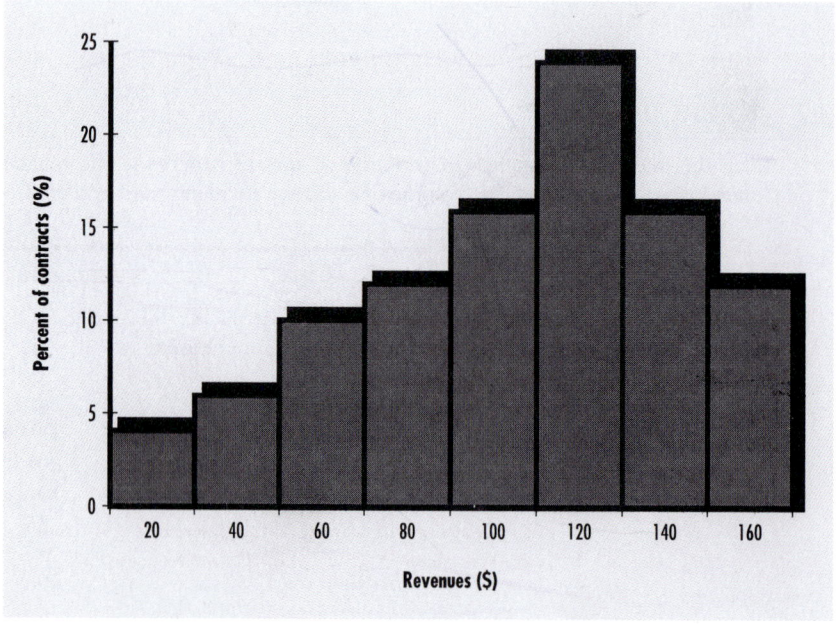

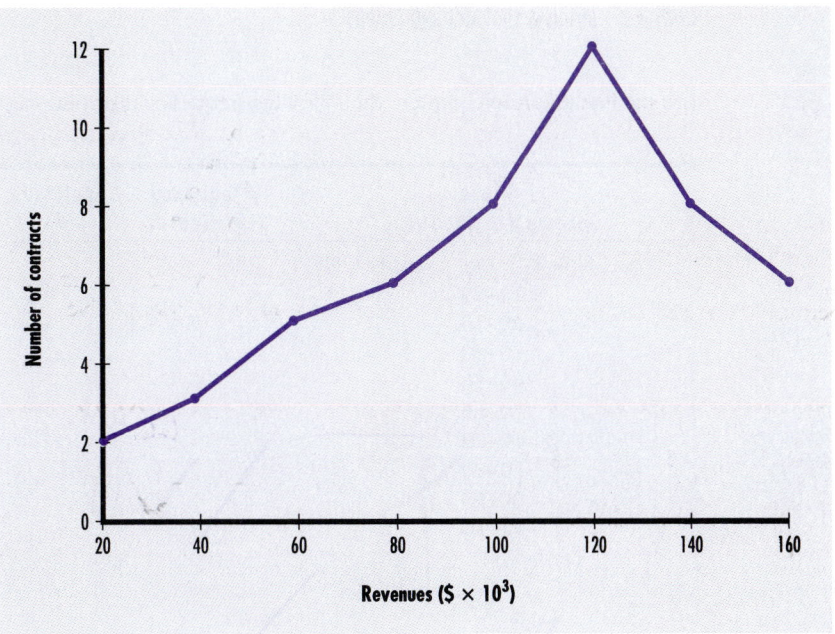

5. Use the frequency distribution of exercise 4 to draw a frequency bar graph.

6. Use the percent frequency distribution in exercise 4 to draw a percent line graph.

7. Gwendolyn Davis operates a personal income tax service out of her home from January 15 through April 15. Last year she formed a frequency distribution of the adjusted gross incomes of her clients. The following percent line graph represents 100 clients.

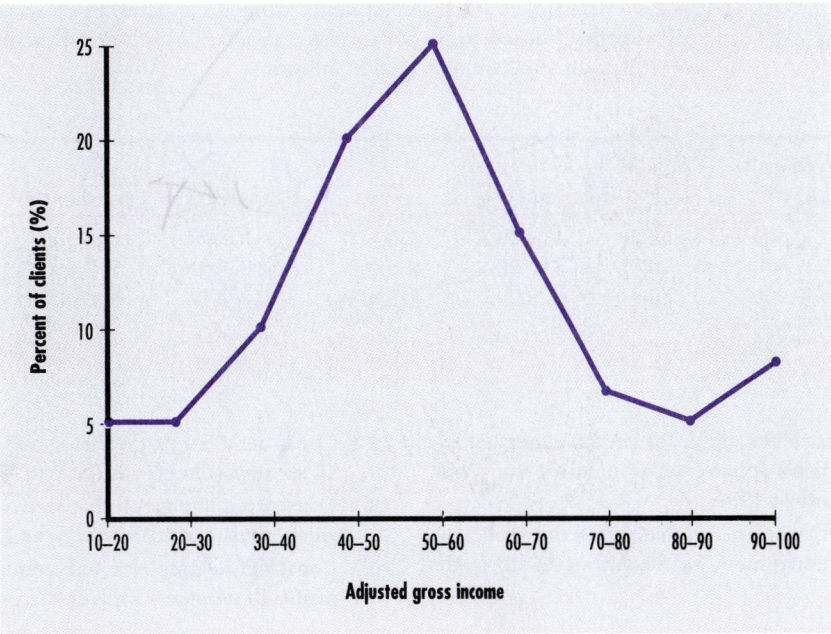

Use the line graph to complete the following frequency distribution.

Class income (× $1,000)	Frequency f(# clients)	Relative f	Percent %f
(lower limit ≤ value < upper limit)			
10–20	5		
20–30	5		
30–40	10		
40–50	20		
50–60	25		
60–70	15		
70–80	7		
80–90	5		
90–100	8		
Totals	100		

8. Use the frequency distribution of exercise 7 to draw
  a. a frequency bar graph
  b. a frequency line graph

9. Use the percent frequency distribution in exercise 7 to draw the following:
  a. a percent bar graph
  b. a pie chart

Use the following information to complete problems 10 through 15.

Easy Mart Store has six departments: women's apparel, men's apparel, children's apparel, toys, housewares, and hardware. Profits for each department for the years 1997 through 1999 are indicated as follows:

Year	Women's apparel	Men's apparel	Children's apparel	Toys	Housewares	Hardware	Total
1997	$30,000	$15,000	$20,000	$10,000	$10,000	$15,000	$100,000
1998	$27,000	$13,000	$17,000	$ 8,000	$ 8,000	$ 7,000	$ 80,000
1999	$40,000	$20,000	$25,000	$15,000	$12,000	$ 8,000	$120,000
							$300,000

10. Construct a bar graph for profits generated in the women's apparel department for the years 1997 through 1999.

11. Construct a line graph that shows the profits of the toy department for the years 1997 through 1999.

12. Construct a bar graph that shows the profits of all six departments for the year 1998.

13. Construct a line graph that shows the profits of all six departments for the year 1999.

14. Construct a composite line graph showing the profits of women's apparel, the profits of men's apparel, and the profits of children's apparel for the 3-year period. Use a solid line for women's apparel, a dotted line for men's apparel, and a dashed line for children's apparel.

15. Construct a pie chart (circle graph) for the profits generated by the six departments during 1997.

16. Construct a bar graph showing the total profits generated by Easy Mart for the 3 years.

17. Construct a line graph showing the total profits generated by Easy Mart for the 3 years.

18. Construct a pie chart showing the total profits generated by Easy Mart for the 3-year period.

3.2 NUMERICAL SUMMARY MEASURES: MEASURES OF CENTRAL TENDENCY

When analyzing business data, we often like to know whether this new set of data is "typical" of previous data collected on the same subjects. If it is not typical, then we wish to know how it differs—are the data greater than or less than what we might consider typical? We use **numerical summary measures** to help us in our analysis. These *summary* measures generally indicate the *center* of a set of data and the *spread* of the data. Measures that indicate the centers of sets of data are called **measures of central tendency,** and measures of spread are called **measures of dispersion.**

Measures of central tendency are statistical measures used to *locate the center of a set of data.* It is much easier to work with a single number that represents a group than to work with the entire group itself. We will consider three such measures: the mean, the median, and the mode.

Mean

The **mean** is the measure of central tendency we usually think of when we mention the word *average.* The mean is easily calculated as follows:

Learning objective
Calculate the mean and trimmed mean from a set of data.

$$\text{mean} = \frac{\text{sum of all values}}{\text{number of values}}$$

Example 11

Let's use the data from the New York applicants' scores to calculate the mean.

Solution

First, add up all of the scores. We will use the letter X to represent the variable "applicant's score." The variable then represents each of the 25 scores.

X
51
59
62
62
66
68
69
70
71
71
73
74
78
80
81
82
84
84
84
84
89
93
94
98
98

sum = 1,925

Next, we count the number of scores in the set of data. There are 25; this is our denominator.

$$\text{mean} = \frac{\text{sum of all values}}{\text{number of values}} = \frac{1925}{25} = 77$$

How do we interpret our answer of 77? Remember, the mean is meant to be the average, or the typical value, of all the values. Our mean of 77 is meant to be the one number that indicates the center of the entire set of scores.

$$\text{mean} = 77$$
↓

51, 59, 62, 62, 66, 68, 69, 70, 71, 71, 73, 74, 78, 80, 81, 82, 84, 84, 84, 84, 89, 93, 94, 98, 98

When working with data, we usually refer to a quantity as a variable and denote it with a letter such as x, y, or z. It is called a variable because it can assume several different values—in this case, the scores of the New York applicants. We use the uppercase Greek letter sigma, Σ, to represent the phrase *sum up,* an *x* with a bar over it, $\bar{x}$, to represent the mean; and *n* to represent the number of values. Our formula for mean then can be written as follows:

$$\text{mean} = \bar{x} = \frac{\Sigma x}{n} = \frac{\text{sum of all values}}{\text{number of values}}$$

The mean usually is a good measure of the center of the data, but it has one fault of which we must be aware. The mean can be affected by very large or very small values. For example, if the score of 52 had been 2 (so that our data total would be 50 less), then the mean would have been

$$\text{mean} = \frac{\text{sum of all values}}{\text{number of values}} = \frac{1875}{25} = 75$$

One number that is much less than any of the other scores can significantly affect the mean. Sometimes we can detect and correct this problem by excluding the largest score and the smallest score and calculating the mean of the remaining 23 scores. This new mean is called the **trimmed mean,** and would be

$$\text{Trimmed mean} = \frac{\Sigma x}{n - 2} \text{ (excluding 2 and 98)} = 77.26$$

This is closer to the original 77 and is a better representation of the typical score in the set of scores.

Median

Learning objective
Calculate the median
from a set of data.

Another measure of central tendency is the **median.** The median is the middle value of all the values. The steps for calculating a median are as follows:

Step 1: Arrange the values in sequential order from highest to lowest, or from lowest to highest.

Step 2: Count the number of values (*n*), add 1 (*n* + 1), and divide by 2.

$$\frac{n + 1}{2}$$

Step 3: Using the value calculated from step 2, count down (or up) the array of values for that number of values.

Example 12

Find the median of the scores for the New York applicants.

Solution

Step 1:   First, we arrange the scores in order from highest score to lowest score:

51
59
62
62
66
68
69
70
71
71
73
74
78
80
81
82
84
84
84
84
89
93
94
98
98

Step 2:   Determine the number of the median term.

$$\frac{n + 1}{2} = \frac{25 + 1}{2} = \frac{26}{2} = 13\text{th score}$$

Note: The number 13 is not the median, it is the *location of the median.*

Step 3:

51		
59		
62		
62		
66		
68		
69	↓	Moving from smallest to largest
70		
71		
71		
73		
74		
78	←	This is the thirteenth number;
80		thus 78 is the median score.
81		
82		
84		
84		
84		
84		
89	↑	Moving from largest to smallest
93		
94		
98		
98		

Example 13

Find the mean and median of the applicants who took the examination in Seattle.

Solution

Since this is a different set of data, let's use y as the variable.

y: 51, 55, 55, 59, 64, 65, 66, 67, 69, 69, 70, 72, 75, 77, 78, 78, 79, 79, 80, 81, 82, 83, 84, 85, 86, 87, 88, 88, 88, 88, 90, 93, 94, 95, 95, 95, 97, 98, 98, 99

$$\text{Mean} = \frac{\Sigma y}{n} = \frac{3200}{40} = 80$$

$$\text{Median:} \quad \frac{n + 1}{2} = \frac{40 + 1}{2} = \frac{41}{2} = 20.5$$

This means that the median is midway between the 20th and 21st values when the data are put in either ascending or descending order.

51, 55, 55, 59, 64, 65, 66, 67, 69, 69, 70, 72, 75, 77, 78, 78, 79, 79, 80, 81, 82, 83, 84, 85, 86, 87, 88, 88, 88, 88, 90, 93, 94, 95, 95, 95, 97, 98, 98, 99

The 20th value is 81 and the 21st value is 82, so the median is the average of the two: $\text{median} = \dfrac{81 + 82}{2} = 81.5$

Mode

Learning objective
Determine the mode
of a set of data.

The **mode** is the third measure of central tendency used to summarize a set of data. The mode is the data value *that occurs most frequently.*

Example 14

Determine the mode of the New York applicants' scores and the Seattle applicants' scores.

Solution

a. The mode of the New York applicants' scores is 84, which occurs four times.

51, 59, 62, 62, 66, 68, 69, 70, 71, 71, 73, 74, 78, 80, 81, 82, 84, 84, 84, 84, 89, 93, 94, 98, 98
↑
mode

b. The mode of the Seattle applicants' scores is 88, which also occurs four times.

51,55,55,59,64,65,66,67,69,69,70,72,75,77,78,78,79,79,80,81,82,83,84,85,86,87,88,88,88,88,90,93,94,95,95,95,98,98,99
↑
mode

Whenever a set of data has one mode, it is called unimodal; if it has two modes it is said to be bimodal; if more than two modes, it is said to be multimodal. Both the data sets of Example 14 are unimodal.

Let us review the three measures of central tendency we calculated for the two data sets.

Example 1: New York applicants	Example 2: Seattle applicants
mean = 77	mean = 80
median = 78	median = 81.5
mode = 84	mode = 88

It would appear that the Seattle applicants performed better on the examination.

Each of these measures (mean, median, and mode) has often been referred to as an *average* for the set of scores. Actually, they each represent a center around which the scores' data are clustered. Notice that even though each is representative of the center of the data, they are quite different from each other. In fact, the overall spread, or *range* between these summary values is $84 - 77 = 7$ in the first set, and $88 - 80 = 8$ in the second set.

In the study of any data set, you should look at all three measures, and take care to know what other people mean when they refer to an average.

Finding the Mean of a Set of Data Represented in a Frequency Distribution

We noted earlier in this section that when data is grouped, we lose sight of the actual values of the scores. Hence, we cannot simply add up the values to get the Σx. So, how do we at least *approximate* this sum so that we can find the mean? Consider the New York applicants' data:

Example 15

Approximate the mean of the New York applicants' scores using the frequency distribution of the New York applicants' scores.

Class scores	Frequency f
(lower limit ≤ score < upper limit)	
50–60	2
60–70	5
70–80	6
80–90	8
90–100	4

Solution

We know there are two scores between 50 and 60, but we do not know what they are. They need to be the subtotal that represents the 50s part of the total sum. Similarly, there are five scores in the 60s that need to be added to give us the necessary subtotal of the 60s for the total sum. There are six scores in the 70s, and so on.

In order to proceed, we make an assumption about the original data represented by the frequencies in each class. We assume that the two scores in the 50s are evenly distributed between 50 and 60; the five are evenly distributed between 70 and 80, and so on. Once we have made that assumption, we can assume that the midpoints, M, of each class (55, 65, 75, 85, 95) are representative of each of the scores in that class. When we multiply *frequency × midpoint*, we will get an approximation of the subtotal in that class. When we add up all subtotals, we have an approximation of the total of all the scores.

Class scores	Frequency f	M	f × M
50–60	2	55	110
60–70	5	65	325
70–80	6	75	450
80–90	8	85	680
90–100	4	95	380
		$\Sigma fM =$	1945

(This approximates the actual sum of 1925 of the data before grouping took place.)

Now we compute our mean of the grouped data as

$$\text{mean} = \bar{x} = \frac{\Sigma fM}{n} = \frac{1945}{25} = 77.8,$$

which is a reasonably good approximation of the true mean, 77, found earlier.

Weighted Mean

Learning objective
Calculate the weighted mean from a set of grouped data.

Finding the mean using midpoints and frequencies is a form of **weighted mean.** In this case, the weighting is provided by the frequencies. The fact that there were eight values in the 80s class meant that 85 was added in eight times, giving the 80s more weight than the 50s, which had only two scores. In fact, the 80s had the greatest weight of any of the classes. Another example of the weighted mean is the way we calculate the grade point averages in colleges.

Example 16

Crampton Community College awards 4 quality points for an A, 3 quality points for a B, 2 quality points for a C, 1 quality point for a D, and 0 quality points for an F. Wally Hurrah earned the following grades during the spring term of his freshman year:

Course	Grade	Credits
English	A	3
Sociology	C	3
Accounting	B	3
Mathematics	A	3
Marketing	A	3

What was his grade point average for the semester?

Solution

Since each course has the same number of credits, the *weighting* is provided by the quality points assigned to the different grades ("A" carries more weight than "B", "B" carries more weight than "C", etc.)

Course	Grade	Credits	QP	Credits × QP
English	A	3	4	12
Sociology	C	3	2	6
Accounting	B	3	3	9
Mathematics	A	3	4	12
Marketing	A	3	4	12
		15		51

$$\text{Term quality point average} = \frac{\Sigma \text{ credits} \times \text{quality pts}}{\Sigma \text{ credits}} = \frac{51}{15} = 3.4$$

If the credits had not been the same for these three courses, both the grade (quality points) and the number of credits would provide different weights.

Example 17

Suzie Markowicz earned the following grades at Crampton:

Course	Grade	Credits
English	B	3
Biology	C	4
Accounting	A	3
Phys. Educ.	A	1
Intro. Computer	B	2
Economics	C	3
		16

What was her grade point average for the term?

Solution

Course	Grade	Credits	QP	$G \times$ QPP
English	B	3	3	9
Biology	C	4	2	8
Accounting	A	3	4	12
Phys. Educ.	A	1	4	4
Intro. Computer	B	2	3	6
Economics	C	3	2	6
		16		45

$$\text{Term quality point average} = \frac{\Sigma \text{credits} \times \text{quality pts}}{\Sigma \text{ credits}} = \frac{45}{16} = 2.81$$

CHECK YOUR KNOWLEDGE

Measures of Central Tendency

Solve the following problems, rounding your answers to the nearest tenth.

1. Ten students in Mr. Garry's Business Mathematics class received the following test scores on their second business mathematics examination: 48, 67, 78, 82, 82, 82, 87, 87, 93, 94.
   a. Determine the mean of the scores.
   b. Determine the median of the scores.
   c. Determine the mode of the scores.

2. A survey was taken asking 20 students how many hours per week they watched television. The results were reported in the following table:

Number of hours	Number of students
7	7
10	2
14	5
20	4
30	2

   a. Use the weighted mean method to determine the mean number of hours that students watched television each week.

    b.   What is the mode of the data?

    c.   What is the median of the data?

3.   A more extensive survey was conducted in the community where the students in problem #2 lived to determine the time people watched television. The survey asked 300 people to indicate how many hours per week they watched television. The results are given in the following frequency distribution.

Number of hours (H)	Number of people (f)	Midpoint (M)	f × M
lower ≤ hours < upper			
0–8	50		
8–16	75		
16–24	125		
24–32	30		
32–40	20		

Calculate the mean number of hours that people in the survey reported watching television per week.

3.2 EXERCISES

Complete the following problems, rounding answers to the nearest one hundredth, or the nearest cent.

1.   Each household on Blueberry Lane was asked, How many pets are in your household? The results were: 1, 2, 3, 2, 1, 1, 1, 1, 2, 3, 0, 3, 0, 0, 2, 1, 3. Find the (a) mean number of pets, (b) median number of pets, and (c) mode number of pets. 00011111112222333

2.   In order to draw up a contract with a leasing company to lease new cars for his sales representatives, Mr. Acme needed to find the average number of miles driven per year. The annual number of miles driven by each sales representative of Acme Company were as follows:

Abrams	18,000
Bartley	21,000
Lions	15,000
Schwinn	20,000
Zikirsky	18,000

Calculate (a) the mean number of miles, (b) the median number of miles, and (c) the mode for the number of miles.

3.   The principal of Oakleaf School wished to know the average number of students using the computer lab per day. If the average was greater than 150 students, he would then allocate money to buy more computers. The data for 2 weeks are as follows:

Monday	200	Monday	190
Tuesday	100	Tuesday	130
Wednesday	10	Wednesday	20
Thursday	220	Thursday	210
Friday	200	Friday	220

Calculate (a) the mean number of students who used the lab per day, (b) the median per day, and (c) the mode per day.

Answers to CYK: *1.* **a.** 80 **b.** 82 **c.** 82   *2.* **a.** 13.95 hours **b.** 14 hours **c.** 14 hours

                           3. 5160/300 = 17.2 hours

4. You are trying to get accepted into the MBA pro-
    gram at Randolph University. Although you are
    uncertain about the acceptance requirements,
    last year you knew six seniors from the state col-
    lege who were accepted into this program. A list
    of their grade point averages follows:

Joshua	95.0
Gabriel	89.0
David	93.2
Bridgette	99.0
Murray	98.8
Trevor	98.8

    You conclude that if your GPA of 95 is at least
    as high as the average GPA of these six individ-
    uals, then you are sure to be accepted. Calculate
    (a) the mean, (b) the median, and (c) the mode,
    to see if you will be accepted.

5. The president of Sunny Realestate Corporation
    asks you to sell real estate for the company. You
    will be paid $1,000 per month plus commissions
    of 5% of the average dollar amount sold per
    month. While having lunch with one of the em-
    ployees of Sunny Realestate, you discover the fol-
    lowing information:

Employee	Amount sold last month
Rebecca	$10,000,000
Maureen	800,000
Steven	800,000
Leslie	800,000

    Calculate (a) the mean, (b) the median, and (c)
    the mode. (d) Which average (mean, median, or
    mode) would you like your 5% commission to
    be based on? Why? (e) Which average would the
    company president like your 5% commission to
    be based on? Why?

6. Use the data from exercise 5 to (a) calculate the
    weighted mean (be sure to show your work!)

and (b) find the median using the columns of
figures you created.

7. A survey was conducted at State University. All
    students and faculty were asked how many cups
    of coffee they drank each day. The results of the
    survey are as follows:

Number of cups of coffee	Number of people
2	600
1	200
6	500
4	400
5	100
3	300

    Calculate (a) the weighted mean (be sure to
    show your work!), (b) the median, and (c) the
    mode. (d) Explain in words your calculation of
    the weighted mean.

8. Mrs. Jensen asked her late afternoon accounting
    class, a class of both full-time and part-time stu-
    dents, how many classes they were enrolled in
    this semester. The results were as follows:

    3, 4, 5, 6, 5, 4, 5, 5, 3, 2, 2, 1, 4, 5, 2, 1, 1, 3, 4, 5.

    Determine (a) the mean, (b) the median, and (c)
    the mode.

9. Jessie Bell decided to start her own floral shop.
    Before establishing the prices for her shop, Jessie
    decided she would look at the prices set by her
    competitors. Ms. Bell decided she would price a
    dozen long-stem roses. The following data are
    the results of her investigation: $60, 70, 75, 60,
    45, 60, 70, 70, 45, 75, 75, 75, 45, 45, 60, 60, 60,
    70, 70, 70. According to these results, what is
    the average price of long-stem roses? (*Hint:* the
    average price may vary, depending on which av-
    erage is calculated.)

10. Edward Blight wished to open a limousine ser-
    vice in Huntwood, Rhode Island. After having
    done preliminary research, Edward discovered
    that a community's average yearly income must
    be at least $1 million in order to support a lim-

ousine service. The following are the incomes of those individuals currently residing at Huntwood:

Annual income	Number of individuals
$2,000,000	400
500,000	600

Should Edward open his limousine service in Huntwood?

11. Phounganh Ho received the following grade report for her first semester at Twin Cities Community College:

Course	Grade	Credits	QP	Credits × QP
English	B	3		
History	A	3		
Marketing	C	3		
Ethics	C	3		
French	D	3		
		15		

Determine her semester average if an A receives 4 quality points, a B receives 3 quality points, a C receives 2 quality points, a D receives 1 quality point, and an F receives 0 quality points.

12. Michael Krubally attends Twin Cities Community College as well, and he received the following grades for his last semester.

Course	Grade	Credits
English	A	3
Chemistry	D	4
Statistics	D	3
Health	F	1
Intro to Internet	A	2
Accounting	B	3
		16

Calculate Michael's semester average.

13. Mikel Lykdel attends Lakeside University, where the grading system includes plus and minus grades on a 7-point scale in which C receives 0 quality points, C+ receives 1 quality point, B− receives 2 quality points, and so on, up to A+ receives 7 quality points. Similarly, C− receives −1 quality point, D+ receives −2, on down to an F−, which receives −7 quality points. Mikel received the following grades for his last term:

Course	Grade	Credits
English	B+	3
Humanities	C−	4
Political Science	A−	3
Phys. Educ.	A+	1
Journalism	B	2
Calculus I	C−	3
		16

Calculate Mikel's grade point average for the term.

14. Herkle Ornamental Manufacturing Enterprises (HOME), a maker of plastic ornamental furnishings for homes, recently did a study of its 200 employees to determine the number of ages of people (employees and their dependents) who were covered by the company's health insurance plan. The results are included in the following table:

Class ages	Frequency f	M	f × M
lower ≤ age < upper			
0–10	20		
10–20	45		
20–30	45		
30–40	80		
40–50	140		
50–60	130		
60–70	40		

Calculate the average (mean) age of the people covered by the plan.

15. Security Plus Insurance Co. wanted to determine the number of its employees who were in certain age brackets. A study of employee records revealed the following:

Class ages	Frequency f	M	f × M
limit ≤ age < upper			
20–30	10		
30–40	15		
40–50	20		
50–60	90		
60–70	45		

Calculate the mean age of the employees.

3.3   NUMERICAL SUMMARY MEASURES: MEASURES OF DATA SPREAD

The measures of central tendency indicate where the data are centered. Another concern we have with the interpretation of data is the spread of the data. How are the data grouped around the center, or centers? What is the overall spread of the data? Is it grouped together somewhat tightly around the center of the data, or is it spread out widely from the center? The spread of the data can be very helpful to us when we are trying to maintain some form of control over a variable or process. Knowing the dispersion (spread) of the data is every bit as important as knowing where its centers are located.

Range

Learning objective
Calculate the range, variance, and standard deviation for a given set of data.

The simplest measure of data spread is known as the range. The **range** of a set of data is defined to be the difference between the largest and the smallest values.

Formula for range

range = largest value − smallest value

Example 18

Consider the following data for tips received by a waiter during the 2-hour lunch period at Chez Paris Restaurant in downtown Syracuse: $2.20, 4.75, 1.00, .75, 5.50, 2.40, 3.20, 4.75, 4.75, 6.00, 3.50, 3.50, 5.50. Determine the mean, median, mode, and range of the tip values.

Solution

The mean of this data is $3.68: $47.80 ÷ 13 = 3.6769.

If we rearrange the data in ascending order, we can see the maximum tip, the minimum tip, and the median tip:

$.75, \$1.00, \$2.20, \$2.40, \$3.20, \$3.50, \$3.50, \$4.75, \$4.75, \$4.75, \$5.50, \$5.50, \$6.00$

We now can see easily that the maximum tip received was $6.00 and the minimum tip was $.75. The median of the tip data is $3.50, and the mode of the data is $4.75. The *range* of the data is:

range = maximum − minimum

$= \$6.00 - \$.75$

$= \$5.25$

We can form a graphical arrangement of the data in Example 18 as in Figure 3.21. Notice that the data are spread out from $.75 to $6.00, and that although the data does group around the centers at $3.50, $3.68, and $4.75, it has a wide range of $5.25. We would like to have a measure that would not only describe the overall spread of the data, but that would also describe the spread within the distribution. The measures of *variance* and *standard deviation* will provide us with such a description.

Figure 3.21

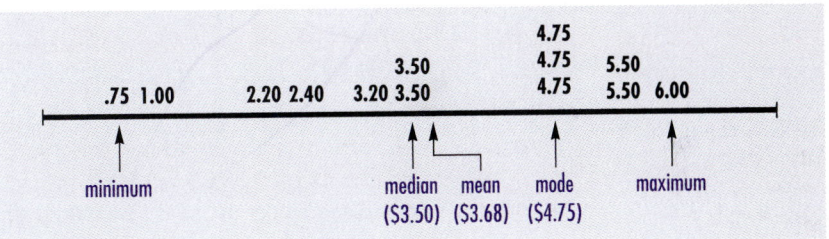

Variance and Standard Deviation

Variance is a form of an average based on the squared difference of each data value with the distribution's mean:

$$\text{variance} = \frac{\Sigma(x - \bar{x})^2}{n - 1}$$

A shorter and easier form of this formula is called the *computational form of variance:*

$$\text{variance} = \frac{\Sigma x^2 - \dfrac{(\Sigma x)^2}{n}}{n - 1}$$

The **standard deviation** is simply the square root of variance:

$$\text{standard deviation} = \sqrt{\text{variance}}$$

Example 19

Determine the mean, variance, and standard deviation for the following set of data: 3, 5, 6, 7, 9.

Solution

Step 1: List the data in a column, and sum up the column:

$$
\begin{array}{l}
3 \\
5 \\
6 \\
7 \\
\underline{9} \\
30
\end{array}
\quad \bar{x} = \frac{\Sigma x}{n} = \frac{30}{5} = 6
$$

Step 2: Make a second column next to the first, and square each piece of the data:

x	x^2
3	9
5	25
6	36
7	49
9	81
sum of data $(\Sigma x) = 30$	sum of squared data $(\Sigma x^2) = 200$

Step 3: Enter these sums into the formula and calculate the variance:

$$
\begin{aligned}
\text{variance} &= \frac{\Sigma x^2 - \dfrac{(\Sigma x)^2}{n}}{n-1} \\[2mm]
&= \frac{200 - (30)^2/5}{4} \\[2mm]
&= \frac{200 - (900/5)}{4} \\[2mm]
&= \frac{200 - 180}{4}
\end{aligned}
$$

$$= \frac{20}{4}$$

$$= 5$$

standard deviation $= \sqrt{\text{variance}} = \sqrt{5} = 2.236$

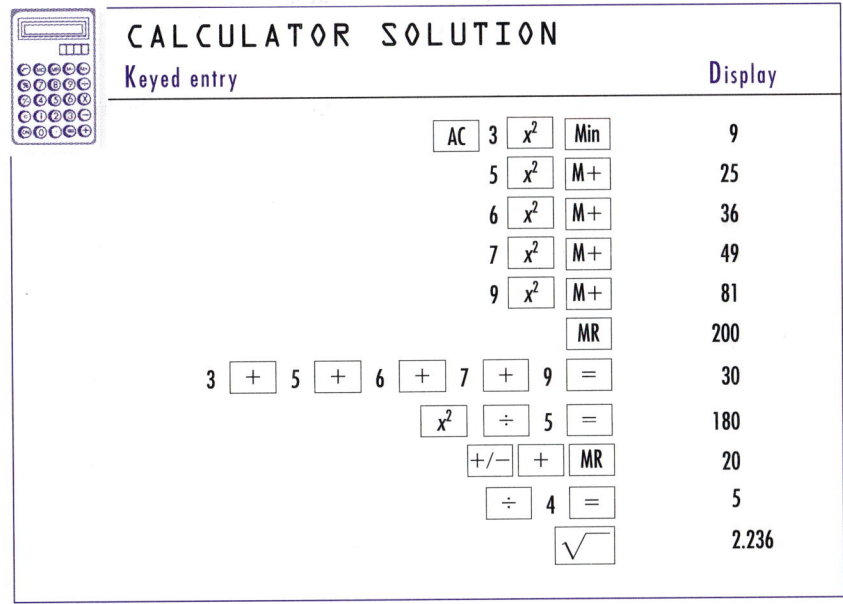

CALCULATOR SOLUTION	
Keyed entry	**Display**
AC 3 x^2 Min	9
5 x^2 M+	25
6 x^2 M+	36
7 x^2 M+	49
9 x^2 M+	81
MR	200
3 + 5 + 6 + 7 + 9 =	30
x^2 ÷ 5 =	180
+/− + MR	20
÷ 4 =	5
√	2.236

Variance is symbolized s^2, and standard deviation is s. So, in Example 8, $s^2 = 5$ and $s = 2.236$.

Let's consider another example that will show how the standard deviation can be helpful in describing the spread of a distribution.

Example 20

Consider the following three sets of data (values ranging from 1 to 9) with their mean, median, mode, and range, and compute the variance and standard deviation of each.

Data set I:

```
                    5
              4     5     6
        3     4     5     6     7
  1     2     3     4     5     6     7     8     9
  |                                         |
```

mean = 90/18 = 5
median = 5
mode = 5
range = 9 − 1 = 8

Data set II:

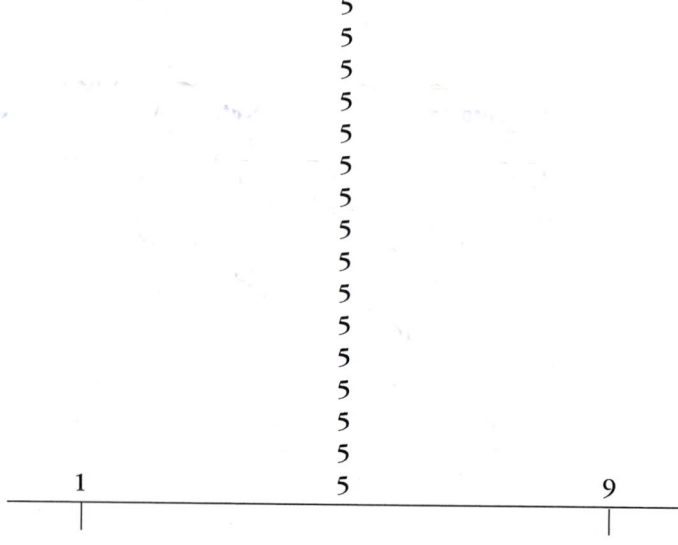

mean = 90/18 = 5
median = 5
mode = 5
range = 9 − 1 = 8

Data set III:

mean = 90/18 = 5
median = 5
mode: bimodal, one at 1 and one at 9
range = 9 − 1 = 8

If we did not have the graphs to look at, and only had the numerical summary measures (mean, median, mode, and range), the first two distributions might seem to be the same, and only the mode would tip us off that the third data set is different from the first two sets. But we can see that all three data sets are significantly different. Determine the variance and standard deviation for each of the data sets:

Solution

Data set I		Data set II		Data set III	
x	x^2	x	x^2	x	x^2
1	1	1	1	1	1
2	4	5	25	1	1
3	9	5	25	1	1
3	9	5	25	1	1
4	16	5	25	1	1
4	16	5	25	1	1
4	16	5	25	1	1
5	25	5	25	1	1
5	25	5	25	5	25
5	25	5	25	5	25
5	25	5	25	9	81
6	36	5	25	9	81
6	36	5	25	9	81
6	36	5	25	9	81
7	49	5	25	9	81
7	49	5	25	9	81
8	64	5	25	9	81
9	81	9	81	9	81
90	522	90	482	90	706
Σx	Σx^2	Σx	Σx^2	Σx	Σx^2

$$s^2 = \text{variance} = \frac{\Sigma x^2 - \dfrac{(\Sigma x)^2}{n}}{n - 1}$$

Data set I:

$$\text{variance } of\ s^2 = \frac{522 - (90)^2/18}{17} = 4.24$$

$$\text{standard deviation} = s = \sqrt{4.20} = 2.06$$

Data set II:

$$\text{variance } of\ s^2 = \frac{482 - (90)^2/18}{17} = 1.88$$

$$\text{standard deviation} = s = \sqrt{1.88} = 1.37$$

Data set III:

$$\text{variance } of\ s^2 = \frac{706 - (90)^2/18}{17} = 15.06$$

$$\text{standard deviation} = s = \sqrt{15.05} = 3.88$$

The largest standard deviation, $s = 3.88$, tells us that the data of the third data set is spread furthest from its center of 5, and the smallest standard deviation, $s = 1.37$, tells us that the data of the second data set is closest to its center of 5.

Quite often, distributions are shaped like data set I, in a mound-like, somewhat symmetrical shape; then we can mentally estimate what our standard deviation should be. Generally, the standard deviation should be equal to about $1/4$ of the range. So we could expect that the standard deviation for a distribution with a range of 8 to be about 2. When our actual calculation is close to this estimate, we would expect our computation to be correct and the shape of the data to be similar to that of data set I. If our actual calculation is substantially different, such as with data set III, we might want to double check our calculations. If the calculations are correct, then we would want to make a closer inspection of the data and note any special characteristics, such as the fact that the data are located at the outer extremities of the distribution, as in data set III.

CHECK YOUR KNOWLEDGE

Measures of Data Spread

Solve the following problems. Round answers to the nearest hundredth.

1. Calculate the mean, range, variance, and standard deviation for 3, 6, 7, 7, 2.

2. Given this set of data: 2, 6, 8, 9, 11, 12, (a) determine the range, (b) estimate the standard deviation by using $1/4$ of the range, and (c) compute the standard deviation using the formula for s.

3. Determine the mean, median, mode, range, and standard deviation for the following set of data: 2, 3, 3, 5, 5, 5, 6, 6, 7, 8.

4. WBXL radio station decided to investigate the cost of unleaded gasoline in its hometown of Beeville. Staff went to the eight self-serve gas stations in town and priced 87-octane unleaded gas at the pump. Their results follow. Determine the mean, median, maximum, minimum, range, and standard deviation of the cost of 87-octane gas on the day of the survey.

Cost per gallon, c	c^2
1.43	_____
1.28	_____
1.55	_____
1.37	_____
1.35	_____
1.33	_____
1.36	_____
1.33	_____

5. Calculate the variance and standard deviation for the tip data in Example 4.

Tip, t	t^2
$ .75	_____
1.00	_____
2.20	_____
2.40	_____
3.20	_____
3.50	_____
3.50	_____
4.75	_____
4.75	_____
5.50	_____
5.50	_____
5.50	_____
6.00	_____

Answers to CYK: 1. mean = 5; range = 5; variance = 5.50; standard deviation = 2.35
2. a. 10; b. 2.5; c. 3.63   3. mean = 5; median = 5; mode = 5; range = 6;
standard deviation = 1.89   4. mean = $1.38; median = $1.36; mode = $1.33;
max = $1.55; min = $1.28; range = $.27; standard deviation = $.08
5. variance = s^2 = 3.12; standard deviation = $1.77

3.3   EXERCISES

Solve the following exercises. Round answers to the nearest hundredth.

1. Calculate the mean, range, variance, and standard deviation for: 7, 4, 9, 11, 13, 17, 21, 14.

2. Calculate the mean, range, variance, and standard deviation for: −3, 5, −7, 4, −6, 2, 5, 11, 7.

3. Calculate the mean, range, variance, and standard deviation for: 20, 21, 25, 29, 30.

4. Given the set of data: 5, 6, 7, 7, 8, 8, 8, 9, 9, 10, 11, (a) determine the range of the data, (b) estimate the standard deviation by using 1/4 of the range, and (c) compute the standard deviation. (d) Was the estimate close to the standard deviation value? If not, then why not?

5. Given the set of data: 1, 7, 1, 7, 1, 7, 1, 7, 1, 7, 1, 7, (a) determine the range of the data, (b) estimate the standard deviation by using 1/4 of the range, and (c) compute the standard deviation. (d) Was the estimate close to the standard deviation value? If not, then why not?

6. Given the set of data: 12, 17, 17, 17, 17, 17, 17, 17, 22, (a) determine the range of the data, (b) estimate the standard deviation by using 1/4 of the range, and (c) compute the standard deviation. (d) Was the estimate close to the standard deviation value? If not, then why not?

Use the following information to complete exercises 7 through 11. Janet Dahl works for a regional power company that employs more than 1,000 people. Jan is one of 100 supervisors who perform similar work and whose jobs are classified at the same level for compensation purposes. Jan felt that male supervisors were paid more than female supervisors. To check her suspicions, she took a random sample of five men's salaries and a second sample of five women's salaries.

Men's salaries (1,000s of dollars)	Women's salaries (1,000s of dollars)
29	30
26	28
31	24
35	22
27	26

7. Determine the mean and median of the men's salaries.

8. Determine the mean and median of the women's salaries.

9. Determine the range and standard deviation of the men's salaries.

10. Determine the range and standard deviation of the women's salaries.

11. Do you believe Jan has a legitimate complaint, based on the data?

Use the following data to complete exercises 12 through 14. Speedy's Pizza delivery service has eight delivery persons. The mileage for each driver on a given day is: 32, 50, 24, 45, 56, 71, 100, 30.

12. Determine the mean of the miles driven for that day.

13. Determine the range of the mileages.

14. Determine the variance and standard deviation.

Use the following information to complete exercises 15 through 20. Elbridge Public Library wanted to determine the use of its computer software during weekends in the fall (September through November) and the spring (April through June). The number of times software was checked out for a weekend in either season follows.

Fall checkouts		**Spring checkouts**	
11	121	13	169
9	81	15	225
6	36	12	144
13	169	16	256
10	100	10	100
8	64	16	256
12	144	9	81
11	121	13	169
12	144	15	225
11	121	8	64
8	64	14	196
12	144	15	225
123	1309	156	2110

15. Determine the mean use during the fall.

16. Determine the mean use during the spring.

17. Determine the range of use during the fall.

18. Determine the range of use during the spring.

19. Determine the standard deviation of the use in the fall.

20. Determine the standard deviation of the use in the spring.

3.4 PROBABILITY AND INFERENCES FROM DATA

In Sections 3.1 through 3.3 we learned how to analyze a set of data by calculating measures of central tendency and dispersion, and by drawing various graphs representing the data. All of these measures and devices help us to form a mental picture of the data and of the situation that data represents. How can we use this information to make business decisions or predictions, called **inferences?** For instance, could we use the information that we gathered and summarized about the people in New York who took the sales aptitude exam to predict how other people who take the same exam at some future date will do on the exam? Can we predict what will happen in trading on the stock market? Can we predict whether people will buy a new product that we have developed? The answer to all these questions is that we cannot predict with 100% certainty what will happen, but we can predict what *probably will happen* in each of these uncertain situations.

Probability: The Mathematics of Uncertainty

Learning objective

Calculate the probability that an event will occur.

The mathematics of uncertainty is known as **probability.** Probability in its simplest form is a *ratio.*

Probability ratio

$$\text{probability that an event will occur} = \frac{\begin{array}{c}\text{number of outcomes}\\\text{in which the event occurs}\end{array}}{\begin{array}{c}\text{total number of outcomes}\\\text{in the experiment}\end{array}},$$

provided all outcomes are equally likely.

Let's explain with two examples.

Example 21

Toss a six-sided die (singular of dice) once and determine the probability that the side showing up will be (a) 3 or (b) 1 or 3.

Solution

There are six different ways the face of the die could turn up; that is, the face showing up could show $1, 2, 3, 4, 5,$ or 6 (each of these is called an **outcome**).

a. The probability the die will show a 3?

$$\text{probability of 3} = \frac{\text{number of ways a 3 could occur}}{\text{total number of ways die could turn up}}$$

$$= \frac{1}{6}$$

$$= .167$$

b. probability of a 1 or 3 occurring $= \frac{2}{6} = \frac{1}{3} = .33$

Example 22

Professor Garcia has an accounting class of 8 men and 17 women, for a total of 25 students. If Professor Garcia randomly picks 1 student from the class (assuming all students have an equal chance to be chosen), what is the probability she will choose: (a) a man? (b) a woman?

Solution

a. $\dfrac{\text{probability}}{\text{of a man}} = \dfrac{\text{number of men in class}}{\text{total number of students in class}} = \dfrac{8}{25}$ or $.32$

b. $\dfrac{\text{probability}}{\text{of a woman}} = \dfrac{\text{number of women in class}}{\text{total number of students in class}} = \dfrac{17}{25}$ or $.68$

The largest value that a probability could be is 1, which is the probability of a *certainty*. For example, the probability that Professor Garcia will select "either a woman or a man" in Example 22 is a certainty, so

Probability of selecting a "woman or a man" $= 25/25 = 1$

Similarly, the probability that Professor Garcia will select one person who is "both a man and a woman" is impossible, since no one person can be both "a man and a woman."

Probability of selecting a woman and a man $= 0/25 = 0$

Hence, the smallest that a probability can be is 0, which is the probability of an impossible event.

Properties of probability

Probability of an event which is certain to happen = 1
Probability of an impossible event = 0
$0 \leq$ probability of any event ≤ 1

How can probability be used in management?

Example 23

Easyflow Conduit Company employs 1,000 workers and keeps daily records showing the number of workers absent from work each day due to illness. The company's records for the last 3 years show that, during the month of October, an average of 22 workers have been absent daily due to illness. Interpret this data in terms of what might be expected (a) for daily absences, and (b) for the likelihood that a particular worker will be absent on a particular day in future Octobers.

Solution

a. We might expect for October of the coming year that a daily average of 22 people will miss work due to illness.
b. We might say that the probability of any one worker being absent on a given day is $^{22}/_{1,000} = .022$.

Example 24

Let's look back at the 25 people who took the sales aptitude examination in New York. In fact, let's look at the composite frequency distribution, which showed frequencies, relative frequencies, and percent frequencies.

Class scores	Frequency f	Relative f	Percent %f
(lower limit ≤ score < upper limit)			
50–60	2	2/25 = .08	8%
60–70	5	5/25 = .20	20%
70–80	6	6/25 = .24	24%
80–90	8	8/25 = .32	32%
90–100	4	4/25 = .16	16%
Totals	25	1.00	100%

We can see that the relative frequency column is made up of ratios, and those ratios could be thought of as probabilities.

Assume that we put each of the scores of the 25 New York applicants on a separate, but otherwise identical, piece of paper and put all 25 pieces of paper in a small box, mixed them up thoroughly, and then randomly drew 1 piece of paper from the box. We could determine several probabilities.

What is the probability we would draw a piece of paper with a score on it that is:

a.   a number in the 70s?
b.   a number in the 90s?
c.   a number 70 or greater?
d.   a number smaller than 80?
e.   a number greater than or equal to 60 but less than 90?

Solution

Clearly, if we review the relative frequencies as probabilities, we can predict the likelihood of each of the outcomes listed.
a.   Probability of a number in the 70s = 6/25 = .24
b.   Probability of a number in the 90s = 4/25 = .16
c.   Probability of a number 70 or greater = P(70s) or P(80s) or P(90s) = (6 + 8 + 4)/25 = .72, or another way is P(70s) + P(80s) + P(90s) = .24 + .32 + .16 = .72
d.   Probability of a number smaller than 80 = .08 + .20 + .24 = .52
e.   Probability of a number greater than or equal to 60 but less than 90 = .20 + .24 + .32 = .76

Chance

We often hear people refer to the "chance" of an event occurring, such as the "chance" of selecting a paper in the experiment of example 24 with a score in the 50s on it. The concepts of **chance** and *probability* are very much the same. The difference between the two is that of using the *decimal form* of a ratio and using the *percent form* of a ratio. When we speak of probability, we use a number between 0 and 1 (i.e., usually a fraction or decimal between 0 and 1).

The *probability* of selecting a score in the 50s = .08

When we speak of the chance of the occurrence of a specific outcome, we use percentages. The chance of an event occurring is between 0% and 100%. Therefore, the chance of selecting a score in the 50s = 8%. Although the difference may seem small, it is an important distinction.

Inferences from Data

Probability, especially as indicated in Examples 23 and 24, brings us to the point of wanting to make *inferences* about the situations represented in these examples. The managers of Easyflow Conduit Company would like to be able to project the last three years of absenteeism into their planning for the future operations of their company. They might plan on about 22 people being absent each day. They realize that since this number is an average, some days there will be more than 22 people absent, and some days fewer than 22 will be absent. They could also look at each employee and predict that there is a probability of .022, or a 2.2% chance, that the employee will be absent on a given day.

Similarly, we might look at the data in Example 24 and try to predict the performance of other applicants who might take the sales aptitude exam in the future. We might predict that 8% of the future applicants will score in the 50s, 20% in the 60s, 24% in the 70s, and so on. But, if we look at the results of the tests of the people in Seattle who took the same exam, those results reflect quite different probabilities, and we recognize that making such inferences is not so simple.

The applied mathematical discipline of *statistics* studies these kinds of problems and provides methods which enable us to make such inferences (predictions and decisions). At this point in our study of statistics, we can legitimately use the data from the scores of the New York applicants and the Seattle applicants to form probabilities about those people who *actually took* the exam and are represented in the frequency distributions. However, we will need a greater knowledge of statistics to be able to predict what will be the performance of other people who have not yet taken the exam.

CHECK YOUR KNOWLEDGE

Probability: The Mathematics of Uncertainty

Solve the following problems. Round probabilities to the nearest hundredth, and chances to the nearest tenth of a percent.

1. A twelve-sided die (each face having a single number from 1 through 12 on it) is tossed once. What is the probability that the side showing "up" is: (a) a side with a 7 on it? (b) a side with a 4 on it? (c) a side with an even number on it? (d) a side with a number less than 4 on it? (e) a side with a number evenly divisible by three on it?

2. The Baldwinsville Optimist Club has a drawing at each of its regular meetings. If the club has three members whose first names are Dave, four Mikes, two Bobs, one Rita, one Bruce, five Eds, and two Cheryls, what is the probability that the winner will have a first name of (a) Bob, (b) Dave, (c) Rita, (d) Ed?

3. The Grandville Chamber of Commerce held a raffle for a $1,000 prize during its Founders' Daze celebration last July and sold 250 tickets for $5 each.
    a. Jeremy D. Luckyone bought ten tickets. What was the probability that Jeremy would win?
    b. Jeremy and seven of his friends bought 50 tickets altogether between them. What is the probability that one of them won the prize?

4. Consider the 50 people's scores on the sales aptitude exam taken in Seattle.

Class scores	Frequency f	Relative f	Percent %f
(lower limit ≤ score < upper limit)			
50–60	5	5/50 = .10	10%
60–70	10	10/50 = .20	20%
70–80	20	20/50 = .40	40%
80–90	12	12/50 = .24	24%
90–100	3	3/50 = .06	6%
Totals	50	1.00	100%

Assume that each of the 50 scores has been written on 50 identical circular discs and put in a container. One disc will be drawn randomly from the container. Use the frequency distribution of the Seattle data to determine the following:

a. Probability that the number drawn is in the 80s
b. Probability that the number drawn is in the 80s or 90s
c. Probability that the number drawn is greater than or equal to 60
d. Probability that the number drawn is less than 80
e. Probability that the number is greater than 69 but less than 90
f. Probability that the number drawn is not in the 80s

5. Repeat problem 4, but determine the "chance" of each of the events a through f occurring.

3.4   EXERCISES

Solve the following problems. Round probabilities to the nearest one hundredth and chances to the nearest tenth of a percent.

1. A four-sided die (each face having a single number from 1 to 4 on it) is tossed once. What is the probability that the face that lands down has a number satisfying the following conditions: (a) the number 3 on it, (b) the number 1 on it, (c) either the number 2 or the number 4 on it, (d) an even number (divisible by 2) on it, (e) a number greater than 3 on it?

2. An eight-sided die (each face having a single number from 1 to 8 on it) is tossed once. What

is the probability that the face that lands down has a number satisfying the following conditions: (a) the number 5 on it, (b) the number 7 on it, (c) either the number 5 or the number 7 on it, (d) a number divisible by 3 on it, (e) a number greater than 3 on it?

3. The "Thespians," a college drama club, has 13 men and 17 women members. Each member's name is to be put in a hat and one name will be drawn; the person whose name is drawn will receive free tickets to a performance of "Les Miserables." What is the probability that the name drawn will be that of (a) a man? (b) a woman?

4. A group of card players who frequent the college snack bar includes two men named Bob, three women named Diane, and one man each named Jack, John, Arnie, and Jose. There are also two women named Michelle and one woman each named Rita, Lynette, and Susan. All the card players agree to meet at the snack bar at 8:00 P.M. on Friday for an evening of cards. If they all have an equal chance of arriving on time for the game, what is the probability that the first person to arrive will be (a) a woman? (b) a man? (c) a woman named Michelle? (d) a man named Bob? (e) a man whose name starts with a J? (f) a woman whose name is either Diane or Michelle?

5. A five-dollar bill, a one-dollar bill, a ten-dollar bill, and a twenty-dollar bill are put into a hat, and one is drawn out randomly. What is the probability that the value of the bill drawn is (a) $10? (b) $20? (c) $5 or $10 or $20? (d) $50?

6. A one-dollar bill, a five-dollar bill, a ten-dollar bill, a twenty-dollar bill, and a fifty-dollar bill are put into a hat, and one bill is drawn out randomly. What is the probability that the value of the bill drawn is (a) $10 or $20? (b) less than $50? (c) $10 or $20 or $50? (d) $50?

7. The "Commuters Club" membership at Good Times College has 7 members who are in their 20s, 10 in their 30s, 5 in their 40s, and 2 in their teens. If one person is chosen at random from the group, what is the probability the person's age is in (a) the 20s? (b) the 40s? (c) the teens? (d) the 50s?

Use the following information to answer problems 8–15.

Given the following frequency distribution of ages of patients in the rehabilitation unit of Community General Hospital, assume that one patient's name will be selected at random.

Class ages	Frequency f	Relative f	Percent %f
(lower limit ≤ age < upper limit)			
5–15	3	.050	5.0
15–25	6	.100	10.0
25–35	15	.250	25.0
35–45	20	.333	33.3
45–55	10	.167	16.7
55–65	3	.050	5.0
65–75	3	.050	5.0
Totals	60		

8. What is the probability that the age of the person selected is greater than or equal to 45 but less than 55?

9. What is the probability that the age of the person selected is at least 25 but less than 65?

10. What is the probability that the age of the person selected is less than 75?

11. What is the probability that the age of the person selected is greater than 75?

12. What is the probability that the age of the person selected is at least 15?

13. What is the probability that the age of the person selected is at least 65 or is less than 25?

14. What is the probability that the person selected has not yet reached her or his 25th birthday?

15. What is the chance that the person selected has not yet reached her or his 55th birthday?

Use the following information to answer problems 16-25.

Well Built Construction Company focuses its business on commercial remodeling and repair work. The management developed the following frequency distribution of anticipated revenues from projects for which they have contracts during the next year. One contract will be selected at random from the pile of contracts for next year's work.

Class revenue (× $1,000)	Frequency f(# contracts)	Relative f	Percent %f
(lower limit ≤ value < upper limit)			
10-30	2		
30-50	3		
50-70	5		
70-90	6		
90-110	8		
110-130	12		
130-150	8		
150-170	6		
Totals	50		

16. Complete the relative frequency and percent frequency columns.

17. What is the probability that the contract selected will have a revenue value of at least $110,000?

18. What is the probability that the contract selected will have a revenue value of at least $50,000 and less than $150,000?

19. What is the probability that the contract selected will have a revenue value of less than $90,000?

20. What is the probability that the contract selected will have a revenue value of less than $170,000?

21. What is the probability that the contract selected will have a revenue value of less than $10,000?

22. What is the chance that the contract selected will have a revenue value of less than $150,000?

23. What is the chance that the contract selected will have a revenue value of at least $30,000 but less than $110,000?

24. What is the chance that the value will be less than $5,000?

25. What is the chance that the value will be at least $10,000?

CHAPTER REVIEW EXERCISES

These review exercises are designed to help you quickly recall and review the primary skills that have been taught in this chapter. Round answers to the nearest hundredth (percents to the nearest tenth of a percent).

1. Complete the last two columns of the following frequency distribution.

Class scores	Frequency f	Relative f	Percent %f
lower ≤ value < upper			
0–10	5		
10–20	10		
20–30	20		
30–40	10		
40–50	5		
	50		

2. Draw a bar graph using the class and frequency columns of the frequency distribution in problem 1.

3. Draw a line graph using the class and frequency columns of the frequency distribution in problem 1.

4. Draw a percent bar graph using the class and percent frequency columns of the frequency distribution in problem 1.

5. Draw a percent line graph using the class and percent frequency columns of the frequency distribution in problem 1.

6. Draw a pie chart using the class and relative frequency columns of the frequency distribution in problem 1.

7. Calculate the mean of the following data set: 1, 1, 1, 2, 2, 2, 3, 3, 3, 3, 3, 4, 4, 4, 5, 5, 5, 6, 6, 17

8. Calculate the median of the following data set: 1, 1, 1, 2, 2, 2, 3, 3, 3, 3, 3, 4, 4, 4, 5, 5, 5, 6, 6, 17

9. Determine the mode of the following data set: 1, 1, 1, 2, 2, 2, 3, 3, 3, 3, 3, 4, 4, 4, 5, 5, 5, 6, 6, 17

10. Of the three measures—mean, median, and mode—of the previous data set, which do you feel is the best indicator of the center of the distribution, and why?

11. Calculate the range of the following data set: 1, 1, 1, 2, 2, 2, 3, 3, 3, 3, 3, 4, 4, 4, 5, 5, 5, 6, 6, 17

12. Calculate the variance of the following data set: 1, 1, 1, 2, 2, 2, 3, 3, 3, 3, 3, 4, 4, 4, 5, 5, 5, 6, 6, 17

13. Calculate the standard deviation of the following data set: 1, 1, 1, 2, 2, 2, 3, 3, 3, 3, 3, 4, 4, 4, 5, 5, 5, 6, 6, 17

14. Calculate the weighted mean for the following data:

Value	Weight
20	3
25	5
35	8
42	6
55	2

15. Calculate the mean of the frequency distribution given in problem 1.

16. A paper bag contains 5 red balls, 3 blue balls, 7 green balls, 4 orange balls, and 6 purple balls. If the balls are all the same size and cannot be seen from outside the bag, and if one ball is picked

at random from the bag, what is the probability that it is blue?

17. What is the chance of drawing an orange ball in problem 16?

18. What is the probability of drawing a brown ball in problem 16?

19. Refer back to problem 1. Assume that all of the values represented in that frequency distribu-

tion were put on identical pieces of paper and put into a container, and one piece of paper was drawn from the container. What is the probability that the piece of paper would have a number with a value of at least 30?

20. What is the chance that the number picked in problem 19 is less than 40?

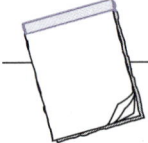

EXPRESS YOUR UNDERSTANDING

Compose one or two well-written sentences to express the requested information in your own words.

1. What do we mean by the word *data?*

2. What is a frequency distribution?

3. Describe how you form a relative frequency distribution from a frequency distribution.

4. How does a percent frequency distribution differ from a relative frequency distribution?

5. Describe how you would build a bar graph representing revenue from departments A, B, and C, if their corresponding percentage of total revenues of $100,000 were 25%, 20%, and 55%.

6. Describe how you would build a pie chart using the information in problem 5.

7. Describe how you would calculate the mean of a set of ungrouped data.

8. Explain what is meant by a weighted mean.

9. Explain how you would determine the median of a set of data.

10. Explain how you would calculate the range of a set of data.

11. What is the mode of a set of data?

12. What is the relationship between variance and standard deviation?

13. Explain how you would determine the probability that an event would occur.

14. What is the difference between probability and chance?

15. Explain what we mean by an inference.

SELF-TEST

A. Terminology

Complete the following items using the key terms presented at the beginning of the chapter. Check your responses against the answer key at the end of the test.

1. When we take ungrouped data and group it into classes and frequencies, we form a _____ distribution.

2. If we use the frequency column and the total of that column to form ratios to replace our frequency column, then the result is a _____ _____ distribution.

3. If we express the ratios just mentioned as percents, and we form a distribution made up of the classes and these percents, we have formed a _____ _____ distribution.

4. The two basic types of graphs that we constructed using our classes and the frequency column are _____ and _____ graphs.

5. A graph that breaks up a circle into proportional
parts of the total is called a _____ graph,
or a pie chart.

6. Statistical measures used to locate the center or
the average of a set of data are called measures
of _____ _____ .

7. The _____ is a measure that indicates the
value that occurs most frequently.

8. The _____ is a measure of central tendency indicating the middle value of a set of values.

9. The _____ for a set of numbers is calculated as the sum of all the values divided by the
number of values in the set.

10. The measuring of dispersion that gives the overall spread of the data is called the _____ .

11. The measure of dispersion that is the square root
of the variance is called the _____
_____ .

12. A ratio of the number of ways that an event can
occur the way we want it to occur compared
to the total number of ways that the event *could*
occur is called the _____ .

13. The percent form of a probability is called
_____ .

14. When we use data described using our de-
scriptive statistics together with probability to
make a prediction or a decision, we are making
an _____ from the data.

B. Calculation

The following concepts and short problems are designed to test your understanding
of the objectives identified at the beginning of the chapter. Answers are provided at
the end of the test. Round your answers to the nearest hundredth, and percents to
the nearest tenth of a percent.

15. Dawn Downes is a personal finance consultant when she is not working at her
full-time job as an accountant at RSVP Health Center. Last year she formed a
frequency distribution of the amount of money initially invested in variable annuities by her clients using her services.

Class amt. inv. ($\times$ \$1,000)	Frequency f(# clients)	Relative f	Percent %f
(lower limit $\leq$ value $<$ upper limit)			
0–20	8		
20–40	10		
40–60	12		
60–80	6		
80–100	4		
Totals	40		

Complete the relative frequency and the percent frequency columns in the frequency distribution.

16. Dawn felt that some graphs would help her get a better visual picture of the small business that she was operating. Help her out by constructing the following graphs for the amounts invested in variable annuities using the information provided in problem 15.

   a.   a bar graph
   b.   a line graph
   c.   a percent bar graph
   d.   a percent line graph
   e.   a pie chart showing the percent of the total that each level of investment represents

17. Andy Riposo works for a car dealership. He receives a weekly salary of $400 plus a commission on each car he sells. During the last quarter his commissions were:

$75, $90, $80, $90, $125, $100, $90, $100, $200, $150, $125, $95

   a.   Calculate the mean of Andy's commissions for the quarter.
   b.   Calculate the median of Andy's commissions for the quarter.
   c.   Determine the mode of Andy's commission for the quarter.
   d.   What is the minimum commission Andy made?
   e.   What is the maximum commission Andy made?
   f.   Calculate the range of Andy's commissions for the quarter.
   g.   Calculate the variance of Andy's commissions using the formula

$$s^2 = [(\Sigma x^2 - (\Sigma x)^2/n](n - 1)$$

given that $\Sigma x^2 = 159,100$ and that $\Sigma x = 1,320$.

   b.   Calculate the standard deviation of Andy's commissions.

18. The men's intramural basketball league has 28 players who meet each week to make up four teams and play basketball for 2 hours. Each week, four names are drawn, one at a time, from a hat. Those four people are team captains for that week. After a name is selected, it is not returned to the hat, since a person can only captain one team at a time. Tom Harkness is in the league.

   a.   What is the probability that Tom will be selected first to be a captain?
   b.   If Tom is not selected first, what is the probability he will be selected for the second captain?
   c.   If Tom is not selected for the first or second captains, what is the probability that he will be selected as the third captain?

19. In problem 18, what is Tom's chance of being selected as captain of the fourth team if he is not selected for the first, second, or third captains?

20. Reconsider problem 15 and assume Dawn randomly pulls one client's file folder from her file drawer, which has one file for each client. What is the probability that she will pull a client's folder who initially invested less than $60,000?

Answers to Self-Test: *1.* frequency   *2.* relative frequency   *3.* percent frequency   *4.* bar, line
   5. circle   *6.* central tendency   *7.* mode   *8.* median   *9.* mean
   10. range   *11.* standard deviation   *12.* probability   *13.* chance
   14. inference

Answers to Self-Test:
(continued)

15.

Class amt. inv. (× $1,000)	Frequency f(# clients)	Relative f	Percent % f
(lower limit ≤ value < upper limit)			
0–20	8	.20	20%
20–40	10	.25	25%
40–60	12	.30	30%
60–80	6	.15	15%
80–100	4	.10	10%
Totals	40		

16.

Investment	# clients	Relative f	%f
0–20	8	0.2	20%
20–40	10	0.25	25%
40–60	12	0.3	30%
60–80	6	0.15	15%
80–100	4	0.1	10%
Totals	40		

a.

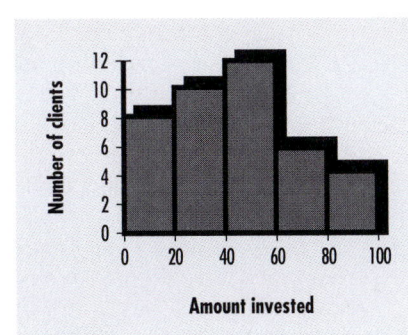

b.

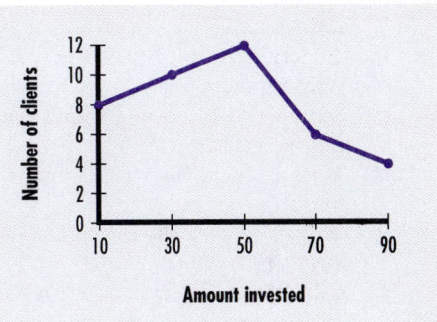

c.

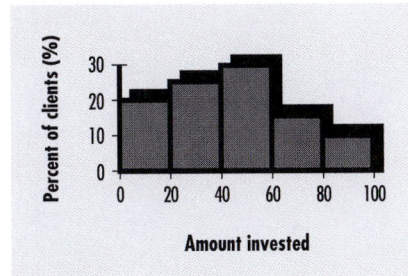

d.

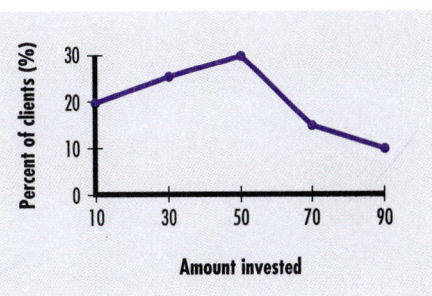

Answers to Self-Test:
(continued)

e.

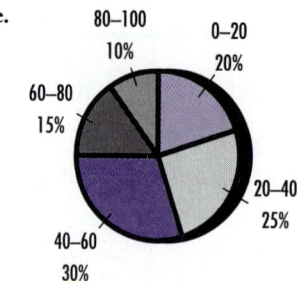

80–100
10%

0–20
20%

60–80
15%

20–40
25%

40–60
30%

17. **a.** $110 **b.** $97.50 **c.** $90 **d.** $75 **e.** $200 **f.** $125 **g.** $1263.64
h. $35.55   *18.* **a.** 1/28 **b.** 1/27 **c.** 1/26   *19.* 4%   *20.* 75

4

Key Terms

balance sheet
assets
liabilities
owner's equity
fundamental accounting
  equation
capital
income statement
cost of goods sold
net income
vertical analysis
gross profit
horizontal analysis
ratio
current ratio
acid test ratio
leverage
debt-to-equity ratio
debt-to-asset ratio
asset turnover ratio
inventory turnover ratio
receivables turnover
  ratio
return on net sales ratio
return on investment
  ratio

FINANCIAL STATEMENTS AND ANALYSIS

Learning Objectives

After completing the exercises in this chapter, you will be able to:

1. Explain the purpose and function of financial statements.

2. Prepare a balance sheet.

3. Prepare an income statement.

4. Complete a vertical analysis of the balance sheet and income statement.

5. Complete a horizontal analysis of the balance sheet and income statement.

6. Calculate primary financial statement ratios.

7. Define the key terms.

INTRODUCTION

Learning objective
Explain the purpose
and function of fi-
nancial statements.

Financial statements are the "output" of the accounting process. They are composed of financial data summarized from business transactions reported in monetary terms. The financial statements of a business provide a format to communicate economic data concerning its financial status to various segments of the financial community. For example, owners and potential investors use the information on the statements to assess the degree of risk and future ability of a company to report profits. Creditors appraise an organization's debt-paying ability before considering making loans or selling goods on credit. Government agencies are interested in the financial position of a firm for the purposes of regulation and taxation. Labor has become increasingly concerned about a company's stability as it relates to job security and wage increases. The individuals who most actively interact with the financial statements are management. Management uses financial information as an integral part of the decision making process when formulating the short-term and long-term plans of the enterprise.

Every business organization, regardless of its size, periodically prepares four primary financial statements: the income statement, the balance sheet, the owner's equity statement, and the cash flow statement. These reports reflect the financial position and the results of operations. The format of these statements may vary slightly from company to company, but their content is standardized by the Financial Standards Accounting Board.

The focus of this chapter is to familiarize you with the content and purpose of the balance sheet and the income statement. You will also be required to apply basic mathematical processes presented in earlier chapters to analyze and interpret the results of business operations reported in these two financial statements. The information contained in this chapter is very important because future chapters will present concepts and quantitative processes related to financial statements.

4.1   THE BALANCE SHEET

The **balance sheet** consists of three sections and it reports the financial position of a business at a specific point in time. **Assets** are anything of value owned by a firm such as its cash, equipment, buildings, and land. **Liabilities,** which are debts, include amounts of money owed by a firm to its creditors as a result of buying assets on credit, or borrowing money. The third section, **owner's equity,** represents the owner's right or claim against the net assets of the firm after the total liabilities are deducted (total assets − total liabilities = owner's equity). The owner's equity consists of the sum of any investments by the owner into the business and the cumulative profitability of the company, less any withdrawals by the owner(s). The relationship between the sections of the balance sheet is expressed in an equation known as the **fundamental accounting equation.**

The fundamental accounting equation

assets	**=**	**liabilities**	**+**	**owner's equity**
items of value owned by the business		amounts owed to creditors		owner's claim or investment in the business

The equation can be used to analyze the effect that various types of business activity can have on a firm's financial position. To illustrate, suppose you wanted to start a small business and you decided to invest $5,000 of your personal savings in the business. The financial position of your business now consists of the following:

$$assets \quad = liabilities + owner's\ equity$$
$$cash\ \$5,000 = \quad \$0 \quad + \quad \$5,000$$

Next, you decide to purchase $8,500 worth of office equipment from a supplier who requires you to pay a $1,500 down payment and who will provide credit for the remainder of the purchase ($8,500 − $1,500 = $7,000). The financial position of your business has changed:

	$assets$	$= liabilities +$	$owner's\ equity$
cash	$3,500		
office equipment	8,500		
totals	$12,000 =	$7,000 +	$5,000

Notice that the owner's equity is still $5,000 but your assets and liabilities have changed in value. The total assets increased to $12,000, of which $8,500 is the value of the office equipment and $3,500 is cash. The reason cash decreased from $5,000 to $3,500 is because you paid a $1,500 down payment at the time of purchase. Your liabilities are now $7,000, which is the amount you owe on the office equipment ($8,500 − $1,500 = $7,000). The residual, your equity in the business at this point, remains at $5,000 (total assets $12,000 − total liabilities $7,000 = owner's equity of $5,000). Creditors' claims to assets have priority over the owner's claims.

The balance sheet contains a variety of accounts in each of the three sections; therefore, the total assets of a company must always be equal to the total of its liabilities and owner's equity. Let's now look at an example that further explains the relationship between the equation and its variables.

Example 1

The Richland Company has total assets of $135,250, total liabilities of $92,430, and owner's equity of $42,820.

Solution

$$assets\ =\ liabilities\ +\ owner's\ equity$$
$$\$135,250\ =\ \$92,430\ +\ \quad\$42,820$$

or

$$assets\ -\ liabilities\ =\ owner's\ equity$$
$$\$135,250\ -\ \$92,430\ =\ \quad\$42,820$$

The report-form balance sheet shown in Figure 4.1 for Rojon Enterprises includes accounts listed under the categories of assets, liabilities, and owner's

Figure 4.1

Balance sheet

Rojon Enterprises
Balance Sheet
December 31, 200X

Short-term

Assets

Current assets:			
Cash		$ 183,250	
Accounts receivable (net)		347,788	
Merchandise inventory		780,000	
Notes receivable (due within current year)		32,400	
Supplies on hand		14,620	
Prepaid insurance		6,975	
Total current assets			$1,365,033
Fixed assets: *long-term*			
Land		$ 135,000	
Machinery and equipment	$ 872,600		
Less: accumulated depreciation	260,780		
		$ 611,820	
Buildings	1,420,000		
Less: accumulated depreciation	646,200		
		$ 773,800	
Total fixed assets			$1,520,620
Total assets			$2,885,653

equity. You should analyze, not memorize, the account location with reference to the definitions provided for each category.

Assets are classified in terms of time. *Current assets* are cash or other items of value owned by the company that will be consumed or converted into cash generally within one year. The current assets in Figure 4.1 are:

- *Cash:* The cash in checking accounts, savings accounts, and any currency on hand as of the balance sheet date. Some companies with long manufacturing cycles will use a longer time period.

- *Receivable Accounts:* Amounts owed to the firm by customers (for purchases of goods and services) sold by the firm. Receivables include accounts receivable and notes receivable. (Notes will be discussed in detail in Chapter 8.)

Figure 4.1 (continued)

Liabilities

Current liabilities:		
Accounts payable	$ 137,285	
Notes payable (payable within one year)	15,200	
Dividends payable	8,300	
Salaries payable	87,495	
Taxes payable	93,470	
Total current liabilities		$ 341,750
Long-term liabilities:		
Mortgage payable, due 2010	265,000	
Debenture 8% bonds payable due December 31, 2015	150,000	
Total long-term liabilities		$ 415,000
Total liabilities		$ 756,750

Stockholders' equity

Contributed capital:		
Preferred 6% stock, $50 par value, 20,000 shares		
authorized and issued	$1,000,000	
Common stock, $5.00 par value, 150,000 shares		
authorized and issued	750,000	
Total contributed capital	$1,750,000	
Retained earnings	378,903	
Total stockholders' equity		$2,128,903
Total liabilities and stockholders' equity		$2,885,653

- *Supply Accounts:* Assets that are used by the firm for operations. Supplies may be broken down into specific-use classifications such as store supplies and office supplies for better control.

- *Prepaid Accounts:* Amount paid in advance by the firm for benefits or services that have not expired or been consumed. Examples of such accounts include prepaid rent and prepaid insurance.

- *Merchandise Inventory:* The cost of all inventory the firm has available for sale on the balance sheet date.

Fixed assets (or *long-term assets,* also referred to as *plant assets*) are assets estimated to be used by the firm for a period longer than one year. They are used in the operation of the business and are not for resale. The cost of fixed assets is charged against income (written off) through a process called *depreciation,* which we will discuss in Chapter 14. In Figure 4.1, the fixed assets include the following:

- *Equipment Accounts:* The *book value* (original cost less accumulated depreciation) of equipment fixtures, machinery, and other such items owned by the firm on the date of the balance sheet.

- *Building Accounts:* The book value of buildings owned by the firm. Depreciation is considered in determining a building's net value.

- *Land Accounts:* The original cost of any land owned by the firm. Land is a fixed asset that does not depreciate.

The liabilities, much like assets, are classified as current or long-term. *Current liabilities* are those debts that must be paid by the firm within a relatively short period of time, generally one year. Current liabilities in Figure 4.1 include:

- *Accounts Payable:* The total of money owed by the firm to creditors for goods and services purchased on credit.

- *Notes Payable:* The amount of money owed by the firm to its creditors in the form of short-term promissory notes. A promissory note is a negotiable financial document that requires the payer to pay the payee an identified amount of money at some specified date, usually less than one year.

- *Salaries Payable:* The amount of money owed by the firm to its employees for wages earned but unpaid as of the date of the statement.

Long-term liabilities are those debts that will not be paid within one year. Long-term liabilities in Figure 4.1 include:

- *Mortgages Payable:* The total balance of all mortgages for which the firm is liable. Mortgages are loans used to finance buildings and property.

- *Bonds Payable:* The obligation to pay the face amount of a bond at maturity by a firm to bondholders. Bonds are issued by firms to generate funds.

Figure 4.2

Presentation of equity: corporation

Stockholders' equity		
Contributed capital:		
Preferred 6% stock, $100 par value,		
20,000 shares authorized and issued	$2,000,000	
Common stock, $25 par value, 250,000		
shares authorized and issued	6,250,000	
Additional paid-in capital, common	1,500,000	
Total contributed capital	$9,750,000	
Retained earnings	4,875,000	
Total stockholder's equity		$14,625,000

Figure 4.3

Presentation of equity: sole proprietorship

Owner's equity		
Owner's equity:		
Janet Adams, Capital, August 1		$35,250
Plus August Net Income	$15,800	
Less: Withdrawals	6,500	
Excess of Income over withdrawals		9,300
Janet Adams, Capital, August 31		$44,550

Owner's Equity: As the accounting equation indicates, the difference between total assets (current assets plus fixed assets) and total liabilities (current liabilities plus long-term liabilities). Therefore, the *owner's equity* section of the balance sheet in Figure 4.1 is determined as follows:

$$owner's\ equity\ =\ total\ assets\ -\ total\ liabilities$$
$$\$2,128,903\ \ =\ \$2,885,653\ -\ \ \ \$756,750$$

The equity section of the balance sheet will vary, depending on how a business is organized. For example, corporations identify ownership as *stockholders' equity,* which is the sum of the *stock value* and *retained earnings* (profits not distributed to shareholders). (See Figure 4.2.) A sole proprietorship (owned by an individual) or a partnership (owned by two or more individuals) identifies ownership as **capital.** Figure 4.3 shows the ownership section of a balance sheet for a sole proprietorship.

Example 2

The following account balances are listed in the books of the Village Diner: cash total $4,650; merchandise inventory $18,250; store supplies $500; office supplies $375; and prepaid insurance $1,240. The Village Diner lists equipment at $30,650 with accumulated depreciation of $2,800; a building at $120,000, with accumulated depreciation of $37,500; and land of $30,000. Liabilities include accounts payable totaling $15,300; a note payable of $4,200; and a mortgage balance of $75,860. The owner's equity is listed at $70,005. Prepare a report form balance sheet as of October 31, 200X, for the Village Diner.

Solution

Village Diner
Balance Sheet
October 31, 200X

Assets

Current assets:			
Cash		$ 4,650	
Merchandise inventory		18,250	
Prepaid insurance		1,240	
Store supplies		500	
Office supplies		375	
Total current assets			$ 25,015
Fixed assets:			
Equipment	$30,650		
Less: accumulated depreciation	2,800	$27,850	
Building	$120,000		
Less: accumulated depreciation	37,500	82,500	
Land		30,000	
Total fixed assets			$140,350
Total assets			$165,365

Liabilities

Current liabilities:		
Accounts payable	$15,300	
Notes payable	4,200	
Total current liabilities		$19,500
Long-term liabilities:		
Mortgage payable		75,860
Total liabilities		$ 95,360

Owner's equity

Village Diner capital	$ 70,005
Total liabilities and owner's equity	$165,365

CHECK YOUR KNOWLEDGE

The Balance Sheet

1. Calculate the value of a company's assets if its liabilities total $37,648 and the equity of its owners amounts to $45,376.

2. If a firm has assets of $1,347,275 and owes creditors $649,589, what amount of the firm's assets can be claimed by the owners?

3. If ownership is $72,965 and assets amount to $118,740, what is the value of the company's debt?

4. Prepare a balance sheet for Cohen's Stationery and Book Shop as of December 31, 200X, from the financial data below.

Cash	$1,252	Accum. depreciation (fixture)	$ 3,600
Accounts receivable	4,748	Building	80,000
Notes receivable	1,250	Accum. depreciation building	6,500
Supplies	875	Land	35,000
Prepaid insurance	2,400	Accounts payable	8,787
Merchandise inventory	16,479	Notes payable (3/03)	2,500
Equipment	4,200	Salaries payable	1,643
Accumulated		Mortgage payable	56,380
depreciation	1,600	H. Cohen capital	77,944
Store fixtures	12,750		

4.1 EXERCISES

Calculate the missing values in the balance sheet data.

	Assets	=	Liabilities	+	Owner's equity
1.	$85,675		_____		$63,987
2.	$372,460		$118,738		_____
3.	_____		$6,875		$24,920
4.	$1,648,250		$410,298		_____
5.	_____		$12,640		$63,875

Classify each of the following accounts as either a current asset, fixed asset, current liability, long-term liability, or owner's equity.

6. cash
7. accounts payable
8. store equipment
9. R. C. Taylor capital

10. store supplies
11. accumulated depreciation
12. mortgage
13. building

14. retained earnings
15. merchandise inventory
16. wages payable
17. land

Answers to CYK: *1.* $83,024 *2.* $697,686 *3.* $45,775 *4.* $147,254 (total assets)

18. Determine total current assets:

Cash	$ 8,460
Accounts receivable	4,370
Store equipment	12,850
Supplies	1,200
Building	42,800
Prepaid insurance	6,750
Merchandise inventory	21,640

19. Calculate total liabilities:

Accounts payable	$ 6,270
Salaries payable	1,430
Advertising expense	850
Mortgage payable	28,270
Interest expense	550
Notes payable	3,600

20. Calculate the owner's equity:

Common stock	$250,000
Cash	85,600
Preferred stock	175,250
Retained earnings	$105,750

21. Arrange the following accounts as they would appear in the assets section of the balance sheet and determine the value of total assets.

Cash	$12,350
Accounts receivable	4,680
Store equipment	15,200
Accumulated depreciation store equipment	3,600
Supplies	875
Prepaid rent	2,500
Office equipment	6,250
Accumulated depreciation office equipment	1,200
Merchandise inventory	23,950

22. The balance sheet of Leung Associates lists the following items and amounts on June 30, 200X: cash $10,250; accounts receivable $3,875; supplies $475; merchandise inventory $18,500; equipment (book value) $7,250; office furniture (book value) $4,800; accounts payable $2,780; salaries payable $2,350. What is the claim on the assets by the owners of Leung Associates?

23. Ellen Grabowski operates a craft shop. Her balance sheet after 3 years of doing business shows her equity as $8,600, which is one-fourth of the value of total assets. What is the amount of total liabilities at this point in time?

24. Last year, Comstock's Pharmacy reported assets of $62,500. This year, assets increased by $15,850 and liabilities decreased by $4,200. If Comstock capital was $52,650 this year, what was the amount of capital reported on last year's balance sheet?

25. The following is a list of account balances taken from the books of the Axton Company on March 31, 200X. Prepare a report form balance sheet for the Axton Company.

Accounts payable	8,325
Merchandise inventory	35,740
Store equipment	15,250
Retained earnings	28,250
Accounts receivable	12,800
Salaries payable	1,200
Accumulated depreciation— store equipment	4,800
Prepaid insurance	550
Common stock	70,000
Mortgage payable	41,200
Land	22,500
Office supplies	825
Cash	14,260
Office equipment	9,570
Building	85,900
Accumulated depreciation— office equipment	1,200
Store supplies	880
Accumulated depreciation— building	43,300

4.2   THE INCOME STATEMENT

Learning objective
Prepare an income statement.

The **income statement** is a summary report that measures the progress of an enterprise by determining its profitability. *Profits* are reported by a business when the inflow of net assets from the sale of goods and services ex-

Figure 4.4

Single-step form income statement

Rojon Enterprises Income Statement December 31, 200X	
Revenues:	
Net sales	$1,661,508
Cost of goods sold	710,815
Gross profit	950,693
Operating expenses	863,555
Net income from operations	87,138
Net other income and expenses	6,525
Net income	$ 80,613

ceeds the outflow of net assets associated with the activities of the period. The income statement and the statement of owner's equity describe how the balance sheet changes from one point in time to another point in time.

The format of the income statement can vary, depending on factors specific to an organization. The two most widely used forms are *single-step* and *multiple-step*. The primary difference in the two forms is the amount of detail presented in the main sections of each of the statements. For example, in the single-step income statement form, the total of all expenses is deducted from the total of all revenues to determine the net income; whereas, in the multiple-step income statement form, several sections and subsections with intermediate balances are used to report net income. Figure 4.4 shows one version of the single-step form. Notice that only the net totals of the main sections are used to determine net income.

The following example illustrates the relationship between the main sections of an income statement.

Example 3

A company reports net sales of $75,000. It cost the company $26,250 to manufacture the products sold at $33,500 in various expenses. What is the net income?

Solution

Net sales	$75,000	
Less cost of goods sold	26,250	
Gross profit	48,750	(gross profit = net sales
		− cost of goods sold)
Less expenses	33,500	
Net income	$15,250	(net income = gross profit
		− operating expenses)

Figure 4.5

Conventional multiple-step income statement

Rojon Enterprises
Income Statement
December 31, 200X

Gross sales:		$1,685,795	
Less: Sales returns and allowances	$ 6,430		
Sales discounts	17,857		
		24,287	
Net sales			$1,661,508
Cost of goods sold:			
Beginning inventory 1/1		$ 635,970	
Purchases	$856,250		
Transportation—in	18,775		
Gross delivered purchases	875,025		
Less: purchase returns and allowances	$ 4,860		
purchase discounts	15,320		
		20,180	
Net purchases		$ 854,845	
Merchandise available for sale		1,490,815	
Less: merchandise inventory 12/31		780,000	
Total cost of goods sold			$ 710,815
Gross profit (net sales − cost of goods sold)			$ 950,693
Operating expenses			
Selling expenses			
Sales salaries expenses	$237,300		
Advertising expenses	85,450		
Store supplies expenses	16,250		
Depreciation expenses—Equipment	4,750		
Depreciation expenses—Buildings	7,465		
Total selling expenses		$ 351,215	

The income statement shown in Figure 4.5 is a conventional multiple-step statement, so called because of its additional sections and the use of interim balances to calculate net income.

The various sections of the multiple-step statement as shown in Figure 4.5 will be discussed in detail in the following paragraphs.

Figure 4.5 (continued)

Conventional multiple-step income statement

General expenses			
Salaries expense	$310,200		
Insurance expense	11,685		
Taxes expense	165,420		
General supplies expense	12,800		
Depreciation expense—equipment	8,940		
Miscellaneous expenses	3,295		
Total general expenses		$ 512,340	
Total operating expenses			$ 863,555
Net income from operations			
(gross profit − total operating expenses)			$ 87,138
Other income and expenses:			
Other income:			
Interest income		350	
Other expenses:			
Interest expenses		6,875	
Net other income and expenses			$ 6,525
Net income			$ 80,613

Note: Net other income and expenses reduces net income from operations in this statement because interest expenses (cost) is greater than interest income (revenues). Therefore, to determine net income we must subtract net other income and expenses ($6,525) from net income from operations ($87,138).

In the *sales* section, *gross sales* to customers for merchandise sold for cash or on account is reported. Sales that are returned and sales discounts (reductions in price) that have been accepted are totaled and deducted from gross sales, to arrive at *net sales*. The net sales in Figure 4.5 is determined as follows:

$$\text{net sales} = \text{gross sales} - (\text{sales returns} + \text{sales discounts})$$
$$\$1,661,508 = \$1,685,795 - (\$6,430 + \$17,857)$$

The section of the income statement that is the most difficult to understand is the **cost of goods sold,** due to the detailed analysis required to determine this important figure. Every merchandising company purchases goods for resale. Therefore, records must be kept of the merchandise that has been purchased (purchases); merchandise that has been purchased but not sold (merchandise inventory); and merchandise that has been purchased and sold (cost of goods sold). In Figure 4.5, *net purchases* represent purchases plus transportation less purchase returns and purchase discounts. The cost of goods sold is determined as follows:

Beginning inventory		$ 635,970
Plus net purchases	+	854,845
Merchandise available for sale	=	1,490,815
Less ending inventory	−	780,000
Cost of goods sold	= $	710,815

The information you will need to determine the valuation of inventory will be presented in Chapter 15. The focus on inventory in this chapter is its effect on cost of goods sold. Cost of goods sold is the actual cost of the inventory that has been sold and when subtracted from net sales generally produces a markup. Markup is often referred to as gross margin or gross profit and will be discussed in detail in Chapter 6.

Operating expenses are the costs incurred directly with the sale of merchandise (selling expense) or in the operations of the business (general expense). Figure 4.5 lists examples of expenses in each of the categories. It is not always necessary to classify expenses by category; they can also be listed under the heading "operating expenses."

If a firm generates income from sources other than operations, an additional section of the income statement is required. Other income or expenses can result from nonoperating activities, such as property rental, investments in other businesses, or the gain or loss on the sale of fixed assets. *Net other income and expenses* is added to (or deducted from) income from operations to determine **net income.**

Example 4

The Gidget Company reported gross sales of $230,500 on December 31 with sales returns of $8,700. Merchandise inventory on January 1 of the year was $32,400. The company purchased $82,750 worth of goods during the year and paid $1,740 in transportation costs. Inventory on December 31 was $46,250. The company's records indicate that it had paid $80,275 in salaries and wages, $2,800 for advertising, $745 for supplies, $12,490 for taxes, $900 for insurance, and depreciated assets of $10,680. Prepare a multiple-step income statement for the Gidget Company for the current year ending December 31.

Solution

Gidget Company
Income Statement
December 31, 200X

Revenue from sales:			
Sales		$230,500	
Less: sales returns and allowances		8,700	
Net sales			$221,800
Cost of goods sold:			
Beginning inventory January 1, 200X		32,400	
Purchases	$82,750		
Transportation—in	1,740		
Net purchases		$ 84,490	
Merchandise available for sale		$116,890	
Less: merchandise inventory (12/31)		46,250	
Total cost of goods sold			$ 70,640
Gross profit			$151,160
Operating expenses:			
Salaries and wages		$ 80,275	
Advertising expense		2,800	
Supplies expense		745	
Taxes expense		12,490	
Insurance expense		900	
Depreciation expense		10,680	
Total operating expenses			$107,890
Net income			$ 43,270

CHECK YOUR KNOWLEDGE

Income Statement Sections

1. If net sales are $350,000, expenses are $168,000, and the cost of goods sold is $140,650, what is the gross profit?

2. Northeast Electrical Supply reported the following financial information for the month ending September 30, 200X: net sales $1,675,430, expenses $823,890, and cost of goods sold $703,668. What is the net income?

3. Determine the value of net sales if a firm reports expenses of $724,500, cost of goods sold $661,500, and net income of $189,000.

4. A business reports total sales for the month at $125,750 of which $4,320 was returned for cash or credit refunds. In addition, the

business recorded $2,560 in discounts on its sales figure. What is the amount of net sales to be reported for the period?

5. Calculate the cost of goods sold based on the following financial data: beginning inventory $65,800; ending inventory $42,225; and net purchases $120,750.

6. The following information was taken from the records of the Calabria Importing Company: beginning inventory $17,950, purchases $46,820, purchase discounts $560, transportation—in $835, and ending inventory $23,935. Determine the cost of goods sold.

7. Prepare a multiple-step income statement for the Southside Beverage Center as of December 31, 200X, from the following financial data: sales $34,280, sales returns and allowances $375, beginning inventory $3,795, purchases $15,250, purchase returns and allowances $460, ending inventory $7,460, rent expense $6,000, salaries expense $12,250, advertising expense $1,500, supplies expense $480, and miscellaneous expense $220.

4.2   EXERCISES

Calculate the missing values from the income statement data.

	Sales	−	Sales returns	=	Net sales	−	Cost of goods sold	=	Gross profit	−	Operating expenses	=	Net income
1.	$ 28,750		1,250		_____		18,350		_____		6,480		_____
2.	_____		3,420		65,800		_____		33,270		_____		11,575
3.	$325,845		_____		321,220		_____		168,570		96,430		_____
4.	$ 7,850		140		_____		2,510		_____		3,670		_____

Calculate the missing values in the cost of goods sold section of the income statement.

	Beginning inventory	+	Net purchases	=	Merchandise available for sale	−	Ending inventory	=	Cost of goods sold
5.	$12,750		38,625		_____		8,750		_____
6.	_____		64,260		82,850		_____		61,518
7.	$46,280		_____		120,250		_____		83,730
8.	$ 3,595		12,845		_____		5,970		_____

Answers to CYK: 1. $209,350   2. $147,872   3. $1,575,000   4. $118,870   5. $144,325
6. $41,110   7. $2,330 (net income)

9. Reynold's Flower Shoppe reported net sales of $182,400. If it cost the shop $94,700 to buy the products it sold and another $52,480 in expenses to sell them, what is the net income?

10. On June 1, the beginning inventory of Video Sounds was valued at $147,650. The firm's records indicated purchases of $120,380, purchase discounts of $3,475, purchase returns and allowances of $2,960, and transportation charges of $4,800. The inventory on June 30 was $82,745. Determine Video Sounds's cost of goods sold for the period.

11. Using the information in problem 10, determine Video Sounds's gross profit for June based on sales of $362,740 and sales discounts amounting to $8,960.

12. The Nine-to-Five Daycare Center incurred the following expenses during the month of August 200X: rent $2,250, utilities $875, supplies $585, salaries $3,600, depreciation—equipment $1,200, and miscellaneous expenses $265. If total expenses are 15% of total revenue, determine (a) total revenue and (b) net income.

13. If a firm's net sales are $48,000 and its operating expenses total $22,500, determine the net income with the cost of goods sold being 40% of net sales.

14. Santana Construction Company purchased a single-family dwelling for $32,500, which it sold the following year for $68,900. During the year, it invested $12,750 in materials and labor for improvements. The real estate firm that sold the house charged a 6½% commission. What is the company's net profit on the sale of the home?

15. Prepare a multiple-step income statement from the following income data taken from the records of Quick-Stop Foods for the month ended September 30, 200X.

Sales	$15,840
Sales returns	260
Merchandise inventory 9/1	3,950
Purchases	5,270
Transportation	835
Purchase discounts	175
Merchandise inventory 9/30	2,485
Salaries	1,250
Rent	850
Advertising	250
Supplies	480
Depreciation equipment	900
Miscellaneous	340
Interest income	150

4.3 PERCENTAGE ANALYSIS OF FINANCIAL STATEMENTS

The solvency and profitability of a business enterprise are reported in the balance sheet and income statement. These two principal statements provide the basic information needed by individuals both inside and outside the organization to make important economic decisions. A firm's managers, creditors, and investors will assess current statement data in terms of past performance to make projections concerning future performance. This process of analysis provides a clearer picture of the firm's performance because it identifies important relationships, developing trends, and the firm's strengths and weaknesses.

Financial statements are often analyzed by preparing a *comparative statement,* which compares and contrasts the results of the current year with the results of prior years. Financial statements are also used to compare the firm's financial data to industry standards and to the financial data of other companies.

The analytical measures generated from the data presented in financial statements are expressed as either percentages or as ratios. In this section we will discuss how financial statements are analyzed both vertically and horizontally.

Vertical Analysis

Vertical analysis of financial statements is used by the analyst when the information needed concerns the percentage relationships of the component parts of the statement to either net sales or total assets for a single period. For example, a manager would want to know the percentage relationship between cost of goods sold and net sales. If the cost of goods sold is 40.5% of net sales, then gross profit is 59.5% (net sales − cost of goods sold = **gross profit**). This percentage figure, when compared to target criteria of 60% gross profit, indicates the effectiveness of management's decisions during the period. A comparative income statement with vertical analysis is shown in Figure 4.6 for Norton Satellite Systems.

Each item of the income statement is stated as a percent of *net sales.* To calculate the percent (rate) we use the rate formula presented in Chapter 2.

$$\text{rate} = \frac{\text{part}}{\text{base}} \qquad \text{or} \qquad B \times R = P$$

Example 5

Using the data presented in Figure 4.6, calculate the percents for (a) sales and (b) total operating expenses for 2000. Round your answer to the nearest tenth of a percent.

Solution

a. $\dfrac{\text{sales}}{\text{net sales}} = \dfrac{\$85,585}{\$84,245} = 1.0159 \times 100, \qquad \text{or} \qquad 101.6\%$

Note that the sales percent is greater than 100%. This is because the *base* is net sales.

b. $\dfrac{\text{total operating expenses}}{\text{net sales}} = \dfrac{\$24,906}{\$84,245} = .2956 \times 100 \qquad \text{or} \qquad 29.6\%$

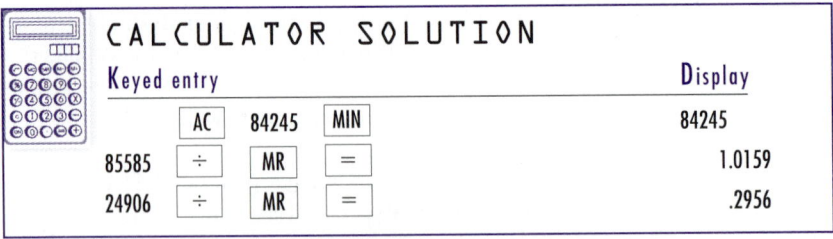

CALCULATOR SOLUTION

Keyed entry				Display
AC 84245 MIN				84245
85585	÷	MR	=	1.0159
24906	÷	MR	=	.2956

Figure 4.6

Vertical analysis:
Comparative in-
come statement

Norton Satellite Systems
Comparative Income Statement
December 31, 1999 and 2000

	2000		1999	
	Amount	**Percent**	**Amount**	**Percent**
Gross sales	$85,585	101.6	$76,187	102.0
Less: sales returns	1,340	1.6	1,480	2.0
Net sales	$84,245	100.0	$74,707	100.0
Cost of goods sold				
Beginning inventory	$12,650	15.0	$14,425	19.3
Net purchases	38,240	45.4	32,500	43.5
Merchandise available	$50,890	60.4	$46,925	62.8
Ending inventory	22,400	26.6	20,200	27.0
Total cost of goods sold	$28,490	33.8	$26,725	35.8
Gross profit	$55,755	66.2	$47,982	64.2
Operating expenses				
Supplies	$ 620	0.8	$ 875	1.1
Wages and salaries	14,356	17.0	12,600	16.9
Advertising	985	1.2	1,400	1.9
Utilities	2,520	3.0	2,535	3.4
Insurance	1,950	2.3	1,500	2.0
Depreciation	4,475	5.3	4,700	6.3
Total operating expenses	$24,906	29.6	$23,610	31.6
Income before taxes	$30,849	36.6	$24,372	32.6
Income taxes	11,105	13.2	8,290	11.1
Net Income	$19,744	23.4	$16,082	21.5

When preparing a vertical analysis of the balance sheet, each item is stated
as a percent of total assets or total liabilities plus owner's equity.

Example 6

Using the data presented in Figure 4.7, calculate for 2000 the percents for
(a) cash (asset) and (b) accounts payable (liability) for Norton Satellite Systems. Round your answer to the nearest tenth of a percent.

Solution

a. $\dfrac{\text{cash}}{\text{total assets}} = \dfrac{\$6,700}{\$90,530} = 0.74 \times 100$ or   7.4%

b. $\dfrac{\text{accounts payable}}{\text{total assets}} = \dfrac{\$4,600}{\$90,530} = 0.51 \times 100$ or   5.1%

Figure 4.7

Vertical analysis: Comparative balance sheet for a sole proprietorship

Norton Satellite Systems
Comparative Balance Sheet
December 31, 1999 and 2000

	2000		1999	
	Amount	**Percent**	**Amount**	**Percent**
Assets				
Current assets:				
Cash	$ 6,700	7.4	$ 4,250	4.8
Accounts receivable	2,980	3.3	3,470	3.9
Merchandise inventory	22,400	24.7	20,200	22.6
Supplies	1,200	1.3	1,275	1.4
Prepaid insurance	700	0.8	1,000	1.1
Total current assets	$33,980	37.5	$30,195	33.8
Fixed assets:				
Land	$12,000	13.3	$12,000	13.4
Building (net)	32,400	35.8	38,650	43.3
Equipment (net)	12,150	13.4	8,375	9.4
Total fixed assets	$56,550	62.5	$59,025	66.2
Total assets	$90,530	100.0	$89,220	100.0
Liabilities				
Current liabilities:				
Accounts payable	$ 4,600	5.1	$ 8,740	9.8
Wages payable	1,320	1.5	960	1.1
Taxes payable	10,260	11.3	8,785	9.8
Total current liabilities	$16,180	17.9	$18,485	20.7
Long-term liabilities:				
Mortgage payable	$43,800	48.4	$52,200	58.5
Total liabilities	$59,980	66.3	$70,685	79.2
Owner's equity				
Norton B, capital	$30,550	33.7	$18,535	20.8
Total liabilities and equity	$90,530	100.0	$89,220	100.0

CHECK YOUR KNOWLEDGE

Vertical Analysis

Complete a vertical analysis of the following partial comparative income statement. Round your answers to the nearest tenth of a percent.

		2000		1999	
		Amount	**Percent**	**Amount**	**Percent**
1.	Sales	$125,500	_____	$110,200	_____
2.	Sales returns	$_____	_____	$_____	_____
	Net sales	122,900	100.0	108,350	100.0
3.	Cost of goods sold	72,750	_____	$_____	_____
4.	Gross profit	50,150	_____	43,850	_____
5.	Operating expenses	45,473	_____	$_____	_____
6.	Net income	$_____	_____	$ 5,200	_____

Complete a vertical analysis of the following balance sheet. Round your answers to the nearest tenth of a percent.

Assets	**Amount**	**Percent**
Current assets:		
7. Cash	$ 2,500	_____
8. Merchandise inventory	8,750	_____
9. Accounts receivable	_____	_____
10. Total current assets	14,850	_____
Fixed assets:		
11. Equipment	_____	_____
12. Buildings	20,750	_____
13. Land	10,000	_____
14. Total fixed assets	36,950	_____
15. Total assets	_____	100.0%

Answers to CYK: *1.* **a.** 102.1; **b.** 101.7   *2.* **a.** 2,600; **b.** 2.1; **c.** 1,850; **d.** 1.7
3. **a.** 59.2; **b.** 64,500; **c.** 59.5   *4.* **a.** 40.8; **b.** 40.5
5. **a.** 37.0; **b.** 38,650; **c.** 35.7   *6.* **a.** 4,677; **b.** 3.8; **c.** 4.8   *7.* 4.8
8. 16.9   *9.* **a.** 3,600; **b.** 6.9   *10.* 28.7   *11.* **a.** 6,200; **b.** 12.0   *12.* 40.1
13. 19.3   *14.* 71.3   *15.* 51,800

For Your Information

FINANCIAL ANALYSIS AT A GLANCE

Toyota, Japan's No. 1 carmaker earned a record pretax operating profit of 3.4 billion on sales of $52 billion during the first half of 1998. However, these figures do not reflect an important fact: Toyota is barely breaking even in its most critical market, Japan. Vehicle sales in Japan have decreased 31% since 1990 and its share of the domestic market has been below 40% since 1995. In 1998, Japan accounted for just 38% of Toyota's worldwide car sales, down from 52% in 1990. Current domestic sales produce approximately 3% of operating income, while North American sales account for 80% of operating income with European sales accounting for the rest. In 1990 Toyota earned as much as 90% of its operating income from domestic sales. A weak yen also helped produce the record profit but, as the yen strengthens against foreign currencies net profit will decrease $68 million for every one-yen increase. If these trends continue, Toyota's financial picture could change drastically.

The illustration shown is based on data taken from Toyota's financial statements and then analyzed using both the vertical and horizontal analysis procedure to generate the percentages contained in the graphs.

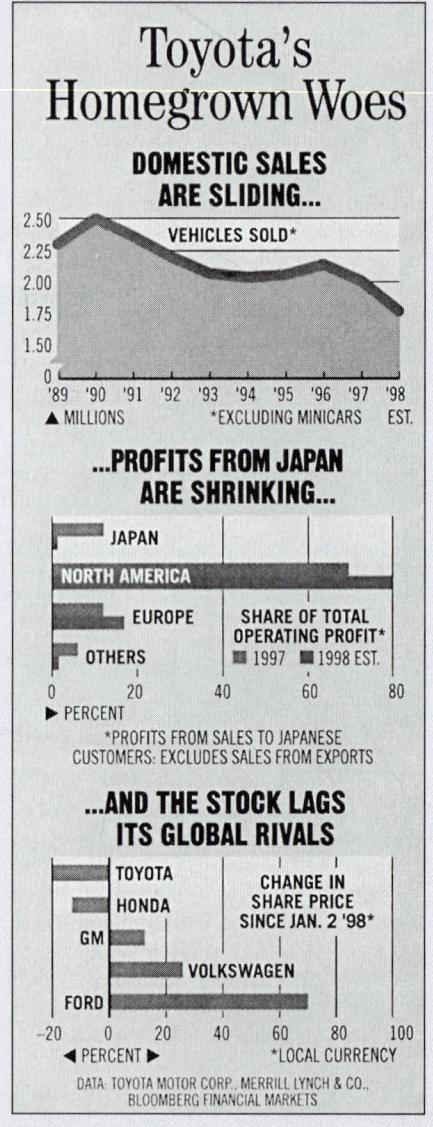

Horizontal Analysis

Financial statements are also analyzed to measure the degree of change (increases or decreases) of a firm's financial data between two time periods. This comparison can reveal important changes that can have either a positive or a negative impact on the firm. For example, the investors of a firm would be pleased to learn that the net income increased 18% over last year while the industry's average increase amounted to 12% and their primary competition reported an increase of 14.5%. On the other hand, management might not be pleased to see an increase in operating expenses at a rate above that considered acceptable in terms of the firm's pricing policies. This percentage analysis of the changes of corresponding items in comparative financial statements is referred to as **horizontal analysis.** Each item of the current period is compared with the corresponding item of one or more earlier periods (base). The increase or decrease in the amount of the item (part) is then listed together with the percent (rate) of increase or decrease. This procedure is illustrated in the comparative balance sheet shown in Figure 4.8 for Norton Satellite Systems.

Learning objective
Complete a horizontal analysis of the balance sheet and income statement.

To calculate the percent increase or decrease for each item we use the rate formula presented in Chapter 2.

$$\text{rate (increase or decrease)} = \frac{\text{amount difference}}{\text{base year (earlier year) amount}} \times 100$$

Example 7

In the illustrated balance sheet (Figure 4.8), the cash account totaled $4,250 in 1999 and increased to $6,700 in 2000. Calculate the percent of change.

Solution

The percent change is determined as follows.

Step 1: Calculate the amount of difference between the 2 years.

$$\text{cash 2000} - \text{cash 1999} = \text{amount difference}$$
$$\$6,700 - \$4,250 \quad = \$2,450 \text{ (increase)}$$

Step 2: Calculate the percent of change.

$$\text{percent change} = \frac{\text{amount difference}}{\text{base year (earlier year) amount}} \times 100$$

$$= \frac{\$2,450}{\$4,250} = .5764 \times 100$$

$$= 57.6\%$$

Figure 4.8

Horizontal analysis: Comparative balance sheet for a sole proprietorship

Norton Satellite Systems
Comparative Balance Sheet
December 31, 1999 and 2000

	2000	1999	Increase/(Decrease) Amount	Increase/(Decrease) Percent
Assets				
Current assets:				
Cash	$ 6,700	$ 4,250	$ 2,450	57.6
Accounts receivable	2,980	3,470	(490)	(14.1)
Merchandise inventory	22,400	20,200	2,200	10.9
Supplies	1,200	1,275	(75)	(5.9)
Prepaid insurance	700	1,000	(300)	(30.0)
Total current assets	$33,980	$30,195	$ 3,785	12.5
Fixed assets:				
Land	$12,000	$12,000	-0-	-0-
Building (net)	32,400	38,650	$ (6,250)	(16.2)
Equipment (net)	12,150	8,375	3,775	45.1
Total fixed assets	$56,550	$59,025	$ (2,475)	(4.2)
Total assets	$90,530	89,220	1,310	1.5
Liabilities				
Current liabilities:				
Accounts payable	$ 4,600	$ 8,740	$ (4,140)	(47.4)
Wages payable	1,320	960	360	37.5
Taxes payable	10,260	8,785	1,475	16.8
Total current liabilities	$16,180	$18,485	$ (2,305)	(12.5)
Long-term liabilities:				
Mortgage payable	$43,800	$52,200	$ (8,400)	(16.1)
Total liabilities	$59,980	$70,685	$(10,705)	(15.1)
Owner's equity				
Norton B, capital	$30,550	$18,535	$ 12,015	64.8
Total liabilities and equity	$90,530	$89,220	$ 1,310	1.5

Note: Figures in ( ) are decreases.

CALCULATOR SOLUTION

Keyed entry					Display
AC	6700	−	4250	=	2450
		÷	4250	=	.576470.

Horizontal analysis can be used to compare the results of business activity reported in any type of financial report, provided the same items are being analyzed over multiple periods of time. Example 8 illustrates the same calculation on data taken from the income statement shown in Figure 4.6.

Example 8

Use the data in Figure 4.6 (comparative income statement) to calculate the percent of change in advertising expenses between 1999 and 2000. Round your answer to the nearest tenth of a percent.

Solution

Step 1: Calculate the amount of difference between two periods.

difference = advertising expenses 2000
− advertising expenses 1999

−$415 = $985 − $1,400

Step 2: Calculate the percent of change.

$$\text{percent change} = \frac{\text{amount difference}}{\text{base year (earlier year) amount}} \times 100$$

$$= \frac{-\$415}{\$1,400} = -.2964 \times 100$$

$$= 29.6\% \text{ (decrease)}$$

When the difference between the two periods results in a negative number (−$415) the percent of change must be identified as a decrease either by placing parentheses ( ) around the number (as shown in Figure 4.8) or by labeling the answer a "decrease."

CHECK YOUR KNOWLEDGE

Horizontal Analysis

Complete a horizontal analysis of the following partial comparative balance sheet. Round your answers to the nearest tenth of a percent.

	Assets	2000	1999	Increase/Decrease Amount	Percent
1.	Current assets	$18,260	$10,475	a. _____	b. _____
2.	Fixed assets	$29,150	$32,640	a. _____	b. _____
3.	Total assets	a. _____	b. _____	c. _____	d. _____
4.	Current liabilities	a. _____	$5,340	b. _____	c. _____
5.	Long-term liabilities	$17,120	a. _____	b. _____	c. _____
6.	Total liabilities	$23,260	$27,635	a. _____	b. _____
7.	Owner's equity	a. _____	$15,480	b. _____	c. _____

Prepare a horizontal analysis of the partial comparative income statement shown below. Round answers to the nearest tenth of a percent.

		2000	1999	Amount	Percent
8.	Gross sales	$278,000	$245,540	a. _____	b. _____
9.	Net sales	276,500	243,300	a. _____	b. _____
10.	Cost of goods sold	134,150	98,750	a. _____	b. _____
11.	Gross profit	142,350	144,550	a. _____	b. _____
12.	Total expenses	114,022	112,825	a. _____	b. _____
13.	Total net income	28,328	31,725	a. _____	b. _____

Answers to CYK: *1.* **a.** $7,785; **b.** 74.3   *2.* **a.** (3,490); **b.** (10.7)
3. **a.** 47,410; **b.** 43,115; **c.** 4,295; **d.** 10   *4.* **a.** 6,140; **b.** 800; **c.** 15
5. **a.** 22,295; **b.** (5,175); **c.** (23.2)   *6.* **a.** (4,375); **b.** (15.8)
7. **a.** 24,150; **b.** $8,670; **c.** 56.0   *8.* **a.** 32,460; **b.** 13.2
9. **a.** 33,200; **b.** 13.6   *10.* **a.** 35,400; **b.** 35.8   *11.* **a.** (2,200); **b.** (1.5)
12. **a.** 1,197; **b.** 1.1   *13.* **a.** (3,397); **b.** (10.7)

4.3   **EXERCISES**

1.   Complete the following comparative income statement analysis for Tri-State Data Systems. Round your calculations to the nearest tenth of a percent.

Tri-State Data Systems, Inc.
Comparative Income Statement
December 31, 1999 and 2000

	2000	Percent	1999	Percent
Net sales	$610,000	_____	$585,000	_____
Cost of goods sold:				
Beginning inventory	$ 30,600	_____	$ 24,500	_____
Net purchases	272,800	_____	260,300	_____
Merchandise available	$303,400	_____	$284,800	_____
Ending inventory	34,300	_____	30,600	_____
Total cost of goods sold	$269,100	_____	$254,200	_____
Gross margin	340,900	_____	330,800	_____
Operating expenses:				
Supplies	$ 5,200	_____	$ 4,850	_____
Wages and salaries	125,400	_____	105,200	_____
Utilities	14,800	_____	12,750	_____
Depreciation	8,000	_____	8,000	_____
Insurance	2,800	_____	2,800	_____
Total operating expenses	$156,200	_____	$133,600	_____
Income before taxes	184,700	_____	197,200	_____
Provision for taxes	92,400	_____	99,050	_____
Net income	$ 92,300	_____	$ 98,150	16.8

2. Fill in the missing information in Northeast Supply Company's balance sheet. Express your answers to the nearest tenth of a percent or dollar.

Northeast Supply Company
Balance Sheet
December 31, 2000

	Amount	Percent
Assets		
Current assets:		
Cash	$ 12,500	_____
Equipment	25,750	_____
Accounts receivable	31,750	12.7
Merchandise inventory	50,500	_____
Total current assets	_____	48.2
Fixed assets:		
Land	40,500	_____
Buildings (Net)	_____	35.6
Total assets	$250,000	
Liabilities		
Current liabilities:		
Accounts payable	26,750	_____
Wages payable	_____	_____
Total current liabilities	$ 36,250	14.5
Long-term liabilities:		
Mortgages payable	85,250	_____
Total liabilities	_____	_____
Owner's equity		
Owner's capital		
Total liabilities and equity	$250,000	100.0

3. Complete the following comparative balance sheet analysis. Round your calculations to the nearest tenth of a percent.

Tri-State Data Systems, Inc.
Comparative Balance Sheet
December 31, 1999 and 2000

| | | | Increase/(Decrease) | |
	2000	1999	Amount	Percent
Assets				
Current assets:				
Cash	$ 12,000	$ 10,500	_____	_____
Accounts receivable	24,500	22,400	_____	_____
Notes receivable	4,800	8,200	_____	_____
Merchandise inventory	30,600	34,300	_____	_____
Total current assets	$ 71,900	$ 75,400	_____	_____
Fixed assets:				
Land	60,000	60,000	_____	_____
Building (net)	77,250	80,750	_____	_____
Equipment (net)	45,650	42,500	_____	_____
Total fixed assets	$182,900	$183,250	_____	_____
Total assets	$254,800	$258,650	_____	_____
Liabilities				
Current liabilities:				
Accounts payable	$ 28,500	$ 25,000	_____	_____
Taxes payable	9,200	12,250	_____	_____
Wages payable	7,600	9,500	_____	_____
Total current liabilities	$ 45,300	$ 46,750	_____	_____
Long-term liabilities:				
Mortgage payable	69,400	65,850	_____	_____
Total liabilities	$114,700	$112,600	_____	_____
Stockholders' equity				
Common stock	85,000	85,000	_____	_____
Retained earnings	55,100	61,050	_____	_____
Total equity	$140,100	$146,050	_____	_____
Total liabilities and equity	$254,800	$258,650	_____	_____

4. Complete the required analysis of Excellent Autoservice Center's condensed comparative income statement.

Excellent Autoservice Center
Comparative Income Statement (condensed)
December 31, 1999 and 2000

	2000	1999	Increase/(Decrease)		Percent of net sales	
			Amount	Percent	2000	1999
Net sales	$240,600	$236,800				
Cost of goods sold	179,900	181,400				
Gross margin	$ 60,700	$ 55,400				
Total operating expenses	30,500	24,250				
Income before taxes	$ 30,200	$ 31,150				
Provision for taxes	13,525	13,925				
Net income	$ 16,675	$ 17,225				

4.4   RATIO ANALYSIS OF FINANCIAL STATEMENTS

Ratio analysis of financial statements provides the managers, creditors, and potential investors of a company with information specific to the type of decisions each must make. **Ratios** provide an assessment of the performance of a company. This assessment is accomplished by comparing a company's current performance either with its own past performance or with the performance of other companies. Ratios expand the vertical and horizontal analysis of financial statements. They are expressed as fractions, such as 6/3, or horizontally using a colon, such as 6:3. They are read as "a ratio of 6 to 3."

Ratios are generally classified as liquidity, leverage, asset utilization ratio, or profit ratios. The most significant and commonly used ratios in each category will be discussed in this section. The examples use information contained in the comparative financial statements presented in Figures 4.6 and 4.7.

Learning objective
Calculate primary financial statement ratios.

Liquidity Assets

The term *liquidity* refers to the ability of a firm to convert its assets into cash to meet its current debts (liabilities). Liquidity ratios are calculated from information contained in the balance sheet and include the current ratio and the acid test ratio. The **current ratio,** sometimes referred to as the *working capital ratio,* is computed by dividing the total of current assets by the total of current liabilities.

$$\text{current ratio} = \frac{\text{current assets}}{\text{current liabilities}}$$

Example 9

Find the current ratio for Norton Satellite Systems (data taken from Figure 4.7).

Solution

2000

$$\text{current ratio} = \frac{33{,}980}{16{,}180} = 2.1, \quad \text{or} \quad 2.1:1$$

1999

$$\text{current ratio} = \frac{30{,}195}{18{,}485} = 1.6, \quad \text{or} \quad 1.6:1$$

These ratios indicate that Norton Satellite Systems' ability to pay its debts has improved (the higher the ratio, the more favorable the position). In 2000, the value of current assets is two times the value of current liabilities, or the company has $2.10 in current assets for every $1.00 that it owes in current liabilities.

The **acid test ratio,** or *quick ratio,* measures the immediate debt-paying ability of a company. *Quick assets* (cash, receivables, and marketable securities) are compared to current liabilities. Inventories are omitted from the calculation because it could take a considerable amount of time to convert those assets into cash. Supplies and prepaid insurance are also omitted because those assets are used in operations and will not be sold for cash to resolve a firm's debts.

$$\text{acid test ratio} = \frac{\text{quick assets}}{\text{current liabilities}}$$

Example 10

Find the acid test ratio for Norton Satellite Systems (data taken from Figure 4.7).

Solution

2000

$$\text{acid test ratio} = \frac{6{,}700 + 2{,}980}{16{,}180} = .60, \quad \text{or} \quad .60:1$$

1999

$$\text{acid test ratio} = \frac{4{,}250 + 3{,}470}{18{,}485} = .42, \quad \text{or} \quad .42:1$$

Norton's creditors may place a greater emphasis on these results than they would on the results of the current ratio. The basis of their concern is the fact that there are only $.60 of liquid assets for every $1.00 of current liabilities. This ratio is below the recommended 1:1 standard used by many creditors when making decisions regarding credit.

Leverage Ratios

Most businesses borrow money to finance their activity. **Leverage** is a term used to describe an investor's ability to use credit to finance investments. The ability to borrow money is based on the current debt position of the company. *Leverage ratios* provide managers and investors with the information they need to determine the way a business is financed. The information required to calculate the **debt-to-equity ratio** and the **debt-to-assets ratio** is found in the balance sheet.

$$\text{debt-to-equity ratio} = \frac{\text{total liabilities}}{\text{total owner's equity}}$$

Example 11

Find the debt-to-equity ratio for Norton Satellite Systems (data taken from Figure 4.7).

Solution

2000

$$\text{debt-to-equity ratio} = \frac{\$59,800}{\$30,550} = 1.96, \quad \text{or} \quad 1.96:1$$

1999

$$\text{debt-to-equity ratio} = \frac{\$70,685}{\$18,535} = 3.81, \quad \text{or} \quad 3.81:1$$

Because this ratio has improved between 1999 (3.81 : 1) and 2000 (1.96 : 1), it may be within the recommended guidelines of creditors. The smaller the ratio the more favorable the margin of safety of the creditors. In 2000, for every $1.00 of ownership, there is almost $2.00 in liabilities. In other words, creditors have a claim on two-thirds of the company's assets. Such a position is not always viewed negatively. There are those who believe a firm's operations should be financed to the extent possible through external sources; therefore, companies will continue to borrow funds provided the earnings generated from the investment of borrowed money is sufficient to meet interest payments.

The relationship between the claim on assets of the owners and creditors is further defined by the debt-to-assets ratio, which compares total liabilities to total assets.

$$\text{debt-to-assets ratio} = \frac{\text{total liabilities}}{\text{total assets}}$$

Example 12

Find the debt-to-assets ratio for Norton Satellite Systems (data taken from Figure 4.7).

Solution

2000

$$\text{debt-to-assets ratio} = \frac{59,980}{90,530} = .66, \quad \text{or} \quad .66:1$$

1999

$$\text{debt-to-assets ratio} = \frac{70,685}{89,220} = .79, \quad \text{or} \quad .79:1$$

These results can be interpreted in two different but related ways. First, for each $1.00 of assets, the company owes $.66 to its creditors. Second, the relationship can be converted to a percentage indicating that the creditors have claim to 66% of the company, compared to the 34% claim held by Norton's owners. A company that uses leverage will have a higher debt-to-assets ratio than one that does not.

CHECK YOUR KNOWLEDGE

Liquidity and Leverage Ratios

The balance sheet for the ICU Corporation on July 31, 200X, contained the following information: cash $25,000, marketable securities $15,000, accounts receivable $85,500, merchandise inventory $130,500, total current assets $256,000, total assets $420,500, total current liabilities $110,250, total liabilities $180,500, and total owner's equity $240,000. From this information, determine: (a) the current ratio, (b) the acid test ratio, (c) the debt-to-equity ratio, and (d) the debt-to-assets ratio. (Express each ratio as a part to 1 and round to the nearest hundredth.)

Answers to CYK: **(a)** 2.32:1 **(b)** 1.14:1 **(c)** .75:1 **(d)** .43:1

Asset Utilization Ratios

Asset utilization ratios analyze the use of assets by a firm. These ratios draw information from both the income statement and the balance sheet because they measure the turnover of assets being utilized to generate sales. Asset utilization ratios are stated in terms of time, which expresses a degree of efficiency. A company that uses its assets efficiently will generate more interest from investors than a company that does not. The **asset turnover ratio** measures the effectiveness with which managers are using assets to produce sales. Assets used to compute the ratio may be the total at the end of the year, the average of the monthly total assets for the year, or the average of the beginning and ending totals of the year. Example 13 is based on total assets reported in Figure 4.6 and net sales reported in Figure 4.7.

$$\text{asset turnover} = \frac{\text{net sales}}{\text{total assets}}$$

Example 13

Find the asset turnover ratio for Norton Satellite Systems (data taken from Figures 4.6 and 4.7).

Solution

2000

$$\text{asset turnover} = \frac{84{,}245}{90{,}530} = .93 \text{ times}$$

1999

$$\text{asset turnover} = \frac{74{,}707}{89{,}220} = .84 \text{ times}$$

The increase in the asset turnover indicates that the assets are being used more efficiently in 2000 than in 1999. Higher ratios suggest better management of assets when compared to activity of other years or other companies within the industry. The ratio also tells us that for each $1.00 invested in assets by management, it returns $.93 in net sales.

The **inventory turnover ratio** analyzes the number of times during the period the inventory has been sold. Since the analysis usually spans a period of time, the average balance of the asset (merchandise inventory) should be used in calculating the ratio. The inventory figures required to calculate the ratio are obtained from the cost of goods sold section of the income statement and from the balance sheet.

$$\text{inventory turnover} = \frac{\text{cost of goods sold}}{\text{average inventory}}$$

Example 14

Find the inventory turnover ratio for Norton Satellite Systems (data taken from Figure 4.6).

Solution

2000

$$\text{inventory turnover} = \frac{28{,}490}{\dfrac{12{,}650 + 22{,}400}{2}} = \frac{28{,}490}{17{,}525} = 1.63 \text{ times}$$

1999

$$\text{inventory turnover} = \frac{26{,}725}{\dfrac{14{,}425 + 20{,}200}{2}} = \frac{26{,}725}{17{,}313} = 1.54 \text{ times}$$

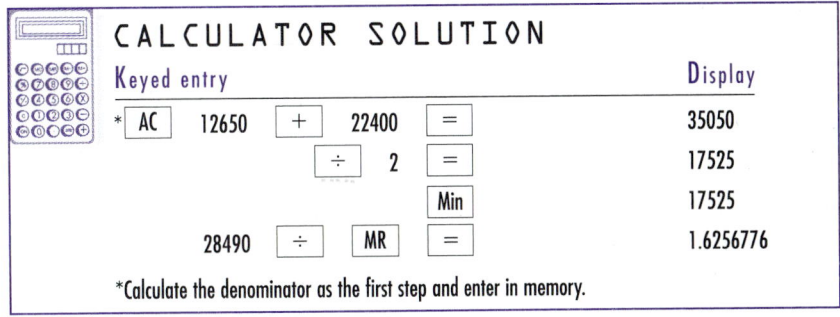

CALCULATOR SOLUTION	
Keyed entry	**Display**
* AC 12650 + 22400 =	35050
÷ 2 =	17525
Min	17525
28490 ÷ MR =	1.6256776

*Calculate the denominator as the first step and enter in memory.

Note: Average inventory was obtained by dividing the sum of beginning inventory plus ending inventory by 2.

The increase in inventory turnover suggests that Norton Satellite Systems may have used its inventories better in 2000 than it did in 1999. However, when comparing the ratio with that of the industry, it may be considered too low. A low ratio indicates Norton may be carrying too much inventory, which increases costs such as insurance, handling, pilferage, and spoilage and obligates amounts of money that could be more effectively utilized. For these reasons many companies have implemented inventory control systems, which by design can increase the inventory turnover ratio. If Norton takes steps to further increase its inventory turnover ratio, the company must be sure it has enough inventory to meet customer demand, as higher ratios usually result in lower inventory levels.

The relationship between credit sales and accounts receivable is referred to as the **receivables turnover ratio.** This ratio indicates the time required by a business to collect money for merchandise sold on credit to its customers. To determine the number of days accounts receivable have been outstanding, we divide the net sales on credit by the average net accounts receivable. The average accounts receivable may be based on the average monthly balances, or the average of the balances at the beginning and the end of the year. In Example 15, we will assume that all sales reported by Norton Satellite Systems in their income statement (Figure 4.6) were credit sales. Otherwise, we would have to identify what percentage of net sales were credit sales versus cash sales.

$$\text{accounts receivable turnover} = \frac{\text{net credit sales}}{\text{average net accounts receivable}}$$

Example 15

Find the accounts receivable turnover for Norton Satellite Systems (data taken from Figures 4.6 and 4.7).

Solution

2000

$$\text{accounts receivable turnover} = \frac{84{,}245}{\dfrac{2{,}980 + 3{,}470}{2}}$$

$$= \frac{84{,}245}{3{,}225} = 26.1 \text{ times}$$

Note: The average accounts receivable figure was found by taking the 2000 account balance ($2,980) and adding to this figure the 1999 account balance ($3,470) and dividing the result by 2. The 1999 balance is, for example purposes, the beginning balance for 2000. The ratio can now be expressed in terms of time by dividing a 360- or 365-day year by the ratio (360 ÷ 26.5 = 13.6). (Many businesses use a 360-day year, which equates to an average of 30 days per month.)

For Norton Satellite Systems, a receivables turnover ratio of 26.5 means that the average credit customer is taking about 14 days to pay its account. A comparison with the credit terms offered by Norton (net payment in 30 days) indicates this average collection period is quite favorable.

$$\text{average collection period} = \frac{\text{number of days in year}}{\text{accounts receivable turnover}} = \frac{360}{26.1}$$

$$= 13.8 \text{ days}$$

Profit Ratios

The focus of attention of most individuals associated with an enterprise is its profitability. If a business is not profitable, it will have difficulty attracting the interest of investors; it will encounter problems buying on credit and borrowing; and most importantly, it will eventually become extinct. *Profit ratios* are a measure of earnings and are developed from information found in both the income statement and the balance sheet.

The **return on net sales ratio** is a measure of earnings that identifies the amount of profit derived for each sales dollar. This ratio involves a vertical analysis of information found in the income statement and can be expressed in amounts or percentages.

$$\text{return on net sales} = \frac{\text{net income after taxes}}{\text{net sales}}$$

Example 16

Find the return on net sales ratio for Norton Satellite Systems (data taken from Figure 4.6).

Solution

2000

$$\text{return on net sales} = \frac{\$19,744}{\$84,245} = .2344$$

1999

$$\text{return on net sales} = \frac{\$16,082}{\$74,707} = .2153$$

In 2000, Norton Satellite Systems made 23.4 cents in profit for each $1.00 of sales. The ratio expressed as a percentage (.2343 = 23.4%) can now be compared to company profit goals and industry averages to measure the degree of achievement.

The actual amount of income is not as clear a measure of earnings as is the relationship between the amount of income earned and the owner's investment. To analyze this more important relationship, we would use the **return on investment ratio,** which is obtained by dividing net income by the average owner's equity. The average owner's equity can be found by adding monthly statement figures or by using equity at the beginning and end of the year.

$$\text{return on investment} = \frac{\text{net income after taxes}}{\text{average owner's equity}}$$

We discussed earlier in this chapter how equity is reported in the balance sheet for both proprietorships and corporations. Example 17 is a presentation for a proprietorship. If you were to prepare a return on investment ratio for a corporation, equity would consist of the sum of common stock, preferred stock, and retained earnings.

Example 17

Find the return on investment for Norton Satellite Systems (data taken from Figures 4.6 and 4.7). For this example, assume owner's equity on January 1, 2000, was $15,750.

Solution

2000

$$\text{return on investment} = \frac{\$19,744}{\dfrac{\$30,550 + \$18,535}{2}}$$

$$= .8044, \quad \text{or} \quad 80.4\%$$

1999

$$\text{return on investment} = \frac{\$16,082}{\dfrac{18,535 + 15,750}{2}}$$

$$= .9381, \quad \text{or} \quad 93.8\%$$

The return on investment for Norton Satellite Systems is quite high. For each dollar invested by its owner, a return of 93.8 cents (1999) and 80.4 cents (2000) results, or a return of 93.8% and 80.4% is realized. The high return on the investment of the owner can be partially explained by the leverage of the firm. Norton has been able to generate good profits by borrowing and purchasing its resale merchandise on credit.

The financial ratios discussed in this section are valuable analytical devices and provide the information necessary to assess an organization's performance and financial position. Careful consideration should be given, however, to the results obtained from these measures; they are not intended to provide their users with specific courses of action. When used with good judgment, industry trends and standards, and other economic indexes, they help to analyze current conditions and interpret future directions of an organization. A summarization of the ratios is presented in Table 4.1 for your convenience when studying or reviewing this chapter.

Table 4.1

Financial ratios

Ratio	Formula	Interpretation
A. Liquidity ratios		
1. Current ratio	$\dfrac{\text{current assets}}{\text{current liabilities}}$	Identifies the amount of current assets available to pay off each \$1 of current debt. A 2 : 1 is favorable.
2. Acid test ratio	$\dfrac{\text{quick assets}}{\text{current liabilities}}$ (Inventories are omitted from current assets to arrive at quick asset figure.)	Identifies the amount of assets that can be quickly liquidated into cash to cover each \$1 of current liabilities. A 1 : 1 is favorable.
B. Leverage ratios		
3. Debt-to-equity ratio	$\dfrac{\text{total liabilities}}{\text{total owner's equity}}$	Identifies the amount of total debt the owners have incurred for each \$1 of equity. Determines the claim on assets by owners. A 1 : 1 is favorable.
4. Debt-to-assets ratio	$\dfrac{\text{total liabilities}}{\text{total assets}}$	Identifies the amount of total debt that can be covered by each \$1 of assets. Determines the claim on assets by the creditors.
C. Asset utilization ratios		
5. Asset turnover ratio	$\dfrac{\text{net sales}}{\text{average total assets}}$	Identifies the amount of sales generated from each \$1 of assets and the number of times assets have turned over.
6. Inventory turnover ratio	$\dfrac{\text{cost of goods sold}}{\text{average inventory}}$	Identifies the number of times during the period the merchandise inventory has been sold. Higher ratios may be considered more favorable.

(continued)

Table 4.1 *(continued)*

Financial ratios

Ratio	Formula	Interpretation
7. Accounts receivable turnover ratio	$$\frac{\text{net credit sales}}{\text{average net accounts receivable}}$$	Indicates the number of times accounts receivable have turned over into sales.
8. Average collection ratio	$$\frac{\text{number of days in year}}{\text{average accounts receivable turnover}}$$	Identifies the average number of days required to collect accounts receivable.
D. Profit ratios		
9. Return on net sales ratio	$$\frac{\text{net income after taxes}}{\text{net sales}}$$	Identifies the amount of profit derived from each $1 of sales. Can also be expressed as a percent.
10. Return on investment ratio	$$\frac{\text{net income after taxes}}{\text{average owner's equity}}$$	Identifies the amount of profit realized from each $1 invested by the owners. Can also be expressed as a percent.

CHECK YOUR KNOWLEDGE

Asset Utilization and Profit Ratios

Use the data obtained from the financial statements of Walsh Industries to prepare the required activity and profit ratios. Round your answers to the nearest hundredth.

1. What is the asset turnover if sales are $1,650,000, sales returns and allowances are $20,500, and assets total $875,000?

2. If beginning inventory was $305,200, and ending inventory was $275,400, what was the inventory turnover on the cost of goods sold amounting to $1,059,500?

3. Eighty percent of the net sales of $1,629,500 are credit sales. What is the average time required to collect accounts from customers if January accounts receivable were $105,250, and December's accounts receivable totaled $110,350?

4. What is the rate of return on net sales if net income after taxes is $120,800 and net sales are reported at $1,629,500?

5. Owner's equity on January 1 was $1,320,500 and $1,400,500 on December 31. What is the return on investment if net income after taxes is $120,800?

4.4   EXERCISES

For questions 1–5 use the financial information provided below taken from the financial statements of the O'Bannion and Grant Paving Company.

Current assets	$ 75,000
Fixed assets	140,500
Total assets	215,550
Current liabilities	27,250
Long-term liabilities	65,175
Total liabilities	92,425
Total owner's equity	123,125
Average owner's equity	89,400
Net sales	932,645
Net income after taxes	75,900

1. Determine the company's current ratio (express answer to the nearest hundredth).

2. Calculate the debt-to-equity ratio (express answer in ratio form).

3. Find the debt-to-assets ratio (express answer to the nearest tenth).

4. Calculate the asset-turnover ratio (express answer in times rounded to the nearest tenth).

5. Determine the return-on investment ratio (express answer to the nearest tenth of a percent).

For the ratios in exercises 6–15, use the comparative financial statements of Tri-State Systems presented in Exercise 4.3, problems 1 and 3. Round answers to hundredths.

6. Current ratio: 1999 _____
                            2000 _____

7. Acid test ratio (quick):   1999 _____
                                   2000 _____

8. Debt-to-equity ratio: 1999 _____
                                   2000 _____

9. Debt-to-assets ratio: 1999 _____
                                   2000 _____

10. Asset turnover ratio: 1999 _____
                                   2000 _____

11. Inventory turnover ratio: 1999 _____
                                   2000 _____

12. Accounts receivable   1999 _____
    turnover (60% of net sales   2000 _____
    are credit sales and 1998
    accounts receivable were
    $21,800):

13. Average collection period   1999 _____
    (365-day year):   2000 _____

14. Return on net sales: 1999 _____
                                   2000 _____

15. Return on investment   1999 _____
    ratio (assume owner's   2000 _____
    equity on January 1, 1999
    of $130,600):

Answers to CYK: **1.** 1.86   **2.** 3.65   **3.** 12.09   **4.** 7.4%   **5.** 8.88%

16. The current ratio of Hamm Music is 2.8:1 based on total current liabilities of $125,000. What is the amount of Hamm's total current assets?

17. Yesterday's Treasures turned over its inventory 5.7 times this past year. If the antique business had an average inventory of $50,000, how much did it cost to purchase the goods that were sold to generate net sales of $475,000?

18. The accounts receivable turnover ratio of the Di-Tung Oriental Supply Co. is 12.5. If average ac-

counts receivable for the period were $30,000, what percent of the $575,000 net sales were credit sales?

19. If Sample Foods's gross margin decreased 12.5% this year to $95,250, what was the gross margin reported on last year's income statement?

20. Rojon Enterprises reported after tax income of $125,000, which represented a 15.7% return on equity. What was the amount of stockholders' equity listed in the firm's balance sheet?

CHAPTER REVIEW EXERCISES

The following problems review all mathematics and business concepts presented in the chapter. If you experience any difficulty solving a problem, go to the appropriate chapter section and review the examples provided. Express all answers as decimals rounded to tenths, or dollars rounded to cents unless otherwise instructed.

1. Determine the value of Ultra-Pure Water Company's total liabilities if the firm's total assets are $2,648,925 and the equity of the owner is $1,642,345.

2. Jill Marskon owns a flower and gift shop, which has the following assets: cash, $2,500; merchandise inventory, $12,400; store supplies, $620; and store equipment, $15,800. If she owes $6,280 for merchandise purchased on credit, how much is her owner's equity?

3. Feldman Fireplace and Stove Centre's March 31, 200X, balance sheet reports owner's equity of $130,696, which is 62% of total assets. What is the amount of (a) total assets, (b) total liabilities?

4. Eaton Yarn and Fabrics reported net sales of $34,600 on February 29. If it cost the store $20,490 to buy the goods that were sold, overhead expenses were $6,125, and taxes paid

amounted to $1,548, what is the net income after taxes for the month?

5. Central Propane Products reported gross sales for the month of $87,645 of which $2,870 in merchandise was returned for credit or allowances. If the company had $5,950 in sales discounts, what was the amount of net sales reported for the period?

6. The following balances were reported on the income statement of Northern Paper Products Inc.: inventory July 1, $265,700; purchases, $425,250; transportation—in, $3,620; and inventory on July 31, was $195,300. Determine the cost of goods sold.

7. A company's net sales for the month are $135,800 and its operating expenses total $56,200. If its cost of goods sold are 45% of net sales, find the net profit before taxes reported on the income statement.

Complete a vertical analysis of the following partial income statement. Round answers to the nearest tenth of a percent.

Midstate Distribution Company
Income Statement
December 31, 200X

		Amount	Percent
8.	Gross sales	$49,000	_____
9.	Sales returns and allowances	1,250	_____
10.	Net sales	47,750	_____
11.	Cost of goods sold	28,420	_____
12.	Gross profit	19,330	_____
13.	Operating expenses	15,070	_____
14.	Net income before taxes	4,260	_____

Complete a horizontal analysis of the following partial balance sheet. Round answers to the nearest tenth of a percent.

Expert Printing Company
Balance Sheet
June 30, 2001

				Increase/Decrease	
		2001	2000	Amount	Percent
15.	Current assets	$15,200	$12,875	$_____	_____
16.	Fixed assets	10,500	18,150	_____	_____
17.	Total assets	25,700	31,025	_____	_____
18.	Current liabilities	7,150	5,900	_____	_____
19.	Total liabilities	13,650	8,200	_____	_____
20.	Owner's equity	12,050	22,825	_____	_____

Use the selected information from the balance sheets and income statement of the Lockwood-Martin Cable Co. for December to calculate the required ratios. Round answers to the nearest tenth of a percent.

Current assets	$40,000
Average accounts receivable	7,400
Merchandise inventory (January 1)	20,150
Merchandise inventory (December 31)	25,640
Average total assets	82,750
Current liabilities	18,200
Total liabilities	32,560
Owner's equity	50,190
Net sales	95,000
Cost of goods sold	36,270
Net income after taxes	22,925

21.   Current ratio: _____

22.   Debt-to-equity ratio: _____

23.   Inventory turnover ratio: _____

24.   Asset turnover ratio: _____

25.   Return on net sales ratio: _____

EXPRESS YOUR UNDERSTANDING

Compose one or two well-written sentences to express the requested information.

1.   Identify and define the three types of accounts found on a balance sheet.

2.   Explain how the ownership section of a balance sheet will vary depending on how a business is organized.

3.   Identify the main sections of an income statement and illustrate the relationship between the various sections.

4.   Discuss the procedure required to determine the cost of merchandise sold at a given amount of net sales.

5.   Describe how you would prepare a vertical analysis of an income statement.

6.   Explain the mathematical procedure required to complete a horizontal analysis of a balance sheet.

7.   Describe how you would interpret the following mathematical expression: 5:4:3

8.   Discuss the primary distinction between the current ratio and the acid test ratio.

9.   What does the term *leverage* mean and how does a firm evaluate its leverage?

10.   Explain the function of profit ratios and discuss how the ROI is determined.

Mind Your Business

Introduction

Beginning with this chapter, you will have an opportunity to apply the business concepts and mathematical procedures presented in each chapter by developing decisions necessary to resolve problems encountered by an owner/manager of a small business enterprise. The exercise will require you to develop both quantitative and written expressions of how you would apply the knowledge acquired in the current and previous chapters to resolve typical day-to-day problems using actual business data of an organization.

Background Information

After graduating from college, Matthew Ryan, Carol Kinney, and yourself formed a partnership to operate a small business enterprise called "Media World." The business originated on January 3, 1997. Each partner invested $10,000 of capital to get the business started, and each agreed to be an active partner in running the day-to-day activities. As a business major, you have assumed the position of business manager and, therefore, are responsible for maintaining the financial records of the business and advising the other partners on all financial matters. Each partner invested $10,000 of capital to get the business started, and the present capital balances represent the partners' original contribution, prior year's earnings, and prior year's withdrawals. The partnership agreement specifies that income and losses are to be shared 40% by you, 30% by Ryan, and 30% by Kinney.

The business occupies 1,200 square feet of space in a strip mall located in the suburbs of a city of 250,000 people. The mall is directly off a major four-lane highway. There is a traffic signal at the mall's entrance, and it contains 500 parking spaces. In addition to your business, the mall contains 20 other businesses that offer a typical mix of businesses providing products and services to local markets. Media World is the only retail business in the mall, or for that matter, the west side of the city, that sells an extensive line of music, books, and videos in one location. Music is the major product line and represents 50% of the total floor space, followed by books, which represent 35% of total floor space, and videos, which have been allocated the remaining floor space. The store offers all kinds of music, sold in both compact disc and cassette formats. Book products offered for sale include paperbacks and magazines, with a focus on specialized personal, professional, and leisure topic markets. The video products offered are limited to family-oriented, full-length motion pictures released on videotape for consumer purchase. Media World also sells accessory products that support each of its three product lines. Since its inception, the business has reported a small but steady increase in sales, providing resources to expand promotion activities and operating hours. In your opinion, the business has not achieved its full potential, and a recent analysis of target markets supports your recommendation to pursue a growth strategy.

As the business manager, you are responsible for all financial operations of the business. Your responsibilities include making all financial decisions that affect the business, evaluating the financial data submitted by employees, preparing reports required to analyze business activity, and advising partners. Your decisions will be guided by the primary goals of the business, which is to remain profitable and expand operations.

On September 30 of the present year, Kate Bradley, the bookkeeper, submitted to you the following report of account balances:

Account title	Amount
Cash	$28,623
Accounts receivable	1,480
Notes Receivable	10,000
Merchandise inventory October 1	34,630
Office supplies	375
Prepaid insurance	1,500
Prepaid taxes	3,725
Prepaid rent	5,000

(continued)

Store equipment	50,000
Accumulated depreciation—store equipment	13,600
Office equipment	12,000
Accumulated depreciation—office equipment	7,590
Accounts payable	15,240
Sales tax payable	11,824
Salaries and wages payable	6,480
Employee benefits payable	3,250
Payroll taxes payable	5,216
Notes payable	6,183
M. Ryan, Capital	26,523
C. Kinney, Capital	30,248
(Your Name), Capital	39,109
Sales (all merchandise)	62,824
Sales returns and allowances	1,282
Sales discounts	4,755
Merchandise inventory October 31	52,560
Purchases	46,387
Transportation—in	2,875
Purchase returns and allowances	1,250
Purchases discounts	5,397
Depreciation expense (sales equipment)	400
Advertising expense	525
Wages and salary expense	6,950
Supplies expense	475
Depreciation expense (office equipment)	147
Insurance expense	138
Legal and accounting expense	750
Rent expense	6,300
Utilities expense	565
Payroll tax expense	1,328
Employee benefits expense	275
Sales tax expense	1,750
Interest expense	500
Interest income	850

Each month this financial data is used to complete the following activities as part of your responsibilities.

Activity 1

Prepare an income statement and a balance sheet for the month ended September 30.

Activity 2

Develop a vertical analysis of both the income statement and the balance sheet.

Activity 3

Prepare specific financial ratios that will provide the following information, based on September's financial statements:

a. The ability of the business to convert its assets into cash to meet its current debt

b. The ability of the business to use credit to finance future expansion

c. The profitability of the business in terms of the investment of the partners if average owner's equity as of September 30 is $62,750.

d. How effective the partners have been using the assets of the business, based on average monthly assets of $118,250

Activity 4

Develop a narrative for the other partners that summarizes your perspective of Media World's performance based on the monthly financial analysis.

SELF-TEST

A. Terminology

Complete the following items using the key terms presented at the beginning of the chapter. Check your responses against the answer key at the end of the test.

1. The _____ details the financial position of a business at a specific point in time.

2. The balance sheet equation states, _____ must always equal _____ plus _____.

3. A claim against the assets of a business by its creditors is called _____.

4. In a sole proprietorship or a partnership, owner's equity is identified as _____.

5. The _____ reflects the results of operations in terms of profits.

6. The relationship between the sections of the balance sheet is expressed mathematically through the _____.

7. In an income statement, _____ is defined as merchandise that has been purchased and sold.

8. When the cost of goods sold is deducted from the value of net sales, the result is _____.

9. The percentage analysis of changes in corresponding items in comparative financial statements is referred to as _____.

10. An analyst will use _____ when the information needed concerns the percentage relationship of the component parts of the financial statement to the total in a single period.

11. A _____ is the comparison of one number to another.

12. The _____ is found by dividing current assets by the current liabilities.

B. Calculations

The following concepts and short problems are designed to test your understanding of the objectives identified at the beginning of the chapter. Answers are provided at the end of the test.

13. If a company's assets total $68,975, and liabilities amount to $24,870, what is the claim against the assets by the owners?

14. The following account balances appeared in the records of the Goodtime Amusement Co. on December 30, 200X: cash $12,265, accounts receivable $18,950, accounts payable $4,290, supplies $1,340, salaries expense $10,240, equipment $67,500, depreciation expense—equipment $2,450, and accumulated depreciation—equipment $18,600. Determine the amount of total assets to be reported on the December 30 balance sheet.

15. From the following—sales $87,260, inventory Jan 1 $17,490, purchases $32,840, purchase returns $4,800, purchase discounts $1,260, inventory Jan 30 $16,820—determine (a) cost of goods sold and (b) gross profit.

16. If sales are $65,000 and operating expenses are 45% of sales, what is the net income if the cost of goods sold is $26,500?

17. Complete the analysis indicated in this partial balance sheet.

	2000	1999	Increase/(Decrease) Amount	Increase/(Decrease) Percent	Percent 2000
Assets					
Current assets:					
Cash	$ 5,000	$ 3,500	a.		
Accounts receivable	8,750	6,800	b.		
Prepaid insurance	1,500	2,400	c.		
Merchandise inventory	27,350	30,250	d.		
Total current assets	$ 42,600	$ 42,950	e.		
Fixed assets:					
Equipment (net)	12,800	7,375	f.		
Building (net)	80,000	85,600	g.		
Total assets	$135,400	$135,925	h.		

18. Charles Lee and Carl Ginter are considering a partnership that requires an investment of $75,000. If Lee invested $25,000 and Ginter invested $50,000, what is the ratio of Lee's investment to (a) the total investment? (b) Ginter's investment?

19. The asset turnover ratio of Smith Supply Company is 5.8:1. If Smith Supply Company's net sales for the period are $464,000, what is the value of Smith's total assets?

20. The following data were abstracted from the financial statements of We-Got-It-All Rental Company.

Current assets	$36,000
Average accounts receivable	8,500
Current liabilities	15,500
Total assets	78,500
Inventory	12,000
Net sales	96,250
Net income after taxes	20,800
Owner's equity (January 1)	40,000
Owner's equity (December 31)	52,500
Total liabilities	38,500

From the data, prepare (a) a current ratio, (b) an acid test ratio, (c) a debt-to-equity ratio, and (d) a return on investment ratio.

Answers to Self-Test: **1.** balance sheet   **2.** assets, liabilities, owner's equity   **3.** liabilities   **4.** capital   **5.** income statement   **6.** fundamental accounting equation   **7.** cost of goods sold   **8.** gross profit   **9.** horizontal analysis   **10.** vertical analysis   **11.** ratio   **12.** current ratio   **13.** $44,105   **14.** $81,445   **15. a.** $27,450; **b.** $59,810   **16.** $9,250   **17. a.** $1,500, 42.9%, 3.7%; **b.** $1,950, 28.7%, 6.5%; **c.** ($900), (37.5)%, 1.1%; **d.** ($2,900), (9.6)%, 20.2%; **e.** $350, .8%, 31.5%; **f.** $5,425, 73.6%, 9.5%; **g.** ($5,600), (6.5)%, 59.1%; **h.** $475, .4%, 100.0%   **18. a.** 1:3; **b.** 1:2   **19.** $80,000   **20. a.** 2.3:1; **b.** 1.54:1; **c.** .73:1; **d.** 45.0%

5

PURCHASING MERCHANDISE

Learning Objectives

After completing the exercises in this chapter, you will be able to:

1. Understand the extent of merchandising distribution networks.

2. Differentiate buying and selling terms and processes.

3. Read and prepare invoices.

4. Compute net price based on single and chain trade discounts.

5. Use the complement method to calculate trade discounts.

6. Find the single-equivalent rate for chain or series discounts.

7. Calculate cash discounts using common terms of payment.

8. Compute adjustment for returned goods and allowances when discounts are offered.

9. Determine freight charges and net amount due on an invoice.

10. Understand the function of commission agents and how to calculate commissions.

11. Calculate the cash discount and the amount of credit when a partial payment is remitted.

12. Define key terms.

INTRODUCTION

Learning objective
Understand the extent of merchandising distribution networks.

Every business engaged in merchandising buys and sells goods for a profit. A business must purchase goods at the lowest possible cost and sell them for a price that will recover all costs and business expenses, and generate an acceptable profit (see Chapter 4). In addition, every business must become part of a distribution network designed to market and transport the goods from the manufacturer to the consumer. Figure 5.1 shows examples of some distribution channels used by manufacturers to market their goods.

The longest channel involves the **manufacturer,** who buys raw materials or parts and produces or assembles them into finished goods that are then sold to wholesalers. The **wholesaler** or **distributor** then stores and promotes the sale of goods to the **retailer,** who then sells the goods to the consumer. The **consumer** is the part of the channel that will use or consume the product.

Notice that the other channels eliminate (to various degrees) the *middlepersons* between the manufacturer and the consumer. These latter channels have become increasingly popular because they reduce the cost of goods, which, in turn, lowers the price charged at each step in the channel. This process is generally referred to as *discounting;* the retailers that participate in these channels, such as Kmart, Target, and Wal-Mart, are called *discount stores.*

Learning objective
Differentiate buying and selling terms and processes.

Because each business in a distribution channel buys and sells goods, it is important that you, the student, understand that the term *cost* refers to *buying* and the term *price* refers to *selling.* These terms are often used interchangeably, because the seller's price is the buyer's cost as the product moves through the distribution channel (Figure 5.2).

Now that you understand these important basic merchandising concepts, we are ready to discuss the focus of this chapter, purchasing merchandise.

5.1   PREPARING INVOICES

An **invoice** is the official record of a business transaction between the seller and the buyer. The buyer prepares a *purchase invoice* that verifies an order has been placed with a specific seller. The seller, in turn, issues a *sales invoice* when the items ordered have been sold and arrangements for deliv-

Figure 5.1

Product distribution channels

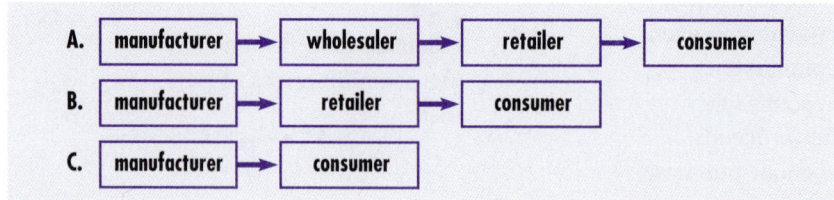

Figure 5.2

Distribution channel: cost versus price analysis

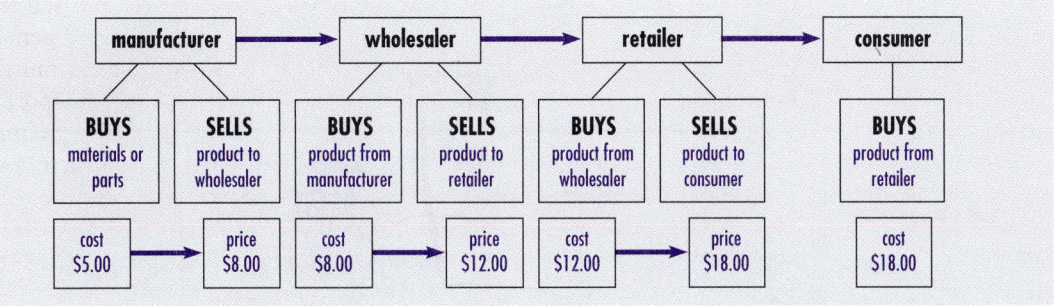

ery are completed. Although invoices vary in appearance, a typical sales invoice will contain the following information:

Learning objective
Read and prepare
invoices.

1. The names and addresses of both the seller and the buyer
2. The invoice number, date, and order reference number
3. A description of the items purchased with quantity, unit price, and extension amount
4. How and who will deliver the merchandise
5. The type and amount of discounts offered
6. The terms of payment
7. Shipping and insurance charges
8. The invoice total (sum of all charges associated with sale)

Figure 5.3 is an example of an invoice. Study the example carefully, as you will be required to prepare invoices for various problems in this chapter. The section of the invoice that identifies the *quantity* (the number of units sold), the *description* (the identification of the item, the stock number, etc.), the **unit price** (the price per measure of unit), and the **extension amount** (the quantity value of items sold under each description and unit price) provides the data necessary to determine the total amount of the invoice. The *invoice total* is the sum of the extension amounts.

The extension amount

The following equation used to calculate the extension amount:

quantity × unit price = extension amount

This equation requires the quantity and price to be stated in the same weight or measure.

Figure 5.3

Sample invoice

Invoice
554

CAPITAL ELECTRICAL SUPPLY INC.
1269 Industrial Drive
Silver Spring MD 20903
(202) 555-4792

Sold to: Midstate Electric Co.
458 Center Street
Mapleview, N.Y. 13107

Ship to: Same

Shipped via Roadway Express Inc.

Customer's Order	Our Order Number	Salesman	Terms	Date Shipped	F.O.B.	Date
XXXXXX		Martin	2/10, n/30	1/4/99	Destination	1/12/99

Quantity Ordered	Quantity Shipped	Description	Unit Price	Amount
6	6	two-pole switches	$ 10	25	$ 61	50
350 ft	350 ft	# 14 wire		035	12	25
12 doz	12 doz	# 6 connectors	1	27	15	24

		Subtotals		$ 88	99

Misc. Information: returned merchandise subject
to a 10 percent restocking fee

Trade Discount
5% of list $ 4 45

Cash discount ▶ **Deduct:** $ ___1.69___
If paid by: ___1/22/99___

Shipping Charge $ 18 20

Invoice Total ▶ $ 102 74

Notification of shortages or damages and requests for proof of
delivery must be made in writing within 30 days of invoice date

Interest of 2 percent per month assessed
on all past due accounts

Look again at Figure 5.3. All quantities are stated in the same weight or
measure as the unit price of the item. The unit price defines how the item
is sold and cannot be changed. To simplify the preparation of the invoice,

sellers generally require the buyer to purchase merchandise in quantities consistent with the unit price measure. When the quantity ordered is not consistent with the unit price measure, the quantity must be adjusted, as shown in the following example.

Example 1

Find the total amount of an invoice if a customer purchased six two-pole switches at $10.25 each; 350 feet of #14 wire at $.035 per foot; and 1 gross of #6 connectors at $1.27 per dozen.

Solution

The quantity of both the switches and wire is multiplied by the unit price to determine the extension amount, as they are quoted in like terms (6 × $10.25 = $61.50 and 350 × .035 = $12.25). The connectors, however, are sold by the dozen. Therefore, one gross must be converted to dozen (12 dozen in one gross), requiring us to adjust the quantity ordered to 12 dozen for calculation (12 × $1.27 = $15.24). We can now sum the extension amounts to find the invoice total.

Quantity	Description	Unit price	Extension amount
6	Two-pole switches	$10.25 ea.	$61.50
350 ft.	#14 wire	$ 0.035/ft.	$12.25
12 doz.	#6 connectors	$ 1.27/doz.	$15.24
		Invoice Total	$88.99

Table 5.1 contains standard abbreviations used on invoices, and Figure 5.4 contains the weights and measures. Refer to both when you prepare your assignments. When sellers use special abbreviations, the notation of their meaning is usually printed on the invoice. (An example of a special abbreviation could be a number code used to identify a choice of color.)

Table 5.1

Invoice abbreviations

Abbreviation	Term	Abbreviation	Term
ea	each	doz	dozen
drm	drum	gro	gross
cs	case	bx	box
ctn	carton	sk	sack
bbl	barrel	qt.	quart
pr	pair	gal.	gallon
C	per hundred (100)	oz.	ounce

(continued)

Table 5.1
(continued)

Invoice
abbreviations

Abbreviation	Term	Abbreviation	Term
M	per thousand (1000)	lb.	pound
cwt.	per hundred weight	ct.	crate
cpm	cost per thousand	ml	milliliter
@	at	cl	centiliter
FOB	free on board	l	liter
COD	cash on delivery	mm	millimeter
ROG	receipt of goods	cm	centimeter
EOM	end of month	m	meter
ex.	extra dating	km	kilometer
FAS	free alongside ship	g	gram
in.	inch	kg	kilogram
ft.	foot	sq. ft.	square feet
yd.	yard	sq. yd.	square yard

Figure 5.4

Units of measure

Quantities

12 units	=	1 dozen
12 dozen	=	1 gross
144 units	=	1 gross
1 ream	=	500 sheets

Weights

16 ounces	=	1 pound
100 pounds	=	1 hundredweight
2,000 pounds	=	1 ton
7,000 grains	=	1 pound

Dry Measure

2 pints	=	1 quart
8 quarts	=	1 peck
4 pecks	=	1 bushel

Liquid Measure

1 cup	=	8 ounces
2 cups	=	1 pint
2 pints	=	1 quart
4 quarts	=	1 gallon
$31\frac{1}{2}$ gallons	=	1 barrel

Linear Measure

12 inches	=	1 foot
36 inches	=	1 yard
3 feet	=	1 yard
$16\frac{1}{2}$ feet	=	1 rod
66 feet	=	1 chain
5280 feet	=	1 mile
320 rods	=	1 mile

Area

length x width	=	square area
144 square inches	=	1 square foot
9 square feet	=	1 square yard
$30\frac{1}{2}$ square yards	=	1 square rod
160 square rods	=	1 acre
640 acres	=	1 square mile

Volume (length x width x height = cubic volume)

1,728 cubic inches	=	1 cubic foot
27 cubic feet	=	1 cubic yard
128 cubic feet	=	1 cord

CHECK YOUR KNOWLEDGE

Preparing Invoice Amounts

Solve each of the following problems. Round dollar amounts to the nearest cent.

1. How much did it cost Sarah Bennett to purchase a gallon of vinegar at a unit price of 29 cents a pint?

2. The Men's Shop purchased 27 shirts for $495.45. What was the unit price if the shirts are sold by the dozen?

3. General Tire Distributors purchased the following tires from a manufacturer: 135, P165/14 for $22.43; 65, P240/15 for $37.25; and 250, 2185/13 for $48.57. What was the total cost of the order?

4. Complete the following invoice calculations:

Quantity	Description	Unit price	Extension amount
10	Helmets, football Model 60, 5 size 6⅞, 5 size 7¼	$62.95 ea	_____
5 doz.	Socks, Athletic, Tube, Style 137	$29.40/doz	_____
12 bx.	Golf Balls Hy-Fly 50's	$18.50/bx	_____
16 doz.	Sweat Shirts, Athletic; 4 doz ea S, M, L, XL; Color Code Z	$87.50/doz	_____
		Invoice Total	_____

5. Identify the following invoice abbreviations:
   a. cs   f. cpm
   b. yd.   g. bbl
   c. cwt.   h. km
   d. gro   i. lb.
   e. l   j. ctn

Answers to CYK: 1. $2.32 per gallon   2. $220.20 per dozen   3. $17,591.80   4. $629.50; $147.00; $222.00; $1,400.00; total $2,398.50   5. a. case; b. yard; c. per 100 lbs.; d. gross; e. liter; f. cost per 1,000 lbs.; g. barrel; h. kilometer; i. pound j. carton

5.1   EXERCISES

Complete the following invoice calculation:

	Quantity	Description	Unit price	Extension price
1.	15 gal	milk	$2.19/gal.	
2.	30 lb.	butter	$4.75/5 lb.	
3.	10 bottles	syrup	$3.25/bottle	
4.	50 lb.	pancake mix	$4.55/5-lb bag	
5.	300	paper plates	$2.75/package of 50	
6.	500	paper napkins	$3.00/package of 100	
7.			Total	

Complete the following invoice calculation:

	Quantity	Description	Unit price	Extension price
8.	1,000	notepads 4″ × 5″ blue	$20/hundred	
9.	13 gross	#2 lead pencils	$47.50/gross	
10.	1,000	convention folders w/assoc. logo	$130/carton (250 folders/ctn)	
11.			Total	

Complete the following invoice calculation:

	Quantity	Description	Unit price	Extension price
12.	10 doz.	adjustable baseball caps—2 doz. ea. of red, blue, green, yellow, and orange	$3.50/cap	
13.	10 doz.	t-shirts with team logo, 2 doz. ea. of red, blue, green, yellow, orange	$3.25/ea.	
14.	30	wood baseball bats, 10 ea. of lengths 26″, 28″, 30″	$12.50 ea.	
15.	5	catcher's chest protector	$27.50 ea.	
16.	5	catcher's mask	$32.25 ea.	
17.	5	catcher's mitt, model S-1138	$43.25 ea.	
18.	6 doz.	Little League approved baseballs	$42/doz.	
19.			Total	

Complete the following invoice calculation:

	Quantity	Description	Unit price	Extension price
20.	5 doz.	yellow canary legal pads	$9.00/doz	
21.	7 bottles	white-out correction fluid	$1.04/bottle	
22.	5 boxes	3½″ computer disks	$15.00/box	
23.	3	desktop staplers	$14.50 ea.	
24.	3 boxes	staples	$2.25/box	
25.	1	staple remover	$3.50	
26.			Total	

Solve the following problems. Round dollar amounts to the nearest cent and rates to the nearest tenth of a percent.

27. Sid Jones purchased 8 six-packs of cola at a unit price of $2.58 per six-pack. What was his total cost?

28. Bob Haskins purchased 1 case of motor oil for $15. If a case contains 12 one-quart cans of motor oil, how much did Bob pay for each quart?

29. While shopping for blank videotapes, Gina Sardino discovered that she could purchase a single tape for $3.99 or she could purchase a pack of four videotapes for $14.50. What was the unit price per tape for the four-pack? How much would she save per tape by purchasing the four-pack if she intended to purchase four tapes anyway?

30. Juan Ramirez priced a 10-pack of 5¹/₄″ computer disks to be $13.90, and a 50-pack to be $59.50.

What is the unit price per disk of the 10-pack? What is the unit price per disk of the 50-pack? If Juan needed to purchase 100 disks, how much would he save by purchasing the 50-packs rather than the 10-packs?

31. Sally Sanchez discovered that during a holiday sale at a local mall, she could purchase one pullover sweater for $22.50, or two of the same style sweater for $35.00. How much would she save per sweater by buying two sweaters?

32. Rachel Mudhamgha purchased 4 boxes of chocolate candy bars at $10.95 per box; 7 boxes of chewing gum at $7.39 per box; 3 canisters of hard candy at $8.49 per canister; and 9 boxes of licorice sticks at $6.79 per box for her high school store. What was the total of Rachel's order?

5.2 TRADE DISCOUNTS

Manufacturers and wholesalers promote the sale of their products through *catalogs,* which contain photos, illustrations, specifications, and descriptions of every product they sell. Catalogs also include the **list price** of each item, which is the suggested selling price the buyers should charge their customers. Sellers may offer buyers trade discounts to encourage buyers to purchase their products. A **trade discount** is a reduction in the list price. The price paid by the buyer after the trade discount is subtracted from the list price is called the **net price** from the seller's perspective, or the **net cost** from the buyer's perspective.

The list price of items listed in catalogs may change and items may be added or deleted before new catalogs can be published. To keep the buyer informed of all changes, the seller will issue a price supplement to the catalog that identifies current trade discounts applicable to the list prices in the catalogs. The price supplement also allows the seller to vary the amount of the trade discount offered to different buyer categories (wholesalers, retailers). If buyers wish to optimize net income, they must be aware of price changes and discount offers of their suppliers so they can purchase the desired quantity of merchandise at the lowest price available.

Trade discounts can be expressed as percentages or as amounts. The size of a discount can vary according to the type of buyer; for example, a wholesaler may receive a 25% discount on an item and a retailer may receive only 15% on the same item. Sellers may also base the amount of the trade discount on the quantity purchased, competitive prices, and product classification. Trade discounts are applied strictly to the list price of an item. Let's see how a trade discount is calculated when a single discount is offered.

Calculating Single Trade Discounts and Net Price

To find the amount of the trade discount and net price when a single discount is offered, we use the two-step procedure shown in Example 2.

Example 2

A refrigerator lists for $1,200 with a trade discount of 20%. Find the amount of trade discount and the net price.

Solution

Step 1:   Find the amount of trade discount.

$$\text{trade discount amount} = \text{list price} \times \text{trade discount rate}$$
$$= \$1,200 \ \times .20$$
$$= \$240$$

Step 2:   Find the net price.

$$\text{net price} = \text{list price} - \text{trade discount amount}$$
$$= \$1,200 \ - \$240$$
$$= \$960$$

As you read the steps, did you recognize the procedure as an application of the percentage formula presented in Chapter 2? The terms would be redefined as follows:

list price	×	trade discount %	=	trade discount amount
base	×	rate	=	part
list price	−	trade discount amount	=	net price
base	−	part	=	complement part
$1,200	−	$240	=	$960

The net price is the **complement** of the trade discount amount (part) because the discount amount is subtracted from the list price (base). In other words, both the net price and trade discount amounts are a part of the base (list price).

The net price for Example 2 can also be found using the *complement method*. The complement of the trade discount expressed as a percent would be 100% − 20% = 80%. A more complete expression of the complement calculation would be:

List price	$1,200	100%
Trade discount	240	20%
Net price	$ 960	80% (complement of the trade discount)

The complement of the trade discount is used as the rate to determine the net price.

net price = list price × trade discount complement rate
= $1,200   × .80
= $960

Calculating Chain Discounts and Net Price

It is quite common for manufacturers and wholesalers to offer retailers additional discounts on merchandise to increase sales, to change prices, to adjust to seasonal trends, and for different types of customers. For example, to increase sales and reduce current inventory, a wholesaler who usually offers a 20% trade discount might offer an additional 15% discount on selected merchandise. The discount would be quoted "20% less 15%" or simply 20/15. This type of discount is called a **chain discount** or *series discount.*

When chain discounts are given, each discount is applied separately. It is important to note that the discounts can never be added together, because each discount is applied to a different base. There are two methods used to calculate chain discounts and net prices. Example 3 illustrates the *declining net price method.* The *complement method* is explained in Example 4.

Example 3

Leisure-Time Distributors lists a sailboard for $1,500 and offers a trade discount of 25% and 20%. Use the declining net price method to determine the net price of the sailboard.

Solution

Step 1:   Multiply list price by first discount rate to find discount.

$1,500 × .25 = $375

Step 2:   Subtract discount from list price to find net amount.

$1,500 − 375 = $1,125

Step 3:   Multiply net amount by second discount rate to find discount.

$1,125 × .20 = $225

Step 4:   Subtract second discount from the net amount found in Step 2. This is the net price.

$1,125 − 225 = $900

If we want to find the amount of the trade discount, we simply subtract the net price from the list price ($1,500 − 900 = $600).

$$\text{trade discount} = \text{list price} - \text{net price}$$
$$= \$1,500 \quad - \$900$$
$$= \$600$$

The net price can also be found by multiplying the list price by the complements of each single discount in a chain.

Example 4

Learning objective
Use the complement method to calculate trade discounts.

The list price of a lawnmower is $525. Use the complement method to determine the net price to the buyer if the seller offers a trade discount of 10/5/10.

Solution

Step 1:   Find the complement of each trade discount.

100% − 10% = 90%
100% − 5% = 95%
100% − 10% = 90%

Step 2:   Convert each percent in step 1 to a decimal and multiply (do not round off). This is the net price equivalent rate.

.90 × .95 × .90 = .7695

Step 3:   Multiply the list price by the net price equivalent rate found in step 2.

$$\text{net price} = \text{list price} \times \text{net price equivalent rate}$$
$$= \$525.00 \ \times .7695$$
$$= \$403.99$$

The trade discount amount is found using the same procedure shown in Example 3.

trade discount = list price − net price
= $525.00 − $403.99
= $121.01

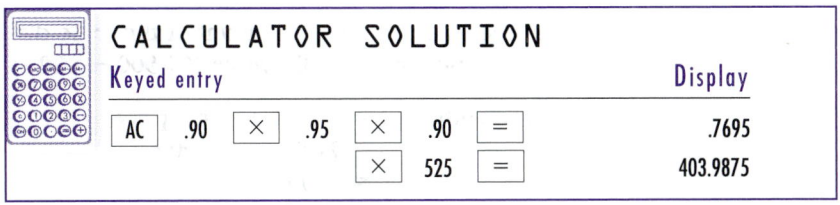

CALCULATOR SOLUTION

Keyed entry							Display
AC	.90	×	.95	×	.90	=	.7695
				×	525	=	403.9875

Learning objective
Find the single-equivalent rate for chain or series discounts.

Calculating the Single-Equivalent Trade Discount Rate

The most convenient way to find the **single-equivalent trade discount rate** is to again use the complement method. To calculate the single-equivalent discount rate for Example 4, we would simply subtract the net price equivalent rate from 1, which gives us its complement—the equivalent discount rate.

single-equivalent discount rate = 1 − (net price equivalent rate)
= 1.000 − .7695
= .2305

The .2305 is the single-equivalent discount rate for a chain discount of 10/5/10. We can also calculate the amount of the trade discount with this rate. Notice that we get the same amount of trade discount shown in Example 4.

amount of trade discount = list price × single-equivalent
discount rate
= $525.00 × .2305
= $121.01

Note: The single-equivalent discount rate can also be calculated with the percentage equation $R = P/B$; where P is the amount of discount and B is the list price ($.2305 = \$121.01/\$525$).

Merchants conducting numerous transactions daily will often use single-equivalent discount tables to simplify the process of preparing or interpreting sales invoices. Example 5 shows how the rates in Table 5.2 are determined.

Table 5.2

Single-equivalent rates of chain discounts

	5%	10%	15%	20%
5	.0975	.145	.1925	.24
10	.145	.19	.235	.28
10/5	.18775	.2305	.27325	.316
10/10	.2305	.271	.3115	.352
15	.1925	.235	.2775	.32
15/10	.27325	.3115	.3475	.388
20	.24	.28	.32	.36
20/15	⟶ .354	⟶ .388 ⟵	.422	.456

Example 5

Find the single-equivalent discount rate for a chain discount of 20/15/10.

Solution

Step 1:   Calculate the complement of the trade discount.

100% − 20% = 80%

100% − 15% = 85%

100% − 10% = 90%

Step 2:   Convert percents to decimals and calculate the net price equivalent rate.

.80 × .85 × .90 = .612 net price equivalent

Step 3:   Calculate the single equivalent discount rate.

single equivalent rate = 1 − (net price equivalent rate)

= 1 − .612

= .388, or 38,8%

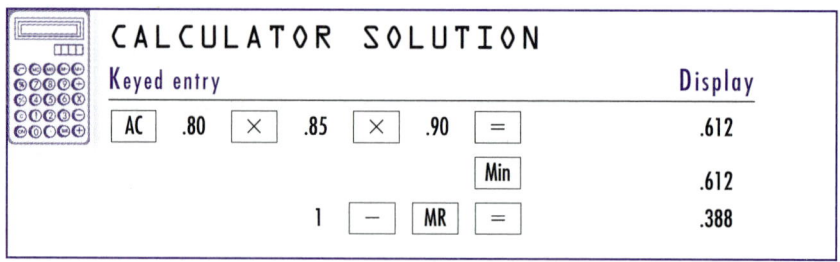

CHECK YOUR KNOWLEDGE

Calculating Trade Discounts

Solve each of the following problems. Round dollar amounts to the nearest cent and rates to the nearest one hundredth of a percent.

1. What is the amount of trade discount and net price of a 10-pound box of high-grade hot dogs that lists for $1.95 per pound and carries a 20% discount per box?

2. Find the net price of a skill saw that lists for $59 with a discount of 35%. (Use the complement method.)

3. The suggested retail price of a 12 foot × 18 foot above-ground pool with filter, ladder, and cover is $2,400. If the dealer offers a trade discount of 30%, 5%, and 10%, what is the retailer's net price?

4. What is the single-equivalent trade discount rate and amount in problem 3?

5. Find the single-equivalent trade discount rate for a chain (series) discount of 40/25/15.

5.2   EXERCISES

Solve the following problems. Round dollar amounts to the nearest cent and rates to the nearest tenth of a percent.

Calculate the following *trade discounts* and *net prices:*

	List price	Trade discount rate	Trade discount amount	Net price
1.	$1,500	15%	225	1275
2.	$2,500	20%		
3.	$1,300	22%		
4.	$5,000	24%		
5.	$7,400	19%		

Answers to CYK: 1. $3.90, $15.60   2. $38.35   3. $1,436.40   4. .4015   5. .6175

Calculate the *trade discount rate* from the given information. Express the discount rate as a percent (%).

	List price	Trade discount amount	Trade discount rate
6.	$1,600	400	_____
7.	$750	250	_____
8.	$2,800	560	_____
9.	$12,000	1,200	_____
10.	$850	170	_____

Determine the *complement percentages* for each of the following percentages, and express them in decimal form.

	Percentage	Complement %	Complement % (decimal form)
11.	15%	_____	_____
12.	45%	_____	_____
13.	22%	_____	_____
14.	37%	_____	_____
15.	17%	_____	_____

Determine the *net price* using the complement method.

	List price	Trade discount rate	Complement trade discount rate	Net price
16.	$2,000	.25	_____	_____
17.	$3,500	.20	_____	_____
18.	$5,400	.15	_____	_____
19.	$720	.30	_____	_____
20.	$85	.10	_____	_____

Calculate the trade discount by the *declining net price* method.

	List price	1st discount rate (%)	1st discount amount	1st net amount
21.	$2,000	15	_____	_____
22.	$3,000	25	_____	_____
23.	$350	20	_____	_____
24.	$90	15	_____	_____
25.	$550	30	_____	_____

2nd discount rate (%)	2nd discount amount	2nd net amount	Total trade discount
10	_____	_____	_____
10	_____	_____	_____
15	_____	_____	_____
12	_____	_____	_____
5	_____	_____	_____

Determine the *net price equivalent rate* for each of the following.

	Discount rate	Complement of discount	Net price equivalent rate
26.	20/20	_____	_____
27.	15/10	_____	_____
28.	30/10/5	_____	_____
29.	40/15/10	_____	_____
30.	18/12/7	_____	_____

Calculate the trade discount by the *net price equivalent rate* method.

	List price	Discount rate (%)	Net price equivalent rate	Net price	Trade discount
31.	$3,000	20/15	_____	_____	_____
32.	$5,000	15/10	_____	_____	_____
33.	$900	25/15/10	_____	_____	_____
34.	$15,500	18/12/6	_____	_____	_____
35.	$9,000	12/8/5	_____	_____	_____

Calculate the *single-equivalent trade discount rate* for each of the following.

	Discount rate	Net price equivalent rate	Single-equivalent trade discount rate
36.	10/20/5	_____	_____
37.	15/10/5	_____	_____
38.	25/12	_____	_____

	Discount rate	Net price equivalent rate	Single-equivalent trade discount rate
39.	20/10/5/3	_____	_____
40.	30/15/10	_____	_____

41. What is the amount of the trade discount and net price of a portable color television set that lists for $675 if the dealer offers a 25% discount on the set?

42. Eberhart's Mills offered a 30% trade discount to buyers on all bedding purchased during December. How much will a customer pay for 8 cartons of assorted sheet and pillow case sets that list for $49.50 a carton?

43. Julio Santo, a buyer for Hometowne Markets, can purchase a crate of melons from one supplier for $18.75. Another supplier offers a list price of $22 per crate. If Julio accepts the lower price, how much will he save if he buys 12 crates?

44. What is the net price for five dozen t-shirts if the list price is $108 per dozen and the seller offers a trade discount of 12% and 8%?

45. Upstate Distributors provides a chain discount of 25/10/5 to retailers on all appliances. Find the amount of trade discount allowed on the sale of three electric ranges that list for $675 each, and 2 refrigerators listed at $1,025 each to City-Wide Appliance Center.

46. Lefkowicz Carpeting purchased a 400-square-yard roll of carpet from a manufacturer at a list price of $6,000. Find the net cost per square yard if the manufacturer offered Lefkowicz Carpeting a series discount of 25/20/15.

47. A manufacturer lists a snowmobile for $3,500. If the manufacturer offers a chain discount 15/10/15 to wholesalers and a 20/15 chain discount to retailers, find the net cost of the snowmobile for (a) the wholesaler and (b) the retailer.

48. Calculate the single equivalent discount rate in problem 47 for (a) the wholesaler and (b) the retailer.

49. If a merchant offers a chain discount of 12/8/3, what is the single equivalent rate of the chain discount?

50. Quality Crafters lists a line of kitchen cabinets at $1,250. If a customer paid a net price of $893.75, what single rate of trade discount was offered?

5.3   CASH DISCOUNTS

A major percent of sales recorded by manufacturers and wholesalers are credit sales. A *credit sale* (sale on account) allows the buyer to pay for the purchased merchandise at a future date agreed upon at the time of the sale. In Chapter 4, we briefly discussed this procedure when we explained why an account receivable is considered a current asset and appears in the balance sheet. The seller who has agreed to sell merchandise on credit has, in essence, agreed to finance that part of the buyer's inventory. This process allows the buyer to (1) carry a larger selection of merchandise, (2) increase sales potential and profits, and (3) have time to sell the merchandise before the payment date. A credit sale can, however, create a *cash flow* problem for the seller because of the time period between the date of the sale and receipt of payment. To encourage buyers to pay invoices before the required date, many sellers offer a **cash discount.** A cash discount is usually expressed as a percent and is always based on the net price of the sales invoice, excluding other charges such as shipping, insurance, trade discounts, and sales tax. The procedure used to calculate the cash discount is another application of the percentage equation.

Learning objective
Calculate cash discounts using common terms of payment.

Calculating cash discount

cash discount = net price × cash discount rate

part = base × rate

The **terms** of the cash discount are listed on the invoice in the section labeled *terms* or *terms of sale.* The terms identify the percent of the cash discount and the time period for which it applies. There are many variations in the terms granted by sellers to buyers. In this section, we will look at the most commonly used expressions of cash discounts.

Cash Discount Dating Methods

The **ordinary dating method** terms on an invoice would be expressed as 2/10, *n*/30. The "2/10" means the buyer may deduct 2% from the net price noted on the invoice if payment is made within 10 days from the invoice date. The "*n*/30" means the buyer must pay the full amount of the invoice within 30 days. After 30 days, payment is considered overdue, and the buyer may be required to pay a late charge. Figure 5.5 illustrates the cash discount time line for this situation.

As you can see, dates are an integral part of the cash discount process. The buyer must be aware of the **discount dates** and the **net payment date** to take advantage of the opportunity to reduce the cost of goods. Both the discount date and the net payment date are counted from the invoice date. Two methods are presented to help you calculate these important dates.

Days-in-a-month-Method.   Thirty days has September, April, June, and November; all the rest have 31 except February, which has 28, and 29 during leap year (2000 and 2004 are leap years).

Figure 5.5

Cash discount time line

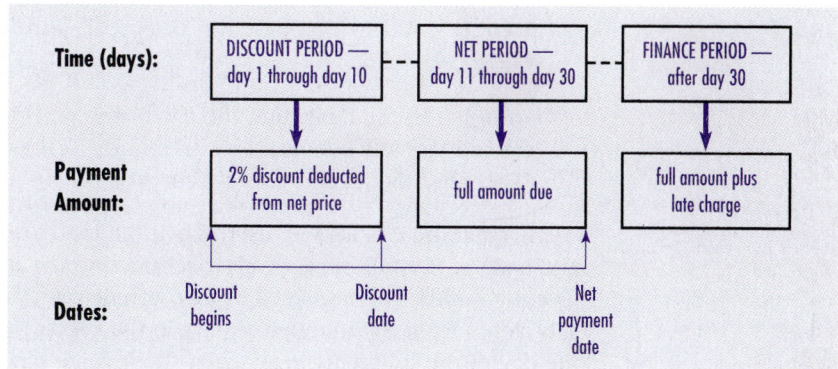

Example 6

An invoice is dated January 12, 1999, with terms 3/10, n/30. (a) Find the discount date; (b) find the net payment date.

Solution

a.   January 12 plus 10 days brings us to January 22, which is the last date to take the cash discount.

b.
$$
\begin{array}{rl}
31 & \text{days in January} \\
-12 & \text{invoice date} \\
\hline
19 & \text{days remaining in January} \\
+11 & \text{days in February net amount due} \\
\hline
30 & \text{days to pay net}
\end{array}
$$

or

$$N + 19 = 30$$
$$N = 30 - 19$$
$$N = 11$$

We first subtract the invoice date from the number of days in the month. Then, we find what number when added to the days remaining in the month equals the net period (30 days). This number is the net payment date (February 11, 1999).

Days-in-a-year Method.   The exact days-in-a-year method requires the use of Table 5.3 to calculate the dates. If we use the information given in Example 3, the invoice due date would be found as follows.

Step 1:   In the table, find the date of the year represented by the invoice date. (January 12 is the 12th day of the year.)

Step 2:   Add to this date the net credit period given on the invoice. (12 + 30 = 42)

Step 3:   In the table, find the day of the year determined in step 2 and read to the left column "Day of month." This is the invoice due date. (The 42nd day of the year is February 11.)

This method can also be used when the credit period begins in one year and ends in the following year by subtracting the invoice date from 365 to give the number of days used in the current year. This figure would then be subtracted from the number of days in the credit period, which would give the day of the following year when the invoice payment is due.

Table 5.3

Exact days-of-the-year calendar

Day of month	31 Jan.	28 Feb.	31 Mar.	30 Apr.	31 May	30 June	31 July	31 Aug.	30 Sept.	31 Oct.	30 Nov.	31 Dec.
1	1	32	60	91	121	152	182	213	244	274	305	335
2	2	33	61	92	122	153	183	214	245	275	306	336
3	3	34	62	93	123	154	184	215	246	276	307	337
4	4	35	63	94	124	155	185	216	247	277	308	338
5	5	36	64	95	125	156	186	217	248	278	309	339
6	6	37	65	96	126	157	187	218	249	279	310	340
7	7	38	66	97	127	158	188	219	250	280	311	341
8	8	39	67	98	128	159	189	220	251	281	312	342
9	9	40	68	99	129	160	190	221	252	282	313	343
10	10	41	69	100	130	161	191	222	253	283	314	344
11	11	42	70	101	131	162	192	223	254	284	315	345
12	12	43	71	102	132	163	193	224	255	285	316	346
13	13	44	72	103	133	164	194	225	256	286	317	347
14	14	45	73	104	134	165	195	226	257	287	318	348
15	15	46	74	105	135	166	196	227	258	288	319	349
16	16	47	75	106	136	167	197	228	259	289	320	350
17	17	48	76	107	137	168	198	229	260	290	321	351
18	18	49	77	108	138	169	199	230	261	291	322	352
19	19	50	78	109	139	170	200	231	262	292	323	353
20	20	51	79	110	140	171	201	232	263	293	324	354
21	21	52	80	111	141	172	202	233	264	294	325	355
22	22	53	81	112	142	173	203	234	265	295	326	356
23	23	54	82	113	143	174	204	235	266	296	327	357
24	24	55	83	114	144	175	205	236	267	297	328	358
25	25	56	84	115	145	176	206	237	268	298	329	359
26	26	57	85	116	146	177	207	238	269	299	330	360
27	27	58	86	117	147	178	208	239	270	300	331	361
28	28	59	87	118	148	179	209	240	271	301	332	362
29	29	—	88	119	149	180	210	241	272	302	333	363
30	30	—	89	120	150	181	211	242	273	303	334	364
31	31	—	90	—	151	—	212	243	—	304	—	365

Example 7

What date is 60 days after November 20?

Solution **Step 1:**
365 days in year
−324 calendar date of November 20
41 days used in current year

> ***Step 2:***   60   credit period
> $\underline{-41}$ days used in current year
>   19   days of following year, or January 19

Ordinary cash discount terms can be extended to give the buyer more time to take advantage of the discount and make payment. For example, a clothing manufacturer might offer a distributor discount terms of 3/30, 2/60, n/90 to purchase its full line in May. The distributor would be able to deduct a 3% discount from the net price if full payment was made within 30 days, or a 2% discount if full payment was made between the 31st and the 60th day. The net or full amount of the invoice would be due 90 days from the date of the invoice. Example 8 illustrates how to apply this discount.

Example 8

If payment was made on July 8, how much money would Reynolds Wholesale receive from the Winton Shop for merchandise invoiced on May 10 that listed for $2,750 with a trade discount of 15% and terms of 4/30, 2/60, n/90?

Solution

Step 1:   The invoice was paid 59 days from the date of invoice; therefore, the 2% cash discount can be deducted from the net price of the invoice.

31	21 May
$\underline{-10}$	30 June
21 days in May	$\underline{8}$ July (payment date)
	59 days

Step 2:   The cash discount is based on the net price, therefore, the trade discount must be deducted from the list price to determine the net price.

$2,750.00 list price
$\underline{-412.50}$ less trade discount (2,750 × .15)
$2,337.50 net price

Step 3:   Calculate the amount of the cash discount on the net price.

cash discount amount = net price × cash discount rate
$$= \$2,337.50 \times .02$$
$$= \$46.75$$

Step 4:   Subtract the cash discount from the net invoice amount to determine the amount due on July 8.

$2,337.50 net invoice amount
$\underline{-46.75}$ less cash discount
$2,290.75 amount due

When merchants use the terms **eom** (*end-of-month*) or **prox** (proximo) on invoices, they are extending their customer's cash discount and periods. For example, when 3/10 eom appears on an invoice, it means the buyer can deduct a 3% cash discount if payment is made 10 days after the end of the month. In other words, the discount period and the credit period do not begin until the first day of the following month. Sellers who use this method will frequently add on an extra month when the date of the invoice is the 26th of the month or later. To illustrate, an invoice dated September 28 with a cash discount of 2/10 eom or 2/10 prox extends the discount period to November 10.

Example 9

Find (a) the discount date and (b) the net payment date for an invoice dated March 12 with terms of 2/15, *n*/30 eom.

Solution

a.   The discount period begins on April 1 and ends on April 15, the discount date.

b.   The net payment date is 15 days after the discount date (April 30) or 30 days after the end of the month.

Example 10

Art's Crafts and Hobby Shoppe received an invoice on June 29 for $290.50 for purchases. How much must the accounts payable clerk remit if payment is made on August 19 and the terms of the invoice are 3/20, *n*/60 prox?

Solution

Step 1:   Determine if the bill is being paid during the discount period. The discount period begins on August 1 and ends on August 20 (end of July plus 20 days comes to August 20). The discount period begins August 1 because the invoice is dated after the 25th of the month.

Step 2:   Calculate the amount of the cash discount.

discount amount = net price × discount rate
= $290.50 × .03
= $8.72

Step 3:   Find the amount due.

$290.50 net invoice amount
    8.72 less cash discount
$281.78 amount due

Manufacturers and wholesalers who produce and distribute seasonal goods to retailers will often extend cash discount periods by using the **extra dating method.** For example, when offering terms of 2/10–90X (or 2/10–90 ex), the seller is allowing the buyer an extra 90 days, or a total of 100 days from the date of the invoice, to take advantage of the cash discount. The net payment date would then be 20 days after the discount date, as each month is considered to have 30 days. The standard discount time, 10 days, is subtracted from 30 days to determine the number of days to the net payment date (30 − 10 = 20). If the terms of the cash discount are 3/5–10X, the discount period would be 15 days (5 + 10 = 15) and the net payment date would be 25 days after the discount date (30 − 5 = 25). For the seller, these terms could induce sales, reduce inventory storage costs, and provide a more manageable shipping period. For the buyers, such terms allow them to purchase goods in the off season and still take the cash discount. The buyers may also be able to sell the goods before the cash discount period expires, which enhances their cash flow position.

Example 11

Solution

If an invoice is dated April 5 with terms 3/15–30X, what is (a) the discount date and (b) the net payment date?

a.
    30 days in April
   −5 date of invoice
    25 days remaining in April
+20 days in May (discount date, May 20)
    45 days in discount period (15 + 30 = 45)

b. May 20 + 15 days after discount date is June 4, net payment date.

Example 12

Solution

An invoice for the purchase of "weed-eaters" totals $945.60 and is dated January 4, 2000, with terms 2/10–60X. If the invoice is paid on March 12, what is the amount of the net payment?

Step 1:   Determine if payment is made by the discount date.

    31 days in January
  −4 date of invoice
    27 days remaining in January
+29 days in February (a leap year)
+14 days in March (discount date)
    70 days in discount period (10 + 60 = 70)

The net payment date is April 3 (March 14 plus 20 days). The discount date is March 14 (January 4 plus 70 days comes to March 14).

Step 2: Calculate amount of cash discount.

$$\text{discount amount} = \text{net price} \times \text{discount rate}$$
$$= \$945.60 \times .02$$
$$= \$18.91$$

Step 3: Find the amount due.

$945.60	net invoice amount
18.91	less cash discount
$926.69	amount due

Notice that in both Example 11 $(15 + 15 = 30)$ and Example 12 $(10 + 20 = 30)$ the net payment period is adjusted to total 30 days based on the number of days in the discount period.

Some merchants may **postdate** invoices to extend the cash discount and net payment periods. A postdated invoice will show the terms of the invoice in the section reserved for terms (2/10, n/30) and the extension of the terms is shown in the date section of the invoice (Date: April 10, *as of* May 15). This postdate means the discount period and the net payment period start on May 15. Therefore, the discount date is May 25 and the net payment date is June 15 (May 15 + 30 days = June 14).

The terms presented thus far are all based on the date of the invoice. However, when it is difficult to determine the shipping or transportation period for merchandise, the seller can use the receipt-of-goods (**ROG**) dating method. This method gives the buyer the opportunity to receive and inspect the goods before the terms of the sale go into effect.

Example 13

Far East Imports ordered merchandise from Orient Export with a net value of $3,500. The invoice for the transaction was dated October 3 with terms of 3/10, n/30 ROG. The order arrived at Far East Imports on November 17. What was the amount of the net payment if payment was made on November 30?

Solution

Step 1: Determine if the payment is made during the discount period.

November	17	date goods received
	+10	cash discount terms in days
November	27	cash discount date

Step 2:   Calculate the amount of cash discount. A cash discount cannot be taken because the invoice was paid on November 30, 3 days after the discount date (November 27).

Step 3:   Find the net amount due.

$3,500   net invoice amount
     0   cash discount amount
$3,500   amount due

Notice that, with this method, the discount period and the credit period begin when the goods are received. The seller will be able to validate this date because the shipping company responsible for delivering the order provides the seller with a shipping invoice that notes the date of delivery.

CHECK YOUR KNOWLEDGE

Calculating Cash Discounts

Solve each of the following problems. Round dollar amounts to the nearest cent.

1.   What is (a) the discount date and (b) the net payment date for an invoice dated June 5 with terms 2/20, *n*/45? (Use Table 5.3.)

2.   What would be the net payment of an invoice dated August 22 for $1,800 with terms 3/10, 2/30, *n*/60, if payment was made on (a) September 1? (b) September 23?

3.   What are the discount dates for invoices dated (a) April 10 and (b) September 26 if both invoices offer terms of 1/15 eom?

4.   General Fasteners purchased merchandise worth $12,500 on May 14 under terms of 3/10–90X. What is (a) the amount of the net payment if the invoice was paid on August 20 and (b) the net payment date?

5.   If an invoice is dated June 5, *as of* August 31 with terms of 2/10, *n*/30, what is (a) the discount date and (b) the net payment date?

6.   Find the net payment for an invoice dated January 15 for $975.60 with terms 2.5/10 ROG if the order was received on March 20 and payment was made on March 29.

Answers to CYK: *1.* **a.** June 25; **b.** July 20   *2.* **a.** $1,746; **b.** $1,800   *3.* **a.** May 15; **b.** November 15 (additional month granted because invoice date is after 25th day of month)   *4.* **a.** $12,125.00; **b.** September 11   *5.* **a.** September 10; **b.** September 30   *6.* $951.21

5.3   **EXERCISES**

Solve the following problems. Round dollar amounts to the nearest cent and rates to the nearest tenth of a percent.

An invoice has the stated terms 4/10, *n*/30 prox. Determine the discount date, and the net payment date.

	Invoice date	Discount date	Net payment date
1.	May 19	_____	_____
2.	Dec. 21	_____	_____
3.	Feb. 26	_____	_____
4.	Oct. 30	_____	_____
5.	Nov. 20	_____	_____

An invoice has the stated terms 3/15, *n*/60 eom. Determine the discount date and the net payment date.

	Invoice date	Discount date	Net payment date
6.	Jan. 20, 2000	_____	_____
7.	March 28, 1999	_____	_____
8.	May 19, 1999	_____	_____
9.	Dec. 27, 2000	_____	_____
10.	June 28, 2001	_____	_____

Determine the discount date and net payment date for an invoice with the terms 4/10–60X if the invoice date is as follows:

	Invoice date	Discount date	Net payment date
11.	Jan. 20, 1999	_____	_____
12.	Sept. 23, 2000	_____	_____
13.	Aug. 11, 2001	_____	_____
14.	Feb. 3, 2000	_____	_____
15.	April 28, 2002	_____	_____

Determine the discount date and the net payment date if the terms of an invoice are 3/15, *n*/60 ROG if the "receipt of goods" dates are as follows:

	ROG date	Discount date	Net payment date
16.	Sept. 21	_____	_____
17.	March 19	_____	_____
18.	May 13	_____	_____
19.	Aug. 15	_____	_____
20.	June 23	_____	_____

Determine the discount date and net payment date for an invoice with the terms 4/10, *n*/30 with these dates:

Invoice date	Discount date	Net payment date
21. March 20 *as of* April 10	_____	_____
22. Feb. 13 *as of* March 14	_____	_____
23. Sept. 16 *as of* Oct. 15	_____	_____
24. July 23 *as of* Aug. 25	_____	_____
25. Nov. 28 *as of* Dec. 15	_____	_____

An invoice is dated March 23, 1999, and the buyer received the goods on April 15, 1999. Determine the discount date and net payment date if the following terms prevailed:

Terms	Discount date	Net payment date
26. 4/10, *n*/60	_____	_____
27. 4/10–90X	_____	_____
28. 4/10, *n*/60 eom	_____	_____
29. 4/10, *n*/60 prox.	_____	_____
30. 4/10, *n*/60 ROG	_____	_____
31. 4/10, *n*/60 *as of* May 15	_____	_____

32. An invoice is dated Feb. 12, 1999, with terms 4/10, *n*/30. Find: (a) the discount date and (b) the net payment date.

33. An invoice is dated September 12 with the terms of payment 4/30, 3/60, *n*/90. Determine (a) the 4% discount date, (b) the 3% discount date, and (c) the net payment date.

34. If payment is made on June 5 on an invoice with payment terms of 5/15, 4/30, 3/45, *n*/60 and dated April 25, is the buyer entitled to any of the stated discounts? If so, which?

35. Hernandez Landscaping purchased merchandise worth $8,500 on June 12 under terms of 5/10, 3/20, *n*/60. What is the amount of the payment if the invoice was paid on (a) June 28? (b) July 29?

36. Melanie's Cards and Gifts made a purchase worth $350 on May 28 with the terms 5/20 ROG. If the goods were received on June 15 and the invoice was paid on July 2, how much did Melanie's have to pay?

37. Oscar's Drive-in Beverage Outlet purchased 1,000 shipping cartons for $975 with the terms 6/15, 4/25, *n*/60, eom. The invoice was dated November 20, the cartons were received on December 10; the invoice was paid on December 12th. How much did Oscar's owe when it paid off the bill?

38. How much should Stevens Electrical Service pay for a $1,345 purchase made on February 14, 1999, with the terms 5/10, 2/20, *n*/30 prox if it receives the goods on February 28 and pays the invoice on March 15?

39. What would Bart's Haberdasherie be required to pay for a purchase of $789 made on January 19, 2000, under the terms of 5/15, 3/25, *n*/40 as of January 25, if the invoice is paid on February 18?

40. Arlene's Beauty Salon purchased $1,234 worth of hair and skin care products on September 28 with the terms of 6/10–60X, and made payment on November 27. What would have been the amount of Arlene's net payment?

41. Marlene's Ice Cream Parlor purchased ice cream, toppings, and supplies on March 29 worth $3,350 with the terms 6/10, 2/20, *n*/45 and paid off the invoice on May 20. What was the net payment due on the 20th?

42. Celebreses' Variety Store purchased assorted merchandise for $1,975 with the terms 5/15, 3/25, *n*/90 eom. The invoice was dated October 17 and was paid on December 12. How much did Celebreses' owe on the invoice when they paid off the bill?

43. How much should Caryl's Plumbing and Sewer Service pay for a $2,568 purchase made on February 24, 2000, with the terms 4/10, 2/20, *n*/30 if it receives the goods on February 28 and pays the invoice on March 6?

44. Guthrie's Pizzeria purchased food and soft drink supplies on October 6 for $567 with the terms 4/10, 2/20, *n*/30 ROG and received the order on November 6. If the invoice was paid on November 15, how much was due on that date?

45. Chevon's Sports Shop purchased $4,500 of baseball uniforms on May 23 with the terms 5/20, 4/30 *as of* June 15 and received the goods on June 6. What was the net payment if it paid the invoice on July 13?

5.4 ADDITIONAL INVOICE CHARGES AND ADJUSTMENTS

In Section 5.1, we explained the invoice in terms of its contents and preparation. In this section, we will increase our understanding of invoices as we learn how discounts and other charges such as transportation and commissions are presented on an invoice. We will also learn how to make adjustments in the amount required to pay invoices that result from partial payments and merchandise being returned for credit or refund by the buyer.

Because the invoice is the official record of the sales transaction, it generally lists the seller's policies regarding methods of payment, price changes, and procedures for returning merchandise for credit or refund. This information is used by the buyer to make important financial decisions after the merchandise has been received.

Returned Goods

Learning objective
Compute adjustment for returned goods and allowances when discounts are offered.

The invoice is either sent to the buyer through the mail or shipped with the merchandise. When the merchandise is received, the buyer checks the merchandise to verify that the quantity listed on the invoice agrees with the quantity delivered. The buyer also checks to make sure the merchandise delivered meets specifications, and that it has not been damaged in transit. Merchandise that does not meet specifications or that is damaged in transit is called **returned goods** when it is sent back to the seller for credit. The seller may offer the buyer an *allowance,* which reduces the cost of the goods if the buyer is willing to keep the merchandise.

Trade and cash discounts are based on the merchandise actually purchased. Therefore, merchandise that has been returned is not subject to any discounts offered and must be deducted before discounts are calculated. Example 14 illustrates how to calculate the adjustment for returned goods when trade and cash discounts are offered by the seller.

Example 14

What is the amount required to pay an invoice for $1,200 with a trade discount of 20%, 10%, 5%, and terms of 2/10 ROG, if the buyer returned $400 worth of merchandise and paid the invoice before the discount date?

Solution

Step 1:   Deduct the amount of returned goods from the invoice list price total.

$1,200   list price
$\underline{-400}$ less returned goods
$ 800   adjusted list price

Step 2:   Calculate the net price.

.80 × .90 × .95 = .684 net price equivalent

$800 × .684 = $547.20 net price amount

Step 3:   Calculate the cash discount.

$547.20   net price
$\underline{\times\ .02}$ cash discount rate
$ 10.94   cash discount amount

Step 4:   Determine the amount due.

$547.20   net price
$\underline{-10.94}$ less cash discount amount

$536.26   amount due

The following calculation summary for Example 14 will help you to better understand the order and basis of the calculations.

$1,200.00   invoice list price total
$\underline{-400.00}$ less returned goods
$ 800.00   adjusted list price
$\underline{-252.80}$ less trade discount (800 × .316)
$ 547.20   net price
$\underline{-10.94}$ less cash discount (547.20 × .02)
$ 536.26   amount due

Freight Charges

Learning objective
Determine freight charges and net amount due on an invoice.

The cost of transporting goods has increased considerably due to the scarcity of natural resources required to produce the fuels needed by the transportation industry and higher labor costs associated with handling transported merchandise. Consequently, manufacturers and wholesalers have had to develop transportation systems that will move shipments in and out of their organizations in a timely manner, at a competitive price, and that accommodate their customers' requirements. Sellers may elect to transport their own merchandise or employ truck, rail, steamship, or airline companies.

When transportation companies are used to ship merchandise, the *freight* charges may be paid by either the seller or the buyer depending on the shipping terms. For example, the term **FOB** (free on board) **shipping point** means that the buyer pays the freight charges directly to the transportation company and that ownership of the merchandise transfers to the buyer before shipment. If the term **FOB destination** is used, the seller prepays the freight charges to the transportation company and retains title to the merchandise until it is delivered. The title distinction is important if the merchandise is lost or damaged in transit.

Figure 5.6

Freight charge terms

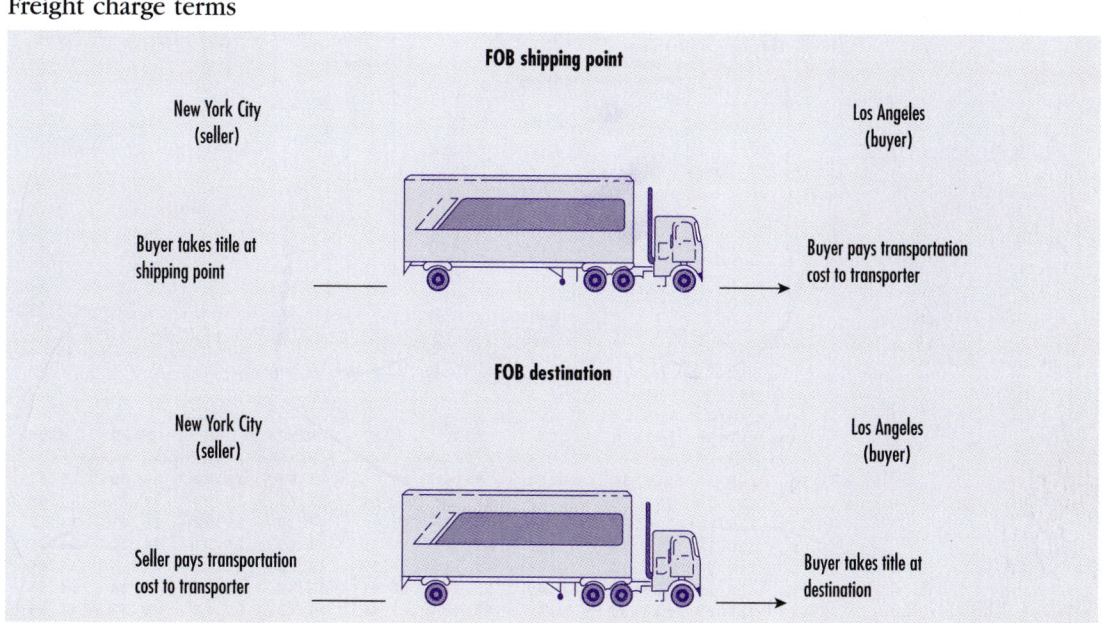

As a matter of practice, the buyer always pays the cost of transportation either directly or indirectly. The buyer pays directly when the goods are shipped FOB shipping point or when the seller prepays freight charges and includes the freight charges on the invoice. The buyer is paying freight charges indirectly when they are included as part of the unit price charged for the merchandise. This procedure is not as popular today because of the constant fluctuation of transportation rates.

FREIGHT RATES MADE EASY

ROADWAY® *Express*

P. O. BOX 471 AKRON, OH 44309-0471
http://www.roadway.com
(RDWY) (EIN 34-0482670)

PRO NUMBER: **511-222500-3**

	INVOICE	PAGE 1 OF 1
INVOICE DATE: 02/08/1999	PICK-UP DATE:	01/29/1999
TARIFF ITEM: 8881.5000		

PAYMENT DUE BY 03-10-1999
A LATE PAYMENT PENALTY MAY APPLY

AMOUNT DUE
545.05

BILL TO:

XYZ Company
987 South St.
Aurora IL 60506

SHIPPER:	CONSIGNEE:	THIRD PARTY:
ABC Company	XYZ Company	
123 North St.	987 South St.	
Greenwood MS 38930	Aurora IL 60506	

COMMENTS: FOR QUESTIONS ABOUT THIS INVOICE, PLEASE CALL (423) 349-6567

PCS	PKG	HM	DESCRIPTION OF ARTICLES	NMFC	CODE	WEIGHT	RATE	AMOUNT
1	SKD		FASTENERS		P70	1450	37.59	545.05
			CL 50 42 CTN	09458000				
			GENERAL SURCHARGE (FUEL/FRT)					
1	TTL							

```
Orig: 38930    Dest: 60506
         M/C      M/C
         <=149#   >=150#
Class                     L5C      M5C      M1M      M2M      M5M      M10M     M20M
========================================================================================
  50    128.38   128.38   48.14    39.34    29.74    23.83    16.72    13.00    12.51
  55    128.38   128.38   51.74    42.29    31.96    25.61    17.97    14.13    13.68
  60    128.38   128.38   54.75    44.75    33.82    27.10    19.02    15.07    14.11
  65    128.38   128.38   57.76    47.21    35.69    28.59    20.07    16.20    14.26
  70    128.38   128.38   60.84    49.73    37.59    30.12    21.13    17.14    14.40
 775    128.38   128.38   65.49    53.53    40.46    32.42    22.75    18.84    14.55
  85    128.38   128.38   70.34    57.49    43.45    34.82    24.43    20.35    14.99
 925    128.38   128.38   75.77    61.93    46.81    37.51    26.32    22.04    15.42
 100    128.38   128.38   80.03    65.41    49.44    39.62    27.80    23.36    16.88
 110    128.38   128.38   89.07    72.80    55.03    44.09    30.94    26.00    18.92
 125    128.38   128.38  101.97    83.35    63.00    50.48    35.42    29.77    21.24
 150    128.38   128.38  119.39    97.59    73.76    59.10    41.47    34.85    34.85
 175    128.38   128.38  138.12   112.89    85.33    68.37    47.98    40.32    40.32
 200    128.38   128.38  152.98   125.05    94.51    75.73    53.14    44.65    44.65
 250    128.38   128.38  196.21   160.38   121.22    97.13    68.16    57.27    57.27
 300    128.38   128.38  222.67   182.00   137.56   110.23    77.35    65.00    65.00
 400    128.38   128.38  315.60   257.96   194.98   156.23   109.63    92.13    92.13
 500    128.38   128.38  390.45   319.15   241.22   193.29   135.64   113.98   113.98
```

Freight Rates Made Easy (continued)

Many sellers transport their goods to buyers by motor freight. The process of determining a freight cost is complex because the rates charged by carriers depend on the classification of the goods being shipped, the weight of the goods, and the distance between point of origin and point of destination. For years, individuals known as "raters" calculated freight charges using a library of rate manuals and a calculator. In 1983, Roadway Express, Inc., introduced "E-Z Rate" the shipping industry's first simplified nationwide, zip-zone computerized freight rating system. The E-Z rate system reduces the older, complex method to a simple procedure. The freight charges shown in the accompanying Roadway Express invoice were determined from the data given and the rate table (shown) as follows:

1. Determine the National Motor Freight Classification class of the freight. Class of freight is C-125.

2. Determine the applicable base rate for the weight bracket and destination postal or zip zone from the rate table. Base rate is $37.59. Shipment weight is 1,450 pounds. Look down column M1M (more than 1,000 pounds but less than 2,000 pounds) and across row C-70.

3. Divide the weight by 100 and multiply the result by the base rate.

$$\frac{1,450 \text{ lbs.}}{100 \text{ lbs.}} \times \$37.59 = \$545.06$$

The rate table includes all classification codes, weight columns ranging from less than 500 pounds to more than 20,000 pounds, applicable base rates for all classes of freight and their respective weights, and the minimum charge for freight shipped between origin zip 38930 and destination zip 60506. A rate table similar to the one shown would be used to calculate freight charges between every zip zone in the United States.

Source: Roadway Express, Inc., Akron, Ohio. Used with permission.

Freight charges are based on a number of factors including weight of the shipment, destination, type of transportation, and the rate charged. The *rate* is the price charged by a transportation company to transport merchandise and is usually quoted per hundred weight (CWT). Rates can also be quoted per thousand pounds (M), or per ton (T). The freight charge is found by multiplying the rate times the number of weight units, as shown in the following example.

Example 15

What is the freight charge for shipping 3,480 pounds if the rate is $12.40 per cwt.?

Solution

Step 1:   Determine the number of weight units (cwt.) in the shipment.

$3,480 \div 100 = 34.8$ weight units

Step 2:   Calculate the freight charge.

freight charge = weight units $\times$ unit rate
$$= 34.8 \times \$12.40$$
$$= \$431.52$$

Rate schedules like the one shown in Table 5.4 are used to determine freight charges and often contain minimum weight requirements. If this rate schedule were used to calculate the freight charge in Example 15, we would have to pay for 5,000 pounds at the $10.20 rate (5,000 lb $\div$ 100 lb = 50 cwt., 50 $\times$ $10.20 = $510) or $510 cwt. Since the 3,480 pounds is less than the lowest minimum weight listed in the rate schedule, we must either ship or pay for 5,000 pounds. When the actual weight shipped is greater than the minimum weight, we multiply the actual weight by the unit rate in that minimum weight class. For example, if we are shipping 12,500 pounds, the freight charge is $1,275.00 (125 $\times$ $10.20).

Notice that the rates in the schedule decrease as the minimum weights increase. This inverse relationship creates a break-even point at which it costs

Table 5.4

Rate schedule

Minimum weight	Unit rate per cwt.
5,000	$10.20
15,000	9.95
30,000	8.24
35,000	6.48
40,000	5.32

less to pay for the greater minimum weight at the lower unit rate price. You will not be required to calculate this break-even point; however, you should understand that it is important to plan shipments so that the greatest amount of merchandise can be shipped at the lowest possible cost whenever possible.

Commissions

Learning objective
Understand the function of commission agents and how to calculate commissions.

Manufacturers and wholesalers often use brokers or **commission agents** to buy or sell their merchandise. Commission agents represent the seller when arranging a sale (**account sale**) or the buyer when arranging a purchase (**account purchase**). Although a commission agent (*consignee*) may accept merchandise "on consignment" from a producer (*consignor*) for sale, the consignee never actually takes title of the merchandise. Commission agents charge a fee for their services based on the net price of the invoice or the gross proceeds of the account sale, and will deduct any expenses associated with selling the merchandise on consignment (see Figure 5.7). Commission agents may be identified on a sales invoice, but the commission charges do not appear on the invoice because the buyer is not responsible for sales commissions. The seller pays the commission agent's **commissions** earned on sales directly to the agent on a monthly basis. Example 16 illustrates how a commission is calculated for an account sale.

Figure 5.7

Account sale

Account Sale

Neuser's Produce Inc.
1376 Commerce Blvd.
Syracuse, NY 13215

Date October 2 19 99

Sales For Account of Farmers Cooperative
Baldwinsville, N.Y. 13027

Date	Description	Price	Amount	Totals
9/10	100 cartons tomatoes	$ 6.50	$650.00	
9/18	50 8-quart boxes plums	8.00	400.00	
9/27	200 50 lb. bags potatoes	4.00	800.00	
				$1,850.00
	Trade Discount 15%			277.50
	Gross Proceeds			$1,572.50
	Charges:			
	Freight		$152.50	
	Storage		65.00	
	Commission 5%		78.63	296.13
		Net Proceeds		$1,276.37

Example 16

John Neuser, a commission agent, sold merchandise on consignment for Farmers Cooperative worth $1,850, with a trade discount of 15%, freight charges of $152.50, and a storage charge of $65.00. If the agent charged a 5% commission, determine (a) the commission earned on the transaction and (b) the net proceeds remitted by the commission agent to Farmers Cooperative.

Solution

a. $1,850.00 list price
  −277.50 less trade discount (1,850 × .15)
$1,572.50 gross proceeds
  ×   .05 commission rate
$ 78.63 commission amount

b. $1,572.50 gross proceeds
           less total charges (freight $152.50, storage $65, and
  −296.13   commission $78.63)
$1,276.37 net proceeds

Commission agents are sometimes used by businesses to purchase merchandise that is not available through regular distribution channels, or that is so specialized that the company's buyers are unable to arrange the transaction. The agent will purchase the merchandise for the buyer and submit an account purchase listing the **prime cost** plus commission and expenses related to the purchase. The amount to be remitted to the commission agent by the purchasing company is called the **gross cost** (gross cost = prime cost + total charges). Figure 5.8 illustrates how an account purchase is prepared from the information provided in Example 17.

Example 17

Lombardi Wholesale Foods Inc., in Watertown, New York, commissioned Global Imports Ltd. in Buffalo, New York, to purchase the following merchandise from a supplier in Naples, Italy. The order was placed on May 10, 20xx for fifteen 20 lb. boxes of Genoa Salami at $4.98 per pound; ten 30 lb. boxes of Provolone cheese at $5.42 per pound; and thirty boxes of assorted types of pasta at $25.50 per box. If Global Imports Ltd. charges an 8% commission for its services, and incurred shipping charges of $268.25; insurance costs of $32.75; and duty taxes of $116.55, determine the gross cost of the account purchase to Lombardi Wholesale Foods Inc.

Now let's consider a transaction involving a sales invoice and the payment of a commission.

Figure 5.8

Account purchase

Account Purchase

Global Imports Ltd.
1547 Northern Blvd.
Buffalo, N.Y. 14203

Date ___June 2___ **20** xx
Account No. ___D173869___

Purchased For Account Of   Lombardi Wholesale Foods Inc.
2765 Arterial Rd.
Watertown, N.Y. 13629

Date	Description	Price	Amount	Total
5/10	15—20lb. boxes Genoa Salami	$4.98 lb.	$1,494.00	
	10—30lb. boxes Provolone Cheese	5.42 lb.	1,626.00	
	30 boxes Assorted Pasta	25.50 bx.	765.00	
		Prime Cost		$3,885.00
	Charges:			
	Insurance		$ 32.75	
	Shipping		268.25	
	Import duty		116.55	
	Commission (8% of $3,885)		310.80	
				728.35
		Gross Cost		$4,613.35

Example 18

A bottling company ships 800 cases of soda to a distributor at a price of $3.75 per carton, with a 20% trade discount and freight charges of $240. How much will the distributor have to pay for the merchandise if a broker is paid a 3% commission for services? The terms of the sale are 2/10, *n*/30, and the goods are sent FOB shipping point. Assume the invoice was paid in full before the discount date.

Solution

$3,000	list price (800 × 3.75)
−600	less trade discount (3,000 × .20)
$2,400	net price
−48	less cash discount ($2,400 × .02)
$2,352	
+240	plus freight charge
$2,592	amount due

Notice that the commission is not included in the calculation. The commission is paid by the bottling company (seller) directly to the commission agent and is not shown on the invoice. The commission received by the broker would be $72.00 ($2,400 net price × .03 commission rate). Commissions are paid on the net amount of the invoice and exclude freight charges.

Partial Payment of Invoice

Learning objective
Calculate the cash discount and the amount of credit when a partial payment is remitted.

Every example involving trade and cash discounts thus far has assumed that the buyer made full payment of the invoice. There are, however, times when the buyer's cash flow position does not permit full payment of the amount due within the discount period. The buyer may instead make a series of **partial payments** on the amount due during the discount period to take advantage of the cash discount.

If it is the seller's policy to grant cash discounts on partial payments, the buyer will receive a discount on the proportionate amount paid. When the full amount due is paid, the buyer deducts the discount and remits a lesser amount that is accepted as full payment by the seller. Conversely, when a partial payment is made during the discount period, the seller must credit the buyer's account with an amount that is greater than the amount of the partial payment.

To determine the amount of the credit, we must first find the percent paid, which is the complement of the discount (1.00 − discount rate = percent paid). This percent is then used to calculate the amount of the credit that is the actual value of the partial payment. The credit can be found using either the equation or formula method.

Equation method for determining credit amount

$$\text{percent paid} \times \text{amount of credit} = \text{amount of partial payment}$$
$$(1 - \text{discount rate}) \times (X) = \text{amount of partial payment}$$

Now we must divide each side by (1 − discount rate).

$$\frac{(1 - \text{discount rate}) \times (X)}{1 - \text{discount rate}} = \frac{\text{amount of partial payment}}{1 - \text{discount rate}}$$

$$\text{amount of credit} = X = \frac{\text{amount of partial payment}}{1 - \text{discount rate}}$$

Formula method for determining credit amount

$$\frac{\text{amount of partial payment}}{1 - \text{cash discount rate}} = \text{amount of credit}$$

To determine the balance due, we have to subtract the amount of the credit from the balance of the account prior to the partial payment (prior balance − amount of credit = balance due) or (invoice amount − amount of credit = balance due).

Example 19

An invoice of $1,250 dated May 18 offers terms of 3/10 eom. A partial payment of $800 is made on June 9. Find (a) the balance due on the invoice and (b) the amount of the cash discount.

Solution

a. $100\% - 3\% = 97\%$ percent paid

$.97X = \$800$

$X = \$800/.97$

$X = \$824.74$ amount of credit

$\$1,250 - \$824.74 = \$425.26$ balance due

b.   amount of credit − partial payment = amount of cash discount

$\$824.74 - \$800.00$ $= \$24.74$

The partial payment is made during the discount period, so the buyer is entitled to a cash discount on the proportionate amount paid. In essence, the buyer is able to reduce $1 of the invoice amount for each $.97 that is paid during the discount period ($800 ÷ .97 = $824.74).

CHECK YOUR KNOWLEDGE

Invoice Charges and Adjustments

Solve each of the following problems. Round dollar amounts to the nearest cent.

1.   On August 8, a merchant purchased goods totaling $865.40 with terms 2/10-50X and a trade discount of 5%. The buyer returned goods worth $124.60 and paid the invoice on October 7. How much money did the merchant remit as full payment of the invoice?

2. In problem 1, what is the net payment date of the invoice?

3. A retailer purchased 1,500 cases of canned goods from a food processing company. If each case weighed 24 pounds and the goods were shipped FOB shipping point, what was the freight charge paid by the retailer? (Use the rate schedule in Table 5.4.)

4. Find the amount due on an invoice dated October 27 for $2,840.50 if terms are 3/10 prox and the invoice is paid on December 10. Freight charges for the shipment amount to $215.40 and the goods were shipped FOB destination.

5. A broker sells merchandise totaling $7,800 on July 14 with a trade discount of 25/10 and freight charges of $462.35 prepaid by the manufacturer. How much will the buyer pay for the order if the terms of the sale are 2/10, 1/20, *n*/30 and payment is made on July 25?

6. If the broker in problem 5 charges a 4% commission, (a) what is the amount of the commission, and (b) who will pay the commission?

7. Develop an account purchase for the following transaction. The Unique Treasures Gift Shoppe commissioned Randall and Martin, international commission merchants in Boston, to purchase 20 pair of English crystal candlestick holders at $135 a pair; ten 20-pound boxes of assorted 1-pound containers of chocolate candies from Switzerland at $8.75 per pound; five 15-pound boxes of assorted 1-pound packages of coffee beans from Columbia at $4.25 per pound; and 50 hand-painted wooden figures from Russia at $15 each. Find the gross cost of the merchandise if Randall and Martin charges include a 5$\frac{1}{2}$ percent commission, insurance of $137, shipping of $85.75, and duty taxes of $62.25.

8. What is the balance due on a $950 invoice dated May 23 with terms 3/20, *n*/60, if a $500 partial payment is made on June 10?

9. If a partial payment of $200 is remitted on September 15 for a $600 invoice dated August 27 with terms 2/10 ROG, what is the balance due? The merchandise was delivered on September 3.

Answers to CYK: *1.* $689.68   *2.* October 27   *3.* $172.80   *4.* $2,755.28   *5.* $5,212.35   *6.* **a.** $210.60;   **b.** seller pays commission   *7.* $6,107.28   *8.* $434.54   *9.* 460.00

5.4 EXERCISES

Solve the following problems. Round dollar amounts to the nearest cent and rates to the nearest tenth of a percent.

Use Table 5.4 to determine the freight charges for the following weights:

1. 2,857 pounds
2. 8,350 pounds
3. 22,640 pounds
4. 34,980 pounds
5. 46,200 pounds
6. What is the cost of freight for 4,890 pounds of goods if the rate is $.75 per cwt?
7. What is the amount required to pay an invoice for $1,700 with a trade discount of 15%, 10%, 5%, and terms of 3/15 ROG if the buyer returned $500 worth of merchandise and paid the invoice in full before the discount date?
8. How much should a buyer pay on an invoice dated August 12 for $980 with a trade discount of 10%, 5%, and terms of 4/10 eom if she returned $270 of the goods and paid the invoice in full on September 9?
9. Industrial Equipment Inc. sold commercial water pumps on consignment for Quality Pumps of Cleveland, Ohio, to various customers during the month of October worth $15,000 with trade discounts of 15%, $18,000 in shipping charges, and $375 in storage charges. If Industrial Equipment Inc. charges a commission of 8%, determine (a) the commission earned on the transactions and (b) the net proceeds reported on the account sale.
10. A petroleum company ships 500 cases of one-quart containers of motor oil (12 quarts per case) to a distributor at a price of $9.60/case with a 30% trade discount and freight charges of $175. How much will the distributor have to pay for the merchandise if a broker is paid a 4% commission for services, and the terms of the sale are 3/10, n/30 FOB shipping point, and the invoice was paid in full before the discount date?
11. An independent grocer orders 200 cases of tomato sauce (12 1-quart jars per case) at $36

per case; 300 cases of vegetables (24 17-oz. cans per case) at $12.00 per case; and 100 cases of canned soup (24 10-oz. cans per case) at $9.00 per case. The grocer is given a trade discount of 20/10/5 and terms of 4/10, 2/20, n/30 eom and freight charges of $43.00 FOB destination point. The invoice is dated July 23 and will be paid in full on August 15. How much will the grocer have to pay?

12. Harden Jewelers commissioned National Gems to purchase rough-cut diamonds, rubies, and emeralds valued at $42,500. During the process of arranging the purchase, National Gems incurred shipping and security charges of $575, insurance costs of $280, and excise taxes amounting to $1,250. Determine the gross cost of the purchase to Harden's Jewelers if National Gems charges $5\frac{3}{4}$% for their services.

13. An invoice of $1,800 dated December 5 offers terms of 4/20 eom. A partial payment of $1,200 is made on December 15. Assuming that the seller does provide cash discounts on partial payments, find (a) the balance due on the invoice and (b) the amount of the cash discount.

14. An invoice of $2,300 dated January 15 offers terms of 5/15 ROG. The shipment was received on February 1, and a partial payment of $1,500 was made on February 12. Assuming that the seller does provide cash discounts on partial payments, find (a) the balance due on the invoice and (b) the amount of the cash discount.

15. A discount store makes a direct purchase of 500 portable television sets from a manufacturer at a price of $250 per set with a trade discount of 30% and terms of 6/20, 4/30, n/60 as of March 15. Shipping charges are $7.00 per set FOB shipping point. The discount store makes a payment of $50,000 on the invoice on March 30, and a second partial payment of $20,000 on April 10. The manufacturer will pay the agent who arranged the sale a commission of 5%. Determine (a) the total amount of money the manufacturer will receive from the discount store, (b) the total amount the transaction will cost

the discount store, and (c) the commission to be paid to the sales agent.

16. A national department store chain purchased 1,000 refrigerators directly from a manufacturer for $650 each with a 20%, 15%, 5% trade discount and terms of 5/15, 2/30 ROG, and FOB shipping point of $35 per refrigerator. The store took delivery of the refrigerators at their central warehouse on April 20 and returned 45 of the refrigerators as damaged goods. The store chain paid $300,000 on April 30, and the remainder of the cost on May 15. Determine (a) the total amount of money the manufacturer will receive from the store and (b) the total amount the transaction will cost the store.

CHAPTER REVIEW EXERCISES

The following problems review all mathematics and business concepts presented in the chapter. If you experience any difficulty solving a problem, go to the appropriate chapter section and review the examples provided. Express all answers as decimals rounded to tenths, or dollars rounded to cents, unless otherwise instructed.

1. A 32 oz. bottle of fruit juice is sold for $3.42. Find the unit price per pint (express answer to the nearest tenth of a cent).

2. The owner of The Bunkhouse purchased a full side of prime beef from Central Provisions for the restaurant. If the side of beef weighed 487 pounds and the supplier's unit price is $72.45 per cwt., what was the list cost of the side of beef?

3. How many gallons of gasoline can Alice Martin purchase for $10.00 if the unit price per gallon is $1.04 9/10. (Round answer to hundredths.)

4. Midstate Appliance Center purchased five electric ranges from a distributor at a list price of $650 less trade discounts of 20% and 10%. What was the net price of the order?

5. A & C Supply sells a particular brand and model video game player at a list price of $149, less trade discounts of 15/5/5. Electro-Mart sells the same brand and model video game player for $169, less trade discounts of 25/15. How much can a retailer save on each player by purchasing from the supplier offering the lowest net price?

6. Find the single-equivalent trade discount rate for a chain discount of $33\frac{1}{3}\%$, $20\frac{3}{4}\%$, and 12%.

7. What is (a) the discount date and (b) the net payment date for an invoice dated May 27 with terms 2/10, n/30 eom?

8. Northwest Compressor Warehouse purchased industrial compressors worth $375,000 on March 15 with terms of 3/10–50X. If the invoice was paid on May 12, what is the amount of the net payment?

9. Determine the amount necessary to pay an invoice for $32,500 with a trade discount of 5%, 3% and terms of 2/10, 1/20, n/30, if the buyer returned $800 worth of merchandise. The invoice is dated July 18 and was paid on August 6.

10. Find the freight charge for shipping 8,250 pounds of boxed dry goods if the rate for the first 2,000 pounds is $2.45; the next 3,000 pounds is $1.85; and the next 5,000 pounds is $1.20. All unit rates are quoted per cwt.

11. Homepatient Care Inc. purchased health care supplies from a distributor on October 20, at a list price of $15,000. How much will Homepatient Care pay the distributor on November 2, if the terms of the sale are 3/15, n/45, the goods are shipped FOB destination, and a sales agent is paid a 10% commission for arranging the transaction?

12. Family Value Foods sold assorted baked goods received on consignment for Gramma's Bakery valued at $4,500 during the month of January. Find the net proceeds listed on the monthly account sale if Family Value Foods charges include a commission of 6.5% and handling and storage charges of $125.

13. A wholesaler commissioned a purchase agent to arrange an account purchase for merchandise to be secured from a manufacturer located in China. The prime cost of the merchandise purchased by the commission agent was $24,800, and the agent's charges included a commission of 10%; shipping charges of $1,500, insurance of $437, and import taxes of $685. Determine the gross cost of the account purchase to the wholesaler.

14. Baldwin Enterprises Inc. sold merchandise to Stetson Motors at a list price of $5,750. The invoice was dated June 22 and offers terms of 4/15, ROG. The merchandise was received on July 10, and a partial payment of $2,500 was made on July 23. If Baldwin Enterprises allows cash discounts on partial payments, find the balance due on the invoice after the acceptance of the partial payment.

15. An industrial wholesaler sold $25,000 worth of grinding fluids and cleaning solvents to a manufacturer on August 12 with a trade discount of 10/5 and terms of 3/15, *n*/45. The merchandise was shipped FOB shipping point at a cost of $215. The sale was arranged by a commission agent, who charged an 8% commission. If the manufacturer makes a partial payment of $15,000 on August 25, determine (a) the total cost of the transaction to the manufacturer, and (b) the balance due to the wholesaler after the partial payment is credited.

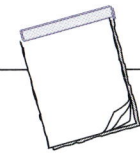 **EXPRESS YOUR UNDERSTANDING**

Compose one or two well-written sentences to express the requested information in your own words.

1. Identify each type of business involved in the longest distribution channel. Describe why each business is considered both a buyer and a seller.

2. Explain the difference between cost and price. Give an example that supports your explanation.

3. Identify the variables found in the equation used to calculate the extension amount. If you have sold 150 items for $3.50 a dozen, what adjustments to the data are necessary to use the equation?

4. Explain how to find the net price when the list price and trade discounts of 10% and 15% are given.

5. Develop a step-by-step description of how you would find the single-equivalent trade discount when given a chain discount of 55%, 10%, and 5%.

6. Explain how the terms 2/10, 1/15, and *n*/30 are applied to determine the amount of cash discount.

7. Describe how you would determine the exact number of days between March 10 and July 18.

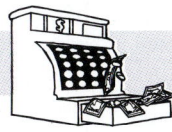 *Mind Your Business*

Media World has a financial policy which requires the payment of all obligations incurred by the business to be paid within the maximum allowed time or the most advantageous discount period. As the business manager, consider the information provided in each of the following activities which occurred during the month of September, and clearly illustrate how you arrived at your decision in each situation.

Activity 1:

The bookkeeper has requested your authorization to write a check for $8,837.59 on September 16th to National Music Warehouse Inc. as total payment of the following invoice. Will you authorize the check? Explain.

Invoice
392547

NATIONAL MUSIC WAREHOUSE
420 Palmer Street
Chicago, Illinois 60615

Sold to: Media World
1425 Commerce Blvd.
Syracuse, N.Y. 13215

Ship to: Same

Shipped via Roadway Express Inc.

Order Number 276384	Salesman Kline	Terms 3/15, n/30	Date Shipped 9/2/2XXX	F.O.B. Shipping Pt.	Date 9/6/200X

Quantity Ordered	Description	Unit Price	Amount
600	30 units of all top 20 albums—CD's	$ 6 50	$ 3,900 00
400	20 units of all top 20 albums—cassette	5 25	2,200 00
200	20 units of all top 10 country albums—CD's	5 90	1,180 00
100	10 units of all top 10 country albums—cassette	4 60	460 00
200	20 units of all top 10 R & B albums—CD's	7 10	1,420 00
150	15 units of all top 10 R & B albums—cassete	5 45	817 50
	Subtotals		$ 9,977 50

Misc. Information:

Trade Discount
3%, 5%, 1% $ 875 18

Cash discount ▶ **Deduct:** $ 299.33 **If paid by:** 10/17/2XXX

Shipping Charge $ 34 60

Invoice Total ▶ $ 8,837 59

Activity 2:

You have commissioned Liverpool Entertainment LTD. in London, England, to purchase various categories of merchandise for resale amounting to $1,725.40. Liverpool Entertainment has informed you that it charges a commission of 12% for its service and will grant a 5% discount on prime cost. In addition, as the buyer, Media World will have to pay shipping charges of $65.30, duty taxes of $108.25, and insurance cost of $28.50. What will be the gross cost amount on the account purchase invoice received by Media World from Liverpool Entertainment LTD.?

Activity 3:

Invoice #3587365, dated August 15, has a total list cost of $2,842.90. The seller, Hardcopy Distributors Inc., granted a trade discount of 15% and terms of 2/10, eom. The merchandise was delivered by UPS on September 18. FOB destination. Your inspection of the merchandise indicates that you must return $184.50 of the magazines included in the order to the distributor for various reasons. What is the total cost of the order to Media World if the invoice was paid on September 9?

Activity 4:

Invoice #136275, consisting of compact discs and cassette tapes purchased from Metropolitan Music Supply Inc., has an invoice total of $8,493.75. The invoice is dated September 3, and offers a trade discount of 10%, 8% and terms of 3/10, *n*/20. After assessing the cash flow position of the business, you decide on September 5 to authorize a partial payment of $4,500 on this invoice.

a. On what date should the bookkeeper send the partial payment check to Metropolitan Music Supply Inc.?

b. What is the unpaid balance of Metropolitan Music Supply's account after the partial payment is credited, assuming the account had no previous balance?

SELF-TEST

A. Terminology

Complete the following items using the key terms presented at the beginning of the chapter. Check your responses against the answer key at the end of the test.

1. The distribution channel of getting goods to a consumer begins with the _____ who buys raw materials or parts and produces and assembles them into finished goods, which are then sold to middle-persons known as wholesalers (or _____) who promote the sale of them to _____ who will sell them to consumers.

2. Store chains that eliminate some or all of the middlepersons in the distribution channel are referred to as _____ store chains.

3. The official record of a business transaction is called an _____.

4. The suggested selling price to the consumer is called the _____ price. A reduction in this price to a buyer is called a _____ discount. The _____ price or cost is determined by subtracting the second of these from the first.

5. The _____ of a 17% trade discount is 83%.

6. A multiple discount expressed as "15% less 10% less 5%" is called a _____ or _____ discount.

7. To determine the net price of an item with the multiple discount listed in problem 6 by the _____ method, one would multiply the list price by .85 × .90 × .95.

8. The discount rate obtained in problem 7 (i.e., .85 × .90 × .95) is called the "_____ rate."

9. Sellers often encourage buyers to pay invoices before the required date by offering _____

_____, which are indicated under terms of the sale using the ordinary dating method such as 2/10, n/30.

10. Proximo and _____ are ways of indicating that the terms of payment of an invoice begin with the first day of the next month, and _____ means that the terms begin when the buyer actually receives the merchandise.

11. Extensions of terms may also be arranged by the _____ dating method (e.g., 3/20–60X) or by the _____ method using the "*as of*" phrase.

12. FOB _____ point means that the buyer pays shipping costs, whereas FOB _____ point means that the seller pays the shipping costs.

13. A _____ is a percentage amount of the net price of a sale paid to the salesperson or agent who arranged for the sale of the goods.

B. Calculations

The following concepts and short problems are designed to test your understanding of the objectives identified at the beginning of the chapter. Answers are provided at the end of the test. Round dollar amounts to the nearest cent and rates to the nearest tenth of a percent.

14. Complete the following invoice calculation:

Quantity	Description	Unit price	Extension price
5 gal.	vanilla ice cream	$2.49/gal.	_____
3 pints	fruit flavored ice cream toppings (1 ea. choc., strawb., blueb.)	$3.25/pint	_____
4 pkg.	4" plastic sauce dishes (10/pkg.)	$1.19/pkg.	_____
1 pkg.	plasticware spoons, 50 spoons/pkg.	$3.00/pkg.	_____
		Total	_____

15. Carmen Juarez was offered a chain discount of 15/20/5 to entice her to purchase a new line of swimwear for her retail outlet store early. The list price of her total purchase was $1,850. Determine (a) the net invoice price of Carmen's purchase, (b) the amount of the trade discount, and (c) the single equivalent discount rate offered to Carmen.

16. Carmen was also offered an incentive to pay her bill for a purchase of women's blouses and sweaters from another seller early with terms of 6/15, 3/25, *n*/60 ROG. The cost of the merchandise was $850 and she paid the entire bill on March 20 after having received her order on March 1. How much cash discount did she receive?

17. How much should Jaoquim's Landscaping Service pay for a $2,175 purchase made on February 10, with the terms 5/10, 2/20, *n*/30 prox if they receive the goods on February 28 and pay the invoice on March 8?

18. What is the amount required to pay an invoice for $2,700 with a trade discount of 20%, 10%, 5%, and terms of 2/10, *n*/30 ROG if the buyer returned $300 worth of merchandise and paid the invoice in full before the discount date?

19. Mueller Farms Nursery in Elgin, Illinois, sold plants and shrubs shipped to them on consignment from the Grower's Cooperative in Toms River, New Jersey, to customers during the month of May. Total sales were $10,500. Determine the net proceeds reported on the monthly account sale if charges include $375 for handling and storage, and $40 for insurance, and a commission of 12% is required for selling the merchandise.

20. An invoice of $7,500 dated January 25 offers terms of 4/20 ROG. The shipment was received on February 16, and a partial payment of $4,000 was made on March 5. Assuming that the seller does provide cash discounts on partial payments, find (a) the balance due on the invoice and (b) the amount of the cash discount.

Answers to Self-Test: **1.** manufacturer, distributors, retailers   **2.** discount   **3.** invoice   **4.** list, trade, net   **5.** complement   **6.** chain   **7.** complement   **8.** net price equivalent   **9.** cash discounts   **10.** eom, ROG   **11.** extra, postdating   **12.** shipping, destination   **13.** commission   **14.** $29.96   **15. a.** $1,195.10; **b.** $654.90; **c.** 35.4%   **16.** $25.40   **17.** $2,066.25   **18.** $1,608.77   **19.** $8,825.00   **20. a.** $3,333.33; **b.** $166.67

6

Key Terms

markup
markup equation
gross margin
markup on cost
markup rate
markup on selling price
perishables
marked price
markdown
reduced price
reduced net profit
break-even point
operating loss
absolute loss
maximum markdown
   percent

PRICING MERCHANDISE

Learning Objectives

After completing the exercises in this chapter, you will be able to:

1. Explain the financial analysis required to price merchandise.

2. Compute the selling price given cost and the markup based on cost.

3. Find the cost when the selling price and markup rate on cost are known.

4. Calculate the amount and percent of markup based on cost.

5. Compute the selling price given the cost and markup based on selling price.

6. Find the cost when selling price and markup rate on price are known.

7. Calculate the amount and percent of markup based on selling price.

8. Convert the markup rate to the opposite base.

9. Calculate list price given a trade discount and markup based on selling price.

10. Determine the selling price required to cover estimated spoilage on perishable merchandise.

11. Compute the markdown and the reduced price amount, given the selling price and markdown rate.

12. Calculate the profit or loss from operations and the absolute loss when a markdown is granted.

13. Find the maximum percent of markdown without incurring an operating loss.

14. Define key terms.

INTRODUCTION

In Chapter 5, we examined the basic elements of purchasing goods from the buyer's perspective. An overview of the merchandising process was provided to help you understand the processes and terminology related to both buying and selling. For instance, Figure 5.2 helped explain the relationship between cost and price by showing you how the seller's price becomes the buyer's cost as goods move through various marketing channels. We learned how manufacturers, wholesalers, and retailers purchase merchandise that they intend to sell to their customers for a profit.

Learning objective
Explain the financial analysis required to price merchandise.

To produce a profit, merchants must price their goods so that a profit is possible. *Profit,* as you recall from our discussion of business operations in Chapter 4, is the amount that remains after the cost of goods and all operating expenses have been recovered. (If you do not clearly understand the analysis below, you should review Chapter 4 before proceeding.)

Formula for computing profit or loss

> revenue (price per unit × units sold)
> − cost of goods sold (cost of units sold)
> = gross margin (markup)
> − operating expenses*
> = net profit or loss
>
> *taxes, supplies, selling expenses, wages, rent, utilities, insurance, etc.

As you can see, the price a merchant charges for goods or services is an economic reflection of cost. A firm must generate sufficient revenue to produce a profit if it is to survive. On the other hand, price is also used by many merchants as a technique to market their products. For example, to stimulate demand, a price lower than the market price may be selected. A price higher than the market price might be used if the price is consistent with the buyer's expectations.

Regardless of the pricing strategy selected, a **markup** will be added to an item's cost to cover operating expenses and profit. To illustrate, if a retailer purchases a personal computer for $850 and sells it for $1,500, the markup is $650. The **markup equation** summarizes the markup calculation.

cost (C) + markup (M) = selling price (S)
$850   + $650   = $1,500

As with any equation, if any two variables are known, the third variable can be found by substituting the given variables into the equation and solving for the unknown variable. Therefore, the following two formulas can be derived from the markup equation.

$$\begin{aligned}
\text{selling price} - \text{markup} &= \text{cost} \\
\$1{,}500 \quad - \quad \$650 \quad &= \$850
\end{aligned}$$

or

$$\begin{aligned}
\text{selling price} - \text{cost} &= \text{markup} \\
\$1{,}500 \quad - \$850 &= \quad \$650
\end{aligned}$$

Markup can be based on the cost of an item or on its selling price. Regardless of which base is used, the markup equation remains as stated: Cost plus markup equals selling price. Markup is also referred to as the **gross margin** or just *margin,* and can be expressed as an amount or as a percent.

In this chapter, you will learn the methods used by merchants to mark up goods, to convert markups to opposite bases, and to apply markdowns under various conditions.

6.1   MARKUP BASED ON COST

Many manufacturers, wholesalers, and retailers use cost systems to analyze the unit costs of their products because the cost approach is more appropriate for their organization. When the **markup on cost** method is used, the cost of the item is considered the base (100%); the operating expenses, selling price, and net profit or loss are considered a part of the product cost. The **markup rate** is expressed as a percent of cost and includes the prorated operating expense rate and the target net profit rate. The amount of the markup, prorated operating expense, and net profit can be found easily through the use of the percent equation ($B \times R = P$) presented in Chapter 2. Let's look at an example that illustrates this important point before we calculate the selling price when the markup is based on cost.

Example 1

An appliance store purchased a television set for $125 and sold it for $200. If the store's operating expenses are 40% of cost, (a) how much markup did the store receive on the sale, (b) what were the store's prorated operating expenses, and (c) what was the amount of net profit on the sale?

Solution

   a. Use the markup equation to calculate the amount of markup.

$$\begin{aligned}
C + M &= S \\
\$125 + M &= \$200 \\
M &= \$200 - \$125 \\
M &= \$75
\end{aligned}$$

Or by formula,

$$S - C = M$$
$$\$200 - \$125 = \$75$$

The markup (gross margin, or margin) on the sale was $75.

b.   Use the percent equation to calculate the prorated operating expense amount.

$$B \times R = P, \quad \text{where } B = \text{cost}$$
$$\$125 \times .40 = \$50 \qquad R = \text{operating expense rate}$$
$$P = \text{operating expense amount}$$

The store's prorated operating expenses on this sale were $50.

c.   The markup includes operating expenses and net profit; therefore:

$$\text{operating expenses} + \text{net profit} = \text{markup}$$
$$\$50 + P = \$75$$
$$P = \$75 - \$50$$
$$P = \$25$$

A net profit of $25 was realized on the sale of the television. Figure 6.1 diagrams the calculations for Example 1.

Figure 6.1

Diagram of calculations in Example 1

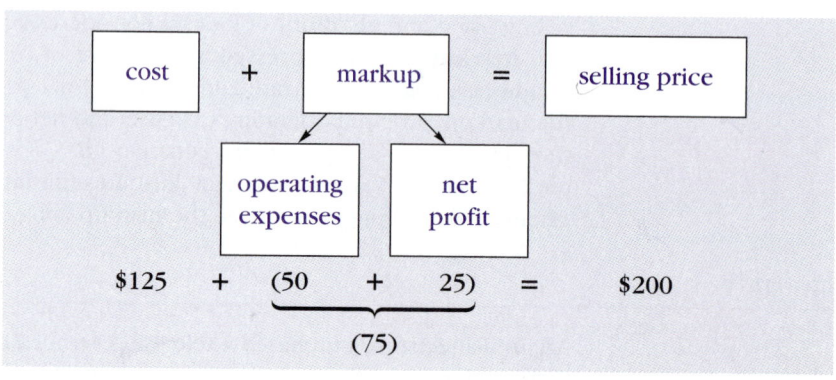

Note that a profit is not always realized when the price of an item is greater than its cost. Merchants must price their merchandise to cover all associated expenses and intended profit when they establish markup rates. We will discuss this concept further in Section 6.3 when we focus on markdowns.

Learning objective
Compute the selling price given cost and the markup based on cost.

Calculating Selling Price Given Cost and Percent Markup Based on Cost

When the cost of an item is known and the rate of markup is based on cost, the selling price is expressed as a percentage of cost as follows:

$$\text{cost} \quad + \quad \text{markup} \quad = \quad \text{selling price}$$
$$100\% \text{ of cost} + 25\% \text{ of cost} = 125\% \text{ of cost}$$

Remember, the markup can be expressed as an amount or as a percent. In either case, when the markup is based on cost, it is added to the cost to determine the selling price.

Example 2

J and B Supply marks up its plumbing supplies 60% of cost. What is the selling price of a bathroom vanity that costs the company $75?

Solution

In the markup equation, express the markup as a percent of cost, convert the percent to a decimal equivalent, and solve.

$$C + M = S$$
$$100\% \text{ of cost} + 60\% \text{ of cost} = 160\% \text{ of cost}$$
$$1.60C = S$$
$$1.60(\$75) = S$$
$$\$120 = S$$

or

$$C + M = S$$
$$\$75 + .60(\$75) = S$$
$$\$75 + \$45 = \$120$$

Did you notice in the two solutions presented that the markup (portion) was found by multiplying the cost (base) times the markup rate? The first solution method expresses the selling price as a percent of cost (160%), and does not produce the amount of the markup. The second solution method produces the amount of markup, which, when added to the cost, gives the selling price.

Calculating Cost Given Selling Price and Markup Based on Cost

Learning objective
Find the cost when the selling price and markup rate on cost are known.

Merchants often find that the cost they are able to pay for merchandise is determined by current market prices of the products they intend to sell as well as their markup policies. For example, if a wholesaler selects a pricing strategy to compete with other wholesalers, the price of its product is determined in part by its price and in part by a markup on cost sufficient to cover operating expenses and required profit. Example 3 shows how to calculate cost when the selling price is known and the markup is based on cost.

Formula for cost

$$\text{cost} = \frac{\text{price}}{1 + \text{markup rate}}$$

Example 3

The buyer for Hanson Shoes wishes to purchase a line of men's walking shoes to sell for $69.95 a pair. How much can the buyer pay for the shoes if the company must realize a 45% markup on cost?

Solution

$$\text{cost} + \text{markup} = \text{selling price}$$
$$100\%C + 45\%C = \$69.95$$
$$145C = \$69.95$$
$$C = \$69.95 \div 1.45$$
$$C = \$48.24$$

or

$$\text{Cost} = \frac{\text{price}}{1 + \text{markup rate}} = \frac{\$69.95}{1 + .45}$$
$$C = \$48.24$$

The cost of the shoes is $48.24; the markup is $21.71 ($48.24 × .45) or as follows:

$$C + M = S$$
$$\$48.24 + \$21.71 = \$69.95$$

Calculating Amount and Rate of Markup Based on Cost

While retailers know the cost of an item and the price they intend to charge for the item when it is sold, they also need to know the percent of markup relative to their cost and price data. The markup percent is important because costs and expenses are constantly changing. In addition, business may have to reduce the price of items to stimulate sales, or to reduce inventory costs. The financial impact of such changes can be measured somewhat by the markup percent. When an item's cost is used as the base, the percent (rate) of markup is found by dividing the markup amount by the cost. Because we are calculating a percent, we can once again use the percentage equation.

Example 4

Karat Jewelers reduced the selling price on an 18-inch solid-gold chain from $285 to $225. If the jewelry store paid $130 for the gold chain, what is the percent markup of cost?

Solution

First, find the amount of markup using the markup equation.

$$\text{cost} + \text{markup} = \text{selling price}$$
$$\$130 + M = \$225$$
$$M = \$225 - 130$$
$$M = \$95$$

or

$$\text{selling price} - \text{cost} = \text{markup}$$
$$\$225 \quad - \$130 = \quad \$95$$

Then, find the markup percent using the percent equation.

$$\text{base} \times \text{rate} = \text{part}$$
$$\$130 \times R = \$95$$
$$130R = \$95$$
$$R = .7307, \text{ or } 73.1\%$$

or

$$\text{rate of markup} = \frac{\text{markup}}{\text{cost}} \left(\text{rate} = \frac{\text{part}}{\text{base}} \right)$$
$$= \frac{\$95}{\$130}$$
$$= .7307, \text{ or } 73.1\%$$

CHECK YOUR KNOWLEDGE

Markup Based on Cost

1. A retailer purchased a microwave for $136 and sells it for $189.95. What is the amount of markup on the sale?

2. Find the cost of an item that sells for $57 and includes a markup of $20.

3. A dealer purchased an above-ground 12- × -24-foot pool for $1,500 and sold it for $2,300. If the company's operating expenses run 30% of cost, what were the expenses related to the sale?

4. In problem 3, what was the amount of (a) markup and (b) net profit on the sale?

5. If a calculator that cost $15 has a markup of 20% of cost, what is the selling price?

6. How much should a wholesaler charge for a mattress that costs $45 if the company's operating expense rate is 50% and the dealer wants a 12% net profit on cost?

7. A clothing store sells a line of jackets for $89.99. What is the maximum price the store can pay for each jacket if its markup on cost is 75%?

8. Find the percent of markup on cost for a radio that costs $18 and sells for $24.30.

6.1 EXERCISES

Fill in the missing amounts.

	Selling price	Cost	Gross margin	Operating expenses	Profit/loss
1.	$85	$52	_____	$28	_____
2.	_____	$35	$13	_____	$3.00
3.	$575	_____	$65	$74	_____
4.	_____	$340	_____	$115	$24
5.	$1,200	$1,500	_____	$325	_____

Calculate the missing numbers. Round dollar amounts to the nearest cent and rates to the nearest tenth of a percent.

	Cost	Markup	% Markup on cost	Selling price
6.	$5.00	_____	30%	_____
7.	_____	$12.50	_____	$25.00
8.	$65.00	$39.00	_____	_____
9.	$350.25	_____	_____	$475.50
10.	_____	$82.35	45%	_____

Solve each of the following problems. Round dollar amounts to the nearest cent, and rates to the nearest tenth of a percent.

11. Upstate Copy Products purchased a desktop copier for 1,200 and sold it for $1,800. If operating expenses run 30% of cost, find (a) the amount of markup, (b) the operating expenses related to the sale, and (c) the amount of profit realized.

12. Family Drugs sells an 8-ounce bottle of cough medicine for $3.69, which includes a markup of $1.05. Find the unit cost of the product.

Answers to CYK: 1. $53.95   2. $37.00   3. $450.00   4. a. $800; b. $350   5. $18
                    6. $72.90   7. $51.42   8. 35%

13. Clark's Men's Shoppe purchased a group of men's topcoats for $89.00 each. If its operating expenses are 20% of cost and the net profit is 15% of cost, find the selling price.

14. Wilkins Jewelers buys a certain style gold bracelet for $450. If the store desires a 150% markup based on cost, determine the selling price of the bracelet.

15. Thornton Bookstore purchases paperback best-sellers from a distributor for $3.40. How much does the bookstore charge customers for the books if it uses a 40% markup on cost?

16. Determine the selling price for a pair of sneakers that cost the retailer $40 if its operating expenses are 30% of its cost, and its net profit margin is 12% of its cost.

17. Sally Steinberg must order 50 fishing rod and reel sets, which her sportshop will sell for $12.50, as part of a promotion for opening day of fishing season. How much can she pay for each set if her markup based on cost is 60%?

18. Find the selling price for an item that costs $149.50 if the seller uses a 50% markup on cost.

19. Raymond Lock Co. sells a door lock for $49 that costs the company $32. What is the percent markup on cost? (Round to the nearest tenth of a percent.)

20. Turnbull Electronics sells a computer system for $2,450. How much does Turnbull pay for each computer system if its markup on cost is 25%?

21. Plainfield Farms Inn offers a complete turkey dinner for $9.95. Find the cost of the dinner if operating expenses are 100% of cost and net profit is 20% of cost.

22. Centervale Hardware purchases interior house paint for $4.25 a gallon, which it sells for $12.95. Find the percent of markup based on cost.

23. A-1 Convenience Store sells a brand of soda pop for $2.59 a six-pack. How much does the store pay for each six-pack if its markup based on cost is 48%?

24. An electrical manufacturer charges distributors $15.25 for switch boxes, which includes a $3.75 markup. Find (a) the cost of the switch boxes and (b) the markup as a percent of cost.

25. A & D Cooling and Heating received an invoice for six hot water heaters with a total list price of $1,275. If the supplier provided a 15% trade discount and charged $20 for delivery, what unit selling price is required to realize a 30% markup on cost?

6.2 MARKUP BASED ON SELLING PRICE

Most retail firms use the **markup on selling price** method to price merchandise because the net sales figure reported in the income statement is the basis for analyzing the efficiency of operations. For example, a firm's cost of goods sold, all operating expenses, and its net profit are all computed as a percent of net sales. (This comparison is referred to as a vertical analysis and was discussed fully in Chapter 4.) In addition, businesses often base payroll commissions, certain taxes, distribution of profit and expenses, and the value of inventory on net sales. These topics will be presented in detail in subsequent chapters.

When merchants mark up their merchandise based on selling price, they are able to determine from the daily sales records the estimated gross profit from operations. To illustrate, assume a business reports sales for a given day at $15,000 and uses a markup based on selling price sufficient to produce a 60% gross margin rate. The merchant is able to determine daily gross profit as follows:

base sales × rate of gross margin = gross margin
$15,000 ×   .60   = $9,000

If the gross margin is $9,000, then the cost of goods sold is $6,000, or 40% of gross sales. The amount and percent relationships can be expressed as an abbreviated income statement.

Net sales revenue	$15,000	100%
Less: cost of goods sold	6,000	40%
Gross margin	$ 9,000	60%
Less: operating expenses	7,500	50%
Net profit	$ 1,500	10%

As explained at the beginning of the chapter, gross margin is also referred to as markup. Therefore, in the preceding analysis, a 60% markup on selling price is used to determine the price of items to cover the operating expenses (50%) and profit (10%). In terms of the markup equation, the example would be presented as follows:

cost   +   markup   =   selling price
40% of selling price + 60% of selling price = 100% of selling price

Now, let's look back at various situations involving calculating markup based on selling price.

Calculating Selling Price Given Cost and Percent Markup Based on Selling Price

Learning objective
Compute the selling price given cost and markup based on selling price.

When the markup is based on the selling price, and the cost is known, the cost is expressed as a percentage of the selling price (as explained in the preceding section). However, in this situation, we do not know the price (base) so we must use either algebra to find the price charged or a formula derived from the algebraic procedure. The algebraic approach utilizes the markup equation as shown in Example 5.

Formula for selling price

$$\text{selling price} = \frac{\text{cost}}{1 - \text{markup rate}}$$

Example 5

Casual Furniture purchased a patio set for $120. What is the selling price of the patio set if the company expects to earn a 25% markup based on price?

Solution

Algebra:

$$cost + markup = selling\ price$$
$$\$120 + .25S = S$$
$$120 = S - .25S$$
$$120 = .75S$$
$$\$160 = S$$

Formula:

$$\frac{cost}{1 - markup\ rate} = selling\ price$$

$$\frac{\$120}{1 - .25} = \frac{\$120}{.75} = \$160$$

In the equation, price (S) is the unknown. Since we know that the markup is 25% of the price, we label it as 25% of the unknown price (.25S). Then we gather terms, subtract like terms as required, and solve for the unknown price by dividing the markup complement into the cost, as shown in the example.

The formula requires memorization of the algebraic procedure. When you divide the cost by 1 minus the markup percentage, you are performing the last two steps of solving the equation ($120 = S - .25S$ and $120 = .75S$, $S = \$160$).

Calculating Cost Given Selling Price and Percent Markup on Selling Price

Learning objective
Find the cost when selling price and markup rate on price are known.

Retailers often stock multiple price lines of each item of merchandise they sell because consumer purchases are, in part, determined by price. For example, a retailer may offer an economy, standard, and deluxe line of merchandise to buyers. When retailers decide to restock a given price line, they must purchase merchandise that will yield the required markup on the target selling price. In other words, the retailers must know the highest price they can afford to pay for the item and realize the required markup.

Example 6

How much can a retailer pay for a product line that will be sold for $40 if a markup of 35% of the selling price is required to cover operating expense and profit?

Solution

Formula:

Step 1: Calculate the cost percent.

selling price   −   markup   =   cost
100% of selling price − 35% of selling price = 65% of selling price

Step 2: Calculate the cost.

selling price × cost rate = cost
$40   ×   .65   = $26

Algebra:

cost + markup = selling price
$C + (.35)(\$40) = \$40$
$C + 14 = \$40$
$C = \$40 - 14$
$C = \$26$

Calculating Amount and Rate of Markup Based on Selling Price

Learning objective
Calculate the amount and percent of markup based on selling price.

In Section 6.1, we explained how to calculate the amount and rate of markup when the markup is based on cost. The same procedure is used to calculate the amount and rate of markup when the markup is based on selling price, except the base is now the selling price. When an item's selling price is used as the base, the percent (rate) of markup is found by dividing the amount of markup by the selling price.

Finding percent of markup

$$\text{percent, or rate, of markup} = \frac{\text{markup}}{\text{selling price}} \quad \left(\text{rate} = \frac{\text{part}}{\text{base}}\right)$$

Example 7

An article that costs $225 was sold for $375. Find (a) the amount of markup and (b) the markup percent based on selling price.

Solution

a. cost + markup = selling price

$$\$225 + M = \$375$$
$$M = \$375 - \$225$$
$$M = \$150$$

b. base × rate = part

$$\$375 \times R = \$150$$
$$\$375R = \$150$$
$$R = \$150/\$375$$
$$R = .40, \text{ or } 40\%$$

or

$$\text{rate} = \frac{\text{markup}}{\text{selling price}} = \frac{\$150}{\$375} = .40, \text{ or } 40\%$$

Converting Markup Rate to Opposite Base

In Section 6.1, we explained how to calculate markup based on cost, a method used primarily by manufacturers. In this section, we explain how to calculate markup based on price, a method used primarily by retailers. Because retailers often purchase merchandise directly from manufacturers, a business may wish to convert its markup rate to the opposite base for purposes of comparison.

Learning objective
Convert the markup rate to the opposite base.

Two methods are used to compare markup percent to the opposite base. The first method requires you simply to compute the markup percent based on cost and then compute the markup percent based on selling price. The second method allows you to convert the markup percent on cost to the markup percent on selling price and vice versa with conversion formulas. Both methods are illustrated in Example 8.

Example 8

If an item that costs $12 is sold for $18, what is the percent of markup based on cost and on the selling price?

Solution

Method 1:   Calculate markup percent for each base.

$$\text{cost} + \text{markup} = \text{selling price}$$
$$\$12 + M = \$18$$
$$M = \$18 - \$12$$
$$M = \$6$$

$$\text{rate} = \frac{\text{markup}}{\text{cost}} = \frac{6}{12} = 50\% \text{ markup on cost}$$

$$\text{rate} = \frac{\text{markup}}{\text{selling price}} = \frac{6}{18} = .333, \text{ or } 33\,\tfrac{1}{3}\% \text{ markup on selling price}$$

Method 2:   Convert to opposite base. To convert percent markup on cost to its equivalent markup percent on selling price, use the following formula:

$$\frac{\text{markup percent on cost}}{1 + \text{markup percent on cost}} = \text{markup percent on selling price}$$

$$\frac{.50}{1 + .50} = \frac{.50}{1.50} = .333, \text{ or } 33\,\tfrac{1}{3}\%$$

To convert percent markup on price to its equivalent markup percent on cost, use the following formula:

$$\frac{\text{markup percent on selling price}}{1 - \text{markup percent on selling price}} = \text{markup percent on cost}$$

$$\frac{.333}{1 - .333} = \frac{.333}{.667} = .50, \text{ or } 50\%$$

CALCULATOR SOLUTION

Keyed entry									Display
AC	.5	÷	(	1	+	.50	)	=	.333333

Calculating List Price Given a Trade Discount and Percent Markup Based on Selling Price

In Chapter 5, we learned that merchants often provide trade discounts to their customers that reduces the list price of an item to its net price (list price − trade discount = net price). Because the list price is not the amount the seller will receive from the buyer as payment, the net price (selling price) must be sufficient to achieve the target markup percent. Example 9 illustrates how to calculate the list price that will guarantee a required rate of markup based on selling price when a trade discount is given.

Example 9

Central City Foods Inc. buys canned vegetables from a producer at a cost of $14 a case. If the company desires a 30% markup on selling price and offers its customers a 15% trade discount, find (a) the selling price (net price), (b) the list price, and (c) the trade discount amount.

Solution

a.   Calculate the selling price (net price) that will provide the required markup.

$$\text{cost} + \text{markup} = \text{selling price (net price)}$$
$$\$14 + \quad .30S = S$$
$$\$14 = S - .30S$$
$$\$14 = .70S$$
$$\$20 = S$$

b.   Calculate the list price required to cover the trade discount and result in the net price (selling price).

$$\text{list price} - \text{trade discount} = \text{net price}$$
$$P \quad - \quad .15P \quad = \quad \$20.00$$
$$.85P \quad = \quad \$20.00$$
$$P \quad = \quad \$23.53$$

c.   Calculate the trade discount amount.

$$\text{list price} \times \text{trade discount rate} = \text{trade discount}$$
$$\$23.53 \ \times \quad .15 \quad = \quad \$3.53$$

or

$$\text{list price} - \text{net price} = \text{trade discount amount}$$
$$\$23.53 - \$20.00 = \quad \$3.53$$

This same procedure is used to calculate the list price when a chain discount is given. However, the chain discount must be converted to its single-equivalent discount rate. If you do not recall how to calculate the single-equivalent discount rate, refer back to Section 5.2.

CHECK YOUR KNOWLEDGE

Markup Based on Selling Price

1. If a retailer uses an average markup of 45% of the selling price and reports $25,000 in gross sales on October 20, what is (a) the amount of gross margin, (b) the percent of cost of goods sold, and (c) the amount of net profit if operating expenses are estimated at 40% of net sales?

2. If an item costs $45 and carries a markup on the selling price of $17.25, what is the selling price?

3. The markup rate by Sounds Unlimited is 30% of the selling price. What is the (a) selling price and (b) markup amount of a television set that cost $179?

4. What is the most a buyer for the SportCenter should pay for a snowboard that will be sold for $39.99 if there is to be a 25% markup based on the retail price?

5. The Garden Shop sells a lawnmower for $250. If the shop's operating expenses are 42% of the selling price and the net profit is 6% of the selling price, how much does the lawnmower cost?

6. If an item is purchased for $20 and is sold for $35, what is the markup percent based on price?

7. What markup percent on selling price is equivalent to a 20% markup on cost?

8. If the markup percent on a computer is 37% of the selling price, what is the markup percent based on cost?

9. A retailer wishes to sell a jacket that costs $43.20 at a list price sufficient to provide a 28% markup on selling price and give the customer a 40% discount. Determine the (a) selling price, (b) the list price, and (c) the discount amount.

Answers to CYK: 1.   a. $11,250;   b.   55%;   c. $1,250   2. $62.25   3.   a. $255.71;   b. $76.71
4. $29.99   5. $130.00   6.   42.9%   7.   16.7%   8.   58.7%
9.   a. $60;   b. $100;   c. $40

6.2   EXERCISES

Calculate the missing numbers. Round dollar amounts to the nearest cent and rates to the nearest tenth of a percent.

	Cost	Markup	% Markup on price	Selling price
1.	_____	_____	20%	$30.00
2.	_____	$50.00	_____	$400.00
3.	$137.00	$27.50	_____	_____
4.	$652.30	_____	_____	$829.50
5.	_____	$94.00	33%	_____

	Cost	Markup	Selling price	% Markup on cost	% Markup on selling price
6.	_____	$20	_____	25%	20%
7.	$12.00	_____	$18.00	_____	_____
8.	_____	$40.65	$135.50	_____	_____
→9.	$240.60	101.65	341.65	_____	42%
10.	_____	$120.00	_____	60%	_____

Convert each of the following markups to the opposite base. Round to the nearest tenth of a percent.

Markup on cost to price

11. 20%   _____
12. 35.5%   _____
13. 50.25%   _____

Markup on price to cost

14. 15%   _____
15. 40⅓%   _____
16. 28%   _____

Determine the missing amounts. Round dollar amounts to the nearest cent.

	Cost	Markup on price	Selling price	Trade discount	List price
17.	$21.60	40%	_____	25%	_____
18.	$240	25%	_____	15%, 10%	_____
19.	$3,500	30%	_____	5%, 15%, 5%	_____

Solve the following word problems. Round rates to nearest whole percent and dollar amounts to the nearest cent.

20. Coffee Merchants reported net sales of $50,245 for July. If the firm uses a markup of 55% of the selling price, find (a) the gross margin amount and (b) the amount of net profit for the month if operating expenses are 48% of net sales.

21. Milton's Farm Supply Co. purchased four garden tillers at $312 each. The store's operating expenses average 42% of the selling price and the net profit is 10% of the selling price. What will be the selling price of the tillers?

22. Herman's Department Store purchased three dozen dress shirts at $72 a dozen. If the markup rate is 25% of the selling price, what is the selling price of each shirt?

23. A dealer pays $129.50 for an article that will be sold at a markup of 37½% on the selling price. What is (a) the selling price and (b) the markup amount?

24. The Toy Shack buys video games for $19.50 and sells them for $39. What is the markup rate based on selling price?

25. Clearwater Beverage Co. buys purified water for resale in 45-gallon barrels at $56.25 per barrel.

The water is sold in 3-gallon plastic bottles. What is the selling price of each bottle if the company uses a 40% markup based on the selling price?

26. The buyer for Paint-n-Paper wishes to buy a line of paint brushes to be sold for $8.50 each. If a markup of 28% based on the selling price is required, how much should the buyer pay for the paint brushes?

27. Milton Restaurant Supply Co. bought four pizza ovens at $312 each. The store's operating expenses average 40% of the selling price, and the net profit is 12% of the selling price. What is the selling price of the ovens?

28. The Discount Furniture Mart purchased 250 student desks from a manufacturer at a total list price of $14,750 less a 12% trade discount. If the Mart sold 140 of the desks at $89 each, 80 at $79, and the remainder at $65 each, determine (a) the total amount of sales from the desks, (b) the total gross margin amount, and (c) the average markup percent on the selling price. (Round to the nearest tenth percent.)

29. The markup rate used by the Village Shop is 20% of the selling price. What is the equivalent markup percent based on cost?

30. A men's clothing store buys ties at $132 per dozen and sells them at $17.50 each. What is the rate of markup (a) based on cost and (b) based on selling price? (Round to the nearest tenth percent.)

31. An item is sold for $59.75, which represents a 27% markup on the cost. What is the equivalent markup percent based on the selling price? (Round to the nearest tenth percent.)

32. Shaheen's Rug Merchants imports 6- × 9-foot area rugs from the Middle East that cost $260 each. Determine the (a) selling price, and (b) the list price of each rug, if retailers are to be given a 20% trade discount and Shaheen desires a 35% markup on selling price.

33. Southwestern Copy Products Inc. purchases photocopiers at $950 each and offers its preferred retailers a trade discount of 10%, 5%, 10%. Find the list price required to realize a gross profit of 20% of the selling price.

6.3   MARKUP OF PERISHABLE GOODS

Many businesses purchase merchandise for resale that will spoil if not sold within a short period of time. Such merchandise is referred to as **perishables** and includes diary products, produce, baked goods, certain canned and packaged goods, and cut flowers. In some cases, merchandise with rapidly changing styles and designs or merchandise that becomes damaged are also considered perishable. Examples that might be considered in this category would be clothing, computers, automobiles, boats, and appliances.

Merchants are able to estimate from past experience the percent or actual amount of their perishable inventories that will be sold at the marked price and at a reduced price. In addition, they often determine the portion that may have to be discarded as spoilage. In the introduction of this chapter, we explained that markup (gross profit) must be sufficient to cover operating expenses and target profit. When pricing perishable items, the price of the items that do sell will generate revenues sufficient to also cover the cost of those items that are expected to spoil (will not sell) or are sold at a reduced price. The examples in this section will illustrate the process of pricing perishable goods under varying conditions.

Learning objective
Determine the selling price required to cover estimated spoilage on perishable merchandise.

Example 10

Bradkey Farms Greenhouse purchased 1,500 4-inch potted geraniums for resale at a cost of $1.25 each. Determine the selling price per plant if a markup of 50% on selling price is required and 30 plants will not be sold due to spoilage.

Solution

Step 1:   Calculate the total cost of the purchase.

quantity $\times$ unit cost = total cost

1,500 $\times$ $1.25   = $1,875.00

Step 2:   Calculate the total selling price of the entire purchase.

$$
\begin{aligned}
\text{cost} \quad + \text{ markup} &= \text{ selling price} \\
\$1,875 + \quad 50\%S \quad &= \quad\quad S \\
\$1,875 + \quad .50S \quad &= \quad\quad S \\
\$1,875 &= \quad S - .50S \\
\$1,875 &= \quad\quad .50S \\
\$3,750 &= \quad\quad\quad S
\end{aligned}
$$

Step 3:   Calculate the selling price per unit required to cover the actual spoilage.

$$
\begin{aligned}
(1,500 - 30)S &= \$3,750 \quad \text{(total selling price)} \\
1,470S &= \$3,750 \\
S &= \$2.55 \quad \text{(selling price per unit)}
\end{aligned}
$$

The next example will illustrate how to price a perishable good when the markup is based on cost and the spoilage is estimated as a percent of the total merchandise purchased.

Example 11

Michelle Ruggiero, a produce buyer for T and C Food Markets, purchased 500 pounds of grapes for 39 cents per pound. She estimates that approximately 5% of the grapes will spoil before they can be sold. If T and C Markets requires a 60% markup on cost, at what price per pound must the grapes be sold?

Solution

Step 1:   Calculate the total cost of the entire purchase.

quantity $\times$ unit cost = total cost

500   $\times$ $.39   = $195.00

Step 2:   Calculate the total selling price of the entire purchase.

cost + markup = selling price
$$C + .6C = S$$
$$1.6C = S$$
$$1.6(195) = S$$
$$\$312 = S$$

Step 3:   Calculate the amount of the purchase lot that is expected to sell.

base × rate = part
$$500 \times .95 = 475 \text{ pounds of grapes are expected to sell}$$

If 5% of the grapes are expected to spoil, then 95% are expected to sell (100% − 5% = 95%).

Step 4:   Calculate the selling price required to cover the estimated spoilage.

$$475S = \$312 \quad \text{(total selling price)}$$
$$S = \$.656, \text{ or } \$.66 \text{ per pound} \quad \text{(rounded to nearest cent)}$$

To determine the price per pound, we divide the total sales in step 2 by the amount of merchandise that is expected to be sold in step 3.

Note: If T and C Markets sells more than 475 pounds of the 500 pounds of grapes purchased before they spoil, each dollar of sales received will be additional profit.

Now, let's look at an example that involves the appropriate markup to cover a reduction in price of perishable goods.

Example 12

Lakeside Bakery made 250 loaves of Italian bread at a cost of 48 cents each. Experience indicates that 8% of the loaves will be sold the following day at a reduced price of 50 cents. Find the marked price if the bakery wishes to obtain a 125% markup on cost.

Solution

Step 1:   Calculate the total cost and total selling price, as we did in steps 1 and 2 for Example 12.

cost + markup = selling price
$$C + 1.25C = S$$
$$2.25C = S$$
$$2.25(\$120) = S$$
$$\$270 = S$$

The loaves of bread cost the bakery $120 to bake (250 × $.48) and the total selling price ($270) will be 225% of the total cost ($120).

Step 2: Calculate the selling price required to cover the estimated reduction in price for day-old merchandise.

$$\frac{\text{sales from}}{\text{original units sold}} + \frac{\text{sales from}}{\text{reduced units sold}} = \text{total sales}$$

$$(250 - 20)S + (20 \times \$.50) = \$270$$
$$230S + \$10 = \$270$$
$$230S = \$270 - 10$$
$$230S = \$260$$
$$S = \$1.13/\text{loaf} \quad \text{(rounded to nearest cent)}$$

The per-unit marked price (selling price) is determined by the total sales required to cover cost and markup (gross margin). Therefore, total sales (revenue) is the number of units sold at the reduced price (250 × .08 = 20; 20 × $.50 = $10) plus the number of units sold at the marked price (250 − 20 = 230 units × S, the unknown selling price).

CHECK YOUR KNOWLEDGE

Markup of Perishable Goods

Solve each of the following word problems. Round amounts to the nearest whole number or cent.

1. Creative Design sells custom sportswear and is licensed to print college logos on various-style sweatshirts and t-shirts. If the store purchased 200 sweatshirts that cost $12.50 each including printing and 10 of the sweatshirts are unsellable due to printing errors, what selling price is required to realize a 30% markup on price?

2. Mother Nature's Basket purchased 125 quarts of strawberries from a local farm at 60 cents per quart. The store estimates that 4% of the berries will spoil before they can be sold. If the store uses a 60% markup on cost, what price should it charge per quart?

3. In-N-Out Convenience Mart bought 50 dozen large eggs at $.75 per dozen. Determine the required selling price per dozen if a markup of 45% on selling price is required, and the manager estimates 4% of the order will not be sold because of damage.

4. The Skate-n-Ski Shop bought 50 pairs of cross-country skis at a cost of $30 a pair. If the shop expects 10% of the skis to be sold on sale at $35

a pair, what price should the skis be marked to make a 40% gross profit on cost?

5. International China Ltd. manufactures a four-place setting of standard chinaware at a cost of $20 per set. If 5% of the 3,000 sets produced were sold in the outlet store for $25 per set, and a markup of 140% on cost is desired, what was the selling price of each set?

6.3 EXERCISES

Find the correct amounts. Round answers to nearest whole number or cent.

	Units purchased	Unit cost	Total cost	Markup on selling price	Total selling price	Units perishable	Units sold
1.	375 lbs.	$0.65	_____	20%	_____	30 lbs.	_____
2.	1,550 ea.	$2.40	_____	12.5%	_____	160 ea.	_____
3.	8,000 gal.	$1.02	_____	42%	_____	295 gal.	_____

Find the selling price per unit of sale. Round amounts to nearest whole cent.

	Total cost	Quantity purchased	Number perishable	Markup on selling price	Total selling price	Selling price per unit
4.	$250	25 ea.	5	40%	_____	_____
5.	$1500	40 doz.	30	15%	_____	_____
6.	$27,625	6500 lbs.	125	30%	_____	_____

Find the correct amounts. Round amounts to the nearest whole number or cent.

	Total cost	Quantity purchased	Percent perishable	Units perished	Units sold	Markup on total cost	Selling price per unit
7.	$750	125 gal.	8%	_____	_____	60%	_____
8.	$2,520	1800 qt.	12%	_____	_____	35%	_____
9.	$190	80 ea.	5%	_____	_____	20%	_____

Find the selling price per unit of sale. Round amounts to nearest whole cent.

	Unit cost	Quantity purchased	Total cost	Percent at reduced price	Reduced price	Markup on total cost	Selling price per unit
10.	$3.00	60 ea.	_____	10%	$2.00	20%	_____
11.	$8.40	1,000 bx.	_____	30%	$6.00	50%	_____
12.	$4.90	150 pr.	_____	2%	$1.75	75%	_____

Answers to CYK: *1.* $18.80 *2.* $1.00 *3.* $1.04 *4.* $42.78 *5.* $49.21

Solve each of the following word problems. Round amounts to nearest whole numbers or cent.

13. Marco's Imported Foods purchased three cases of Italian Cappicolla ham from a supplier at a cost of $2.75 a pound. Each case contains six 5-lb. pieces. Determine (a) the total cost of the purchase, (b) the total selling price if a markup of 60% on selling price is required, and (c) the selling price per pound if 3 pounds of the merchandise is unsellable.

14. Bryant's Dairy purchased 200 cases of 2% milk from Eastern Dairy Cooperative at a cost of $.95 per half gallon. Each case contains six half-gallon glass bottles. Find (a) the total cost of the purchase, (b) the total selling price if a markup of 25% on cost is desired, and (3) the selling price per half-gallon bottle if $1\frac{1}{2}\%$ of the bottles purchased are estimated to be recorded as breakage.

15. The produce manager of Hometowne Markets purchased five boxes of white mushrooms that contained 20 pounds each at a cost of $.655 per pound. The manager estimates 15% of the merchandise will perish before it can be sold. If the market's markup on produce is 120% on cost, what is the mushroom's selling price per pound?

16. Claudette's Fruit and Floral Designs bought 75 dozen long-stem roses at $18 per dozen. Claudette estimates 8% of the roses will not be sold for various reasons. At what price per dozen should the roses be marked to realize a 50% markup on selling price?

17. Silver Mountain Ski Shop purchases 60 pair of downhill skis at a cost of $150 each. Past experience suggests 25% of the skis purchased will have to be sold on sale at $175 each. What marked price is required to provide an 80% markup on the cost of the skis?

18. The Media Boutique bought 150 video games of various titles at a total cost of $3,750. The management expects 20% of the games will have to be sold at a reduced price of $29.99 each. What selling price is necessary to obtain a gross profit of 30% on selling price for the order?

19. General Filter Inc. manufactures oil and air filters for industrial and commercial markets. A production run of 15,000 WD 120, oil filters cost the company $19,500 to manufacture. If $1\frac{1}{2}\%$ of the filters were rejected by quality control as unsellable, what selling price is required to generate a 35% markup on cost?

20. Wilbur's Market purchased 120 pounds of bananas at 20 cents per pound. If 5% of the purchase will spoil, what price per pound must be charged to realize a gross margin of 15% on selling price?

21. Casual Furniture Gallery purchased 300 outdoor patio sets at a cost of $125 per set. Past records indicate that the store will likely have to sell 20 sets at a reduced price of $169. If the store wants a markup on cost of 115%, what is the selling price per set?

22. The buyer at Boardwalk Beach and Surf Shop purchased 50 bikini swimsuits at a total cost of $2,950. The shop plans to sell 30% of the order at $69.99 each. Determine the selling price necessary to obtain a gross profit of 38% on selling price.

6.4   MARKDOWNS

In the preceding section we learned how to price merchandise to realize a required percent markup on price or cost knowing that a percent of the merchandise purchased would have to be sold at a reduced price. This procedure guarantees an acceptable gross profit because the reduced price was considered as part of the merchandise's original markup. However, sellers are not always able to accurately estimate the percent of purchases that will be marked down or to anticipate market conditions that force them to

Learning objective
Compute the
markdown and the
reduced price
amount, given the
selling price and
markdown rate.

reduce their prices. Merchants may reduce the selling price of their products for any one of the following reasons:

1. To reduce excessive inventories
2. To adjust for seasonal and fashion changes
3. To move slightly damaged goods
4. To be competitive with prices offered by other merchants
5. To stimulate the sale of unsold merchandise because of size or color
6. To accommodate a reduction in the cost of the merchandise.

When merchants decide to reduce the **marked price** of their merchandise, they usually announce the decision to their market in a promotion referred to as a *sale* in the local media. The reduction in the marked price is called a **markdown,** and is expressed as a percent or an amount of the marked price. The price of the item after the markdown has been applied is called the **reduced price.** These terms are considered merchandising terminology, but are also accounting terms. For example, in Chapter 4, we learned that gross sales − sales discounts = net sales. Therefore, in retailing an item's marked price is equivalent to its gross sale amount, its markdown amount is equivalent to its sales discount, and its reduced price is equivalent to its net price.

$$
\begin{aligned}
\text{Marked price} &= \text{Gross sales} \\
\underline{-\ \text{Markdown}} &= \underline{-\ \text{Sales discount}} \\
= \text{Reduced price} &= = \text{Net sales}
\end{aligned}
$$

This procedure is the basis of the markdown equation, which can be expressed as follows:

marked price − markdown = reduced price

Because the markdown is expressed as a percentage of the marked price, we once again have a percent equation application where

marked price = base

percent of markdown = rate

markdown amount = part

Now, let's look at a few examples involving markdowns that can be solved with the percent equation.

Example 13

Alter's Shoes reduced the price of its entire stock of shoes 30% for one week. Find the reduced price of a pair of boots with a marked price of $78.99.

Solution

Step 1: Find the amount of the markdown.

base × rate = part

$78.99 × .30 = $23.70 (markdown amount)

Step 2: Find the reduced price.

marked price − markdown = reduced price

$78.99 − $23.70 = $55.29

Note: The percent of markdown is always based on the marked price.

Example 14

Center City Auto Parts sells a case of 10-W-40 motor oil for $18. To reduce an overstock of inventory, the motor oil was marked down to a sale price of $13.50 per case. What is the percent of markdown?

Solution

Step 1: Find the amount of markdown.

marked price − reduced price = markdown

$18.00 − $13.50 = $4.50

Step 2: Find the percent (rate) of markdown.

$$\text{rate} = \frac{\text{part}}{\text{base}} = \frac{\$4.50}{\$18.00} = .25 \text{ or } 25\%$$

On some occasions, a single markdown may not be sufficient to promote the sale of an item. Consequently, a retailer may decide to continue to reduce the price until the item is sold. When a series of markdowns are applied, each markdown is based on the previous sale price, as illustrated in Example 15.

Example 15

Clara's Boutique displayed a sweater for $95 on September 10. The sweater was marked down 15% on October 8 and another 20% on November 12. Find the reduced price if the sweater was sold on November 15.

Solution

Step 1: Calculate the reduced price on October 8.

$95 × .15 = $14.25 markdown amount

$95 − $14.25 = $80.75 reduced price October 8

Step 2:   Calculate the reduced price on November 12.

$80.75 × .20 = $16.15 markdown amount

$80.75 − $16.15 = $64.60 reduced price November 15

A series markdown can also be calculated with the complement method.

Complement Method

Step 1:   Determine the complement of each markdown.

  100% marked price

 −15% markdown percent on October 8

  85% reduced price complement on October 8

 100% reduced price on October 8

 −20% markdown percent on November 12

  80% reduced price complement on November 12

Step 2:   Multiply the marked price by the complements determined in step 1.

$95.00 × .85 × .80 = $64.60 reduced price on November 15

(You may have noticed the complement method is the same method we used in Chapter 5 to calculate the net price when a chain trade discount was offered.)

When merchants reduce the prices of their goods, the selected markdowns should be sufficient to promote the sale of the merchandise and provide profits. Under extreme circumstances, merchants may mark down goods to cost and below cost. Obviously, markdowns of this nature must be implemented with a clear understanding of their impact on the financial condition of the business. Figures 6.2 through 6.5 illustrate the various financial effects of markdowns. A **reduced net profit** results when the reduced price is sufficient to cover cost and operating expenses yet provide some profit (Figure 6.2). The **break-even point** occurs when the reduced price is sufficient to cover only cost and operating expenses (Figure 6.3). An **operating loss** results when the reduced price is sufficient to cover cost but not all of the operating expenses (Figure 6.4). The amount of loss is the difference between the break-even point and the sales price. An **absolute loss** occurs when the reduced price is below the actual cost of the merchandise (Figure 6.5). The amount of the absolute or gross loss is the difference between the reduced price and the cost.

Let's look at a couple of examples that will help you gain a better understanding of these financial concepts.

Figure 6.2

Reduced net profit

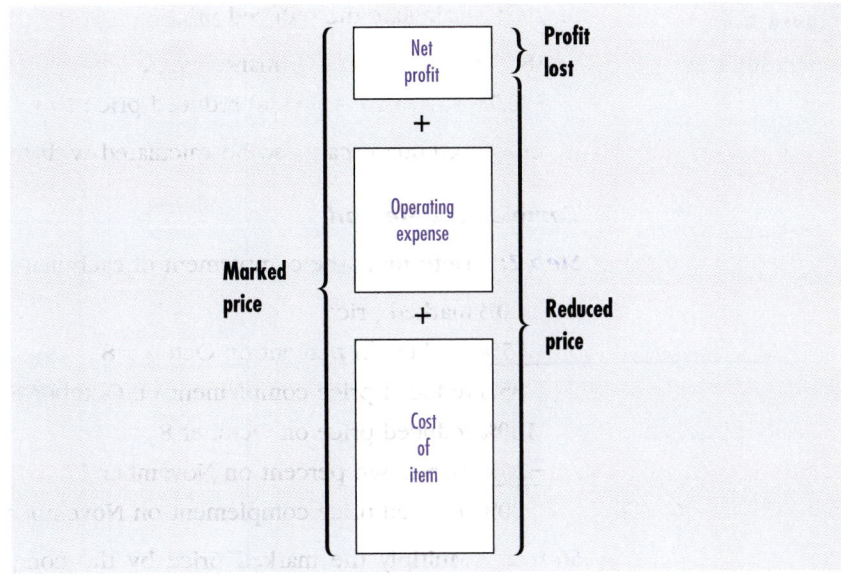

Figure 6.3

Break-even point

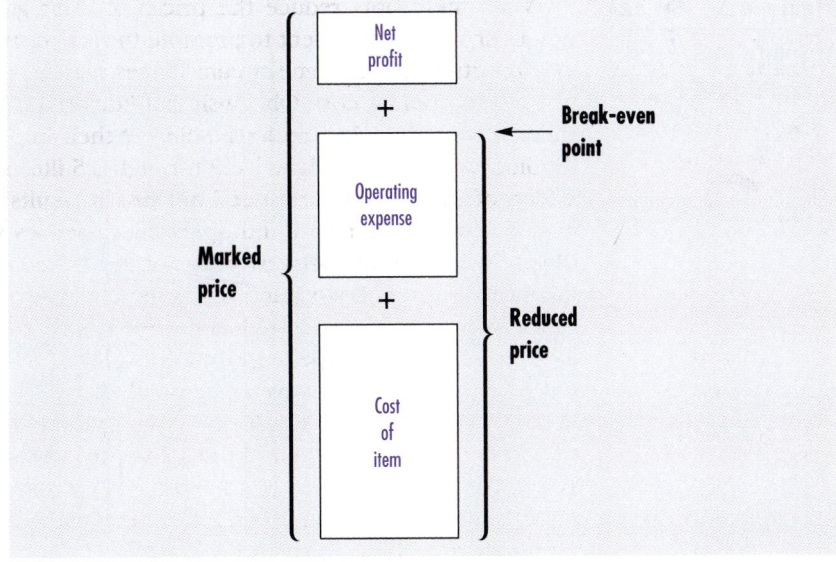

Figure 6.4

Operating loss

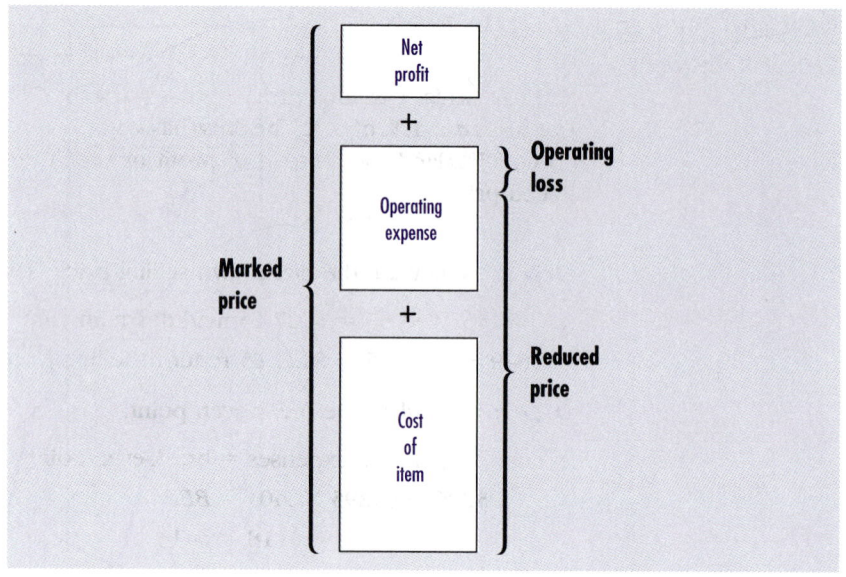

Figure 6.5

Absolute loss

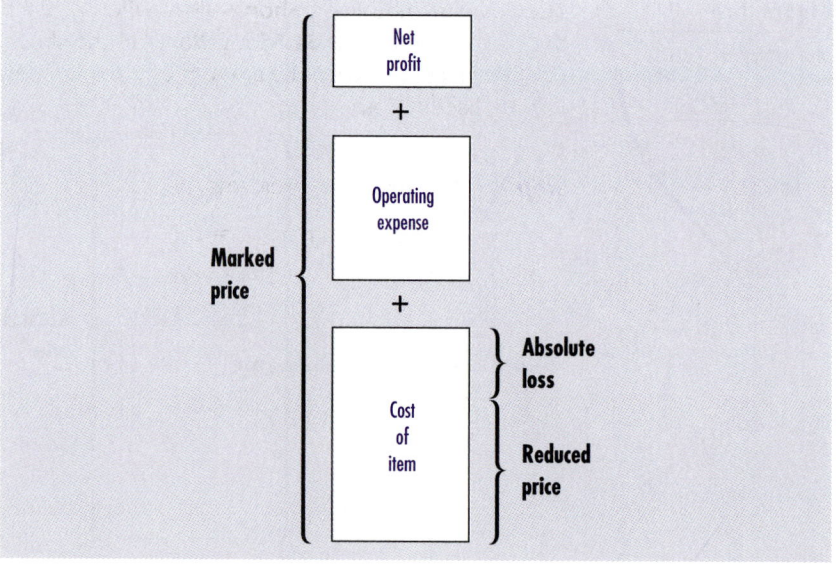

Example 16

Yvette's Bridal Fashions purchased a gown for $295. The store's operating expenses run 40% of cost. The dress has a marked price of $649, but is marked down 35%. Find the amount of profit or loss if the dress is sold at the reduced price.

Solution

Step 1: Calculate the markdown selling price (sale price).

$$\$649 \times .35 = \$227.15 \text{ markdown amount}$$
$$\$649 - \$227.15 = \$421.85 \text{ reduced selling price}$$

Step 2: Calculate the break-even point.

$$\text{cost} + \text{operating expenses} = \text{break-even point}$$
$$\$295 + (\$295 \times .40) = BEP$$
$$\$295 + \$118 = \$413$$

Step 3: Determine if a profit or a loss resulted from the transaction.

$$\text{reduced selling price} - \text{break-even point} = \text{profit/loss}$$
$$\$421.85 - \$413.00 = \$8.85 \text{ profit}$$

Example 17

The Convenience Food Shop sells a gallon of 2% milk on sale for $1.49. If the cost of the milk is $1.55 a gallon and the shop's operating expenses are 30% of the cost, find (a) the amount of the operating loss and (b) the amount of the absolute loss.

Solution

Step 1: Calculate the break-even point.

$$\text{cost} + \text{operating expenses} = BE$$
$$\$1.55 + (1.55 \times .30) = BE$$
$$\$1.55 + .47 = \$2.02$$

Step 2: Calculate the amount of the operating loss and the absolute loss.

a. reduced selling price − break-even point = operating profit/loss
$$\$1.49 - 2.02 = (\$.53) \text{ operating loss/gallon}$$

b. reduced selling price − cost = absolute loss
$$\$1.49 - \$1.55 = -\$.06 \text{ absolute loss/gallon}$$

For Your Information

ECONOMIC STATISTICS INFLUENCE RETAIL PRICES

Retail sales data are collected by various federal and state governmental agencies from retail establishments of all sizes and types throughout the country. The data are compiled and published on a regular basis to provide valuable information on sales. For example, the Department of Commerce collects data using a monthly survey of retail establishments and reports the composition of retail sales as shown in the circle graph. Retail sales are broken down into two major categories, durables and nondurables, which are then subdivided by the type of retail establishment. The data reported represent sales of merchandise for cash or credit by organizations engaged in retail trade. They do not include sales of manufacturers, wholesalers, or service establishments.

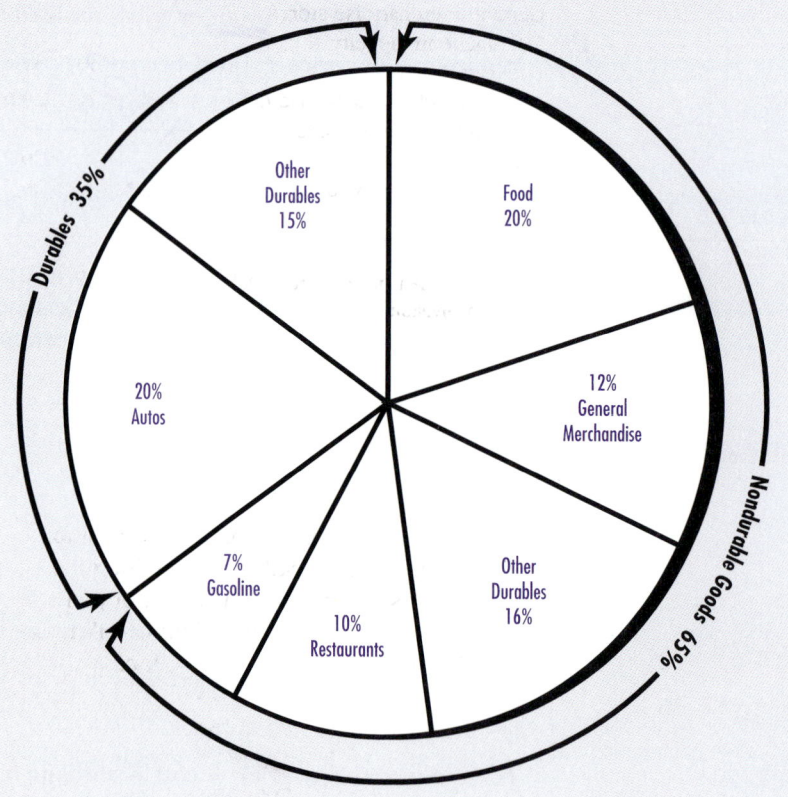

The U.S. Treasury Department collects retail sales data from income tax returns reported by retail businesses to the Internal Revenue Service. The Statistics Division of the IRS compiles the data and publishes numerous reports utilized by business and government as the basis of financial decision making. The following table is one example of the retail sales data published by the U.S. Treasury Department.

Retail store average markup (Based on selling price)

Type of establishment	Markup percent
Apparel and accessories	37.6
Automotive dealers (new)	12.8
Diners and restaurants	56.4
Drug and proprietary stores	30.8
Furniture and home furnishings	35.8
Gasoline service stations	14.5
General merchandise stores	30.0
Gift, souvenir, novelty shops	41.9
Grocery stores	22.1
Motor vehicle dealers (used)	29.6
Convenience food stores	27.3
Liquor stores	20.2
Sporting goods stores	29.7
Taverns, pubs, etc.	52.5

Source: U.S. Treasury Department, Internal Revenue Service, Statistics Division

Markdowns can have both a positive and negative impact on a firm's financial condition. Markdowns may increase volume, which in turn can increase profits, or markdowns may not generate sales sufficient to recover costs, resulting in some type of loss. To illustrate this point, Figure 6.6 compares the financial impacts of reducing the per unit marked price of an item that has a unit cost of $4.00 and a marked price of $10.00 per unit. The item's marked price is based on operating expenses of 50% of selling price, and net profit of 10% of selling price. Examine the figure carefully. Once again, financial statement concepts are the basis of a business application.

Figure 6.6

Financial impact
of markdowns

Original price		Reduced profit	
marked price	$10.00	reduced price	$9.50
cost of item	4.00	cost of item	4.00
gross profit	6.00	gross profit	5.50
operating expense	5.00	operating expense	5.00
net profit(loss)	$ 1.00	net profit(loss)	$ .50
Breakeven		**Operating loss**	
reduced price	$9.00	reduced price	$8.00
cost of item	4.00	cost of item	4.00
gross profit	5.00	gross profit	4.00
operating expense	5.00	operating expense	5.00
net profit(loss)	$0.00	net profit(loss)	($1.00)

As you can see, if markdowns do not increase the volume of sales suffi-
ciently to cover operating expenses, a loss will occur. Therefore, merchants
use this type of profit analysis to determine the maximum percent of markdown they can offer their customers without incurring an operating loss.
Example 16 will show you how to calculate the **maximum percent of
markdown** based on a required percent markup on selling price.

Example 18

Learning objective
Find the maximum
percent of markdown
without incurring an
operating loss.

The Bookmark buys paperback novels at a cost of $4.50 each. The store's
operating expenses are 30% of selling price and net profit is 10% of selling
price. Determine (a) the selling price of a book and, (b) the maximum percent of markdown allowed without an operating loss.

Solution

a. Calculate the selling price (marked price) using the markup equation.

$$\text{cost} + \quad \text{markup} \quad = \text{selling price}$$
$$\$4.50 + (.30S + .10S) = \quad S$$
$$\$4.50 + \quad .40S \quad = \quad S$$
$$\$4.50 \quad = \quad S - .40S$$
$$\$4.50 \quad = \quad .60S$$
$$\$7.50 \quad = \quad S$$

b. Calculate maximum percent of markdown.

Step 1: Find the breakeven point.

$$\text{cost} + \text{operating expenses} = \text{break-even point}$$
$$\$4.50 + (\$7.50 \times .30) = \quad \text{BEP}$$
$$\$4.50 + \quad \$2.25 \quad = \quad \$6.75$$

Step 2: Find the maximum markdown amount allowed without an operating loss.

$$\text{selling price} - \text{break-even point} = \text{maximum markdown}$$
$$\$7.50 \quad - \quad \$6.75 \quad = \quad \$.75$$

Step 3: Convert maximum markdown amount to a percent.

$$\text{part} = \text{base} \times \text{rate}$$
$$\$.75 = \$7.50 \times R$$
$$\$.75 = \$7.50R$$
$$.10 = R$$
$$or$$
$$10\% = R$$

CHECK YOUR KNOWLEDGE

Markdowns

Calculate the markdown and the sale price.

	Marked price	Markdown percent	Markdown amount	Sale price
1.	$55	12%	_____	_____
2.	$480	8%	_____	_____
3.	$6,590	24.5%	_____	_____

Calculate the markdown amount and the percent of markdown.

	Marked price	Sale price	Markdown amount	Markdown percent
4.	$4.50	$3.75	_____	_____
5.	$239.99	$167.99	_____	_____
6.	$3,785	$2,865	_____	_____

Solve each of the following word problems. Round dollar amounts to the nearest cent and rates to the nearest tenth of a percent.

7. A clothing store sells a men's overcoat for $225. The coat was marked down 20% on February 7 and an additional 25% on February 21. Find

the sale price if the coat was sold on February 25. (Use the complement method.)

8. Using the information in problem 7, calculate the percent of markdown.

9. Harding Furniture Annex plans to sell a certain style of bedroom set at its inventory clearance sale for $995. If the cost of the set was $800 and the operating expenses were 30% of the cost, find the amount of profit or loss on the sale of each set.

10. A stereo system cost a retailer $750. The retailer's regular marked price was $1,250, but the system was marked down 40% to reduce inventory. If operating expenses are 32% of cost, find (a) the operating loss and (b) the absolute loss.

11. The Pro Shop at Cedervale Gold Club buys Distance King golf clubs at a cost of $299 per set. If the shop's operating expenses are 25% of selling price and net profit is 15% of selling price, find (a) the selling price per set, and (b) the maximum percent of markdown allowed without incurring an operating loss.

6.4   EXERCISES

Fill in the blanks with the correct amount. Round amounts to the nearest whole cent.

	Marked price	% Markdown	Markdown amount	Sale price
1.	$19.50	30%	_____	_____
2.	_____	15%	_____	$85.00
3.	$.80	_____	_____	$.60
4.	$1,250	_____	$62.50	_____
5.	_____	33⅓%	$3.16	_____
6.	$367.80	20%	_____	_____
7.	$24.00	_____	_____	$14.40

Complete each calculation using the information provided. If there is no operating loss or absolute loss, place a 0 in the blank space.

	Cost	Operating expense	Break-even amount	Sale price	Operating loss	Absolute loss
8.	$12.00	$4.00		$14.00		
9.	$65.00		$80.00	$62.50		
10.	$135.70	$60.40		$189.25		
11.	$8.50	$3.40		$8.65	$3.25	
12.	$525.00		$705	$510		

Solve each of the following word problems. Round dollar amounts to the nearest cent, and rates to the nearest tenth of a percent.

13. Heritage Spas marked down its standard outdoor hot tub 40% to be competitive with other area retailers. Find the reduced price if the marked price of the tub was $3,500.

14. The Vision Center sells wire-rimmed eyeglass frames for $139. If the marked price is reduced to $95, what is the percent of markdown?

15. Sal's Hobby Supply sells a radio-controlled car for $420. The cars were marked down 25% on March 15 and another 15% on April 7. What was the reduced price on April 7?

16. A 3-drawer steel filing cabinet has a marked price of $49.99. If the cabinet is marked down 40%, what is the reduced price?

17. Acme Hardware stores marked price for a 52-inch ceiling fan on June 1 was $69.99. The fan was marked down to $59.99 on August 1 and to $39.99 on November 1. Find the percent of markdown on the units sold as of November 1.

18. Family Apparel's marked price on girls jeans on March 15 was $19.99. If the jeans were marked down 20% on April 10 and another 10% on April 30, find the reduced price of a pair of jeans sold on May 2.

19. Fischer Interiors purchased stain resistant nylon carpeting for $10 per square yard. The store's operating expenses are 30% of cost, and the carpet is sold at 25% off the marked price of $18 per square yard. Find the amount of profit or loss if the store sold 3,000 square yards of carpeting during the month.

20. A gas-powered chain saw selling for $225 is marked down 40%. If the cost of the chain saw is $140 and the operating expenses are 20% of the cost, find (a) the amount of the operating loss and (b) the amount of the absolute loss.

21. A retailer buys a 25-inch color TV with remote from a distributor for $175, which it plans to sell for $350. If the retailer's operating expenses are 45% of cost and the TV's are marked down 25%, find the amount of net profit or loss of each set sold at the reduced price. (Use the profit analysis shown in Figure 6.6 to solve this problem.)

22. Beauty Supply Warehouse buys reclining cutting chairs at a cost of $350 each. The supply store's operating expenses are 28% of cost, and net profit is planned at 15% of cost. Determine (a) the selling price of a cutting chair and (b) the maximum percent of markdown the store can offer without incurring an operating loss.

23. The marked price of a dozen eggs is $1.40. The eggs cost $1.10, and the retailer's related operating expenses are 40% of cost. If the retailer marks the eggs down 25% to promote business, what is (a) the operating loss and (b) the absolute loss?

24. Quality Footwear purchases leather sandals for $24 per pair. The sandals have a marked price of $59.99, but are reduced $20 for a sale. If the store's operating expenses are 35% of cost, find the amount of profit or loss at the sale price.

25. An upright vacuum cleaner costs a retailer $148. The retailer's operating expenses are 42% of selling price and net profit is 8% of selling price. Find (a) the selling price of the vacuum cleaner and (b) the maximum percent of markdown allowed without an operating loss.

CHAPTER REVIEW EXERCISES

The following problems review all mathematics and business concepts presented in the chapter. If you experience any difficulty solving a problem, go to the appropriate chapter section and review the examples provided. Express all answers as decimals rounded to tenths, or dollars rounded to cents, unless otherwise instructed.

1. Village Auto Mart bought 8 new automobiles from a manufacturer for $12,500 per unit. If they sell the vehicles during a special promotion for $16,265 each, and their operating expenses are 25% of their cost, what will be the amount of net profit realized from the sale of the 8 automobiles?

2. Commercial Security Systems purchases office safes at a cost of $1,500 each. Find the selling price of a safe if the company's operating expenses are 20% of its cost and its net profit must be 10% of cost.

3. What percent markup on cost is a retailer using if they buy a 36-bit color flatbed scanner for $95 and sell it for $149.99?

4. The Home Center sells an 18.2 cu. ft. refrigerator for $569. How much can the store pay the supplier for each unit if it must realize a 65% markup on cost?

5. Harmon's sells a seven-piece set of concentric air cookware for $89.99, which is 130% of cost. Find (a) the cost of each set and (b) the amount of markup.

6. After-Market Installation buys remote control car starter kits for $71.50. What is the selling price of the kits if the store's markup is 45% of the selling price?

7. Kimball's Drugs purchases its store brand deodorant from a supplier at $25.20 per case. Find the selling price per unit if each case contains 24 units and the store desires a 40% markup on selling price.

8. Determine the cost of a multimedia notebook computer that is sold for $1,299 if the selling price includes a 25% markup on selling price.

9. Power Discount Jewelry sells a gemstone bracelet for $120, which it buys for $40. Determine (a) the markup percent based on price and (b) the equivalent markup percent based on cost.

10. Arnold's Lighting buys kitchen ceiling lights from a supplier at a cost of $20 per fixture. If it desires a 40% markup on selling price and offers contractors a 10% trade discount on this model, find (a) the net selling price and (b) the list price required to cover the discount.

11. The Northside Girls and Boys Club bought 100 Christmas trees for resale as a fund raiser at a cost of $12 each. They estimate that 7 trees will not be sold due to various types of damage. Determine the selling price per tree if the club desires a 50% markup on selling price.

12. Electronics Unlimited bought 60 pagers at a cost of $20 each. The store plans to sell 40% of the pagers at a sale price of $25. Find the selling price of the pagers if the store needs a 75% markup on cost.

13. A used car was advertised in a local newspaper for $4,200 on March 20. The following week the same car was offered for sale at $3,800. The vehicle was sold on April 8 for $3,150. Calculate the percent of markdown on April 8.

14. Total Fitness buys a 2 HP treadmill for $115. The regular marked price of the treadmill is $179.99, but the store recently marked down the product 20% to reduce inventory. If the business's operating expenses are 35% of cost, find the amount of profit or loss on the sale of each treadmill.

15. Abbott's Hardware buys national-brand circular saws at a cost of $24.50 each. The hardware's operating expenses are 40% of selling price, and net profit is 15% of selling price. Determine (a) the selling price of the saw and (b) the maximum percent of markdown allowed without incurring an operating loss.

✍ EXPRESS YOUR UNDERSTANDING

Compose one or two well-written sentences to express the requested information in your own words.

1. Describe the effect markup has on an organization's ability to generate profits.

2. Identify the variables of the markup equation. Create an example to illustrate the relationship between each variable.

3. Describe in detail the two methods used to mark up merchandise.

4. Explain how you would determine the selling price of an item that was purchased for $5.00 and is marked up 30% on cost.

5. Describe how you would find the cost of an item that is sold for $29.95 and carries a markup of 50% on cost.

6. Explain how to determine the rate of markup based on cost if both the cost and selling price are known.

7. How would you find the cost of an item if you were given both the markup percent based on selling price and the selling price?

8. If you know the markup percent based on cost, how would you find the equivalent markup percent based on selling price?

9. Explain how a business would price an item when a portion of the items purchased is expected to spoil and a specified markup on cost must be realized.

10. Describe the financial impact of a reduced price that results in an absolute loss.

Mind Your Business

Media World's pricing strategy is based on a number of economic considerations. The prices the company sets for its merchandise are determined primarily by the products' cost, the company's overhead costs, profit criteria, and competition. You decided when the business was started to use the markup on selling price method to price all merchandise. Each product line has been assigned a different markup percent, and the company's financial plan requires an average gross margin of 55%.

Activity 1:

It is your responsibility to price all merchandise for sale. You are to prepare a pricing sheet that a sales associate will use to mark the prices on all merchandise received, as noted on National Music Warehouse's Invoice No. 392547 ("Mind Your Business" Chapter 5). The pricing sheet should include a brief description of the merchandise and the required marked price for each product category. Media World's markup is 60% on all compact discs and 45% on all cassette tapes. All prices are marked to the nearest $.99 of a dollar. For example, a calculated price of $5.69 would require a marked price of $5.99 and a calculated price of $7.49 would require a marked price of $6.99.

Activity 2:

You have made arrangements to attend a trade show in Boston for the purpose of identifying new products to sell at Media World, assessing the competition in the multimedia markets, and selecting more competitive suppliers of your product lines. As part of your preparation for the business trip, you want to develop an analysis that will determine the maximum purchase price you can pay for the following product line based on their respective target selling prices.

a. Full-length videotape movies that will be sold for $29.99 based on a markup of 30% of the selling price

b. Soft-cover "bestseller" novels, which will be sold for $9.95 based on a markup of 42% of the selling price

c. 10-pack, type II blank cassette tapes to be sold for $14.99 based on a markup of 27% of the selling price

d. Video games, CD-ROM, which will sell for $59.99 based on a markup of 55% of selling price

Activity 3:

A number of products purchased during the month of September will be sold at a reduced price as part of the store's merchandising strategy. The marked price of these products must be sufficient to obtain the store's required markup percent and at the same time allow a predetermined amount of the inventory to be sold at a reduced price.

a. 500 CDs are purchased at a cost of $5.50 each, of which 200 will be sold at a reduced price of $10.99 each. What marked price is required to realize a 60% markup on selling price?

b. Magazines are purchased from a supplier at a cost of $1.25 each. Media World's markup on this product is 40% of selling price, and customers receive a 30% sales discount. Determine the required marked price and reduced price based on this pricing strategy.

c. Video games are purchased at a cost of $26.80 ea. What would be the maximum percent of markdown without incurring an operating loss if Media World's operating expenses are 40% of selling price and net profit is 15% of selling price?

Activity 4:

You have made a number of important decisions involving the pricing of merchandise this month. The effect of these decisions on the financial position of the business needs to be analyzed, as you must advise your partners of any necessary changes. You are to analyze September's financial statements (Mind Your Business, Chapter 4) and report all information to your partners that resulted directly or indirectly from your pricing strategies.

SELF-TEST

A. Terminology

Complete the following items using the key terms presented at the beginning of the chapter. Check your responses against the answer key at the end of the test.

1. A _____ is added to an item's cost to cover operating expenses and profit.

2. Markup is also referred to as _____ and can be expressed as an amount or a percent.

3. When the markup amount is added to the cost, the result is referred to as the _____.

4. The rate of markup based on cost is found by dividing the _____ by the _____.

5. When we divide the percent markup on cost by 1 plus the markup percent on cost, we are converting the percent markup on _____ to its equivalent markup percent on _____.

6. A reduction in the marked price is called a _____.

7. _____ are merchandise that will spoil if not sold within a short period of time.

8. The _____ occurs when the reduced price is sufficient to cover cost and operating expenses.

9. If the reduced price is below the actual cost of the merchandise, _____ will occur.

10. The _____, when applied to the marked price, determines a reduced price that is sufficient to cover all costs.

B. Calculations

The following concepts and short problems are designed to test your understanding of the objectives identified at the beginning of the chapter. Answers are provided at the end of the test.

11. Plaza Photo and Camera Shop buys a camera for $85 that it sells for $123.25. If the shop's operating expenses are 35% of cost, find (a) the amount of the markup and (b) profit from the sale of a camera.

12. Charles Simari, International Rug Merchant, purchased four oriental rugs of the same value for a total cost of $4,800. Find the selling price per rug if the merchant is to realize a 40% markup based on cost.

13. Country Pride Pizza must reprice a two-topping pizza because of recent price changes. Per unit costs are as follows: 1 pound of dough $.85; 1 cup of sauce $.40; 8 ounces of cheese $1.25; toppings $.75 each. If the shop's operating expenses are estimated at 60% of the cost, and the required profit is 15% of the cost, what is the selling price of each pizza?

14. What is the most a retailer should pay for an item that will be sold for $350 if a markup rate of 60% on cost is required?

15. The Lighting Gallery reported net sales of $40,580 for June. If the store averaged a 30% markup on the selling price, what was (a) the gross margin amount and (b) the amount of net profit if operating expenses were 24% of net sales?

16. A department store buys men's dress shirts at $96 a dozen. If the store's markup rate is 20% based on selling price, what is the selling price of each shirt?

17. The Triangle Shoe Store wants to sell a line of boys' shoes for $54. How much should the store pay for the shoes if its operating expenses are 32% of the selling price and the profit is 8% of the selling price?

18. Raymond's Discount Store carries a line of ladies' watches that cost $35 each. The price tag on the watch case lists two different prices, a manufacturer suggested retail price of $85 and a discount price of $60. What is the markup percent based on (a) the manufacturer suggested retail price and (b) the discount price?

19. The markup percent on a gallon of interior latex paint is 20% of the selling price. What is the equivalent markup percent based on cost?

20. Koss Home Supply buys 10 × 10 oak kitchen cabinet sets from Reliable Cabinets at a cost of

$1,200 per set. All contractors are offered a contractors' discount of 30% off list. If Home Supply requires a 36% markup on selling price, what list price should be printed in their upcoming contractors' catalog?

21. A & B Markets bought six flats of blueberries at $7.60 per flat. If each flat contains 10-quart baskets of berries and approximately 5% of the purchase will spoil before it can be sold, what price should be marked on each quart to realize a markup of 50% on the cost?

22. The Teen Scene purchased six dozen Rugby shirts of assorted sizes at a cost of $264 per dozen. The store expects 25% of the total shirts purchased will be sold on sale for $28 each. What marked price should be placed on the shirt in order to make a 65% markup on total cost?

23. Green Acres Farms planted 5,000 six-plant flats of various types of perennial flowers to be sold to retail stores during the spring planting pe-

riod. Each flat of flowers cost $1.05 to grow. Past experience indicates that 4% of the flats will not be sold due to weather and growing conditions. What selling price per flat is required to realize a 20% markup on selling price?

24. During a clearance sale, Hartin's Furniture Store marked down its entire inventory 30%. Find the reduced price of a traditional living room set if the marked price was $1,499.

25. A camcorder with a marked price of $1,000 is marked down 40%. If the cost of the camcorder is $650, and the operating expenses are 28% of the cost, find (a) the operating loss and (b) the absolute loss.

26. A retailer purchases a product at a cost of $15.00 each. If the retailer's operating expenses are 40% of the selling price and net profit is 8% of the selling price, what is (a) the selling price of the product, and (b) the maximum percent of markdown allowed without incurring an operating loss?

7

Key Terms

interest
principal
interest rate
maturity value
exact interest
ordinary interest
future value
present value
finance charge
sinking fund

SIMPLE INTEREST

Learning Objectives

After completing the exercises in this chapter, you will be able to:

1. Calculate the dollar amount of interest using the simple interest formula.

2. Calculate the principal, rate, time, or interest when the other three quantities are known.

3. Understand the difference between ordinary interest and exact interest, and be able to calculate both.

4. Calculate the maturity value of a simple interest investment.

5. Calculate the future value of a simple interest investment.

6. Determine the interest earned when future value and principal are known.

7. Calculate the present value of a simple interest investment at a given time within the term of the investment.

8. Determine the finance charge for a simple interest loan.

9. Define key terms.

INTRODUCTION

It is unlikely that you will go through life without having either to borrow money or to lend it. Perhaps you have already borrowed money in order to go to college or to buy a car. As the owner of a retail business, you may have needed to borrow money to make improvements or to expand your business. When you borrow money from a bank or lending institution, the bank charges you a sum of money (this is called *interest* or a *finance charge*) for the use of the money for the length of time that you use it.

Interest may be either an income item or an expense item. If a business borrows money and *pays* interest for the use of it, then the interest paid is an *expense*. If the business loans money for use by someone else and *receives* interest, then, as we learned in Chapter 4, that interest is an *income* item.

There are two basic types of interest: simple interest and compound interest. In this chapter, we will study *simple interest*, which is interest earned on only the original principal invested. *Compound interest*, which is interested earned on principal and past interest, will be discussed in Chapter 9. Chapter 11 will familiarize you with some day-to-day situations in which individuals and businesses must involve themselves with interest.

7.1   SIMPLE INTEREST

Learning objective
Calculate the dollar amount of interest using the simple interest formula.

Interest is a sum of money paid or charged for the use of money; that is, for borrowing someone else's money. Simple interest is usually calculated for short-term loans rather than for long-term investments and loans. Simple interest is calculated using an expanded form of the "base, rate, and part" formula ($P = B \times R$). The formula for simple interest follows.

Formula for simple interest

interest = principal $\times$ rate $\times$ time

$I = P \times R \times T$

$I =$ interest, the amount charged for the use of money

$P =$ **principal,** the initial amount of money invested or borrowed

$R =$ annual **interest rate,** expressed as a percent

$T =$ length of time the principal is invested or borrowed, expressed in terms of years or a fraction of 1 year

Example 1

Kevin Robinson borrowed $1,200 for 1 year at a simple interest rate of 11.3%. How much interest did he pay?

Solution

P = amount borrowed = $1,200

R = rate = 11.3%, or .113 (decimal form)

T = time in years = 1 year

$I = P \times R \times T$

$I = \$1,200 \times .113 \times 1$

$I = \$135.60$

At the end of the year, Kevin was required to pay $135.60 interest.

Example 2

Sorensen's Bargain Books borrowed $50,000 from its banker at 12.5% simple interest for 3 years. How much interest will Sorensen's pay at the end of 3 years?

Solution

$P = \$50,000$

$R = .125$

$T = 3$ years

$I = P \times R \times T$

$I = \$50,000 \times .125 \times 3$

$I = \$18,750$

$18,750 interest is due at the end of 3 years.

Both examples express time in terms of full years. Terms of loans also occur in fractional parts of a year. To determine the fractional part of a year, T is divided by 1 for years, 12 for months, and 52 for weeks (T = number of months/12, T = number of weeks/52, etc.).

Example 3

The Social Services Department of Minnehaha County has received funding for a special project 9 months prior to when they need to expend the monies. They decide to invest the funds, $130,000, in a short-term

investment for 8 months at 12% simple interest. How much will the department earn on the investment?

Solution

$P = \$130{,}000$

$R = 12\%, \text{ or } .12$

$T = \dfrac{8 \text{ months}}{12 \text{ months}} = \dfrac{8}{12} \leftarrow \text{fractional part of 1 year}$

$I = P \times R \times T$

$I = \$130{,}000 \times .12 \times \dfrac{8}{12}$

$I = \dfrac{\$130{,}000}{1} \times \dfrac{.12}{1} \times \dfrac{8}{12}$

$I = \dfrac{\$130{,}000 \times .12 \times 8}{12}$

$I = \$10{,}400$

The interest earned is $10,400.

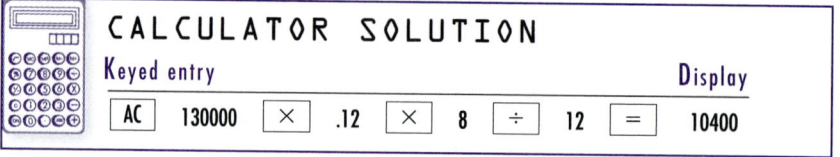

CALCULATOR   SOLUTION

Keyed entry   Display

| AC | 130000 | × | .12 | × | 8 | ÷ | 12 | = | 10400 |

At the end of the term (time period) for a simple interest investment or
loan, the total amount to be paid to the lender is the original principal *plus*
interest. The **maturity value** of a simple interest investment or loan is the
sum of the principal and interest as depicted in Figure 7.1. This maturity
value is also referred to as the *future value* of the investment or loan.

Figure 7.1

Maturity value of a
loan

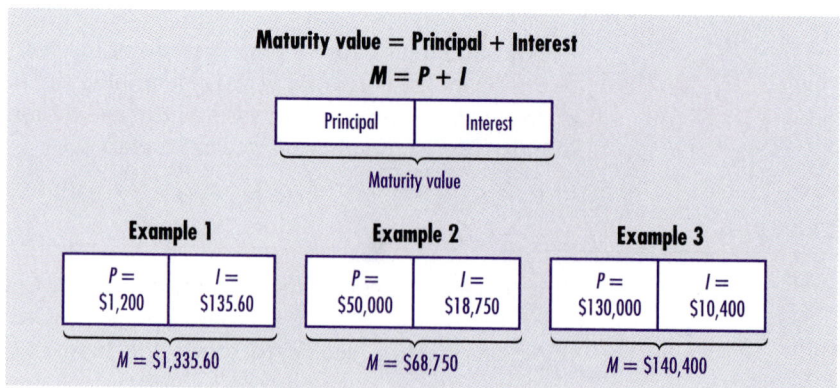

Maturity value = Principal + Interest
$M = P + I$

| Principal | Interest |

Maturity value

Example 1

| $P =$ | $I =$ |
| $1,200 | $135.60 |

$M = \$1{,}335.60$

Example 2

| $P =$ | $I =$ |
| $50,000 | $18,750 |

$M = \$68{,}750$

Example 3

| $P =$ | $I =$ |
| $130,000 | $10,400 |

$M = \$140{,}400$

Maturity value of a loan

Maturity value = principal + interest

$$M = P + I$$
$$M = P + (PRT)$$
$$M = P(1 + RT)$$

You may also see this formula written as $Fv = Pv(1 + RT)$, where Fv represents the *future value* (or maturity value) of the investment or loan, and Pv represents the *present value* (or principal) of the future value. The present value of money invested is its value at the time it is invested, and with time it grows to a future value, or maturity value. These are all terms that are commonly used in financial transactions.

Example 4

Jones and Johnson, Attorneys at Law, have received a $100,000 retainer fee to represent R & J Plumbing Supplies in any litigation that might develop in the next 3 years. Jones and Johnson invest the $100,000 at 11% simple interest for 1½ years. What will be the total value of the investment at the end of 1½ years?

Solution

total value = maturity value
maturity value = $P + (P \times R \times T)$

$P = \$100,000$

$R = .11$

$T = 1.5$ years

$M = P + (P \times R \times T)$
$M = \$100,000 + (\$100,000 \times .11 \times 1.5)$
$M = \$100,000 + \$16,500$
$M = \$116,500$

CALCULATOR SOLUTION

Keyed entry	Display
AC 100000 + (100000 × .11	
× 1.5) =	116500

CHECK YOUR KNOWLEDGE

Simple Interest

Determine the interest for each of the following (round answers to the nearest cent).

	Principal	Rate	Time	Interest
1.	$10,000	12%	5 years	_____
2.	$750	9%	6 months	_____
3.	$1,300	11.5%	13 weeks	_____
4.	$2,500	7¾%	78 weeks	_____

Determine the maturity value of each of the following:

	Principal	Rate	Time	Maturity value
5.	$7,500	10¼%	2 years	_____
6.	$3,000	8.75%	9 months	_____
7.	$4,000	12.68%	26 weeks	_____

Solve the following word problems. Round dollar amounts to the nearest cent and rates to the nearest hundredth of a percent.

8. Santiago Rubbish Removal borrowed $75,000 at 16% simple interest for 10 months. (a) How much interest will Santiago owe at the end of the term of the loan? (b) Find the total amount Santiago will have to pay at the end of the loan term.

9. Marla Bryant purchased a $7,000 certificate of deposit that paid 9% simple interest for 3 years. What will the value of the CD be at the end of the 3-year period?

10. J. M. Merrill invested $30,000 in a 5-month simple interest investment paying 13% per annum (per year). At the end of the 5-month term, she invested the entire maturity value in an account paying 10% simple interest for 7 months. What was the total value of her investments at the end of the 12-month period?

Answers to CYK: *1.* $6,000 *2.* $33.75 *3.* $37.38 *4.* $290.63 *5.* $9,037.50 *6.* $3,196.88 *7.* $4,253.60 *8.* a. $10,000; b. $85,000 *9.* $8,890 *10.* $33,469.79

7.1 EXERCISES

Find the simple interest (round to the nearest cent).

	Principal	**Rate**	**Time**	**Interest**
1.	$12,000	12%	8 months	_____
2.	$8,000	10.5%	10 months	_____
3.	$2,000	12.75%	1½ years	_____
4.	$4,500	11.5%	1 year	_____
5.	$10,000	13%	6 months	_____
6.	$2,350	7%	5 months	_____
7.	$7,750	9.5%	15 months	_____
8.	$12,450	7.75%	26 weeks	_____
9.	$3,345	8.75%	39 weeks	_____
10.	$17,875	13.8%	4 years	_____

Find the maturity value.

	Principal	**Rate**	**Time**	**Maturity value**
11.	$12,000	12%	8 months	_____
12.	$8,000	10.5%	13 weeks	_____
13.	$2,000	12.75%	2½ years	_____
14.	$4,500	11.5%	5 years	_____
15.	$10,000	13%	15 months	_____

Solve the following word problems. Round dollar amounts to the nearest cent and rate values to the nearest hundredth percent.

16. Cecilie Morganstein invested $5,000 in a simple interest account paying 9% per year. How much interest will she earn if she leaves her money in the account for 2½ years?

17. Jerry Moss wishes to take out a $7,500 loan at 14% simple interest for 7 months. (a) What is the interest that Jerry must pay at the end of 7 months? (b) What is the total amount that Jerry must repay to the bank at the end of 7 months?

18. Karen Arch wishes to take out a $17,200 loan at 7.9% simple interest for 1½ years. (a) What is the interest that Karen must pay at the end of the term of the loan? (b) What is the total amount that Karen must pay at the end of the term of the loan?

19. Jacksonville Technical College received $3,445,553 in state aid on September 15 for the fall academic semester. The vice-president for finance decided to invest $2,000,000 in a 2-month investment that pays 11.5% simple interest. How much interest will the college earn on the investment?

20. Serendipity County's Planning and Development Office received a $1.5 million grant for urban development. Realizing they would expend the funds somewhat uniformly throughout the year, the agency set aside 50% of the funds for the first 6 months' expenditures. The remainder of the funds were invested in short-term investments paying 15% simple interest per year. One half of these funds were invested for 4 months, and the remainder were invested for 8 months. (a) How much was set aside for the 8-month investment? (b) How much interest was earned on the 8-month investment? (c) What was the maturity value of the 4-month investment? (d) What was the total amount of interest earned on the two investments?

7.2   SOLVING FOR PRINCIPAL, RATE, AND TIME IN THE SIMPLE INTEREST FORMULA

Learning objective
Calculate the princi-
pal, rate, time, or in-
terest when the
other three quanti-
ties are known.

In Section 7.1, we used the simple interest formula to solve for the variable I, interest. We can use algebra to solve for the principal, rate, and time variables as well.

Example 5

If Sumi Komoto paid $42 in interest for a 7-month loan at a rate of 12%, what principal was borrowed?

Solution

$$I = P \times R \times T$$

$$\$42 = P \times .12 \times \frac{7 \text{ months}}{12 \text{ months}}$$

$$\$42 = .07P$$

$$\$\frac{42}{.07} = \frac{.07P}{.07}$$

$$\$600 = P$$

$$\text{principal} = \$600$$

If we solve P in the formula for simple interest, we can develop a formula for principal, too. Since the variable P is multiplied by ($R \times T$) we will use division, the opposite operation of multiplication, to isolate P on one side of the equal sign.

Formula for
principal

$$I = PRT$$

$$\frac{I}{RT} = \frac{P\cancel{R}\cancel{T}}{\cancel{R}\cancel{T}}$$

$$\frac{I}{RT} = P$$

or

$$P = \frac{I}{RT}$$

Principal equals interest divided by the product of rate and time; this formula supplied with the information in Example 5 would give us the same result:

$$P = \frac{I}{RT} = \frac{42}{(.12)\left(\frac{7}{12}\right)} = \$600$$

Example 6

Harrison Supply Co. invested $10,000 in a 3-month simple interest account that paid $400. What was the simple interest rate?

Solution

$$I = PRT$$

$$\$400 = \$10,000(R)\frac{3 \text{ months}}{12 \text{ months}}$$

$$\$400 = \$2,500R$$

$$\frac{\$400}{\$2,500} = \frac{\$2,500R}{\$2,500}$$

$$.16 = R$$

$$\text{rate} = 16\%$$

We can develop the formula for R (rate) the same way we developed the formula for P.

Formula for rate

$$I = PRT$$

$$\frac{I}{PT} = \frac{PRT}{PT}$$

$$\frac{I}{PT} = R$$

or

$$R = \frac{I}{PT}$$

Rate is equal to the interest divided by the product of principal and time. In Example 6, our solution by the rate formula would then be:

$$R = \frac{I}{PT} = \frac{\$400}{\$10,000(3/12)} = \frac{\$400}{\$2,500} = .16$$

Example 7

Jody Richardson paid $76.32 on an $1,100 loan at $9\frac{1}{4}\%$ simple interest. What was the length of the loan?

Solution

$$I = PRT$$
$$\$76.32 = \$1,100(.0925)T$$
$$\$76.32 = \$101.75T$$
$$\frac{\$76.32}{\$101.75} = \frac{\$101.75T}{\$101.75}$$
$$.75 = T$$
$$\text{Time} = \sqrt[3]{4} \text{ of 1 year}$$

We develop the formula for T (time) as follows:

Formula for term

$$I = PRT$$
$$\frac{I}{PR} = \frac{PRT}{PR}$$
$$\frac{I}{PR} = T$$

or

$$T = \frac{I}{PR}$$

The solution to Example 7 using the term formula then would be:

$$T = \frac{I}{PR} = \frac{\$76.32}{\$1,100(.0925)} = \frac{\$76.32}{\$101.75} = .75$$

Solving for $P, R,$ or T can be accomplished most easily by memorizing one formula, $I = PRT$, and then using algebra. However, you can also memorize all four formulas and simply choose the proper one for the task at hand.

$$I = PRT \qquad P = \frac{I}{RT} \qquad R = \frac{I}{PT} \qquad T = \frac{I}{PR}$$

 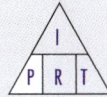 

The triangular figures are a pictorial device that can help you to remember these formulas. The shaded quantity is equal to the remaining products and quotients.

CHECK YOUR KNOWLEDGE

Solving for Principal, Rate, and Time

Solve for the remaining quantity in each (round dollar amounts to the nearest cent and rates to the nearest hundredth of a percent).

	Interest	Principal	Rate	Time
1.	_____	$1,500	13%	5 months
2.	$1010.63	_____	11%	15 months
3.	$200	$625	_____	4 years
4.	$262.50	$8,750	12%	_____

5. What percent simple interest is the borrower paying on a 2-year loan of $5,000, of which $990 interest is paid?

6. Martha's Flower Shoppe invested $2,000 in a simple interest investment paying 12½%, and it received $1,000 in interest. How long was the Shoppe's money invested?

7. Frank Farkwilder is retired and receives $4,950 every 6 months from an investment paying 11% simple interest. Frank receives all the interest generated by the account each 6 months and never takes any of the principal. What is the principal Frank has invested?

Answers to CYK: *1.* $81.25   *2.* $7,350.04   *3.*   8%   *4.*   .25 year, or 3 months   *5.*   9.9%
                 6.   4 years   *7.* $90,000

7.2   EXERCISES

Fill in the missing entries (round dollar amounts to the nearest cent and rates to the nearest hundredth percent).

	Principal	Rate	Time	Interest
1.	_____	13.1%	6 months	$393.00
2.	_____	14.0%	1 year	$2,478
3.	_____	4.8%	2 years	$216
4.	_____	9.5%	1½ years	$712.50
5.	_____	8.6%	9 months	$838.50
6.	$4,800	_____	11 months	$440.00
7.	$5,200	_____	1¼ years	$487.50
8.	$9,750	_____	12 months	$565.50
9.	$11,000	_____	13 months	$1,251.25
10.	$20,000	_____	1 year	$2,240.00
11.	$30,000	12%	_____	$7,200.00
12.	$33,000	14.1%	_____	$2,326.50
13.	$4,400	13.5%	_____	$594.00
14.	$7,800	11.7%	_____	$608.40
15.	$10,000	6.9%	_____	$1,035.00
16.	$4,000	5.6%	3 months	_____
17.	$6,750	7.8%	14 months	_____
18.	$3,000	10%	2 years	_____
19.	$5,500	11%	4 months	_____
20.	$9,000	7.0%	1 year	_____

Solve the following word problems. Round dollar amounts to the nearest cent and rates to the nearest hundredth of a percent.

21. If the simple interest on $15,000 was $555 for 6 months, what rate of interest was charged?

22. Ray Grey needs to borrow $10,000. The simple interest rate is 12.15%. If Ray can't afford to pay more than $1,200 for interest, what is the length of time that Ray can borrow the money?

23. Ricardo Rivera loaned his brother Carlos some money at 13% simple interest per year. At the end of 5 months, Ricardo had earned $1,110 in interest from the loan. How much did Carlos borrow from Ricardo?

24. Barney Casey borrowed $40,000 from his parents for 2 years. He paid them a total of $45,000 at the end of the 2-year term of the simple interest loan. What rate of interest did he pay his folks?

25. Carla Swerzik confided in a friend that she had invested some money in an investment paying 7.5% simple interest per year, and that she had earned $200 interest in just 1 year. The friend wanted to know how much was invested, but Carla would not reveal the amount. How much did Carla invest?

7.3 EXACT AND ORDINARY INTEREST

Up until now, time (T) in our simple interest formula has been represented in years, months, or weeks, which means that the denominator in the fraction used to determine T as a part of 1 year has been 1, 12, or 52. When time (T) is given in days, the denominator is either 365 or 360. **Exact interest** is the simple interest determined by dividing the number of days by 365. Exact interest using exact time is:

$$T = \frac{\text{number of days}}{365}$$

Ordinary interest is the simple interest determined by dividing the number of days by 360. Ordinary interest using ordinary time is:

$$T = \frac{\text{number of days}}{360}$$

Use of 360 as the denominator is often referred to as the *banker's rule,* because each month is considered to have 30 days ($30 \times 12 = 360$). Let's look at specific examples of each.

Example 8

Find the exact interest for a loan whose length is 120 days; the amount borrowed is $18,000; and the rate of simple interest is 8%.

Solution

$$I = PRT$$

$$I = \$18{,}000 \times .08 \times \frac{120 \text{ days}}{365 \text{ days}}$$

$$I = \frac{\$18{,}000 \times .08 \times 120}{365}$$

$$I = \$473.42$$

Note: Be careful in these calculations to multiply the numerator values together, then multiply the denominator values together, and last complete the division. If you choose to first find the decimal equivalent for 120/365 and substitute it into the equation, do not round the decimal or any other part of the calculation until the final step. Here, the decimal equivalent for the time fraction would be expressed in the following calculation:

$$I = \$18{,}000 \times .08 \times (120/365)$$

$$I = \$18{,}000 \times .08 \times (.3287671 \dots)$$

$$I = \$473.42466 \dots = \$473.42$$

Example 9

Find the ordinary interest for a loan whose length is 120 days; the amount borrowed is $18,000; and the rate of simple interest is 8%.

Solution

$$I = P \times R \times T$$

$$I = \$18,000 \times .08 \times \frac{120 \text{ days}}{360 \text{ days}}$$

$$I = \$480$$

When we compare Examples 8 and 9, we can see that the dollar amount of interest is higher for ordinary interest than it is for exact interest. Unless a problem specifically states that exact interest is to be used, you should calculate ordinary interest.

Calculating the Number of Days and Due Date

In the previous examples, the number of days of the loan was given. In some instances, the *dates* of a loan are given, and then you have to calculate the number of days of the loan. Either of the two methods we discussed in Section 5.3, the days-of-the-month method or the days-in-a-year method, can be used to determine the exact number of days.

Example 10

If a loan is issued on March 4 and is due on July 15, find the length of this loan.

Solution

First let's use the days-of-the-month method:

March	27 days of loan in March $(31 - 4 = 27)$
April	30 days of loan in April
May	31 days of loan in May
June	30 days of loan in June
July	15 days of loan in July
total days of loan	133 days

Second, we will use Table 5.3 on page 207, which tells the numerical day of each date of the year. For this example, we see that July 15 is the 196th day of the year and that March 4 is the 63rd day of the year. The difference, $196 - 63$, is 133 days. When using Table 5.3, be sure to note whether the calculation includes the month of February during a leap year; this would require 1 additional day to be added to the calculation.

For Your Information

THE INFLUENCE OF INTEREST RATES ON INVESTMENT

Economists study the relationships between different types of investments. Banks and other lending institutions loan money and charge interest. They in turn pay interest to their depositors for the use of their money. That is one way that consumers can invest their money and make it grow for them. Another way people can invest is to buy stock in a company and earn interest on their investment in the form of dividends, or make money on the appreciated value of the stock when they sell it. When interest rates on loans are high, consumers have little money to spend, either on consumer goods or on investments. When interest rates on loans decrease, there is more money in the pockets of consumers, businesses, and the government to be spent in the commercial marketplace. The accompanying graph shows an economic model of the increase in investment, which can be stimulated by the lowering of interest rates.

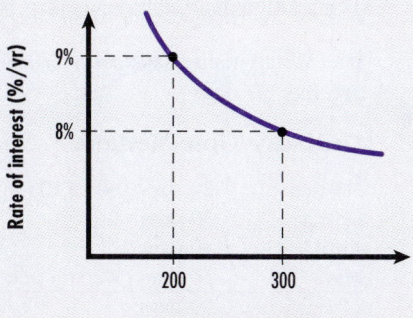

If interest rates decrease, then consumers have more money to spend because they are not spending it on their own loan debt. If they are not satisfied with the lower interest that they are earning in their bank, then they will look for other investments to purchase, such as stocks and bonds. Classical economic theory states that lower interest rates will cause an increase in investments, and higher interest rates will result in less money invested in stocks.

Example 11

Calculate the due date (a) using exact time for a 90-day loan made on February 10, 2000, and (b) using ordinary time.

Solution

a. Since 4 divides into 2000 evenly with no remainder, 2000/4 = 500, it follows that 2000 was a leap year and that February 2000 had 29 days.

Days-of-the-Month Method:

February	19 days (29 − 10 = 19)
March	31 days (50 days to March 31)
April	30 days (80 days to April 30)
May	10 days (we need 10 days in May)
total	90 days

Therefore, May 10 is the due date.

Days-of-a-Year Method:

February 10 is the 41st day.
Add 90 to 41 to arrive at the 131st day.
Note that February has 29 days in this leap year so we subtract 1 more day, to give us the 130th day of the year.
The 130th day in Table 5.3 is May 10, the due date.

b. When using *ordinary time,* we assume each month, including February, has 30 days.

Ordinary Time Method:

February	20 days (30 − 10)
March	30 days
April	30 days
May	10 days (We still need 10 days in May)
	90 days

The due date is May 10.

CHECK YOUR KNOWLEDGE

Exact and Ordinary Interest

Solve the following problems. Round dollar amounts to the nearest cent and rates to the nearest hundredth of a percent.

	Principal	Rate	Time	Exact interest	Ordinary interest
1.	$2,500	12%	90 days	_____	_____
2.	$13,000	9%	250 days	_____	_____
3.	$48,000	10%	400 days	_____	_____

	Date issued	Date due	Number of days
4.	June 7, 2000	November 19, 2000	_____
5.	January 23, 2000	March 30, 2000	_____
6.	February 6, 1999	April 6, 1999	_____

	Date issued	Number of days	Due date (exact time)
7.	November 12, 2000	60	_____
8.	July 23, 1999	120	_____
9.	January 17, 2000	240	_____

10. A loan of $4,400 was issued at a simple interest rate of 11½% on June 14 and is due on September 2. Determine (a) the ordinary interest, (b) the exact interest, and (c) the difference of the ordinary minus the exact interest.

7.3 EXERCISES

Find the exact interest (round answers to the nearest cent).

	Principal	Rate	Time	Exact interest
1.	$12,500	10.4%	120 days	_____
2.	$1,200	11.6%	45 days	_____
3.	$4,700	8.8%	360 days	_____
4.	$6,600	14.5%	365 days	_____
5.	$800	16.0%	280 days	_____

Answers to CYK: **1.** $73.97; $75 **2.** $801.37; $812.50 **3.** $5,260.27; $5,333.33 **4.** 165 **5.** 67 **6.** 59 **7.** Jan. 11, 2001 **8.** Nov. 20, 1999 **9.** Sept. 13, 2000 **10.** **a.** $109.63; **b.** $110.90; **c.** $1.27

Find the ordinary interest (round answers to the nearest cent).

	Principal	Rate	Time	Ordinary interest
6.	$6,600	14.5%	365 days	_____
7.	$800	16.0%	280 days	_____
8.	$2,000	9.9%	70 days	_____
9.	$8,000	10.5%	100 days	_____
10.	$11,750	12.7%	60 days	_____

Determine the length of the loan that is issued on each given date and due on the date indicated, using exact time.

	Date issued	Date due	Number of days
11.	January 12, 1999	April 12, 1999	_____
12.	August 23, 1999	November 15, 1999	_____
13.	March 30, 2000	September 30, 2000	_____
14.	December 15, 1999	April 15, 1999	_____
15.	January 25, 2000	May 1, 2000	_____

Calculate the due date using exact time for the following loan dates and terms.

	Date issued	Number of days	Due date
16.	January 22, 1999	60	_____
17.	August 25, 2000	120	_____
18.	May 10, 2000	240	_____
19.	November 15, 1999	30	_____
20.	February 5, 2000	45	_____

Determine the exact interest. (Assume the year is *not* a leap year. Round answers to the nearest hundredth of a percent.)

	Principal	Rate	Dates of loan	Interest
21.	$100,000	8.9%	Jan. 1 to May 1	_____
22.	$1,500	12.3%	June 11 to Aug. 2	_____
23.	$20,000	14.0%	Feb. 15 to July 31	_____
24.	$7,500	14.7%	Apr. 4 to Nov. 3	_____
25.	$15,000	11.0%	May 16 to Dec. 27	_____

Solve the following word problems. Round dollar amounts to the nearest cent and rates to a hundredth of a percent.

26. Sarai Sherman agreed to deposit $4,450 in an account paying 16% simple interest per year for 60 days. If she made the deposit on February 25,

determine (a) the date of the end of the term of the investment and (b) the ordinary interest Sarai will earn.

27. Marshall Peters borrowed $3,000 at 20% simple interest per year for 90 days. If he borrowed the money on June 1, determine (a) the due date of his loan based on exact interest and (b) the amount he must repay.

28. Elvis Jones has $10,000 he wishes to invest for 18 months (548 days). He could invest the money in Alpha Bank for 13% exact simple interest or in Beta Bank for 13.5% ordinary interest. (a) Determine how much interest he would earn at Alpha Bank. (b) Determine how much interest he would earn at Beta Bank.

7.4 FUTURE VALUE AND PRESENT VALUE OF SIMPLE INTEREST

Learning objective
Calculate the maturity value of a simple interest investment.

Earlier in this chapter we learned that if we invested an amount of money P into a simple interest account at a rate R for T years, then the interest earned would be $I = PRT$ and the total value of the investment would be

> Maturity Value = Principal + Interest

We also referred to the maturity value as the future value of the investment and the principal as the present value of the investment. These two terms, *future value* and *present value,* have more extensive meanings when we consider investments and loans.

If we consider a simple interest investment of $1,000 at 10% simple interest for three years, we can calculate the maturity value of the investment at the end of three years to be:

$$M = P(1 + RT) = \$1,000 \times .10 \times 3 = \$1,300$$

We also realize that the investment is earning $100 each year. So as each year passes in the investment, the investment grows from $1,000 to $1,100 at the end of the first year, to $1,200 at the end of the second year, to $1,300 at the end of the third year. The future value of this investment is $1,300. **Present value** indicates the value of the investment at different times throughout the 3-year time frame of the investment. The present value of the investment at the beginning is $1,000. At the end of one year, the present value is $1,100. At the end of two years the present value is $1,200, and finally at the end of three years the full future value of $1,300 is realized. **Future value** is the value of an investment at the end of the investment time frame. We will use the symbols Pv and Fv to indicate present value and future value, respectively.

A time line of investment might look like this:

0 years	1 year	2 years	3 years
$1,000	$1,100	$1,200	$1,300
↑	↑	↑	↑
present value at the time of investment	present value at the end of 1 year	present value at the end of 2 years	present value at the end of 3 years = future value

Learning objective
Calculate the future value of a simple interest investment.

The value of money invested changes as time passes. This is generally referred to as the *time value of money.*

Example 12

Arnie McElroy has invested $1,700 for 3 years in a certificate of deposit that pays 11% simple interest. What will be the future value of Arnie's investment at the end of three years?

Solution

The present value of Arnie's investment is $1,700, and the future value will be $1,700 plus the interest earned.

$$Fv = \text{Maturity value} = P(1 + RT)$$
$$Fv = \$1{,}700(1 + (.11 \times 3)) = \$1{,}700 \times 1.33 = \$2{,}261$$

The interest earned on Arnie's investment can be determined two ways:

1. $I = PRT = \$1{,}700 \times .11 \times 3 = \$561$

or

2. Since $Fv = P + I$, solving this equation for I gives us

$$I = Fv - P$$
$$I = \$2{,}261 - \$1{,}700 = \$561$$

Learning objective
Determine the interest earned when future value and principal are known.

This approach to determining interest, Interest earned equals future value minus the principal invested, is a very useful method.

Example 13

Consider Arnie McElroy's investment in Example 12, and determine the present value of the investment at the end of (a) one year, (b) one and one-half years, and (c) two years.

Solution

a. We can use the formula $M = P(1 + RT)$ to determine the present value at the end of one year:

$$M = P(1 + RT)$$

Present value = $Pv = M = \$1{,}700 \times (1 + (.11 \times 1)) = \$1{,}887$
The present value of the investment at the end of one year is $1,887.

Learning objective
Calculate the present value of a simple interest investment at a given time within the term of the investment.

b. Similarly, the present value of the investment at the end of one and one-half years is
Present value = $Pv = M = P(1 + RT) = \$1{,}700(1 + (.11 \times 1.5))$
                                         $= \$1{,}980.50$

c. The present value of the investment at the end of two years would be
Present value = $Pv = M = \$1{,}700(1 + (2 \times .11)) = \$2074$

Loans can be viewed in a similar manner, since loans are simply an investment made by the lender of the loan.

Example 14

Learning objective
Determine the finance charge for a simple interest loan.

Michelle Dieu borrowed $1,500 to buy a new computer at 9% simple interest for the 18 month loan. (a) What is the future value of Michelle's loan? (b) How much finance charge will she pay?

Solution

a.   The $1,500 that Michelle borrowed is the present value of the loan, and the future value is $1,500 plus interest that she will pay.

$$\text{Future value} = M = P(1 + RT)$$
$$Fv = \$1,500(1 + (.09 \times 1.5))$$
$$Fv = \$1,500(1 + .135)$$
$$Fv = \$1,702.50$$

b.   **Finance charge** means the *amount of interest* that Michelle will pay for her loan.

$$\text{Finance charge} = I = PRT$$
$$\text{Finance charge} = \$1,500 \times .09 \times 1.5 = \$202.50$$

or

$$\text{Finance charge} = I = Fv - P$$
$$\text{Finance charge} = \$1,702.50 - \$1,500 = \$202.50$$

Example 15

Viet Ho expects to purchase a new refrigerator in two years to replace and upgrade the current refrigerator in his restaurant. He expects the new refrigerator to cost about $2,000. How much must Viet invest *now* (present value) in a simple interest account paying 10% for two years so that the investment will be worth $2,000?

Solution

In this case, Viet is saving ahead for a future purchase of $2,000. When someone creates an investment such as this, it is referred to as a **sinking fund.** The $2,000 is the future value of Viet's investment fund, and we need to determine the present value of that $2,000 in this investment. The present value is simply the principal that Viet must invest in order to have the desired future value of $2,000.

$$M = P(1 + RT)$$

or in this case, we could use the notation

$$Fv = Pv(1 + RT)$$

Solve for Pv:

$$\frac{Fv}{(1 + RT)} = \frac{Pv\cancel{(1 + RT)}}{\cancel{(1 + RT)}}$$

$$Pv = \frac{Fv}{(1 + RT)}$$

$$Pv = \frac{\$2,000}{(1 + (.10 \times 2))} = \frac{\$2,000}{1 + .2} = \frac{\$2,000}{1.2} = \$1,666.67$$

Example 16

In Chapter 2 we referred to the Consumer Price Index (CPI), which tracks the rate of inflation in our economy on a year-by-year basis. Assume the cost of living increased at a rate of 3% from 1997 to 1998.

a.   How much would you expect to have to pay for an item in 1998 that cost $100 to purchase in 1997?

b.   How much would you have expected to pay in 1997 for an item that had a cost of $100 in 1998?

Solution

a.   In 1997 the present value of the item was $100, and we wish to determine the future value in 1998.

$$M = P(1 + RT)$$
$$Fv = Pv(1 + RT)$$
$$Fv = \$100(1 + (.03 \times 1))$$
$$Fv = \$103$$

b.   In 1998 the cost of the item was $100, and we wish to determine the value a year earlier, so the Fv is $100 and the Pv is the value in 1997.

$$M = P(1 + RT)$$
$$Fv = Pv(1 + RT)$$
$$\$100 = Pv(1 + (.03 \times 1))$$
$$Pv = \frac{\$100}{1.03} = \$97.09$$

So an item valued at $100 in 1997 would have been expected to cost $103 in 1998, just due to inflation. Inflation, as indicated in Example 16, has the effect on us individually that we can purchase less with the same amount of money. Consumers in 1998 needed $3 more for every $100 to have the same purchasing power that they had in 1997. We can put this in different terms by considering a person's salary from one year to the next.

Example 17

Maggie Provost earned $20,000 in 1997 and received a 2% raise in 1998. If the cost of living increased by 3% from 1997 to 1998, determine (a) how much Maggie needed to earn in 1998 so that she would have the same purchasing power, (b) how much she did earn in 1998, and (c) how much loss of purchasing power she realized from 1997 to 1998.

Solution

a.   Maggie would need a 3% increase in earnings just to stay even with inflation, so the future value of her $20,000 earnings needed to be

$$Fv = M = P(1 + RT) = \$20,000(1 + (.03 \times 1)) = \$20,600$$

b.   Maggie's 2% increase in earnings resulted in earnings of

$$M = P(1 + RT) = \$20,000(1 + (.02 \times 1)) = \$20,400$$

c.   Maggie's loss of purchase power is $20,600 − $20,400 = $200

As we continue our study of business finance, we will consider both the investment of money and the borrowing of money, as indicated by the previous examples. Generally, the simple interest investment involves depositing a "lump sum" of money into an account, and letting it earn interest without adding any other money to the principal (present value). The sum of the principal and interest (future value) is a larger "lump sum" of money.

In general terms, we can think of two different forms of a simple interest investment or a simple interest loan. In each case, a lump sum investment or loan is made, and it results in a new and larger lump sum, which is the principal plus interest.

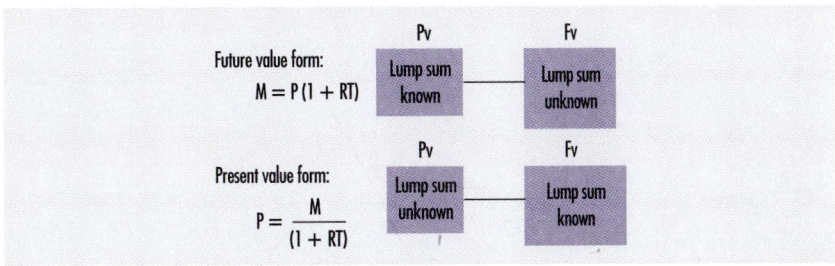

We have given these phrases a very straightforward and simple meaning in this section. We will enhance that meaning as we proceed with our study of finance. Notice that there is no single formula for future value or present value for simple interest. The context of the problem determines the formula to be used in that problem.

CHECK YOUR KNOWLEDGE

Future Value and Present Value of Simple Interest

1. What is the future value of a simple interest investment of $3,000 for 5 years at 9%?

2. How much interest is earned in an investment with a principal of $4,000 and a future value of $5,200?

3. What is the present value of a simple interest investment you have just made at 10% for 4 years, if its future value is $4,480?

4. How much is the finance charge for a simple interest loan that has a principal of $2,500 and a maturity value of $3,000?

5. How much is the interest or finance charge for a simple interest loan of $2,000 at 7% for 18 months?

6. What is the value of a $6,000, 4-year simple interest investment at a rate of 8% at the end of the first year?

7. What is the value of a $5,000, 6-year simple interest investment at a rate of 7% at the end of the second year?

8. What is the value of a $3,000, 5-year simple interest investment at a rate of 9% at the end of the fourth year?

9. What is the present value of a 10-year simple interest investment at a rate of 12% that will have a total value of $10,000?

10. How much interest is earned on a $6,000, 4-year simple interest investment at a rate of 8%?

Answers to CYK: *1.* $4,350   *2.* $1,200   *3.* $3,200   *4.* $500   *5.* $210   *6.* $6,480
7. $5,700   *8.* $4,080   *9.* $4,545.45   *10.* $1,920

7.4   EXERCISES

Complete the following chart by filling in the blanks:

	Present value	Rate	Time	Interest	Future value
1.	$1,000	5%	2 years	_____	_____
2.	$3,000	8%	15 months	_____	_____
3.	$2,700	11.5%	4.5 years	_____	_____
4.	$10,000	10%	_____	$3,000	_____
5.	$9,000	8%s	_____	_____	$11,880
6.	$2,500	6%	_____	_____	$2,875
7.	$4,000	_____	2 years	_____	$4,800
8.	$20,000	_____	3 years	_____	$24,800
9.	$500	_____	1/2 year	_____	$520
10.	_____	10%	3 years	$600	_____
11.	_____	9%	3.5 years	$1,260	_____
12.	$3,500	8%	_____	_____	$4,900
13.	$7,000	_____	8 years	_____	$13,720

Solve each of the following problems.

14. Anna Cavanaugh loaned her friend Jason $1,000 for 6 months at 6% simple interest. What is the future value of the loan, and how much finance charge will Jason pay?

15. Tionn Washington invested $500 in a simple interest investment that pays 10% for 4 years. Tionn likes to keep track of the value of the investment every six months. What is the value of Tionn's investment at the end of 18 months?

16. Arky Johnstone is saving to purchase a new all-terrain vehicle in 3 years. If the ATV is expected to have a total cost of $2,000, how much must Arky invest now in a simple interest account paying 12% in order to have the necessary amount?

17. Sid Soukey's parents invested $4,000 in a simple interest account paying 8% 10 years ago. Sid expects to go to a community college 2 years from now. How much is the account worth now? How much will it be worth in 2 years? If Sid decides to take only the interest earned out

of the account 2 years from now, how much will he get?

18. Kara Krazewski borrowed $6,000 from her parents to buy a car for commuting to Seneca Community College. Her parents agreed that she could pay interest only on the 6% simple interest loan each month while she was a college student. How much must she pay her parents each month? If she sells the car for $6,000 and repays her parents in full at the end of $2\frac{1}{2}$ years of college, how much finance charge will she have paid?

19. Frank Beckworth invested $2,000 2 years ago in a 4-year certificate paying 8% simple interest, and 1 year ago he invested $1,500 in a separate 5-year certificate paying 10% simple interest. What is the current total value of his investments in the two certificates?

20. Carla Morris currently earns $40,000 as a marketing representative in an advertising agency. She suspects that the cost of living will increase by 4% next year. What salary will she have to earn to stay even with inflation?

CHAPTER REVIEW EXERCISES

The following problems review all mathematics and business concepts presented in the chapter. If you experience any difficulty solving a problem, go to the appropriate chapter section and review the examples provided. Express all answers as decimals rounded to the nearest hundredth, or dollars rounded to the nearest cent, unless otherwise instructed.

1. Determine the interest earned on a simple interest account of $700 at 15% for 30 weeks.

2. Determine the simple interest rate if the interest is $243 on a principal of $1,800 for 18 months.

3. Determine the time necessary for $3,200 to earn $1,600 interest at a 10% simple interest rate.

4. Determine the principal that earns $243.75 simple interest at 13% in 30 months.

5. How long will it take for $100 investment to double in value at 9% simple interest?

6. Determine the ordinary interest on a simple interest investment of $1,500 at 13% for 4 months, assuming each month has 30 days.

7. Determine the exact interest on a simple interest investment of $1,500 at 13% interest, invested from June 1 through September 30.

8. Determine the interest paid on a loan principal of $500 if the total payoff of the loan was $580.

9. Determine the future value of a simple interest investment of $35,000 at 15% for 4 years.

10. Determine the value at the end of the fourth year of a $10,000 investment at 20% simple interest for 6 years.

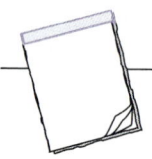

EXPRESS YOUR UNDERSTANDING

Compose one or two well-written sentences to express the requested information in your own words.

1. Explain how you would determine the interest rate in a simple interest problem wherein $320 was earned on a principal of $4,000 in 8 months.

2. Explain what is meant by the *future value* in a simple interest problem.

3. Describe the most important difference between *ordinary interest* and *exact interest.*

4. Describe the steps necessary to determine the exact number of days from January 15 to March 1, inclusive.

5. Explain what is meant by *present value* in a simple interest problem.

6. Explain what is meant by the *finance charge* in a loan.

7. Explain how you would determine the finance charge of a loan for which you knew the principal borrowed and the maturity value.

8. Explain how you would determine the principal of a simple interest loan of 10% for 2 years if you knew its maturity value was $3,600.

9. Explain how you would determine the length of a simple interest investment if you knew the interest earned, the principal invested, and the interest rate.

10. Explain how you would determine your increased buying power if you earned 10% more this year than last year, and the cost of living increased 4% in that same time period.

Mind Your Business

About midway through the month of September, Matt Ryan, one of your partners in Media World, asked you and your other partner Carol Kinney if he could borrow $10,000 from the business for about 2 months. He suggested the possibility of a simple interest loan at 9% and paying the loan off (principal plus interest) at the end of the term. He explained his request by stating that the 9% would be more than the business could earn on other types of short-term interest, and it would be cheaper for him than the 12% personal loan rate that he would be charged at a bank—or an even higher rate if he borrowed through his credit card.

Activity 1

Write up an agreement between the business and Matt Ryan that clearly delineates the principal amount, interest rate, the maturity value of the loan, and the maturity date of the loan (assuming it is a 60-day loan beginning on September 10).

Activity 2

Develop an analysis that describes the impact that the loan to Matt Ryan would have on the business presented in the September balance sheet and income statement.

SELF-TEST

A. Terminology

Complete the following items using the key terms presented at the beginning of the chapter. Check your responses against the answer key at the end of the test.

1. The amount of money initially invested or borrowed is called the _____.

2. Money charged for the use of money is called _____.

3. Principal plus interest is the _____, denoted M.

4. If you use a 360-day calendar year when calculating interest, you are calculating _____ interest.

5. Using a 365-day calendar year for the calculation of interest results in _____ interest.

6. Principal plus interest in an investment is sometimes called the _____ of the investment.

7. The interest charged for a loan is also called the _____ of the loan.

8. An expression for value of an investment at some specific time during the term of the investment in a simple interest investment is called the _____ of the investment.

9. An investment designed to have a future value sufficient to purchase a specific item in the future is called a _____ fund.

B. Calculation

Solve each of the following problems. Round dollar amounts to the nearest cent and rates to the nearest hundredth of a percent. Answers are provided at the end of the test.

10. What is the exact interest for a $1,000 loan from January 14 to June 28, 1999, at a 9³/₄% rate of simple interest?

11. What is the principal of an investment if the interest received on a 6-month 8% simple interest loan is $89?

12. What percent interest is the investor earning on a 2-year loan of $5,000 if $990 is the interest earned? (Round to the nearest tenth of a percent.)

13. Calculate the ordinary interest on a simple interest investment paying 10.5% interest on a principal of $1,200 from January 12, 2000, to May 12, 2000.

14. Jacob S. Latter borrowed $2,500 from Abe Long at 12% simple interest for 2 years and 6 months. If Jake paid back the loan in full at the end of the loan period, how much did he pay Abe in total?

15. In Problem 14, how much finance charge did Jake pay for his loan from Abe?

16. Sally Potts borrowed $800 at 8% and paid $96 simple interest on the loan. How long did she have the loan?

17. What is the future value of a $20,000 simple interest investment at 11% for 4 years?

18. What is the present value of a $20,000 simple interest investment at 11% for 4 years at the end of 3 years?

19. Bobby Mandally borrowed $450 for 3 months and paid back a total of $475. What was the finance charge on the loan?

20. What was the simple interest rate on the loan made to Bobby Mandally in Problem 19?

Answers to Self-Test: 1. principal   2. interest   3. maturity value   4. ordinary   5. exact
6. future value   7. finance charge   8. present value   9. sinking
10. $44.08   11. $2,225   12. 9.9%   13. $52.50   14. $3,250   15. $750
16. 1.5 years   17. $28,800   18. $26,600   19. $25   20. 22.2%

Key Terms

simple discount
discounting
discount rate
proceeds
bank discount
maturity value
true rate of interest
promissory note
simple interest note
face value
simple discount note
partial payment of a
   loan
U.S. rule

SIMPLE DISCOUNT

Learning Objectives

After completing the exercises in this chapter, you will be able to:

1. Calculate the simple discount using the simple discount formula, figure the amount of the proceeds, and determine the maturity value of a discounted loan.

2. Calculate the true interest rate for a loan that has been discounted.

3. Compare a simple interest loan and a simple discount loan to determine which is better.

4. Determine the face value, interest, and maturity value of a simple interest note.

5. Determine the face value, discount, proceeds, and maturity value of a simple discount note.

6. Determine the proceeds of a discounted simple interest note.

7. Use the U.S. Rule to calculate the balance of the principal of a loan when a partial payment has been made on the loan.

8. Define key terms.

INTRODUCTION

Through our study of simple interest in Chapter 7, we learned that when you take out a loan the initial *amount borrowed is called the principal* and the *amount due is called the maturity value,* which is the principal plus interest.

The concept of interest, money paid for the use of someone else's money, is an essential concept of the process of conducting many business transactions. People need money in order to do business. If they do not have the money they need, then they need to borrow someone else's money. People who have money to loan—*lenders*—are in the business of having their money work for them to earn more money. Lenders want to earn the highest interest in the shortest time that they can. Borrowers want to use other people's money for as long as they can and pay as little interest as possible. Consequently, people are continually trying to negotiate different ways to work with interest.

In this chapter, we will look at the concept of the *simple discount* and the process of *discounting.* We will build on the concepts that we learned in Chapter 7 to compare simple interest and simple discount, and loans involving both of these concepts. We will learn about promissory notes and the process of one lender selling a note to another lender at a discounted price. We will calculate the earnings that each of the lenders will receive, and the interest rates that their discounted loans have generated for them. Finally, we will study partial payments on loans, and the effect that a partial payment has on the amount of interest that a borrower should pay for that loan.

8.1   SIMPLE DISCOUNT AND TRUE INTEREST RATE

Learning objective
Calculate the simple discount using the simple discount formula, figure the amount of the proceeds, and determine the maturity value of a discounted loan.

The concept of simple discount is similar to the concept of simple interest, except that the terminology is different. We will look at a list of definitions before exploring the formula for discounting.

Proceeds: The actual dollar amount of money that the borrower receives or takes home from the lender

Bank discount: The interest that is collected at the *beginning* of the loan

Maturity value: The proceeds plus the bank discount

Sometimes banks or other lending institutions collect the interest from a loan at the beginning of the loan rather than at the end of the loan. The process of deducting interest at the beginning of a loan is called **discounting.**

Formulas for simple discount

discount = maturity value $\times$ discount rate $\times$ time

$$D = M \times R \times T$$

where

M = maturity value

R = simple discount rate

T = time, or length of loan in terms of years

D = discount or interest

proceeds = maturity value − discount

$$P' = M - D$$
$$P' = M - (MRT)$$
$$P' = M(1 - RT)$$
$$P' = \text{proceeds}$$

The symbol P' is read "P prime."

Example 1

Lynette Cushing needed a loan before she could go to college. Lynette went into the bank and asked to borrow $7,000. The rate was 11% and the length of the loan was for 1 year. Assuming the bank discounted the loan, how much money will Lynette actually receive from the bank?

Solution

Lynette has asked for $7,000, which will be the maturity value. We calculate the simple discount on $7,000 to be

$D = MRT = \$7,000 \times .11 \times 1$ (which is the same form as the simple interest formula)

$D = MRT = \$770$

Now this discount (interest) is *subtracted* from the maturity value to give us the proceeds.

$P' = M - (MRT) = \$7,000 - \$770 = \$6,230$

The maturity date is 1 year after Lynette borrowed the money. At that time, Lynette must pay the bank $7,000 (proceeds plus discount).

Figure 8.1

Figure 8.1

Maturity value —
simple discount

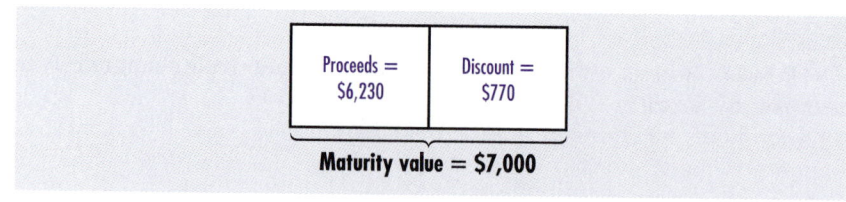

| Proceeds = $6,230 | Discount = $770 |

Maturity value = $7,000

Notice the similarity between Figure 8.1 and Figure 7.1 in Chapter 7. Since a bank discount is simple interest deducted from the maturity value to arrive at the proceeds, $M = P + I$ and $M = P' + D$ are very similar.

True Interest Rate

Learning objective
Calculate the true interest rate for a loan that has been discounted.

When working with simple discount, we need to be careful not to confuse the discount rate with the simple interest rate. In Example 1, the bank discount of $770 was based on $7,000 maturity value multiplied by the rate of 11%. However, we must remember that Lynette did not have the use of $7,000 but, rather, $6,230. So, to find the **true rate of interest** we must compare the proceeds ($6,230, the actual amount of money borrowed) with the dollar amount of the bank discount ($770). We can compute the simple interest rate that this discount of $770 represents of the actual amount borrowed (proceeds) by using the simple interest formula:

$I = PRT$

$I = \text{interest} = \text{discount amount} = \$770$

$P = \text{principal} = \text{proceeds} = \$6,230$

$T = \text{time} = 1 \text{ year}$

$\$770 = (\$6,230)R(1)$

$\dfrac{\$770}{\$6,230} = R$

or

$R = \dfrac{\$770}{\$6,230} = .1235955 \ldots = .1236 \text{ or } 12.36\%$

This rate of interest is called the true rate of interest of a simple discount loan. We can develop the formula for a true rate of interest as follows: $I = PRT$; for a discount loan,

$I = \text{discount interest} = D$

$P' = \text{proceeds}$

$T = \text{time}$

$$R = \text{true interest rate}$$

$$D = P'RT$$

$$\frac{D}{P'T} = \frac{P'RT}{P'T}$$

$$R = \frac{D}{P'T}$$

Formula for true rate of interest

$$\text{true rate of interest} = \frac{\text{discount interest}}{\text{proceeds} \times \text{time}}$$

To solve Example 1 using the formula, we would start with the formula for true rate of interest and end with the same answer.

$$R = \frac{D}{P'T}$$

$$R = \frac{\$770}{(\$6,230)(1)} = .1236$$

It is not uncommon to think of the "simple discount rate" as the "simple interest rate" of the loan. However, we must note that they are not the same, even though they are used in a similar way. Remember that the base for the simple discount rate is the *maturity value* of the loan, whereas the base of the simple interest rate is the *principal* of the loan.

CHECK YOUR KNOWLEDGE

Simple Discounts and True Interest Rate

Determine the discount for the following simple discount loans (round rates to the nearest hundredth of a percent):

	Maturity value	Rate	Time	Discount
1.	$10,000	9%	1 year	_____
2.	$3,850	11.5%	8 months	_____
3.	$17,500	8%	40 weeks	_____

Determine the proceeds for the following simple discount loans:

	Maturity value	Rate	Time	Proceeds
4.	$6,700	9%	2 years	_____
5.	$21,500	10.5%	78 weeks	_____
6.	$73,450	8%	2.5 months	_____

Determine the true interest for each of the following simple discount loans.

	Maturity value	Rate	Time	Discount	Proceeds	True interest rate
7.	$2,350	15%	1 year	$352.50	$1,997.50	_____
8.	$875	11.5%	8 months	$75.47	$799.53	_____
9.	$3,940	13%	18 months	$768.30	$3,171.70	_____

Solve the following problems. Round dollar amounts to the nearest cent and rates to the nearest hundredth of a percent.

10. Sammy Phong borrowed $14,000 at 10% for 10 months. Assuming the bank discounted the loan, determine: (a) the dollar amount of the discount, (b) the dollar amount of the proceeds, and (c) the dollar amount of the maturity value.

11. Determine the true interest rate (i.e., the simple interest rate) that Sammy paid for the loan.

12. Brenda Lemke borrowed $6,700 on January 7, 1999, at a simple discount rate of 11.5% for 6 months. Assuming the loan was a simple discount loan, determine (a) the dollar amount of the discount, (b) the dollar amount of the proceeds, (c) the maturity value, (d) the maturity date, and (e) the true interest rate.

8.1 EXERCISES

Determine the dollar amount of the discount and the proceeds (round answers to the nearest cent).

	Maturity value	Discount rate	Time	Discount amount	Proceeds
1.	$11,000	15.9%	9 months	_____	_____
2.	$15,000	13.9%	15 months	_____	_____
3.	$900	12.0%	4 months	_____	_____
4.	$7,600	8.0%	11 months	_____	_____

Answers to CYK: *1.* $900   *2.* $295.17   *3.* $1,076.92   *4.* $5,494   *5.* $18,113.75   *6.* $72,225.83   *7.* 17.65%   *8.* 14.16%   *9.* 16.15%   *10.* a. $1,166.67; b. $12,833.33; c. $14,000   *11.* 10.91%   *12.* a. $385.25; b. $6,314.75; c. $6,700; d. July 7, 1999; e. 12.20%

	Maturity value	Discount rate	Time	Discount amount	Proceeds
5.	$12,000	7.7%	2 years	_____	_____
6.	$4,100	4.8%	2.5 years	_____	_____
7.	$2,200	10.0%	30 months	_____	_____
8.	$1,300	9.9%	48 weeks	_____	_____
9.	$6,300	9.0%	39 weeks	_____	_____
10.	$10,900	10.0%	15 months	_____	_____

Determine the maturity value that will produce the proceeds as shown (round to the nearest cent).

	Maturity value	Discount rate	Time	Proceeds
11.	_____	10.0%	15 months	$10,000
12.	_____	13.0%	20 months	$40,000
13.	_____	14.0%	24 months	$5,000
14.	_____	15.0%	11 months	$20,000
15.	_____	16.0%	1.25 years	$7,000
16.	_____	16.0%	1.5 years	$40,000
17.	_____	12.0%	5 months	$33,000
18.	_____	13.0%	40 weeks	$16,800

Find the proceeds to the nearest cent, and the true interest rate to the nearest hundredth of a percent.

	Maturity value	Discount rate	Time	Proceeds	True rate
19.	$40,000	16.0%	36 months	_____	_____
20.	$35,000	11.1%	25 months	_____	_____
21.	$28,000	12.2%	8 months	_____	_____
22.	$25,000	18.0%	2 years	_____	_____
23.	$22,000	14.4%	7 months	_____	_____
24.	$19,000	13.3%	10 months	_____	_____
25.	$17,000	15.5%	4 months	_____	_____
26.	$5,650	9.5%	2.25 years	_____	_____

Solve the following word problems. Round dollar amounts to the nearest cent and rates to the nearest hundredth of a percent.

27. Carrie Lewis borrowed $23,500 at 13% for 9 months. Assuming the bank discounted the loan, determine (a) the amount of the discount, (b) the amount of the proceeds, and (c) the maturity value of the loan.

28. Penny Jenkins borrowed $34,765 on January 30, 1999, at a simple discount rate of 10% for 18 months. For the loan determine (a) the proceeds, (b) the maturity date, and (c) the true interest rate of the loan.

29. Syd Sentlowicz was offered a discount loan with a maturity value of $15,000 for 2 years. He would actually receive proceeds of $12,000. What

discount rate would Syd have accepted if he accepted the loan?

30. New York Trade Shows needs to borrow $15,000 for 6 months in order to set up its next trade show. How much must the company ask to borrow from a bank that offers a simple discount loan at 9.5% in order to actually receive $15,000?

8.2   COMPARING SIMPLE INTEREST AND SIMPLE DISCOUNT LOANS

We can now compare the two types of loans we have studied thus far, simple interest loans and simple discount loans. Recall that *interest is money paid for the use of someone else's money.* The three factors that are important to calculating how much interest is to be paid are principal, interest rate, and time. The following illustration compares two loans from two different banks and provides a means of analyzing the better offer.

Let's consider Jake Monroe, a beauty shop owner, who needs a $5,000 loan for 1 year. Jake asks two different banks, Ace Bank and Barnaby Bank, to consider loaning him money.

Ace Bank's Offer.   Ace Bank offers Jake a simple interest loan of $5,000 at 12% per annum for 1 year. The interest would be

Learning objective
Compare a simple interest loan to a simple discount loan to determine which is better.

$$I = \$5,000 \times .12 \times 1 = \$600$$

and the maturity value would be

$$M = P + I = \$5,600.$$

Barnaby Bank's Offer.   Barnaby Bank offers Jake a discount loan with maturity value of $5,000, discount rate of 12%, and a time period of 1 year. The discount amount would be:

$$D = MRT = \$5,000 \times .12 \times 1 = \$600$$

and the proceeds would be

$$P' = M - D = \$4,400.$$

In each case, Jake asks for a $5,000 loan, but in the discount case, he receives less than requested. We also know that because Jake really would have only $4,400 in hand from Barnaby Bank, the true interest rate that he would pay on the loan would be:

$$\text{true interest rate} = \frac{D}{P'T} = \frac{\$600}{\$4,400 \times 1} = 13.64\%$$

What if Jake needs the full $5,000?

Example 2

If Jake Monroe decided that he preferred to do business with Barnaby Bank, and that he really needed the full $5,000 to work with for the year, how much should he request from Barnaby Bank to have proceeds of $5,000?

Solution

We can use algebra to answer this question.

$$\text{proceeds} = M - D = M - (MRT)$$
$$\text{proceeds} = M(1 - RT)$$
$$\$5,000 = M(1 - .12 \times 1) = .88M$$

or

$$.88M = \$5,000$$
$$M = \frac{\$5,000}{.88} = \$5,681.82$$

Check:

$$\text{Proceeds} = M - D = \$5,681.82 - \$5,681.82 \times .12 \times 1$$
$$\text{Proceeds} = \$5,681.82 - \$681.82 = \$5,000.$$

Consequently, Jake should ask for a $5,681.82 loan at a discount rate of 12% in order to receive the needed $5,000.

It would seem that considering the higher interest rate (13.64% rather than 12%) and higher interest payment (the discounted loan would cost $81.82 more), a discounted loan would not be what Jake would want in this case.

Example 3

Sharilyn Wysocki needs to borrow $8,000 for six months to help her pay some unexpected operating costs that she incurred in her Fantastic Nails Salon. Crystal City Finance Corp. offered to loan her the money at 9% simple interest for 6 months, and Business First Finance Corp. offered her a loan of $8,000 discounted at 8% for 6 months. Compare the two loan opportunities.

Solution

Crystal City offer: This is a simple interest loan, so

$$I = PRT = \$8,000 \times .09 \times .5 = \$360$$
$$M = P + I = \$8,360$$

Business First offer: This is a simple discount loan, so

$$D = MRT = \$8,000 \times .08 \times .5 = \$320$$
$$P' = M - D = \$7,680$$
$$\text{True interest rate} = D/P'T = .0833 \text{ or } 8.33\%$$

In order to actually receive $8,000 Sharilyn would need to borrow a maturity value of

$$M = P'/(1 - RT) = \$8,000/(1 - .08 \times .5) = \$8,333.33$$

So, in this case, Sharilyn would pay a lower true interest rate with the discounted loan, and would pay $26.67 less in interest than if she took the simple interest loan.

Clearly each loan situation needs to be analyzed before any decision can be made as to which is the better business decision.

Obviously the simple interest loan is the more straightforward loan and most easily understood. Consequently, you may not be confronted with having to make a decision between which kind of loan to choose. Discounting, on the other hand, is a common business process.

Have you ever wondered how a company can stay in business when it offers you the opportunity to buy merchandise with no money down and no interest paid for one full year?

Example 4

Randall and Farnsdale Furniture Company decided to offer its customers the opportunity to purchase furniture and pay nothing down and no interest for one year. Michael and Shandra Jones purchased $1,000 worth of furniture from Randall and Farnsdale, and filled out an application for financing that would allow them to pay for the furniture with monthly payments over a 3-year period, beginning 1 year from the date of purchase of their furniture. The interest, also called the *finance charge*, on the agreement is $300 for the 3-year loan. Their application was approved, they signed a loan contract, and they had the furniture delivered to their home. How is it possible for Randall and Farnsdale to have so much of their money tied up in furniture that is sitting in its customers' homes for a full year?

Solution

Obviously, unless Randall and Farnsdale are incredibly overburdened with excess cash, the furniture company cannot afford to have its money out of circulation for that long.

Randall and Farnsdale have a contract from Michael and Shandra in which they promise to pay for their furniture in the future. If they pay off the entire debt 1 year from purchase, they will only pay the purchase price of the furniture. If they pay off the debt according to the loan agreement, they will pay the purchase price plus the interest on the loan.

Therefore, the contract that Randall and Farnsdale has with Michael and Shandra has monetary value of at least $1,000, and could have a value of $1,300. Wright Investors Inc. offers to purchase the finance contract from Randall and Farnsdale at a 15% discount on the $1,000 value. So Wright agrees to purchase the contract for $850.00.

$$D = MRT = \$1,000 \times .15 \times 1 = \$150$$
$$P' = M - D = \$1,000 - \$150 = \$850$$

Now Michael and Shandra have their furniture for 1 year at no cost. Wright Investors can earn at least $150 on its investment, and perhaps $450. Randall and Farnsdale has received $850 for the furniture that it sold to Michael and Shandra. Presumably, Randall and Farnsdale paid less than $850 for the furniture, so it has made money on the sale as well. Everybody appears to be a winner in this sale.

Transactions such as the one in Example 4 are common in the business community, so understanding the mathematics of the situation is very important. We will explore these kinds of business transactions more carefully in the next section.

CHECK YOUR KNOWLEDGE

Comparison of Simple Interest and Simple Discount Loans

Viet Nguyen wished to borrow $10,000 for 1 year to purchase new equipment for Southeast Asian foods specialty market. Merchants Bank offered him a 9% simple interest loan for 1 year, and Faldo Commercial Bank offered him a 1 year simple discount loan at a 9% discount.

1. Calculate the maturity value of the loan offered by Merchants Bank.

2. What is the maturity value of the loan offered by Faldo Commercial Bank?

3. How much interest will be paid on each loan?

4. Calculate the proceeds that Viet will receive from Faldo Commercial Bank.

5. Calculate the true interest that Viet would pay on the Faldo loan.

6. If Viet needs the full $10,000, calculate the maturity value of the loan from Faldo that would provide him with the full proceeds of $10,000.

Anna Garcia needed to borrow $8,000 for 15 months. She approached two banks inquiring about loans. Aldo National Bank offered her a simple interest loan at a rate of 13%. Bonzai International Bank offered her a simple discount loan with a 12% simple discount rate.

7. Determine the interest and maturity value of the Aldo Bank offer.

8. Determine the discount amount and proceeds of the Bonzai offer assuming a maturity value of $8,000.

9. Determine the true interest rate of the Bonzai loan.

10. Determine how much Anna would have to borrow from Bonzai in order to have $8,000 in proceeds.

Answers to CYK: *1.* $10,900 *2.* $10,000 *3.* $900 *4.* $9,100 *5.* 9.89% *6.* $10,989.01
7. $1,300; $9,300 *8.* D = $1,200 P = $6,800 *9.* 14.12% *10.* $9,411.76

8.2   EXERCISES

Sven Anderson inquired about a $15,000 loan for 1 year to repair a heating and cooling system in his small manufacturing facility. Sydney Savings and Loan offered him a 10% simple interest loan for 1 year, and Magnuson Canal Trust Bank offered him a 1-year simple discount loan at a 10% discount.

1. Calculate the maturity value of the loan offered by Sydney Savings.

2. What is the maturity value of the loan offered by Magnuson Trust?

3. How much interest will be paid on each loan?

4. Calculate the proceeds that Sven will receive from Magnuson Trust.

5. Calculate the true interest rate that Sven would pay on the Magnuson loan.

6. If Sven needs the full $15,000, calculate the maturity value of the loan from Magnuson that would provide him with the full proceeds of $15,000.

Javier Castro wanted to borrow $4,000 for 2 years. He called two banks about loans. El Paso Savings Bank stated that they have a simple interest loan at a rate of 15%. The Lone Star Commercial Bank offered him a simple discount loan with a 13% simple discount rate.

7. Determine the interest and maturity value of the El Paso Bank loan.

8. Determine the discount amount and proceeds of the Lone Star offer.

9. Determine the true interest rate of the Lone Star bank loan.

10. How much would Javier have to borrow from Loan Star in order to have $4,000 in proceeds?

Samuel Smart wants to borrow $3,800 for 10 months. First National Bank offers him a 12% simple interest loan. Second International Bank offers him an 11% simple discount loan.

11. Determine the maturity value of the loan at First National.

12. Determine the proceeds of the loan at Second International.

13. Determine the true interest rate of the loan at Second International.

14. What amount must Samuel request to borrow from Second International Bank in order to actually receive $3,800 in proceeds?

15. What will be the true rate of interest for the loan in exercise 14?

Mountain Road Auto Parts needs $50,000 to redesign its store front and showroom. Mountain Road's bank offers a simple discount loan at an 11% discount for 2 years, or it will also offer a 13% simple interest loan for 2 years.

16. How much must Mountain Road borrow in order to actually receive $50,000 if it opts for the discounted loan?

17. How much interest would Mountain Road pay on the discounted loan at the maturity value that you determined in exercise 16?

18. What would the true interest rate be on the discounted loan with the maturity value that you determined in exercise 16?

19. How much interest would Mountain Road pay on the simple interest option offered by the bank?

20. Which loan would seem to you to be the better value?

8.3   PROMISSORY NOTES

A **promissory note** is a written promise by a borrower (*maker*) to pay a sum of money (*face value*) to a designated person or bearer (*payee*) of the note, at a set time or *on demand*. A promissory note is similar to a personal check. When a person—let's call him Abe—writes a personal check to a second person—let's call her Barb—the following transactions occur. First, Abe deposits cash in his bank (called Abe's bank) where he has a checking ac-

Figure 8.2

Two-party check transaction

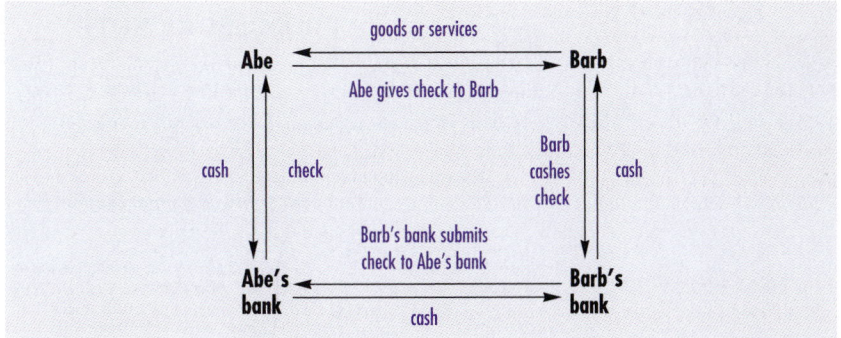

count. Abe then writes a check to Barb on an official check supplied by his bank for part or all of the money he deposited. Barb takes the check, which she accepted instead of cash, to her bank, which gives her the designated amount of cash on the check. Barb's bank now sends the check to Abe's bank and receives the cash from that bank. Figure 8.2 depicts this transaction. A promissory note is a *negotiable instrument* similar to a check, except that it cannot be cashed until the designated date on the note, and it usually earns interest.

For Your Information

PROMISSORY NOTES

Many legal forms, including negotiable instruments such as promissory notes, can be purchased in office supply stores. A sample of such a form for a promissory note is shown below. In many states these blank promissory notes can be filled out by the parties involved in a loan, and the completed form becomes a legal document, and perhaps even a negotiable instrument that can be purchased and sold. The parts of this promissory note are listed below, with a letter designation that corresponds to the designation in the note shown on page 316.

a. Amount, date, and location of the loan agreement between the lender(s) and borrower(s)

b. Lender, amount of loan, interest rate, and any specified terms of the loan

c. Terms if payments are late, or if the borrower defaults on the loan

d. The promise and any conditions associated with it

e. Borrower signatures witnessed by two other people

f. Witnessed signatures of anyone else guaranteeing that the loan will be paid

A293-10
R293-04
BB240

PROMISSORY NOTE

$ 3,000   Dated: July the first   2000   (year)

Principal Amount three thousand dollars State of Euphoria

a.

FOR VALUE RECEIVED, the undersigned hereby jointly and severally promise to pay to the order of

Jessica Sweet Enterprises, Inc.

the sum of three thousand

Dollars ($3,000.00   ), together with interest thereon at the rate of 8 % per annum on the unpaid balance. Said sum shall be paid in the manner following: 1/8 of the principal plus the interest on that portion of the principal accrued to the date of that payment shall be paid every three months from the date above for a term of two years.
    Payment shal be made by cash, certified bank check, or cashier's check payable to Jessica Sweet Enterprises, Inc.

b.

    All payments shall be first applied to interest and the balance to principal. This note may be prepaid, at any time, in whole or in part, without penalty. All prepayments shall be applied in reverse order of maturity.

    This note shall at the option of any holder hereof be immediately due and payable upon the failure to make any payment due hereunder within   10   days of its due date.

c.

    In the event this note shall be in default, and placed with an attorney for collection, then the undersigned agree to pay all reasonable attorney fees and costs of collection. Payments not made within five (5) days of due date shall be subject to a late charge of   15 % of said payment. All payments hereunder shall be made to such address as may from time to time be designated by any holder hereof.

    The undersigned and all other parties to this note, whether as endorsers, guarantors or sureties, agree to remain fully bound hereunder until this note shall be fully paid and waive demand, presentment and protest and all notices thereto and further agree to remain bound, notwithstanding any extension, renewal, modification, waiver, or other indulgence by any holder or upon the discharge or release of any obligor hereunder or to this note, or upon the exchange, substitution, or release of any collateral granted as security for this note. No modification or indulgence by any holder hereof shall be binding unless in writing; and any indulgence on any one occasion shall not be an indulgence for any other or future occasion. Any modification or change of terms, hereunder granted by any holder hereof, shall be valid and binding upon each of the undersigned, notwithstanding the acknowledgment of any of the undersigned, and each of the undersigned does hereby irrevocably grant to each of the others a power of attorney to enter into any such modification on their behalf. The rights of any holder hereof shall be cumulative and not necessarily successive. This note shall take effect as a sealed instrument and shall be construed, governed and enforced in accordance with the laws of the State first appearing at the head of this note. The undersigned hereby execute this note as principals and not as sureties.

d.

Signed in the presence of:

Witness *Jerod Minskt*
Witness Jerod Minskt

Witness *Jacob Arno*
Witness Jacob Arno

Borrower *Symantha Lynx*
Borrower Samantha Lynx

Borrower *Symantha Lynx*
Borrower Samantha Lynx

e.

GUARANTY

    We the undersigned jointly and severally guaranty the prompt and punctual payment of all moneys due under the aforesaid note and agree to remain bound until fully paid.

In the presence of:

Witness *Sylvia Soffietti*
Witness Sylvia Soffietti

Witness

Guarantor *Michael Lynx*
Guarantor Michael Lynx

Guarantor

f.

Source: Courtesy of E-Z Legal Forms.

Figure 8.3

A simple interest note

```
$4,050.00                    Brocksburg, OR 97215        Oct. 1, 1999
60 days                                    AFTER DATE  I  PROMISE TO PAY TO
THE ORDER OF _____ E. R. Hawley Construction Co. _____
         Four thousand fifty and 00/100 _____ DOLLARS
PAYABLE AT    Brocksburg First National Bank
VALUE RECEIVED WITH INTEREST AT 10.5% per annum.

No. 35                        Signed  Arn Svedman
```

Learning objective
Determine the face value, interest, and maturity value of a simple interest note.

Simple Interest Note

A promissory note that includes interest added to the face value is called a **simple interest note.** Figure 8.3 shows a typical format for a simple interest note.

Example 5

With the information given on the note in Figure 8.3, calculate (a) the due date, (b) the ordinary interest, and (c) the maturity value at the due date.

Solution

a. The term of the note is 60 days. From October 2 through October 31 is 30 days; from November 1 through November 30 is 30 days; therefore, November 30 is the due date. By the *days of the year method,* October 1 is the 274th day plus 60 days is the 334th day, which is November 30.

b. $I = PRT$
$I = \$4{,}050 \times .105 \times 60/360$
$I = \$70.88$

c. $M = P + I$
$M = \$4{,}050 + \$70.88 = \$4{,}120.88$

The **face value** of this note is its present value of $4,050, the interest is $70.88, and the **maturity value** or future value of this note is $4,120.88. Notice that the maturity value of a simple interest note is *not* the same as the face value. Maturity value equals face value plus interest.

Learning objective
Determine the face value, discount, proceeds, and maturity value of a simple discount note.

Simple Discount Note

Next, let's consider a **simple discount note** in which the face value and maturity value are the same.

Example 6

Tom Anderson received a discounted loan from Solvay Bank for $9,000 at a 13% discount for 8 months. He agreed to sign a simple discount note for a face value of $9,000 due in 8 months (Figure 8.4). Determine (a) the proceeds Tom received and (b) the due date of the note.

Solution

a. The face value of the simple discount note is also the maturity value.

Step 1: Determine the discount amount.

$$\text{discount} = MRT$$
$$= 9{,}000 \times .13 \times 8/12$$
$$= \$780$$

Step 2: Determine the proceeds.

$$\text{proceeds} = \text{maturity value} - \text{discount}$$
$$= \$9{,}000 - \$780$$
$$= \$8{,}220$$

b. The due date of the note will be 8 months after May 15, 1999. Since May is the fifth month of the year, and $5 + 8 = 13$, the due date will be in the first month of 2000. The due date will be January 15, 2000.

We refer to notes that have interest added (such as the simple interest note in Example 5) as *interest bearing notes*. The maturity value of an interest bearing note is the face value plus interest. Notes like the simple discount note in Example 6 are called *non-interest bearing notes*. The maturity value of a non-interest bearing note is the same as the face value of the note. Non-interest bearing notes may have been issued as simple discount

Figure 8.4

A simple discount note

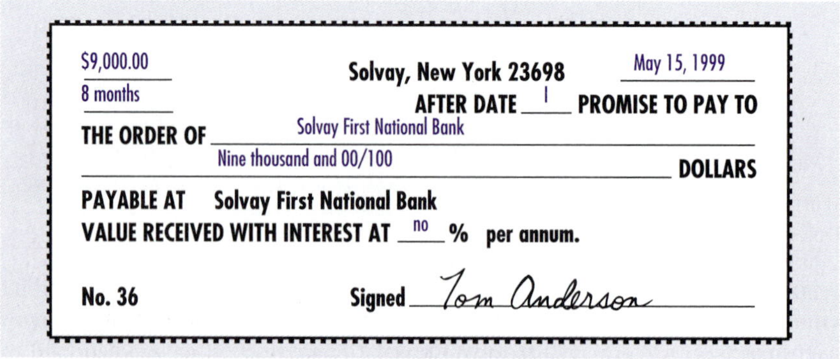

notes, or they may have been issued for the full principal with no interest charged either in the form of simple interest or a simple discount.

Discounting a Note

Promissory notes are negotiable instruments that have a specific monetary value at a specified future date. When a note is sold or traded prior to its maturity date, the note's value is usually discounted. Consider the following illustration.

E. T. Steele Wholesalers sold $10,000 worth of school supplies to Norton's Notions and Books with a cash discount of 5/10, *n*/30. After 20 days, Norton realizes that he cannot pay the invoice within the 30-day limit, and he asks Steele for an extension. Steele considers Norton to be a preferred customer and wants to continue to do business with him, but is concerned about his inability to pay on time. Rather than offer Norton a simple extension, Steele asks Norton for a promissory note due in 90 days for the full $10,000. To show good faith, Steele offers to charge no interest for this short-term loan. Norton goes to his bank and obtains a promissory note for $10,000 at 0% interest that is due in 90 days, and presents it to Steele. (See Figure 8.5).

Figure 8.5

A promissory note

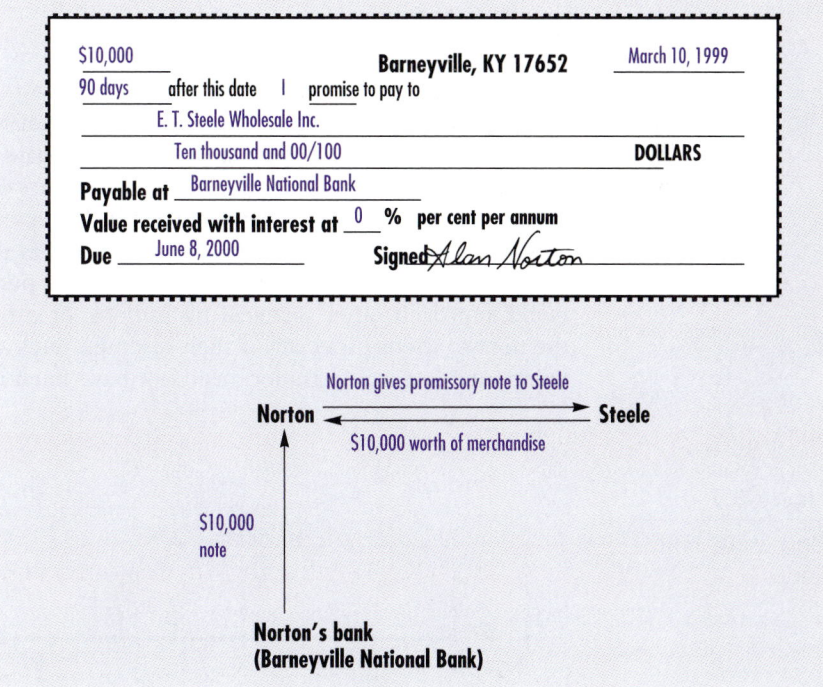

Figure 8.6

Discount period
time line

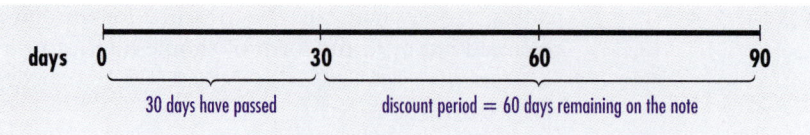

Learning objective
Determine the
proceeds of a dis-
counted simple
interest note.

After 30 days have passed, Steele discovers that she is short of cash and needs to liquidate some of her assets. Steele takes the promissory note to her bank and asks the bank to purchase the note, which has a $10,000 face value. Since the bank wants to make money on the transaction, it negotiates with Steele to buy the note from Steele at a 15% discounted rate; that is, Steele's bank offers to buy the note from her for

$$\text{proceeds} = M - MRT = \$10,000 - \$10,000 \times .15 \times 60/360$$
$$P = \$10,000 - \$250 = \$9,750$$

Notice that the discount time here is 60 days, as shown in Figure 8.6. Steele's bank has now purchased a note that will be worth $10,000 for $9,750. This transaction is shown in Figure 8.7.

Steele's bank will receive $10,000 from Norton 60 days later, at the end of the original 90-day period; her bank will earn $250 on the transaction, which is a true interest rate of

$$R = \frac{250}{9,750 \times 60/360} = .1538, \text{ or } 15.38\%$$

Figures 8.8 and 8.9 show diagrammatically what has transpired.

Finally, Norton makes sure at the end of 90 days that there are sufficient funds in his bank to cover the note, and Steele's bank cashes in the note with Norton's bank and receives $10,000.

The use of the discount method here provides the bank and Steele with a definite method for negotiation of the sale and purchase of the note. Steele could approach other financial institutions, or other individuals, and offer the note to them, especially if their discount rates are lower. The two banks mentioned in this illustration need not have been involved at all in the ex-

Figure 8.7

Note value time
line

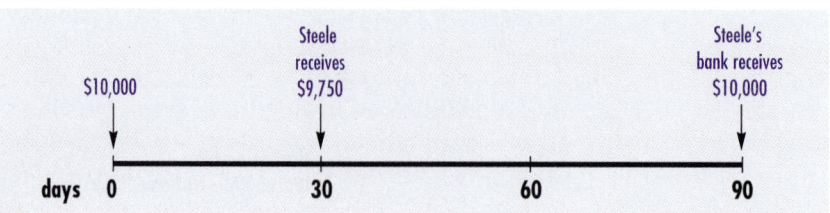

Figure 8.8

Transaction between Steele and her bank

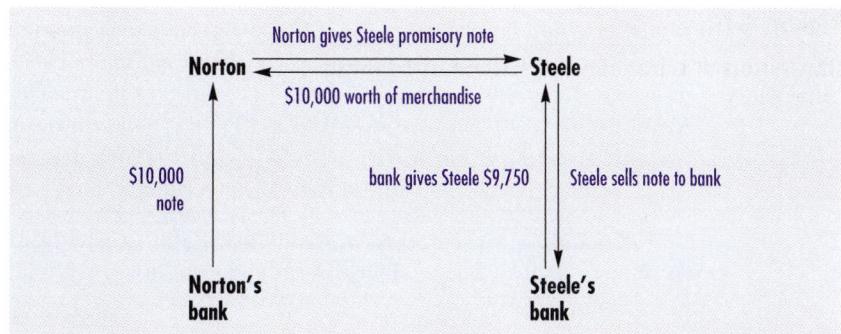

Figure 8.9

Transaction between Steele's bank and Norton's bank

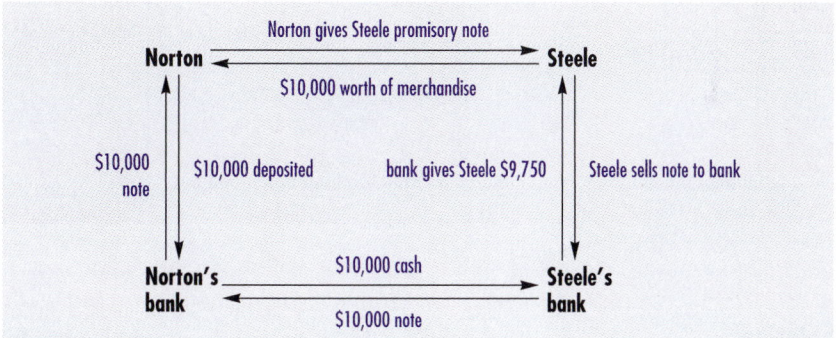

change of money. Once the promissory note was obtained from the bank, the remaining transactions could have involved only the two parties, Norton and Steele.

Now let's look at an example of discounting a simple interest note.

Example 7

Emilio Johnson borrowed $50,000 from Ben Simpson. Emilio gave Ben a promissory note stating that he would pay Ben $50,000 plus 13% simple interest 9 months from the date of the note. After holding the note for 5 months, Ben sold the note to Janet Whitaker at a 16% discount. Determine: (a) the discounted amount Ben received for the note, (b) Ben's net gain or loss, (c) the amount Janet received for the note, (d) Janet's net gain or loss, (e) the true interest rate Ben earned, (f) the true interest rate Janet earned, (g) the amount Emilio paid, and (h) the value of the note at the time Janet bought it. Refer to Figure 8.10, which illustrates these transactions.

Figure 8.10

Illustration of transaction described in Example 7

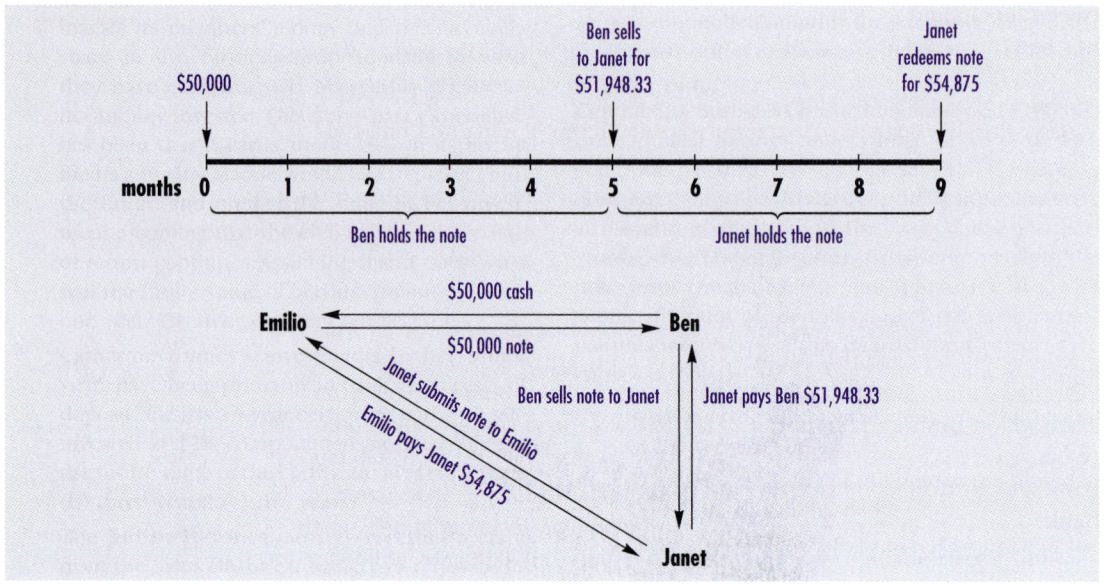

Solution

a.   To determine the discounted amount (proceeds), we must complete two steps:

   1.   Determine the maturity value of the note.

   2.   Discount the total maturity value to determine the proceeds.

Step 1:   The maturity value of the simple interest note is

$$\text{maturity value} = P + I$$
$$= P + PRT$$
$$= \$50{,}000 + (\$50{,}000)(.13)(9/12)$$
$$= \$54{,}875$$

Step 2:   Ben received the proceeds when the note was discounted on the maturity value of $54,875. Ben has held the note for 5 months, so there are 4 months remaining in the term of the loan. Hence, the discount period is 4 months, and $T = 4/12$.

$$\text{proceeds} = M - D$$
$$= \$54{,}875 - (\$54{,}875)(.16)(4/12)$$
$$= \$51{,}948.33$$

CALCULATOR SOLUTION

Keyed entry											Display

AC 50000 `+` 50000 `×` .13 `×`

9 `÷` 12 `=`   54875

`−` 54875 `×` .16 `=`

4 `÷` 12 `=`   51948.3333

b. Ben's net gain is the difference between the amount Janet paid him for the note ($51,948.33), and the amount he loaned Emilio ($50,000). The net gain for Ben is $1,948.33.

c. Janet received the face value of $54,875.

d. Janet's net gain is the difference between what she received from Emilio and what she paid Ben.

Janet's net gain = $54,875 − $51,948.33 = $2,926.67

e. Ben's true interest rate $= \dfrac{\$1,948.33}{\$50,000(5/12)} = .0935198$ or 9.35%

f. Janet's true interest rate $= \dfrac{\$2,926.67}{\$51,948.33(4/12)} = .1690143$ or 16.9%

g. Emilio paid $54,875.

h. The value of the note when Janet bought it would be the original principal ($50,000) plus 5 months' interest earned on that principal at 13% per annum.

$$\text{value after 5 months} = M = P + I$$
$$= \$50,000 + (\$50,000)(.13)(5/12)$$
$$= \$52,708.33$$

Both Ben and Janet should know this value before they enter into negotiations. Ben wants to receive an amount as close as possible to this figure. Janet wants to pay something less than this figure. They would negotiate the discount rate to be used in the sale of the note to Janet. Since Ben received less than $52,708.33, Janet was more successful in the negotiations.

The Steele illustration and the illustration in Example 7 are but two of the many complex business transactions that could occur with negotiable business instruments.

CHECK YOUR KNOWLEDGE

Promissory Notes

Solve the following word problems.

1. Susan Humphreys received a loan from First National Bank. She signed a promissory note with a face value of $9,500, agreeing to pay off the loan in 6 months with interest at 11.5% per annum. Determine the maturity value of the note.

2. Harry Hanson signed a 120-day simple discount note with a face value of $23,500 on March 3, 1999, with a discount rate of 12%. Determine the proceeds Harry actually received. (Assume 360 days per year.)

3. Sharon Tasker borrowed $40,000 from Jackson Finance Company. She signed a promissory note on June 2, 2000, with a face value of $40,000 plus 10% interest due in 18 months from the date of the note. Determine the maturity value of the note.

4. Refer to problem 3. Jackson Finance Co. sold Sharon's note (after holding it for 6 months) to Ferguson Funding Corp. at a 15% discount. Determine the proceeds received by Jackson from Ferguson.

5. Now refer to your results in problems 3 and 4 and determine (a) Jackson's net gain or loss and (b) Ferguson's net gain or loss.

Problems 6 through 8 refer to the promissory note below.

6. Determine the due date of the note.

7. Determine the maturity value of the note (1 year = 360 days).

8. Assuming that Haskins Loan Co. sells the note to Quick Dollar Funding Corp. at a 17% discount on May 26, 1999, determine the proceeds received by Haskins from Dollar Funding.

Answers to CYK: 1. $9,500 + $546.25 = $10,046.25   2. $P = $23,500 − $940 = $22,560$
3. $40,000 + $6,000 = $46,000   4. $39,100   5. $900; $6,900   6. June 4, 1999
7. $28,750 + $1,245.83 = $29,995.83   8. $29,868.35

8.3 EXERCISES

Solve the following word problems. Round dollar amounts to the nearest cent and rates to the nearest hundredth of a percent.

1. Arnold Jensen received a loan from Chemung Savings and Loan. He signed a promissory note with a face value of $11,500 agreeing to pay off the loan in 15 months with simple interest at 13% per annum. Determine the maturity value of the note.

2. Consider the following promissory note:

$4,750   Harrowville, Va. 58349   March 12, 1999

240 days   **after this date ___|___ promise to pay to**
            Signet Loan Co

            Four thousand seven hundred fifty and 00/100   **DOLLARS**

Payable at ___First Bank of Harrowville___

Value received with interest at ___14.5%___ per cent per annum

                            Signed ___Marla Key___

Determine the date that the loan is due, and the maturity value of the loan.

3. Consider the following promissory note:

$23,450   Lodi, New York 58349   April 15, 1998

300 days   **after this date ___|___ promise to pay to**
            Kermet Loan Co. Inc.

            Twenty-three thousand four hundred fifty and 00/100   **DOLLARS**

Payable at ___Lodi First National Bank___

Value received with interest at ___0___ per cent per annum

                            Signed ___Janet Hamlin___

Assume that this loan was discounted at 12% and determine (a) the proceeds and (b) the true interest rate of the loan.

4. Consider the following promissory note signed by Rooney Construction Co. for money to purchase a new backhoe:

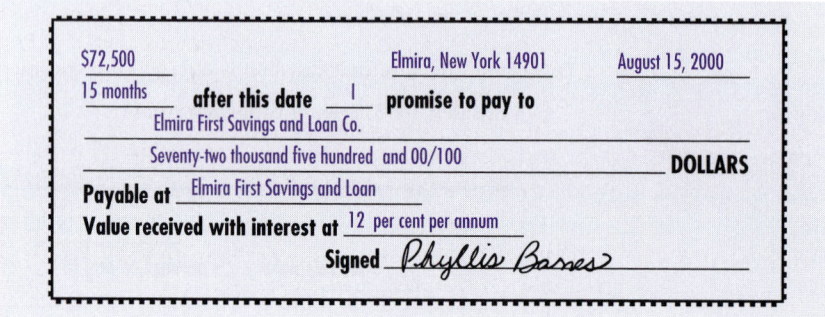

$38,500	Valoise, New York 14980	September 1, 1999

18 months **after this date** I **promise to pay to**
Valois Savings and Loan Co.

Thirty-eight thousand five hundred and 00/100 _____ **DOLLARS**

Payable at Valois Savings and Loan

Value received with interest at 10.5 per cent per annum

Signed *John R. Rooney*

Determine the due date of the loan, and the maturity value of the loan.

5. Refer to the promissory note in exercise 4. Assume that the Valois Savings and Loan sold the note to Nelson Funding Group on September 1, 1999, at a discount rate of 15%. What amount did the Valois Savings and Loan receive from Nelson Funding?

6. Consider the following promissory note signed by Barnes Construction Co. for money to purchase a new utility truck:

$72,500	Elmira, New York 14901	August 15, 2000

15 months **after this date** I **promise to pay to**
Elmira First Savings and Loan Co.

Seventy-two thousand five hundred and 00/100 _____ **DOLLARS**

Payable at Elmira First Savings and Loan

Value received with interest at 12 per cent per annum

Signed *Phyllis Barnes*

Determine the due date of the loan, and the maturity value of the loan.

7. Refer to the promissory note in exercise 6. Assume that the Elmira Savings and Loan sold the note to Johnson Funding Group on January 15, 2000, at a discount rate of 17%. What amount did the Elmira Savings and Loan receive from Johnson Funding?

8. Refer to exercise 7 and determine Elmira Saving's net gain.

9. Refer to exercise 7. Assume that the Johnson Funding Group sells the promissory note to Fast Funds on June 15, 2000, at a 20% discount. Determine the proceeds Johnson received for the note.

10. Refer to exercise 9 and determine the net gain or loss (a) for Johnson and (b) for Fast Funds.

8.4   PARTIAL PAYMENTS ON A SIMPLE INTEREST LOAN

Recall that interest is a sum of money charged for the use of the principal or proceeds of a loan. The longer the borrower has the principal, the greater the interest cost for the loan. If all or part of the principal could be paid off early, the borrower could save on interest costs. When the borrower pays part of the principal prior to the maturity date, the payment is referred to as the **partial payment of a loan.** The U.S. rule provides guidelines for partial payments made on loans prior to their due dates.

The U.S. Rule

Learning objective
Use the U.S. rule to calculate the balance of the principal of a loan when a partial payment is made.

The **U.S. rule** states that in the United States, a partial payment must be divided into two parts, principal and interest, just as the loan is made up of those two parts. First, the partial payment must pay any interest accrued to date on the loan, then the remainder of the partial payment is subtracted from the principal to reduce it. The U.S. rule is so named because a U.S. Supreme Court ruling in the nineteenth century ruled that this procedure is a valid method of applying partial payments to a loan and its interest.

Example 8

Bill Lewkowicz borrowed $3,000 at 12% simple interest for 2 years on March 10, 1998. On March 10, 1999, he made a partial payment of $1,000. Bill made a second partial payment of $500 on June 10, 1999. How much does Bill still owe the loan after each partial payment?

Solution

First, let's look at a time line on this loan (Figure 8.11).

Step 1:   Determine the amount of interest to be paid on the principal from the beginning of the loan until the first partial payment; that is, for 1 year. This is the simple interest earned at 12% on $3,000 for 1 year.

$$I = PRT$$
$$I = \$3{,}000 \times .12 \times 1 = \$360$$

Figure 8.11

Partial payment time line for Example 8

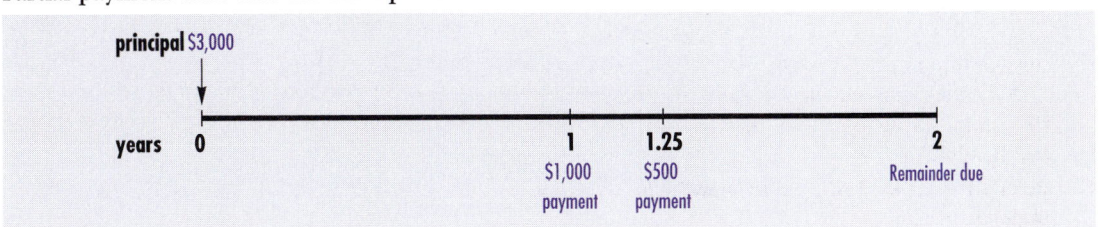

Step 2:   We must first pay off all of this accrued interest before any of the principal can be reduced. Subtract the interest, $360, from the partial payment to determine the remaining amount to be credited to the principal.

partial payment − interest = balance of partial payment
       $1,000   − $360 = $640

Step 3:   Subtract the amount to be used to reduce the principal from the principal to determine the unpaid principal, or balance, at the end of 1 year.

principal − balance of partial payment = unpaid principal
    $3,000 − $640   = $2,360

Let's update our time line now (Figure 8.12).

Now let's look at the second partial payment of $500, which is made $1/4$ year after the first partial payment. We apply the same steps to determine the new (reduced principal) balance.

Step 4:   Determine the interest from the 1-year to the $1\,1/4$-year time periods.

$$I = PRT$$
$$P = \text{current balance (\$2,360)}$$
$$R = .12$$
$$T = 1/4 \text{ year from 1 year to } 1\,1/4 \text{ year}$$
$$I = \$2,360 \times .12 \times 1/4 = \$70.80$$

Step 5:   Subtract the interest from the partial payment:

amount to reduce principal = $500 − $70.80 = $429.20

Figure 8.12

First partial payment calculation for Example 8, U.S. rule

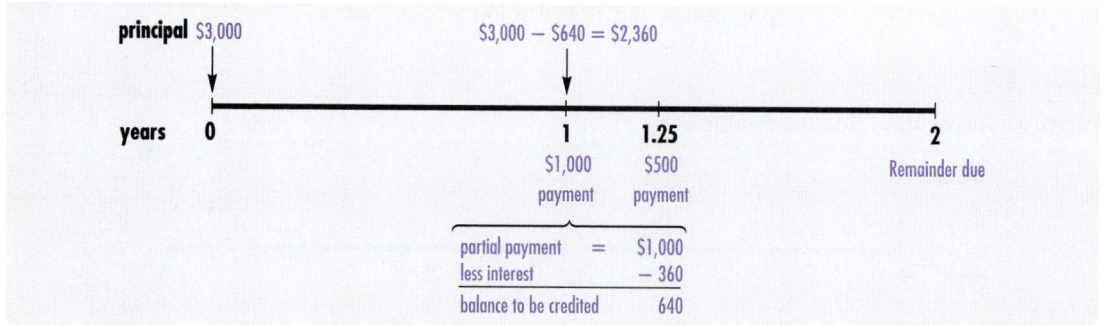

Figure 8.13

Second partial payment calculation for Example 8, U.S. rule

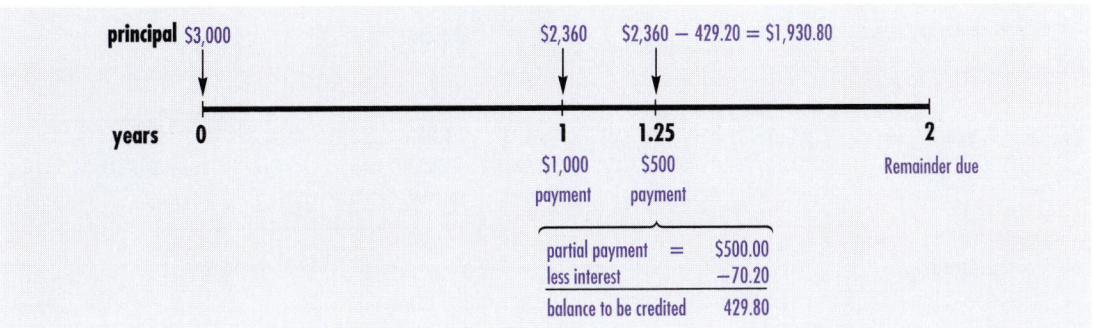

Step 6: Determine the balance of the principal and update the time line again, as shown in Figure 8.13:

balance = $2,360 − $429.20 = $1,930.80

The interest for the remaining ¾ of a year will be calculated on the principal of $1,930.80:

$$I = PRT = \$1{,}930.80 \times .12 \times \tfrac{3}{4} = \$173.77$$

The amount of the last payment will be $1,930.80 (unpaid principal) plus $173.77 (interest for the last ¾ year) = $2,104.57. (See Figure 8.14.)

Assuming Bill makes no more partial payments, and pays off the loan when it is due, he will have paid a total interest of $604.57 (see Figure 8.15):

total interest paid = $360 + $70.80 + $173.77 = $604.57

Figure 8.14

Final partial payment calculation for Example 8, U.S. rule

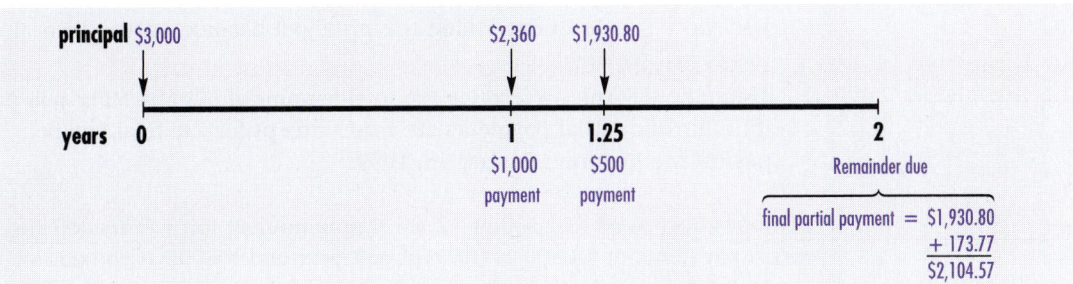

Figure 8.15

Interest with partial payments

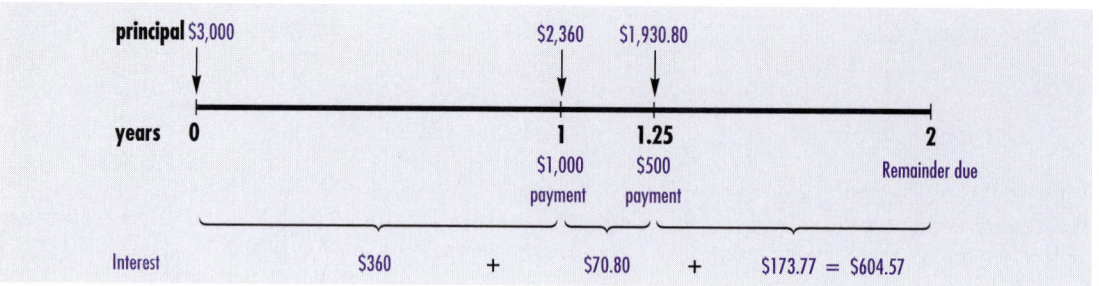

If Bill had made no partial payments, and had simply paid off the principal and interest at the end of the 2-year period, he would have paid $720.00:

$$I = PRT = \$3,000 \times .12 \times 2 = \$720.00$$

So, the partial payments saved him $115.43 in interest ($720 − $604.57 = $115.43).

CHECK YOUR KNOWLEDGE

Partial Payments of Simple Interest Loans

Solve the following problems. Round dollar amounts to the nearest cent and rates to the nearest hundredth of a percent.

Mary Meston borrowed $8,000 at 10% simple interest for $1\frac{1}{2}$ years on April 15, 1998. Mary made a $4,000 partial payment on April 15, 1999. Use this information to answer problems 1 and 2.

1.   Use the U.S. rule to determine the principal balance on the loan after the partial payment.

2.   Using the U.S. rule, determine the total amount of interest Mary will pay if no further partial payments are made except for the final payment to pay off the loan on October 15, 1999.

Jeff Jones borrowed $14,300 at 12.5% simple interest for 3 years. Jeff made partial payments of $8,500 at the end of 1 year, and $5,000 at the end of $1\frac{1}{2}$ years, and then paid off the loan at the end of the term. Use this information to answer problems 3 and 4.

3. Use the U.S. rule to determine the principal balance on the loan after the first partial payment.

4. Use the U.S. rule to determine the principal balance on the loan after the second partial payment.

8.4 EXERCISES

Solve each of the following problems. Round dollar amounts to the nearest cent.

Bart Barlow borrowed $15,000 at 12% simple interest for 2 years on June 15, 1997. He made a partial payment of $9,000 on June 15, 1998. Use this information to answer exercises 1 through 4.

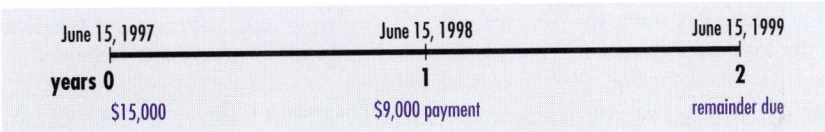

1. Use the U.S. rule to determine the principal balance on the loan after the partial payment.

2. Calculate the interest that will accrue on the remaining principal that you determined in exercise 1 if no further partial payments are made before the loan's due date of June 15, 1999.

3. Using the U.S. rule, determine the total amount of interest Bart paid if no further partial payments were made except for the final payment to pay off the loan on June 15, 1999.

4. How much interest will Bart save as a result of his partial payment?

Lou's Martville Car Care service station borrowed $15,000 to purchase a new hoist for his tow truck. Lou borrowed the money on January 15 at 11.5% simple interest for 9 months. After 6 months, Lou made a partial payment of $9,000. Use this information to answer exercises 5 through 8.

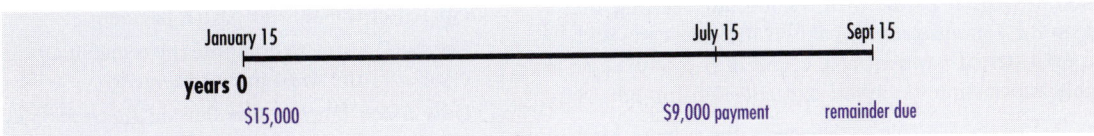

5. Use the U.S. rule to determine the principal balance on the loan after the partial payment.

6. How much interest will accrue on the remaining principal that you determined in exercise 5 if no further partial payments are made before the loan's due date?

7. Using the U.S. rule, determine the total amount of interest Lou will pay if no further partial payments are made except for the final payment.

8. How much interest will Lou's save as a result of the partial payment?

Answers to CYK: *1.* $4,800.00   *2.* $1,040.00   *3.* $7,587.50   *4.* $3,061.72

Dr. Janice Marlow, DDS, borrowed $9,600 at 13.25% simple interest on March 20, 1998, for 18 months to purchase a new x-ray machine for her dental practice. She made a partial payment of $3,000 on June 20, 1998, and a second partial payment of $5,000 on April 20, 1999. Use this information to answer exercises 9 through 13.

March 20, 1998	June 20, 1998	April 20, 1999	Sept. 20, 1999
years 0			
$9,600	$3,000 payment	$5,000	remainder due

9. Use the U.S. rule to determine the principal balance on the loan after the first partial payment.

10. Use the U.S. rule to determine the principal balance on the loan after the second partial payment.

11. Determine the interest that will accrue on the remaining principal that you determined in problem 10 if no further partial payments are made before the loan's due date.

12. Calculate the total interest Dr. Marlow paid for the loan.

13. How much interest did Dr. Marlow save as a result of the partial payment?

Edwardson's Hog Farm needed to borrow $6,500 to purchase corn seed. The owners signed a 6-month promissory note with 10% simple interest added, due August 15. Edwardson's made a partial payment of $1,500 after having the loan for 2 months. It made a second partial payment of $3,500 after having the loan for 4 months, and it paid off the loan early with a final partial payment at the end of 5 months. Use this information to answer exercises 14 through 18.

14. Use the U.S. rule to determine the principal balance on the loan after the first partial payment.

15. Use the U.S. rule to determine the principal balance on the loan after the second partial payment.

16. Determine the interest that Edwardson's had to pay for the remaining principal calculated in exercise 15, having paid off the loan earlier than its due date.

17. Determine the total amount of interest Edwardson's paid as of the final loan payment.

18. How much interest did Edwardson's save as a result of the partial payments?

Williamson's Craft Store borrowed $30,000 at 10% simple interest for three years to purchase new display shelving in a minor remodeling project. At the end of 1 year, Williamson's made a partial payment of $10,000 on the loan. At the end of the second year, it made a second partial payment of $10,000 on the loan. At the end of $2\frac{1}{2}$ years, it made a third partial payment of $5,000 on the loan. At the end of the third year it paid off the remainder of the loan in full.

19. At the beginning of the loan, what was the total anticipated interest for the 3-year loan?

20. Use the U.S. rule to compute the remaining principal after the first partial payment.

21. Use the U.S. rule to compute the remaining principal after the second partial payment.

22. Use the U.S. rule to compute the remaining principal after the third partial payment.

23. How much interest was due on the remaining principal after the third partial payment through the final payment?

24. How much did Williamson's pay in total interest on the loan?

25. How much interest cost did Williamson's save by making the partial payments?

CHAPTER REVIEW EXERCISES

The following problems review all mathematics and business concepts presented in the chapter. If you experience any difficulty solving a problem, go to the appropriate chapter section and review the examples provided. Express all answers as decimals rounded to the nearest one hundredth, or dollars rounded to the nearest cent, unless otherwise instructed.

Determine the unknown quantities in the following simple discount problems.

	Maturity value	Discount rate	Time	Discount amount	Proceeds
1.	$13,000	11%	8 months		
2.		9%	15 months	$2,250	$17,750
3.	$19,000		2 years		$15,960
4.	$30,000	6%		$5,400	
5.	$10,000	12%	18 months		$8,200

Marcille Bourdeau borrowed $18,500 at 11% for 18 months. Assuming the bank discounted the loan, use this information to answer problems 6 through 11.

6. Determine the amount of the discount on Marcille's loan.

7. Determine the amount of the proceeds that Marcille received.

8. What is the maturity value of Marcille's loan?

9. Determine the true interest rate of Marcille's loan.

10. If Marcille really needed her proceeds to be $18,500, then how much would the maturity value of her loan need to be?

11. How much interest would Marcille pay on the loan if it did have the maturity value you determined in problem 10?

Whipple Lumber and Home Supply wants to borrow $38,000 for 2 years. Columbia Bank offers the company a 10% simple interest loan. Chenango National Bank offers it an 8% simple discount loan.

12. Determine the maturity value of the loan at Chenango National.

13. Determine the proceeds of the loan at Chenango National.

14. Determine the true interest rate of the loan at Chenango National.

15. What amount must Whipple Lumber request to borrow from Chenango National Bank in order to actually receive $38,000 in proceeds?

16. What will be the true rate of interest for the loan in problem 15?

17. What is the maturity value of the simple interest loan offered by Columbia Bank?

18. Which loan would give Whipple Lumber the full $38,000 needed and also cost the least amount of interest?

Asha Shaffer borrowed $8,000 at 7% simple interest for one year to begin a cottage industry in her home manufacturing Snuggles Teddy Bears, which she marketed and sold at craft shows in the summer. She borrowed the money on January 15, 1998, made a partial payment of $5,000 on August 15, and paid off the remainder on January 15, 1999. Use this information to answer problems 19 through 25.

19. Calculate the original interest due on the loan if no partial payments had been made.

20. Use the U.S. rule to determine the interest accrued on the loan from the beginning of the loan until the partial payment.

21. Determine the amount of the partial payment that should be applied to reduce the principal by the terms of the U.S. rule.

22. By the U.S. rule, what is the principal balance on the loan after the partial payment?

23. Determine the interest that Asha had to pay for the remaining principal.

24. Determine the total amount of interest Asha paid as of the final loan payment.

25. How much interest did Asha save by making the partial payment on the loan?

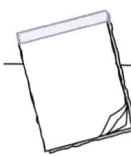

EXPRESS YOUR UNDERSTANDING

Compose one or two well-written sentences to express the requested information in your own words.

1. Explain how you would calculate the amount of discount to be deducted from the maturity value of a discount loan.

2. Explain how a simple discount loan is different from a simple interest loan.

3. Explain why it is important to know the true interest rate of a simple discount loan.

4. In what way does the true interest rate differ from the discount rate of a discount loan?

5. Explain why a simple interest loan of $1,000 at the simple interest rate of 10% is better than a simple discount loan of $1,000 at a simple discount rate of 10% if both loans are for the same time period.

6. What is a promissory note?

7. Who is the payee of a promissory note?

8. Who is the maker of a promissory note?

9. Explain what the U.S. rule says about early partial payments of a loan.

10. Explain how the U.S. rule would be applied to a simple interest loan of $1,000 at 10% for 1 year if one partial payment of $500 were made halfway through the term of the loan.

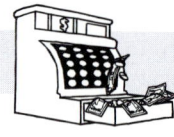

Mind Your Business

Media World purchased $10,000 of display cases from Metropolitan Commercial Furnishings on October 10. Metropolitan Furnishings agreed to accept a 6-month promissory note for the full purchase price of the equipment, at a simple interest rate of 12% per year.

Activity 1

You have decided to finance the equipment with the terms provided by Metropolitan and you have signed the promissory note on October 15. Based on the terms, determine (a) the maturity date of the loan; (b) the maturity value of the simple interest loan; and (c) the total amount of interest due at maturity.

Activity 2

Assume in the future that you receive a notice that Metropolitan has sold the promissory note to Consolidated Funding Inc. on January 15 at a discount rate of 15%. Determine the following:

(a) the discounted price that Consolidated paid for the note

(b) how much Consolidated will earn on the note if it holds the note to maturity

(c) the amount of interest Metropolitan would earn on the loan

(d) whether Metropolitan would lose any earnings by selling the note early, and if so, how much it would lose

(e) the amount Media World would owe at maturity and to whom it would be owed

SELF-TEST

A. Terminology

Complete the following items using the key terms presented at the beginning of the chapter. Check your responses against the answer key at the end of the test.

1. The amount of money a borrower actually receives on a discounted loan is called the _____.

2. The interest rate used to determine the amount to be deducted from the maturity value of a simple discount loan is called the _____.

3. The rate of interest calculated using simple interest methods on a discount loan is called the _____ rate of interest.

4. A _____ note is a written pledge by the borrower to pay a sum of money to the bearer of the note.

5. The _____ value of a simple interest note is the principal plus simple interest on the principal.

6. The _____ value of a simple interest note is the same as the principal of the simple interest loan.

7. A _____ _____ note means that the interest is deducted in advance.

8. When a payment is made before the simple interest loan is due, and it does not pay off the entire loan early, it is called a _____ payment.

9. When a payment is made early on a loan and the payment is used to pay interest accrued on the loan up to that point in time first, and the remainder of the payment is used to decrease the principal, the _____ Rule is being applied.

B. Calculation

Formbees Electronics Inc. borrowed $30,000 from a bank for 3 years at 9% simple discount rate. Use this information to answer problems 9 through 11.

10. Calculate the amount of the discount that will be applied to the loan.

11. Calculate the proceeds of the discounted loan.

12. Calculate the true rate of interest on the loan.

E. Z. Draw Artistic Paper needs to borrow $10,000. The company can obtain a 12% simple interest loan for 2 years from Seneca Lake Bank and Trust, or an 11% simple discount loan for 2 years from Hammondsport Bank of Commerce. Use this information to answer problems 13 through 15.

13. Calculate the maturity value of the simple interest loan.

14. What maturity value will E. Z. Draw need to request on the simple discount loan in order to actually receive the needed $10,000?

15. Which loan will cost less in interest, and how much interest will be saved?

Use the following information to complete problems 16 through 18. Eberhard's Card Shoppe signed a promissory note with a face value of $8,000 for 18 months, agreeing to pay the face value plus 10% simple interest to Bronxtown Citizens' Bank.

16.   Determine the maturity value of the note.

17.   Assume Citizens' Bank sells the note to Hometown investors at a 14% discount rate after Citizens' had held the note 8 months. How much did Citizens' receive for the note?

18.   How much will Hometown Investors earn when the note comes due and is paid in full?

Use the following information to solve problems 19 and 20.

Martha Miller's Home Aids borrowed $4,500 for 6 months at 9% simple interest. A partial payment of $2,500 was made after 4 months.

19.   Use the U.S. rule to determine the principal balance after the partial payment.

20.   Calculate the final payment on the loan, so that the remaining principal and interest will be paid in full.

Answers to Self-Test: *1.* proceeds   *2.* discount rate   *3.* true   *4.* promissory   *5.* maturity   *6.* face   *7.* simple discount   *8.* partial   *9.* U.S.   *10.* $8,100   *11.* $21,900   *12.* 12.33%   *13.* $12,400   *14.* $12,820.51   *15.* Simple interest loan will save $420.51   *16.* $9,200   *17.* $8,126.67   *18.* $1,073.33   *19.* $2,135   *20.* $2,167.03

9

Key Terms

compound interest
compound amount
future value of a
    compound amount
periodic interest rate
present value of a
    compound amount
time value of money
effective rate
nominal rate

COMPOUND INTEREST

Learning Objectives

After completing the exercises in this chapter, you will be able to:

1. Explain the difference between simple interest and compound interest.

2. Calculate the compound interest and compound amount using the multistep method.

3. Calculate the compound amount by using the compound interest formula.

4. Calculate the compound amount by using the compound interest table.

5. Calculate the future value of a compound amount by using the formula or tables.

6. Calculate the present value of a compound amount using the present value formula or tables.

7. Identify *nominal rate* and *effective rate* and be able to distinguish between them.

8. Determine the effective rate of a compound interest problem by using a formula or a table.

9. Define key terms.

INTRODUCTION

As we mentioned earlier in this text, there are different methods used to calculate interest on investments and loans. In Chapters 7 and 8 you learned about the simple interest and simple discount methods for computing interest. The next step in our study of interest is to learn about compound interest.

In this chapter we will discuss the concept of compound interest as interest on the original principal *plus interest on past interest earned.* Most savings accounts in banks and other investment institutions are compound interest accounts. Our discussions will consider a single lump sum deposited into a compound interest account, which will grow to be a larger lump sum, principal plus interest, much the same as with simple interest. The concepts of simple interest apply, but result in a slightly different formulation that produces a larger amount of earned interest.

9.1 COMPOUND INTEREST

Compound interest differs from simple interest in the following way. *The principal amount in a simple interest calculation always remains the same,* no matter how many years, months, weeks, or days are in the term of the investment or loan. In a **compound interest** calculation the *principal amount is increased periodically* by adding the past interest earned to the previous principal. Thus, the principal increases periodically throughout the duration of the investment or loan.

Learning objective
Explain the difference between simple interest and compound interest.

Compound interest is calculated using the interest formula, $I = PRT$, which multiplies the principal, P, times the rate, R, times the time, T (where the principal includes previous interest already earned). Figure 9.1 illustrates how the compound interest, when added to the principal (present value), accumulates up to the future value.

P here is the principal that will grow to be the compound amount, A, also called the future value.

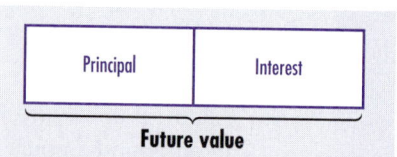

Principal	Interest

Future value

Figure 9.1

Future value of a compound amount

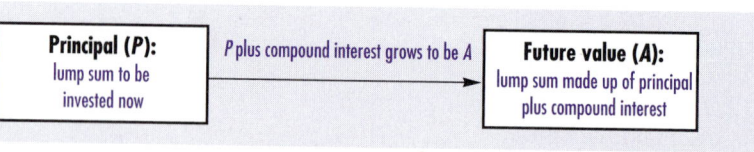

| **Principal (P):** lump sum to be invested now | *P* plus compound interest grows to be *A* → | **Future value (A):** lump sum made up of principal plus compound interest |

In order to better understand the difference between simple interest and compound interest, let us examine the following two examples, with $P = \$1,000$; $R = 6\frac{1}{2}\%$, and $T = 3$ years.

Example 1

Ian Simon deposited $1,000 in a savings account that offered $6\frac{1}{2}\%$ simple interest for 3 years. How much interest did Ian earn in the 3 years, and what was the total value of his account at the end of 3 years?

Solution

At the end of 3 years, Ian had earned in interest

$$I = PRT = \$1,000 \times .065 \times 3 = \$195$$

The total value of Ian's account would be the maturity value of the simple interest account:

$$M = P + I = \$1,000 + \$195 = \$1,195$$

Example 2

Ian Simon deposited $1,000 in a savings account that offered $6\frac{1}{2}\%$ interest compounded (or calculated) *yearly*. Ian left the original principal plus all of the accumulated interest in his savings account for 3 years. What was the total value of his account at the end of 3 years? How much interest did the account earn in the 3-year period?

Solution

Step 1: Calculate the maturity value for the first year.

$$M = P + (PRT)$$
$$M = \$1,000 + (\$1,000 \times .065 \times 1) = \$1,065$$

The total amount at the end of the first year is $1,065

Step 2: Calculate the maturity value for the second year. The principal here is the *sum* of the previous principal and the interest earned during the first year.

$$M = P + (PRT) = \$1,065 + (\$1,065 \times .065 \times 1)$$
$$M = \$1,065 + 69.23 = \$1,134.23$$

The total amount at the end of the second year is $1,134.23

Step 3: Calculate the maturity value for the third year. The principal here is the *sum* of the previous principal and the interest earned during the first and second years.

$$M = P + (PRT) = \$1,134.23 + (\$1,134.23 \times .065 \times 1)$$
$$M = \$1,134.23 + 73.73 = \$1,207.95$$

The total amount at the end of the third year is $1,207.95. The total interest earned in the 3-year term is

$$\text{total interest} = \$1,207.95 - \$1,000 = \$207.95$$

Let us now examine more closely the solution to Example 2. The **compound amount** is the principal plus interest, where the principal includes previous interest already earned. Notice that the compound amount at the end of the first year ($1,065) is substituted into the formula as P at the beginning of the second year. This principal amount of $1,065 includes the previous interest of $65 earned during the first year. The principal amount of $1,065 at the beginning of the second year will now be earning interest throughout the second year. This process continues and the compound amount of $1,134.23 at the end of the second year is the principal amount of $1,134.23 at the beginning of the third year. Figure 9.2 illustrates the comparison of compound versus simple interest from Examples 1 and 2.

Notice that Examples 1 and 2 are virtually the same *except* that in Example 1 interest was calculated only on the original principal of $1,000 each year, and for Example 2 interest was calculated on the principal plus interest accumulated at the end of the first and second year. The simple interest

Figure 9.2

Time line comparison of simple interest and compound interest

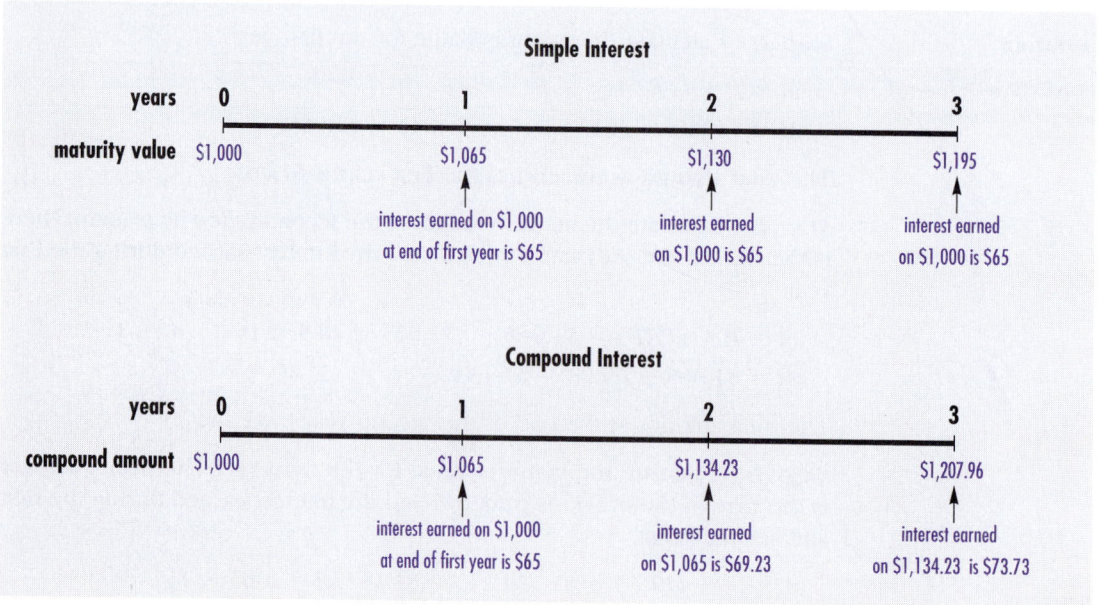

produced \$195 (\$65 each year) interest, whereas interest compounded yearly produced the greater amount of interest, \$207.96 (\$65 + \$69.23 + \$73.73 = \$207.96).

Let's step back for a moment to remind ourselves what we are trying to accomplish here. We know the exact amount of money that we are investing *today*. What we are trying to find is an accumulated *future value*. The **future value of a compound amount** includes the original principal plus the interest.

Multistep Method

Whenever interest is compounded more than once a year (e.g., semiannually, quarterly, monthly, daily), the formula $I = PRT$ must be calculated for each *period* that interest is compounded during the length of the investment. If interest is compounded annually, then we calculate the interest for 1 period per year. If interest is compounded semiannually, quarterly, monthly, or weekly, there are 2, 4, 12, and 52 periods per year, respectively.

When the interest rate is given, it is given as an annual rate. Whenever interest is compounded more than once a year, the rate R must be adjusted for the number of times interest is compounded. We will let i represent the **periodic interest rate,** which is the annual interest rate divided by the total number of compounding periods in 1 year. We will need to know the total number of compounding periods that occur in the full term of the loan. We will let n represent that total number of compounding periods.

Example 3

Consider an investment of 12% per year for 5 years, and determine i and n for compounding annually, semiannually, quarterly, monthly, weekly, and daily.

Solution

Compounding	Periods per year	i = %/periods per year	$n = \left(\dfrac{\text{number of periods}}{\text{per year}}\right) \times \left(\dfrac{\text{number}}{\text{of years}}\right)$
Annually	1	12%/1 = 12%, or .12	$n = 1 \times 5 = 5$
Semiannually	2	12%/2 = 6%, or .06	$n = 2 \times 5 = 10$
Quarterly	4	12%/4 = 3%, or .03	$n = 4 \times 5 = 20$
Monthly	12	12%/12 = 1%, or .01	$n = 12 \times 5 = 60$
Weekly	52	12%/52 = .23%, or .0023	$n = 52 \times 5 = 260$
Daily	365	12%/365 = .03288%, or .0003288	$n = 365 \times 5 = 1,825$

Example 4

Joe Schulle deposited $100 into a savings account that offered 8% interest compounded semiannually. Joe kept the original deposit plus the accumulated interest in the savings account for 3 years. What was the accumulated amount at the end of 3 years?

Solution

The method we are using here is called the *multistep method,* because we calculate the compound amount for each period, one step at a time, until we have the future amount. The first step is to calculate the interest for the first period, which is for the first half-year. Recall that the yearly rate is 8% and this is compounded semiannually (2 times per year), so the periodic rate is

$$i = \frac{R}{2} = \frac{.08}{2} = .04$$

Also note that T for compound interest will always equal 1, meaning "one" period.

	P + (P × i × T) =	Compound amount
End of 1st period	100 + (100 × .04 × 1)	
	100 + 4 =	$104.00
End of 2nd period	104 + (104 × .04 × 1)	
	104 + 4.16 =	$108.16
End of 3rd period	108.16 + (108.16 × .04 × 1)	
	108.16 + 4.33 =	$112.49
End of 4th period	112.49 + (112.49 × .04 × 1)	
	112.49 + 4.50 =	$116.99
End of 5th period	116.99 + (116.99 × .04 × 1)	
	116.99 + 4.68 =	$121.67
End of 6th period	121.67 + (121.67 × .04 × 1)	
	121.67 + 4.86 =	$126.53

Figure 9.3 illustrates how the periodic interest amount is added to the previous principal to determine the new principal at the end of each period. The compound amount at the end of 3 years is $126.53.

Learning objective
Calculate the compound amount by using the compound interest formula.

Calculating Compound Interest Using a Formula

You can see from the previous examples how long and tedious the multistep process of calculating compound interest can be. This can be avoided

Figure 9.3

Compound interest time line

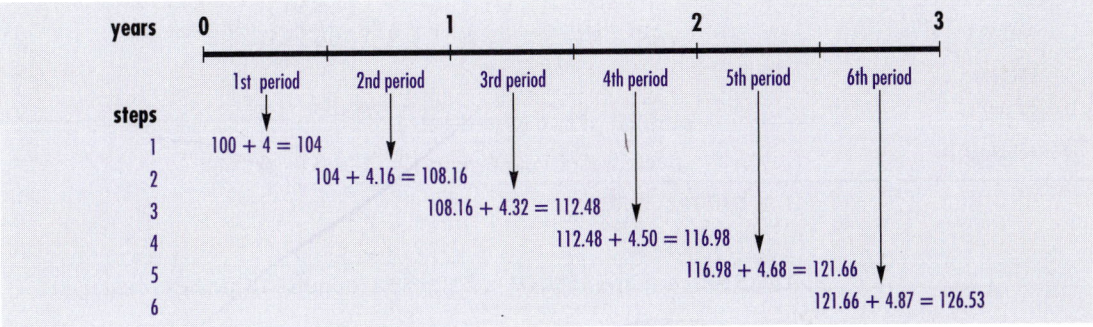

Figure 9.4

Compound interest formula

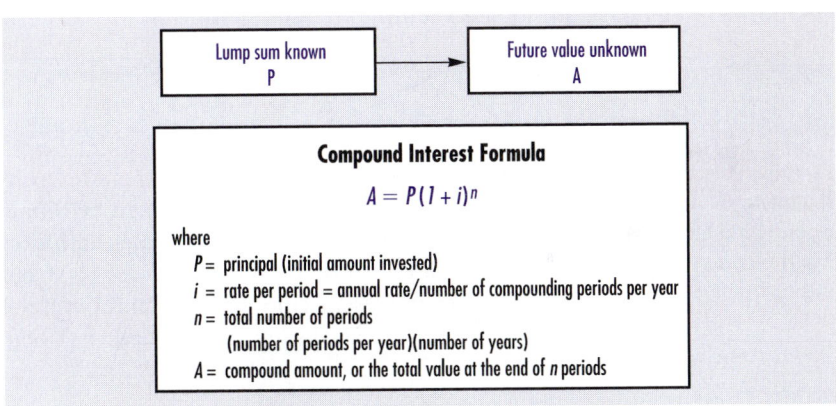

by using the formula in Figure 9.4. Let's use the formula to calculate the compound amount from Example 4.

Example 5

Joe Schulle deposited $100 into a savings account that offered 8% interest compounded semiannually. Joe kept the original deposit plus the accumulated interested in the savings account for 3 years. What was the accumulated amount at the end of 3 years?

Solution

Refer to Figure 9.4. We know the lump sum is $100. We are trying to determine the unknown future value.

$$A = P(1 + i)^n$$

$$P = \$100$$

$$i = \frac{R}{\text{periods per year}} = \frac{R}{2} = \frac{.08}{2} = .04$$

$$n = (\text{periods per year})(\text{years}) = 2 \times 3 = 6$$

$$A = \$100(1 + .04)^6$$

$$A = \$100(1.04)^6$$

$$A = \$100 \times 1.265319018 = \$126.53 \text{ (rounded to nearest cent)}$$

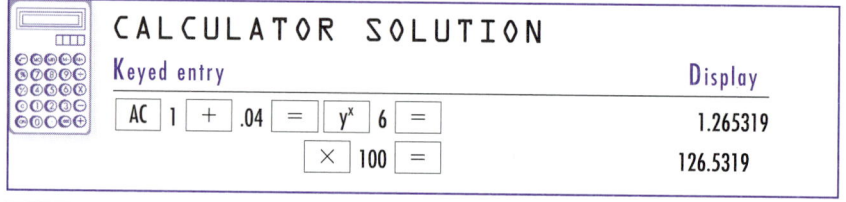

CALCULATOR SOLUTION	
Keyed entry	Display
AC 1 + .04 = y^x 6 =	1.265319
× 100 =	126.5319

Calculating Compound Interest Using Tables

Learning objective
Calculate the compound amount by using the compound interest table.

Another method for computing compound interest is to use *compound interest tables.* A sample of these tables can be found in Tables 9.1 and 9.2 as well as Appendix B. Column A in the tables in this text is titled "Future Value." In other sources, it may be referred to as "$1 at compound interest," or "future value of a present amount." The tables will act in lieu of the y^x key on your calculator. Let us rework Example 4 to demonstrate how the tables work for us.

Example 6

Joe Schulle deposited $100 into a savings account that offered 8% interest compounded semiannually. Joe kept the original deposit plus the accumulated interest in the savings account for 3 years. What was the accumulated amount at the end of 3 years?

Solution

Refer to Figure 9.4. The lump sum (known) is $100. We are trying to determine the unknown future value.

Step 1: Determine i in terms of %, and n.

$$i = \frac{\%}{\text{compounding periods per year}} = \frac{8\%}{2} = 4\%$$

$$n = (\text{compounding periods per year})(\text{years}) = (2)(3) = 6$$

Table 9.1

Sample compound interest table at 4%

N	A Future Value	B Present Value	C Ordinary Annuity	D Sinking Fund	E Present Annuity	F Amortization
1	1.040000	0.961539	0.999999	1.000001	0.961538	1.040001
2	1.081600	0.924556	2.039999	0.490196	1.886094	0.530196
3	1.124864	0.888997	3.121597	0.320349	2.775088	0.360349
4	1.169859	0.854804	4.246462	0.235490	3.629893	0.275490
5	1.216653	0.821927	5.416319	0.184627	4.451821	0.224627
6	1.265319	0.790315	6.632972	0.150762	5.242135	0.190762
7	1.315932	0.759918	7.898289	0.126610	6.002053	0.166610
8	1.368569	0.730690	9.214220	0.108528	6.732742	0.148528
9	1.423312	0.702587	10.582790	0.094493	7.435328	0.134493
10	1.480244	0.675564	12.006100	0.083291	8.110891	0.123291
11	1.539454	0.649581	13.486340	0.074149	8.760472	0.114149
12	1.601032	0.624597	15.025800	0.066552	9.385071	0.106552
13	1.665073	0.600574	16.626820	0.060144	9.985642	0.100144
14	1.731676	0.577475	18.291900	0.054669	10.563120	0.094669
15	1.800943	0.555265	20.023570	0.049941	11.118380	0.089941
16	1.872981	0.533908	21.824510	0.045820	11.652290	0.085820
17	1.947900	0.513374	23.697490	0.042199	12.165660	0.082199
18	2.025816	0.493628	25.645390	0.038993	12.659290	0.078993
19	2.106848	0.474643	27.671200	0.036139	13.133930	0.076139

Rate = 4 %

Table 9.2

Sample compound interest table at 1.5%

N	A Future Value	B Present Value	C Ordinary Annuity	D Sinking Fund	E Present Annuity	F Amortization
1	1.015000	0.985222	0.999999	1.000001	0.985221	1.015001
2	1.030225	0.970662	2.014995	0.496279	1.955879	0.511279
3	1.045678	0.956317	3.045217	0.328384	2.912192	0.343384
4	1.061363	0.942184	4.090890	0.244446	3.854374	0.259446
5	1.077284	0.928261	5.152250	0.194090	4.782629	0.209090
6	1.093443	0.914543	6.229528	0.160526	5.697167	0.175526
7	1.109845	0.901027	7.322972	0.136557	6.598195	0.151557
8	1.126492	0.887711	8.432810	0.118584	7.485903	0.133584
9	1.143390	0.874593	9.559298	0.104610	8.360494	0.119610
10	1.160540	0.861668	10.702680	0.093435	9.222154	0.103435
11	1.177948	0.848934	11.863220	0.084294	10.071090	0.099294
12	1.195617	0.836388	13.041160	0.076680	10.907470	0.091680
13	1.213552	0.824028	14.236780	0.070241	11.731500	0.085241
14	1.231755	0.811850	15.450320	0.064724	12.543340	0.079724
15	1.250231	0.799852	16.682080	0.059945	13.343200	0.074945
16	1.268985	0.788032	17.932310	0.055765	14.131230	0.070765
17	1.288019	0.776386	19.201290	0.052080	14.907610	0.067080
18	1.307340	0.764912	20.489300	0.048806	15.672520	0.063806
19	1.326950	0.753608	21.796640	0.045879	16.426130	0.060879

Rate = 1.5 %

Step 2: Look in the future value for compound interest table in the column for 4% and the row for $n = 6$. Go to the table for rate = 4%. Now go to the compound amount table column *A*. Next, go down the *n* column until you reach the number 6, then move to the right until you reach the compound amount column. The number at the intersection of these two columns is 1.265319.

Step 3:   Multiply the principal ($100) times this table number.

$A = 100 \times$ (table value for $i = 4\%$ and $n = 6$)

$A = \$100 \times (1.265319) = \$126.5319$, or $\$126.53$ (rounded)

We are still working with our formula for compound amount.

$A = P(1 + i)^n$

$A = 100(1.04)^6$. The compound interest tables will give us the value of $(1.04)^6$; that is $(1.04)^6 = 1.265319$

$A = 100(1.265319) = \$126.53$ (rounded)

Joe's compound amount is $126.53, and he has earned $26.53 in interest during the 3 years he invested it. Notice that multiplying the number in the table (1.265319) by the principal ($100) was exactly what we did when using the formula in Example 5.

At this time, it is important to understand exactly what the numbers in the table are telling us. Let's refer to the number that we just looked up, 1.265319. What exactly is this number telling us? Essentially, it indicates that if we had invested $1 at 8% interest compounded semiannually, at the end of 3 years we would have a compound amount of $1.27 (this amount was obtained by rounding 1.265319). However, we did not invest $1, but rather we invested $100, which is why we multiplied the $100 principal by the number in the table to obtain the compound amount of $126.53.

CHECK YOUR KNOWLEDGE

Compound Interest

Solve the following problems. Round dollar amounts to the nearest cent and rates to the nearest hundredth of a percent.

Use the multistep method for problems 1 and 2.

1. Determine the compound amount at the end of 4 years for $20,000 deposited at 10% interest compounded annually.

2. If $2,500 is deposited for 3 months into an account paying 18% interest compounded monthly, determine: (a) the compound amount at the end of 3 months and (b) the amount of interest earned during this time.

3. Identify i and n for each of the following: (a) 6% compounded quarterly for 4 years, (b) 10.5% compounded semiannually for 5 years, (c) 12% compounded monthly for 6 years, (d) 10% compounded weekly for 3 years, and (e) 15% compounded daily for 5 years. (Use 360 days per year.)

Use the compound interest formula to complete problems 4 through 6.

4. Determine the compounded amount of $30,000 invested at 6% compounded semiannually for 4 years.

5. Find the compound interest for a $13,500 investment at 12% compounded quarterly for 2 years.

6. If $7,000 is invested at 18% compounded monthly for 1 year, determine (a) the compound amount and (b) the total interest earned.

Use the compound interest tables to complete problems 7 and 8.

7. Fred Rooney invested $25,000 in a compound interest account paying 6% compounded monthly for 5 years. Determine the value of (a) Fred's investment at the end of 5 years and (b) the interest earned.

8. Emily Broadbent has a certificate of deposit that pays 12% compounded quarterly. If Emily's initial investment was $17,500 and it is a 6-year certificate, what will the value of the certificate be at the end of the 6-year term?

9.1   EXERCISES

Solve the following exercises. Round dollar amounts to the nearest cent and rates to the nearest hundredth of a percent.

Determine i and n.

	Rate/ year	Compounding	Number of years	i	n
1.	8%	semiannually	9		
2.	21%	quarterly	6		
3.	6%	monthly	10		
4.	9½%	yearly	30		
5.	10%	quarterly	15		

Find (a) the compound amount and (b) the compound interest using the multistep method for each of the following:

6. $3,000 at 6% compounded quarterly for 1 year

7. $7,000 at 11% compounded semiannually for 2 years

8. $300,000 at 9½% compounded yearly for 3 years

9. $1,000 at 11.5% compounded semiannually for 2 years

10. $10,000 at 4% compounded quarterly for 1 year

11. $10,000 at 5% compounded quarterly for 12 months

Determine the compound amount (future value) using the compound interest formula for each of the following:

12. $5,000 at 6% compounded monthly for 10 years
13. $5,000 at 6% compounded quarterly for 10 years
14. $5,000 at 6% compounded semiannually for 10 years
15. $5,000 at 6% compounded annually for 10 years
16. $35,000 at 12% compounded monthly for 9 years

In exercises 17–23, determine the compound amount using the compound interest tables found in Appendix A.

17. $5,000 at 6% compounded monthly for 10 years
18. $5,000 at 6% compounded quarterly for 10 years
19. $5,000 at 6% compounded semiannually for 10 years
20. $5,000 at 6% compounded annually for 10 years
21. $10,000 at 7% compounded semiannually for 8 years
22. $35,000 at 12% compounded monthly for 7 years
23. $11,000 at 9% compounded semiannually for 4 years

Solve the following exercises using either the compound interest formula or the compound interest tables.

24. Connors Realty Corp. enjoyed a fruitful month of income last month and found that it had $30,000 that it could invest in a compound interest investment paying 12% compounded monthly. If the firm leaves the money in the account for 2½ years, what will be the value of the account at that time?

25. Bianco Big "B" Markets made a $13,000 investment in a compound interest account paying 18% compounded monthly. What was the value of its investment at the end of 8 months?

26. Hazelmyer Heirloom Plastic Furniture borrowed $125,000 from a bank to make improvements in its showroom. If it borrowed the money at 10% compounded quarterly for 3 years, how much will it owe at the end of the 3-year period?

27. Sally Betson invested $300 at 6% compounded monthly for a total of 3 years. Sally took the total value of her investment at the end of 3 years, and added $500 to it and invested the total at 8% compounded quarterly for 4 years. What was her total investment worth at the end of the 7-year period?

Use the following information to complete exercises 28 through 30. Rita Menkins recently inherited $43,500. She has chosen to invest it in an insurance policy that guarantees to pay 10% compounded quarterly for the first 5 years, and a minimum of 6% compounded quarterly for the remaining 3 years of the policy.

28. Determine the value of Rita's investment at the end of the first 5-year period.

29. Determine the value of Rita's investment at the end of the last 3-year period, assuming she only receives the minimum rate and that she leaves all her money in the investment account.

30. Determine the total interest Rita will earn over the 8-year term of the policy assuming she receives only the minimum rate for the last 3 years.

31. Jack LaMonti borrowed $10,000 and agreed to pay interest at 8% compounded quarterly. How much will Jake have to pay at the end of the term of the loan if he keeps the loan for 2 years?

32. Herbert Farnsley deposited $7,850 into an accounting paying 10% compounded twice yearly. Determine (a) the compound amount and (b) the interest earned if he leaves all of the money in the account for 8 years.

33. Charlene Jenkins borrowed $40,000 to purchase new equipment for her Copies-Plus Shoppe. If she agreed to pay 12% compounded monthly for the use of the money for 2 years, how much will she have to have to pay off the loan at the end of the 2-year period?

9.2   FUTURE VALUE OF A COMPOUND AMOUNT

The concept of **time value of money** has been suggested in our work in our study of simple interest and simple interest notes. In the context of our studies thus far, *time value of money* simply means that as time passes, the value of a monetary investment changes. A simple interest investment that has a maturity value determined at the end of the term of the investment also has an increasing value throughout the investment. We used the phrase *future value of a simple interest investment or loan* as a means of expressing the time value of the investment. We used the formula for finding the maturity value of a simple interest investment or loan to determine the future value at any future point in time. We also used the phrase *present value of a simple interest investment* to express the value of the investment right now, or presently. Present value of an investment is another way to look at the value of the investment at a given point in time. Both of these concepts refer to the time value of money, and both play an important role in the decision-making processes involved in making important financial decisions, such as buying and selling promissory notes.

The time value of money also applies to compound interest investments and loans, as we have suggested in Section 9.1, when we referred to the compound amount as the future value.

The **future value of a compound interest investment** is the compound amount of the investment at a specific time within the term of the investment.

Example 7

Jane Sylvester has $35,000 invested in a 5-year certificate of deposit that pays 12% compounded monthly. Jane doesn't want to withdraw any money from the investment early because there is a substantial penalty for doing so, but she likes to track her net worth year by year at the end of each year. Assuming that she just made the investment, what will be its value at the end of (a) the first year, (b) the end of the third year, and (c) the end of the fifth year?

Solution

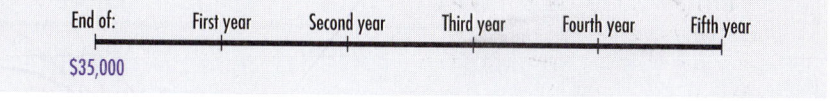

This is a compound interest problem in which we are seeking the future value of Jane's investment at the end of the first, third, and fifth years. We can use the compound interest formula, $A = P(1 + i)^n$, with $P = \$35,000$, $i = .12/12 = .01$, and n will vary for each part of the problem.

a. At the end of the first year, $n = 12$.

$A = \$35,000(1 + .01)^{12} = \$35,000(1.126825)$

$Fv = \$39,438.88$ or, by the tables,

$A = \$35,000 \times$ (table value for $i = 1\%$ and $n = 12$)

$A = \$35,000 \times 1.126825$

$Fv = \$39,438.88$

End of:	First year	Second year	Third year	Fourth year	Fifth year
$35,000	$39,438.88				

b. At the end of the third year, $i = .01$, or $i = 1\%$, and $n = 12 \times 3 = 36$

$A = \$35,000(1 + .01)^{36} = \$35,000(1.430768)$

$Fv = \$50,076.91$ or, by the tables,

$A = \$35,000 \times$ (table value for $i = 1\%$ and $n = 12$)

$Fv = \$35,000 \times 1.430768 = \$50,076.91$

End of:	First year	Second year	Third year	Fourth year	Fifth year
$35,000	$39,438.88		$50,076.91		

c. At the end of the fifth year, $i = .01$ or $i = 1\%$, and $n = 12 \times 5 = 60$.

$A = \$35,000(1 + .01)^{60} = \$35,000(1.816697)$

$Fv = \$63,584.38$ or, by the tables,

$A = \$35,000 \times$ (table value for $i = 1\%$ and $n = 60$)

$A = \$35,000 \times 1.816697$

$Fv = \$63,584.38$

End of:	First year	Second year	Third year	Fourth year	Fifth year
$35,000	$39,438.88		$50,076.91		$63,584.38

Example 8

Eddie Koslowski owns and operates a commercial sign business in his home town of Liverpool. Three years ago he needed to make some improvements to his business, so he borrowed $10,000 from Merchants National Bank. Eddie signed a promissory note agreeing to repay the loan principal plus 10% interest compounded quarterly at the end of three years. Merchants National sold the promissory note to East Coast investors at the end of 2 years, agreeing to accept a 10% simple discount on the final value of the note at the end of the term of the loan. Determine (a) the future value of the note at the end of the term of the loan, (b) the value of the note at the time of sale to East Coast Investors, and (c) the amount that Merchants National received for the note.

Solution

a.   The compound interest formula can be used directly to determine the future value of the note at the end of 3 years.

$$A = Fv = P(1 + i)^n$$
$$Fv = \$10,000(1 + .10/4)^{3 \times 4} = \$10,000(1.344889) = \$13,448.89$$

b.   Similarly, the same formula can be used to calculate the value of the note at the time of sale to East Coast Investors, 2 years into the term of the note.

$$A = Fv = P(1 + i)^n$$
$$Fv = \$10,000(1 + .10/4)^{2 \times 4} = \$10,000(1.21840) = \$12,184.03$$

c.   The discounted price paid to Merchants National can be determined by using our simple discount formulas, with 1 year left on the loan.

$$D = MRT = \$13,448.89 \times .10 \times 1 = \$1,344.89$$

Amount Merchants National received = $\$13,448.89 - \$1,344.89 =$ $12,104.00

So, Merchants National will make $2,104.00 on the loan, and East Coast Investors will make $1,344.89.

Obviously a good comprehension of simple interest, simple discount, compound interest, future value, and present value is essential to understanding and working with a variety of financial and business transactions.

CHECK YOUR KNOWLEDGE

Future Value of a Compound Amount

Solve each of the following problems. Round your answers to the nearest cent.

Cissy Washington has $10,000 in a savings account that pays 6% compounded quarterly. She hopes to be able to let the account grow from interest it earns for several years without withdrawing any money or adding any other money

For Your Information

THE POWER OF TAX DEFERRED GROWTH

The Rule of 72

The **Rule of 72** is a rule of thumb for quickly estimating how many years it will take for the principal of a compound investment to double in value. It is a technique of mental estimation and is therefore only an approximation. To estimate by the Rule of 72, we simply divide 72 by the yearly interest rate, and the quotient is the estimated number of years for the principal to double.

For example, at 12% the principal would double in about $72/12 = 6$ years. Indeed, if we invested $50,000 at 12% compounded annually for 6 years,

$$Fv = \$50,000(1.12)^6 = \$98,691.13 \text{ which is almost doubled.}$$

Taxable Investment versus a _Tax Deferred Investment_

Some investments, such as Individual Retirement Accounts (IRAs) are *tax-deferred* investments, which means that you do not pay taxes on the investment's earnings each year, but that you can defer paying those taxes until you withdraw the money from the account in which it is invested. Other investments are taxable investments, which means you pay taxes on investment's earnings each year at the end of that year.

Assume that Holly Green invests $50,000 in a tax-deferred account and Matt Green invests $50,000 in a taxable account, both with an

to the account. Cissy would like to know what the future value of the account will be in

1. two years
2. three years
3. five years

Jacob Jensen has been talking to Cissy and trying to convince her that he could invest her money in other ways that would earn her an average of 12% per year compounded annually on her money. Assuming Jake is accurate in his prediction, use his average rate of interest to determine the future value of Cissy's $10,000 in

4. two years
5. three years
6. five years

expected growth rate of 12%. Also assume that Matt and Holly are in the 33% federal tax bracket. Each year Matt will have to pay one-third of his investment's earnings to federal taxes, so only two-thirds of his earnings are reinvested. The effect of this ⅓ loss each year is to reduce his 12% growth to ⅔ × 12% = 8%. Holly will pay no taxes on her earnings each year, so she will receive the full 12% growth rate.

Holly's investment will almost double every 72/12 = 6 years

Matt's investment will almost double every 72/8 = 9 years

Number of Years	Matt's taxable investment (approx.)	Holly's tax deferred investment (approx.)
0	$50,000	$50,000
3		
6		$100,000
9	$100,000	
12		$200,000
15		
18	$200,000	$400,000

Essentially, the tax dollars that Holly is able to keep in her investment allow her investment to grow almost twice as fast as Matt's investment.

You might want to compute the exact time values of Matt and Holly's investments to see that the approximations are pretty good estimates.

9.2 EXERCISES

Solve each of the following, round answers to the neatest cent.

1. Frank Fitzgibbons loaned Art MacArthur $1,500 for 1 ½ years at 14% compounded semiannually. What is the future value of Frank's money invested in this loan?

2. Aiesha Thompson purchased a 2 ½-year $7,000 certificate of deposit, which pays 8% compounded quarterly. What is the future value of

the certificate at the end of (a) one year? (b) two years? (c) the term of the certificate?

3. Sid Belafonte's parents gave him a $1,000 savings bond which pays 6% compounded quarterly. If Sid doesn't cash in the bond for 10 years, what is the future value of the bond in (a) five years? (b) eight years? (c) ten years?

4. How many years would it take for the future value of Sid Belafonte's bond to at least double in value? (Think about this problem and how

Answers to CYK: 1. $11,264.93 2. $11,956.19 3. $13,468.55 4. $12,544.00
5. $14,049.28 6. $17,623.41

you could reason out the answers using the future value tables, or a calculator.)

5. Marcie Baker belongs to an investment club that invests its members' money and the members share in the earnings proportionally to what they have each invested. Marcie has $5,000 of her money invested. The club's past experience has been that it earns about 23% on its investments annually. Marcie would like to project into the future and predict the value of her investment assuming that the club's 23% average rate of return continues. Assuming that it does, what will the future value of her investment be in (a) one year? (b) five years? (c) ten years?

6. Catherine Kinney's investments in her senior years have been primarily in bank certificates of deposit. She has a 5-year certificate with $40,000 invested at 12% compounded monthly. What is the future value of the certificate in (a) one year? (b) three years? (c) five years?

7. The Jordan Historical Society received a grant from the State Historical Society to renovate its museum building, and will not need to use the $45,000 grant money until all work on the project is complete 10 months from now. The board of directors decided to invest the money at 12% compounded monthly for 9 months. What is the future value of their investment?

Central City Business Consortium borrowed $100,000 from Omaha Savings and Commerce Bank at 10% compounded quarterly for 5 years. CCBC signed a note agreeing to pay back the principal plus interest at the end of the term of the loan. At the end of 4 years, Consolidated Investors Corporation bought the note from Omaha Savings and Commerce at a 12% simple discount of the full value of the note (principal plus interest). Use this information to answer exercises 8 through 10.

8. What was the future value of the note at the end of the term of the loan?

9. What was the value of the note at the time that Consolidated purchased it?

10. What was Consolidated's purchase price of the note?

9.3   PRESENT VALUE OF A COMPOUND AMOUNT

Learning objective
Calculate the present value of a compound amount using the present value formula or tables.

In Sections 9.1 and 9.2, we learned how to find a compound amount (future value) when given the amount of the principal plus the terms of interest. In this section, the problems presented will be just the opposite; that is, we will be given the compound amount that we need to attain in the future and our unknown variable will be the principal, which we will call the present value. The **present value of a compound amount** is the amount of principal that we need to invest *today* to attain our desired future value, as depicted in Figure 9.5.

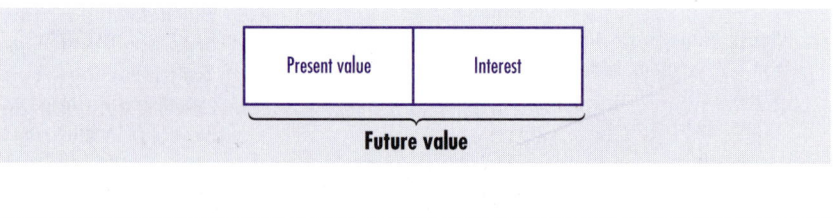

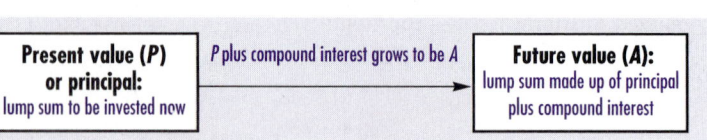

Figure 9.5

Present value of a compound amount

P here is the principal, which will grow to be the compound amount, A, also called the future value. P in this case is also called the present value of the compound amount.

Example 9

Cheryl Terry wishes to have $12,000 2 years from now in order to buy the car she has her eye on. What amount of money does she need to invest today at an interest rate of 6% compounded quarterly in order to attain her goal?

Solution

This is a compound interest problem in which we know the future value A ($12,000), and we need to determine P (present value) if it is to grow at an interest rate of 6% compounded quarterly for 2 years.

$$A = P(1 + i)^n$$

$$A = P(\text{table value for } i = 1.5\% \text{ and } n = 8)$$

$$A = \$12,000$$

$$i = \frac{6\%}{4} = 1.5\%$$

$$n = 2 \text{ years} \times 4 \text{ quarters per year} = 8$$

$$\$12,000 = P(1.126492)$$

Next, solve for P by dividing both sides by 1.126492.

$$P = \frac{\$12,000}{1.126492} = \$10,652.539 = \$10,652.44 \text{ (rounded)}$$

The above calculation for P is the same as

$$P = \$12,000 \times \frac{1}{1.126492}$$

This fraction, 1/1.126492, is the same as .8877111, and the calculation is the same as

$$P = \$12,000 \times \frac{1}{1.126492} = \$12,000 \times (.8877111)$$

This value can be found in the present value tables.

Refer back to Table 9.2 (page 345), which has a sample of the present value tables for a compound amount at the 1.5% rate. Looking down column B, present value, to where the n column equals 8, the intersection of that row and column is the value .8877111.

We can check our solution by solving the compound interest formula for A, using $P = \$10,652.53$, $i = .015$, and $n = 8$. If A is the desired $12,000, then we have correctly found the present value.

$$A = \$10,652.54 \times (1.015)^8 = \$12,000.007$$

The slight discrepancy here is due to the rounding of $10,652.539 to $10,652.54.

What does our answer of $10,652.54 mean to us? It means that at 6% interest compounded quarterly, Cheryl Terry must deposit or invest $10,652.54 *today* so that this amount plus interest will accumulate to $12,000 in 2 years time and Cheryl may buy her car. The $12,000 amount is the *future value* of her account. The *present value* is the amount that must be in the account *at the present time* to generate the necessary interest to achieve that future value in 2 years.

Once again, it is important to understand exactly what the numbers in the table are telling us. Let us refer to the number .8877111 that we obtained from the present value table. If Cheryl Terry had wished to have $1 two years from now, at 6% interest compounded quarterly, she would have had to invest approximately $.89 today. (The $.89 came from rounding .88771112.) However, Cheryl wished to have $12,000, not $1, so we needed to multiply .88771112 by $12,000.

It is also important to understand that the present value of a compound amount uses the same compound interest formula, $A = P(1 + i)^n$, that we used in Section 9.1. The difference is that we solve the formula for P, called our principal or present value.

$$A = P(1 + i)^n$$

$$\frac{A}{(1 + i)^n} = P\frac{\cancel{(1 + i)^n}}{\cancel{(1 + i)^n}}$$

$$P = A\frac{1}{(1 + i)^n} \quad \text{present value formula}$$

$$P = \$12,000 \times \frac{1}{(1 + .06/4)^8}$$

$$= \$12,000 \times .88771112$$

$$P = \$10,652.54 \text{ (rounded)}$$

CALCULATOR SOLUTION	
Keyed entry	Display
AC 0.6 ÷ 4 + 1 =	1.015
x^y 8 =	1.1264926
1/x	.8877111
× 12000 =	10652.533

Let's look at another problem in which we use only the tables to solve the problem. Please note that although tables are a helpful tool in making our work a bit easier, tables are not available for all situations. Consequently, you should be familiar with the formula and be able to use it if necessary.

Example 10

H. R. Jason anticipates the need to replace an office copier in 5 years. How much should the company invest in a compound interest account paying 18% compounded monthly so that it will have $5,000 in 5 years?

Solution

Refer to Figure 9.6. The future value (known) is $5,000. We are trying to determine the lump sum.

$$P = A \times (\text{present value table value for } n\ i)$$

$$A = \$5,000$$

$$i = \frac{18\%}{12} = 1.5\%$$

$$n = 5 \text{ years} \times 12 \text{ periods per year} = 60$$

$$P = (\$5,000)(.409296) \text{ (see Table 9.2, page 345)}$$
$$= \$2,046.4798, \text{ or } \$2,046.48$$

We can check our answers by calculating

$$A = P(1 + i)^n = \$2,046.48(2.4432198) = \$5,000$$

Figure 9.6

Present value formula

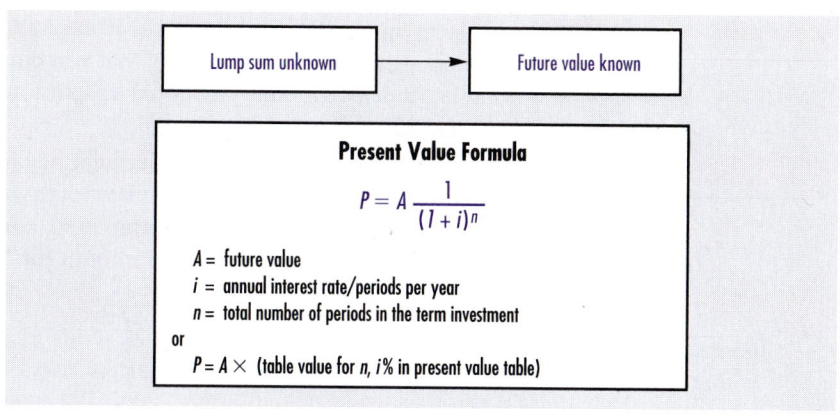

Present Value Formula

$$P = A \frac{1}{(1 + i)^n}$$

A = future value
i = annual interest rate/periods per year
n = total number of periods in the term investment
or
$P = A \times$ (table value for n, $i\%$ in present value table)

If H. R. Jason invests $2,046.48 in a compound interest account paying 18% compounded monthly and leaves it there untouched for 5 years, the account will be worth $5,000.

CHECK YOUR KNOWLEDGE

Present Value of a Compound Amount

Determine the present value of the compound amount for each set of conditions. Round your answers to the nearest cent.

	Future value	Rate per year	Compounded	Number of years	Present value
1.	$7,000	10%	annually	5	_____
2.	$7,000	10%	semiannually	5	_____
3.	$7,000	10%	quarterly	5	_____
4.	$8,000	6%	monthly	3	_____
5.	$3,750	8%	semiannually	8	_____
6.	$14,900	12%	monthly	7	_____

Solve the following word problems. Round dollar amounts to the nearest cent and rates to the nearest hundredth of a percent.

7. Your company needs to have $40,465 five years from now. If the interest rate at your bank is 6% compounded monthly, how much will you need to deposit into your account today in order to have the desired amount in 5 years?

8. If a compound amount in 1 year is $575,000, what is the present value of this amount if the interest terms are 8% compounded quarterly?

9. The interest rate at your bank is 12% compounded quarterly. If someone gave you a choice of either receiving $27,443 six years from now or receiving a lesser amount today that you could deposit in your bank, what is the least amount you could accept for payment today and still have the $27,443 value in 6 years?

10. Nancy Wilkinson won a contest that will pay her $100,000 three years from now. How much must the contest organizers deposit today into a compound interest account paying 10% compounded quarterly in order to have the required payoff amount for Nancy in 3 years?

Answers to CYK: 1. $4,346.45   2. $4,297.40   3. $4,271.90   4. $6,685.16   5. $2,002.16
6. $6,459.39   7. $29,999.62   8. $531,211.45   9. $13,500.14
10. $74,355.60

9.3   **EXERCISES**

Solve the following exercises. Round each dollar amount to the nearest cent and each rate to the nearest hundredth of a percent.

In exercises 1 through 8, determine the present value of the compound amount.

	Future value	*Rate*	*Compounded*	*Time*	*Present* value
1.	$1,000	7.5%	annually	3 years	_____
2.	$10,500	9%	semiannually	2 years	_____
3.	$6,300	6%	monthly	1 ½ years	_____
4.	$23,450	14%	quarterly	4 years	_____
5.	$81,540	8%	annually	24 months	_____

6. Determine the principal that must be deposited at 10% compounded annually to have an amount of $5,000 in 5 years.

7. MacGregor Sports Shoppe in Kermittown agrees to pay a $5,000 scholarship to the outstanding junior high school athletes, one male and one female, at the end of successful completion of their first year of college (i.e., 5 years after they receive the award). How much must MacGregor invest in a compound interest account paying 12% compounded quarterly in each winner's name in order to have the correct amount to award them at the end of 5 years?

8. Bob Farley wants to have $11,500 to purchase a new auto 2 years from now. How much must he invest in a compound interest account paying 8% compounded semiannually in order to achieve his goal?

Use the following information to solve exercises 9 through 12. Mary Walton's parents know that Mary is going to attend a state college where the tuition, room and board, and other costs for each year for the next 4 years will be approximately $6,000. Mary's parents have saved for her education and they decide to invest money in four separate certificates of deposit (compound interest accounts), each of which will mature at the beginning of a new academic year (i.e., the first will mature at the beginning of her freshman year, the second at the beginning of her sophomore year, and so on). Mary will begin college 1 year from

now. Each certificate of deposit will earn 10% compounded quarterly.

9. How much must be invested in the first CD in order for Mary to have $5,800 at the beginning of her freshman year?

10. How much must be invested in the second CD in order for Mary to have $5,800 at the beginning of her sophomore year?

11. How much must be invested in the third CD in order for Mary to have $5,800 at the beginning of her junior year?

12. How much must be invested in the fourth CD in order for Mary to have $5,800 at the beginning of her senior year?

Use the following information to solve exercises 13 through 17. Marcia Jankowski worked for 5 years after graduation from college and saved her money in order to pay for 3 years of law school, which she will begin 1 year from now. Edinborough Bank will offer Marcia the following rates on certificates of deposit:

1 year	4% compounded quarterly
2 year	6% compounded quarterly
3 year	8% compounded quarterly

13. How much must Marcia deposit in the 1-year CD in order to have $15,000 for her first-year costs?

14. How much must she deposit in a 2-year CD in order to have $15,750 for her second-year costs?

15. How much must she deposit in a 3-year certificate in order to have $16,500 for her third-year costs?

16. Assuming Marcia does follow this plan, how much will she have invested for her law degree?

17. How much interest will she have earned if she follows this plan?

Solve the following problems. Round each dollar amount to the nearest cent and each rate to the nearest hundredth of a percent.

	Principal	R	Period	Total time	Future value
18.	$19,300	12%	quarterly	5 years	_____
19.	_____	18%	monthly	20 months	$450
20.	$3,250	10%	semiannually	6 years	_____
21.	_____	8%	quarterly	10 quarters	$2,780
22.	$6,500	15%	daily	2 years	_____
23.	$19,300	12%	quarterly	5 years	_____
24.	_____	18%	monthly	20 months	$1,000
25.	$3,250	10%	semiannually	6 years	_____
26.	$2,870	8%	quarterly	10 quarters	_____
27.	_____	15%	daily	2 years	$6,550

28. Hartnett Heating and Air Conditioning borrowed $75,000 at 12% interest compounded monthly for 1 year. At the end of the loan term, the company paid off its total debt (principal and interest). How much did Hartnett pay back totally, and how much interest did it pay?

29. The owners of Happy Hideaway Campground anticipate that an expenditure of $16,500 will be necessary to renovate the lavatory and laundromat areas of the campground. They wish to set aside the money now for the renovation project, which will begin in 1 year. If the owners can invest the money they set aside at 10% interest compounded quarterly, what is the minimum they should invest in order to meet their goals?

30. Anslow Furniture Company deposited $53,000 into an account paying 6% interest compounded monthly and left it there for 8 months. How much interest did the company earn on the investment?

31. Hamilton Warehousing must replace two forklift trucks 5 years from now. If each truck will cost $19,500, what is the present value of the trucks if money can be invested at 6% interest compounded monthly?

32. The interest rate at your bank is 8%, compounded quarterly. If someone gave you a choice of either receiving $15,500 8 years from now or receiving a lesser amount today that you could deposit in your bank, what is the least amount you could accept for payment today and still have the $15,500 value in 8 years?

9.4 EFFECTIVE RATE

Savings institutions today offer a wide variety of opportunities for people to safely invest money in compound interest accounts. Such accounts offer a variety of annual interest rates and compounding periods. Investors need a

Learning objective
Identify *nominal rate* and *effective rate* and be able to distinguish between them.

means by which these different forms of compound interest opportunities can be compared so that they can choose wisely. The effective rate is a tool we can use to make comparisons between different compound interest opportunities.

Effective rate is the rate of simple interest you would need to receive from a bank so that the simple interest account will earn the same amount of interest in 1 year as a specific compound interest account would earn in 1 year. **Nominal rate** is the compound interest rate for 1 year.

Example 11

The Whimsical National Bank of Condolenceville offers a compound interest account with a nominal (yearly) rate of 8% compounded quarterly. What annual simple interest rate (effective rate) would yield the same **earnings**?

Solution

Simple interest:

$$A = P + (PRT)$$
$$A = P(1 + RT), \qquad T = 1 \text{ year}$$
$$A = P(1 + R)$$

Compound interest:

$$A = P(1 + i)^n$$

where

$$i = \frac{\text{nominal (yearly rate)}}{\text{number of compounding periods in 1 year}}$$

In this case,

$$A = P\left(1 + \frac{.08}{4}\right)^4$$

Let's let our principal P be $1, then each dollar we invest ($P = \$1$) in the compound account will accumulate to a future value of $A = 1(1 + .02)^4 = 1.0824322$. Each $1 we invest in the simple interest account will accumulate to $A = 1(1 + R)$.

Simple interest:

$$A = P(1 + PR), \qquad P = \$1$$
$$A = 1(1 + 1R)$$
$$A = 1 + R \quad \text{(This } R \text{ will be our effective rate, so we will label it } R_{\text{eff}}.)$$
$$A = 1 + R_{\text{eff}}$$

Compound interest:

$$A = P(1 + i)^n$$

$$P = \$1$$

$$i = \frac{.08}{4} = .02$$

$$n = 4 \text{ periods in 1 year}$$

$$A = 1(1 + .02)^4$$

$$A = 1.0824322$$

Now the A value for simple interest must *equal* the A value of compound interest, so let's set them equal to each other.

simple interest = compound interest

$$1 + R_{eff} = 1.0824322$$

$$R_{eff} = 1.0824322 - 1$$

$$R_{eff} = .0824322$$

or

$$R_{eff} = 8.24\% \text{ (rounded)}$$

Effective rate formula

Learning objective
Determine the effective rate of a compound interest problem by using a formula or a table.

$$R_{eff} = (1 + i)^m - 1$$

$$i = \frac{(\text{nominal rate})}{m}$$

m = number of compounding periods in 1 year

We can determine effective rate either by using a calculator alone, or by using compound interest tables.

Three-step method for finding the effective rate using compound interest tables

Step 1: Look up table value for m and $i\%$ in a compound interest table.

Step 2: Subtract 1 from this table value. This is the effective rate in decimal form.

Step 3: Multiply the result of step 2 by 100 to give effective rate in % form.

Example 12

Determine the effective rate of a compound interest account paying 6% interest compounded monthly.

Solution

First we will use the three-step method:

Step 1:   Determine m and $i\%$ and look up the table value $m = 12$ compounding periods in one year:

$$i\% = \frac{6\%}{12} = .5\% \text{ per month}$$

The table value for $m = 12$; $i\% = .5\% = 1.061678$.

Step 2:   Subtract 1 from table value:

$$R_{\text{eff}} = 1.061678 - 1 = .0616787 \text{ (decimal form)}$$

Step 3:   Multiply by 100 to put in % form:

$$R_{\text{eff}} = .061678 \times 100 = 6.17\% \text{ (rounded)}$$

Second, let's look at the calculator solution for this problem:

$$R_{\text{eff}} = (1 + i)^m - 1$$

$$= \left(1 + \frac{.06}{12}\right)^{12} - 1$$

$$= 1.06168 - 1 \text{ (value from compound interest table)}$$
$$= .06168, \text{ or } 6.17\% \text{ (rounded)}$$

CALCULATOR SOLUTION	
Keyed entry	**Display**
AC .06 ÷ 12 + 1 =	1.005
y^x 12 =	1.0616778
− 1 =	.0616778

Example 13

Emilio Fernandez has an opportunity to invest his money in an account paying 12% compounded quarterly or into a second account paying 12% compounded monthly. Which account has the better yield?

Solution

To determine the better yield, we compare the effective rates of the two opportunities. The higher effective rate will be the better yield. (The *yield* on an investment is the same as the *effective rate* of the investment: The two expressions are used interchangeably.) First we use the three-step method. Determine the effective rate for the 12% quarterly:

Step 1: $m = 4$; $i\% = 3\%$ table value $= 1.125509$

Step 2: $R_{eff} = .125509$

Step 3:   or 12.55%

Determine the effective rate for 12% monthly:

Step 1: $m = 12$, $i\% = 1\%$ table value $= 1.126825$

Step 2: $R_{eff} = .126825$

Step 3:   or 12.68%

    The second rate has a better yield, so Emilio would earn .13% more interest by investing at 12% monthly.

    If you were to use a calculator the solution for this problem would be as follows:

12% interest compounded quarterly

$$R_{eff} = \left(1 + \frac{.12}{4}\right)^4 - 1 = .1255, \text{ or } 12.55\%$$

12% interest compounded monthly

$$R_{eff} = \left(1 + \frac{.12}{12}\right)^{12} - 1 = .1268, \text{ or } 12.68\%$$

Again, we can see that the first rate has a better yield, so Emilio would earn .13% more interest by investing at 12% monthly.

CHECK YOUR KNOWLEDGE

Effective Rate

Solve the following problems. Round rates to a hundredth of a percent.

A credit union offers 12.0% compounded quarterly. Use this information to answer the following.

1. What is the nominal rate?
2. What is the periodic rate?

3. What is the compounding period?

4. Determine the effective rate for deposits made in this credit union.

Compute the effective rate (R_{eff}) for each of the following:

Nominal rate (yearly rate)	Compounding periods	Number of compounding periods per year	R_{eff}
5. 12%	semiannually	_____	_____
6. 10%	quarterly	_____	_____
7. 6%	monthly	_____	_____

Tara Buffa can earn 8% compounded quarterly at Blarneyville Bank, or she can earn 9% compounded semiannually at Morristown First National Bank. Use this information to answer the following.

8. Determine the effective yield at Blarneyville Bank.

9. Determine the effective yield at Morristown First National.

10. Which bank has the higher yield, and what is the difference?

9.4   EXERCISES

Solve the following exercises (round rates to the nearest hundredth of a percent).

Compute the effective rate (R_{eff}) for each of the following. *Note:* You will need a scientific calculator to complete exercises 7 and 8.

Nominal rate (yearly rate)	Compounding periods	Number of compounding periods per year	R_{eff}
1. 18%	semiannually	_____2_____	_____
2. 18%	quarterly	_____	_____
3. 18%	monthly	_____12____	_____
4. 15%	semiannually	_____	_____
5. 15%	quarterly	_____4_____	_____
6. 12%	monthly	_____	_____
7. 18%	weekly	_____52____	_____
8. 18%	daily (360 days)	_____	_____

Answers to CYK: *1.* 12%   *2.* 3%   *3.* quarter   *4.* 12.55%   *5.* 12.36%   *6.* 10.38%
7. 6.17%   *8.* 8.24%   *9.* 9.20%   *10.* Morristown First Nat'l; .96% difference

9. Chemung Bank and Trust offered a compound interest account paying 10% compounded quarterly. What should it advertise as the effective yield on the account?

Use the following information to answer exercises 10 through 12. Juan Carlos is wondering whether it would be wiser to invest in a compound interest investment paying 10% compounded quarterly or one paying 11% semiannually.

10. What is the effective rate of the 10% investment?
11. What is the effective rate of the 11% investment?
12. Determine the difference of the two yields (higher yield − lower yield), and determine how much more interest would be earned in 1 year on a $1,000 investment.

13. Merchants State Bank advertises that its regular savings account pays 6% interest compounded monthly for an annual yield of 6.2%. Is its claim correct?

14. First Trust and Deposit advertises that its effective yield on 10% interest compounded quarterly is 10.5%. Determine the magnitude of the error in this claim.

15. (Scientific calculator required) Jackson Investment offers a money market account for 8% compounded daily (365 days). If you invested $10,000 with Jackson for exactly 1 year, (a) what would be the effective rate of the account and (b) how much interest would you earn?

CHAPTER REVIEW EXERCISES

The following problems review all mathematics and business concepts presented in the chapter. If you experience any difficulty solving a problem, go to the appropriate chapter section and review the examples provided. Express all answers as decimals rounded to the nearest one hundredth, or dollars rounded to the nearest cent, unless otherwise instructed.

Given the following interest rates per year, the type of compounding periods per year, and the number of years, determine i and n for problems 1 through 5.

	Rate/ year	Compounding	Number of years	i	n
1.	10%	annually	5		
2.	12%	monthly	4		
3.	4%	semiannually	7		
4.	6%	quarterly	12		
5.	13%	weekly	8		

6. Determine the compound amount on $1,000 at 6% interest per year compounded semiannually for 2 years using the multistep method.
7. How much interest is earned on the investment in problem 6?

8. Determine the compound amount and compound interest earned on $2,000 at 6% interest per year compounded semiannually for 2 years using the compound interest formula.
9. How much interest is earned on the investment in problem 8?
10. Determine the compound amount on $3,000 at 6% interest per year compounded semiannually for 2 years using the compound interest tables.
11. How much interest is earned on the investment in problem 10?
12. Fipps Landscaping Company invested $20,000 in a compound interest account paying 12% compounded monthly, and left the money in the account for 3 years without making any other deposits or withdrawals on the account. What was the future value of the account at the end of the 3 years?
13. How much interest is earned on the investment in problem 12?

Kimberly Marsallis won $30,000 in her state's lottery and chose to invest $15,000 of it in an insurance investment that guaranteed her 10% per year compounded quarterly for the first 5 years and a minimum of 4% per year compounded quarterly for the next 5 years. Use this information to answer problems 14 through 16.

14. What is the future value of the investment at the end of the first 5 years?

15. What is the future value of the investment at the end of 10 years, assuming that she only earns the minimum 4% per year?

16. What is the total amount of interest earned on the investment as of the end of the tenth year, assuming the conditions of problem 15?

17. Frank Stankowicz owns a dump truck for hauling away construction debris from construction sites that his crews are working on for his New Homes from Old Remodeling and Renovation Company. Frank anticipates that he will need to replace the truck body 5 years from now at a cost of approximately $18,000. How much must Frank invest in his credit union account, which pays 10% compounded quarterly, so that he will have the $18,000 needed in 5 years?

18. How much interest will Frank earn on the investment in problem 17?

Tawanda Jackson has the opportunity to invest some money in her credit union at 10% compounded quarterly, or she can go to her local bank and purchase a savings certificate of deposit that pays 9% compounded monthly. Use this information to answer problems 19 and 20.

19. What is the effective rate at Tawanda's credit union?

20. What is the effective rate at Tawanda's local bank?

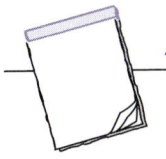

EXPRESS YOUR UNDERSTANDING

Compose one or two well-written sentences to express the requested information in your own words.

1. Describe the difference between compound interest and simple interest.

2. Explain how the future value in a compound interest problem differs from the future value in a simple interest problem.

3. Explain why $1,000 compounded daily at 12% would yield a higher interest in one year than $1,000 compounded weekly at 12%.

4. Describe how we use the concept of future value in compound interest loans.

5. Describe the steps you would use to calculate the future value of $1,000 at 12% compounded monthly for 3 years.

6. Describe what we mean by the present value of a compound amount.

7. Describe the steps you would use to calculate the present value of a compound amount of $1,000 at 12% compounded daily.

8. Explain the difference between present value and future value in a compound interest problem.

9. Explain what is meant by the nominal rate of a compound interest problem.

10. Explain what is meant by the effective rate of a compound interest rate.

Mind Your Business

Now that you have served as business manager of Media World for over 2 years, you have noticed that for the last 12 months the business has regularly had cash assets of $20,000 or more at the end of each month. You have found a 6-month certificate of deposit that pays 6% compounded monthly. To obtain this rate of interest you must invest a minimum of $2,000. You also have found a high interest savings account that pays 3% compounded daily. Based on the cash position of the business at this time, assume that you decide to invest $4,000.

Activity 1

Assume that you will invest the full amount in a certificate of deposit.

a. What would be the future value of the CD at the end of the investment term?

b. How much interest would the investment earn for the period?

c. What would be the effective rate of the investment?

Activity 2

Assume that you decide to invest the $4,000 in the high-interest savings account.

a. What future value would you expect to receive at the end of 6 months?

b. How much interest would the investment earn for the period?

c. What would be the effective rate of the investment?

Activity 3

Write a recommendation to the partners justifying a short-term investment of business funds at this time, recommending one of these investments. Include your analysis from activities 1 and 2 in your recommendation.

SELF-TEST

A. Terminology

Complete each of the following items using the key terms presented at the beginning of the chapter. Check your responses against the answer key at the end of the test.

1. _____ _____ is produced when interest is earned on principal plus previously earned interest.

2. The future value of the principal plus compound interest is called the _____ _____.

3. The interest rate per compounding period is called the _____ interest rate.

4. The amount of money you must invest today (or the principal) in a compound interest account to produce a known compound amount sometime later is also called the _____ _____ of a compound amount.

5. The yearly interest rate is also called the _____ rate.

6. The simple interest rate necessary to generate the same amount of interest in 1 year as a particular compound interest rate is called the _____ _____.

B. Calculation

The following concepts and short problems are designed to test your understanding of the objectives identified at the beginning of the chapter. Answers are provided at the end of the test.

Emilio Santiago's father, Hernando, has invested $50,000 in a long-term investment account that guarantees to pay 12% interest compounded quarterly for 10 years. Hernando cannot make any more deposits to the account, and there will be a severe penalty if he withdraws any money from the account before the designated end of the term of the investment. Use this information to answer questions 7 through 13.

7. What will be the value of the account at the end of the 10-year period?

8. How much interest will have been earned at the end of the 10-year investment?

9. Since Emilio had studied simple interest in school, he thought that the account would earn just half the interest in 5 years that it would in 10 years. Calculate the future value of the account at the end of the first 5 years.

10. How much interest did the account earn in the first 5 years?

11. How much interest did the investment earn in the second 5 years?

12. Emilio now knows that he was wrong because his dad's account was a compound interest account. How much more interest was earned in the last 5 years than in the first 5 years?

13. Emilio's interest about his dad's investment was growing now, and he wondered how much the account would be worth in $7\frac{1}{2}$ years. Help him out by calculating the future value of the investment for that time.

14. You wish to make a one-time deposit into a compound interest account paying 8% interest compounded quarterly, and you wish to have the value of the account be $5,000 at the end of 5 years. What is the present value of the $5,000 (i.e., How much must you deposit now to meet your goal)?

15. What is the effective yield of a certificate of deposit paying 11% compounded semiannually?

Answers to Self-Test

1. Compound interest 2. compound amount 3. periodic 4. present value
5. nominal 6. effective rate 7. $163,101.89 8. $113,101.89
9. $90,305.56 10. $40,305.56 11. $113,101.89 − $40,305.56 = $72,796.33
12. $72,796.33 − $40,305.56 = $32,490.72 13. $121,363.12 14. $3,364.86
15. 11.30%

10

Key Terms

annuity
future value annuity
future value (sum) of an
   ordinary annuity
future value (sum) of an
   annuity due
present value annuity
sinking fund
amortization

ANNUITIES

Learning Objectives

After completing the exercises in this chapter, you will be able to:

1. Explain the difference between a compound amount and an annuity.

2. Calculate the future value of an ordinary annuity by using the formula or the table for annuities.

3. Calculate the future value of an annuity due using the formula or the table for annuities.

4. Calculate the amount necessary to establish a present value annuity by using the formula or a table for annuities.

5. Calculate the regular payment necessary for establishing a sinking fund by using the formula or a table for annuities.

6. Calculate the equal regular payments that will amortize a specific sum of money over a set period of time.

7. Define key terms.

INTRODUCTION

In this chapter we complete the study of the basic building blocks that form the solid foundation on which is built the mathematics of finance, and the study of finance in general. Finance in its many sophisticated forms is founded on the mathematical models of simple interest, simple discount, compound interest, and annuities, which we began studying in Chapter 7 and will complete in this chapter.

An **annuity** is a special type of compound interest account. Thus far, we have presented compound interest, investments in terms of one lump sum—the *principal*—that remains in the account and earns interest. The principal increases only as the interest earned is added to it. An annuity provides a series of periodic deposits into the account or periodic payments out of the account, which add to or subtract from the account principal. In Section 10.1 we will examine future value annuities, and in Section 10.2 we will learn about present value annuities.

10.1   FUTURE VALUE ANNUITIES

Learning objective
Explain the difference between a compound amount and an annuity.

A **future value annuity** is an application of compound interest in which regular deposits are made into a lump sum account, adding to the principal in that account. This differs from the compound interest that we studied in Chapter 9, in that *the principal is increased by the interest earned, and by the regular payments,* as indicated in Figure 10.1.

p here is the regular equal payments that will each earn interest and accumulate to be a lump sum amount we will denote as *S,* also called the future value.

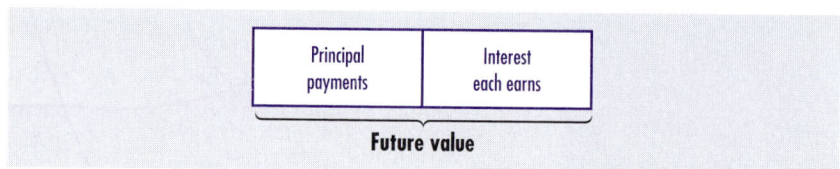

Ordinary Annuities

Suppose we agree to pay $100 per year, at the end of each year, into an account paying 8% compounded annually for 4 years. This type of account is called an *ordinary annuity,* which means that the payments are made at

Figure 10.1

Future value of an annuity

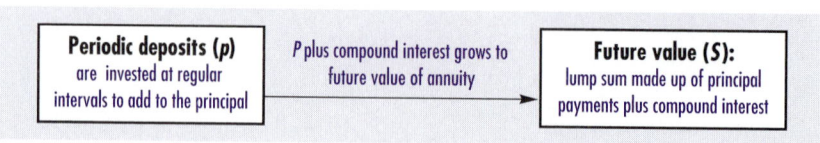

Figure 10.2

Time line for the future value of an ordinary annuity

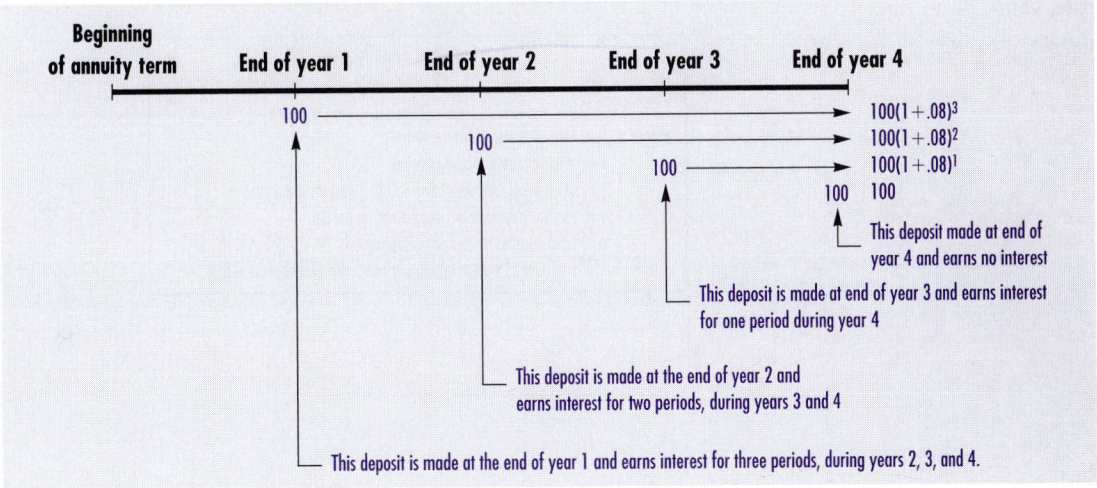

the end of each period. Consider the diagram in Figure 10.2, which illustrates the activity in this account.

The first $100 deposit grows to be $125.97; the second $100 grows to be $116.64; the third $100 grows to be $108; and the last $100 deposit earns no interest. The total value of the annuity (future value) is the sum of these parts.

$$\text{future value} = 100(1.08)^3 + 100(1.08)^2 + 100(1.08)^1 + 100$$
$$= 125.97 + 116.64 + 108 + 10$$
$$= 450.61$$

Learning objective

Calculate the future value of an ordinary annuity by using a formula or the table for annuities.

Notice that the value of the annuity increased as a result of regular deposits, which increased the principal and earned interest on that changing principal.

The formula diagrammed in Figure 10.3 allows us to compute the **future value** (or future sum) **of an ordinary annuity.**

Appendix B at the end of the book provides us with *amount of annuity* values in column C, which we can use instead of the formula to find the value of a future value annuity.

Example 1

Joel Bradd deposited $100 per year into an ordinary annuity paying 8% compounded annually for 4 years. What was the total value of the annuity at the end of the fourth year?

Figure 10.3

Formula for the future value of an ordinary annuity

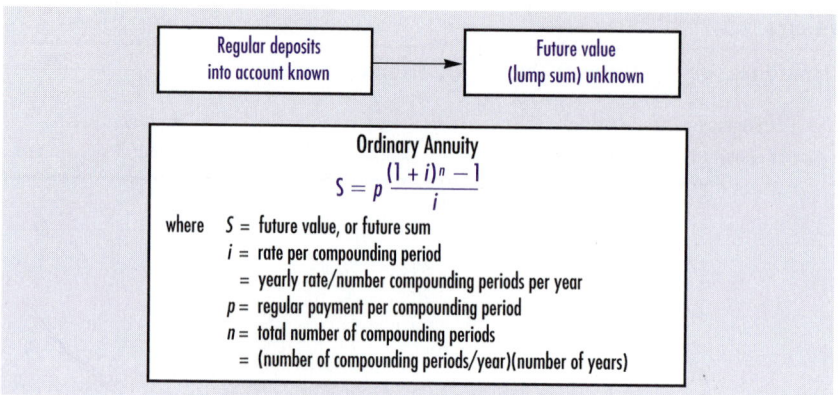

Solution

Refer to Figure 10.3. Regular deposits into the account (known) are $100 per year. We must determine the future value of the account.

$$p = \$100$$
$$i = 8\%, \text{ or } .08$$
$$n = 4$$

Table:

Table value for 8% page,
$n = 4$ in column C = 4.506114
S = (payment)(table value)
$S = 100 \times (4.506114)^*$
$S = \$450.61$

Formula:

$$S = p\frac{(1 + i)^n - 1}{i}$$

$$S = 100\frac{[(1 + .08)^4 - 1]}{.08}$$

$$S = 100(4.506112)$$

$$S = \$450.61$$

*This value can be calculated on a calculator, or it can be found in the table in Appendix B. In order to use the table you need only know that $n = 4$, $r = 8\%$, and that we are working with an annuity. Column C, entitled *Ordinary Annuity*, will provide the value to be multiplied by the regular payment to determine the future value. Find the page for the 8% rate, move down column C to the row opposite $n = 4$, and read 4.506112, which is the entire value of

$$\frac{(1 + .08)^4 - 1}{.08}$$

CALCULATOR SOLUTION

Keyed entry					Display
AC 1.08 y^x 4 =					1.360489
− 1 =					.360489
÷ .08 =					4.5061125
= 100 =					450.61125

Example 2

Sophia Warner decides she will deposit $2,000 a year into her savings account, which pays 6½% compounded annually. What amount of money will Sophia have in her savings account at the end of 5 years?

Solution

Refer to Figure 10.3. Regular deposits into the account (known) are $2,000 per year. We must determine the future value of the account.

$p = \$2,000$

$i = .065$ or $6\frac{1}{2}\%$

$n = 5$

Table:

Table value = 5.69364098*

$S = $ (payment)(table value)

$S = \$2,000 \times 5.69364098$

$S = \$11,387.28$

Formula:

$$S = p\frac{(1 + i)^n - 1}{i}$$

$$S = \$2,000\frac{(1 + .065)^5 - 1}{.065}$$

$S = \$2,000 \times 5.69364098$

$S = \$11,387.28$

*Turn to the Appendix for $i = 6.4\%$, column C, and look up the number where $n = 5$. This number is 5.69364098, which is then multiplied by $2,000 to give us

$S = 2,000(5.69364098) = \$11,387.28$ (rounded)

Example 3

Raenel Jones decides to deposit $50 per month into an annuity paying 12% compounded monthly for 3 years. What will the annuity be worth at the end of 3 years? How much interest will be earned?

Solution

Refer to Figure 10.3. Regular deposits into the account (known) are $50 per month. We must determine the future value of the account.

$$p = \$50$$

$$i = \frac{.12}{12} = .01 \text{ or } 1\%$$

$$n = \frac{12 \text{ periods}}{\text{year}} \times 3 \text{ years} = 36 \text{ periods}$$

Table:

Table value for $i = 1\%$, column C, and $n = 36$ is 43.076878

S = payment $\times$ table value

S = 50 $\times$ 43.076878

S = $2,153.84

Formula:

$$S = p\frac{(1 + i)^n - 1}{i}$$

$$S = \$50\frac{(1 + .01)^{36} - 1}{.01}$$

$$S = 50(43.076878)$$

$$S = \$2,153.84$$

Raenel's deposits total 50×36 payments = $1,800.00. The difference, $2,153.84 - $1,800.00 = $353.84, is the interest she earned.

Annuity Due

Learning objective
Calculate the future value of an annuity due using the formula or the table for annuities.

When the payments are made at the beginning of each payment period rather than at the end, the annuity is called an *annuity due.*

We can see that when compared to the amount earned by the ordinary annuity, the annuity due earned $36.05 more, which is 8% more interest earned. The effect of putting the money to work at the beginning of the period is to earn an extra period's interest.

Figure 10.4

Formula for the future value of an annuity due

$$Sd = p\frac{(1 + i)^n - 1}{i} \times (1 + i)$$

where

Sd = future value, or future sum, of annuity due

i = rate per compounding period

= yearly rate/number of compounding periods per year

p = regular payment per compounding period

n = total number of compounding periods

= (number of compounding periods/year)(number of years)

Figure 10.5

Time line for future value of annuity due

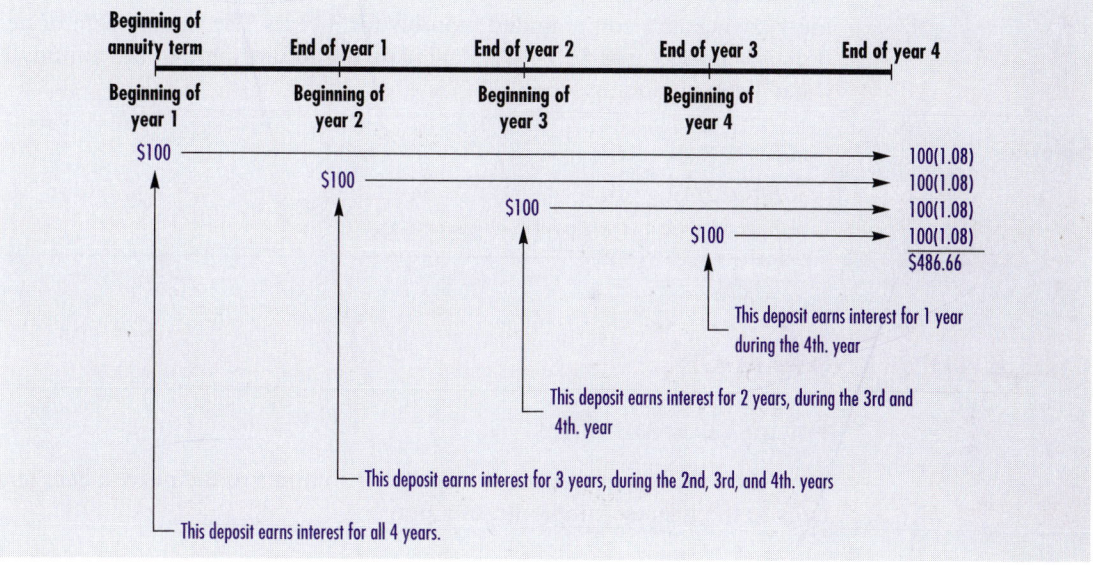

Example 4

Reconsider Example 2 in which Sonia Warner would deposit $2,000 each year into a savings account paying 6.5% for 5 years. Calculate the future value of her account if the deposits are made at the beginning of each year.

Solution

Step 1:   Calculate the *ordinary annuity:*

Table:

Table value for 6.5%, page
$n = 5$ in column C = 5.69364098
S = payment × table value
$S = \$2,000 \times 5.69354098$
$S = \$11,387.28195$

Formula:

$$S = \$2,000 \frac{(1.065)^5 - 1}{.065}$$

$S = \$11,387.28195$

Step 2:   Multiply the ordinary annuity amount by $(1 + i)$, or 1.065 in this case.

$Sd = (1 + i) \times S = (1.065)(\$11,387.28195) = \$12,127.45528$
$Sd = \$12,127.46$

Example 5

In Example 3, Raenel Jones deposited $50 per month into an ordinary annuity paying 12% compounded monthly for 3 years. The future value of her ordinary annuity was $2,153.84. Calculate the future value of her annuity if it had been an annuity due.

Solution

$$Sd = S \times (1 + i)$$
$$I = .12/12 = .01$$
$$Sd = \$2,153.84 \times (1.01) = \$2,175.38$$

CHECK YOUR KNOWLEDGE

Future Value Annuities

Solve the following problems. Round dollar amounts to the nearest cent and rates to the nearest hundredth of a percent.

For each of the compounding periods and time spans listed in problems 1 through 5, determine i and n:

	Compounded	Yearly rate	Number of years	i	n
1.	annually	7.25%	5		
2.	quarterly	12%	3		
3.	semiannually	10%	4		
4.	monthly	6%	7		
5.	daily (365 days)	15%	10		

6. Determine the future value of an ordinary annuity paying 12% interest compounded monthly for 5 years if a regular deposit of $100 is made monthly.

7. Sally Martin's parents began a savings program in which they deposited $200 quarterly into an annuity paying 8% interest compounded quarterly. They began this program when she was born and gave her the total 15 years later. How much was the ordinary annuity worth when she assumed ownership of it?

8. Molly Rainsford, attorney, has established an ordinary annuity retirement account into which she will deposit $3,000 semiannually until she is 65 years old. How much will Molly accrue in the account if she establishes it at age 29 and the account earns 11% interest compounded semiannually?

9. In problem 8, calculate the total amount of the payments Molly will have deposited.

10. Use the results of problems 8 and 9 to determine the total interest Molly's retirement account will earn.

11. Determine the future value of an annuity due paying 12% interest compounded monthly for 5 years if a regular deposit of $100 is made monthly.

12. If Sally Martin's parents had established an annuity due savings plan instead of the ordinary annuity (problem 7), how much would it have been worth when she assumed ownership?

10.1 EXERCISES

Solve the following exercises. Round dollar amounts to the nearest cent and rates to the nearest hundredth of a percent. Each annuity problem is an ordinary annuity unless it is otherwise stated as an annuity due (i.e., the payment is made at the end of each period unless otherwise stated).

Complete problems 1 through 5 by determining the value of the ordinary annuity at the end of the specified term.

	Regular payment	Rate per year	Payments made	Number of years	Future value
1.	$200	9%	annually	10	_____
2.	$300	10%	semiannually	20	_____
3.	$600	8%	quarterly	5	_____
4.	$1,000	12%	quarterly	7	_____
5.	$150	6%	monthly	3	_____

6. Determine the amount of an ordinary annuity for annual deposits of $1,000 at 9% compounded annually at the end of the fourth deposit.

7. What is the accumulated amount for annual deposits of $1,000 invested in an account paying 11% compounded semiannually following the eighth deposit?

8. Three years ago, Marley Symanski's parents began putting $75 each month into a savings account. If the terms of the interest were 12% compounded monthly, how much money has been accumulated?

9. What is the value at the end of 15 years of quarterly deposits of $2,000 with terms of 8% compounded quarterly?

Answers to CYK: 1. $i = .0725; n = 5$,   2. $i = .03; n = 12$ 3. $i = .05; n = 8$
4. $i = .005; n = 84$ 5. $i = .00041; n = 3,650$ 6. $8,166.96   7. $22,810.20
8. $2,521,377.90   9. $216,000   10. $2,305,377.90   11. $8,248.64
12. $23,266.51

10. Jason M. Goldman established his own retirement account 10 years ago by making semiannual deposits of $6,000 into an ordinary annuity that pays 10% compounded semiannually. (a) What is the value of his retirement account today? (b) How much interest has been earned?

11. Jason Goldman has discovered that he can obtain a better rate for the next 10 years at 11% interest compounded semiannually. Consequently, Jason established a new ordinary annuity account (beginning amount $0) and he will contribute $7,000 semiannually into it for the next 10 years. (a) What will the value of this account be at the end of the 10-year period? (b) How much interest will the account earn?

12. Jennifer Jones is an Army reservist as well as an accountant for a local auditing firm. Jennifer regards the salary she earns from her reserve activities as extra income that she can invest at this time. If Jennifer invests $500 every 3 months into an ordinary annuity paying 8% interest compounded quarterly, how much will she have accumulated at the end of 5½ years?

13. Arnie Mankewicz owns five video rental stores in the Springfield area, and they are realizing a good profit. Arnie decides to invest part of the profits in an annuity offered by United Life Insurance. United Life will guarantee Arnie 10% interest compounded quarterly for up to 5 years as long as he deposits $10,000 every quarter of the term of the guaranteed rate. Assuming that Arnie does fulfill the obligations of the investment, (a) what will be the value of his investment at the end of the 5-year term? (b) how much interest will he earn?

14. Everette Aldo receives $330 in dividends every quarter from a stock investment he holds. If Everette deposits his dividend into an ordinary annuity paying 14% interest compounded quarterly, how much will he earn in interest in an 11-year period?

15. Estelle Fareweathers plays drums in a dance band on weekends in addition to her full-time job at the Marcyville Savings and Loan. Estelle decided on her 35th birthday to establish her own retirement savings account by investing $400 of her weekend earnings every month into an ordinary annuity paying 12% interest compounded monthly. If Estelle makes these regular deposits until her 65th birthday, how much will this retirement account by worth?

16. Art and Janey Longly's son Jim has a disability that will likely prevent him from earning a full-time income as an adult. Art and Janey want to assure that Jim will be able to financially support himself after reaching the age of 21 years. Art and Janey have invested $1,000 monthly into an ordinary annuity earning 12% interest compounded monthly ever since Jim's first birthday. How much will they have accumulated for Jim on his 21st birthday?

17. Frank Hofstedder was injured while employed at Martin Industries and will receive disability pay since he will not be able to return to work. Martin has agreed to establish an education fund for Frank's 14-year-old son Ryan. For the next 4 years, Martin will deposit $2,000 per quarter into an ordinary annuity paying 8% interest compounded quarterly. If Martin fulfills its commitment for 4 years, how much will be available for Ryan's education when he goes to college in 4 years?

Use the following information to answer problems 18 through 20. Jan Greeley has deposited $250 per month in Fastgrowth Investments monthly for the last 10 years. Fastgrowth is an ordinary annuity investment that has offered 12% interest compounded monthly. At the end of the 10-year term, Jan withdrew her total investment and deposited it into a compound interest account paying 9% interest compounded yearly. Jan began a new annuity savings program with Fastgrowth, but due to changing economic conditions, the company could only guarantee her 10% compounded semiannually for the next 5 years. Jan will now deposit $1,500 semiannually into this new annuity.

18. What was the value of Jan's 10-year investment with Fastgrowth at the end of the 10-year term?

19. Recall that at the end of the 10-year plan, Jan put her accumulated savings in a straight compound interest account. If Jan leaves her investment in the compound account and allows it to grow in interest only for the next 5 years, what will its value be at the end of the 5-year term?

20. If Jan fulfills her commitment to the new annuity, what will its value be at the end of the 5-year period?

21. Tionn Wellman just agreed to begin an annuity plan with his insurance agent in which he will deposit $1,000 each quarter for 20 years. Tionn's annuity is an annuity due. Calculate the value of Tionn's annuity at the end of the 20 years, if he is guaranteed 6% compounded quarterly.

22. Wanda Seligman has been depositing $250 per month into an annuity due that guarantees a rate of 6% compounded monthly for the past 20 years. How much is her annuity worth now?

23. PhoungAnh Ho works for a company that makes surgical needles. She enrolled in a 401(k) program in which she deposits $100 each month and her company contributes an additional $50

each month. The company's 401(k) is an annuity due that pays 12% compounded monthly. (a) If PhoungAnh continues in this program for 10 years, what will be the value of her 401(k)? (b) If PhoungAnh continues for 20 years what will be the value of her 401(k)?

24. Ahmad Al-Mudhamga is ready to retire after contributing to an annuity due investment for 40 years. The investment has averaged a 10% rate throughout the 40 years that he has made payments of $550 quarterly. (a) How much is Ahmad's investment worth now? (b) How much interest has the investment earned?

25. Marcus Dangler invested money in a fixed-rate annuity due that paid a rate of 8% compounded quarterly for 10 years. He paid in $350 each quarter for 10 years, and earned a total interest of $7,563.50. What was the total value of his investment?

10.2   PRESENT VALUE ANNUITIES

A **present value annuity** is a form of a compound interest account that is established with a single lump sum, which periodically pays out of the account a specified number of equal payments that liquidate the account, as shown in Figure 10.6.

The lump sum initially invested earns interest, and therefore, the lump increases by the interest it earns. However, regular payments are deducted from the lump sum, so the lump is reduced in size by the regular payments. Each period these two changes occur to the original lump sum—interest added and payment deducted—until the lump is reduced to zero dollars (liquidated).

Figure 10.6

Present value annuity

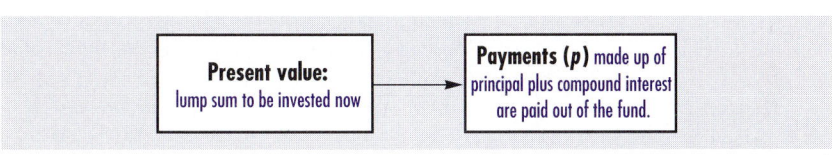

Present value: lump sum to be invested now → Payments (**p**) made up of principal plus compound interest are paid out of the fund.

For Your Information

RETIRE A MILLIONAIRE: IT PAYS TO START EARLY

More and more people are finding that they must plan and save for their retirement. Social Security may not provide sufficient income when you are eligible to receive benefits, and there is no guarantee that you will work for any one company long enough to receive retirement benefits that will guarantee you a comfortable retirement. People are finding that they must plan for and take control of their own financial futures through IRAs and other investments. Annuities are one way to build for the future. But the sooner a person invests in an annuity, the easier it will be to meet a specific goal. Let's look at how much you would need to save each year in an account paying 8% compounded annually in order to have $1 million when you are 65 years old.

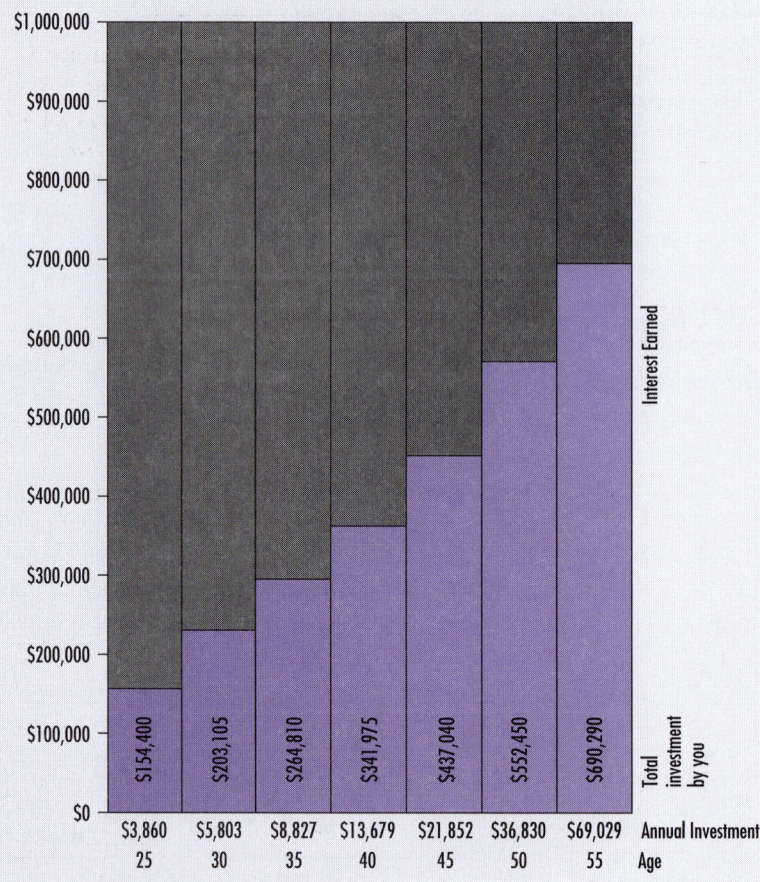

In this case, time is money. The graph shows that a person who begins investing at the age of 25 can save $3,860 annually, for a total of $154,400 by age 65, to be a millionaire at age 65. By contrast, a person who begins investing at the age of 45 has to save $13,679 annually, for a total of $341,975, to be a millionaire at age 65. The earlier that you start saving, the less it will cost you to meet your goal. Postponing your savings program causes you to lose time, and with it, the opportunity for compounding returns that can perform much of the work in reaching your goal.

Example 4

Sherry Moss's grandfather wants to provide her with $5,000 at the end of each of her four college years to help her pay off any extraneous expenses and to help her have a pleasant summer. Grandpa Moss intends to deposit a lump sum of money into a present value annuity, which pays 10% interest compounded annually and which will pay Sherry $5,000 at the end of each year from principal and accumulated interest, and which will liquidate itself with the last payment (that is, after the last payment the account will have exactly $0.00 in it). How much must Grandpa Moss invest in the present value annuity to accomplish his stated goals?

Solution

Figure 10.7 shows diagrammatically what action is taking place in the account. Note that the original principal continues to earn interest throughout the 4 years, but that each payment contains part of the principal and

Figure 10.7

Time line for a present value annuity

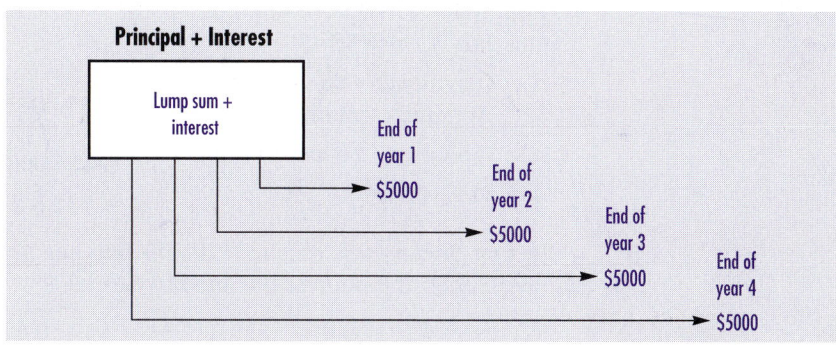

Figure 10.8

Formula for a present value annuity

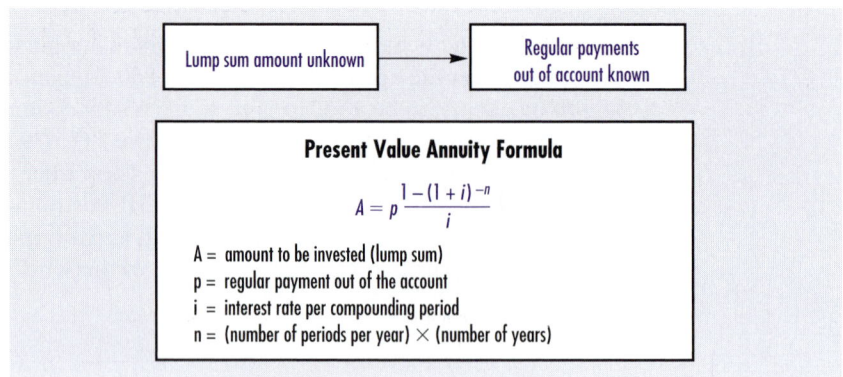

part of the interest earned. Altogether, the principal and interest total $20,000, which Sherry will receive. More principal will be needed in the first year, when less interest will have been earned, than in the fourth year when 4 years' interest has been earned.

Learning objective

Calculate the amount necessary to establish a present value annuity by using a formula or by using a table.

Figure 10.8 gives us a formula that allows us to compute the present value of an annuity. It is not always necessary to use this formula since there are tables for present value annuities, similar to the compound interest and ordinary annuities, that make our work easier. These tables are found in Appendix B.

Referring to Figure 10.8, the regular payments out of the account (known) are $5,000 per year. We must determine the lump sum amount (unknown).

$p = \$5,000$

$i = 10\%, \text{ or } .10$

$n = 4$

Table:

Table value for 10%, $n = 4$, colume E is 3.1698655

$A = (\text{payment})(\text{value from the Appendix, column E})$

$A = (\$5,000)(3.1698655)$

$A = \$15,849.33$

Formula:

$$A = p\frac{1 - (1 + i)^{-n}}{i}$$

$$A = \$5,000\frac{1 - (1 + .10)^{-4}}{.10}$$

$$A = (\$5,000)(3.1698655)$$

$$A = \$15,849.33$$

Note that Sherry will receive $20,000 but her grandfather only invested $15,849.33; so the interest the annuity earned is

interest $= \$20,000.00 - \$15,849.33 = \$4,150.67$

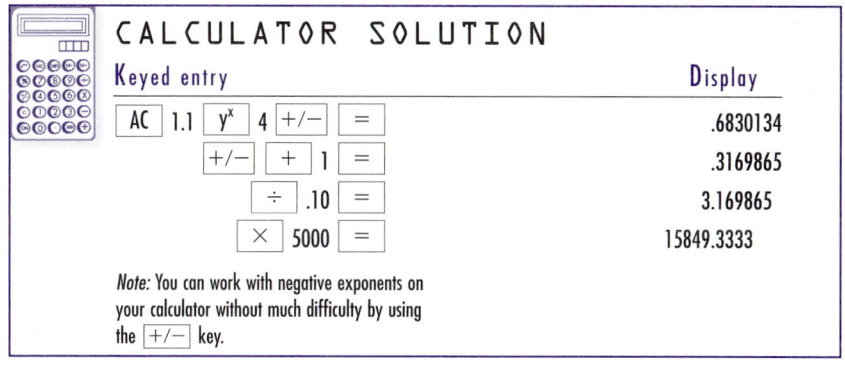

Note: You can work with negative exponents on your calculator without much difficulty by using the $\boxed{+/-}$ key.

Example 7

Suppose you had to pay your ex-spouse $10,000 per half-year in alimony payments over the next 5 years. How much would you need to put into a savings account paying 8% interest compounded semiannually so that you would have enough money to cover each alimony payment?

Solution

Refer to Figure 10.8. The regular payments out of the account (known) are $10,000 per half-year. We must determine the lump sum amount (unknown).

$$p = \$10,000$$

$$i = \frac{.08}{2} = .04 \text{ or } 4\%$$

$$n = \left(\frac{2 \text{ periods}}{\text{year}}\right)(5 \text{ years}) = 10$$

Table:

Table value for 4% page, $n = 10$ in column E = 8.11089578

$A = $ (payment)(table value)

$A = (\$10,000)(8.11089578)$

$A = \$81,108.96$

Formula:

$$A = p\frac{1 - (1 + i)^{-n}}{i}$$

$$A = \$10,000\frac{1 - (1 + .04)^{-10}}{.04}$$

$$A = (\$10,000)(8.11089578)$$

$$A = \$81,108.96$$

CHECK YOUR KNOWLEDGE

Present Value Annuities

Solve the following problems. Round dollar values to the nearest cent and rates to the nearest hundredth of a percent.

1. Mari and Lou Hencle intend to retire next year and they want to establish a present value annuity that will pay out $8,000 each year to cover the costs of their real estate taxes, home insurance costs, and other incidental home costs for the next 3 years while they travel. How much must they invest at 7% interest compounded annually to accomplish their goals?

2. Farley Davidson has decided to take a year off from work without pay to sail around the world on a cruise ship. He wants to establish a present value annuity that will pay him $2,000 per month for 1 year. How much must he invest at 12% interest compounded monthly to meet his needs?

3. Home Owner's Digest is running a $1 million lottery as a sales promotion. The winner will receive $50,000 per year for the next 20 years (i.e., a total of $1,000,000). What is the minimum amount the Digest must invest in an account paying 8% interest compounded annually in order to meet the commitment to the winner?

4. Faraday Construction Co. was judged to have been liable for the death of Ben Arnold in one of its construction projects. Ben's relatives sued Faraday for negligence and settled out of court for $130,000 per year for 15 years. The family agreed to accept payment in the form of four quarterly payments of $32,500 each year. The court ordered Faraday to establish a present value fund that would guarantee the family received their award. How much must Faraday deposit into the fund if it pays 10% interest compounded quarterly in order to settle the judgment and satisfy the requirements of the court?

Answers to CYK: *1.* $20,994.54   *2.* $22,510.14   *3.* $490,907.45   *4.* $1,004,531.10

10.2   EXERCISES

Solve the following problems. Round dollar amounts to the nearest cent and rates to the nearest hundredth of a percent.

Complete exercises 1 through 5 by determining the present value amount.

	Regular payment	Rate per year	Payments made	Number of years	Present value
1.	$200	9%	annually	5	_____
2.	$300	10%	semiannually	11	_____

Regular payment	Rate per year	Payments made	Number of years	Present value
3. $600	8%	quarterly	6	_____
4. $1,000	12%	quarterly	7	_____
5. $150	6%	monthly	4	_____

6. Determine the present value of an annuity for which the fund pays out $10,000 per year at 9% interest compounded annually for 30 years.

7. Jackson and Parkins Greenhouses needs to have $25,000 from investments each year to make improvements in its business. How much must the company invest at 7½% interest compounded annually in order to achieve its goal for the next 5 years?

8. What is the minimum amount of money you could accept today in place of receiving quarterly payments of $4,000 each at 8% interest compounded quarterly for the next 12 years?

9. What is the present value of semiannual payments of $15,500 at 11% interest compounded semiannually over the next 9 years?

10. What is the current value of $999 semiannual payments made over the next 9 years if the interest rate is 9% interest compounded semiannually?

11. Linda Kowalski will begin attending Barry Town College 1 year from now and will attend for the next 4 years; her annual costs will be $14,000 which her parents have agreed to pay. Linda's parents have saved enough money to meet Linda's needs. How much must they invest in an account paying 10% interest compounded annually in order to receive $14,000 at the end of each of the next 4 years, and have nothing left in the account at the end of that time?

12. Ramon's Plumbing and Heating is a very successful business developed by José Ramon over the last 30 years. José will retire this year and has sold his business for $1,250,000. José will invest part of this money in an account paying 12% interest compounded monthly, which will pay him $10,000 per month for the next 20 years. What is the minimum amount he must invest to accomplish his goal?

13. The Super Sweepstakes lottery was won by Barbara Thompson when the winnings were $500,000 to be paid in quarterly payments of $12,500 each over a 10-year period. How much must Super Sweepstakes deposit in an account paying 8% interest quarterly for 10 years in order to meet its commitment to Barbara?

14. Security Life Insurance Co. offers an income disability policy that will pay $2,000 per month to an insured claimant until age 65 if a full disability occurs. Julie Hanson suffered such a disability at age 55 and filed a claim that was approved for her to receive the full benefit of $2,000 per month for 9½ years until her 65th birthday. What is the minimum amount Security Life must invest in an account paying 6% interest compounded monthly in order to meet its commitment to Julie?

15. Mrs. Christianson found it necessary to sell her home and move into a senior citizens housing complex. She agreed to receive a $65,000 down payment, plus the buyers will pay her $500 per month for the next 20 years. If the terms she and the buyers agreed to were at 12% interest compounded monthly for the next 20 years, what is the present value of their agreement? (*Note:* Figure this out as a present value annuity for $500 per month at 12% interest compounded monthly for 20 years.)

16. Refer to exercise 15. What was the selling price of the house? (*Note:* Selling price = down payment + present value of financing.)

17. Refer to exercise 15. How much will Mrs. Christianson realize from the sale of her house if we add the down payment and the total amount she will receive in the financing arrangement?

18. Assume you sell a business for $35,000 in cash plus monthly payments of $750 for 3 years. If today's interest terms are 12% compounded

monthly, what is the equivalent cash selling price of your business?

19. Li Tung bought a car and financed it with a 4-year loan at a 12% nominal interest rate compounded monthly. Li is to pay off the loan and interest with 48 payments of $200 each. How much did Li borrow? (*Hint:* This is a present value problem; i.e., what is the current value of $200 monthly for 48 months at 12% interest compounded monthly?)

20. John Cooper purchased a car with a 7-year loan at 18% interest compounded monthly, and is paying $300 per month for 84 months. John recently inherited a large sum of money and wants to pay off his loan early. He has 36 payments left. How much does he still owe? (*Hint:* What is the current value of $300 monthly for 36 months at 18% interest compounded monthly?)

10.3   SINKING FUNDS

Often in our business and personal lives we are able to anticipate future expenses. A business may anticipate the replacement of a piece of equipment or a parent may anticipate the need for a lump sum of money to assist a child in obtaining a college education. Sometimes it is possible to plan ahead financially for anticipated expenses and either set aside a sum of money immediately to meet later expenditures, or plan to set aside smaller sums of money on a regular basis that together with interest will accumulate to the desired amount of money. An annuity **sinking fund** is an annuity into which regular equal periodic deposits are made so that a specific lump sum of money is accumulated (payments plus interest on them) at the end of a set time.

Example 8

Expresso Printing anticipates the need to replace a printing machine in 5 years. The anticipated cost of the replacement machine is $9,000. The company can earn 8% interest compounded quarterly and wishes to invest the funds prior to the need so they will not have to borrow money to replace the machine.

Solution

a. One option is for Expresso to make a one-time lump-sum investment into the compound interest account and allow that sum to accumulate interest until the interest and initial principal total the $9,000 needed. How much should be invested?

$$A = P(1 + i)^n \qquad \text{compound interest formula}$$

$$\$9,000 = P\left(1 + \frac{.08}{4}\right)^{20}$$

$$P = \frac{\$9{,}000}{\left(1 + \dfrac{.08}{4}\right)^{20}}$$

$$P = \frac{\$9{,}000}{1.485946} \qquad \text{value from column A in the Appendix}$$

$$P = \$6{,}056.74$$

This is the present value of a compound interest problem with $i = .02$ and $n = 20$. A single lump sum investment of $6,056.74 at 8% interest compounded quarterly will yield the $9,000 in 5 years.

b. But what if Expresso cannot spare $6,056.74 to be set aside immediately? A second option would be to make regular payments into a sinking fund, which will accumulate, with interest, to $9,000 at the end of 5 years. Since regular payments are made into a compound interest fund, we are working with an annuity, and we will assume it to be an ordinary annuity for all of our annuity sinking funds.

Using the ordinary annuity formula,

$$S = p\frac{(1 + i)^n - 1}{i}$$

in which we know $S = \$9{,}000$, $n = 20$, $i = .08/4 = .02$, we need to determine the regular payment p.

$$\$9{,}000 = p\frac{(1 + .08/4)^{20} - 1}{.02}$$

$$\$9{,}000 = p(24.29737)$$

$$p = \frac{\$9{,}000}{24.29737}$$

$$= \$370.41$$

CALCULATOR SOLUTION

Keyed entry	Display
AC .08 ÷ 4 + 1 =	1.02
y^x 20 =	1.4859474
− 1 =	.4859474
÷ .02 =	24.29729
1/x × 9,000 =	370.41046

Figure 10.9

Formula for a sinking fund

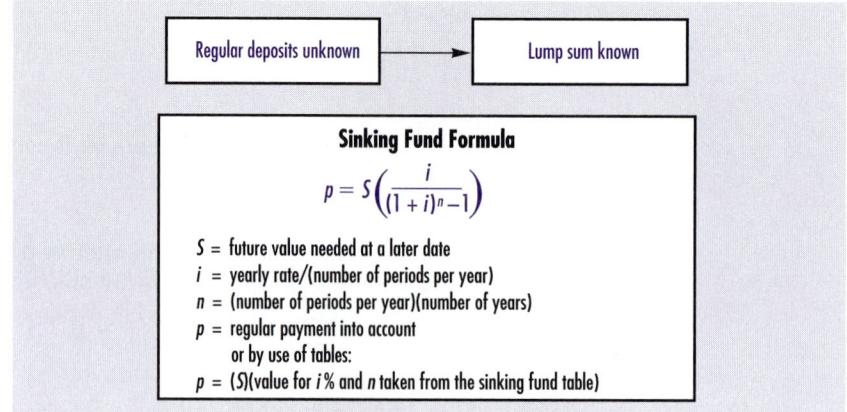

Either of the options in Example 8 may be thought of as a sinking fund, which is a fund accumulated to pay off a long-term expenditure. However, the second option, which uses the future sum of an ordinary annuity, is formally regarded as a sinking fund. The formula for a sinking fund is depicted in Figure 10.9.

Learning objective

Calculate the regular payment necessary for establishing a sinking fund by using the formula or a table for annuities.

We can use the tables in Appendix B to determine the regular payment for the sinking fund. Column D, labeled "sinking fund," provides us with a simple means of determining p.

In Example 8, we would look to column D on the page for $i = 2\%$ and $n = 20$. Moving down the column and across the row for $n = 20$ we would find .041157. If we multiplied this factor by $9,000 we would obtain the regular payment of $370.41.

Example 9

Sara Delmore plans to vacation in Europe 3 years from now. She wants to have $6,000 available for the trip, and plans to make regular semiannual payments into an account paying 10% compounded semiannually. How much should her semiannual deposits be to achieve her goal?

Solution

Since regular deposits are being made into the fund, we have a future sum annuity; and since we know the future sum of the annuity ($6,000) but not the size of the regular payments, it is a sinking fund.

Refer to Figure 10.9. The lump sum (known) is $6,000. We must determine the regular deposit to be made.

$S = \$6,000$

$i = \dfrac{10\%}{2} = 5\%$

$n = (3 \text{ years})(2 \text{ periods per year}) = 6$

Table:

Table value for 5%, $n = 6$, column D is .1470174

$p = (\text{lump sum})(\text{table value})$

$p = (\$6,000)(.1470174)$

$p = \$882.10$

Formula:

$S = p\dfrac{(1 + i)^n - 1}{i}$

$\$6,000 = p\dfrac{(1 + .10/2)^6 - 1}{.05}$

$\$6,000 = p(6.801912)$ value from the future value annuity column C, $i = 5\%$, $n = 6$

$p = \dfrac{\$6,000}{6.801912} = \$882.10$

Sara will make six payments of $882.10 for a total of $5,292.60, which means that the interest she will earn is the difference of $6,000 and $5,292.60; that is,

interest $= \$6,000 - \$5,292.60 = \$707.40.$

Remember that we can check our solution by substituting the information into the ordinary annuity formula.

Check: Ordinary annuity formula

$S = p \times (\text{future annuity table value for } i = 5\%, n = 6)$

$\quad = \$882.10 \times (6.801912)$

$\quad = \$5,999.96$

Or with a calculator,

$S = p\dfrac{(1 + i)^n - 1}{i}$

$\quad = \$882.10 \times \dfrac{(1 + .05)^6 - 1}{.05}$

$\quad = \$882.10 \times (6.801912)$

$\quad = \$5,999.96$

(The 4-cent discrepancy is due to rounding error.)

CHECK YOUR KNOWLEDGE

Sinking Funds

Solve the following problems. Round dollar amounts to the nearest cent.

For each of the following sinking fund annuities in problems 1 through 5, determine the regular payments.

	Future value	Rate/year	Compounded	Number of years	Regular payment
1.	$10,000	9%	annually	5	_____
2.	$22,500	10%	semiannually	3	_____
3.	$8,750	16%	quarterly	4	_____
4.	$43,250	18%	monthly	4	_____
5.	$875	8%	quarterly	6	_____

6. Sureway Machine Shop will need to replace a drill press 3 years from now. The projected cost of a new press is $10,000. How much should be deposited quarterly into an annuity paying 8% interest compounded quarterly so that SMS will have $10,000 in 3 years?

7. Acme Vending Co. needs to purchase a new van to be used in servicing its machines every 4 years. With the purchase of each new van, Acme establishes a sinking fund to save for the next purchase. Acme anticipates the cost in 4 years to be $17,500. How much should Acme invest monthly into an account paying 12% interest compounded monthly so it will have $17,500 at the end of 4 years?

8. Refer to problem 7 and determine the following for Acme: (a) the total in payments Acme will make and (b) the total interest that will be earned by the sinking fund.

9. Matt Brewster Oil Distributors decides to purchase a motor home to transport company personnel to business meetings, and to use for a variety of sales promotions. Brewster decides to establish a sinking fund paying 10% interest compounded quarterly to accrue a total of $50,000 in 5 years to make the purchase. How much must Matt Brewster deposit into the fund each quarter to achieve his goal?

10. Refer to problem 9 and determine for Brewster (a) the total amount that will be deposited into the account, and (b) the total interest the sinking fund will earn.

Answers to CYK: *1.* $1,670.92   *2.* $3,307.91   *3.* $400.93   *4.* $621.72   *5.* $28.76
6. $745.60   *7.* $285.85   *8.* **a.** $13,720.80;   **b.** $3,779.20   *9.* $1,957.35
10. **a.** $39,147;   **b.** $10,853

10.3 EXERCISES

Solve the following exercises. Round dollar amounts to the nearest cent and rates to the nearest hundredth of a percent.

For each of the sinking fund annuities in exercises 1 through 5, determine the regular payments.

	Future value	Rate/year	Compounded	Number of years	Regular payment
1.	$5,000	8%	quarterly	8	_____
2.	$13,000	9%	semiannually	11	_____
3.	$7,000	10%	annually	12	_____
4.	$28,000	12%	monthly	4	_____
5.	$4,350	14%	quarterly	5	_____
6.	$3,275	11%	semiannually	3.5	_____

7. You wish to establish a sinking fund that will be worth $3,000 in 4 years. If you establish such a fund in an account paying 8% interest compounded quarterly, how much must you deposit quarterly to meet your goal?

8. Jack Smythe will need $6,000 accumulated in 8 years in order to replace an office copy machine for his business. How much must he deposit semiannually into a sinking fund paying 7% interest compounded semiannually in order to have the required amount?

9. E. Z. Fixit is a small repair business run by its owner, Ed Zambino. Ed's main asset is his own ability and a good truck. He needs to replace his truck every 4 years, and saves ahead so that he can pay cash for the new vehicle. Ed just bought a new truck for $19,500. He believes inflation will be about 4% per year for the next 4 years. Using the compound interest formula, he has calculated that the same truck will cost $22,812.24 then. How much will Ed need to deposit monthly into a sinking fund paying 12% interest compounded quarterly in order to have the desired amount?

10. Tim and Sally Fredette gave birth to a baby girl this year and want to establish a college fund for her higher education. They estimate that even a public education will cost at least $10,000 per year when she attends college. How much must they invest quarterly into a fund pay-

ing 8% interest quarterly in order to have $40,000 in 18 years?

11. Sharon Barrister is a professor at a small college. She has decided to take a leave of absence 3 years from now and travel throughout Eastern Europe for a semester and the summer that follows it. She will need $12,000 to accomplish her goal. How much must she deposit annually into an annuity paying 9% interest compounded annually in order to meet her needs?

Use the following situation to answer exercises 12 through 15. Art Barrett has signed a promissory note that will be due in 30 months, at which time he will have to pay a total of $9,000. Since the note was a discounted note, $2,025 is interest.

12. If Art establishes a sinking fund paying 12% interest compounded monthly, how much must he invest monthly in order to have the $9,000 in 30 months' time?

13. How much did Art actually deposit in the fund?

14. How much interest did the fund earn him?

15. In effect, Art's plan retrieved some of the interest cost on the original promissory note. What was the net cost of the original note if we credit this interest earned against that interest paid?

10.4   **AMORTIZATION OF AN AMOUNT OF MONEY**

Learning objective

Calculate the equal regular payments that will amortize a specific sum of money over a set period of time.

Amortization of a lump sum of money is an application of the present value annuity. When we worked with the present value annuity, we asked the question, *How large a sum of money must be invested* in an account paying $R\%$ interest per year compounded m times per year in order to receive n equal payments of $\$p$? We knew p, R, i, and n, and we were looking for A.

With an "amortization," we start out with a known lump sum of $\$A$ and we ask, *How large will our equal payments be* if we invest $\$A$ in an account paying $R\%$ per year compounded m times per year for a total of n payments? Here we know A, R, i, and n, and we are looking for p.

Amortization is used whenever we have an account with a lump sum of money that is earning interest and that we wish to liquidate (decrease the value of principal + interest to $\$0.00$) with n equal payments. Typical examples of amortizations are retirement funds and loans (such as auto loans and mortgages on property).

Amortization is a present value annuity in which we are trying to determine the value of the regular payment, p. Consequently, if we begin with our present value annuity formula and solve for p, we can develop a formula for amortization (see Figure 10.10).

Present value annuity formula

$$A = p\left[\frac{1 - (1 + i)^{-n}}{i}\right]$$

If we solve this equation for p, we will have

$$p = A\frac{1}{\left[\dfrac{1 - (1 + i)^{-n}}{i}\right]}$$

The value of the entire quantity that is multiplied by A is given to us in the Appendix column F, which will assist us in our work.

Figure 10.10

Amortization
formula

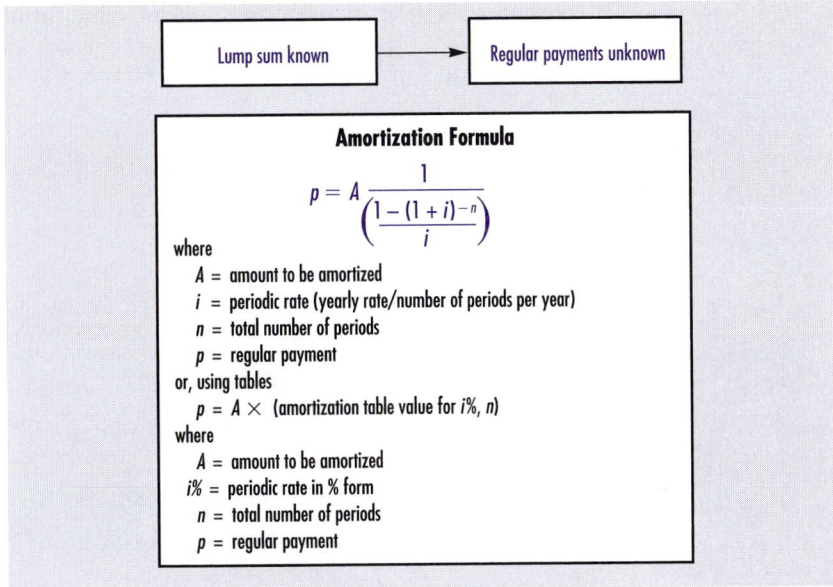

Example 10

Martha McGrath, a sculptor of miniature wildlife scenes, has worked as an independent craftswoman for 15 years. She created a retirement account and has saved a total of $200,000 from the sales of her work. She will retire this month and wishes to establish a fund with her savings ($200,000), which will make equal monthly payments to her for the next 20 years. If she sets up the retirement fund with an institution paying 12% compounded monthly, how much will she receive monthly?

Solution

Martha is establishing a present value annuity since regular payments will be made out of the fund to her. Since we know the amount of money with which she will establish the fund, and we need to determine p, this is an amortization problem. Refer to Figure 10.10. The lump sum (known) is $200,000; we are trying to determine the regular payments (unknown).

First, let's solve this problem by using the tables:

$$p = A \times \text{(amortization value for } i\%, n)$$

$$A = \$200,000$$

$$i\% = \frac{.12}{12} = .01 \text{ or } 1\%$$

$$n = (12 \text{ periods per year})(20 \text{ years}) = 240$$

Table value for 1%, $n = 240$, column F is .0110109

$$p = \$200,000 \times (.0110109) = \$2,202.17 \text{ per month}$$

Second, let's solve by using the present value formula:

$$A = p\frac{1 - (1 + i)^{-n}}{i}$$

$$A = \$200,000$$

$$i = \frac{.12}{12} = .01$$

$$n = 12 \times 20 = 240$$

$$\$200,000 = p\frac{1 - (1.01)^{-240}}{.01}$$

$$\$200,000 = p(90.819410)$$

$$p = \frac{\$200,000}{90.819410} = \$2,202.17 \text{ per month}$$

Notice that the actual amount Martha will receive is

total = 240 × \$2,202.17 = \$528,520.80

which means that the account will earn \$328,520.80 in interest as it liquidates itself.

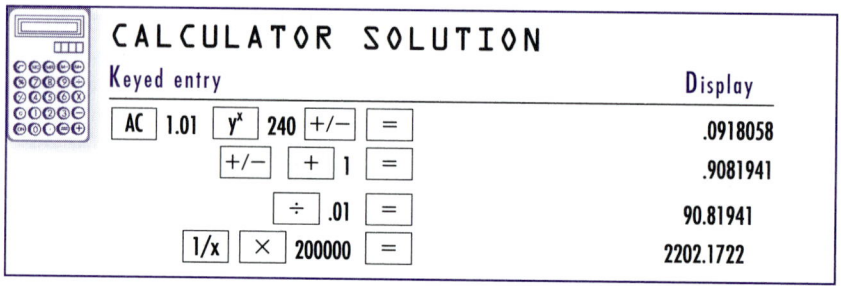

CALCULATOR SOLUTION

Keyed entry	Display
AC 1.01 y^x 240 +/− =	.0918058
+/− + 1 =	.9081941
÷ .01 =	90.81941
1/x × 200000 =	2202.1722

Loans that are amortized work in this same way; that is, they earn interest as they liquidate themselves. The interest they earn is the finance charge the borrower pays for the use of the money.

Example 11

Faith MacArthur bought a new car and financed \$17,000 at 18%. The loan was amortized monthly for 5 years. Determine Faith's monthly payment and the total interest she will pay for the loan.

Solution

Refer to Figure 10.10. The lump sum (known) is \$17,000. We are trying to determine the regular payments (unknown).

The amount, A, to be amortized here is the $17,000 she borrowed. The periodic rate is $i = .18/12 = .015$ per month. The loan is for 5 years, so there will be 60 periods in the loan.

$p = A$(amortization table value for $i\% = 1.5\%$, $n = 60$)

$\quad = \$17,000(.025393) = \$431.688$, or $431.69 per month.

The total of Faith's payments will be

total $= 60$ payments $\times \$431.69 = \$25,901.40$

The interest Faith will pay will be

interest $=$ total payments $-$ amount borrowed

$\quad = \$25,901.40 - \$17,000$

$\quad = \$8,901.40$

Amortized loans provide the benefits of equal payments while truly charging interest on the unpaid balance only of the loan.

CHECK YOUR KNOWLEDGE

Amortization of an Amount of Money

Solve each of the following problems. Round dollar amounts to the nearest cent.

Determine the regular payment p in the amortization of the amount in problems 1 through 5.

	Amount	Rate/year	Compounding	Number of years	Payment
1.	$4,000	8%	semiannually	6	_____
2.	$12,500	10%	quarterly	5	_____
3.	$3,250	6%	monthly	3	_____
4.	$22,250	9%	annually	13	_____
5.	$50,000	14%	quarterly	8	_____

6. Jake McDougald won a $500,000 lottery prize in his state's lottery. His winnings were reduced to $400,000 after taxes were deducted, and he decided to amortize the winnings by investing them in a fund that would make quarterly payments to him for 15 years. If the fund pays 10% compounded quarterly, how much will he receive each quarter?

7. Refer to problem 6. Determine the total amount that Jake will receive from the fund. Also determine the total interest the fund will earn.

8. An application of amortization is the paying down of a loan, such as an auto loan or a home loan. Mike Bisbee bought a new Dodge Dakota truck and borrowed $12,000, which he will amortize with regular monthly payments over the next 5 years. If the interest rate on the loan is 9%, how much will be his regular monthly payments?

9. Angie's parents borrowed $10,000 to put new siding on their home. The loan will be amortized at 12% compounded monthly for the next 10 years. How much will Angie's parents have to pay in equal monthly payments to pay off the loan and interest?

10. Etan Burgan borrowed $8,000 to help him get through his graduate program in public policy. He will not be charged interest until June, when he finishes up his program. At that time, he has agreed to amortize the debt at 8% compounded quarterly with equal regular quarterly payments for 3 years. How much will he have to pay each quarter?

10.4   EXERCISES

Solve the following problems. Round dollar amounts to the nearest cent and rates to the nearest hundredth of a percent.

Determine the regular payment for the amortization of the given amount of money at the given rate for the stated time in exercises 1 through 5.

	Amount	Rate/year	Compounding	Number of years	Payment
1.	$20,000	5%	annually	10	_____
2.	$12,000	6%	monthly	5	_____
3.	$200,000	8%	quarterly	15	_____
4.	$75,000	10%	semiannually	7	_____
5.	$17,000	12%	monthly	5	_____

6. Jean LaFleur recently inherited $100,000 and wants to receive regular quarterly payments from this sum to supplement her income for the next 10 years. If she invests the money in an account paying 10% compounded quarterly that will liquidate itself at the end of 10 years, how much will she receive each quarter?

7. Mark Green bought a new power boat for $75,000. He paid a $30,000 down payment and financed the remainder with an amortized loan at 12% compounded monthly for 5 years. What will be his monthly payment?

8. Sally Markham has saved $180,000 over the last 15 years to supplement her income during retirement. Now that she has retired, Sally wants to amortize this amount so that she will receive regular payments over the next 30 years and liquidate the account. She arranges to do this with an investment that pays 10% compounded quarterly. How much will she receive each quarter?

Answers to CYK: 1. $426.21   2. $801.84   3. $98.87   4. $2,971.87   5. $2,622.10
6. $12,941.20   7. a. $776,472; b. $376,472   8. $249.10   9. $143.47
10. $756.48

Use the following information to complete problems 9 through 11. Janet Ordway won the Pennsylvania lottery and was given a prize of $500,000. She decided to invest this amount in an account that would amortize it over a 10-year period at 12% interest compounded monthly.

9. How much will she receive each month?

10. What will be the total amount of the payments that she will have received as of the end of the 10-year period?

11. How much interest will the account have earned?

Use the following information to answer problems 12 through 15. Jack and Marge Janson purchased a home for $170,000 and put a 20% down payment on the purchase. They financed the remainder with a loan that was amortized at 12% interest compounded monthly for 20 years.

12. How much did Jack and Marge finance?

13. What will be Jack and Marge's monthly payment?

14. How much will their total payment be to pay off the loan?

15. How much interest will they pay for the loan?

CHAPTER REVIEW EXERCISES

The following problems review all mathematics and business concepts presented in the chapter. If you experience any difficulty solving a problem, go to the appropriate chapter section and review the examples provided. Express all answers as decimals rounded to the nearest one hundredth, or dollars rounded to the nearest cent, unless otherwise instructed.

1. Marley Stonecipher deposited $2,000 per year into an IRA that was an ordinary annuity paying 6% compounded annually for 25 years. How much was the annuity worth at the end of the 25 years?

2. Monique Anderson deposited $550 semiannually into an annuity due account that paid 8% compounded semiannually for 15 years. What was the value of her account at the end of the 15-year period?

3. Huong Viet anticipates that he will need $5,000 in 6 years, so he is depositing $150 per quarter into an ordinary annuity that pays 10% compounded quarterly. How much will he have at the end of the 6-year period?

4. In problem 3, how much should Houng invest each quarter into the annuity so that he would have the desired amount in 6 years?

5. During her 4 years in the military, Shanda Lee saved $300 per month in an ordinary annuity which paid 6% compounded monthly. How much did Shanda have saved at the end of the 4 years? How much interest did Shanda earn?

6. In problem 5, how much should Shanda have saved monthly if she wanted a total of $20,000 at the end of the 4-year period?

7. Rob Knight won $100,000 in his state's lottery. The state rules require that Rob be paid $10,000 per year for 10 years. The state will deposit a lump sum of money into a bank account that will pay Rob his yearly sum. How much must the state deposit into an account that will pay 10% compounded annually for 10 years so that Rob will receive the money due to him?

8. Sid Jackson was injured while working as an electrician on a construction job. Sid's employer agreed to deposit a lump sum of money into an account paying 8% compounded quarterly, which will make payments of $600 per quarter to Sid for 2 years while he recuperates. How much must the employer deposit into the account so that Sid will receive his money and the account will liquidate itself?

9. Andrea Wiggins purchased a new entertainment center for her home for $3,500. She financed the entire amount of the purchase and agreed to an

amortization of the loan at 18% compounded monthly with equal monthly payments for 4 years. (a) How much will each of her payments be? (b) How much will Andrea pay in interest if she completes the financing agreement on schedule?

10. Withro Washington borrowed $8,000 to buy a car. He agreed to monthly payments of $250 for 4 years. Was he charged the correct amount if the loan was amortized at 12% compounded monthly for 4 years? If not, how much per month was he overcharged or undercharged?

EXPRESS YOUR UNDERSTANDING

1. Explain how an annuity differs from the basic compound interest problems that we studied in Chapter 9.

2. Describe the circumstances that make an annuity an *ordinary annuity*.

3. Explain in what way an *annuity due* differs from an ordinary annuity.

4. If you know that the future value of an ordinary annuity is $1,000 and that the interest rate per payment period of the annuity is 3%, explain how you could calculate the future value of an annuity due that has the same terms as the ordinary annuity.

5. Give an example of a situation in which you might create a *sinking fund*.

6. How is an ordinary annuity related to a sinking fund?

7. Explain how a present value annuity differs from an ordinary annuity.

8. How is amortization related to present value annuity?

9. Give three examples of uses of amortization of a sum of money.

10. Explain the kinds of annuities someone might use to build a retirement fund until the person is 65 years old and then liquidate the fund with regular payments until age 85.

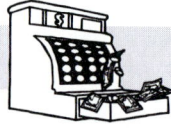

Mind Your Business

Annuities

The partners of Media World have decided to offer their employees the opportunity to invest in an individual retirement account, a 401(k) plan. You have decided to invest in a company called Integrated Mutual Funds Inc. Although the interest earned by ITF varies with the stock market to some degree, ITF has produced a steady growth rate of about 6% per year in the value of its shares. The partners have agreed to deposit $25 each pay period for each employee, and the employee can contribute as well. Kate Bradley, your bookkeeper, has signed up and will put 3% of her own money from each paycheck into the fund. There are 52 pay periods per year, but the accumulated money for each month will be sent in at the end of the month.

Activity 1

Write up a summary for Kate, who will turn 23 in January, stating (a) the total that will be deducted from her paycheck each month, (b) the total per month that will be invested in ITF on her behalf, (c) the expected value of her investment in 10 years (assume the 9% rate is the average rate per year), (d) the expected value of her investment in 20 years, and (e) the expected value of the investment when she reaches the age of 65.

Activity 2

To further help Kate to understand the potential value of this benefit, indicate that she may be able to deduct her contributions to her retirement fund from her gross earnings, and reduce her income tax liability. Use an example of some specific number of years of investing and provide her with the following information, assuming the annuity grows at an average yearly rate of 6% compounded monthly: (a) the total contributions that the business would have made to Kate's retirement at the end of 20 years; (b) the total of the contributions that Kate will have made at the end of 20 years; and (c) the total interest Kate's retirement account will have earned at the end of 20 years.

A SUMMARY OF THE MATHEMATICS OF FINANCE CHAPTERS

Your study of simple interest, simple discount, compound interest, future value annuities, present value annuities, sinking funds and amortizations, gives you the foundation for understanding and working with the many variations of these formulas that make up our financial world today. Although our current world of finance may seem very complex and at times mind boggling, its logic and acceptability by people doing business with each other is based on the concepts that you have learned in Chapters 7, 8, 9, and 10.

Taken altogether, the concepts and formulas developed in Chapters 7 through 10 make up a mathematical model for finance. The diagram that follows represents that basic financial model using a decision tree that can help you choose the right formulas to solve a mathematics of finance problem.

When we first read a problem, how do we know where to begin and which formulas to use? Clearly, if the problem is a simple interest or simple discount problem, the wording of the problem will indicate *simple interest* or *simple discount,* and we will know to use those formulas. But what if it is a compound interest problem. How do we know if it is an annuity or a basic compound interest problem? If periodic deposits or payments are involved, then it is an annuity. If it is an annuity, how do we know if it is an ordinary annuity, annuity due, or a present value annuity? If regular deposits are made into an account that will grow to a lump sum, it is an ordinary annuity or an annuity due. If, on the other hand, regular payments are made from the account with the goal of liquidating the account, then it is a present value annuity.

Using this information, we can diagram a means of analyzing a problem after we first read it, as follows:

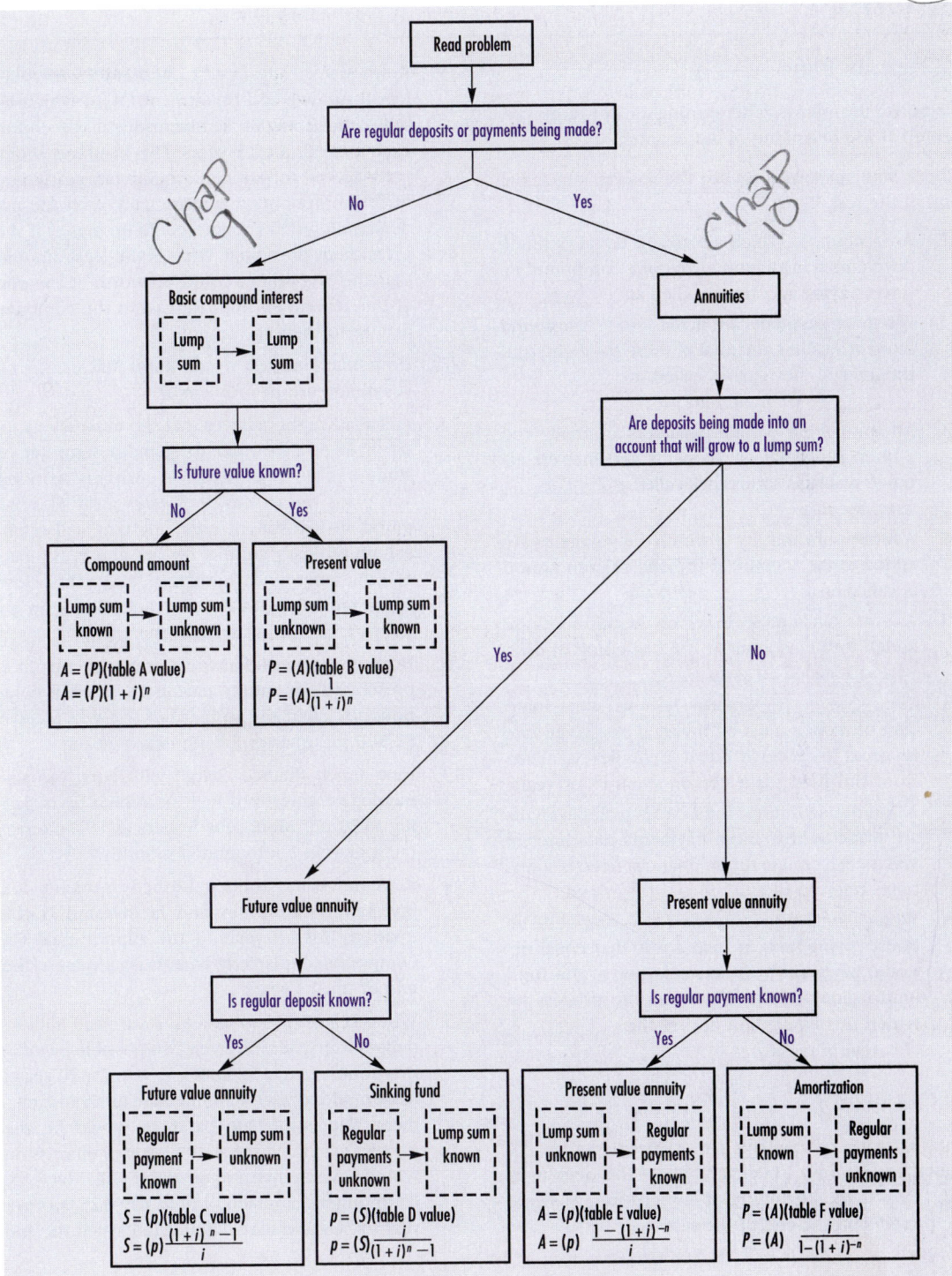

SELF-TEST

A. Terminology

Complete the following items using the key terms presented at the beginning of the chapter.

Check your responses against the answer key at the end of the test.

1. An account in which a series of equal periodic payments is made into or out of a compound interest paying account is called an _____.

2. If regular payments are made into a compound interest account, the total of these payments and the interest they earn is called the _____ _____ of an annuity.

3. An annuity into which payments are made and a lump sum total of payments and interest accrued results is sometimes called a _____ *sum annuity.*

4. A future sum annuity in which the payments are added to the account at the end of each period is called an _____ *annuity.*

5. A future sum annuity in which the payments are added to the account at the beginning of each period is called an *annuity* _____.

6. The _____ annuity tells us what lump sum of money must be invested at a given rate in order to receive equal periodic payments from the fund for a certain number of years.

7. A compound interest fund into which payments are made so that those payments plus interest will meet a certain future financial need is sometimes referred to as a _____ *fund.*

8. When a lump sum is invested in a compound interest paying fund in such a way that equal periodic payments will be made out of the fund until it liquidates itself, we call the process by which this liquidation occurs the _____ of a sum of money.

B. Calculation

The following concepts and short problems are designed to test your understanding of the objectives identified at the beginning of the chapter. Answers are provided at the end of the test.

9. Rufus MacSnead agreed to a contract with Ouitless Insurance and Investment Co. in which he will deposit $3,000 semiannually, at the end of each period, for 15 years. The insurance company agreed to pay him a minimum yearly rate of 8% compounded semiannually, with the understanding that the rate could be higher if the investment does well. What is the least amount that the investment could be worth at the end of the 15 years if both Rufus and the company live up to their agreement?

10. How much interest would Rufus MacSnead's investment earn in problem 9?

11. Priscilla Knight believes that she must save a total of $8,500 in order to spend a semester of study in Cairo, Egypt. She is currently working and plans the trip to be 3 years away. She has found an investment company that will establish an ordinary annuity for her at a yearly rate of 12% compounded monthly for 3 years. How much must Priscilla deposit each month in order to meet her goal under this plan?

12. How much must Emelde Hernandez put in a present value annuity paying 8% interest compounded quarterly so that her son will receive $2,500 per quarter for the next 5 years?

13. How much finance charge will Trung Nguyen pay for an amortized loan on which he is paying $150 per month for 5 years at 18% interest rate per year compounded monthly?

14. Sadie Hawker's father established an annuity due investment for her in which he invested $1,000 quarterly for 18 years. If the annuity paid 8% compounded quarterly, what was its value at the end of the 18 years?

15. When Zeb MacAdoo won his state's half-million-dollar lottery, he was given two possible payoffs, Zeb could receive $25,000 per year for 20 years, or he could receive an immediate lump sum payment. The immediate payment would be the present value that the state's lottery commission would have to invest in an annuity that pays 9% compounded annually, would pay $25,000 per year to Zeb, and would liquidate itself at the end

of the 20 years. How much would the one-time lump-sum payment be?

16. Consider problem 15. (a) If the lottery commission sold 200,000 two-dollar tickets, how much did they make on the lottery Zeb won? (b) If Zeb took the option to receive the $25,000 per year, how much interest would the annuity account earn?

Use the following information to answer questions 17 through 20. Joe Armani purchased a new car for $18,000 and made a down payment of $8,000. Joe financed the remaining amount with an amortized loan at 12% interest compounded monthly for 5 years.

17. How much did Joe finance?

18. What are Joe's regular payments?

19. How much will Joe pay back in total for the loan?

20. How much interest will Joe pay for the loan?

Answers to Self-Test: **1.** annuity   **2.** future value   **3.** future   **4.** ordinary   **5.** due   **6.** present value   **7.** sinking fund   **8.** amortization   **9.** $168,254.81   **10.** $78,254.81   **11.** $197.32   **12.** $40,878.58   **13.** $3,092.96   **14.** $161,218.16   **15.** $228,213.64   **16.** **a.** $171,786.36; **b.** $271,786.36   **17.** $10,000   **18.** $222.44   **19.** $13,346.40   **20.** $3,346.40

11

Key Terms

collateral
unsecured loan
secured loan
open-ended credit
closed-end credit
finance charge
revolving charge
    account
average daily balance
daily balance
installment loan
mortgage loan
annual percentage rate
amortization
interest rebate
sum of digits
rule of 78s
mortgage

BUSINESS AND CONSUMER CREDIT

Learning Objectives

After completing the exercises in this chapter, you will be able to:

1. Determine the daily balance for a revolving credit account.

2. Calculate the average daily balance for a revolving credit account.

3. Calculate the monthly finance charge for a revolving credit account.

4. Calculate the amount of the regular payments of an installment contract.

5. Using tables, calculate the equal regular payments of an amortized loan.

6. Calculate the regular payment for a mortgage and determine total interest paid.

7. Create loan repayment tables for amortized loans.

8. Calculate the outstanding principal of an amortized loan.

9. Calculate the interest rebate for a loan paid off early.

10. Calculate the approximate interest rebate using the sum of digits and rule of 78s methods.

11. Determine the annual percentage rate for a loan.

12. Define key terms.

INTRODUCTION

Whenever a person or a company purchases goods or services, the purchaser must decide whether to pay cash or buy on credit (i.e., purchase now and pay later). If the only choice possible were to pay cash, our economy as we know it would not exist. The economies of countries and of the world depend on the ability of individuals and companies to buy now and pay later. Our goal in this chapter will be to explore the basics of consumer credit.

A merchant's goal is to sell merchandise to a customer. If the customer has the cash to buy the merchandise, the merchant may need only a quality product and a competitive price to make the sale. If the customer does not have the cash available to buy, then the merchant may need to provide the customer with one more incentive, *financing*.

Whenever a merchant offers financing to a customer, that merchant assumes increased responsibilities, risks, and costs. Merchants may choose to finance the customer's purchase themselves, or they may seek out financing to offer the customer from an additional source. Merchant-financed loans are those the seller makes directly to the buyer, for example, the short-term loan, such as when a furniture store offers "90 days is the same as cash." In this instance, the seller agrees to give the buyer up to 90 days to pay for merchandise without a finance charge. If the buyer holds the item during this 90-day period, the seller has effectively loaned the value of that merchandise to the buyer for 90 days, interest free.

The merchant may also finance a loan by offering an installment contract over a period of time with a finance charge included. Revolving credit accounts and charge cards offered by local department stores or national chain stores—such as Sears, Radio Shack, or Kmart—are another form of merchant financing.

A merchant may prefer to provide financing through a third party. For example, an automobile dealer may initiate an installment contract for a customer to buy a car, but the contract is really between the customer and a financial institution, such as a bank. The car dealer may receive a commission for completing the initial paperwork; however, he or she may be required to pay the institution a fee for a lower interest rate necessary to entice the customer to buy.

Many merchants, especially those with small businesses, prefer to use the services of national or international credit cards such as MasterCard, Visa, American Express, or Discover, which reduce the merchant's responsibilities and risks. Whenever the borrower requests to use such credit, the merchant obtains an electronic verification from the lender and extends the credit accordingly. The merchant does incur costs of approximately 3% to 5% of the average ticket value for the benefit of using national credit card services.

A lender may reduce the risk of making a loan by requiring some form of security in the form of collateral from the borrower. **Collateral** is simply something of value that is used as security for a loan. If the borrower cannot pay off the loan, the borrower agrees to give up all or part of the ownership of the item used as collateral to the lender. **Unsecured loans** are loans that have no specific collateral provided to the lender; the lender is relying on the borrower's ability and willingness to pay. **Secured loans** generally provide lower interest rates than unsecured loans because they represent less risk to the lender.

We will consider two basic categories of consumer credit: **open-ended credit,** which includes various types of charge accounts and credit cards, and **closed-ended credit,** which includes installment buying and mortgages. We will also briefly consider consumer credit from the seller's (merchant's) point of view.

Perhaps the following illustration will help explain the basis on which open-ended credit is offered.

> Arlene Slater has lived in the village of Johnsville for most of her life and she has purchased her groceries at Sigsbee's Market for the last 30 years. Each week she shops, and the owner, Jay Sigsbee, keeps a record of how much Arlene owes. At the end of each month, after Arlene receives her retirement check, she pays Jay the entire amount that she owes for the month.

Arlene's arrangement with Jay Sigsbee is, perhaps, one of the simplest forms of open-ended credit. Her credit arrangement is based on "trust." Jay trusts that Arlene will pay her bill in an honest and timely manner. Arlene may or may not be required to pay a finance charge for the privilege of doing her grocery shopping in this way. A **finance charge** is interest paid for using someone else's money for a period of time. The finance charge is usually based on the amount of money used, principal, P; a percentage rate, R; and the time, T, the money is used (finance charge = interest = $P \times R \times T$). With this model, all credit advanced to either an individual or to an organization is based on *trust* that the loan, plus interest, will be paid back fully and in proper time.

Today's highly mobile and transient society has resulted in fewer financial arrangements like the small town grocery arrangement of Arlene Slater. However, credit accounts offered by local department stores or national store chains such as Sears or JCPenney are very similar. These stores accept our application for a credit line, evaluate our ability and willingness to pay our debts (trust), and issue us a line of credit that is open-ended up to some upper limit such as $500, $1,000, or $5,000. We, then, are able to purchase goods and services openly up to our upper limit just by saying "Charge it."

11.1 REVOLVING CREDIT

Revolving credit is the most common type of open credit. Revolving credit as we know it is really a form of two types of charge accounts. With a *regular charge account,* we take the goods and agree to pay for them fully within a set time period, say 30, 60, or 90 days. Furniture stores often offer this type of credit with no finance charge as an inducement to convince a customer to buy now. A **revolving charge account** is an open account in which we are allowed to charge up to a certain limit and we agree to pay at least a percentage, say 10%, 15%, or 20%, of our outstanding balance at the end of a set time period (called a *billing period* or *billing cycle*), which is often set at 25 or 30 days. A finance charge is assessed on the outstanding balance of the account each month.

Revolving credit is credit automatically available up to a specified limit while payments are periodically made. If the entire balance due at the end of a billing period is paid, then no finance charge is assessed. If only the minimum amount due, or some amount less than the entire balance, is paid, then the finance charge will be assessed on the amount *carried over* into the next billing period. Typically, a statement describing the customer's purchases and payments is sent to the customer each month, as shown in Figure 11.1. If a customer pays the balance in full each month, then there is no finance charge added to the balance. If, however, the customer pays less than the balance in full, then a finance charge must be paid each month a balance is left. The finance charge is calculated as a percentage of either the *adjusted balance* or the *average daily balance*. We will demonstrate both of these methods of calculation in this section.

In the adjusted balance method, the finance charge is calculated as a percentage of the unpaid balance as of the end of the previous month.

Example 1

José Chingo has a revolving charge account for which the terms of payment are as follows: "The customer must pay a minimum of $25 and a finance charge of $1\frac{1}{2}\%$ per month. To avoid paying a finance charge, the customer must pay the entire balance amount." Assume that José pays the minimum payment of $25 each month for the next 3 months and that his previous balance was $654. Find (a) his new balance and (b) the total amount in finance charges he has paid during these 3 months.

Solution

Month	Previous balance	+	Finance charge	+	Current purchases	−	Payments	=	New balance
1	$654.00		$9.81 (654 × .015)		0		$25.00		$638.81
2	$638.81		$9.58 (638.81 × .015)		0		$25.00		$623.39
3	$623.39		$9.35 (623.39 × .015)		0		$25.00		$607.74

a. The new balance is $607.74.

b. The total amount of finance charges is $28.74 (9.81 + 9.58 + 9.35 = 28.74).

Figure 11.1

Billing statement

Anita's Department Stores
Statement of Account

**The Customer Is Always
Our First Concern**

ACCOUNT NUMBER **000-00-00000**

CURRENT BILLING DATE **02-27-XX**

PAGE 01 OF 01   To Avoid Additional Finance Charge, Pay New Balance By ▶ **03-24-XX**

Date	Store/Reference Number	Balance Type	Item Description	Charges	Payments and Credits
02-13	6501-0025375	R	PYMT -THANK YOU		60.00

THANK YOU!

BALANCE SUMMARY	PREVIOUS BALANCE	− PAYMENTS AND CREDITS	+ FINANCE CHARGE	+ CHARGES	= NEW BALANCE	MINIMUM PAYMENT
REG	**136.36**	**60.00**	**1.91**	**.00**	**78.27**	**20.00**

Your finance charge rates are: (See reverse side for important information and explanation of your finance charge method)

BAL TYPE		ON BALANCE OF	AS OF	ON BALANCE THRU $	MONTHLY PERIODIC RATE(S) %	ANNUAL PERCENTAGE RATE(S) %	ON BALANCE OVER $	MONTHLY PERIODIC RATE(S) %	ANNUAL PERCENTAGE RATE(S) %
R	**A**	**109.08**	**02-98**	**ALL**	**1.75**	**21.0**			

△ Your Finance Charge Method is (see reverse)

MINIMUM *FINANCE CHARGE*   **$ .50**

Example 2

Stacy's previous balance on her charge account is $780.21. This month she has purchased goods worth $35 and has made a payment of $40. What is the new balance in her account if the finance charge is 1.75%?

Solution

Month	Previous balance	+	Finance charge		+	Purchases	−	Payments	=	New balance
1	$780.21		$13.65	(780.21 × .0175)		$35		$40.00		$788.86

Learning objective
Determine the daily balance for a revolving credit account.

Although the methods for calculating finance charges vary for different institutions that issue credit cards, many use the average daily balance method. The **average daily balance** is the sum of daily balances divided by the number of days in a billing cycle. The **daily balance** is equal to the previous balance plus new purchases and cash advances, minus payments and credits.

Calculating daily balance

daily balance = previous balance + new purchases + cash
advances − payment − credits

Example 3

The following information represents the monthly statement for Joan Gansk's credit card account.

6/17	previous balance	$420
6/28	purchase	$ 50
7/2	payment	$100
7/9	purchase	$ 30

Find the daily balance for Joan's account on (a) June 28, (b) July 2, (c) July 7, and (d) July 12.

Solution

a. previous balance − payments + purchases = daily balance on 6/28
$420   + $50   = $470

b.   previous balance − payments + purchases = daily balance on 7/2

   $470   − $100   = $370

c.   previous balance − payments + purchases = daily balance on 7/7

   $370   = $370

d.   previous balance − payments + purchases = daily balance on 7/12

   $370   + $30   = $400

Example 4

Learning objective
Calculate the average daily balance for a revolving credit account.

The following information represents the monthly statement for Kerry Milhouse's credit card account:

3/25	previous balance	$580
3/31	purchase	$ 80
4/4	payment	$ 50
4/17	purchase	$ 25

Learning objective
Calculate the monthly finance charge for a revolving credit account.

Assume the billing cycle has 31 days beginning on March 25, and ending on April 25.

Find the daily balance for Kerry's account as of (a) March 31, (b) April 4, and (c) April 25. (d) Find the average daily balance as of April 25. (e) Find the finance charge as of April 25 if it's 1½% of the daily average balance.

Solution

a.   previous balance − payments + purchases = daily balance on 3/31

   $580   + $80   = $660

b.   previous balance − payments + purchases = daily balance on 4/4

   $660   − $50   = $610

c.   previous balance − payments + purchases = daily balance on 4/25

   $610   + $25   = $635

d.   daily balance × number of days

$580 × 5   (3/26–3/30) = $ 2,900
660 × 4   (3/31–4/3) =   2,640
610 × 13   (4/4–4/16) =   7,930
635 × 9   (4/17–4/25) =   5,715
         31                     $19,185

$$\frac{\$19,185}{31} = \$618.87 \text{ average daily balance}$$

e.   .015 × $618.87 = $9.28 amount of finance charge

CHECK YOUR KNOWLEDGE

Revolving Credit

Use the following information to answer problems 1 and 2. Maggie Green's previous balance on her revolving charge account was $453. One month ago she made a payment of $150 and had new purchases of $47. This month she has made a payment of $100 and purchased goods worth $75. The finance charge on the account is 1.5% per month on the unpaid balance.

1. Complete the following table to determine the new balance on Maggie's account at the end of the 2-month period.

Previous balance	Finance charge	Current purchases	Payments	New balance
$453	_____	$47	$150	_____
$	_____	$75	$100	_____

2. What is the total finance charge that Maggie paid over the 2 months?

Refer to the following monthly statement for Randall Stewart's revolving charge account for problems 3 through 5.

11/23	previous balance	$890
11/30	payment	100
12/5	purchase	210
12/12	purchase	75

3. Find the daily balance for Randall's account as of (a) November 30, (b) December 5, and (c) December 15.

4. Find the average daily balance for the period November 24 through December 23.

5. What is the amount of the finance charge as of December 23 if it is 1.75% of the average daily balance?

Refer to the following monthly statement for Shannon Rossi's revolving charge account to answer problems 6 through 8:

1/28	previous balance	$570
2/10	payment	150
2/15	purchase	120
2/17	cash advance	250
2/20	credit for returned merchandise	325

6. Find the daily balance for Shannon's account as of (a) February 10, (b) February 15, (c) February 17, and (d) February 20.

7. Find the average daily balance for the period January 29 through February 28.

8. What is the amount of the finance charge as of February 28 if it is 1.25% of the average daily balance?

11.1   EXERCISES

Solve the following problems. Round dollar amounts to the nearest cent.

Use the following information for exercises 1 and 2.

Mark Pogat's previous balance on his revolving charge account was $750. One month ago he made a payment of $250 and had new purchases of $124. This month he has made a payment of $300 and purchased goods worth $95. The finance charge on the account is 1.5% per month on the unpaid balance.

1. Complete the table below to determine the new balance on Mark's account at the end of the 2-month period.

Previous balance	Finance charge	Current purchases	Payments	New balance
$750	_____	$124	$250	_____
$	_____	$ 95	$300	_____

2. What is the total finance charge that Mark paid over the 2 months?

Use the following information for exercises 3 through 6.

Millie Farmer's use of her Iszard's Department Store charge card for the last three months is listed in the table below. Iszard's finance charge rate is 1.25% per month.

Month	Previous balance	Finance charge	Current purchases	Payments	New balance
1	$185.60	_____	$64.50	$85	167.42
2	$ 167.42	2.09	$27	$100	96.57
3	$ 96.57	1.21	$43.75	$141.47	_____

3. Calculate the finance charge and new balance at the end of the first month.

4. Calculate the finance charge and new balance for the second month.

5. Calculate the finance charge and new balance for the third month.

6. How much did the finance charge total for the three months?

Answers to CYK: *1.* $337.15 *2.* $12.15 *3.* a. $790.00; b. $1,000.00; c. $1,075.00
4. $973 *5.* $17.03 *6.* a. $420; b. $540; c. $790; d. $465
7. $534.68 *8.* $6.68

Refer to the following monthly statement for Leah Johnson's revolving charge account to answer exercises 7 through 9.

Aug. 3	previous balance	$326.70
Aug. 10	payment	125.00
Aug. 24	purchase	73.20
Aug. 30	purchase	89.60

7. Find the daily balance for Leah's account as of (a) August 10, (b) August 24, and (c) August 30.
8. Find the average daily balance as of August 30.
9. What is the finance charge as of August 30 if it is 1.5% of the average daily balance?

Refer to the following monthly statement for Sean Humphrey's revolving charge account to answer exercises 10 through 12.

May 12	previous balance	$435.67
May 15	purchase	143.50
May 23	payment	213.45
May 30	payment	67.00

10. Find the daily balance for Sean's account as of (a) May 15, (b) May 23, and (c) May 30.
11. Find the average daily balance as of May 30.
12. What is the finance charge as of May 30 if is 1.5% of the average daily balance?

Refer to the following monthly statement for Karen Sweet's revolving charge account for exercises 13 through 15.

July 1	previous balance	$745.98
July 12	payment	145.00
July 17	purchase	111.65
July 22	purchase	47.34
July 31	purchase	223.54

13. Find the daily balance for Karen's account as of (a) July 12, (b) July 17, and (c) July 31.
14. Find the average daily balance as of July 31.
15. What is the finance charge as of July 31 if it is 1.75% of the average daily balance?

Refer to the following monthly statement for Andy Frederick's revolving charge account for exercises 16 through 18.

March 2	previous balance	$543.24
March 8	payment	543.24
March 15	purchase	258.96
March 27	purchase	87.43
March 30	purchase	275.87

16. Find the daily balance for Andy's account as of (a) March 8, (b) March 27, and (c) March 30.
17. Find the average daily balance as of March 31.
18. What is the finance charge as of March 31 if it is 1.85% of the average daily balance?

11.2 INSTALLMENT CREDIT

Closed-ended credit differs from open-ended credit in that the fixed principal borrowed in closed-ended credit is set at the time of the loan. Two types of closed-ended credit that we will consider are installment loans and mortgages.

Installment loans are generally used when purchasing items that can be paid off with monthly income on a short-term basis. Automobiles, boats, motorhomes, motorcycles, and appliances are often purchased on an installment plan. **Mortgage loans** are generally used for long-term loans of larger principals. Homes, summer cottages, and businesses are purchased with mortgage loans.

Example 5

Jake Shuron is about to purchase a new car for $10,300. Jake will put $3,000 down and finance the remainder at 12% interest for 48 months with monthly payments of $192.24 per month. (a) How much is Jake actually financing, assuming that all other fees and charges (license fee, tax, preparation charges, etc.) will be paid for separately? (b) How much will he actually pay back if he completes the terms of the loan as stated? (c) How much will the finance charge for this loan be?

Solution

a.   amount to be financed = (cost of car) − (down payment)
$$= \$10,300 - \$3,000$$
$$= \$7,300$$

b.   actually paid = (number of payments) × (amount each payment)
$$= 48 \times \$192.24$$
$$= \$9,227.52$$

c.   amount of finance charge = amount paid − amount financed
$$= \$9,227.52 - \$7,300$$
$$= \$1,927.52$$

Example 6

Barney Bluff and his wife have just purchased an entirely new kitchen for their home for $13,000. They made a down payment of $4,000 and agreed to finance the remainder at an annual percentage rate of 18% interest for 36 months. The finance charge for their installment plan is $2,713.32. (a) How much did Barney actually finance? (b) How much will they pay back totally? (c) How much will each of the 36 equal payments be?

Solution

a.   amount financed = (total cost) − (down payment)
$$= \$13,000 - \$4,000$$
$$= \$9,000$$

b.   total paid back = (amount financed) + (finance charge)
$$= \$9,000 + \$2,713.32$$
$$= \$11,713.32$$

c.   amount of each payment $= \dfrac{\text{total paid back}}{\text{number of payments}}$

$$= \frac{\$11,713.32}{36}$$

$$= \$325.37$$

Example 7

Jethro and Kay Sullivan have agreed to buy a home for $115,000 with a down payment of $45,000 and a $70,000 mortgage at an annual percentage rate of 9% interest for 30 years. Their monthly payments will be $563.24. Assuming Jethro and Kay pay off their mortgage in 30 years, (a) how much will they pay back totally? (b) how much will their interest costs total?

Solution

a.   There are 360 months in 30 years, so
total paid back = 360 × $563.24 = $202,766.40.

b.   total interest paid = total paid back − amount financed
= $202,766.40 − $70,000
= $132,766.40

Determination of Regular Payment Amount

How do we determine the amount of each regular payment? Although there are several ways to determine payment, we will consider three simple examples.

First, consider an example in which simple interest is added to the principal.

Example 8

Tom Evans borrowed $1,200 at 9% simple interest for 1 year. Thrifty Finance Company, Tom's lender, and Tom agreed that the simple interest would be calculated for 1 year and added to the principal. Tom's payment each month would be $1/12$th of the total. Determine (a) the simple interest, (b) the total Tom will owe, and (c) Tom's monthly payment.

Solution

a.   Thrifty Finance Company calculated the interest using the simple interest formula:

$I = PRT = \$1,200 \times .09 \times 1 = \$108$

b.   Thrifty then added the interest to the principal for a total of $1,308.
c.   Thrifty and Ton agreed that he would pay the total in 12 monthly installments of $1,308/12 = $109 each.

Now we will consider a simple discount loan handled in a similar way.

Example 9

Alyssa Meyers borrowed $1,200 from Thrifty Finance Company as well. If her loan was discounted at 9%, determine the amount she actually received and the amount she must pay back.

Solution

$$\text{proceeds} = M - MRT$$
$$= \$1{,}200 - 1{,}200 \times .09 \times 1$$
$$= \$1{,}200 - 108$$
$$= \$1{,}092$$

So Alyssa actually received $1,092 in proceeds but agreed to pay back the full $1,200 since it was a discount loan. She paid Thrifty $100 each month for 1 year to pay off her loan.

We shall show later, in Section 11.5, that although Tom and Alyssa may believe they have obtained 9% interest loans, their true percentage rates as determined by federal standards are actually higher.

A third example of regular payment determination leaves the borrower with regular but unequal payments.

Example 10

Nancy Green borrowed $1,200 from Integrity Finance and agreed to pay $1/12$th of the principal plus 1% interest on the unpaid principal balance each month for 1 year. What will Nancy's payments be for the year?

Solution

Nancy's payments will be as follows:

Month	Principal balance	Interest on unpaid balance	Payment
1	$1,200	$1{,}200 \times .01 = 12$	$100 + 12 = \$112$
2	1,100	$1{,}100 \times .01 = 11$	$100 + 11 = 111$
3	1,000	$1{,}000 \times .01 = 10$	$100 + 10 = 110$
4	900	$900 \times .01 = 9$	$100 + 9 = 109$
5	800	$800 \times .01 = 8$	$100 + 8 = 108$
6	700	$700 \times .01 = 7$	$100 + 7 = 107$
7	600	$600 \times .01 = 6$	$100 + 6 = 106$
8	500	$500 \times .01 = 5$	$100 + 5 = 105$
9	400	$400 \times .01 = 4$	$100 + 4 = 104$
10	300	$300 \times .01 = 3$	$100 + 3 = 103$
11	200	$200 \times .01 = 2$	$100 + 2 = 102$
12	100	$100 \times .01 = 1$	$100 + 1 = 101$

Nancy's loan at 1% per month would be considered a 12% per year rate of interest. Tom and Alyssa's loans provide the benefit of equal regular payments, but they have been led to believe that their loan rates are lower than Nancy's. This occurs because the simple interest computed on Tom's and

Alyssa's loans assumes that they have the entire principal amount for the full year, which is not true. After each payment they have less of the principal in their possession. In truth, their interest rates are considerably higher than Thrifty Finance has suggested. Federal laws on "Truth in Lending" require that Thrifty inform its borrowers of the true annual percentage rate. The **annual percentage rate** is the yearly rate of interest that is being charged on the *unpaid balance* after each payment.

In the next section, we will examine *amortization of a loan,* which provides us with both the benefits of equal regular periodic payments and interest on the unpaid balance after each payment.

CHECK YOUR KNOWLEDGE

Installment Credit

1. For each of the following, determine the amount to be financed.

	Total cost	Down payment	Amount to be financed
a.	$8,000	$2,500	_____
b.	$3,000	$700	_____
c.	$6,000	$1,500	_____
d.	$750	10% down	_____

2. For each of the following, determine the amount of the finance charge.

	Amount to be financed	Monthly payments	Interest rate	Total number of months	Finance charge
a.	$10,000	$332.14	12%	36	_____
b.	$10,000	$263.14	12%	48	_____
c.	$10,000	$222.44	12%	60	_____
d.	$10,000	$195.50	12%	72	_____
e.	$10,000	$176.53	12%	84	_____

3. Determine for each of the following the total amount to be paid and the amount of each equal installment.

	Amount to be financed	Finance charge	Total amount to be paid	Number of months	Amount of each payment
a.	$8,000	$80	_____	12	_____
b.	$10,000	$1,297	_____	24	_____

c.	$6,000	$435.29	_____	6 _____
d.	$15,000	$3,960.48	_____	48 _____

4. Debby Ross purchased a $3,000 piano and borrowed $2,000 at 8% simple interest add-on for 3 years. (a) Determine the interest (finance charge) Debby agreed to pay. (b) Determine the amount of the equal payments if Debby agreed to pay off the principal and interest in 12 equal quarterly payments.

5. Kevin Mark borrowed $200 from his father and agreed to repay the loan in four quarterly payments of $50 each plus 2% interest on the unpaid portion of the loan for that quarter. Construct a payback table for Kevin's loan. (See Example 10.)

	Number	Principal	Interest	Payment
a.	1	_____	_____	_____
b.	2	_____	_____	_____
c.	3	_____	_____	_____
d.	4	_____	_____	_____

11.2   EXERCISES

Determine the finance charge in each of the following problem sets. Round dollar amounts to the nearest cent.

	Amount to be financed	Monthly payment	Total number of payments	Finance charge
1.	$11,000	$325	36	_____
2.	$3,000	$130	24	_____
3.	$780	$75	12	_____
4.	$1,500	$35	48	_____
5.	$23,450	$1,200	20	_____

Determine the total amount to be paid and the amount of each equal installment in each of the following problem sets. Round dollar amounts to the nearest cent.

	Amount to be financed	Finance charge	Total amount to be paid	Number of months	Amount of each payment
6.	$7,000	$84	_____	12	_____
7.	$3,900	$780	_____	24	_____

Answers to CYK: *1.* **a.** $5,500; **b.** $2,300; **c.** $4,500; **d.** $675   *2.* **a.** $1,957.04; **b.** $2,630.72; **c.** $3,346.40; **d.** $4,076.00; **e.** $4,828.52   *3.* **a.** $8,080; $673.33; **b.** $11,297; $470.71; **c.** $6,435.29; $1,072.55; **d.** $18,960.48; $395.01   *4.* **a.** $I = \$2,000(.08)(3) = \$480.00$; **b.** $206.67   *5.* **a.** $200; 4; 54; **b.** $150; 3; 53; **c.** $100; 2; 52; **d.** $50; 1; 51

Amount to be financed	Finance charge	Total amount to be paid	Number of months	Amount of each payment
8. $16,500	$32,000	_____	360	_____
9. $4,700	$1,380	_____	36	_____
10. $9,500	$3,040	_____	48	_____

11. Erin Cole purchased a new refrigerator for $990 and after making a 10% down payment on her purchase, financed the remainder at a 10% add-on simple interest for 1 year. (a) How much did Erin finance? (b) How much interest will Erin pay? (c) If Erin agreed to pay the total (principal + interest) that she owed in four equal quarterly installments, how much will each installment be?

12. Lynette Harvey purchased a new computer for $1,650 and financed the total cost with a 12% per year add-on simple interest for 2 years. (a) How much interest will Lynette pay? (b) If Lynette agreed to pay the total (principal + interest) with eight equal quarterly installments, how much will each installment be?

13. Carl Fignon purchased a mountain bike for $650. He made a 15% down payment and financed the remainder at 12% yearly interest rate for 36 months. If Carl's monthly payments are $18.35, how much finance charge will he pay?

14. Carry Marshall purchased a new large-screen television set for $5,500. After making a 20% down payment, she financed the remainder at 10% interest per year to be paid in quarterly installments for 3 years. Her finance charge was $747.31. How much will she pay in each of her equal quarterly installments?

15. Don Browne borrowed $4,000 to make some improvements to his home. Don agreed to repay the loan in four quarterly payments of $1,000 per quarter plus 3% interest on the unpaid portion of the loan for that quarter. Construct a payback table for Don's loan. How much interest will Don pay for the loan?

16. Singh Po borrowed $1,000 from her father to cover a downpayment on a used car. She agreed to repay the money with four semiannual payments of $250 each plus 2% interest per period on the unpaid balance of the loan for that semi-annual period. Complete the following payback table, and determine the total amount of interest Singh will have to pay her dad.

Unpaid principal	Principal payment	Interest for period	Total payment
$1,000	$250	_____	_____
$	$250	_____	_____
$	$250	_____	_____
$	$250	_____	_____

17. The new drum set purchased by the Hard Luck Six rock band's drummer Earl Pott's cost $8,500. Earl made a downpayment of $2,100 and agreed to pay off the remainder of the cost with eight quarterly payments of $800 each plus 2.5% interest on the unpaid balance of the loan for that quarter. Complete the following payback table for Earl's loan, and determine the total amount of interest he will pay.

Unpaid principal	Principal payment	Interest for quarter	Total payment
$6,400	$800	_____	_____
$	$800	_____	_____
$	$800	_____	_____
$	$800	_____	_____
$	$800	_____	_____
$	$800	_____	_____
$	$800	_____	_____
$	$800	_____	_____

18. Jan Barrett borrowed $8,000 to pay for a trip to Europe for her family. Jan agreed to repay the loan in eight semiannual payments of $1,000 each plus 4% interest on the unpaid portion of the loan for that period. Construct a payback table for Jan's loan. How much interest will Jan pay for the loan?

19. Jeff Jones borrowed $9,000 to add a small room to his summer cottage on the lake. Jeff agreed to repay the loan in six quarterly payments of $1,500 per quarter plus 2.5% interest on the un- paid portion of the loan for that quarter. Con- struct a payback table for Jeff's loan. How much interest will Jeff pay for the loan?

20. Betty Brunno borrowed $3,600 to repair her mo- tor home. She agreed to repay the loan in twelve quarterly payments of $300 per quarter plus 3% interest on the unpaid portion of the loan for that quarter. Construct a payback table for Betty's loan. How much interest will Betty pay for the loan?

11.3   AMORTIZATION OF A LOAN

Learning objective
Using tables, calcu- late the equal regular payments of an amortized loan.

The most common method to determine the regular monthly payment for a loan is called **amortization.** Generally, to amortize a debt is to liquidate the debt with a series of regular payments or installments. In the case of a loan, we will liquidate the total of the principal and finance charge of the loan with regular equal payments. How do we determine the amount of each payment? We use the *present value of an annuity.* Recall that present value annuity problems asked the question, "How much (lump sum) must we de- posit into a fund so that the fund will pay out regular payments of a certain amount to us?"

Example 11

How much must be invested in a fund paying 12% compounded monthly to receive $100 per month for 48 months?

Solution

$$A = p\frac{1 - (1 + i)^{-n}}{i} = p \times \text{(value for } n, r \text{ in present value table, Appendix A, column E)}$$

$$A = 100\frac{1 - (1.01)^{-48}}{.01} = 100 \times (37.97395)$$

$$= 3{,}797.395, \text{ or } \$3{,}797.40$$

From the individual consumer's point of view, a deposit of $3,797.40 into the fund will generate $100 per month (principal and interest) for 48 months and the fund will liquidate itself to $0.

We use the same concept with a small modification for loans. Picture the role of the bank and the consumer reversed. The bank invests a lump sum of money (the loan principal) in the consumer at a rate of interest. The consumer will pay back to the bank the principal plus interest with regular equal payments. The consumer now becomes the fund generating regular payments, which are in part repayment of the principal and in part payment of the interest. The difference is that we know the size of the lump sum (loan principal) and we need to determine the size of the regular payment.

Amortization formula

$$A = p\frac{1 - (1 + i)^{-n}}{i}$$

A = loan amount

p = regular payment

$$p = A\frac{1}{\dfrac{1 - (1 + i)^{-n}}{i}} \qquad \text{Solve for } p.$$

or

$$p = A\frac{i}{1 - (1 + i)^{-n}}$$

Fortunately, we have *amortization tables* in Appendix B, column F that assist us in this process.

Example 12

Determine the regular monthly payment for a loan of $4,800 borrowed at a 12% annual percentage interest rate for 48 months.

Solution

By tables and calculator:

$A = \$4,800$

$n = 48$

$i = \dfrac{.12}{12} = .01 \text{ or } 1\%$

$p = A(\text{table value in column F}$
  for $1\%, n = 48)$

$p = \$4,800(.026334)$

$p = \$126.40$

By formula and calculator:

$A = \$4,800$

$n = 48$

$i = \dfrac{.12}{12} = .01$

$$p = A\frac{1}{\dfrac{1 - (1 + i)^{-n}}{i}}$$

$$p = A\frac{i}{1 - (1 + i)^{-n}}$$

$$p = \$4,800\frac{.01}{1 - (1.01)^{-48}}$$

$$p = \$4,800\frac{.01}{.3797396}$$

$$p = \$4,800(.026334)$$

$$p = \$126.40$$

Now let's consider a car buying situation in which the same interest rate is applied monthly over two different time spans, 4 years and 5 years. We can observe the effect of the time periods on the monthly payments, as well as on the total finance charge.

E x a m p l e 1 3

Everett Michels has decided to purchase a new car for $17,000. The dealer will give him a trade-in allowance of $5,000 on his old car and Everett will put an additional $3,000 down on the car. He can finance the remaining $9,000 at an annual percentage interest rate of 12% for either 48 months or 60 months. (a) How much will his monthly payments be in each of the two plans? (b) How much finance charge will he pay in each of the two plans?

Solution

48 Months:	60 Months:
$A = \$9,000$	$A = \$9,000$
$n = 48$	$n = 60$
$i = \dfrac{.12}{12} = .01$ or 1.0%	$i = \dfrac{.12}{12} = .010$ or 1.0%

(a) $p = A$(table value in column F for 1.0%, $n = 48$)

$p = 9,000(.026334)$

$p = \$237.00$

(a) $p = A \times$ (table value in column F for 1.0%, $n = 60$)

$p = 9,000(.022244)$

$p = \$200.20$

(b) Total payments:

$48 \times 237 = \$11,376.00$

Finance charge

$\$11,376.00 - 9,000 = \$2,376.00$

(b) Total payments:

$60 \times \$200.20 = \$12,012.00$

Finance charge

$\$12,012 - 9,000 = \$3,012.00$

Notice that Everett can reduce his monthly payment by $36.80 ($237 − 200.20) per month by electing the 60-month plan; however, it will cost him an additional $636 ($3,012 − $2,736) in finance charges.

For Your Information

SHOULD YOU BUY OR SHOULD YOU LEASE?

Historically, consumers have generally purchased their automobiles for personal and business use. Over the past few years, however, dealers have actively marketed the lease option to their customers. The advertisement shown here is a typical presentation of these options. To determine which option is most appropriate, a potential buyer would have to consider a number of factors, many of which are mathematical in nature.

The primary distinction between the two options is ownership. If the buyer elects the lease option, title will not transfer to the leasee (customer) at the end of the lease. The monthly payments made by the leasee include interest and the declining value of the vehicle during the lease period. For this reason, the monthly lease payment is always less than the monthly payment under the purchase option for the same time period. In addition, the lease option usually requires the leasee to pay special fees and a mileage penalty if the vehicle is driven more than a certain number of miles per year. Both options require the buyer to pay taxes, registration, insurance, and all service and repair costs. A mathematical analysis of the accompanying ad is given on the next page.

Source: Dick Haskins, Olds GMC Truck Inc., North Syracuse, New York. Used with permission.

Buy option		Lease option	
$33,867	Cost of buy or lease	$10,389	($33,867 − $23,478)
2,598	Less: Down payment	2,598	
31,629	Amount to be financed	7,791	
2,371	Plus: Sales Tax	1,265	
33,640		9,056	
4,674	Plus: Finance charge (8.0%)	7,275	(lease charge)
38,314		16,331	
2,598	Add $2,598 back in for total	2,998	($2,598 + document fee)
$40,912	Total cost to buy or lease	$19,329	
$23,478	Projected value of car after 39 months to investor	$0	
$982	Monthly payment for 39 months	$398	

Mortgages

A **mortgage** is a *pledge of property* to a creditor to secure the repayment of a debt. The most common form of a mortgage by consumers is to secure a loan to purchase real estate. Mortgage loans usually involve a larger loan principal and a longer term to maturity, 15 to 30 years, than other types of installment loans. The property pledged provides the lender with security for the loan. If the borrower defaults on the loan, then he or she may be required to forfeit the property to the lender.

Mortgage loans are commonly used by consumers to buy homes and by businesses to finance fixed assets. The simplest type of mortgage loan is called a *fixed-rate mortgage loan*. Fixed-rate mortgages are calculated in the same manner as the other amortized loans presented in this section.

Example 14

Bob and Marlene Harvey have purchased a new home for $90,000. Bob and Marlene have been granted a 6% interest rate mortgage for $60,000 to be amortized monthly for 30 years. Determine (a) Bob and Marlene's monthly payment, (b) the total amount they will repay, and (c) the total interest they will pay, if they pay off the loan in 30 years.

a.   Using the amortization table in Appendix A column F, for $i = .06/12 = .005$ or .5%,

$$n = 360$$

$$p = A \times i/(1 - (1 + i)^{-n})$$

$$p = 60{,}000 \times (.005)/(1 - (1.005)^{-360})$$

$$p = 60{,}000 \times (.005996)^* = \$359.76 \text{ per month}$$

$$^*\text{value in column F for } i = .5\%, n = 360$$

b.   Total repaid = 360 months × \$359.76/month = \$129,513.60

c.   Total interest = \$129,513.60 − \$60,000 = \$69,513.60

A loan repayment table can be generated for Bob and Marlene using the techniques in Section 11.2 and 11.4—and preferably a computer with a high-speed printer, as well.

Example 15

How much would Bob and Marlene (Example 14) save if they took a 20-year mortgage rather than a 30-year mortgage?

Solution

$$\text{monthly payment} = p = A \times (\text{value in column F for } i = .5\%, n = 240)$$

$$p = \$60{,}000 \times (.007164)$$

$$= \$429.84 \text{ per month}$$

total payments = 240 × \$429.84 = \$103,161.60

total interest = \$103,161.60 − \$60,000 = \$43,161.60

total interest savings = \$69,513.60 − \$43,161.60

$$= \$26{,}352.00$$

So, if Bob and Marlene can find a way to pay \$70.08 more each month for 20 years, they can save \$26,352.00 in interest.

Bob and Marlene could also consider a *biweekly mortgage* plan. Under this plan they would divide their \$359.76 monthly payment in half (\$359.76/2 = \$179.88) and pay \$179.88 every 2 weeks. Since they will be paying off part of the principal 2 weeks earlier each month, this plan results in substantial savings of time and money.

Increasing costs of homes and rising interest rates have resulted in lending institutions' creating new variations of conventional mortgage financing. *Adjustable-rate mortgages (ARMS)* were introduced to make it possible for first-time homeowners to purchase homes with smaller down payments and/or lower monthly payments in the earlier portion of the term of the mortgage. Adjustable-rate mortgages have variable interest rates that may

change every year, or every 3 years. The interest rate is determined by the interest rate of U.S. treasury bonds during the month that the ARM's term ends and a new term begins. The rate is then fixed for the new term. Usually, there is a limit on the amount of change that can occur in the interest rate (e.g., 2% increase or decrease) and there is an upper *cap* on the rate (e.g., 15%). Adjustable-rate mortgages usually carry lower interest rates than fixed-rate mortgages, but provide less stability because of the potential for rate fluctuation.

Adjustments during the term of the mortgage loan can be made in interest rates, unpaid balance, and overall length of the term of the loan. As these adjustments are made, there are corresponding changes in the monthly payments and amount of interest (finance charge) paid.

Additionally, loans can be offered by the builders and/or owners of homes to help move both older and new homes on the real estate market. These innovative forms of mortgage loans have become necessary due to higher costs of real estate and of borrowing money. The use of computers has made such innovations possible because the computer can be programmed to perform many tasks very quickly. While we might hesitate to redo the computations necessary to calculate a new mortgage payment, new interest charge, or mortgage repayment table, the computer can complete such tasks in minutes.

CHECK YOUR KNOWLEDGE

Amortization of a Loan

For problems 1 through 5, use the tables in column F of Appendix A to determine the monthly payment.

	Loan principal (A)	i	n	p
1.	$1,000	1.5%	36	_____
2.	$4,000	.5%	24	_____
3.	$8,000	1.0%	48	_____
4.	$20,000	1.5%	30	_____
5.	$100,000	1.0%	360	_____

6. Determine the quarterly payments of a loan of $40,000 taken for 5 years at an annual percentage interest rate of 12%.

7. Marcia Wilkins has obtained an auto loan for $15,000 for 5 years at 12% annual percentage interest rate per year. (a) Determine Marcia's monthly payments. (b) Determine the total of the payments Marcia will make if she makes all scheduled payments. (c) Determine the finance charge.

8. Melanie Archer recently purchased a home for $90,000. She made a down payment of $25,000 and financed the remainder with a 6% interest rate mortgage to be paid monthly for 30 years. Determine: (a) how much Melanie mortgaged, (b) Melanie's monthly payment, and (c) the total interest Melanie will have paid if she makes *all* the payments over the 30-year period.

11.3 EXERCISES

Solve the following problems. Round dollar amounts to the nearest cent.

For exercises 1 through 4, use the tables in column F of Appendix A to determine the monthly payment.

Loan (A) principal	*i*	*n*	*p*
1. $3,000	1.0%	36	_____
2. $5,000	1.5%	24	_____
3. $1,000	2.0%	30	_____
4. $2,000	0.5%	48	_____

For exercises 5 through 8, use the tables in column F of Appendix A to determine the periodic payment.

Loan (A) principal	*i*	Compounding	Number of years	*p*
5. $4,200	3.0%	quarterly	4	_____
6. $15,000	3.5%	semiannually	10	_____
7. $2,500	9.0%	annually	5	_____
8. $6,850	0.5%	monthly	3	_____

9. Bruce Bodkin recently borrowed $10,000 to purchase a new motorcycle. The loan is to be amortized monthly over a 5-year term and 12% interest per year. Determine: (a) the monthly payment and (b) the total finance charge for the loan.

10. Francine Martin borrowed $7,500 to invest in a small business venture. Her loan is to be amortized on a quarterly basis for 6 years at 10% interest per year. Determine: (a) Francine's quarterly payments, (b) the total amount she will pay

on the loan, and (c) the total finance charge for her loan.

11. Marla Kex purchased a new audio–video entertainment center for her home for $4,200. She made a down payment of $1,200 and borrowed the remainder, which will be amortized monthly over a 2-year period at 18% interest. Determine: (a) the monthly payment and (b) the finance charge.

12. Bruce and Nancy Hanson purchased a new van. They wanted to finance the van with a loan of

$10,000 so they shopped around for financing. Fultown Trust offered them 8% amortized monthly for 4 years. What will be the amount of their monthly payments, and what will be their total finance charge? (Formula required)

13. Fred and Ann Karpel recently agreed to buy a home for $95,000. After a down payment of $20,000, they financed $75,000 with a 30-year fixed-rate mortgage of 6% interest per year to be paid monthly. (a) Determine the monthly payments of the loan. (b) Determine the total interest charge if the loan is paid off according to the mortgage agreement in 30 years.

14. Mary and Jack Kronnen recently agreed to buy a home for $105,000. After a down payment of $25,000, they financed $80,000 with a 30-year fixed-rate mortgage of 7.5% interest per year to be paid monthly. (a) Determine the monthly payments of the loan using the formula. (b) Determine the total interest charge if the loan is paid off according to the mortgage agreement in 30 years.

15. Amie McElroy recently agreed to buy a home for $88,000. After a down payment of $20,000, she financed $68,000 with a 20-year fixed-rate mortgage of 6% interest per year to be paid monthly. (a) Determine the monthly payments of the loan. (b) Determine the total interest charge if the loan is paid off according to the mortgage agreement in 20 years.

16. Ingvar Borg recently agreed to buy a cottage for $60,000. After a down payment of $20,000, he financed $40,000 with a 20-year fixed-rate mortgage of 6% interest per year to be paid monthly. (a) Determine the monthly payments of the loan. (b) Determine the total interest charge if the loan is paid off according to the mortgage agreement in 20 years.

17. Sue Marshall purchased a townhouse for $125,000 in 1985. After a down payment of $30,000, she financed the remainder with a 30-year fixed-rate mortgage of 12% interest per year to be paid monthly. (a) Determine the principal financed. (b) Determine the monthly payments of the loan. (c) Determine the total interest charge if the loan is paid off according to the mortgage agreement in 30 years.

18. Cindy Jacobi purchased a two-family home for $185,000. After a down payment of $55,000, she financed the remainder with a 30-year fixed-rate mortgage of 6% interest per year to be paid monthly. (a) Determine the principal financed. (b) Determine the monthly payments of the loan. (c) Determine the total interest charge if the loan is paid off according to the mortgage agreement in 30 years.

19. Pat Avery purchased a townhouse for $140,000. After a down payment of $30,000, she financed the remainder with a 20-year fixed-rate mortgage of 8% interest per year to be paid monthly. (a) Determine the principal financed. (b) Determine the monthly payments of the loan. (c) Determine the total interest charge if the loan is paid off according to the mortgage agreement in 20 years. (Formula required)

20. Any and Bob Harkhess purchased a small camp for $25,000. After a 20% down payment, they financed the remainder with a 10-year fixed-rate mortgage of 6% interest per year to be paid monthly. (a) Determine the principal financed. (b) Determine the monthly payments of the loan. (c) Determine the total interest charge if the loan is paid off according to the mortgage agreement in 10 years.

Use the following information to answer questions 21 through 25. Jonathan Marx recently purchased a home and sought a mortgage for $50,000. Solvay Trust & Deposit offered to amortize the loan monthly at 12% interest per year for either 20 or 30 years. Determine:

21. Jonathan's monthly payments on the 20-year loan.

22. The total interest Jonathan would pay on the 20-year loan.

23. Jonathan's monthly payments on the 30-year loan.

24. The total interest Jonathan would pay on the 30-year loan.

25. The interest savings of the 20-year loan over the 30-year loan.

Marcie Metzler purchased a summer home for $65,000. She had a choice of either a 10-year, 20-year, or 30-year mortgage amortized monthly at 6% interest per year. She was required to make a 20% downpayment on the home. Use this information for exercises 26 through 30.

26. Calculate (a) Marcie's monthly payment on the 30-year loan and (b) the total interest she would pay for the loan if she paid if off in 30 years.

27. Calculate (a) Marcie's monthly payment on the 20-year loan and (b) the total interest she would pay.

28. Calculate (a) Marcie's monthly payment on the 10-year loan and (b) the total interest she would pay.

29. Calculate the difference in interest Marcie would pay between (a) the 30-year and the 20-year loans, (b) the 30-year and the 10-year loans, and (c) the 20-year and the 10-year loans.

30. Calculate the difference in monthly payments Marcie would pay between (a) the 30-year and the 20-year loans, (b) the 30-year and the 10-year loans, and (c) the 20-year and the 10-year loans.

11.4   LOAN REPAYMENT TABLES AND EARLY LIQUIDATION OF A LOAN

Learning objective
Create loan repayment tables for amortized loans.

Amortization as explained in Section 11.3 is a fair and honest form of consumer loan, for both lender and borrower. By "fair" we mean simply that in the borrowing and repayment process the borrower pays the greatest amount of interest when he or she is holding the most amount of the loan principal, and the least amount of interest when holding the least amount of the loan principal.

Example 16

Consider a loan of $1,000 borrowed at a 12% annual percentage interest rate (APR), which will be paid back in 1 year with equal monthly payments. Determine (a) the monthly payments and (b) construct a loan repayment table.

Solution

a.   Using amortization, we can determine the monthly payments and finance charge ($n = 12$, $i = .12/12 = .01$ or 1%).

$$p = A\left(\frac{i}{1 - (1 + i)^{-n}}\right) = 1{,}000(.088848)^* = \$88.85$$

*Appendix A, column F, $i = 1\%$, $n = 12$
Note that the finance charge is:

   $12(88.85) - 1{,}000 = \$66.20$

b.   We will build our table by (1) computing the interest (1%) on each month's outstanding principal balance, (2) subtracting that interest from that month's regular payment to determine how much of the payment will be used to reduce the principal, and (3) determining the next month's principal by subtracting that principal part of the payment from that month's original principal (see Table 11.1).

Table 11.1

A loan repayment table for an amortized loan

Month	Principal	Payment number	Interest on unpaid balance	Regular payment	Amount of payment of principal	New principal
1	$1,000.00	1	$1,000 × .01 = $10.00	$88.85	$88.85 − 10 = $78.85	$1,000 − 78.85 = $921.15
2	921.15	2	921.15 × .01 = 9.21	88.85	88.85 − 9.21 = 79.64	921.15 − 79.64 = 841.51
3	841.51	3	841.51 × .01 = 8.42	88.85	88.85 − 8.42 = 80.43	841.51 − 80.43 = 761.08
4	761.08	4	7.61	88.85	81.24	679.84
5	679.84	5	6.80	88.85	82.05	597.79
6	597.79	6	5.98	88.85	82.87	514.92
7	514.92	7	5.15	88.85	83.70	431.22
8	431.22	8	4.31	88.85	84.54	346.68
9	346.68	9	3.47	88.85	85.38	261.30
10	261.30	10	2.61	88.85	86.24	175.06
11	175.07	11	1.75	88.85	87.10	87.96
12	87.97	12	.88	88.85	87.97	0*

*Occasionally this final balance is *not exactly* $0. Rounding numbers may result in small errors that cause the discrepancy in the final balance.

Notice that the largest amount of interest ($10) is paid when the borrower possesses the largest amount of the lender's money ($1,000). The least amount of interest ($.88) is paid when the borrower holds the least amount of the lender's money ($87.95). Notice also that after six payments, the outstanding principal is $514.92, and more than half ($1,000/2 = $500) of the loan is still outstanding.

Example 17

Construct a loan repayment table for a loan of $5,000 borrowed at 8% interest per year to be paid off quarterly over a 2-year period.

Solution

$A = \$5,000$

$i = .8/4 = .02$ or 2% per quarter

$n = 2 \text{ years} \times \dfrac{4 \text{ quarters}}{\text{year}} = 8 \text{ quarters}$

regular payment = $5,000 × (column F value for $n = 8$, $i = 2$)

$p = \$5,000(.136510) = \$682.55$

Quarter	Principal	Interest-on unpaid-balance		p	Principal part	New principal
1	$5,000.00	$5,000 × .02 =	$100.00	$682.55	$582.55	$4,417.45
2	4,417.45	4,417.45 × .02 =	88.35	682.55	594.20	3,823.25
3	3,823.25	3,823.25 × .02 =	76.46	682.55	606.09	3,217.16
4	3,217.16		64.34	682.55	618.21	2,598.96
5	2,598.96		51.98	682.55	630.57	1,968.39
6	1,968.39		39.37	682.55	643.18	1,325.20
7	1,325.20		26.50	682.55	656.05	669.16
8	669.16		13.38	682.55	669.16	0.00

CALCULATOR SOLUTION

Keyed entry	Display
AC 5000 Min × .02 =	100
− 682.55 =	−582.55
M+ MR	4417.45

First quarter principal balance

	Display
× .02 =	88.349
− 682.55 =	−594.201
M+ MR	3823.249

Second quarter principal balance

Learning objective
Calculate the outstanding principal of an amortized loan.

Determining Outstanding Principal of a Loan

Recall that the present value annuity formula was used to establish the amortization formula.

Amortization formula

present value $\quad A = p\dfrac{1 - (1 + i)^{-n}}{i}$

amortization $\quad p = A\dfrac{1}{\dfrac{1 - (1 + i)^{-n}}{i}}$

$\qquad\qquad\quad = A\dfrac{i}{1 - (1 + i)^{-n}}$

The present value annuity can be used to determine the outstanding principal of a loan at any stage in the loan.

Example 18

Determine the amount of outstanding loan principal for a loan of $1,000 at 12% interest per year to be amortized monthly over 1 year after (a) six payments, (b) three payments, and (c) nine payments have been made. (*Note:* Please consult the loan repayment table of Example 16.)

Solution

a. If we had the loan repayment table available, we could see that the outstanding principal is $514.93. If we did not have the loan repayment table, we could use the present value annuity.

$$A = p\frac{1 - (1 + i)^{-m}}{i} \qquad \text{(Appendix, column E)}$$

for $p = \$88.85$, $i = 1\%$, and $m = 6$ payments yet to be made. The present value formula tells us what lump sum of money would be necessary to generate six more payments of $88.85 each at 1% interest per month. That same amount is also the amount of the unpaid principal of the loan.

$$A = \$88.85(5.795473)$$
$$= \$514.93$$

Note: We are replacing n with m in the formula simply to point out that the power in the formula for computing the outstanding principal is:

m = **the number of *remaining payments* in the loan**

b. After three payments, there will be $12 - 3 = 9$ payments left, so

$m = 9$

$i = .01$ or 1%

$p = 88.85$

$$A = p\frac{1 - (1 + i)^{-m}}{i} = \$88.85(8.566011) = \$761.09$$

Checking the repayment table, we can see that this is correct with a minor rounding error.

c. After eight payments, there will be $12 - 8 = 4$ payments left so,

$m = 4$

$i = .01$ or 1%

$p = 88.85$

$$A = p\frac{1 - (1 + i)^{-m}}{i} = 88.85\frac{1 - (1.01)^{-4}}{.01}$$

$$= 88.85(3.90165) = 346.69$$

This form of consumer loan is very common because it provides the borrower with the convenience of equal regular payments, and because it distributes the interest portion of the payments fairly.

CHECK YOUR KNOWLEDGE

Loan Repayment and Early Liquidation of a Loan

Solve the following problems. Round dollar amounts to the nearest cent.

1. Construct an amortized loan repayment table for a principal of $1,000 borrowed at 12% interest per year and repaid quarterly over 2 years. (*Note:* $p = 1,000(.142456) = \$142.46$ per quarter.)

n	Principal	Interest	Payment	Principal payment	New principal
1	$1,000.00	30	$142.46	$112.46	$887.54
2	887.54	____	142.46	____	____
3	____	____	142.46	____	____
4	____	____	142.46	____	____
5	____	____	142.46	____	____
6	____	____	142.46	____	____
7	____	____	142.46	____	____
8	____	____	142.46	____	____

2.   Construct an amortized loan repayment table for the first four payments of a $600 loan at 18% interest to be paid monthly for 1 year. (*Note: p =* 600(.091680) = $55.01 with rounding.)

n	Principal	Interest	Payment	Principal payment	New principal
1	$600	_____	$55	_____	_____
2	_____	_____	55	_____	_____
3	_____	_____	55	_____	_____
4	_____	_____	55	_____	_____

3.   Use the present value annuity formula to determine the outstanding principal amounts in problem 1 after (a) three payments, (b) five payments, and (c) seven payments. (Check your answers with the table you developed in problem 1.)

4.   Use the present value annuity formula to determine the outstanding principal amounts in problem 2 after (a) two payments, (b) five payments, and (c) nine payments.

5.   Brian Paul bought a new car for $13,000. After paying all incidental costs and putting a down payment on the car, Brian agreed to finance the remaining costs with an $11,000 loan at 12% APR to be paid monthly for 4 years. After 3 years of payments (36 months), Brian wants to see if he can afford to pay off the remaining principal of the loan. How much does Brian owe on the principal? (Note: You must first determine Brian's regular payment, and then determine his outstanding principal.)

Answers to CYK: *1.*   principal: 771.71, 652.40, 529.51, 402.94, 272.57, 138.29; interest: 26.63, 23.15, 19.57, 15.89, 12.09, 8.18, 4.15; principal payment: 115.83, 119.31, 122.89, 126.57, 130.37, 134.28, 138.31; new principal: 771.71, 652.40, 529.51, 402.93, 272.56, 138.28, 0   *2.*   principal: 554.00, 507.31, 459.92; interest: 9.00, 8.31, 7.61, 6.90; principal payment: 46.00, 46.69, 47.39, 48.10; new principal: 554.00, 507.31, 459.92, 411.82   *3.*   **a.** $652.42; **b.** $402.96; **c.** $138.31   *4.* **a.** $507.22; **b.** $362.90; **c.** $160.17   *5.*   payment = $289.67, remaining after 36 months = $3,260.26.

11.4   EXERCISES

1. Construct a loan repayment table for a loan of $4,500 borrowed at 8% interest per year amortized quarterly for 2 years.

n	Principal	Interest	Payment	Principal payment	New principal
1	$4,500	90	$614.30	$524.30	$3,975.70
2	3,975.70	_____	614.30	_____	_____
3	_____	_____	614.30	_____	_____
4	_____	_____	614.30	_____	_____
5	_____	_____	614.30	_____	_____
6	_____	_____	614.30	_____	_____
7	_____	_____	614.30	_____	_____
8	_____	_____	614.30	_____	_____

2. Construct a loan repayment table for a loan of $8,000 borrowed at 12% interest per year amortized monthly for 1 year.

n	Principal	Interest	Payment	Principal payment	New principal
1	$8,000	_____	$710.79	_____	_____
2	_____	_____	710.79	_____	_____
3	_____	_____	710.79	_____	_____
4	_____	_____	710.79	_____	_____
5	_____	_____	710.79	_____	_____
6	_____	_____	710.79	_____	_____
7	_____	_____	710.79	_____	_____
8	_____	_____	710.79	_____	_____
9	_____	_____	710.79	_____	_____
10	_____	_____	710.79	_____	_____
11	_____	_____	710.79	_____	_____
12	_____	_____	710.79	_____	_____

3. Construct a loan repayment table for a loan of $2,400 borrowed at 10% interest per year amortized semiannually for 3 years.

4. Construct a loan repayment table for a loan of $3,000 borrowed at 18% interest per year amortized monthly for 1 year.

5. Construct a loan repayment table for a loan of $12,000 borrowed at 8% interest per year amortized quarterly for 2 years.

6. Consider a $9,000 loan amortized monthly at 12% interest per year for 4 years. If the monthly payments are $237, determine the outstanding principal on this loan after (a) 12 payments, (b) 24 payments, (c) 36 payments, and (d) 44 payments.

7. Consider a $12,000 loan amortized at 12% interest per year for 4 years. If the monthly payments are $316, determine the outstanding prin-

cipal on this loan after (a) 15 payments, (b) 24 payments, (c) 36 payments, and (d) 40 payments.

8. Consider a $19,000 loan amortized monthly at 12% interest per year for 5 years. If the monthly payments are $422.64, determine the outstanding principal on this loan after (a) 12 payments, (b) 24 payments, (c) 36 payments, and (d) 48 payments.

9. Consider a $15,000 loan amortized monthly at 12% interest per year for 3 years. If the monthly payments are $498.21, determine the outstanding principal on this loan after (a) 12 payments, (b) 24 payments, and (c) 36 payments.

10. Consider a $2,000 loan amortized monthly at 18% interest per year for 2 years. If the monthly payments are $99.85, determine the outstanding principal on this loan after (a) 6 payments, (b) 12 payments, (c) 18 payments, and (d) 20 payments.

11. Maryanne Leo purchased a new silver tea set for her parents' 50th wedding anniversary for $1,800. Maryanne agreed to an installment contract for the entire $1,800 to be amortized monthly for 2 years at 12% interest per year. She has made 12 payments on the installment contract and now wishes to pay off the loan. How much does Maryanne still owe on the original principal?

12. Denise Farmer purchased a new boat for $6,700 and borrowed $5,000 to be amortized quarterly for 4 years at 10% interest per year. She has made 12 payments on the installment contract and now wishes to pay off the loan. How much does Denise still owe on the original principal?

13. Jerry Schurfun purchased a new jetski and borrowed $4,100 to be amortized monthly for 2 years at 6% interest per year. He has made 15 payments on the installment contract and now wishes to pay off the loan. How much does Jerry still owe on the original principal?

14. Lloyd and Sue Darryl purchased matching snowmobiles last winter. The snow machines cost $11,000, $7,500 of which was amortized monthly for 2 1/2 years at 12% interest per year. They have made 24 payments on the installment contract and now want to pay off the loan. How much do they still owe?

15. Brian and Sonia Pepoy purchased a home 20 years ago with a 30-year 6% interest rate mortgage on which they have been paying $98 monthly. They have recently inherited some money and would like to pay off their mortgage. What is their outstanding principal over 240 payments?

11.5 REBATE OF INTEREST ON A LOAN PAID OFF EARLY

Learning objective
Calculate the interest rebate for a loan paid off early.

Whenever a borrower pays off a loan principal sooner than the loan contract requires, the borrower expects to save on interest costs. The amount of interest saved is called an **interest rebate.** After all, if the borrower is no longer holding the loan principal, no interest should be charged.

Determining the outstanding principal on either an *amortized loan* (equal regular payments) or a *principal plus interest on the unpaid balance loan* can be accomplished either by a loan repayment table or the techniques we discussed in Section 11.4. Consequently, if a borrower chooses to liquidate a loan early, either the individual or the lending institution can determine the outstanding principal that will then be paid to the lender and the contract will be brought to closure. Some lending institutions prefer that the repayment terms of the contract be met fully, and then they offer a rebate of interest to the borrower.

Next we will use some of the methods and examples developed earlier in this chapter to determine the amount of rebate a borrower might receive on a loan paid off early.

Present Value Method for Determining Interest Rebate

Example 19 is based on the information we developed earlier in Example 16. Table 11.1 should be helpful.

Example 19

Craig Barrett borrowed $1,000 at 12% annual percentage interest rate to be amortized monthly over a 1-year period. Craig's contract stated that he would make 12 monthly payments of $88.85 each and that the finance charge would be $66.20. Craig made six payments as the contract specified and then decided he would pay off the loan. Since Craig had studied business in college, he knew his outstanding balance to be $514.92. Primal Loan Company, Craig's lender, insisted that Craig make the final six payments, after which it would grant him an interest rebate. How much should Craig be rebated?

Solution

Step 1:   Determine the amount of the unpaid principal. We also know Craig still owes $514.92 on the principal (Example 18) of the loan, so, thus far, he has paid $1,000 − 514.92 = $485.08 of the principal.

Step 2:   Determine the amount of interest paid to date. The difference between the total payments made and the total principal paid is the amount of interest Craig should pay for six months' use of Primal Loan's money.

We know that the monthly payments are $88.85 and that Craig has made six payments for a total of 6 × $88.85 = $533.10.

interest = $533.10 − 485.08 = $48.02

Step 3:   Calculate the amount of the rebate. Craig's rebate can be calculated by subtracting this amount from the full finance charge of $66.20.

interest rebate = full finance charge − 6 months' interest
$$= (\$66.20 - 48.02)$$
$$= \$18.18$$

The next example will demonstrate quite clearly the benefit of paying off a loan as early as possible. It also reinforces the concept mentioned earlier that less interest is paid when less principal is held by the borrower.

Example 20

Eileen Simpson borrowed $2,000 at a yearly percentage rate of 18% interest amortized monthly for 2 years. The installment contract indicates a finance charge of $396.35 and monthly installments of $99.85. What interest rebate should Eileen receive if she pays off her loan after (a) 18 months, (b) 12 months, or (c) 6 months?

Solution

a.   *Step 1:* Determine the outstanding principal after 18 months. Present value formula:

$$A = p\frac{1 - (1 + i)^{-m}}{i}$$

$m = 24 - 18 = 6$

$i = .18/12 = .015$ or 1.5%

$A = \$99.85(5.697167) = \$568.86$

Step 2: Determine the amount of interest paid thus far.

interest paid = sum of 18 payments − principal paid to date

$= (18)(99.85) - (\$2,000 - 568.86)$

$= \$1,797.30 - \$1,431.14$

$= \$366.16$

Step 3: Determine the rebate.

rebate of interest = finance charge − interest paid

$= \$396.35 - \$366.16$

$= \$30.19$

b.   *Step 1:* Determine the outstanding principal and the principal paid to date.

$p = \$99.85$

$m = 24 - 12 = 12$

$i = .015$ or 1.5%

$A = \$99.85(10.907470) = \$1,089.11$

principal paid to date $= \$2,000 - \$1,089.11 = \$910.89$

Step 2: Determine the interet paid to date.

interest paid = 12(99.85) − $910.89

$= \$1,198.20 - \$910.89 = \$287.31$

Step 3: Determine the interest rebate.

interest rebate = finance charge − interest paid to date

= $396.35 − $287.31

= $109.04

c. **Step 1:** Determine the principal paid to date.

$$p = \$99.85$$

$$m = 24 - 6 = 18$$

$$i = .015 \text{ or } 1.5\%$$

$$A = \$99.85(15.672520) = \$1{,}564.90$$

principal outstanding = $1,564.90

principal paid to date = $2,000 − $1,564.90 = $435.10

Step 2: Determine interest paid to date.

interest paid = 6(99.85) − $435.10

= $599.10 − $435.10

= $164

Step 3: Determine the interest rebate.

interest rebate = $396.35 − $164 = $232.35

Sum of Digits Method of Determining Interest Rebate

Another method for computing the amount of interest rebate a borrower is entitled to on an early pay loan is based on a mathematical formula that quickly adds up a sequence of digits from 1 to n that differ from each other by 1. This formula is the **sum of digits** formula.

$$S = \text{sum of digits} = \frac{n(n + 1)}{2}$$

Example 21

Determine the sum of the digits (a) 1 through 10, (b) 1 through 12, (c) 1 through 24, and (d) 1 through 36.

Solution

a. $n = 10$ because there are 10 digits, and

$$\text{sum} = \frac{10(10 + 1)}{2} = \frac{(10)(11)}{2} = 55$$

b. $n = 12$

$$\text{sum} = \frac{12(12 + 1)}{2} = 78$$

c. $n = 24$

$$\text{sum} = \frac{24(25)}{2} = 300$$

d. $n = 36$

$$\text{sum} = \frac{36(37)}{2} = 666$$

Learning objective
Calculate the approximate interest rebate using the sum of digits and rule of 78s methods.

The **rule of 78s** is used to *approximate* the amount of interest contained in each regular payment of a 12-installment loan. The rule of 78s is a very common example of the general sum of digits concept. This method makes use of the fact that the sum of the 12 digits $1 + 2 + 3 + 4 + \cdots + 12 = 78$ and that, if we form the fractions $1/78,\ 2/78,\ 3/78, \cdots, 12/78$, the sum of these fractions is 1; that is,

$$\frac{1}{78} + \frac{2}{78} + \frac{3}{78} + \cdots + \frac{12}{78} = \frac{78}{78} = 1$$

Now we take the total finance charge and break it up into parts using these fractions. For the 12 payments of a loan, these fractions represent the amount of the finance charge that corresponds to each payment; that is,

$12/78$ of finance charge is included in payment 1

$11/78$ of finance charge is included in payment 2

$10/78$ of finance charge is included in payment 3

$9/78$ of finance charge is included in payment 4

$8/78$ of finance charge is included in payment 5

$7/78$ of finance charge is included in payment 6

$6/78$ of finance charge is included in payment 7

$5/78$ of finance charge is included in payment 8

$4/78$ of finance charge is included in payment 9

$3/78$ of finance charge is included in payment 10

$2/78$ of finance charge is included in payment 11

$1/78$ of finance charge is included in payment 12

Again, note that if you add up the fractions $12/78 + 11/78 + 10/78 + \cdots + 1/78 = 78/78 = 1$, which means that all of the finance charges will be collected over the 12-installment period. If the loan has fewer or more than 12 payment periods, then a different rule must be developed using the sum of digits technique.

Example 22

Consider the loan in Example 19. $A = \$1,000$, $p = \$88.85$, $i = 1\%$, $n = 12$, finance charge $= \$66.20$, and Craig chose to pay off the loan after 6 months of payments. Determine Craig's interest rebate using the rule of 78s.

Solution

The rule of 78s would distribute the finance charge of $66.20 over the life of the loan in the following way:

Payment number	Rule of 78s part of finance charge	Amount of finance charge
1	$12/78$	$12/78 \times 66.20 = 10.18$
2	$11/78$	$11/78 \times 66.20 = 9.34$
3	$10/78$	$10/78 \times 66.20 = 8.49$
4	$9/78$	$9/78 \times 66.20 = 7.64$
5	$8/78$	$8/78 \times 66.20 = 6.79$
6	$7/78$	$7/78 \times 66.20 = 5.94$
7	$6/78$	$6/78 \times 66.20 = 5.09$
8	$5/78$	$5/78 \times 66.20 = 4.24$
9	$4/78$	$4/78 \times 66.20 = 3.39$
10	$3/78$	$3/78 \times 66.20 = 2.55$
11	$2/78$	$2/78 \times 66.20 = 1.70$
12	$1/78$	$1/78 \times 66.20 = .85$
	$78/78 = 1$	66.20

CALCULATOR SOLUTION

Keyed entry	Display
AC 66.20 ÷ 78 =	.8487179
Min. MR × 12 =	10.184615
MR × 11 =	9.3358974
MR × 10 =	8.4871795
MR × 9 =	7.6384615
. . . .	
. . . .	
. . . .	
MR × 1 =	.84871

Now adding the last six interest amounts gives us $17.82 rebate of interest. Our earlier calculation in Example 19 yielded a higher rebate of $18.18 to the borrower. The $17.82 yielded by the rule of 78s is an *approximation* to the correct rebate of $18.18. You might want to compare Table 11.1 to the table in Example 22 to see the assignment of interest to each payment and to observe where the differences occur. A shorter and quicker means of computing the approximate interest rebate by the rule of 78s follows:

1. Note that after 6 months of payments on a 12-month loan, 6 payments remain. Use the sum of digits formula to add the digits 1 through 6.

$$\text{sum} = \frac{6(7)}{2} = 21$$

2. Form the ratio 21/78; this is the ratio of the finance charge remaining in those 6 payments, or the approximate interest rebate due to the borrower.

$$\frac{21}{78} \times \$66.20 = \$17.82$$

Example 23

Use the rule of 78s to approximate the interest rebate due on a loan amortized over 12 months with a finance charge of $65 if the loan is paid off after the following number of payments: (a) 10, (b) 7, and (c) 4.

Solution

a. $n = 12 - 10 = 2$ $\text{sum} = \dfrac{2(3)}{2} = 3$

approximate interest rebate $= \dfrac{3}{78} \times 65 = \$2.50$

b. $n = 12 - 7 = 5$ $\text{sum} = \dfrac{(5)(6)}{2} = 15$

approximate interest rebate $= \dfrac{15}{78} \times 65 = \$12.50$

c. $n = 12 - 4 = 8$ $\text{sum} = \dfrac{(8)(9)}{2} = 36$

approximate interest rebate $= \dfrac{36}{78} = 65 = \$30$

The sum of digits method described above for the rule of 78s (12 installment payments) also may be applied to loans with different numbers of payments, for example, 24 or 36.

Example 24

Use the sum of digits method to approximate the interest rebate on a loan amortized over 24 months with a finance charge of $120 if the loan still has the remaining number of payments: (a) 18, (b) 12, and (c) 9.

Solution

The denominator of our fraction is determined by the sum of the digits 1 through 24 because there are 24 installments in the entire loan.

$$\text{denominator} = \text{sum} = n\frac{(n + 1)}{2} = \frac{24(25)}{2} = 300$$

a. $n = 18$ payments remain

$$\text{sum} = \frac{18(19)}{2} = 171$$

$$\text{approximate interest rebate} = \frac{171}{300} \times \$120 = \$68.40$$

b. $n = 12$ payments remain

$$\text{sum} = \frac{12(13)}{2} = 78$$

$$\text{approximate interest rebate} = \frac{78}{300} \times \$120 = \$31.20$$

c. $n = 9$ payments remain

$$\text{sum} = \frac{9(10)}{2} = 45$$

$$\text{approximate interest rebate} = \frac{45}{300} \times \$120 = \$18.00$$

Although the difference in the amount of interest rebate calculated using present value methods and sum of digits methods may have seemed minor in Example 22, in other situations, the differences could be significant.

Example 25

Carin Sward obtained an auto loan for $10,000 at 12% APR that was amortized monthly over a 4-year period (48 months). Carin paid off the loan after 24 monthly payments. Determine (a) Carin's monthly payment, (b) Carin's total finance charge under the original agreement, and (c) Carin's interest rebate after 24 months of payments by: (1) the present value method, and (2) the sum of digits method.

Solution

a. $A = \$10,000$

$n = 48$

$i = .12/12 = .01$ or 1%

Amortization formula:

$$p = A\frac{i}{1 - (1 + i)^{-n}} = 10,000 \times \frac{.01}{1 - (1.01)^{-48}}$$

$p = 10,000 \times (.026334)^* = 263.34$ per month

*Appendix A, column F, 1%, $n = 48$

b. total payments $= 48$ months $\times 263.34 = \$12,640.32$
finance charge $=$ total payments $-$ loan principal $= \$2,640.32$

c. Interest rebate:
1. present value method:
rebate $=$ finance charge $-$ interest paid to date

principal remaining: $A = p\dfrac{1 - (1 + i)^{-m}}{i}$

$m = 48 - 24 = 24$

$i = .01$ or 1%

$p = 263.34$

$A = \$263.34 \times \dfrac{1 - (1.01)^{-24}}{.01}$

$A = \$263.34 \times (21.24338)^*$
*Appendix, column E, $n = 24$, $i = 1\%$

principal remaining $= \$5,594.23$

principal paid to date $= \$10,000 - \$5,594.23 = \$4,405.77$

interest paid to date $=$ total payments to date $-$ principal to date

$= 24(\$263.34) - \$4,405.77$

$= \$6,320.16 - \$4,405.77$

$= \$1,914.39$

interest rebate $= \$2,640.32 - \$1,914.39 = \$725.93$

2. sum of digits method:
First determine the sum of the digits 1 through 48 (denominator).

$$\text{sum} = \frac{n(n + 1)}{2} = \frac{(48)(49)}{2} = 1,176$$

Second, determine the sum of digits for the remaining payments (numerator). 24 payments remain, so

$$\text{sum} = \frac{(24)(25)}{2} = 300$$

$$\text{interest rebate} = \frac{300}{1,176} \times \$2,640.32 = \$673.55$$

Notice that the sum of digits method would rebate $52.38 less than the borrower should receive. Some states have prohibited the use of the sum of digits method because it does not accurately determine the interest rebate due to the borrower.

CHECK YOUR KNOWLEDGE

Interest Rebate

1. Use the present value annuity method to determine the interest rebate on a $20,000 loan at 6% APR amortized monthly ($608.44 per month) over a period of 3 years, if the loan is paid off after (a) 10 payments, (b) 20 payments, and (c) 30 payments.

2. Simon Florcyk borrowed $30,000 for 1 year and agreed to pay a finance charge of $1,200. He agreed to pay back the loan with twelve monthly installments of $2,600 each. Use the rule of 78s to determine the amount of the finance charge assigned to each of the twelve payments.

Payment	Rule of 78s fraction	Amount of finance charge
1	_____	_____
2	_____	_____
3	_____	_____
4	_____	_____
5	_____	_____
6	_____	_____
7	_____	_____
8	_____	_____
9	_____	_____
10	_____	_____
11	_____	_____
12	_____	_____

3. Use the sum of digits method to approximate the interest rebate for a $15,000 loan amortized monthly over a 5-year period with a finance

charge of $2,595.00 if the loan is paid off after (a) 50 payments, (b) 30 payments, and (c) 15 payments.

4. Harry Orwell purchased new appliances for his kitchen and signed an installment contract for a loan of $3,000. The loan will be amortized over 24 months at an annual percentage interest rate of 18%. The contract calls for 24 monthly payments of $149.77 each, which means the finance charge is $594.48. If Harry pays off the loan after 18 months, what will his interest rebate be by (a) the present value method? (b) the sum of digits method?

11.5 EXERCISES

Solve the following. Round dollar amounts to the nearest cent.

1. Consider a $9,000 loan amortized monthly at 12% interest per year for 4 years. If the monthly payments are $237, and the finance charge is $2,376, determine the interest rebate due if the loan is paid off after (a) 12 payments, (b) 24 payments, and (c) 36 payments.

2. Consider a $12,000 loan amortized at 12% interest per year for 4 years. If the monthly payments are $316, and the finance charge is $3,168, determine the interest rebate due if this loan is paid off after (a) 15 payments and (b) 24 payments.

3. Consider a $19,000 loan amortized monthly at 12% interest per year for 5 years. If the monthly payments are $422.64 and the finance charge is $6,358.40, determine the interest rebate due if the loan is paid off after (a) 12 payments, (b) 24 payments, (c) 36 payments, and (d) 48 payments.

4. Consider a $15,000 loan amortized monthly at 12% interest per year for 3 years. If the monthly payments are $498.21, and the finance charge is $2,935.56, determine the interest rebate due if the loan is paid off after (a) 12 payments, (b) 24 payments, (c) 36 payments, and (d) 48 payments.

5. Consider a $2,000 loan amortized monthly at 18% interest per year for 2 years. If the monthly payments are $99.85 and the finance charge is $396.40, determine the interest rebate if the loan is paid off after (a) 6 payments, (b) 12 payments, and (c) 18 payments.

6. Consider a $3,000 loan amortized quarterly at 12% interest per year for 4 years. If the quarterly payments are $238.33 and the finance charge is $821.33, determine the interest rebate due if the loan is paid off after (a) 4 payments, (b) 8 payments, and (c) 12 payments.

7. Consider a $9,000 loan amortized quarterly at 8% interest per year for 5 years. If the quarterly payments are $550.41, and the finance charge is $2,008.20, determine the interest rebate if the loan is paid off after (a) 5 payments, (b) 10 payments, (c) 12 payments, and (d) 15 payments.

8. Consider a $6,000 loan amortized semiannually at 12% interest per year for 4 years. If the semiannual payments are $966.22, and the finance charge is $1,729.76, determine the interest rebate if the loan is paid off after (a) 2 payments, (b) 4 payments, and (c) 6 payments.

9. Sherry Lynne purchased new appliances for her kitchen, borrowing $4,500 to be amortized for 2 years at 12% interest per year. She has made 18 payments on the installment contract and now wishes to pay off the loan. How much would Sherry save in interest if she paid off the loan after 18 payments?

10. Marybeth Leo purchased six place settings of flatware for her granddaughter's wedding present for $1,800. Marybeth agreed to an installment contract for the entire $1,800 to be amortized monthly for 2 years at 12% interest per year. She has made 12 payments on the installment contract and now wishes to pay off the loan. How much will Marybeth save in interest is she does pay off the loan?

11. Denise Farmer purchased a new boat for $6,700; she borrowed $5,000 to be amortized quarterly for 4 years at 10% interest per year. She has made 12 payments on the installment contract and now wishes to pay off the loan. How much does Denise expect to save in interest if she does pay off the loan?

12. Jerry Schurfun purchased a new Jetski and borrowed $4,100 to be amortized monthly for 2 years at 6% per year. He has made 15 payments on the installment contract and now wishes to pay off the loan. How much will Jerry save in interest if he does pay off the principal?

13. The Andersons purchased matching custom saddle sets for horseback riding costing $11,000; $7,500 was amortized monthly for 2½ years at 12% interest per year. They have made 24 payments on the installment contract and now want to pay off the loan. How much interest will they save?

14. Use the formula for the sum of digits to determine the sum of the digits from 1 to (a) 20, (b) 30, (c) 40, (d) 50, (e) 100, and (f) 1,000.

15. Use the rule of 78s to approximate the interest rebate on a 1-year loan paid monthly with a finance charge of $132.38 if the loan is paid off after (a) 4 months, (b) 6 months, (c) 8 months, and (d) 11 months.

16. Use the rule of 78s to approximate the interest rebate on a 1-year loan paid monthly with a finance charge of $198.56 if the loan is paid off after (a) 3 months, (b) 6 months, (c) 9 months, and (d) 10 months.

17. Use the rule of 78s to approximate the interest rebate on a 1-year loan paid monthly with a finance charge of $330.94 if the loan is paid off after (a) 3 months, (b) 6 months, and (c) 9 months.

18. Use the rule of 78s to approximate the interest rebate on a 1-year loan paid monthly with a finance charge of $529.50 if the loan is paid off after (a) 4 months, (b) 6 months, (c) 8 months, and (d) 11 months.

19. Cheryl Martin purchased her home 20 years ago with a 30-year fixed-rate mortgage of 6% interest to be paid monthly on a $25,000 principal. She has made 240 monthly payments of $150 per month. (a) Determine how much is still owed on the original principal. (b) Determine how much interest will be saved if the loan is paid off at this time.

20. Dave Jansen purchased a home 20 years ago with a 30-year fixed-rate mortgage of 6% interest to be paid monthly on a $35,000 principal. He has made 240 monthly payments of $210 per month. (a) Determine how much is still owed on the original principal. (b) Determine how much interest will be saved if the loan is paid off at this time.

11.6 ANNUAL PERCENTAGE RATE

Learning objective
Determine the annual percentage rate for a loan.

The Truth in Lending Act passed by Congress in 1969 is designed to protect consumers against dishonest lenders. It also provides consumers with standardized interest rate terms that must be given to them in writing and that provide them with a means by which they can compare different loan offers.

The *annual percentage rate (APR)* is the rate of interest stated on a yearly basis and calculated on the unpaid principal balance of each period of loan payments. For a loan with equal periodic payments, it is the yearly interest stated in the amortized loan methods studied in Section 11.3. For loans like the one in Example 10, with an unequal payment based on a part of the principal plus a percentage of the unpaid balance, APR is that interest multiplied by the number of payments in 1 year.

In this section, we will use the APR tables found in Appendix C to show how APR is determined. The procedure requires that we follow four steps:

Step 1: Determine the total number of payments, n.

Step 2: Calculate M:

$$M = \left(\frac{\text{finance charge}}{\text{loan principal}}\right) \times 100$$

Step 3: Consult the APR table in the Appendix (of this text) and find M in row n.

Step 4: Determine the annual percentage rate by looking to the top of the column in which M appears.

Example 26

Determine the APR for the loan Tom Evans obtained in Example 8. Tom borrowed $1,200 at 9% simple interest add-on for 1 year. The finance charge was $108, and Tom was to pay $109 monthly for 1 year.

Solution

$$n = 12$$

$$M = \frac{\$108}{\$1,200} \times 100 = 9$$

Find M in row n at the top of the column APR = 16.25% (nearest value). Notice that the true APR based on interest on the unpaid balance is much greater than the suggested 9%.

Example 27

Reconsider Alyssa Meyer's discounted loan of $1,200 with a $108 (Example 9) finance charge for 1 year to be paid monthly. Determine the true APR charged Alyssa.

Solution

Recall that Alyssa received only $1,092 since she accepted a discounted loan.

loan principal = proceeds = $1,200 − $108 = $1,092

$n = 12$

$$M = \frac{\$108}{\$1,092} \times 100 = 9.89$$

APR = 17.75%

Now let's consider the last of our three original examples of installment loans, that of Nancy Green. Nancy borrowed $1,000 and paid back the loan in twelve equal payments plus 1% of the unpaid balance.

Example 28

Determine the APR being charged Nancy Green in Example 10. Nancy borrowed $1,200 for 1 year and paid a finance charge of $78. The terms of her loan were that she would pay 1% interest per month on the unpaid balance as well as $1/12$ of the principal each month. The 1% interest per month suggests a yearly rate of 12%. The terms of her agreement are in the true spirit of APR, and thus we should find her APR to be 12%, or at least very close to it.

Solution

$n = 12$

loan principal = $1,200

finance charge = $78

$$M = \frac{\$78}{\$1,200} \times 100 = 6.5$$

APR = 11.75% (nearest value)

Let's look at one last example that will represent a situation we almost all find ourselves in at one time or another, that of buying a vehicle.

Example 29

Andy Jones purchased a motorcycle and borrowed $8,000 for 4 years with a finance charge of $2,687.00. What is the APR charged Andy, if his payments were made monthly?

Solution

$$n = 48$$

$$M = \frac{\$2,687}{\$8,000} \times 100 = 33.59$$

$$\text{APR} = 15\%$$

A formula exists for calculating APR as well, but it provides only an approximation to APR, so we have not included it here.

CHECK YOUR KNOWLEDGE

Annual Percentage Rate

Solve the following problems. Round dollar amounts to the nearest cent and rates to the nearest one-hundredth of a percent.

1. Determine the annual percentage rate for each of the following loans.

Principal	Interest finance charge	Total number monthly payments	APR
$3,000	$1,649.75	60	_____
$7,000	$1,859.59	36	_____
$9,000	$2,165.23	48	_____
$20,000	$5,496.54	60	_____
$12,000	$3,595.90	48	_____
$15,000	$3,673.23	42	_____

2. Sean Gillis borrowed $3,000 from Sure and Quick Loan Finance Company for 2 years. The finance charge method S&QL used was a 12% simple interest add-on that was calculated to be $2 \times .12 \times \$3,000 = \$720$, with Sean making 24 payments of $155 each. What is the APR that S&QL is charging Sean?

3. Marty Bickoff needed $1,500 in a hurry and borrowed it from his uncle. Marty agreed to pay his uncle $300 interest, and to pay back the loan plus interest in monthly installments over a 24-month period. What APR did Marty agree to?

Answers to CYK: *1.*   18.75%; 16%; 11%; 10%; 13.5%; 12.75%   *2.*   21.5%   *3.*   18.25%

11.6 EXERCISES

For each of the following problem sets, determine the annual percentage rate for the loans. Round dollar amounts to the nearest cent and rates to the nearest one-hundredth of a percent.

	Principal	Interest	Number of monthly payments	APR
1.	$5,000	$1,465.00	24	_____
2.	$8,000	$3,912.80	36	_____
3.	$3,000	$317.70	12	_____
4.	$6,000	$1,495.20	30	_____
5.	$2,000	$434.80	48	_____

	Principal	Interest	Number of monthly payments	APR
6.	$7,000	$908.60	24	_____
7.	$4,000	$714.40	36	_____
8.	$8,000	$2,064.80	48	_____
9.	$5,000	$1,930.00	30	_____
10.	$3,000	$1,141.80	60	_____

Solve the following problems. Round dollar amounts to the nearest cent and rates to the nearest one-hundredth of a percent.

11. Les Jackson borrowed $3,900 from his local E-Z Loan office and agreed to pay a finance charge of $955.89. He also agreed to pay the total $4,855.89 (loan + interest) in 24 monthly installments. Les thought he had a 12.16% (955.89/2 = 477.95, 477.95/3,900 = .1226) interest loan rate. What was Les's APR?

12. Maggie Jenks borrowed $7,000 from her local loan company and agreed to pay a finance charge of $2,754.50 to make a total of $9,754.50 that she would pay back in 36 equal monthly installments of $270.96. What annual percentage rate did Maggie agree to pay?

13. Jake and Joanna Jefferson were a bit short on cash when they needed $2,100 worth of repairs done on their car. They borrowed the $2,100 from a local loan company and agreed to pay off the loan and interest in 15 monthly installments of $167.54 each. How much will they pay

back totally? How much finance charge will they pay? What is their true annual percentage rate?

14. Janna Millken bought a new living room set of couches, tables, and chairs from a local furniture company for $4,500. She made a downpayment of $500 and financed the remaining amount at 8% simple interest for 3 years. The furniture company calculated the simple interest for the 3-year period, added it to the principal of the loan, and divided by 36 to determine the amount of Janna's 36 equal monthly payments. Janna thought that the 8% rate was a good rate. (a) How much did Janna finance? (b) Calculate the simple interest Janna will pay. (c) Determine the actual APR.

15. Farley Fonda recently purchased a used motor bike for $6,500. Farley agreed to pay 20% down and make regular monthly payments of $130 per month for 5 years to pay off the loan and fully own the bike. (a) How much did Farley finance? (b) How much interest did Farley agree to pay? (c) Determine the actual APR of the loan.

CHAPTER REVIEW EXERCISES

Stacy MacDonald's revolving charge account's monthly statement is given below, with a billing cycle of March 6 through April 5. The account has a 1.5% finance charge imposed on the average daily balance for the billing period. Use this information to answer questions 1 through 3.

3/5	previous balance	$420
3/12	payment	$200
3/26	shoes purchased	$ 75
4/3	casual wear purchased	$100

1. Find the daily balance of Stacy's account on April 1.

2. Find the average daily balance of Stacy's account for the billing period.

3. Find the finance charge for the billing period.

Abdul Mohammed bought a new commercial freezer for $10,000. He made a 20% down payment and financed the remainder of the cost for 2 years with a 10% simple interest add-on finance charge. Use this information to answer problems 4 through 6.

4. How much did Abdul finance?

5. How much interest was charged as the finance charge?

6. If Abdul agreed to pay back the total amount (principal + finance charge) in eight equal quarterly installments, how much would each installment be?

7. In need of $2,000 to help furnish her new house, Siri Shazahd borrowed the money from her grandmother and agreed to repay the loan in four quarterly installments of $500 each plus 2% interest on the unpaid balance of the loan for that quarter. Complete the following table, and determine the total interest paid.

Unpaid principal	Principal payment	Interest for period	Total payment
$2,000	$500	_____	_____
$1,500	$500	_____	_____
$1,000	$500	_____	_____
$ 500	$500	_____	_____
Total Interest =			

Sierra Jones borrowed $1,000 for 1 year. The loan was amortized at 8% compounded quarterly with a quarterly payment of $262.62. Use this information to answer problems 8 through 10.

8. How much will Sierra pay back totally on the loan?

9. How much interest will Sierra pay on the loan?

10. Complete the following payback table to show the interest payments each quarter on Sierra's loan.

Unpaid principal	Interest on principal	Quarterly payment	Principal payment	Remaining principal
$1,000	_____	$262.62	_____	_____
	_____	$262.62	_____	_____
	_____	$262.62	_____	_____
	_____	$262.62	_____	_____

Kara and Al Jenkins purchased a lakefront cottage for $45,000. After a 10% down payment, they financed the remainder with a 10-year fixed-rate mortgage of 6% interest per year to be paid monthly. Use this information to answer questions 11 through 16.

11. Determine the principal financed.

12. Determine the monthly payments of the loan.

13. Determine the total interest charge if the loan is paid off according to the mortgage agreement in 10 years.

14. If Kara and Al Jenkins decided to pay off their mortgage on the cottage after 5 years of payments, how much principal would they still owe on the mortgage?

15. How much interet would Al and Kara Jenkins have paid at the end of the 5-year period?

16. If Al and Kara paid off the loan after 5 years of payments, how much interest would they save on the loan?

17. Determine the sum of the digits 1 through 25.

18. Use the Rule of 78s to approximate the amount of interest that Abdel Al-Mudamgh will save if he pays off a 1-year $750 loan after 4 months of payments if the 1-year's interest on the loan is $78.

19. Determine the annual percentage rate that Abdel is paying on a 1-year $750 loan if the monthly payment on the loan is $69 per month.

20. What is the APR on a loan of $5,000 with an interest of $1,200 over 4 years if monthly payments are made?

EXPRESS YOUR UNDERSTANDING

Compose one or two well-written sentences to express the requested information that follows in your own words.

1. What is the daily balance of a revolving credit account?

2. Explain the difference between an installment loan and a mortgage loan.

3. What is interest on the unpaid balance?

4. Explain what it means to amortize a $10,000 loan over a 2-year period with monthly payments.

5. Explain how you would determine the outstanding principal of a loan that has 20 more monthly payments of $98 each, if the annual interest rate is 12% compounded monthly.

6. What is the rebate of interest on a loan paid off early?

7. Explain how the rule of 78s distributes a $156 yearly interest over the 12 months of a year.

8. Describe a quick way to determine the sum of the digits in the first four hundred numbers, i.e., $1 + 2 + 3 + \cdots + 400$.

9. Describe the kinds of purchases generally financed by mortgage loans.

10. What is the annual percentage rate (APR) of a loan?

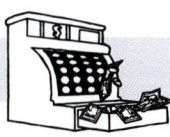

Mind Your Business

Media World has prospered and is beginning to outgrow its space in a small strip mall. The partners have considered either moving to a larger store in the mall, should one become available, or purchasing and renovating a building. A stand-alone building adjacent to the mall has just been put up for sale for $250,000. It is 40 feet by 60 feet, and would require about $50,000 in renovations to make it fully usable by Media World.

Activity 1

Assume that the renovations could be paid out of operating expenses and accumulated funds, and that you could obtain a 20-year mortgage for the purchase of the building with a 10% down payment at 7% compounded monthly. Determine the monthly payments you would have to make, and the total interest the business would pay on the loan, if you paid it off in 20 years according to the terms of the loan.

Activity 2

Develop an amortization table for the first 12 months of the loan.

Period	Principal	Interest	Payment	Remaining principal
Month 1				
Month 2				
Month 3				
Month 4				
Month 5				
Month 6				
Month 7				
Month 8				
Month 9				
Month 10				
Month 11				
Month 12				

Activity 3

Assume that Media World does purchase the building and assumes the mortgage just described. Calculate the remaining principal at the end of 5 years, 10 years, and 15 years and the amounts of interest that would be saved on the loan if the mortgage were paid off at the end of each of those terms.

SELF-TEST

A. Terminology

Complete the following items using the key terms presented at the beginning of the chapter. Check your responses against the answer key at the end of the test.

1. The interest paid for using someone else's money for a period of time is called a _____ _____.

2. Two categories of consumer credit are: (1) _____ _____ credit, such as credit cards, and (2) _____ _____ credit, such as installment loans and mortgages.

3. The _____ _____ of a revolving charge account is computed as:

   ___ ___ = previous balance + purchases
         + cash advances − payments
         − credits.

4. The _____ _____ _____ is computed by adding together the daily balances of each of the days in the billing period and dividing by the number of days in the billing cycle.

5. _____ loans are generally used to purchase items such as appliances, cars, and boats, while _____ loans are used for longer term loans to purchase homes.

6. An _____ _____ is a loan in which the principal and interest are liquidated with a series of equal regular payments.

7. The _____ _____ rate is the rate of interest calculated on the unpaid principal of each period of the loan payments.

8. Something of value that is used to secure a loan is called _____.

9. The _____ is a sum of digits method used to approximate the amount of interest contained in each regular payment of a 12-month installment loan.

B. Calculation

The following concepts and short problems are designed to test your understanding of the objectives identified at the beginning of the chapter. Answers are provided at the end of the test.

10. Consider the following monthly statement for Leah Ficker's revolving charge account:

Date	Activity	Amount
Feb. 28	previous balance	$256.38
March 12	purchase	$128.75
March 23	purchase	$73.18
March 31	payment	$150.00

(a) Find the daily balance for Leah's account as of March 12, March 23, and March 31. (b) Find the average daily balance for the period March 1 through March 31. (c) What is the finance charge as of March 31 if it is 1.5% of the average daily balance?

11. Chanel Maxson purchased a new rack stereo system with CD player for $3,750. After making a 20% down payment she financed the remain-

der at 10% interest per year to be paid in quarterly installments for 3 years. Her finance charge was $509.53. How much will she pay in each of her equal quarterly installments? How much

will she pay in total to complete her financial contract?

12. Use the tables in Appendix B column F to determine the periodic payments.

Loan (A) principal	r	Compounding period	Years	P
$5,300	12.0%	quarterly	5	_____
$16,000	7.0%	semiannually	8	_____
$4,700	9.0%	annually	4	_____
$7,850	6.0%	monthly	6	_____

13. Jeffry Sussex recently bought a video camera, VCR, and TV for $3,500. He made a down payment of $1,200 and borrowed the remainder, which will be amortized monthly over a 3-year period at 18%. Determine (a) the monthly payment and (b) the finance charge.

14. Use the rule of 78s to approximate the interest rebate on a 1-year loan paid monthly with a finance charge of $147.78 if the loan is paid off after (a) 4 months, (b) 7 months, and (c) 9 months.

15. Use the sum of digits rule to approximate the interest rebate on a 2-year loan paid monthly with a finance charge of $400 if the loan is paid off after (a) 10 months and (b) 18 months.

16. Alvin Johnson borrowed $4,500 from the Quick and Slick Loan Company and agreed to pay a finance charge of $902.70. He also agreed to pay the total $5,402.70 (loan + interest) in 18 monthly installments. Alvin thought he had a 13.37% (902.70/1.5 = 601.80, 601.80/4,500 = .1337) interest loan rate. What was Alvin's APR?

17. Mark and Sandy Barker agreed to buy a home for $175,000. After a down payment of $50,000, they financed $125,000 with a 30-year fixed-rate mortgage at 6% interest per year to be paid with monthly payments of $749.44. Determine the total amount Mark and Sandy will pay back if they pay off the mortgage according to the contract.

18. How much will Mark and Sandy pay in interest on the mortgage if the loan is paid off according to the mortgage agreement in 30 years?

19. If Mark and Sandy take a 20-year mortgage at 6%, their monthly payments will be $895.54. How much interest will they pay on this loan if they pay it off in 20 years according to the mortgage agreement?

20. If Mark and Sandy pay off the 30-year loan (problem 17) after 20 years (240 payments), they will still owe $67,504.48 on the principal (i.e., they will have paid off $57,485.52). How much interest would they save by paying off the loan at the end of 20 years?

Answers to Self-Test: **1.** finance charge **2.** open-ended, closed-ended **3.** daily balance **4.** average daily balance **5.** installment, mortgage **6.** amortized loan **7.** annual percentage **8.** collateral **9.** liquid **10. a.** (1) $385.13 (2) $458.31 (3) $308.31; **b.** $355.85; **c.** $5.34 **12.** $356.24; $1,322.96; $1,450.74; $130.10 **13. a.** $83.15; **b.** $693.39 **14. a.** $68.21; **b.** $28.42; **c.** $11.37 **15. a.** $140; **b.** $28 **16.** 24% **17.** $269,798.40 **18.** $144,798.40 **19.** $89,929.60 **20.** $22,418.32

12

Key Terms

share account
check
signature card
service charge
simple interest
deposit
restrictive endorsement
blank endorsement
special endorsement
check stub
check register
certified check
returned check (NSF)
stop payment order
merchant's deposit
   summary
discount fee
bank statement
cancelled check
reconciliation
account form

BANK SERVICES AND RECORDS

Learning Objectives

After completing the exercises in this chapter, you will be able to:

1. Perform all activities necessary to establish a personal or business checking account.

2. Describe the different types of checks and endorsements.

3. Prepare the required calculations to maintain checking account records.

4. Explain checking account services provided by banks.

5. Complete deposit records of credit card transactions.

6. Reconcile bank statements and check registers by preparing bank reconciliation statements.

7. Define key terms.

INTRODUCTION

Banking services can be traced as far back as 352 B.C. with the invention of the check by the Romans. However, the earliest printed checks appeared in England in the latter part of the seventeenth century to allow people to receive checks from different people drawn on different banks. Although this service was somewhat limited, its convenience and safety proved to be invaluable.

Today, due to dynamic changes in the marketplace, banks offer customers many services besides checking accounts and a safe place to deposit their savings. For example, in addition to a complete line of long- and short-term loan programs, banks offer financial management, electronic banking, credit cards, investment counseling, and insurance. These services have resulted because of changes in federal banking laws designed to protect and better serve the depositor. To illustrate this point a change in banking regulations allowed savings and loan associations, credit unions, and other similar financial institutions to offer their customers a service comparable to the checking account called a **share account.** This change made these institutions more competitive with full-service banks and, in turn, forced full-service banks to reevaluate their scope of services and associated fee schedules.

This chapter examines bank services associated with checking accounts. It will instruct you on how to maintain both personal and business checking accounts records, deposits credit card transactions, and prepare bank reconciliation statements. All of the other banking services noted above will be presented in later chapters.

12.1   CHECKING ACCOUNTS AND RECORDS

Banks provide many services to individuals and businesses, the most notable being the checking account. This services enables us to conduct all types of business transactions, such as purchasing goods and services or paying debts, without having to use cash. A check is a safe, convenient method of payment. In addition, a check provides an accurate record of the transaction for future reference. It is for these reasons that most individuals and businesses use the checking account service provided by financial institutions including commercial banks, savings and loan associations, and credit unions.

Learning objective
Perform all activities necessary to establish a personal or business checking account.

A **check** is a written order from an individual or a company (drawer) to their banking institution instructing it to pay a designated party (payee) a specified amount of funds kept on deposit. Checks may involve more than two parties, and are often sent great distances. Therefore, it is important to understand how to fill out a check correctly. The various parts of a check are explained in Figure 12.1.

When filling out a check, it is important to follow these guidelines so that the check cannot be altered, and errors can be detected easily.

Figure 12.1

How to write a check

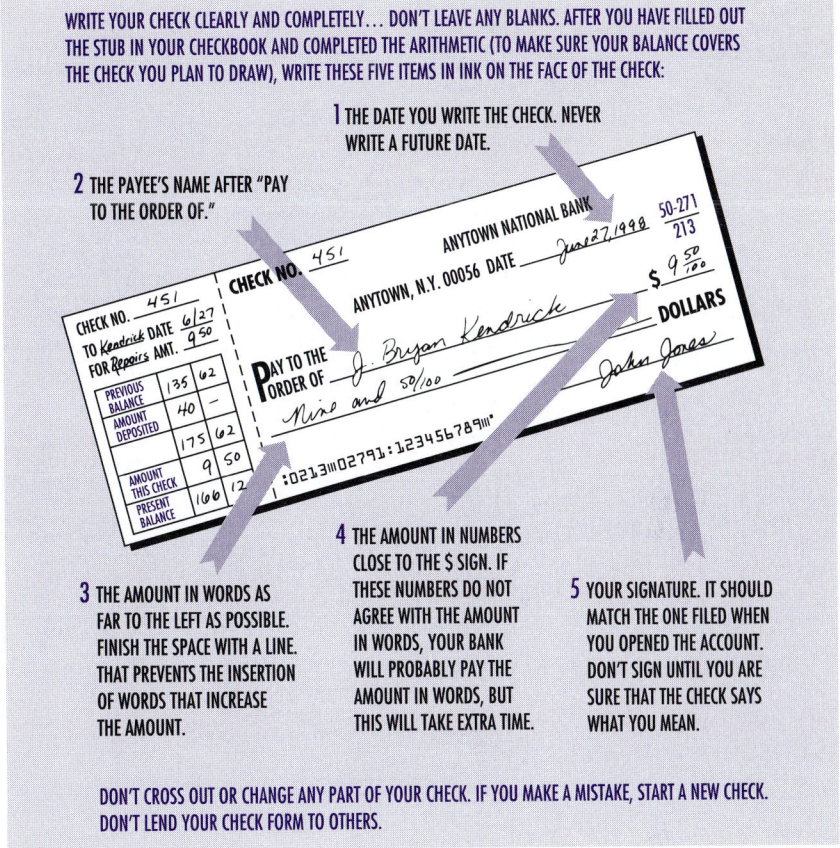

WRITE YOUR CHECK CLEARLY AND COMPLETELY... DON'T LEAVE ANY BLANKS. AFTER YOU HAVE FILLED OUT THE STUB IN YOUR CHECKBOOK AND COMPLETED THE ARITHMETIC (TO MAKE SURE YOUR BALANCE COVERS THE CHECK YOU PLAN TO DRAW), WRITE THESE FIVE ITEMS IN INK ON THE FACE OF THE CHECK:

1 THE DATE YOU WRITE THE CHECK. NEVER WRITE A FUTURE DATE.

2 THE PAYEE'S NAME AFTER "PAY TO THE ORDER OF."

3 THE AMOUNT IN WORDS AS FAR TO THE LEFT AS POSSIBLE. FINISH THE SPACE WITH A LINE. THAT PREVENTS THE INSERTION OF WORDS THAT INCREASE THE AMOUNT.

4 THE AMOUNT IN NUMBERS CLOSE TO THE $ SIGN. IF THESE NUMBERS DO NOT AGREE WITH THE AMOUNT IN WORDS, YOUR BANK WILL PROBABLY PAY THE AMOUNT IN WORDS, BUT THIS WILL TAKE EXTRA TIME.

5 YOUR SIGNATURE. IT SHOULD MATCH THE ONE FILED WHEN YOU OPENED THE ACCOUNT. DON'T SIGN UNTIL YOU ARE SURE THAT THE CHECK SAYS WHAT YOU MEAN.

DON'T CROSS OUT OR CHANGE ANY PART OF YOUR CHECK. IF YOU MAKE A MISTAKE, START A NEW CHECK. DON'T LEND YOUR CHECK FORM TO OTHERS.

1. Write legibly in ink. If typewritten, complete all parts but the signature line.

2. Begin the amount to be paid in words at the far left of the line provided. Express any part of a dollar as a numerator of a fraction with a denominator of 100. Fill any space between the end of the written amount and the printed word *dollars* with a solid line, as shown in Figure 12.1.

3. Position the numerals of the check amount close to the printed dollar sign to prevent any adjustment in the amount. (By law, the bank will honor the check at the amount that is to be paid in words.)

4. Sign all checks with your legal signature as noted on your signature card (Figure 12.2).

Figure 12.2

Signature card

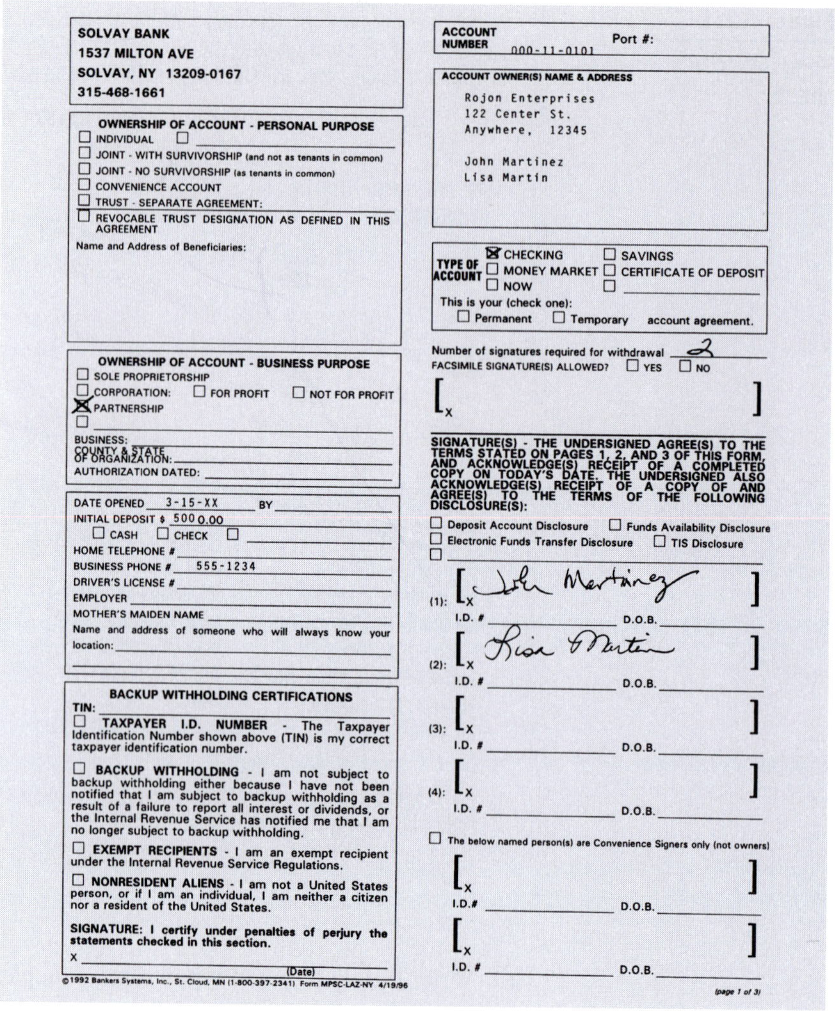

SOLVAY BANK
1537 MILTON AVE
SOLVAY, NY 13209-0167
315-468-1661

ACCOUNT NUMBER 000-11-0101 Port #:

ACCOUNT OWNER(S) NAME & ADDRESS
Rojon Enterprises
122 Center St.
Anywhere, 12345

John Martinez
Lisa Martin

OWNERSHIP OF ACCOUNT - PERSONAL PURPOSE
☐ INDIVIDUAL ☐ ________
☐ JOINT - WITH SURVIVORSHIP (and not as tenants in common)
☐ JOINT - NO SURVIVORSHIP (as tenants in common)
☐ CONVENIENCE ACCOUNT
☐ TRUST - SEPARATE AGREEMENT:
☐ REVOCABLE TRUST DESIGNATION AS DEFINED IN THIS AGREEMENT
Name and Address of Beneficiaries:

TYPE OF ACCOUNT ☒ CHECKING ☐ SAVINGS ☐ MONEY MARKET ☐ CERTIFICATE OF DEPOSIT ☐ NOW ☐ ________
This is your (check one): ☐ Permanent ☐ Temporary account agreement.

Number of signatures required for withdrawal 2
FACSIMILE SIGNATURE(S) ALLOWED? ☐ YES ☐ NO
[x]

OWNERSHIP OF ACCOUNT - BUSINESS PURPOSE
☐ SOLE PROPRIETORSHIP
☐ CORPORATION: ☐ FOR PROFIT ☐ NOT FOR PROFIT
☒ PARTNERSHIP
☐ ________
BUSINESS: ________
COUNTY & STATE OF ORGANIZATION: ________
AUTHORIZATION DATED: ________

SIGNATURE(S) - THE UNDERSIGNED AGREE(S) TO THE TERMS STATED ON PAGES 1, 2, AND 3 OF THIS FORM, AND ACKNOWLEDGE(S) RECEIPT OF A COMPLETED COPY ON TODAY'S DATE. THE UNDERSIGNED ALSO ACKNOWLEDGE(S) RECEIPT OF A COPY OF AND AGREE(S) TO THE TERMS OF THE FOLLOWING DISCLOSURE(S):
☐ Deposit Account Disclosure ☐ Funds Availability Disclosure
☐ Electronic Funds Transfer Disclosure ☐ TIS Disclosure
☐ ________

DATE OPENED 3-15-XX BY ________
INITIAL DEPOSIT $ 500 0.00
☐ CASH ☐ CHECK ☐ ________
HOME TELEPHONE # ________
BUSINESS PHONE # 555-1234
DRIVER'S LICENSE # ________
EMPLOYER ________
MOTHER'S MAIDEN NAME ________
Name and address of someone who will always know your location: ________

(1): [x John Martinez] I.D. # ________ D.O.B. ________
(2): [x Lisa Martin] I.D. # ________ D.O.B. ________
(3): [x] I.D. # ________ D.O.B. ________
(4): [x] I.D. # ________ D.O.B. ________

BACKUP WITHHOLDING CERTIFICATIONS
TIN: ________
☐ TAXPAYER I.D. NUMBER - The Taxpayer Identification Number shown above (TIN) is my correct taxpayer identification number.
☐ BACKUP WITHHOLDING - I am not subject to backup withholding either because I have not been notified that I am subject to backup withholding as a result of a failure to report all interest or dividends, or the Internal Revenue Service has notified me that I am no longer subject to backup withholding.
☐ EXEMPT RECIPIENTS - I am an exempt recipient under the Internal Revenue Service Regulations.
☐ NONRESIDENT ALIENS - I am not a United States person, or if I am an individual, I am neither a citizen nor a resident of the United States.
SIGNATURE: I certify under penalties of perjury the statements checked in this section.
X ________ (Date)

☐ The below named person(s) are Convenience Signers only (not owners)
[x] I.D.# ________ D.O.B. ________
[x] I.D. # ________ D.O.B. ________

©1992 Bankers Systems, Inc., St. Cloud, MN (1-800-397-2341) Form MPSC-LAZ-NY 4/19/96

(page 1 of 3)

Source: Solvay Bank, Solvay, NY. Used with permission.

Before you open a checking account, you should investigate the scope of service and fees charged for services of various financial institutions in your area. The federal government has modified banking regulations which has increased competition and resulted in a variation of fees and services.

All banks require checking account applicants to complete a **signature card** similar to the one shown in Figure 12.2 that lists personal information and identifies the signature you will use to sign checks. An initial deposit is usually required to open the account, and the bank will provide some blank checks with your account number and a checkbook record.

Table 12.1
Service charge
schedule

Average balance	Monthly charge
$0–$200	$6.50
$201–$400	$4.00
$401–$599	$2.50
$600 or more	free
plus $.20 per check paid	

Calculating Service Charges on Checking Accounts

There are two basic types of checking accounts, personal and business. *Personal checking accounts* are used by individuals. The bank provides personalized printed checks for a fee, a checkbook cover, and a selected checkbook record.

The *regular account plan* requires a monthly service charge to cover the cost of maintaining the account. The **service charge** is usually based on the average balance and the number of checks written during the month. Many banks will eliminate or prorate the monthly service charge based on a required minimum monthly balance. Table 12.1 is a service charge schedule that illustrates how the service charge in Example 1 is determined.

Example 1

Rose Heckla has an average balance of $375 in her account. What is the amount of her service charge if she wrote 34 checks during the month?

Solution

Monthly charge	$4.00	($201 − $400)
Checks paid charge	6.80	(34 checks × $.20)
	$10.80	

Most banks will provide *special checking accounts* for individuals who write only a few checks each month and do not want to leave a minimum balance in their account. The service charge may consist of a charge for blank checks, a charge for each check processed, or a monthly maintenance fee.

Example 2

A bank charges a per-check charge of $.30 and a monthly maintenance fee of $2.50. Determine the service charge for a customer who had 12 checks processed this month.

Solution

Check fee	$3.60	(12 checks × $.30)
Maintenance fee	2.50	
Total service charge	$6.10	

Learning objective
Describe the different types of checks and endorsements.

Many banks offer *flat fee checking accounts.* Under this plan, a fixed charge per month covers the cost of checks, the checking account, bank charge card, and other bank services.

Another type of personal checking account that has become popular is the *interest-bearing account.* This account pays interest on the average daily balance or on the lowest balance recorded during the month. The interest-bearing account also provides free checking if the required minimum balance is maintained during the month. If the balance drops below the stipulated balance, a monthly fee and a per check fee will be charged. The interest earned on checking accounts is **simple interest,** which means the rate of interest paid is applied to the principal (lowest or average account balance) on an annual or per annum basis. Because banks provide bank statements to their customers each month, the time period of the interest equation is adjusted to $\frac{1}{12}$ (one-twelfth) of a year, as shown in the solution to Example 3.

Example 3

Community Bank offers interest-bearing checking, which provides free checking based on the schedule shown in Table 12.1. It also pays $2\frac{1}{2}\%$ interest on the lowest monthly balance. A customer's account record indicates an account balance ranging from $457.85 to $1,365.90, with an average balance of $682.50 for the month. What charges (debits) or payments (credits) are there on the customer's account for the period if the bank paid 32 checks?

Solution

P = lowest monthly balance

r = annual rate of interest paid

t = time period based on 1 full year or a part of 1 year

I = interest earned

interest payment: $I = P \times r \times t$ (interest earned on lowest balance)

$$\$0.95 = \$457.85 \times .025 \times \frac{1}{12}$$

There is no service charge because the customer's average balance exceeded the minimum $600 average balance for free checking. The customer will, however, collect 95 cents in interest based on the lowest balance of $457.85 recorded in the account during the month, and will pay a $6.40 checks paid fee.

Business checking accounts, sometimes referred to as commercial accounts, have different rate schedules and charges than personal accounts because of the financial services required to control the business cash flow. For example, banks will collect receipts and make payments for a business. A typical rate schedule, as shown in Table 12.2, might include

1. a fee for checks paid on behalf of the business
2. a fee for each deposit made by the business during the month
3. a charge for each check received by the business included with deposits
4. a monthly maintenance fee for maintaining the account
5. an earned credit, which is annual interest based on the average daily balance

Example 4 illustrates the application of charges and payments to a business checking account.

Example 4

Northside Auto Supply made 12 deposits, which included 287 customer checks, had 130 checks paid by its bank, and had an average balance for September of $5,392.60. What are the monthly charges (debits) and payments (credits) to the account? (Use Table 12.2 schedule of charges.)

Solution

Maintenance fee	$10.00	
Checks paid	11.70	(130 × $.09)
Deposit fee	3.60	(12 × $.30)
Checks deposited	17.22	(287 × $.06)
total charges (debit)	$42.52	

Interest payment (credit) $23.59 \left(\$5,\!392.60 \times .0525 \times \dfrac{1}{12} \right)$

Table 12.2
Service charge
schedule

Service	Charge
1. Maintenance fee	$10.00
2. Checks paid	.09 each
3. Deposit fee	.30 per deposit
4. Checks deposited	.06 each
5. Interest credit	5.25 percent per annum based on average balance

Maintaining Checking Account Records

At this point, we have explained the parts of a check, how to calculate service charges on different types of checking accounts, and how to open a checking account. We are now ready to learn how to deposit money in an account and to record various types of transactions in the checkbook register.

A **deposit** increases the amount of money in a checking account and can include checks as well as cash. When checks are deposited in an account, they must be properly endorsed or they will not be accepted by the bank.

An *endorsement* is a signature or instructions with a signature written on the back of the check legally transferring ownership of the check. There are three methods of endorsing a check, each serving a specific purpose. A **restrictive endorsement** limits the ability to cash the check. It includes the words "for deposit only" and the signature of the depositor. Most businesses will use this type of endorsement when depositing customer checks. The **blank endorsement** requires only the signature of the person or firm appearing on the check as the payee. Consequently, if the check is lost or stolen, it could be further endorsed and cashed. The final endorsement is called a **special endorsement.** This method allows the payee to transfer the check amount to a third party (another person or company). However, only the party named in the endorsement can cash the check. Sample endorsements are illustrated in Figure 12.3.

The *deposit slip* is the official record of the transaction and is provided by the bank. The slip contains personalized printed information and space for the amounts to be deposited in cash and checks for both personal and business accounts. In Figure 12.4 are copies of actual deposit slips for personal and business accounts.

Learning objective
Prepare the required calculations to maintain checking account records.

Figure 12.3

Check endorsements

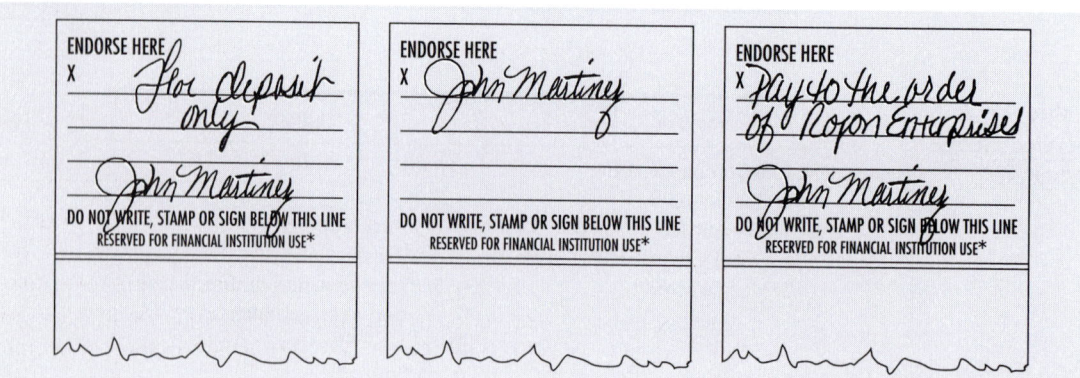

Figure 12.4

Deposit slips

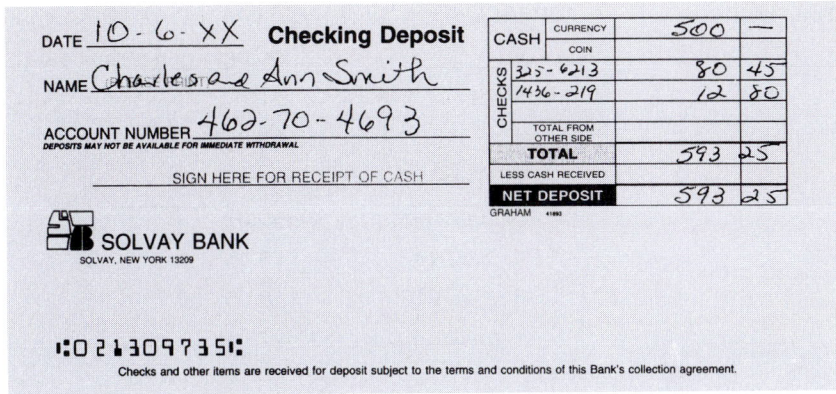

(a) Personal account deposit slip

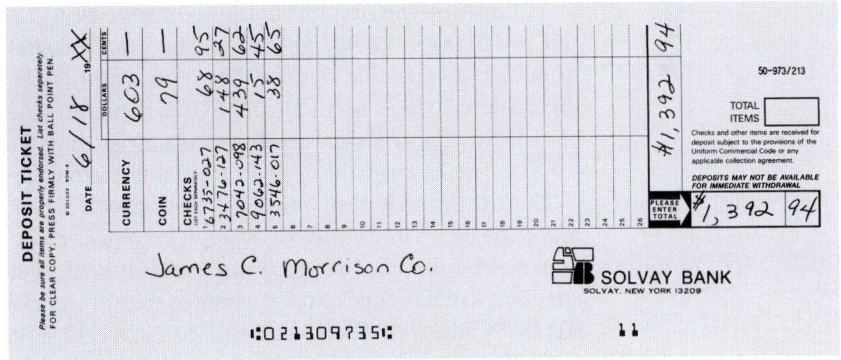

(b) Commercial deposit slip

Source: Solvay Bank, Solvay, NY. Used with permission.

Note that when you deposit checks, you must list, in addition to the amount, each check's bank identification number or check serial number. With each deposit, you will receive a copy for your records so that you can verify the transaction. The reconciliation procedure will be explained later in the chapter.

Example 5

Using the following financial data, prepare the daily deposit for the Country Pride Gift Shoppe: Currency: 15 twenty-dollar bills, 20 ten-dollar bills, 12 five-dollar bills, 43 one-dollar bills; coins: 5 rolls of quarters, 3 rolls of dimes, 6 rolls of nickels, 4 rolls of pennies; checks: no. 6735-027 for $68.95, no. 3746-127 for $148.27, no. 7042-098 for $439.62, no. 9062-143 for $15.45, and no. 3546-017 for $38.65.

Solution

Currency	$603.00	($300 + 200 + 60 + 43)
Coin	79.00	($50 + 15 + 12 + 2)
Checks		
1. 6735-027	68.95	
2. 3746-127	148.27	
3. 7042-098	439.62	
4. 9062-143	15.45	
5. 3546-017	38.65	
total	$1,392.94	

Banks require currency and coins to be separated by denomination and either banded or rolled. Therefore, in Example 5, the total currency was found by multiplying the number of bills by their value ($15 \times \$20.00 = \$300$) plus ($20 \times \$10.00 = \$200$), and so on. The same procedure was followed to determine the total coin deposit. The coin wrapper for each coin size indicates the required value of each roll: $10 for quarters, $5 for dimes, $2 for nickels, and 50 cents for pennies. In the example, the quarters totaled $50 ($10 \times$ 5 rolls). The completed deposit slip is illustrated in Figure 12.4(b).

Each time a deposit is made or a check is written, the transaction is recorded on a check stub or in a check register. The type of record selected depends on the preference of the individual or business. The **check stub** is attached to each check and remains as part of the checkbook once the check has been removed. The check stub in Figure 12.5 illustrates the information and procedure necessary to complete the record. Notice that the check number is printed on the check stub. To complete the check stub, start at the top and fill in each line as required. The information shown in Figure 12.5 is developed in Example 6.

Figure 12.5

Check stub

Source: Solvay Bank, Solvay, NY. Used with permission.

Example 6

Amanda Chambers, an accounts payable clerk for County Airport, wrote check no. 991 on October 10 to General Supply in the amount of $386.95 for a file cabinet. If the balance in the account is $4,735.57, and a deposit of $1,320 had been made since the last check was written, what is the balance forward on the check stub?

Solution

Step 1: Complete the date, to whom the check was written, and for what purpose.

Step 2: Record "balance brought forward" $4,735.57.

Step 3: Add deposit to balance brought forward for total ($6,055.57).

Step 4: Enter the amount of this check, no. 1001, and subtract from the total amount ($5,668.62).

Step 5: Record the difference on the balance forward line and on the balance brought forward line of the check stub in no. 1002 (next check stub).

#1001	October 10, 1999	
To: General Office Supply		
For: file cabinet		
Balance brought forward		$4,735.57
Amount deposited	+	1,320.00
Total	=	6,055.57
Amount this check	−	386.95
Balance forward	=	5,668.62

A **check register** may be preferred by some depositors because it is usually in book form and can be separated from the checkbook. The check register also allows depositors to evaluate multiple transactions at one time and provides a column to check off each check as it is returned from the bank. The same information is recorded in the check register that is recorded on the check stub. The process of entering transactions is shown in Figure 12.6. Although check registers may vary slightly, all require that you add deposits (credits) and deduct payments (debits) from the balance forward. Whatever method is used to record checkbook transactions, it is important that you record all required information for each check when the check is written.

Figure 12.6

Checkbook register

NUMBER	DATE	DESCRIPTION OF TRANSACTION	PAYMENT/DEBIT (-)	√ T	FEE (IF ANY) (-)	DEPOSIT/CREDIT (+)	BALANCE
					BALANCE BROUGHT FORWARD →		$ 689 47
172	11/25	Central States Electric	$ 49 75	$	$		639 72
173	11/28	Community Bank	235 62				404 10
	12/2	Deposit				850 —	1,254 10
174	12/4	Food Center	148 27				1,105 83

Example 7

Enter the following transactions in the checkbook register and determine the balance forward: balance brought forward, $689.47; payment on November 25 of $49.75, check no. 172 to Central States Electric Company for monthly electric bill; payment on November 28 of $235.62, check no. 173 to Community Bank for November car payment; deposit December 2 of $850; payment on December 4 of $148.27, check no. 174 to Food Center for weekly groceries. (Refer to Figure 12.6.)

Solution

Step 1:   Enter balance brought forward from preceding check register page ($689.47).

Step 2:   Record the first transaction, which is a payment of $49.75 as indicated. This reduces the balance brought forward to $639.72.

Step 3:   Enter the second transaction—a payment of $235.62 reducing the balance forward to $404.10.

Step 4:   Record the next required transaction—a deposit of $850 on December 2, which is added to the balance forward. The balance forward is now $1,254.10.

Step 5:   Enter the final transaction—another payment of $148.27, which reduces the balance forward to $1,105.83.

CHECK YOUR KNOWLEDGE

Checking Account Records and Service Charges

1. What is the monthly service charge for Sam's Hobby Shop if the average balance is $527.89 and 64 checks were paid during the month? (Use Table 12.1.)

2. Gladys Raymond has a special checking account. She is charged 20 cents for each check, 30 cents for each check paid by her bank, plus a monthly maintenance fee of $4.50. If Gladys wrote nine checks during the month and five were paid by the bank, how much would the bank charge her account for checking service?

3. You have your checking account with a bank that pays $4\frac{3}{4}\%$ interest on the lowest monthly balance. The bank also offers free checking under the terms shown in Table 12.1. Your account balance range this month was $289.45 to $726.84, with an average balance of $482.60. If you wrote 28 checks during the month, what would be the net cost (charges minus payments) of your account for the month?

4. Homemaid Cleaning Service had an average balance for January of $3,692.37. During the month, it made 8 deposits, which included a total of 124 checks, and the bank paid 82 checks on the company's behalf. Using the rate schedule in Table 12.2, calculate Homemaid's monthly service charge.

5. Using the following data, complete a sketch of the deposit slip provided: 9 fifty-dollar bills, 25 twenty-dollar bills, 47 ten-dollar bills, 87 one-dollar bills, 19 rolls of quarters, 13 rolls of dimes, 26 rolls of nickels, 6 rolls of pennies; checks: no. 291-32 for $289.22, no. 432-61 for $542.87, no. 1036 for $56.99, no. 148 for $136.45, no. 273-10 for $198.65.

DATE _____ **Checking Deposit**	CASH	CURRENCY		
		COIN		
NAME _(PLEASE PRINT)_____	CHECKS			
ACCOUNT NUMBER_____		TOTAL FROM OTHER SIDE		
DEPOSITS MAY NOT BE AVAILABLE FOR IMMEDIATE WITHDRAWAL		**TOTAL**		
_____ SIGN HERE FOR RECEIPT OF CASH	LESS CASH RECEIVED			
	NET DEPOSIT			
	GRAHAM 55873			

SOLVAY BANK
SOLVAY, NEW YORK 13209

⑈021309735⑈

Checks and other items are received for deposit subject to the terms and conditions of this Bank's collection agreement.

6. Sketch a blank check and check stub like the one shown and complete it. Make the check out to the campus bookstore in the amount of $126.89 for textbooks. Use the current date. The balance brought forward in the account is $267.45 and a deposit of $300 was made prior to writing this check.

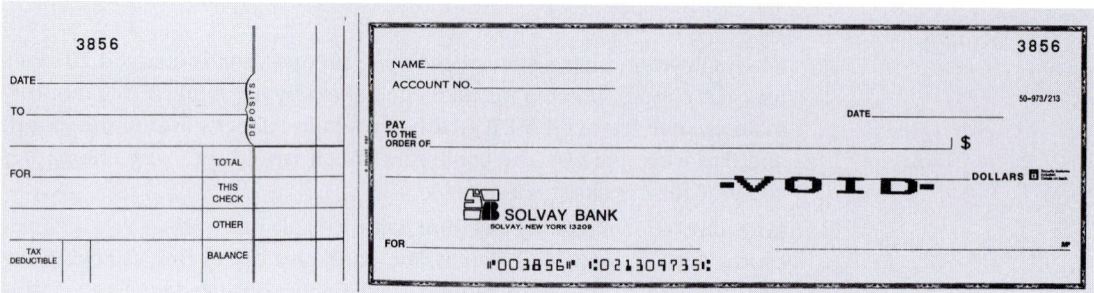

7. Complete the balance column for each transaction.

NUMBER	DATE	DESCRIPTION OF TRANSACTION	PAYMENT/DEBIT (-)	√T	FEE (IF ANY) (-)	DEPOSIT/CREDIT (+)	BALANCE
					BALANCE BROUGHT FORWARD →		647 80
67	3/8	Adams Cleaners	35 60				
68	3/9	Hillman Dept. Store	49 32				
69	3/12	Central Mortgage	462 50				
	3/15	Deposit				800 00	
70	3/14	Janet Sullivan	45 00				
71	3/20	Food-Mart	79 30				
72	3/22	Mike's Repair	29 80				
74	3/25	Value Hardware	146 28				
75	3/27	The Spent-Shop	387 57				
	3/29	Deposit				950 00	
76	4/2	Bill Charles	65 00				

Answers to CYK: **1.** $15.30   **2.** $7.80   **3.** $6.95   **4.** $11.07   **5.** $3,041.18   **6.** $440.56
7. $1,097.43

12.2 CHECKING ACCOUNT SERVICES AND CREDIT CARD TRANSACTIONS

Checking Account Services

Learning objective
Explain checking account services provided by banks.

Financial institutions are constantly analyzing their markets to improve existing services and develop new services. As in any competitive industry, banks must strive to offer the highest quality service at the lowest possible cost within federal and state regulations. To attract depositors, some banks currently offer the following additional checking account services to personal and business accounts. For these services, they may charge a fee that will appear on the bank statement as a *debit memo* (DM). Figure 12.7 illustrates typical charges for the miscellaneous checking account services described in this section. These charges must be deducted from the depositor's checkbook balance at the time of reconciliation. The reconciliation process will be presented in Section 12.3 of this chapter.

A **certified check** is a personal or business check that is certified by the bank, guaranteeing the payee payment. A certified check eliminates the possibility of having regular checks returned due to non-sufficient funds. The bank will hold a sufficient amount of money in the depositor's account to cover the check amount until the check is paid.

Overdraft protection requires the bank to pay all checks presented for payment even though there may be insufficient funds in the account. Instead of being charged for each overdraft the bank automatically transfers funds from the depositor's savings account or issues a minimum loan. The charge for this service is the current interest charge for the type of loan provided, together with a transfer fee. A fee for a **returned check** is usually charged to a depositor's account for checks deposited and returned to the bank due to nonsufficient funds (NSF) in the account of the party issuing the check.

Commercial banks will perform *collections* of funds for business depositors that usually result from transactions financed with negotiable orders of payment called *notes.* (See Chapter 8, Section 3.) When a note is collected by a bank, the proceeds are added to the depositor's account and they appear on the bank statement as a *credit memo* (CM). Such credits are added to the checkbook balance when the account is reconciled.

Many banks offer *NOW* (negotiable orders of withdrawal) *accounts,* which are savings accounts with terms and conditions similar to interest-bearing checking accounts. The depositor prepares an order of withdrawal that works like a check.

Figure 12.7

Checking account service charge schedule

SOLVAY BANK

Schedule of Fees and Charges

Solvay Bank ATM	$.00
Foreign ATM	$.75
Lost ATM Card Replacement	$5.00
Certified Check/Depositor	$10.00
Certified Check/Non-Depositor	$10.00
Cashiers Check/Depositor	$3.00
Cashiers Check/Non-Depositor	$5.00
Counter Check	$.50
Travelers Checks per 100	$1.00
Money Order/Depositor	$.75
Money Order/Non-Depositor	$2.00
Stop Payment	$15.00
Overdraft/Returned Check	$15.00
Charged Back Check/Personal	$5.00
Charged Back Check/Business	$10.00
Protest Check	$15.00
Wire Transfer/Outgoing	$20.00
Wire Transfer/Incoming - Customer	$.00
Wire Transfer/Incoming - Non-Customer	$20.00
Bond Coupon	$5.00
Bond Coupon	$5.00
Collection of Foreign Check	$5.00
Collection/Incoming	$5.00

Collection/Outgoing	$10.00
Telephone Transfer less than $200	$2.00
Telephone Transfer $200 or More	$.00
Statement Request	$2.00
Loan or Deposit Verification	$10.00
Research per Hour (Min - $5.00)	$20.00
Photocopy of Deposit	$1.00
Photocopy of Withdrawal	$1.00
Photocopy of Paid Check	$1.00
Photocopy	$.50
Legal Restraint	$60.00
Duplicate Passbook	$1.00
Duplicate Safe Deposit Box Key	$10.00
Drilling of Safe Deposit Box	$100.00
Night Deposit Bag	$20.00
Amortization Schedule	$5.00
Fax/Long Distance	$4.00
Fax/Local	$2.00
Coin Counting per Bag	$5.00

FOR BANK USE ONLY

PERSONAL ACCOUNTS

Checking Accounts

All checking accounts have monthly statement
- Checks are listed in date paid order
- Checks are listed in number order
- All checks are returned with statement
- ATM available - Visa Debit or Moneycard 24
- Overdraft Line of Credit

ECONOMY CHECKING Class Code 01
- Maintenance - $ 2.00 monthly
- Activity - $ 15 items paid
- Credit Back - none
- Balance $ 1.00

BASIC BANKING Class Code 08
- Maintenance - $ 3.00 monthly
- Activity - $ 8 free checks
- Activity - $ 9 and over $.50 each
- Credit Back - none
- Balance $ 1.00

PERSONAL CHECKING Class Code 02
- Maintenance - $ 4.00
- Activity - $ .20 paid item
- Credit Back - none
- Balance - $ 400.00 average
(If average balance is above $ 400 no charges)

NOW - PERSONAL Class Code 03 (Negotiable Order of Withdrawal)
- INTEREST - variable (2.32 APY) compound monthly
- Maintenance - $ 5.00
- Activity - $ .15 item paid
- Credit Back - None
- Balance - $ 500.00 average
(If average balance is above $ 500 no charges)

MONEY MARKET Class Code 06
- INTEREST - variable (2.78 APY) compound monthly
- Balance $ 2,500 to earn interest and avoid SC
- Checks - limit 6
- Maintenance - $ 5.00
- Activity - $ .15 paid debit
- Credit Back - None

LIQUID PRIME Class Code 07
- All same as Money Market, as well as
- over $ 10,000 balance earns 3.82 APY

Savings Accounts

STATEMENT SAVINGS Class Code 20
- Service Charge - None
- Interest - variable (3.10 APY)
  Compounded daily/ Posted monthly

PASSBOOK SAVINGS Class Code 05
- NO Statements, only passbook
- Interest - variable (3.10 APY)
  Compounded daily/Posted quarterly

CHRISTMAS CLUBS Class Code 80 October 15 / 2.89 APY Close early $10.00
VACATION CLUBS Class Code 70 May 15 / 2.89 APY Close early $10.00

Certificates of Deposit
- Term:
  - 91 days to 60 months
  - Minimum $ 500
  - SEE MESSAGE SCREEN FOR RATES

BUSINESS ACCOUNTS

Checking Accounts

All account have monthly statements
- Checks are listed in date paid order
- Checks are listed in check number order
- Checks are returned to customer

BUSINESS Class Code 11
- Maintenance - $ 6.00
- Activity - $.17 paid debit
- Activity - $.12 item deposited
- Overdraft Line of Credit

NOW - Non Personal (Business)Class Code 13
- INTEREST - 2.32 APY if balance over $ 1,000
- Maintenance - $ 6.00
- Activity - $.17 paid debit
- Activity - $.12 item deposited
- Credit Back - annual rate of 25% of prime
  applied on minimum balance (.0020)
  to offset SC only/does not carry to next month

MONEY MARKET -Non Personal Class Code 16
- INTEREST - 2.78 APY if balance over $ 2,500
- Maintenance - $ 6.00
- Activity - $.17 paid debit
- Activity - $.12 item de bited
- Credit Back - None
- Checks - limit 6

(If $ 2500 balance no Maintenance fee, but other charges apply)

LIQUID PRIME-Non-Personal Class Code 17
- Same as Money Market, including
- over $ 10,000 balance 3.82 APY

Savings Accounts

STATEMENT SAVINGS Class Code 25
- Service Charge - None
- Interest - variable (3.10 APY)
  Compound daily/Post monthly

PASSBOOK SAVINGS Class Code 05
- NO Statements, only passbook
- Interest - variable (3.10 APY)
  Compound daily/Post quarterly

Certificates of Deposit
- Term:
  - 91 days to 60 months
  - Minimum $ 500
  - SEE MESSAGE SCREEN FOR RATES

(If $1000 balance no Maintenance fee, but other charges apply)
Only sole proprietors and non-profits
Corporations and partnerships are NOT allowed

revised 4/98

Source: Solvay Bank, Solvay, NY. Used with permission.

Stop payment orders are provided by banks when a depositor requests that the bank not pay a check that has been written.

Banks will provide *duplicate bank statements* and passbooks when requested by the depositor.

Cancelled checks are often needed to verify payments of debts. As most banks no longer return cancelled checks with monthly bank statements, photocopies are made available upon request.

Electronic banking is a service made available to depositors that uses computer technology to transfer funds into and out of checking accounts without the presence of the depositor or without the traditional documents required by the transaction. Funds can be transferred anywhere, for any purpose, wherever the telecommunications networks are available. An example of electronic banking offered by financial institutions is the *automatic teller machine* (ATM). These computerized bank machines can be found outside financial institutions and places of business with high traffic patterns. Depositors can conveniently withdraw cash, make deposits, or transfer funds to other accounts any time day or night.

Banks will also provide *account reconcilement* if the depositor is unable to reconcile their own account. The fee will vary depending on the type of account and the number of items involved. Many banks set a minimum fee for this type of service.

For Your Information

BANK CREDIT CARDS: UNDERSTANDING THE MONTHLY STATEMENT

Credit cards can provide a convenient and effective means of paying for goods and services. Some 83 million Americans hold credit cards, for an average of 2.68 cards per person. But many people who use credit cards do not fully understand their monthly statement and therefore end up paying increased costs. Understanding how fees and interest are determined can decrease credit car costs and enable consumers to choose the least costly cards. The explanations and suggestions that follow are keyed numerically to the monthly bank credit card statement shown.

1. **Transaction/posting date:** The transaction date is the date on which the card was used by the cardholder for a purchase or cash advance. A record of the transaction is sent to the cardholder's bank and is entered in the cardholder's account on the posting date. If the cardholder does not pay the balance on time, interest is charged on the average daily balance as of the posting date.

2. **Annual percentage rate:** This is the yearly interest rate used to determine finance charges to the customer for card use. Rates average about 18.6% per year, but may vary from around 11% to around 22%. If a bank charges variable rates, the rates run approximately 6 to 10 percent higher than the current prime lending rate.

3. **Days in the billing cycle:** This is the number of days between one month's payment due date and the next month's payment due date. The more days in the cycle, the higher the finance charge if the card holder carries a balance from month to month.

4. **Payment due date:** This is the date on which the full amount of the account balance or the minimum payment is due. If the cardholder pays nothing, some banks impose a late fee. Such fees are illegal in some states.

5. **Minimum payment due:** This is the smallest amount that the cardholder is required to pay on his or her account. Most banks require a minimum payment of $20 on account balances from $20 to $720. On balances of more than $720, the minimum payments can range from 2.5% to 5% of the average daily balance, plus any amount the cardholder has charged in excess of the credit line.

6. **Finance charge:** This is the interest charged on the average daily balance when the entire balance is not paid within a grace period (which is usually 25 to 30 days). The charge is calculated by multiplying the average daily balance by the daily periodic rate by the number of days in the billing period. It may be calculated as $1/12$ of the bank's annual interest rate if a monthly periodic rate is used. A minimum finance charge may be imposed when the actual charge is less than the minimum.

7. **Debit adjustments:** If a cardholder asks the bank to correct a mistake involving a charge to her or his account, the bank will delete the charge from the cardholder's account while the error is being investigated. The adjustment will appear in this section of the monthly statement. If no error has occurred, the bank will rebill the cardholder's account for the amount and will also add interest for the period.

8. **Cashback bonus:** To encourage use of their card, some financial institutions, such as Discover Card, provide a periodic cash-back bonus based on qualified purchases. This bonus reduces the cost of the card.

account number	**6011 0008 0000 0576**
new balance	**$ 1,762.50**
payment due date	**December 18, 1998**
minimum payment due	**$ 37.00**
amount enclosed	$

Please make check payable to Discover Card.

```
IlbIllmIllmnbIblImdbIbIdImnbIdbd
      stan t260
      123 PARK
      LOS ANGELES CA 90046-2505
```

PO BOX 30395
SALT LK CITY UT 84130-0395
IlmIdmIlmdbIbIImmdbIdmdbIbmbmdb

Address or telephone change? Please print change in the space above.

00000601100008000005760176250000000000003700

Closing Date: November 23, 1998   page 1 of ??PN??

Cashback Bonus®Award	this period	to date
qualified purchases	$750.00	$750.00
Cashback Bonus award earned	$1.88	$1.88
Cashback Bonus anniversary date:	October 23	

Discover Card Account Summary

account number	6011 0008 0000 0576
payment due date	December 18, 1998
minimum payment due	$37.00
credit limit	$3,000.00
credit available	$1,237.00
cash credit limit	$1,500.00
cash credit available	$1,000.00

previous balance		$0.00
payments and credits	−	0.00
purchases	+	750.00
cash advances	+	500.00
balance transfers	+	500.00
FINANCE CHARGES	+	12.50
new balance	=	$1,762.50

To avoid additional finance charges, pay your entire new balance by December 18, 1998.

Transactions

Gas/Automotive	Nov 11	SALES WELLBORN FL	$ 750.00
Cash Advances	Nov 11	CASH SHELBY NC	500.00
		CASH ADVANCE TRANS FEE FINANCE CHARGE	12.50
Balance Transfers	Nov 11	BALANCE TRANSFER COLUMBUS OH	500.00
		BALANCE TRANSFERRED TO DISCOVER	

	Average Daily Balances	Daily Periodic Rates	ANNUAL PERCENTAGE RATES	Periodic FINANCE CHARGES	Transaction Fee FINANCE CHARGES
current billing period: 25 days					
Purchases	$0	0.04792%	17.49%	$0	none
Cash Advances	$0	0.05751%	30.00%*	$0	$12.50
Balance Transfers	$0	0.01890%	6.90%	$0	none
previous billing period: 0 days					
Purchases	$0	N/A	N/A	$0	none
Cash Advances	$0	N/A	N/A	$0	none

* Rate reflects transaction fees. Rate corresponding to daily periodic rate is 20.99%.

Questions? Call 1-800-DISCOVER (1-800-347-2683) . For TDD (Telecommunication Device for the Deaf) assistance, see reverse side. Send billing error notice to: Discover Card; P.O. Box 52164; Phoenix, AZ 85072-2164.

Source: Discover Financial Services. Used with permission.

Credit Card Transactions.

Learning objective
Complete deposit
records of credit
card transactions.

A large portion of today's retail transactions are made with credit cards. Have you ever wondered how a business collects payment for merchandise you purchased with either a Visa or MasterCard?

The process begins when you present your card to the cashier, who transfers required information on the card to the credit card sales slip using a stamping machine. You must sign the sales slip (Figure 12.8) verifying the purchase information and authorizing payment. The business then deposits the credit card transactions into its checking account using the **merchant's deposit summary** shown in Figure 12.9. Credit card sales and cash sales can be deposited at the same time using separate deposit slips.

The merchant's deposit summary lists both sales slips (charges) and credit slips (refunds). The net deposit (total sales slips recorded minus total credit slips) increases the balance in the check register just like a regular deposit. The bank charges the business a **discount fee** of 2–5% of the net amount

Figure 12.8

Credit card sales slip

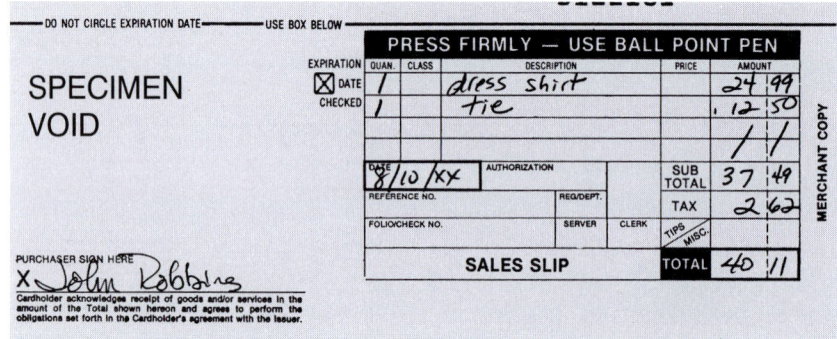

Source: General Credit Forms, Inc., St. Louis, MO. Used with permission.

Figure 12.9

Merchant's deposit ticket

Source: General Credit Forms, Inc., St. Louis, MO. Used with permission.

deposited during the month. The discount fee covers the cost of immediately providing funds to the business while the bank awaits payment from the customer through the interbank fund transfer process. This fee in effect reduces the net deposit as it becomes part of the total bank charges appearing on the monthly bank statement to be reconciled.

Example 8

On March 10, 200X, Northshore Marine Supply had credit card sales of $29.85, $162.49, $8.35, $57.14, $237.99, and credit card refunds of $71.45 and $14.62. (a) Complete the deposit summary. (b) Calculate the discount fee that will appear on Northshore's bank statement if the bank charges 4¾%. (Refer to Figure 12.9.)

Solution

a. **Deposit Summary—March 10, 200X**

Credit slips	Sales slips
1. $71.45	1. $ 29.85
2. 14.62	2. 162.49
3. _____	3. 8.35
4. _____	4. 57.14
5. _____	5. 237.99
Total $86.07	$495.82

Total amounts: Sales slips   495.82
                Less credit slip   86.07
                Net deposit $409.75

b. $409.75 × .0475 = $19.46 discount fee

CHECK YOUR KNOWLEDGE

Checking Account Services and Credit Card Transactions

The sales records of Carl's Men's Shoppe indicate the following credit card transactions for April 12. Deposits are made daily.

Sales		Refunds
$139.69	$76.49	$ 9.34
10.42	29.87	34.69
49.99	52.12	116.50
259.78	187.06	

1. What are the total credit card sales?

2. What are the total credits (refunds)?

3. What is the net deposit to be recorded on the merchant's deposit summary?

4. If the bank charges a 3½% discount fee, what amount will be charged to Carl's account for credit card transactions?

5. Sketch a deposit slip like the one shown and complete it.

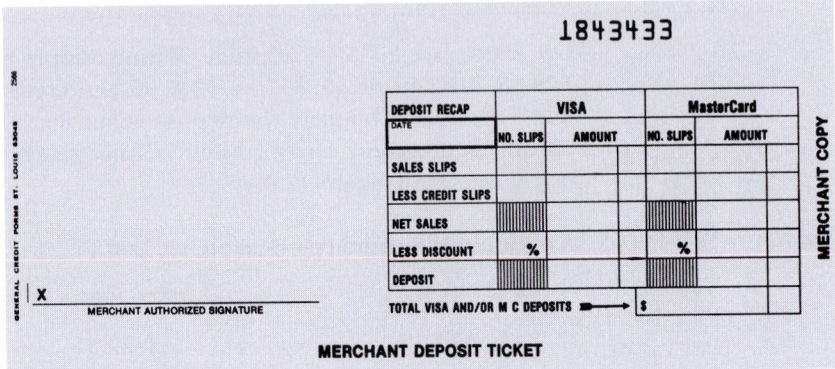

Source: General Credit Forms, Inc., St. Louis, MO. Used with permission.

12.2 EXERCISES

Use Table 12.2 to find the monthly service charge for the following personal checking accounts based on the regular account plan.

Name	Checks written	Average balance	Service charge
1. Charles Adams	15	$1,451	_____
2. Cynthia Johnson	28	$4,622	_____
3. Elizabeth Cruz	62	$2,103	_____
4. Matthew Koski	46	$7,154	_____
5. Mavis Tuttle	85	$3,875	_____

Use Table 12.2 to find the monthly service charge for the following business checking accounts.

Company	Checks paid	Deposits made	Checks deposited	Average balance	Service charge
6. Pasta Pantry	52	8	12	$ 874.90	_____
7. Eastside Sports Center	127	12	293	$5,346.75	_____
8. Goldman Furniture	243	20	461	$8,790.50	_____

Answers to CYK: **1.** $805.42 **2.** $160.53 **3.** $644.89 **4.** $22.57 **5.** $622.32 total deposits

Use the deposit information below to complete each deposit slip on the dates indicated for Perfection Cleaners.

| | | Bills (quantity) | | | | Coin (rolls) | | | | Checks | |
	Date	20s	10s	5s	1s	25	10	5	1	Number	Amount
9.	Nov. 7	15	26	12	43	10	3	8	6	2146-126	$127.10
										5621-287	56.45
										1874-521	365.10
10.	Nov. 15	28	47	54	185	22	12	18	9	7364-432	212.75
										4267-893	62.20
										3642-127	436.82
										2983-622	38.46
										6876-539	123.95

9. **Deposit slip**

currency _____

coin _____

checks _____

_____ _____

_____ _____

_____ _____

_____ _____

_____ _____

_____ _____

Total _____

10. **Deposit slip**

currency _____

coin _____

checks _____

_____ _____

_____ _____

_____ _____

_____ _____

_____ _____

Total _____

For exercises 11–13, sketch checks like the one shown, and endorse each with your signature as required using the information provided.

11. You want to transfer the amount of this check to Sutter's Hardware to be applied to your account balance.

ENDORSE HERE
X _____

_____

_____

DO NOT WRITE, STAMP OR SIGN BELOW THIS LINE
RESERVED FOR FINANCIAL INSTITUTION USE*

12. This check has been made out to you and you are about to cash it at your bank.

ENDORSE HERE
X _____

_____

_____

DO NOT WRITE, STAMP OR SIGN BELOW THIS LINE
RESERVED FOR FINANCIAL INSTITUTION USE*

13. You received this check from one of your customers and you want to deposit the amount in your checking account.

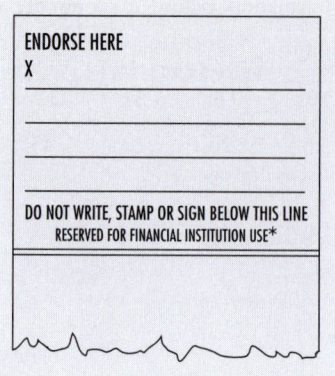

ENDORSE HERE
X

DO NOT WRITE, STAMP OR SIGN BELOW THIS LINE
RESERVED FOR FINANCIAL INSTITUTION USE*

For exercises 14–16, sketch check stubs like the ones shown, and complete each for We-Fix-It Appliance Repair using the following information.

Stub number	Date	To	For	Amount
14. 136	Jan 6	Adams Electric Company	Electric Motor	$562.15
15. 137	Jan. 8	Spectrum Business Supply	Office Supply	83.28
16. 138	Jan. 9	Millie's Auto Center	Truck Repair	249.50

Balance brought forward on stub #136: $2,892.40

Deposits made: Jan 7 $135.25
                        Jan 9 $379.70

DATE _____

TO _____

DEPOSITS

FOR _____

		TOTAL		
		THIS CHECK		
		OTHER		
TAX DEDUCTIBLE		BALANCE		

DATE _____

TO _____

DEPOSITS

FOR _____

TOTAL		
THIS CHECK		
OTHER		
BALANCE		

TAX DEDUCTIBLE

DATE _____

TO _____

DEPOSITS

FOR _____

TOTAL		
THIS CHECK		
OTHER		
BALANCE		

TAX DEDUCTIBLE

17. Complete the required computations, arranged as in the check register.

NUMBER	DATE	DESCRIPTION OF TRANSACTION	PAYMENT/DEBIT (-)		√ T	FEE (IF ANY) (-)	DEPOSIT/CREDIT (+)		BALANCE	
		BALANCE BROUGHT FORWARD →							$ 147	60
124	3/12	Dr Wilson	$ 35	50		$	$			
	3/15	Deposit					450	—		
125	3/16	Paper Cutter	18	70						
126	3/18	Solvay Bank (loan)	146	20						
127	3/20	West End Pharmacy	26	80						
128	3/21	P & C Foods	59	90						
129	3/23	Niagra Mohawk	169	40						
	3/23	Deposit					500	—		
130	3/25	Rent	425	00						
131	3/26	Addis & Dey's	48	72						
132	3/28	Byrne Dairy	8	40						
133	4/2	Deluxe Checks	4	60						
134	4/5	Chuck's Service Center	29	75						
135	4/8	Wegman's Market	47	35						

Use the following credit card transactions to answer problems 18–22. Sales codes (V) Visa; (M) MasterCard.

Sales slips	Credit slips	Sales slips	Credit slips
$ 27.38(V)	$15.95(M)	167.32(M)	
52.45(V)	38.64(M)	48.50(V)	
129.36(M)	87.12(V)	109.25(M)	
82.47(V)	9.45(M)	21.89(M)	
15.95(M)		65.75(M)	

18. What are the total sales slips for the month?

19. What are the total credit slips for the month?

20. What is the amount of the net deposits?

21. If the bank charges a 3½% discount fee, what is the amount of the discount fee to appear on the customer's bank statement?

22. Complete a sketch of the merchant's deposit summary shown using the calculations from exercises 18–21.

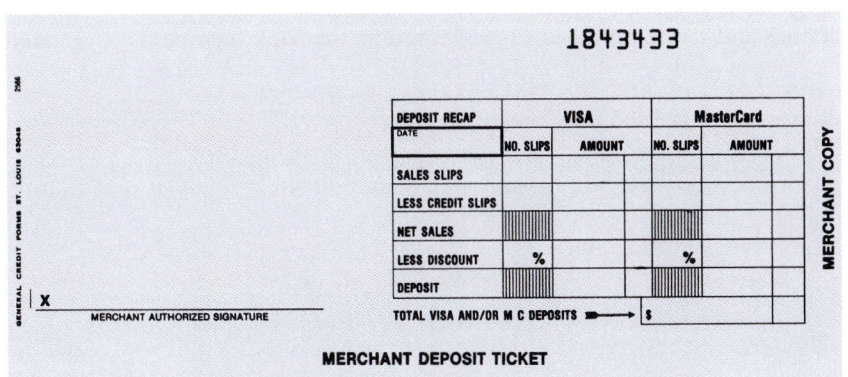

Source: General Credit Forms, Inc., St. Louis, MO. Used with permission.

23. Find the monthly service charge for Lisa Williams if her average balance is $387.50 and 27 checks were written during the month of October. (Use Table 12.1.)

24. Harold Cohen has a special checking account with Community Bank. The bank charges his account 10 cents for each check, 20 cents for each check paid, plus a monthly maintenance fee of $6.25. If Harold wrote 17 checks during the month, how much would Community Bank charge his account for checking service?

25. Christine Plavocus has a checking account with a bank that pays 3¼% interest on the lowest monthly balance. Using Table 12.1, determine the net service charge (charges minus payments) to her account if the bank paid 23 checks during the month and her account balance range was $137.45 to $489.62 with an average balance of $368.15.

26. Determine the charges and payments to a checking account if the bank pays 4¼% interest on the lowest monthly balance. The lowest account balance is $592.65; the average balance for the month is $725.90; and 44 checks were paid by the bank. (Use Table 12.1.)

27. Panther Lake Inn made 15 deposits during July, which included 392 customer checks. A total of 64 checks were paid by the bank and the average balance for the month was $3,689.45. Determine the total debits and credits to the account based on the service charge schedule, Table 12.2.

28. Determine the total amount to be deposited by Frank's Pizzeria based on the following: 10 twenty-dollar bills, 30 ten-dollar bills, 25 five-dollar bills, 87 one-dollar bills; 6 rolls of quarters, 8 rolls of dimes, 5 rolls of nickels, 2 rolls of pennies; checks amounting to $689.50.

29. Alice Newser bought a desk for $329.62 and paid with check no. 108. If the balance prior to the purchase of the desk was $586.48 and she made a deposit of $250.00 since the last check was written, what is her balance forward on check stub number 108?

30. Fred's Cycle Shop had credit card sales that were deposited on December 15 of $79.95, $149.57, $356.92, and $89.75, and credit refunds of $59.95 and $24.69. Determine (a) the amount of the net deposit recorded on the merchant's deposit summary and (b) the discount fee charged to Fred's account if the bank charges 3½%.

12.3   THE RECONCILIATION PROCESS

Learning objective
Reconcile bank statements and check registers by preparing bank reconciliation statements.

Now that we understand how to maintain the required checking account records and how the bank determines the monthly service charge, we are ready to discuss the reconciliation process. Each month the bank will send depositors a **bank statement** (Figure 12.10) that lists deposits made during the period and checks paid by the bank, as well as a record of other transactions and service charges. Checks listed on the bank statement are referred

Figure 12.10

A sample bank statement

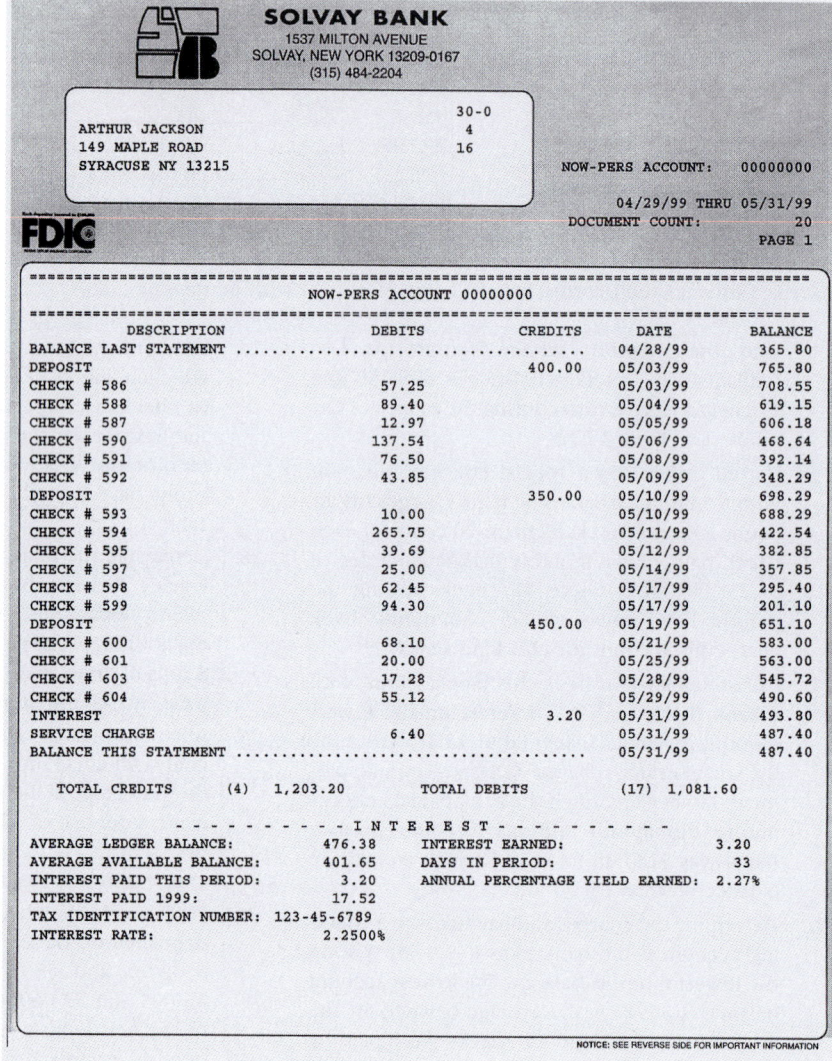

Source: Solvay Bank, Solvay, NY. Used with permission.

Figure 12.11

Check cancellation process

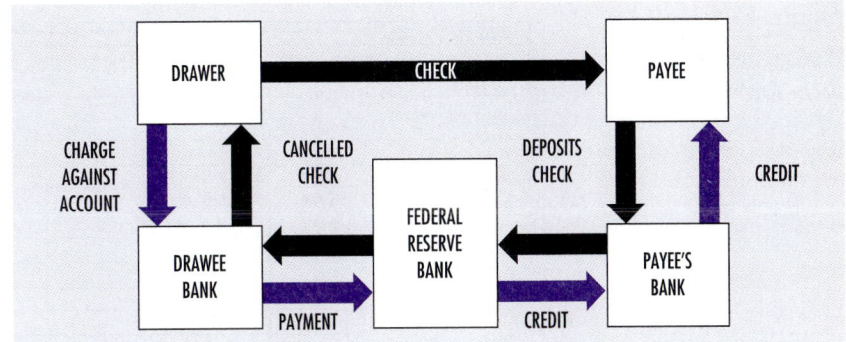

to as **cancelled checks.** Figure 12.11 shows how a check written by a depositor in New York to a payee in California is processed by the federal reserve system. Notice that the check is physically returned to the depositor's bank for payment. Some banks still return the cancelled checks to the depositor with the bank statement.

The ending balance on the bank statement will most likely differ from the balance on the depositor's checkbook record. The process of analyzing these two records and making the required adjustments to bring the balance into agreement is called **reconciliation.** The fact that the two balances do not agree does not mean that either record is in error. What the difference does suggest is that amounts have been added to or deducted from the balance in one record but not the other. These amounts and their purposes become clear when the information on the bank statement is compared to the checkbook record. Ordinarily the adjustments to the records are due to the following reasons:

1. *Deposits in transit.* These are deposits made to the checking account that do not appear on the bank statement. These deposits have been included in the checkbook balance but are not received by the bank in time to be recorded on the bank statement.

2. *Collections.* The bank will often collect funds on the depositor's behalf. Such funds are reflected in the bank balance but seldom in the checkbook balance.

3. *Outstanding checks.* These are checks that have been written and recorded in the checkbook register but have not been paid by the bank as of the statement date.

4. *Bank charges.* The bank will deduct from the bank balance all charges for the services provided and/or requested during the period. These fees may or may not be included in the checkbook balance.

Figure 12.12

Bank reconciliation form for Example 9

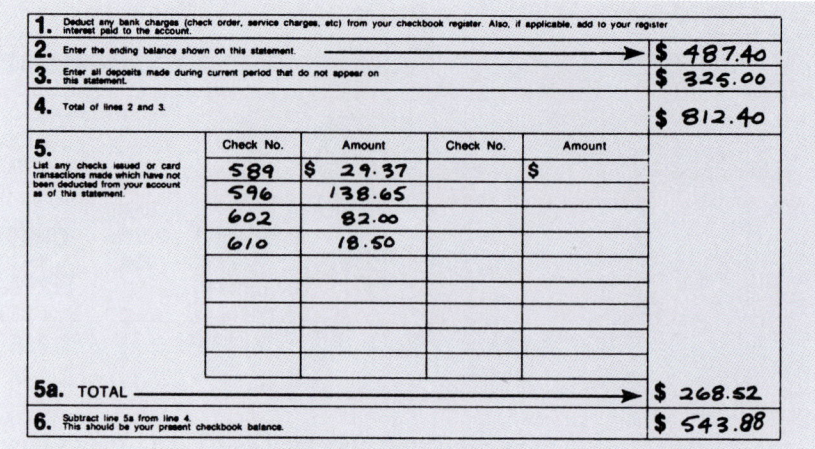

Source: Solvay Bank, Solvay, NY. Used with permission.

5. *Errors.* Both depositor and bank make errors. Errors must be ana-
    lyzed carefully and corrected. If the error is a bank error, the de-
    positor should notify the bank so that the proper adjustment can
    be made to the bank record.

The bank statement can be reconciled by using the *bank form* (Figure
12.12) printed on the reverse side of the statement or by preparing the **account form** shown in Figure 12.13. The account form is often preferred by
businesses when preparing the monthly bank reconciliation statement.

Figure 12.13

Account form bank reconciliation for Example 9

Art Jackson
Bank Reconciliation
May 31, 1993

Bank balance	$487.40	Book balance	$547.08
Add:		*Add:*	
Deposits in transit	325.00	Interest credit	3.20
Total	$812.40	Total	$550.28
Deduct:		*Deduct:*	$ 6.40
Outstanding checks		Service charge	
589 $ 29.37			
596 138.65			
602 82.00			
605 18.50	268.52		
Adjusted Bank balance	$543.88	Adjusted Book balance	$543.88

To illustrate the steps involved in reconciliation, Example 9 uses the information on the bank statement provided in Figure 12.10 to complete the bank form (Figure 12.12) on the reverse side of the statement and the account form (Figure 12.13).

Example 9

The following checkbook register of Art Jackson shows a checkbook balance of $547.08. After comparing the entries in the checkbook register with

NUMBER	DATE	DESCRIPTION OF TRANSACTION	PAYMENT/DEBIT (-)	√T	FEE (IF ANY) (-)	DEPOSIT/CREDIT (+)	BALANCE
					BALANCE BROUGHT FORWARD →	$	405 32
584	4/18	Telephone Co.	24 50	✓			380 82
585	4/20	Cash	15 00	✓			365 82
586	4/28	Sports Outfit	57 25				308 57
587	4/29	Fred's Drugs	12 97				295 60
588	4/30	P&C Markets	89 40				206 20
	4/30	Deposit				400 00	606 20
589	5/4	Cable TV	29 37				576 83
590	5/4	Niagara Mohawk	137 54				439 29
591	5/5	Raymond's Furniture	76 50				362 79
592	5/6	Northshore Lumber	43 85				318 94
	5/7	Deposit				350 00	668 94
593	5/7	Youth Center	10 00				658 94
594	5/7	Car Payment	265 75				393 19
595	5/9	Chappels	39 69				353 50
596	5/10	Credit Union	138 65				214 85
597	5/11	Sears	25 00				189 85
598	5/12	Snyder's Autocenter	62 45				127 40
599	5/13	Sibley's	94 30				33 10
	5/16	Deposit				450 00	483 10
600	5/17	Gary's TV Repair	68 10				415 00
601	5/19	Marie's Gifts	20 00				395 00
602	5/22	William's Market	82 00				313 00
603	5/26	Village Water	17 30				295 70
604	5/27	Healthcenter East	55 12				240 58
	5/30	Deposit				325 00	565 58
605	5/30	Byrne Dairy	18 50				547 08

NUMBER	DATE	DESCRIPTION OF TRANSACTION	PAYMENT/DEBIT (-)	√T	FEE (IF ANY) (-)	DEPOSIT/CREDIT (+)	BALANCE
					BALANCE BROUGHT FORWARD →	$	547 08
	6/3	Service Charge	6 40				540 68
		Interest			(Adjusted Balance)	3 20	543 88

the bank statement, we find that check nos. 589 for $29.37, 596 for $138.65, 602 for $82.00, and 610 for $18.50 have not been paid, and that the bank paid $3.20 in interest and charged $6.40 for its services. Also the register lists a deposit of $325 that was not included in the bank statement balance of $487.40. Use the bank form (Figure 12.13) to reconcile Art Jackson's checking account.

Solution

Step 1: First, add to the checkbook balance any bank payments (credit memos) such as interest earned on the account. The interest paid by the bank appears in the bank statement. (Checkbook balance, $547.08, plus interest, $3.20, totals $550.28.)

Second, deduct all bank charges (debit memos) from the total obtained above ($550.28 − $6.40 = $543.88). This total is the *adjusted checkbook balance* and should agree with the adjusted bank balance found by completing steps two through six of the bank reconciliation form.

Step 2: Enter statement balance ($487.40) on line 2 of the reconciliation form.

Step 3: List all deposits recorded in the check register that do not appear in the bank statement. (A deposit of $325 is not yet recorded on the bank statement.)

Step 4: Add the total deposits in transit to the bank statement balance ($487.40 + $325 = $812.40).

Step 5: Compare the checks written in the register with the cancelled checks listed in the bank statement. Checks listed in the register and not in the bank statement are outstanding. Record these outstanding checks in the space provided on the form and total.

#589	$29.37
596	138.65
602	82.00
610	18.50
Total	$268.52

Step 6: Deduct the total outstanding checks ($268.52) from the step 4 total ($812.40). This total ($543.88) is called the *adjusted bank balance.*

If the adjusted balances do not agree, check arithmetic on the reconciliation statement and recheck each step of the process to make sure you did not omit any adjustments. When rechecking, make sure the adjustment is made to the correct record. Other common errors to look for include: recording the amount of the check incorrectly, addition and subtraction errors in the check register, and the omission of transactions by both depositor and bank.

Example 10 shows how to prepare a bank reconciliation statement for a business account using the *account form*.

Example 10

Spenser's Travel Agency has a checkbook balance of $4,396.63. The balance shown on the November bank statement is $3,684.97. A comparison of the checking account records indicates the following checks have not yet been paid by the bank: no. 689 for $237.85, no. 704 for $50.35, no. 712 for $382.72, no. 713 for $129.68, and no. 716 for $672.35. In addition, Spenser's had deposits of $825.50 and $1,345.00, which were recorded in the checkbook register but did not appear in the bank statement. The bank statement also shows a check charge of $12.50, a monthly service charge of $6.40, and an interest payment of $4,79. Prepare a bank reconciliation statement for Spenser's Travel Agency.

Solution

Spenser's Travel Agency
Bank Reconciliation
November 1999

Bank balance	$3,684.97	Checkbook balance	$4,396.63
Add:		*Add:*	
Deposits in transit		Interest credit	4.79
$ 825.50			
1,345.00	2,170.50		
Total	$5,855.47	Total	$4,401.42
Deduct:			
Outstanding checks		Service charge 6.40	
no. 689 $237.85		Check charge 12.50	
no. 705 50.35			18.90
no. 712 382.72			
no. 713 129.68			
no. 716 672.35	1,472.95		
Adjusted bank balance	$4,382.52	Adjusted book balance	$4,382.52

```
CALCULATOR  SOLUTION
Keyed entry                                        Display

                    AC  3684.97  Min                3684.97
                   825.5  +  1345  =                2170.50
                           M+   MR                  5855.47
         237.85  +  50.35  +  382.72  +
                129.68  +  672.35  =                1472.95
                                  +/−              −1472.95
                           M+   MR                  4382.52
```
Follow same procedure to find checkbook balance

The steps involved in preparing the report form reconciliation statement are the same as those used to prepare the bank form. In both presentations, you are adjusting the records to reflect any increases or decreases to the balances that have not been recorded in the records as of that date.

CHECK YOUR KNOWLEDGE

The Reconciliation Process

1. Find the adjusted bank balance using the following information. Deposits in transit: $262.50 and $379.62. Checks outstanding: no 1031 for $60.00, no. 1046 for $187.50, no 1051 for $42.95, and no. 1056 for $13.50. Bank statement balance: $429.34.

2. Determine the adjusted checkbook balance if the bank statement showed an interest payment to the customer's account of $3.21, an NSF (nonsufficient funds) charge of $15.00, a check printing charge of $6.75, and a monthly service charge of $4.50. The balance in the check register is $279.54.

3. Use the report form to reconcile Bob's Lawn and Garden Store account. The bank statement shows a balance of $3,742.90, a service charge of $4.50, an NSF charge of $20.00, and an interest credit of $5.74. Checks outstanding are: $69.30, $127.48, $642.25, $312.60, and $258.75. The checkbook register balance is $3,926.78, which includes a deposit of $1,575.50 not recorded on the bank statement.

Answers to CYK: *1.* $767.51   *2.* $256.50   *3.* $3,908.02

12.3 EXERCISES

Find the adjusted bank balance for each of the following accounts.

	Opening bank balance	Deposits in transit	Outstanding checks	Adjusted bank balance
1.	$648.72	$287.36 129.43	$24.37 186.25 62.94	_____
2.	$7,246.25	$1,436.80	$572.12 78.52 168.97	_____
3.	$32,568.47	$2,876.25 4,165.97	$12,145.10 1,215.50 861.43 79.43	_____

Calculate the adjusted checkbook balance for each of the following business accounts.

	Opening checkbook balance	Check charges	Service charge	Notes collected	Interest payment	Adjusted checkbook balance
4.	$457.90	$9.00	$6.50		$3.45	_____
5.	$1,648.32	$18.20	$12.15		$8.97	_____
6.	$12,366.15	$54.90		$1,578.40	$29.17	_____

7. From the following information, prepare an account form bank reconciliation for Nu-Wave Coiffures.

Bank balance	$4,246.25
Checkbook balance	2,931.05
Deposits in transit	1,650.00
Service charge	12.50
Outstanding checks	2,972.50
Interest credit	14.20
Check charge	9.00

Nu-Wave Coiffures
Bank Reconciliation Statement
July 200X

Bank Balance $_____	Book balance $_____
Add:	*Add:*
Total _____	Total _____
Deduct:	*Deduct:*
Total _____	Total _____
Adjusted bank balance _____	Adjusted book balance _____

8. Use the following information to complete a bank form like the one provided. Verify bank balance against checkbook balance.

Bank balance		$682.40
Deposits in transit		350.00
		175.00

Outstanding checks	
#243	29.62
#257	245.20
#253	119.75

Checkbook balance	$828.43
Service charge	9.00
Interest credit	6.25
Check charge	12.85

		Check No	Amount	Check No.	Amount	
1.	Deduct any bank charges (check order, service charges, etc.) from your checkbook register. Also, if applicable, add to your register interest paid to the account.					
2.	Enter the ending balance shown on this statement.					$
3.	Enter all deposits made during current period that do not appear on this statement.					$
4.	Total of lines 2 and 3.					$
5. List any checks issued or card transactions made which have not been deducted from your account as of this statement.			$		$	
5.a TOTAL						$
6.	Subtract line 5a from line 4. This should be your present checkbook balance.					$

Source: Solvay Bank, Solvay, NY. Used with permission.

9. Janet Green received her bank statement showing a balance of $197.14, a service charge of $12.50, an interest credit of $2.50, and a returned check amounting to $35.25. When Janet compared her checkbook register to the bank statement, she noted a checkbook balance of $465.20, a deposit of $350.00 that was not recorded on the bank statement, and that check no. 136 for $74.65, check no. 138 for $12.55, and check no. 143 for $39.99 had not been processed by her bank for payment. Prepare a bank reconciliation for Janet organized like the account form shown.

Bank balance	_____	Checkbook balance	_____
Add:	_____	*Add:*	_____
Deduct:	_____	*Deduct:*	_____
Adjusted bank balance	_____	Adjusted checkbook balance	_____

10. Central Trust charges "economy" checking account customers 20 cents for each check written and a monthly maintenance fee of $4.50. Reconcile an "economy" statement showing a balance of $369.42 with twelve checks having been paid by the bank. The customer's register indicates a deposit of $200 is not shown on the statement, a balance of $110.82, and that check no. 78 for $15.50 and check no. 82 for $450 are outstanding. Reconcile the customer's checking account, organizing the information according to the blank bank form shown.

1. Deduct any bank charges (check order, service charges, etc.) from your checkbook register. Also, if applicable, add to your register interest paid to the account.						
2. Enter the ending balance shown on this statement.						$
3. Enter all deposits made during current period that do not appear on this statement.						$
4. Total of lines 2 and 3.						$

	Check No	Amount	Check No.	Amount	
5. List any checks issued or card transactions made which have not been deducted from your account as of this statement.		$		$	

5.a TOTAL		$
6. Subtract line 5a from line 4. This should be your present checkbook balance.		$

Source: Solvay Bank, Solvay, NY. Used with permission.

11. New-Release Video received its November bank statement showing a balance of $1,247.25, check print-ing charges of $23.40, a service charge of $16.61, and an interest credit of $9.30. The checkbook register balance was $1,509.41. The bank statement indicated deposits in transit of $850, $1,200, and $575, and that checks of $98.50, $470.25, $689.20, and $1,135.60 had not been paid by the bank. Use the account form to reconcile the accounts.

Bank balance   _____   Checkbook balance   _____
Add: _____   *Add:*   _____
Deduct: _____   *Deduct:*   _____
Adjusted bank balance   _____   Adjusted checkbook balance   _____

12. Reconcile the business checking account for Rojon Enterprises for the month of March. All debits and credits to Rojon's account were computed using rates from Table 12.2. The bank statement shows a bal-ance of $2,194.70, a service charge including all charges of $58.52, a discount fee for credit card sales of $22.50, and an interest credit of $5.58 based on an average balance of $1,275.50. The bank statement does not show a deposit of $875 or check nos. 246, 252, 257, 263, and 265 for $45.50, $163.20, $239.60, $87.35, and $19.15, respectively, having been paid by the bank. Rojon's checking records indicate a balance of $2,590.34; 20 deposits were made during the month (which included 356 customer checks), 124 checks were paid by the bank, and the checkbook balance is $2,590.34. (Note: Verify all bank charges.)

Bank balance   _____   Checkbook balance   _____
Add: _____   *Add:*   _____
Deduct: _____   *Deduct:*   _____
Adjusted bank balance   _____   Adjusted checkbook balance   _____

CHAPTER REVIEW EXERCISE

The following problems review all mathematics and business concepts presented in the chapter. If you experience any difficulty solving a problem, go to the appropriate chapter section and review the examples provided. Express all answers as decimals rounded to tenths, or dollars rounded to cents, unless otherwise instructed.

1. Art Sanders's bank charges his checking account 25 cents for each check, 15 cents for each check paid, and a monthly maintenance fee of $6.75. Determine Art's monthly service charge if he wrote 23 checks and 14 checks were paid by this bank this past month.

2. Using the Table 12.1 service charge schedule, find the monthly service charge for an account with an average balance of $198 and for which the bank paid 17 checks.

3. Carrie's Nail Salon checking account record indicates the following activity: 10 deposits, which included 143 customer checks; 62 checks paid by the bank; and an average daily balance of $3,675.50. What are the monthly charges and payments to the account? (Use Table 12.2 service charge schedule.)

4. Moore's Insurance Company wrote check No. 1146 for $1,200 to Sidney Johnson for "payment of coverage." On the same date the check was written, a deposit of $4,000 was made. If the balance brought forward on the check stub from check no. 1145 was $20,762.25, what is the balance brought forward amount on check stub no. 1147?

5. Determine the final balance that would appear in the check register based on the following transactions:

Number	Date	Description	Payment	Deposit	Balance
					$562.00
101	5/2	Federal Finance	$250.00		
102	5/3	Mike's Plumbing	87.50		
	5/5	Deposit		$950	
103	5/7	General Mortgage	652.30		
104	5/10	City Travel	163.40		
105	5/12	Quality Foods	124.87		
106	5/16	Visa card	50.00		
107	5/18	Dr. Fallon	42.00		
	5/19	Deposit		875	
108	5/23	Walmart	64.18		
109	5/27	Patterson's Market	159.73		

6. Prepare the daily deposit for Village Cleaners; Currency: 3 fifty-dollar bills, 17 twenty-dollar bills, 30 ten-dollar bills, 24 five-dollar bills, 58 one-dollar bills; Coins: 10 rolls of quarters, 8 rolls of dimes, 4 rolls of nickels; Checks: No. 2642-136 for $15.95, No. 5832-096 for $24.35, No. 9237-1064 for $18.20, No. 7149-632 for $37.90 and No. 4845-721 for $44.50.

7. Heyden Jewelers had credit card sales of $36.98, $149.25, $52.45, $532.20, $265.40, and credit refunds of $85.60 and $28.35. Determine (a) the net deposit amount, and (b) the discount fee if the bank charges 3 1/2%.

8. Determine the adjusted bank balance from the following information. Deposits in transit of $380.60 and $525.00. Checks outstanding: No.

246 for $39.25, No. 248 for $145.50 and No. 249 for $63.40. The bank statement indicated that check No. 225 for $35.00 had been deducted from the bank balance twice by error. The bank statement balance is $847.62.

9. What would be the adjusted checkbook balance if the bank statement shows the following transactions. An interest payment of $5.40, check printing charges of $12.25, and NSF charge of $20.00, and a monthly service charge of $8.50. The balance in the check register is $538.70 and check No. 315 for $22.50 had been deducted from the balance as $25.20.

10. Tri-State Heating and Cooling has a checkbook balance of $3,480.25. The balance on the October bank statement is $4,986.40. An analysis of the checking account records provides the following information. Checks No. 421 for $565.80, No. 423 for $1,312.50 and No. 425 for $2,483.10 have not been paid by the bank. The statement indicates a deposit of $2,800 was in transit. In addition, the bank statement lists a check charge of $15.50, a monthly service charge of $26.35, a note collection charge of $25.00, and an interest payment of $11.60. Prepare a report from bank reconciliation statement for Tri-State Heating and Cooling.

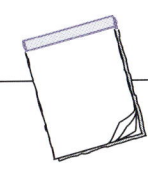

EXPRESS YOUR UNDERSTANDING

Compose one or two well-written sentences to express the requested information in your own words.

1. Explain the procedure required to properly write a check. Why should this procedure be followed?

2. How would you endorse a check that you received from a customer that is to be transferred to one of your suppliers as payment for goods purchased? Why would you use this type of endorsement?

3. Describe how you would prepare a commercial deposit if you planned to deposit $500 in bills of various denominations, $125 in coin of various denominations, and 20 checks of various amounts.

4. Identify the types of information required to enter a transaction in a checkbook register.

5. Describe fully how a business collects payment for sold merchandise that was purchased by the buyer with a credit card.

6. Explain how a check written by you for merchandise moves through the cancellation process. Be sure to use appropriate terminology in your explanation.

7. Why is it necessary for a business or an individual to prepare a periodic bank reconciliation?

8. Identify the information on a bank statement that is required to determine the adjusted book balance portion of a reconciliation statement. Explain how the information affects this balance.

9. Explain how information found in both the checkbook register and the bank statement is used to determine the adjusted bank balance portion of the reconciliation statement.

10. What would you do if, after completing a bank reconciliation, you discovered that the adjusted balances did not agree?

Mind Your Business

Banking services for Media World are provided by Solvay Bank. Kate Bradley, the bookkeeper, is responsible for maintaining all records pertaining to the cash control system. Therefore, she records all transactions involving receipt of cash from both cash and credit sales, writes all checks involving the disbursement of cash, prepares all bank deposits, and at the end of each month prepares the bank reconciliation statement. During the month of September, Kate completed a number of activities, which she submitted to you on various dates to keep you informed of the cash position of the business.

Activity 1

A branch office of Solvay Bank is located in the same mall as Media World, which allows Kate Bradley to make bank deposits as needed. Based on the following information, what would be the total amount of the September 13 bank deposit?

	Currency	Checks	
		No.	Amount
5	fifty-dollar bills	6423-015	32.07
25	twenty-dollar bills	8146-132	53.48
34	ten-dollar bills	2197-837	74.89
120	one-dollar bills	7619-1053	13.90
8	rolls of quarters	5974-621	83.05
5	rolls of dimes	2732-496	139.08
10	rolls of nickels	4368-258	26.50
		6130-720	46.94
		3761-582	10.09

Activity 2

Media World accepts only Visa and MasterCard for credit sales. On September 14, Kate Bradley made a bank deposit of credit card transactions. If Visa charges a 3.5% discount fee and MasterCard charges a 3.0% discount fee, what was the total amount of the deposit recorded on the merchant deposit ticket? The deposit was based on the following financial data.

Visa Sales:

Sales slip amounts	Credit slip amounts
1. $15.09	1. $32.08
2. 32.48	2. 17.10
3. 21.10	
4. 64.18	
5. 27.29	
6. 48.12	
7. 34.21	
8. 160.49	

MasterCard Sales:

Sales slip amounts	Credit slip amounts
1. $34.21	1. $11.76
2. 26.73	2. 20.32
3. 17.09	3. 19.24
4. 128.38	
5. 44.90	
6. 16.03	

Activity 3

On September 27, Kate Bradley received the monthly bank statement from Solvay Bank. In accordance with financial policy, she is required to prepare the monthly bank reconciliation statement on the last day of each month and forward a copy to you for your assessment. Use the data in the following financial records to determine (a) the adjusted checkbook balance amount and (b) to prepare the bank reconciliation statement for September.

SOLVAY BANK
1537 MILTON AVENUE
SOLVAY, NEW YORK 13209-0167
(315) 484-2204

```
MEDIA WORLD                        49-0
1425 COMMERCE BLVD                 31
SYRACUSE NY 13215                  53
                                        BUSINESS ACCOUNT:   00000000

                                        08/24/99 THRU 09/23/99
FDIC                                    DOCUMENT COUNT:        89
                                                          PAGE 3
```

```
===================================================================
                     BUSINESS ACCOUNT 00000000
===================================================================
         DESCRIPTION        DEBITS      CREDITS    DATE      BALANCE
DEPOSIT                                    520.00  09/16/99  43,188.06
DEPOSIT                                  1,940.00  09/16/99  45,128.06
DEPOSIT                                  2,500.00  09/17/99  47,628.06
CHECK # 1568                 199.75                09/17/99  47,428.31
DEPOSIT                                  1,425.00  09/20/99  48,853.31
DEPOSIT                                  2,150.00  09/20/99  51,003.31
CHECK # 1576                 110.82                09/20/99  50,892.49
CHECK # 1578                 131.52                09/20/99  50,760.97
CHECK # 1572                 394.75                09/20/99  50,366.22
CHECK # 1567               4,500.00                09/20/99  45,866.22
DEPOSIT                                  1,650.00  09/21/99  47,516.22
CHECK # 1575                 352.07                09/21/99  47,164.15
CHECK # 1573                 369.91                09/21/99  46,794.24
CHECK # 1574                 478.94                09/21/99  46,315.30
DEPOSIT                                    480.00  09/22/99  46,795.30
DEPOSIT                                  1,875.00  09/22/99  48,670.30
CHECK # 1577                 321.87                09/22/99  48,348.43
CHECK # 1569                 672.35                09/22/99  47,676.08
CHECK # 1570               1,243.60                09/22/99  46,432.48
DEPOSIT                                  2,000.00  09/23/99  48,432.48
CHECK # 1580                 325.00                09/23/99  48,107.48
INTEREST                                    15.01  09/23/99  48,122.49
SERVICE CHARGE                46.25                09/23/99  48,076.24
BALANCE THIS STATEMENT......................................09/23/99  48,076.24

   TOTAL CREDITS    (31) 51,035.01    TOTAL DEBITS    (54) 48,334.14

           - - - - - - - - - I N T E R E S T - - - - - - - - - -
AVERAGE LEDGER BALANCE:       45,579.72  INTEREST EARNED:            15.01
AVERAGE AVAILABLE BALANCE:    38,716.69  DAYS IN PERIOD:                31
INTEREST PAID THIS PERIOD:        15.01  ANNUAL PERCENTAGE YIELD EARNED: 2.27%
INTEREST PAID 1999:              138.14
TAX IDENTIFICATION NUMBER:  123-45-6789
INTEREST RATE:                  2.2500%
```

Note: The bank statement is page 3 of 3 and lists checks and deposits paid or received by the bank from September 16 to the 23. Consider all checks and deposits not listed on this page of the statement or checked (✓) on the checkbook register prior to September 13 as account activity that appears on pages 1 and 2 of the statement. The checkbook register balance of $39,368.06 is a correct balance as of that date, and there are no errors on the bank statement.

NUMBER	DATE	DESCRIPTION OF TRANSACTION	PAYMENT/DEBIT (-)	√	FEE (if any)	DEPOSIT/CREDIT (+)	BALANCE
		BALANCE BROUGHT FORWARD →					35,982 27
1564	9/10	Luis Mendez	110 82	√			110 82 / 35,871 45
1565	9/10	Susan Evans	321 87	√			321 87 / 35,549 58
1566	9/10	Shelia McCaffery	131 52	√			131 52 / 35,418 06
	9/10	Deposit		√		2,000	2,000 00 / 37,418 06
	9/11	Deposit		√		1950	1,950 00 / 39,368 06
1567	9/13	Metropolitan Music Supply	4,500 00				
	9/13	Deposit				1,800	
1568	9/14	Herald Tribune	199 75				
	9/14	Deposit		√		1,500	
1569	9/15	Mercantile Supply Inc.	672 35				
1570	9/16	Upstate Magazine Dist.	1,243 60				
1571	9/16	Vision Communications Media	5,380 75				
	9/16	Deposit				1,940	
	9/16	Deposit				520	
1572	9/17	Kate Bradley	394 75				
1573	9/17	Carol Kanz	369 91				

NUMBER	DATE	DESCRIPTION OF TRANSACTION	PAYMENT/DEBIT (-)	√	FEE (if any)	DEPOSIT/CREDIT (+)	BALANCE
		BALANCE BROUGHT FORWARD →					
	9/22	Deposit				1,875	
1584	9/23	J&S Music Wholesalers	3,748 25				
1585	9/23	General Insurance	138 00				
1586	9/23	Mid-State Gas + Electric	480 50				
	9/23	Deposit				2,000	
	9/23	Deposit				575	
1587	9/24	Kate Bradley	394 75				
1588	9/24	Carol Kane	369 91				
1589	9/24	Matthew Welch	478 94				
1590	9/24	Gordon Stephens	352 07				
1591	9/24	Luis Mendez	110 82				
1592	9/24	Susan Evans	321 87				
1593	9/24	Shelia McCaffery	131 52				
	9/24	Deposit				2,500	
	9/25	Deposit				2375	

NUMBER	DATE	DESCRIPTION OF TRANSACTION	PAYMENT/DEBIT (-)	√	FEE (if any)	DEPOSIT/CREDIT (+)	BALANCE
		BALANCE BROUGHT FORWARD →					
1574	9/17	Matthew Welch	478 94				
1575	9/17	Gordon Stephens	352 07				
1576	9/17	Luis Mendez	110 82				
1577	9/17	Susan Evans	321 87				
1578	9/17	Shelia McCaffery	131 52				
	9/17	Deposit				2,500	
	9/18	Deposit				2,150	
1579	9/20	County Sales tax Bureau	1,750 00				
1580	9/20	Central Office Supply	325 00				
	9/20	Deposit				1,425	
1581	9/21	National Music Warehouse	8,775 37				
	9/21	Deposit				1,650	
	9/21	Deposit				480	
1582	9/22	Herald Tribune	375 00				
1583	9/22	United Freight Services	530 00				

NUMBER	DATE	DESCRIPTION OF TRANSACTION	PAYMENT/DEBIT (-)	√	FEE (if any)	DEPOSIT/CREDIT (+)	BALANCE
		BALANCE BROUGHT FORWARD →					
1594	9/27	Liverpool Entertainment	2,037 88				
1595	9/27	Bell Atlantic Co.	84 50				
	9/27	Deposit				1,600	
1596	9/28	Infotech Computer Services	267 40				
1597	9/28	State Income Tax Bureau	1,328 00				
1598	9/28	Roadway Express Inc.	435 00				
	9/28	Deposit				2,400	
1599	9/29	Carter Plumbing Co	136 50				
1600	9/29	Audio-Visual Supply Co	5,450 75				
	9/29	Deposit				1,750	
	9/29	Deposit				350	
1601	9/30	Noble Accounting Services	750 00				
	9/30	Deposit				2,350	
	9/30	Deposit				475	

SELF-TEST

A. Terminology

Complete the following items using the key terms presented at the beginning of the chapter. Check your response against the answer key at the end of the test.

1. A _____ is a written order from an individual or company to their bank instructing it to pay a designated party a specified amount of funds.

2. Banks deduct a _____ to cover the cost of maintaining the customer's account.

3. A _____ increases the amount of money in a checking account and can include checks as well as cash.

4. A _____ limits the ability to cash the check as it includes the words "for deposit only" and the signature of the depositor.

5. When a person or business wishes to transfer the amount of a check to a third party they must use a _____.

6. The record maintained by the depositor listing all transactions to their checking account is called a _____.

7. Banks charge businesses a _____ for credit card sales indicated on the merchant's deposit summary.

8. A check that is guaranteed by a bank is referred to as a _____.

9. Checks that have been deposited but not paid by the bank because there was not enough money in the account to cover the check are called _____.

10. A _____ is a check that has been paid by the depositor's bank.

11. When a depositor requests the bank not to pay a check, they have the bank issue a _____.

12. _____ are provided to depositors by banks and list all charges (debits) and payments (credits) to the checking account during a specified period.

13. The process of analyzing the check register and bank statement to bring their balances into agreement is called _____.

B. Calculation

The following concepts and short problems are designed to test your understanding of the objectives identified at the beginning of the chapter. Answers are provided at the end of the test.

14. Centerville Bank's personal checking account carries service charges of 15 cents for each check written; 30 cents for each check paid; and a monthly maintenance fee of $3.50. What is the monthly service charge if a depositor wrote 18 checks, 12 of which were paid by the bank during the month?

15. Oak Corners Grocery had an average balance for May of $2,875.50. During the month, the store made 10 deposits, which included a total of 192 checks, and the bank paid 63 checks. Using the rate schedule provided, calculate the store's monthly service charge.

Service	Charge
$4.00	Maintenance fee
$0.07	Per check paid
$0.10	Per check deposited
$0.50	Per deposit fee

16. Howard Dolan banks with First Federal Savings and Loan, which pays $5\frac{1}{4}\%$ interest on its NOW accounts if the average monthly balance is at least $500. How much interest should be credited to Mr. Dolan's account if his average monthly balance is $892.70?

17. What amount of money would be deposited based on the following: 3 fifty-dollar bills, 18 twenty-dollar bills, 32 ten-dollar bills, 13 five-dollar bills, 108 one-dollar bills; 7 rolls of quarters, 9 rolls of dimes, 12 rolls of nickels; and checks of $115.47, $73.69, $328.62, $39.99, and $12.50?

18. Determine the amount to be brought forward on check stub no. 153 if the amount brought forward on check stub no. 152 was $1,467.29, check no. 152 was written for $215.95, and a deposit of $375.00 was made before check no. 152 was written.

19. Jean's Unlimited had the following credit card transactions during a recent period. Sales: $39.95, $20.42, $69.55, $103.26, $9.67, $54.32, and $71.77. Refunds: $48.52, $20.42, and $19.25. Determine (a) the net deposit for the merchant deposit summary and (b) the amount of the discount fee to be charged to Jean's Unlimited account if the bank charges a $4\frac{1}{4}\%$ discount fee.

20. On October 8, J. T. Ryan and Sons received a bank statement showing a balance of $1,874.30.

The checking account balance was $2,139.18. An inspection of the check register and bank statement revealed the following: There were deposits in transit of $850 and $525, and outstanding checks of $267.42, $98.25, $143.27, $364.18, and $56.00. There were check printing charges of $12.50, a service charge of $8.40, an interest credit of $5.20, a discount fee of $18.30, and a note collected by the bank for $250. A check from Charles Williams for $35 that had been deposited was returned due to nonsufficient funds. Prepare a bank reconciliation for J. T. Ryan and Sons.

13

SECURITIES AND DISTRIBUTION OF INCOME AND EXPENSES

Learning Objectives

After completing the exercises in this chapter, you will be able to:

1. Explain the advantages and disadvantges an investor has in owning stocks and bonds.

2. Explain the advantages and disadvantages a company has in issuing stocks and bonds.

3. Interpret stock and bond quotations.

4. Calculate a capital gain or loss from the sale of stock.

5. Prorate bond interest.

6. Interpret mutual fund quotations.

7. Calculate the net asset value (NAV) of a mutual fund.

8. Calculate the profit or loss on the trading of mutual fund shares.

9. Distribute company profits to partners and stockholders.

10. Distribute the expenses of a company on the basis of the number of employees, the square footage, and the amount of sales.

11. Define key terms.

INTRODUCTION

Perhaps you currently own a business or maybe you plan to own a business someday. The day will probably come when you will need a substantial amount of extra money to expand or improve your business. This chapter discusses two ways in which a business generates extra cash:

1. By issuing stocks
2. By issuing bonds

We often hear references made to stocks and bonds, but what are they and how do they affect us? A **stock** is a share or portion of ownership of a particular company or corporation. A **bond** is an interest-bearing loan that the investor makes to a company. Stocks and bonds are referred to as **securities.**

To illustrate, let us suppose you have some extra cash that you would like to invest. You decide that you would like to purchase some shares of American Telephone and Telegraph (AT&T) stock. (At this point, the term *stock* shall refer to common stock rather than preferred stock, each of which will be clearly explained later.) In order to do this, you go to a **stockbroker,** (a broker is a person who buys and sells stocks and bonds). The stockbroker, in turn, contacts the **stock exchange** where your shares of AT&T stock will be purchased, makes the purchase for you, and charges you a fee for the services provided. The stock exchange, sometimes called the **stock market,** is a place where securities that are registered with the exchange are bought and sold. If each share of AT&T stock is selling for $40 per share, and if you had $400 to invest, then you would be able to purchase 10 shares of stock ($400/$40 = 10 shares). After purchasing this stock, you would receive a **stock certificate,** which is a certificate indicating ownership of 10 shares of AT&T stock. Since purchasing this stock means that you would now be part owner of AT&T and would be considered a **stockholder** in its corporation (a stockholder is a holder or owner, of stock in a corporation), you would be entitled to dividends. **Dividends** are a portion of a company's profits, usually paid to its stockholders in the form of cash. As an illustration, let us suppose that AT&T has $70,000 in profits that it wishes to distribute among 200,000 shares of outstanding common stock shareholders. This means that for each share of AT&T stock, you would receive $.35 in dividends ($70,000/200,000 = $.35). Since you have just purchased 10 shares of stock, you would receive $3.50 in dividends ($.35 × 10 shares).

A second way for corporations to acquire cash is by issuing bonds. A bond is an interest-bearing certificate issued by a government or business, redeemable on a specified date. In other words, the investor is lending money to the corporation for a certain length of time. In turn, the corporation must pay the investor a predetermined rate of interest during the life of the bond and it must repay the **principal** (the original amount borrowed) at some future specified date.

Most corporate bonds are in $1,000 denominations, although bonds of $5,000 and $10,000 are also issued. Each bond bears a stated rate of interest, such as 7%, 9⅛%, or 12⅜%. Usually, corporations pay interest to the bondholder semiannually. The life of a bond begins the day the bond is purchased and ends on the **maturity date,** that is, the date on which the principal of the bond is to be repaid. Most bonds have a fairly long life such as 10, 20, or 30 years, although the life of a bond can be as short as 2 years.

From the investor's point of view, there are both advantages and disadvantages to owning securities. The advantages to the investor of owning stocks in a company are as follows:

Learning objective
Explain the advantages and disadvantages an investor has in owning stocks and bonds.

- The dollar value of the stock may increase.
- Dividends may be paid to the stockholder by the issuing company.
- Ownership of stocks can easily be sold or traded.
- The stockholder, as an owner of the corporation, has voting privileges in major company decisions and therefore has some control over business decisions.
- Stocks are negotiable; therefore, they can be bought and sold.

Some disadvantages of owning stock are as follows:

- The dollar value of the original shares may decrease.
- The corporation is not required to issue dividends.

The advantages to the investor who purchases bonds are as follows:

- The investor receives regular, semiannual interest payments during the life of the bond.
- Bonds are considered a low-risk investment and therefore a good investment for the risk avoider.
- Bonds are negotiable, which means that the original purchaser of the bond can sell it to someone else.

Disadvantages of owning bonds include:

- The investor has no voting privileges in the issuing corporation and therefore has no input into the operation of the company.
- Bond prices may decrease.

Since we have just looked at some reasons why an investor would want to purchase securities, let us take the opposite perspective and look at the reasons why corporations issue securities.

When a company needs money to expand its business, one way to get that necessary money is to issue shares of stock. To illustrate, let us suppose that 5 years ago, three people started a business. This company now needs $50,000 to expand its business. If the three people could find one person to invest $50,000 then there would be four owners (three original plus one

Learning objective
Explain the advantages and disadvantages a company has in issuing stocks and bonds.

Advantages of Issuing Stock to the Company

1. There is an increase in its assets and capital.
2. The principal amount invested does not have to be repaid to the investor as it would be if a bond were issued.
3. There is no requirement that a dividend be paid to the investor.

Disadvantages of Issuing Stock

1. There is some loss of control over business decisions.

Advantages to a Company of Issuing Bonds

1. The company still maintains control over business decisions since bondholders have no voting privileges.

Disadvantages to the Company

1. The company must repay the principal amount to the bondholders at some future date.
2. The company must pay interest to bondholders during the life of the bond.

---

new owner). However, it might be easier to get five people to each invest $10,000. In this case, there would be eight owners (three original plus five new owners). It might even be easier to get 2,000 people to each invest $25. Then, of course, there would be a total of 2,003 owners. Remembering that each stockholder receives one vote for each share of stock held, this company would now have 2,003 people voting rather than just the original three. The original owners would now probably lose some control over the business decisons affecting the company. Loss of control would be one reason why a company might not want to issue stock. On the other hand, an advantage of issuing stocks, rather than issuing bonds, is that the company's capital, or assets, increases. This means that there is no debt to repay as there would be with bonds. Another advantage is that the issuing company does not have to pay out dividends to stockholders on a regular basis. These advantages and disadvantages are listed in Figure 13.1.

If you have neither the desire nor the opportunity to own a business, this chapter will still be of interest to you since we will discuss practical calculations such as calculating profits and losses from the sale of securities (stocks, bonds and mutual funds) and distributing income and expense amounts in a business. Each of these topics will help you with your financial investments and your future success in the business world.

13.1 STOCK QUOTATIONS AND CAPITAL GAINS

People who wish to invest their money in stocks need information about the value, performance, and costs of the stocks available for purchase. Such information is provided in a format called a *stock quotation.* When investors want to sell stocks, they need to know the profit or loss that they have realized with their investments. This profit or loss is referred to as a *capital gain* or *capital loss.*

Stock Quotations

Learning objective
Interpret stock and bond quotations.

As an investor, it is important that you understand how to read stock and bond quotations. Stock and bond quotations are printed in local, daily papers as well as in financial publications such as the *Wall Street Journal.* Figure 13.2 represents a stock quotation and the following paragraphs explain this figure.

The first two columnns of numbers in Figure 13.2 labeled "52 Weeks High-Low" represent the highest price and lowest price, respectively, that this stock sold for within the past year. To convert these mixed numbers to dollar and cents amounts, simply convert the fractions to their decimal equivalents. Therefore, 130⅞ becomes $130.875 (⅞ = .875) and 106¼ becomes $106.25. The heading "Stock" refers to the name of the corporation that is issuing the stock. Usually, it is necessary to abbreviate the name of the company. The fourth column, "Div.," approximates the annual dividends per share paid to the stockholders. In this case, $4.84 was paid in annual dividends. The next column, "Yld %," shows that 4.2% is the percent of dividends you would receive if you bought this stock at the current closing price. In other words, divide the dividend by the closing price (4.84/115.50 = .0419, or 4.2%).

$$\text{Current Yield} = \frac{\text{Annual dividend per share}}{\text{Current price per share}}$$

Figure 13.2

Sample stock quotation

| 52 Weeks | | Stock | Div. | Yld % | P-E Ratio | Sales 100s | High | Low | Close | Net Change |
High	Low									
130⅞	106¼	**XYZ**	4.84	4.2	12	96543	119⅜	114⅞	115½	− ¼
highest and lowest prices over the last 52 weeks		company name	annual dividend	% yield	price-earnings ratio	number of shares sold at current price	highest and lowest trading prices on this day		closing price on this day	change in the closing price from previous day

The yield is expressed as a percentage (rounded to the nearest tenth) and is used by investors to compare the dividends paid on stocks selling at various prices.

The "PE Ratio" column is the price–earnings ratio. This ratio compares the closing price with the corporation's earnings per share for the past year. A low PE ratio could suggest the stock is undervalued or that investors do not believe the company has a promising financial future. A high PE ratio could indicate that the current selling price is overpriced in relation to the company's earnings or that investors believe that the earnings will increase in the near future. The PE ratio is found with the following formula:

$$\text{PE ratio} = \frac{\text{Price per share}}{\text{Annual net earnings per share}}$$

In Figure 13.2 the PE ratio is 12, which means that the current selling price is 12 times the earnings per share.

Notice that the company's earnings are not stated on a stock quotation. The next column, "Sales 100s," shows the number of shares (in hundreds) traded on this one day. Therefore, 9,654,300 shares of XYZ stock were traded on this day. The eighth, ninth, and tenth columns, labeled "High," "Low," and "Close," represent the highest price the stock traded for on this day, the lowest price the stock traded for on this day, and the closing price for the day. In this case, the high price was $119.375, the low price was $114.875, and the closing price at the end of the day was $115.50.

The last column, "Net Change," shows the net change in closing price from the previous day with that of today's closing price. In other words, today's closing price is $-\frac{1}{4}$, or $0.25, less than yesterday's closing price. Therefore, yesterday's closing price was $115.75.

Capital Gains

Learning objective
Calculate a capital gain or loss from the sale of stock.

As an investor, when you determine that it is time to sell your shares of stock, you need to know how to calculate capital gains. A **capital gain** is the profit earned from the sale of assets, such as securities or real estate.

It is important for you to know that each time you buy or sell securities, you must pay the broker a commission. Calculating the broker's commission is confusing because each brokerage house (e.g., Merrill Lynch, Prudential-Bache, Dean Witter Reynolds) has its own set of rules for calculating commissions. Adding to the confusion is the fact that there is a different commission structure for the over-the-counter market (where lesser-known stocks that are not registered with a stock market are traded) than there is for the stock exchange, such as the New York Stock Exchange (where larger, well-known stocks are traded). The phrase "over-the-counter market" is used to describe a variety of ways that stocks and bonds not registered with a stock exchange—and therefore not traded through a stock exchange—can be bought and sold. Usually, you will pay a larger comission for a smaller stock. In order to simplify this process, we will assume in our examples that a broker's commission is 3% of the selling price of the stock.

When it comes time to sell your stock, you will need to compute your *capital gain,* or profit, from the sale. There are three steps in computing the capital gain. Step 1 is to determine the total cost of buying the shares. This includes the purchase price plus broker's commission. Step 2 determines the total amount received from the sale of stock. The total amount received equals the selling price minus the broker's commission. Step 3 is the difference between Step 1 and Step 2. Example 1 illustrates how to calculate a capital gain.

Example 1

If you buy ten shares of a stock at $40 each and then later sell them at $57, what is your capital gain? Assume the commission is 3% of the market price.

Solution

Step 1: Determine the cost to buy shares.

10 shares at $40 per share	$400.00
+ commission (.03 × 400)	12.00
total cost	$412.00

Step 2: Determine the amount received from sale of the stock.

10 shares at $57 per share	$570.00
− commission (.03 × $570)	17.10
total received	$552.90

Step 3: Determine the difference.

total received	$552.90
− total cost	412.00
capital gain	$140.90

Note: If the total cost is greater than the total received, then this would be a capital loss rather than a capital gain.

Long-term capital gains (a gain on an asset that has been held longer than 12 months) receive tax-favored treatment. Instead of paying taxes on 100% of the capital gain, taxes are generally subject to a 20% tax rate, or 10% if you are in a lower regular tax bracket. The tax reforms initiated in 1998 changed this so that long-term capital gains could receive more favored treatment.

CHECK YOUR KNOWLEDGE

Stocks, Stock Quotations, and Capital Gains

1. If XYZ Corporation wishes to distribute $110,000 in dividends among its 400,000 shareholders' shares of outstanding stock, how much will you receive in dividends if you own 10 shares of XYZ?

2. From the investor's point of view, list one advantage of owning bonds.

3. From the corporation's point of view, list one advantage of issuing stocks.

Refer to the following stock quotation when answering questions 4 through 7.

52 Weeks High	52 Weeks Low	Stock	Div.	Yld %	P–E Ratio	Sales 100s	High	Low	Close	Net Change
30¾	24⅝	ABC	.88	3.2	9	112	28	27¾	27⅞	+ ¼

4. What was the low trading price for the day?

5. What was the low price for the year?

6. How many shares of stock were traded on this day?

7. Did the stock increase or decrease in price from the previous day, and by how much did it change?

8. If you bought 300 shares at 108½, and the broker received 3% commission, what would have been the total cost to you?

9. If you sold 300 shares at 105¾, and the broker received a 3% commission, what would have been the total amount you received?

10. Compute the gain or loss using the information from problems 8 and 9, assuming that it was the same stock traded in each case.

13.1 EXERCISES

Find the dividends per share and total dividends received.

	Profits to distribute	Number of shareholders	Dividends per share	Number of shares owned	Total dividend
Example	$ 60,000	100,000	$.60	25	$15.00
1.	$250,000	100,000	_____	25	_____

Answers to CYK: *1.* $2.75   *2.* steady (periodic) interest income   *3.* not a debt, therefore, nothing to repay   *4.* $27.75   *5.* $24.625   *6.* 11,200   *7.* increased by ¼, or $.25   *8.* $33,526.50   *9.* $30,773.25   *10.* $2,753.25 capital loss

2.	$500,000	554,900	_____	500	_____
3.	$400,000	998,000	_____	100	_____
4.	$300,000	400,000	_____	250	_____
5.	$200,000	50,000	_____	20	_____

Refer to the following stock quotation to complete the missing information below. Convert all fractions to decimals.

52 Weeks				Yld	P–E	Sales				
High	Low	Stock	Div.	%	Ratio	100s	High	Low	Close	Net Change
37⅞	31	BBD	2.56	7.0	11	12955	37¾	36⅜	36½	−1
10½	6⅞	IBN	20	2.0	7	2803	10½	8⅜	10⅛	+1⅜
117	106⅞	ATD	15.25	13.3	—	3080	115½	114	114½	−5
99⅜	87⅓	GMK	9.25	9.7	—	420	98	98	98	−1⅜
35¼	22⅞	BMZ	1.00	3.0	6	4112	35¼	33⅜	33⅝	−½
60⅛	41½	NBO	1.20	2.2	22	1173	59¼	51¼	55¾	+4¾
23⅞	14¾	TRL	.60	2.6	8	2871	23⅞	22⅛	23½	+1½
73¼	52½	BVD2	.04	2.8	19	22934	73½	64⅛	72⅝	+8
25¾	21⅜	PGE	1.07	4.2	14	15788	25⅝	24½	25⅜	+⅛
17¼	12⅞	GTS	.32	1.9	18	58431	17¼	15⅞	16½	+½

		52-week high	Yearly dividend	Number of shares sold	Close	Net change
6.	BBD	_____	_____	_____	_____	_____
7.	IBN	_____	_____	_____	_____	_____
8.	ATD	_____	_____	_____	_____	_____
9.	GMK	_____	_____	_____	_____	_____
10.	BMZ	_____	_____	_____	_____	_____
11.	NBO	_____	_____	_____	_____	_____
12.	TRL	_____	_____	_____	_____	_____
13.	BVD2	_____	_____	_____	_____	_____
14.	PGE	_____	_____	_____	_____	_____
15.	GTS	_____	_____	_____	_____	_____

Calculate the total cost to the buyer. Assume the broker's commission is 3% of the market price per share.

		Stock cost	Commission	Total cost
16.	100 shares at 63	_____	_____	_____
17.	100 shares at 24⅞	_____	_____	_____
18.	500 shares at 78	_____	_____	_____
19.	200 shares at 182¾	_____	_____	_____
20.	250 shares at 33¼	_____	_____	_____

Calculate the total amount received if the shares were sold at the prices listed below. Assume the broker's commission is 3% of the total market price.

		Stock revenue	Commission	Total received
21.	100 shares at 63	_____	_____	_____
22.	500 shares at $101\frac{7}{8}$	_____	_____	_____
23.	250 shares at $88\frac{1}{8}$	_____	_____	_____
24.	200 shares at $11\frac{5}{8}$	_____	_____	_____
25.	400 shares at $36\frac{3}{4}$	_____	_____	_____

Find the capital gain or loss. Assume a broker's commission of 3%.

26. Bought 100 shares at $14\frac{1}{4}$; sold 100 shares at $15\frac{1}{4}$.

27. Bought 200 shares at $21\frac{1}{4}$; sold 200 shares at 22.

28. Bought 200 shares at $21\frac{1}{4}$; sold 100 shares at 22.

29. Bought 500 shares at 108; sold 500 shares at 138.

30. Bought 500 shares at $76\frac{5}{8}$; sold 350 shares at $80\frac{3}{8}$.

31. Bought 400 shares at $26\frac{1}{4}$; sold 400 shares at $22\frac{1}{8}$.

32. Bought 150 shares at $120\frac{7}{8}$; sold 150 shares at $80\frac{7}{8}$.

33. Bought 60 shares at 52; sold 60 shares at $36\frac{7}{8}$.

34. Bought 800 shares at $8\frac{3}{8}$; sold 800 shares at $64\frac{7}{8}$.

35. Bought 360 shares at $11\frac{3}{8}$; sold 360 shares at $7\frac{5}{8}$.

Refer to the stock quotation provided for problems 6 through 15.

36. Yesterday you bought 100 shares of BVD2 stock at the quoted price; you sold all 100 shares at today's quoted price and paid a broker's fee of 2% for each transaction. (a) What was the total selling price of the stock when you bought it? (b) What was the total you paid for the stock, including broker's fees? (c) What was the total selling price of the stock when you sold it? (d) What was the total you received for the stock? (e) What were the capital gains or losses?

37. Yesterday you bought 2,100 shares of NBO stock at the quoted price; you sold all 2,100 shares at today's quoted price and paid a broker's fee of 1% for each transaction. (a) What was the total selling price of the stock when you bought it? (b) What was the total you paid for the stock, including broker's fees? (c) What was the total selling price of the stock when you sold it? (d) What was the total you received for the stock? (e) What were the capital gains or losses?

38. Yesterday you bought 500 shares of GMK stock at the quoted price; you sold all 500 shares at today's quoted price and paid a broker's fee of 3% for each transaction. (a) What was the total selling price of the stock when you bought it? (b) What was the total you paid for the stock, including broker's fees? (c) What was the total selling price of the stock when you sold it? (d) What was the total you received for the stock? (e) What were the capital gains or losses?

39. Jack Mangan bought 250 shares of ATD stock yesterday and must sell some of it today. He has decided to sell 100 shares. The broker's fees were 3% in both transactions. What is Jack's capital loss?

40. Bob Burnet bought 1,000 shares of GTS stock yesterday; he sold it at today's high price, with broker's fees of 2%. What is his capital loss or gain?

13.2 BOND QUOTATIONS AND ACCRUED INTEREST

When investing in bonds, information about value, performance, and costs of bonds is contained in a *bond quotation*. The interest earned while an investor owns a bond is called *accrued interest*.

Bond Quotations

As demonstrated earlier, bonds are issued as a way for corporations to borrow money. The denomination of a bond is referred to as the **face value of a bond.** In this chapter, we will assume that each bond has a face value of $1,000. If you were interested in buying or selling bonds, you would go to a stockbroker just as you would if you were buying or selling stocks. Bonds are traded at a stock exchange or on the over-the-counter market.

Even though a bond has a face value of $1,000, it can be purchased at a price higher or lower than $1,000. When the market price (or selling price) of a bond is lower than its face value, the price is called a **discount price** and the bond is said to be selling at a *discount*. When the market price is higher than its face value, the price is called a **premium price** and the bond is said to be selling at a *premium*. The market price of a bond fluctuates from day to day to reflect the current state of the economy. This fluctuation allows bonds to compete with the current interest rates of other investments. The market price of a bond can also reflect the current status of the issuing corporation (e.g., whether it's growing, stable, innovative, and so on).

The market price of a bond is stated as a percent of its face value. To illustrate, suppose a bond is listed as 99. This means that the current market price is 99% of the $1,000 face value or $990 (.99 × $1,000). Suppose a second bond is listed as 105. Then the current market price is 105% of $1,000, or $1,050 (1.05 × $1,000).

The current market price of a bond can be found in a bond quotation. As with stock quotations, bond quotations can be found in daily newspapers or in financial periodicals. Figure 13.3 represents a bond quotation. The following paragraphs explain this figure.

Figure 13.3

Sample bond quotation

Bond			Cur. yld.	Vol.	High	Low	Close	Net chg.
Btmm Corp	7 5/8	20	8.0	20	95	95	95	−1 5/8
name of company issuing the bond	interest rate	year bond matures (2020)	current yield	volume of bonds sold on this day	highest and lowest prices bond sold for on this day		closing price on this day	net change from previous day

Under the heading of "bonds," first the abbreviated name of the company is listed; next, the $7\frac{5}{8}$ represents the $7\frac{5}{8}\%$ stated interest rate on the bond. The large "20" stands for 2020, the year in which the bond matures. If, instead of "20" the number was "05," it would mean that the maturity date was in 2005.

The data in the column labeled "Cur. Yld." means the bond has a current yield of 8.0%. Current yield is determined by dividing the dollar amount of interest you expect to make on the $1,000 bond by today's closing price of $950; your return on investment would be 8% here.

$$\text{expected interest} = 7\frac{5}{8} \times 1{,}000 = \$76.25$$

$$\text{current yield} = \frac{76.25}{950} = .0802 = 8\%$$

The "Vol." heading stands for the volume of bonds that were traded on this 1 trading day. In this case, 20 bonds were traded. The column labeled "High" represents the highest price that one of these bonds traded for on this day. Notice that the market price is stated as a percent of the face value of the bond. That is, the current market price is $950 (.95 $\times$ $1,000).

The next two columns represent the lowest price that one of these bonds traded for on this day and the closing price, respectively.

The last column is the net change from the previous day's closing price compared with the closing price of today. The number listed in this column is stated as a percent of the face value. Therefore, a net change of $-1\frac{5}{8}$ means that today's closing price is lower than yesterday's closing price by $16.25.

$$-1\frac{5}{8}\% = -1.625\% \qquad -.01625 \times \$1{,}000 = -\$16.25$$

Accrued Interest on a Bond

Learning objective
Prorate bond interest.

Two important points were mentioned earlier that bear repeating at this time. First, bonds are negotiable, which means that the original purchaser of the bond can sell it to someone else. Second, interest is paid to the bondholder semiannually. In this chapter, we will assume that interest is paid to the bondholder on January 1 and July 1. If bondholders decide to sell their bonds on a date other than January 1 or July 1, then they must know how to prorate (to divide proportionately) bond interest. Example 2 illustrates this concept and procedure.

Example 2

Alice currently owns a $1,000 bond with a stated interest rate of 10%. On June 1, she will sell it to Benny. The commission is $20. On July 1, the issuing corporation will send the interest payment, for the past 6 months, to Benny, who will be the current bondholder. In order for Alice to receive

For Your Information

WHAT IS A BOND AND HOW DOES IT WORK?

A bond is essentially a loan that pays a set interest rate for a set period. The interest rate is called a *coupon* and the period is called the *maturity*. Individual bonds usually have a face (or par) value of $1,000. If you, as an individual, buy a 10-year bond at par paying 6%, you will receive $60 a year in interest; after 10 years, at maturity, you will recoup your principal of $1,000.

Bond prices are interest-rate sensitive. Generally speaking, if interest rates rise—to 7% for instance—your bond will be worth less if you sell it, because it pays $60 per year while newer bonds are paying $70. To make up for this difference, you will have to sell at a discount, receiving less than $1,000. However if you hold the bond to maturity, you will receive the full $1,000 and will not take a loss regardless of prevailing interest rates.

If interest rates drop to 5 percent, your bond will be worth more if you sell it because it pays $60 per year rather than the new rate of $50. You can sell the bond at a premium and receive more than $1,000, or you can hold it to maturity, continuing to receive a higher-than-market interest rate, and receive $1,000.

History of Bond Performance 1925 - 1997

This image illustrates the hypothetical growth of a $1 investment in stocks, corporate bonds, government bonds, and Treasury bills over the period December 31, 1925, through December 31, 1997.

In accordance with their higher risk (due to risk of default), corporate bonds outperformed government bonds and Treasury bills. However, when compared to the ending wealth of stocks, corporate bonds fell short.

Historically, bonds have not proven to be the best investment vehicle for long-term growth. However, there have been short periods of time when bonds have outperformed stocks, such as the mid-'70s recession. For this reason, bonds can provide excellent diversification benefits.

Note: The data assumes reinvestment of income and does not account for taxes or transaction costs. Unlike Treasury bonds and bills, stocks and corporate bonds are not backed by the full faith and credit of the United States. Diversification does not eliminate the risk of experiencing investment losses. Underlying data is from the 1998 Stocks, Bonds, Bills, and Inflation (SBBI) Yearbook, by Roger G. Ibbotson and Rex Sinquefield. Updated Annually. Past performance is no guarantee of future results.

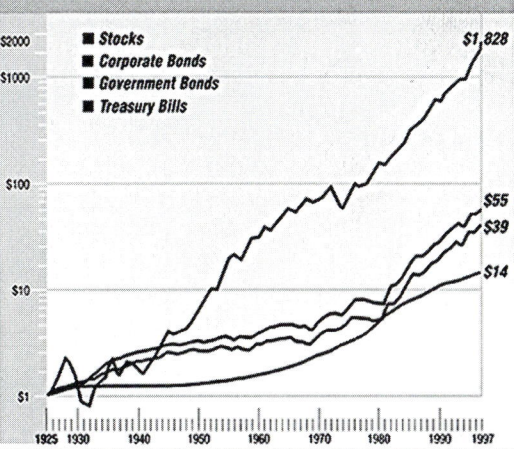

Source: *Stocks, Bonds, Bills and Inflation® 1999 Yearbook*, © 1999 Ibbotson Associates, Inc. Based on copyrighted works by Ibbotson and Sinquefield. All rights reserved. Used with permission.

interest payments for the 5 months during which she held the bond, the buyer must pay the seller for any interest that has accrued since the last interest payment. Determine how much Alice should receive from Benny.

Solution

The formula for determining how much the buyer pays, and the seller receives, is

buyer pays = market price + accrued interest + commission

seller receives = market price + accrued interest − commission

Determine the accrued interest amount. The interest for 5 months, or 150 days of 360 days in a year, is 150/360 of the total interest.

$$I = prt$$
$$= 1,000 \times .10 \times \frac{150}{360}$$
$$= \$41.67$$

Benny pays = market price + accrued interest + commission
$$= \$1,000 + \$41.67 + \$20$$
$$= \$1,061.67$$

Alice receives = market price + accrued interest − commission
$$= \$1,000 + \$41.67 − \$20$$
$$= \$1,021.67$$

Let's look at a similar transaction except in which the market price will not be exactly \$1,000 and more than one bond will be purchased.

Example 3

Mark Van Kuren owned five bonds with face values of \$1,000 each when he purchased them. He sold all five bonds on May 1 to Sharon Martin. The bond quotation for May 1 was 11½ quoted at 96, and the commission for each bond was \$8. Find (a) the amount that the buyer paid and (b) the amount the seller received.

Solution

January 1 to May 1 is 4 months, or 120 days/360.

$$I = prt$$
$$= 1,000 \times .115 \times \frac{120}{360}$$
$$= \$38.33$$

a. buyer pays = market price + accrued interest + commission

= $960 + $38.33 + $8

= $1,006.33 for one bond

buyer pays $5,031.65 total for five bonds (1,006.33 × 5)

b. seller receives = market price + accrued interest − commission

= $960 + $38.33 − $8

= $990.33 for one bond

Seller receives $4,951.65 total for five bonds (990.33 × 5)

CHECK YOUR KNOWLEDGE

Bond Quotations and Accrued Interest

Refer to the following bond quotation when answering questions 1 through 4.

Bond			Cur. yld.	Vol.	High	Low	Close	Net chg
Woody corp	8 ¼	01	7.8	30	105	103	105	+ ¼

1. What is the year of maturity?

2. What is the rate of return?

3. What is the stated rate of interest?

4. How much would you pay for this bond if you bought it at the closing price?

5. Today is August 1, and on this date, Shelley decides to sell her bond. The stated rate of interest is 9¼%, while the bond is quoted at 106. The commission is $20. How much interest has accrued since July 1?

6. If Edna buys the bond from Shelley under the conditions stated in problem 5, what is the total amount that Edna will pay?

7. What is the total amount that Shelley will receive?

Answers to CYK: *1.* 2001 *2.* 7.8% *3.* 8.25% *4.* $1,050.00 *5.* $7.71 *6.* $1,087.71 *7.* $1,047.71

13.2   EXERCISES

Refer to the following bond quotation when answering questions 1 through 20.

Bond			Vol.	Close	Net chg
Lavender C	9⅜	99	10	103	+6
Teal Corp	8¾	01	200	94	−¾
Lime C	9⅜	09	13	105	−⅜
Pink Pw	7⅛	98	20	97	+¼
Blue Ed	7¼	05	14	62½	−⅝
Green Pw	14	00	32	117½	−3½
Yellow Mills	5¾	11	12	85	+2⅛
Orange Ed	11	09	8	99	−1
Red Dyn	12⅛	15	24	101	+1⅞
Peach Cp	8¼	20	26	124½	−2¼

For each bond, list the year of maturity and calculate the annual interest income.

	Corporation	Year of maturity	Interest income
1.	Lavender C	_____	_____
2.	Teal Corp	_____	_____
3.	Lime C	_____	_____
4.	Pink Pw	_____	_____
5.	Blue Ed	_____	_____
6.	Green Pw	_____	_____
7.	Yellow Mills	_____	_____
8.	Orange Ed	_____	_____
9.	Red Dyn	_____	_____
10.	Peach Cp	_____	_____

For each bond, calculate the current yield. Remember that the current yield is the relationship between the interest income (already calculated in exercises 1–10) and the closing market price.

	Company	Interest income	Closing market price	Current yield
11.	Lavender C	_____	_____	_____
12.	Teal Corp.	_____	_____	_____
13.	Lime C	_____	_____	_____
14.	Pink Pw	_____	_____	_____
15.	Blue Ed	_____	_____	_____
16.	Green Pw	_____	_____	_____
17.	Yellow Mills	_____	_____	_____
18.	Orange Ed	_____	_____	_____
19.	Red Dyn	_____	_____	_____
20.	Peach Cp	_____	_____	_____

For exercises 21 through 30, calculate (a) the accrued interest, (b) the amount the buyer pays, and (c) the amount the seller receives. Assume that each bond is of $1,000 denomination, the commission is $20 per bond, and that interest is paid to the bondholder on January 1 and July 1.

21. One ABE 8 1/4, closing at 98, date of sale is February 1.

22. Two XYZ 9 5/8, closing at 100, date of sale is March 1.

23. Five BCD 6 1/4, closing at 102, date of sale is October 1.

24. Four LMN 7 3/8, closing at 92, date of sale is November 1.

25. Eight STU 11 1/2, closing at 101, date of sale is May 1.

26. Ten QRS 10, closing at 99, date of sale is April 1.

27. Twelve TUV 7 7/8, closing at 105, date of sale is February 1.

28. One CDE 8 3/8, closing at 88, date of sale is December 1.

29. One FGH 8 3/8, closing at 100, date of sale is December 1.

30. Five NOP 9 3/4, closing at 106, date of sale is March 1.

Solve the following word problems.

31. Sara Johnson owned ten bonds whose face value was $1,000 each when she purchased them. She sold all ten bonds on April 1 to Joyce Birch. The bond quotation for April 1 was 9 1/2 quoted at 98, and the comission for each bond is $10. Find (a) the amount that the buyer paid and (b) the amount the seller received.

32. Alan Swarde purchased 20 bonds with a face value of $1,000 each. When he purchased them, they were being sold at a discount quoted at 97. He sold all 20 bonds on June 1 to Carole Harvey at a premium of 103. The interest rate on the bonds was 10 3/8 and the broker's fee was $15 per bond. Find (a) the amount that the buyer paid and (b) the amount the seller received.

33. J. C. Roppel purchased 40 bonds with a face value of $1,000 each. When she purchased them, they were being sold at a premium quoted at 102. She sold all 40 bonds on March 1 to Sharon Dall at a premium of 105. The interest rate on the bonds was 9 5/8 and the broker's fee was $13 per bond. Find (a) the amount that the buyer paid and (b) the amount the seller received.

34. Harmon Ziminski purchased 100 bonds with a face value of $1,000 each. When he purchased them they were being sold at a premium quoted at 102. He sold all 100 bonds on March 1 to Steve Malii at a discount of 95. The interest rate on the bonds was 12 7/8 and the broker's fee was 4% of the face value per bond. Find (a) the amount that the buyer paid and (b) the amount the seller received. (c) If Steve had held the bonds for 4 years and received eight interest payments prior to this sale, how much interest income did Steve realize in this investment?

35. Ann Warwick purchased five bonds with a face value of $1,000 each. When she purchased them they were being sold at a premium quoted at 105. She then sold all of the bonds on Feb. 1 to Stan James at a premium of 107. The interest rate on the bonds was 14 7/8 and the broker's fees were 3% of the face value per bond. Find (a) the amount that the buyer paid and (b) the amount the seller received. (c) If Ann had held the bonds for 2 years and received four interest payments prior to this sale, how much interest income did she realize in this investment?

13.3 MUTUAL FUNDS

A **mutual fund** is an investment company that pools monies of individuals and organizations and invests in stocks, bonds, and short-term financial securities. Investors purchase fund shares and are considered partial owners of the fund portfolio. Mutual funds attract billions of investment dollars because they are designed to achieve a wide range of investment objectives. With more than 7,000 mutual funds available today, investors can select a high risk/reward potential, capital growth fund; a moderate risk/reward potential, growth and income fund; to a low risk/reward potential, capital preservation and income fund.

Mutual funds offer several investment advantages:

- **Financial management.** Each fund has numerous financial experts who select and monitor the investments in the fund portfolio.

- **Transaction costs.** The costs of buying and selling investments in the fund portfolio are generally lower because the funds are traded in large numbers and costs are shared by thousands of shareholders.

- **Diversification.** Mutual funds invest shareholder dollars in different types of investments depending on the funds classification. The selection of investments is designed to protect the shareholders against the potential poor performance of any one investment.

- **Liquidity.** Pursuant to fund regulations, the investors can sell their shares back to the fund (redeem) for cash.

Mutual Fund Quotations

Learning objective
Interpret mutual fund quotations.

Like stocks and bonds, mutual fund quotations are listed in the financial section of most daily newspapers. However, the presentation of information will vary, depending on the publication. Figure 13.4 is the presentation format listed in most financial publications like the *Wall Street Journal* and the *New York Times*. Figure 13.5, on the other hand, is a more complete presentation, as it lists more investment data pertaining to the fund's past and present performance. For this reason, the exercises in this chapter are based on this type of mutual fund quotation.

The figure contains a partial listing of a mutual fund, the variety of investments in the funds portfolio, and the relevant financial data necessary to monitor the funds activity. The first column identifies the name of the fund investment. The second column "Inv. Obj." is the investment objective as stated in the funds *prospectus* (capital growth, growth and income, etc.) The "NAV" heading lists the closing price quote per fund share on that date. The "offer price" column is the price that each fund share will cost that day when a load charge (sales commission) is charged. When a sales commission is charged for either buying or selling fund shares, the offer price (NAV plus commission) will be more than the net asset value. The amount of difference is the actual commission charge. The "NAV Chg." heading shows the increase or decrease in the previous day's NAV. The change is quoted in cents. The next three columns listed under "Total Return" heading identifies the percent return for varying time periods. YTD means January 1 to current date, 1 wk. indicates current past weeks, and 3 yrs. means over the past 36 months. The last column heading labeled with an "R" identifies the fund's performance compared to all funds categorized in the same investment objective. An A ranking indicates top 20%; B, next 20%; C, middle 20%; D, next

Figure 13.4

Mutual fund quotation: Daily financial publication

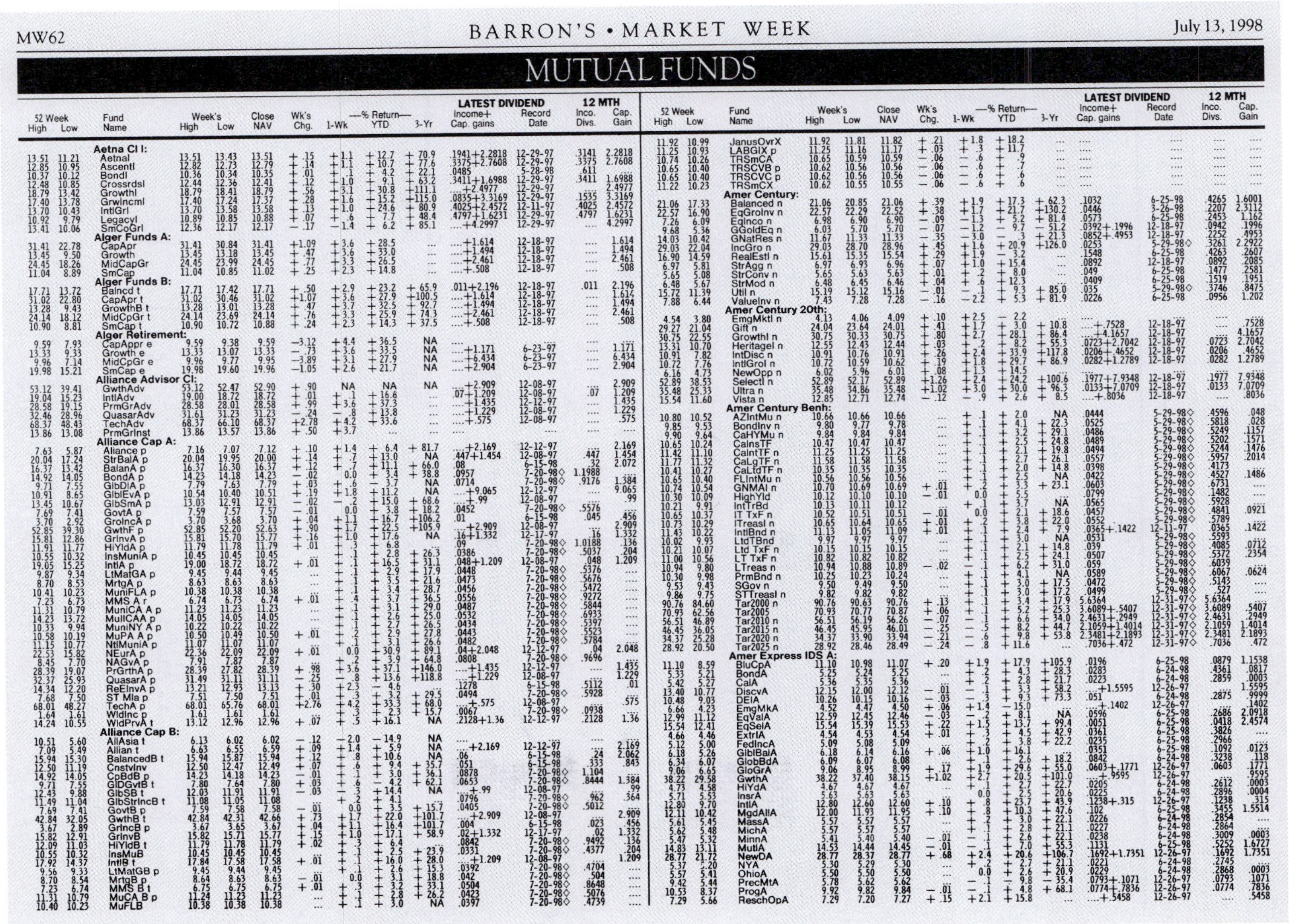

BARRON'S • MARKET WEEK

MUTUAL FUNDS

MW62   July 13, 1998

Source: Republished by permission of Dow Jones, Inc. via Copyright Clearance Center, Inc. © 1998, Dow Jones and Company, Inc. All Rights Reserved Worldwide.

Figure 13.5

Mutual fund quotation: General presentation

	Inv. obj.	NAV	Offer Price	NAV Chg.	YTD	Total return 1 wk.	3 yrs.	R
First Security Fund A								
Bal Inst	G&I	12.37	12.85	+0.08	+10.7	+0.6	+70.9	C
Mrtg.	INC	8.63	8.74	+0.03	+ 3.5	+0.1	+21.6	B
EqGr p	GRO	49.51	52.63	+0.05	+15.2	+1.3	+68.5	A
ATK p	BND	11.26	11.89	+0.02	+ 1.4	+0.2	+04.5	C
INT C p	EQI	27.59	27.59	+0.25	+30.5	+2.8	+58.3	B
EuGr p	I T L	52.35	56.02	+0.07	+ 6.8	+3.7	+84.1	A
IntBd p	INC	10.35	11.03	−0.04	+ 1.3	−0.5	+12.7	C
General Fund B								
HiYd p	S&B	11.78	12.37	+0.10	+ 8.4	+0.4	+32.6	B
Bal	G&I	15.63	15.96	−0.02	+ 0.1	−0.1	+48.5	B
Fed B p	MTG	9.81	10.05	+0.04	+ 3.6	+0.2	+16.9	C

20%; and E, bottom 20%. Some quotations will rank a fund numerically from 1 to 10. When numbers are used, a 1 ranking indicates top 10%; 2, next 10%, and so forth.

Example 4

The following information was published in the mutual funds quotations for the MFG fund.

	Inv. obj.	NAV	Offer price	NAV chg.	YTD	Total return 1 wk.	3 yrs.	R
MFG	gro	9.72	10.21	.05	+6.5	+0.2	+8.3	C

Determine (a) NAV, (b) offer price, (c) NAV Chg., (d) year-to-date change, (e) rating, and (f) the amount and rate of sales commission.

Solution

a. $9.72
b. $10.21
c. increase of $0.05
d. 6.5% increase
e. middle 20%
f. Sales commission amount $0.49 ($10.21 − $9.72 = $0.49)
    Sales commission rate 5% ($0.49 ÷ $9.72 = .05 or 5%)

Now that you understand how to read a mutual fund quotation, we can discuss the procedures involved in buying and selling mutual funds.

Trading Mutual Fund Shares

When you invest in a mutual fund the cost of your investment is determined by the funds **net asset value (NAV),** or the dollar value of one mutual fund share, the number of shares you purchase, and the type of fund you select. Investors in **no-load funds** are not charged sales commissions when they buy or sell fund shares. However, investors in **load funds** will be required to pay a sales commission ranging from 2% to 8%. The commission can be paid when the share is purchased, **(front-end load)** or when the shares are sold **(back-end load).** A load charge may not be the only expense an investor incurs when trading mutual fund shares. Many funds will charge fees sufficient to cover the costs of marketing and/or managing the fund.

Learning objective
Calculate the net asset value (NAV) of a mutual fund.

The net asset value of a mutual fund share is used by the investor to (1) determine the cost of purchasing shares in the mutual fund, (2) determine the price to sell back mutual fund shares, and (3) track the value of their investment in the mutual fund. For these reasons, investors should know how the NAV is calculated. At the end of each business day, fund personnel subtract the fund's current liabilities from the market value of the investments in the fund, and then divide this difference by the number of fund shares outstanding. This procedure is illustrated in the following equation:

$$NAV = \frac{\text{Market value of fund investments} - \text{Current liabilities}}{\text{Number of fund shares outstanding}}$$

Example 5

ITD Funds reports the current market value of its fund investments at $18,275,650 and current fund liabilities of $2,540,000, and it has 1,250,000 fund shares outstanding. Determine the fund's net asset value (NAV).

Solution

$$NAV = \frac{\text{Market value of fund investments} - \text{Current liabilities}}{\text{Number of fund shares outstanding}}$$

$$= \frac{\$18,275,650 - \$2,540,000}{1,250,000}$$

$$= \frac{\$15,735,650}{1,250,000}$$

$$= \$12.588 \quad \text{or} \quad \$12.59$$

Learning objective
Calculate the profit or loss on the trading of mutual fund shares.

We are now ready to analyze a transaction involving the purchase and sale of mutual fund shares. Example 6 will illustrate this process for a load fund. Remember a load fund has a sales commission assessed at the time the fund shares are either purchased or sold.

Example 6

Pat O'Brien purchased 150 shares of the Franklin Growth and Income Fund. The NAV of the fund at the time of purchase was $12.80 and the fund has a front-end load charge of 6%. Five years later Pat sold his 150 shares. If the NAV at the time of the sale was $25.50, what is the amount of Pat's profit or loss on the trade?

Solution

Step 1:   Determine the total cost of the fund shares purchased.

$$\begin{aligned}
\text{cost of} &= \text{number of shares purchased} \times \text{offer price} \\
\text{shares} &= 150 \times (\text{NAV} + \text{commission}) \\
&= 150 \times [\$12.80 + (\$12.80 \times .06)] \\
&= 150 \times [\$12.80 + \$0.77] \\
&= 150 \times \$13.57 \\
&= \$2,035.50
\end{aligned}$$

Step 2:   Determine the total selling price of the fund shares.

$$\begin{aligned}
\text{selling} &= \text{number of shares sold} \times \text{net asset value at sale} \\
\text{price} &= 150 \times \$25.50 \\
&= \$3,825
\end{aligned}$$

Step 3:   Determine the profit or loss on the trade of fund shares.

$$\begin{aligned}
\text{profit} &= \text{total selling price} - \text{total cost} \\
\text{or loss} &= \$3,825.00 - \$2,035.50 \\
&= \$1,789.50 \text{ profit}
\end{aligned}$$

Note: The sales commission must be added to the NAV to determine the offer price as the fund has a front-end load charge of 6%.

CHECK YOUR KNOWLEDGE

Mutual Funds

Use the information in Figure 13.5 for the mutual fund identified as EqGr to answer questions 1–5.

1.   What is the closing price quote per fund share?

2.   What is the price per fund share when a sales commission is charged?

3. What is the net asset value change expressed in dollars?

4. What is the year-to-date total return?

5. What is the fund rating?

6. Northstar Fund has current fund liabilities of $580,400 and the current market value of all fund investment is $7,450. If there are 1,200,000 fund shares outstanding, what is the net asset value of the fund?

7. Marcia Ortega purchased 200 shares of the Hamilton Fund with an offer price of $14.32 on the day of the trade. If the fund has an NAV of $13.75, what did Marcia pay for the fund shares?

8. The Strong Capital Growth fund has an NAV of $21.67. The fund lists a front-end load charge of $7\frac{1}{4}\%$. Determine (a) the fund's offer price and (b) the cost of 500 shares.

9. In today's mutual fund quotations, Highmark Income Fund has a NAV of $17.95 and an offer price of $18.76. What is the load charge percent? (Express answer to nearest tenth of a percent.)

10. Fadi Abhallah purchased 50 shares of the Fidelity Advisor Fund. The fund's NAV at the time of purchase was $45.49. Three years later Fadi sold all shares of the fund. The NAV and offer price at the time of the sale was $56.32 and $58.29, respectively. What is the amount of profit or loss on the transaction if the fund has a back-end load charge?

13.3 EXERCISES

Complete questions 1–5 by using the following mutual fund quotation.

	Inv. obj.	NAV	Offer price	NAV chg.	YTD	Total return 1 wk.	3 yrs.	R
DG Equity	GI	$23.77	$24.65	+.26	+7.3	+3.4	+10.2	B

1. What is the closing price quote per fund share?

2. What is the price per fund share with front-end load charge?

3. What is the NAV change expressed in dollars?

4. What is the annaul return as of quotation date?

5. What is the fund rating?

Determine the net asset value (NAV) for each of the following (round answers nearest cent):

Current market value of fund shares	Current fund liabilities	Fund shares outstanding	NAV ($)
6. $ 4,750,000	$ 365,000	$ 825,000	_____
7. 12,642,500	1,212,300	1,460,000	_____
8. 48,390,125	3,546,270	3,165,200	_____

Fill in the blanks with the correct amounts or percents (round amounts to nearest cent and percents to nearest tenth).

NAV	Load charge %	Load charge ($)	Offer price ($)
9. $ 6.25	5%	_____	_____
10. 13.44	_____	$0.51	_____
11. 17.92	_____		$18.30
12. _____	6 1/5%	$0.27	_____
13. _____	_____	$0.45	$62.80

14. Tracy Howe purchased 100 shares of the State Farm no-load Growth and Income Fund. The NAV of the fund was $41.90. If the offer price was $43.70, how much did Tracy pay for the shares?

15. On July 15, TDY Income Fund had investments with a current market value totaling $5,750,000. The fund's current liabilities on that date was $430,500 and the number of shares outstanding was 865,240. What was the fund's NAV on July 15?

16. Determine the offer price of a mutual fund with a net asset value of $38.74 and a back-end load charge of 4 3/4 %.

17. How much would you have to pay for 250 shares of a front-end load fund with a current NAV of $57.22? The fund has an "NAV change" quote of +.05 and an offer price of $58.94.

18. Giles Morgan bought 175 shares of the Fidelity Balance Fund. The NAV of the fund was $20.18 and the offer price was $21.39. Determine the fund's front-end load charge percent.

19. The Preferred Growth Fund has fund investments totaling $25,780,000, current liabilities of $2,350,000 and 4,850,000 fund shares outstanding. The NAV was listed in the mutual fund quotes that day at $4.38. Was the NAV reported in the mutual fund quotations correct?

20. Carol Nurse sold 500 shares of the Oakmark International Fund, which has a back-end load. The NAV on the day of the sale was $12.72. If the load charge was 5%, what amount did she receive from the sale of her shares?

21. An investor purchased 80 shares of a no-load mutual fund. The NAV when the shares were purchased was $15.08. Four years later the investor sold all shares in the fund. The NAV at the time of the sale was $14.75 and the offer price was $15.78. Determine the amount of profit or loss on the sale.

22. Gary DeBlasie bought 150 shares of American Century Income and Growth Fund with a net asset value of $27.17. The fund had a front-end load charge of 3% when the shares were purchased. What is the amount of profit or loss if Gary sold the shares recently with an NAV of $34.21 and an offer price of $35.24?

13.4 DISTRIBUTION OF INCOME AND EXPENSES

There are three basic types of ownership of a business: sole proprietorship, a partnership, and a corporation. The sole proprietor of a business receives all of the income from the business and also is responsible for all expenses. Because a partnership and a corporation have more than one owner, a procedure for distribution of income and expenses is necessary.

Learning objective
Distribute company
profits to partners
and stockholders.

Distribution of Profits to Partners of a Business

First, let's consider a business that is owned by four partners who have made equal investments in the company and, therefore, who own an equal part of the business. Whenever the business makes a profit, the partners must decide how to divide this income. There are two usual ways of doing this. The first is simply to divide the profit equally among all partners. Table 13.1 demonstrates the equal distribution of income assuming the company has a profit of $100,000. Since there are four partners, each receives one-fourth of the income.

An alternative method of dividing income is to divide it according to a predetermined ratio. For instance, suppose a group of partners did not make equal investments in a business and they decided to divide future profits by the percent of their individual investments in the business. Income or profit would then be multiplied by this percent to determine the amount of money each partner would receive.

E x a m p l e 7

Three years ago, the following partners invested money in a business: Abbey, $35,000; Burt, $15,000; Caroline, $40,000; and Darla, $60,000. They now have a total of $142,000 profit, which they want to divide according to the percent that each initially invested. What amount will each partner receive?

Table 13.1

Example of equal
distribution of
income in a
partnership

Partner	Income		Fraction		Portion of income
Abbey	$100,000	×	1/4	=	$25,000
Burt	100,000	×	1/4	=	25,000
Caroline	100,000	×	1/4	=	25,000
Darla	100,000	×	1/4	=	25,000
					$100,000

Solution

Partner	Initial investment
Abbey	$ 35,000
Burt	15,000
Caroline	40,000
Darla	60,000
total investment	$150,000

Partner	Initial investment / total investment	=	Rate each invested
Abbey	35,000/150,000		.233
Burt	15,000/150,000		.100
Caroline	40,000/150,000		.267
Darla	60,000/150,000		.400
			1.000

Partner	Rate invested	×	Income	=	Portion of income
Abbey	.233		142,000	=	$ 33,086
Burt	.100		142,000	=	14,200
Caroline	.267		142,000	=	37,914
Darla	.400		142,000	=	56,800
					$142,000

CALCULATOR SOLUTION

Keyed entry	Display
AC 35000 ÷ 150000 =	.2333
× 142000 =	33086
Abbey's share of income	
AC 15000 ÷ 150000 =	.10
× 142000 =	14200
Burt's share of income	

Use the same procedure to find other partners' shares.

　The partners of a company share in the assets of their business, and they share in any liabilities that the company may incur. They are personally responsible for the company. A corporation on the other hand is a legal entity that is set up in such a way that the stockholders own the business and are responsible for it rather than the people who work for the company and operate it.

Distribution of Profits to Stockholders of a Corporation

Earlier we mentioned two types of stock that could be purchased in a corporation, preferred stock and common stock. An investor who purchases **preferred stock** is entitled to some preferential treatment in terms of receiving dividends (a share of company profits) and of having an early claim on any assets the company may have if it is dissolved as a company. An investor who purchases **common stock** receives his/her share of the dividends and company assets after the preferred stockholder. Neither the preferred nor the common stockholder has a guarantee that the company will ever pay a dividend on the stock, but if it does, the preferred stockholders receive their share first. Both types of investors hope that their stock will increase in value before they sell it, and that the company will make a profit that will be shared with the investors in the form of dividends.

Investors who purchase preferred stock do not received a guarantee that their stock will appreciate in value, but they are guaranteed a yearly dividend at a fixed rate of interest based on the purchase value of the stock, if the company pays a dividend that particular year. *Cumulative preferred stockholders* also enjoy the privilege that while the company may defer payment of these dividends, the company is responsible for paying them, whether or not it makes a profit during a given year. Whenever the company does declare its intent to pay a dividend to its investors, cumulative preferred stockholders receive their share of the profits first and the remainder of the profits are then distributed to the common stockholders.

Example 8

Furbush Electric Components Company has 10,000 stockholders, of which 4,000 hold preferred stock and 6,000 hold common stock. The preferred stock was sold for $100 per share (called the *face value* or *par value* of the stock) with a guaranteed yearly dividend of 5%. Each share of common stock was sold for $50. The company has made a profit of $40,000. Determine the amount of dividend per share that will be paid for preferred stock (a) and for common stock (b), and the common dividend rate (c).

Solution

a. *Step 1:* Determine the amount to be paid per share for the preferred stock. Each share of preferred stock must receive 5% of the $100 cost of the stock.

dividend for the preferred stock = .05 × $100 = $5.00/share

Step 2: Determine the total to be paid to preferred stockholders.

total = dividend per share × number of shares

= $5.00 × 4,000 = $20,000

b. *Step 3:*   Determine the amount of dividends to be distributed to common stockholders.

$$\begin{array}{c}\text{amount to common} \\ \text{stockholders}\end{array} = \text{total dividends} - \begin{array}{c}\text{amount to preferred} \\ \text{stockholders}\end{array}$$

$$= \$40{,}000 - \$20{,}000$$

$$= \$20{,}000$$

Step 4:   Determine the amount of dividends to be distributed to each share of common stock.

$$\frac{\text{dividend/share}}{\text{common stock}} = \frac{\text{Amount to be distributed to common stock}}{\text{Number of common shares outstanding}}$$

$$= \$20{,}000 \div 6{,}000$$

$$= \$3.33/\text{share}$$

c. *Step 5:*   Determine the percentage dividend to each common stockholder.

$$\text{percentage dividend} = \frac{\text{dividend per share}}{\text{cost per share}} \times 100$$

$$= \frac{\$3.33}{\$50} \times 100$$

$$= 6.66\%$$

In Example 8, the preferred stockholders received a larger dividend in dollars per share of stock they owned than the common stockholders; however, the common stockholders received a greater percentage return on their investment.

Finally, let's consider a situation in which the company does not declare a dividend during a given year, and defers payment of its cumulative preferred stock dividend as well.

Example 9

Furbush Electric had a bad year during 1999 and chose not to pay any dividends to its investors. During 2000, the company declared a profit of $100,000 to be distributed to its stockholders. The cumulative preferred stock was purchased at $100 per share and pays 5% per share. There are 6,000 shares of common stock and 4,000 shares of cumulative preferred stock that will receive dividends in 2000.

Solution

As determined in Example 8, the company owes $20,000 in dividends to its cumulative preferred stockholders each year.

amount due cumulative preferred stockholders, 1999	$20,000
amount due cumulative preferred stockholders, 2000	$20,000
total to be paid to cumulative preferred stockholders	$40,000

amount to be distributed to common
   stockholders = $100,000 − $40,000 = $60,000
amount per share of common stock = $60,000 ÷ 6,000 = $10/share

In these past examples, we assumed that the preferred stockholders were entitled only to their 5% of the dividends, which classifies them as nonparticipating preferred stockholders. Participating preferred stockholders also are entitled to a share of the dividends left over after they have received their normal distribution. Consequently, they receive not only their guaranteed dividend amount, but also an additional share of the remaining dividends.

Distribution of Expenses

Learning objective
Distribute the expenses of a company on the basis of the number of employees, the square footage, and the amount of sales.

Just as it is important to understand how to divide income, it is equally important to be able to divide costs and expenses. For instance, if your business has two departments, the accounting department and the sales department, how will you divide overhead expenses (such as gas and electric, rent, or garbage pickup)? There are three common ways to divide expenses. One is according to the number of employees in each department, another is according to the square footage of each department, and a third is determined by the proportion of sales made by each department. First, let's consider the distribution of expenses by the proportion of employees working in each department.

Example 10

The accounting department of a small department store has 3 employees and the sales department has 12 employees. Divide a $500 utility bill according to the percentage of employees in each department.

Solution

Department	Number of employees	÷	Total employees	=	Rate
Accounting	3		15		20%
Sales	$\frac{12}{15}$		15		$\frac{80\%}{100\%}$

Department	Rate	×	Expense	=	Department expense
Accounting	.20		$500		$100
Sales	.80		$500		$400
					$500

Next, consider the distribution of these same expenses by the number of square feet of space each occupies within the company.

Example 11

The accounting department, with all of its computer equipment, uses 1,680 square feet of office space, whereas the sales department uses only 720 square feet. Distribute a $500 utility bill according to the percent of square feet used by each department.

Solution

Department	Square feet	÷	Total square feet	=	Rate
Accounting	1,680		2,400		.70
Sales	720		2,400		.30
	2,400				1.00

Department	Rate	×	Expense	=	Department expense
Accounting	.70		$500		$350
Sales	.30		$500		$150
					$500

A third method of distribution of expenses could be used in a business that is primarily involved in sales of different products, such as a department store.

Example 12

Harmon's Department Store reported department sales for October as follows: Housewares, $15,000; Women's Apparel, $26,000; Men's Clothing, $18,000; Children's Clothing, $22,000; Jewelry, $11,000; and Footware, $8,000. The store's overhead expenses for the month totaled $80,000. Distribute the expenses on the bases of sales per department.

Solution

Step 1:   Determine total sales for the month.

Total sales = $15,000 + $26,000 + $18,000 + $22,000

+ $11,000 + $8,000

= $100,000

Step 2:   Determine each department's rate of total sales.

$$\text{rate of total sales} = \frac{\text{sales amount for department}}{\text{total sales}}$$

Department	Sales amount	Total sales	Rate
Housewares	$15,000	$100,000	.15
Women's Apparel	$26,000	$100,000	.26
Men's Clothing	$18,000	$100,000	.18
Children's	$22,000	$100,000	.22
Jewelry	$11,000	$100,000	.11
Footware	$ 8,000	$100,000	.08

Step 3:   Determine the amount of the expenses to be distributed to each department.

amount = rate × total expenses

Department	Rate	× Total expenses	= Department share
Housewares	.15	$80,000	$12,000
Women's Apparel	.26	$80,000	$20,800
Men's Clothing	.18	$80,000	$14,400
Children's	.22	$80,000	$17,600
Jewelry	.11	$80,000	$ 8,800
Footware	.08	$80,000	$ 6,400
			$80,000

These are three basic ways that expenses could be distributed, but obviously they cannot be applied to every business. Variations on these basic concepts are often used to find the most appropriate way to distribute expenses fairly within any particular business operation.

CHECK YOUR KNOWLEDGE

Distributing Income and Expense

1.   Three people start a business together. Divide a $90,000 income equally among all partners.

2. Three partners invest the following amounts of money when they initially began their business: Robinson, $80,000; Cripe, $110,000; and Short, $130,000. Distribute a profit of $60,000 among them according to the percent that each initially invested.

3. Thompson Office Supplies has 7,000 shares of stock. 3,000 are preferred stock and 4,000 are common stock. The preferred stock was sold for $100.00 per share with a guaranteed yearly dividend of 6%. If the company declared $60,000 for the year's profits for dividends, determine the amount of dividend per share that will be paid for preferred stock and for common stock.

4. A company has 35 employees located on the first floor and 50 employees on the second. Distribute a $27,000 installation charge between the two departments according to the number of employees in each department.

13.4 EXERCISES

Solve the following word problems. Round answers to the nearest cent and interest rates to the nearest tenth of a percent.

1. Three years ago, the following partners invested money in a business as follows: Abner, $60,000; Twila, $75,000; and Kari, $40,000. They now have a total of $600,000 profit that they will divide according to the percent that each initially invested. What amount of profit does each partner receive?

Partner	Amount invested	Rate of total	Share of profit
Abner	_____	_____	_____
Twila	_____	_____	_____
Karl	_____	_____	_____

2. Four partners invest the following amounts of money when they start their business: Larry, $200,000; Bobby, $8,000; Johnny, $50,000; and Colleen, $150,000. Distribute a profit of $500,000 among them according to the percent that each initially invested.

Partner	Amount invested	Rate of total	Share of profit
Larry	_____	_____	_____
Bobby	_____	_____	_____
Johnny	_____	_____	_____
Colleen	_____	_____	_____

Answers to CYK: *1.* $30,000 each *2.* Robinson's share = $15,000; Cripe's share = $20,625.00; Short's share = $24,375 *3.* preferred stock = $6.00 per share; common stock = $10.50 *4.* first floor = $11,117.65; second floor = $15,882.35

3. Distribute a $15,000 dividend among 300 preferred stockholders and 700 common stockholders if the preferred stockholders are guaranteed 6% on each share of stock whose per value is $200.

Amount of dividend for each share of preferred stock   _____

total amount to be paid preferred stockholders   _____

total to be distributed to common stockholders

($15,000 − _____)   _____

amount of dividend per share of common stock   _____

4. Distribute a $215,000 dividend among 5,000 preferred stockholders and 15,000 common stockholders if the preferred stockholders are guaranteed 7% on each share of stock whose per value is $300.

Amount of dividend for each share of preferred stock   _____

total amount to be paid preferred stockholders   _____

total to be distributed to common stockholders

($215,000 − _____)   _____

amount of dividend per share of common stock   _____

5. Distribute an expense of $2,500 to four departments by percent of employees if each department has the following number of employees:

Department	Number of employees	Rate of total	Amount of expense share
A	5	_____	_____
B	2	_____	_____
C	3	_____	_____
D	10	_____	_____

20

6. Distribute an expense of $4,000 to five departments by percent of employees if each department has the following number of employees:

Department	Number of employees	Rate of total	Amount of expense share
E	4	_____	_____
F	3	_____	_____
G	2	_____	_____
H	6	_____	_____
I	5	_____	_____

7. Distribute an expense of $8,000 to four departments by amount of floor space each department occupies:

Department	Floor space	Rate of total	Amount of expense share
J	300	_____	_____
K	200	_____	_____
L	100	_____	_____
M	400	_____	_____

8. Distribute an expense of $7,000 to five departments by amount of floor space each department occupies:

Department	Floor space	Rate of total	Amount of expense share
N	700	_____	_____
O	550	_____	_____
P	320	_____	_____
Q	610	_____	_____
R	120	_____	_____

9. Distribute an expense of $10,000 to four departments by amount of sales that each department made for a set period of time.

Department	Sales	Rate of total	Amount of expense share
S	$50,000	_____	_____
T	$25,000	_____	_____
U	$15,000	_____	_____
V	$20,000	_____	_____

10. Distribute an expense of $3,500 to five departments by amount of sales for each department.

Department	Sales	Rate of total	Amount of expense share
W	$2,300	_____	_____
X	$4,300	_____	_____
Y	$6,500	_____	_____
Z	$1,100	_____	_____
A	$ 900	_____	_____

11. Five sorority sisters start a dating service. Each invested the following amounts: Felicia, $2,000; Muriel, $3,000; Aubrey, $4,400; Ingrid, $3,078; and Guinevere, $2,050. Divide their $500,000 profits from their first-quarter earnings according to the percent that each initially invested.

12. The O'Malley Brothers opened a tune-up garage. Each brother contributed the following amount: Hank, $15,000; Henry, $1,000; Harold, $2,000; Harvey, $2,300; and Homer, $1,700. Distribute their $50,000 loss according to the percent that each invested.

13. Three business math instructors quit their jobs in order to start a dairy farm. Each invested the following amounts: Turner, $8,000; Babcock, $5,000; Connor, $6,000. Distribute the $22,000 loss among the three ex-instructors according to the percent that each initially invested.

14. Levinson and Esposito Company has 10,000 stockholders of which 2,000 hold preferred stock and 6,000 hold common stock. The preferred stock was sold for $100.00 per share with a guaranteed yearly dividend of 7%. Each share of common stock was sold for $50.00. The com-

pany has made a profit of $40,000. Determine the amount of dividend per share that will be paid for preferred stock and for common stock.

15. Delorme and Schurman Real Estate Corporation has 1,000 preferred stockholders and 8,000 common stockholders. The preferred stock was sold for $150 per share with a guaranteed yearly dividend of 5.5%. Each share of common stock was sold for $75. The corporation has made a profit of $80,000. Determine the amount of dividend per share that will be paid for preferred stock and for common stock.

16. The Popsicle Recording Company has three departments. Divide the $980 utility bill among the departments according to the percentage of employees in each department. The legal department has 43 employees, the accounting department has 5 employees, and the music department has 30 employees.

17. The Bee-Bop Art Supply Store wishes to divide a $500 bonus between its two departments according to the percentage of employees in each department. The paint department has two employees while the brush department has one.

18. Wings is a pet store in Gotham City. Management wishes to divide a $4,100 maintenance bill among the three departments on the basis of the number of birds (or flying mammals) in each department. There are 23 parrots, 41 canaries, and 198 bats.

19. The largest muscial instrument store in Grant City needs to divide a $1,001 utility bill among its departments according to the percent of employees in each department. The string instrument department has 404 employees, the wind instrument department has 606 employees, the percussion instrument department has 808 employees, and the brass instrument department has 909 employees.

20. The largest musical instrument store in Grant City has received another bill—this one is for cleaning services that were rendered last month. This time the store wishes to divide the $1,150 bill among its departments according to the square footage in each department. The strings have 1,700 square feet, the winds have 800 square feet, the percussions have 990 square feet, and the brass have 1,000 square feet.

21. Cunningham's plumbing store wishes to divide a $67 bill between its two departments according to the square footage in each department. The chrome department has 2,300 square feet, while the copper department has 300 square feet.

22. The Mahatma Jewelry Store needs to divide a $7,000 bill between its two departments based on the square footage used by each. The yellow gold department has 30 square feet, and the white gold department has 20 square feet.

23. Addis's Department Store reported department sales for October as follows: Housewares, $5,000; Women's Apparel, $15,000; Men's Clothing, $8,000; Children's Clothing, $12,000; Jewelry, $1,000; and Footware, $700. The store's overhead expenses for the month totaled $21,000. Distribute the expenses on the basis of sales per department.

24. Bee Town Mini Mall has four shops within it that have agreed to share any real estate tax bills the building may have each year. Distribute this year's tax bill of $16,000 by sales if each shop had the following sales: Jan's Maternity Shoppe, $120,000; Fredd's Hobby and Crafts, $80,000; Sally's Cafe, $75,000; and Marti's Music Box Shoppe, $90,000.

25. The Germantown Oktoberfest Committee rents large tents to house several nonprofit community groups that sell food and other items, as well as provide game concessions for the weekend festivities. The committee distributes its costs for rental, electric, plumbing, etc. by the amount of sales reported by the groups. Distribute an expense of $2,200 to the following groups: Grace Church, $3,500; St. Luke's Church, $1,400; Optimist Club, $890; Rotary, $1,300; Boy Scout Troop #119, $1,700; Pop Warner Football Club, $1,200; Lacross Support Group, $2,300; Project Children, $450; Marching Band Support Group, $1,700; and Theater Guild, $1,300.

CHAPTER REVIEW EXERCISES

The following problems review all mathematics and business concepts presented in the chapter. If you experience any difficulty solving a problem, go to the appropriate chapter section and review the examples provided. Express all answers as decimals rounded to tenths, or dollars rounded to cents, unless otherwise instructed.

1. Northern Paperboard Inc. allocated $250,000 of its current quarter's profit to common stock dividends. If the firm has 50,000 common stock shares outstanding, how much would a shareholder receive who owns 400 shares at the time of payment?

2. If you bought 500 shares at 67⅝ and the broker received a 2½% commission for the brokerage service, what was your total cost?

3. Assume you sold the 500 shares in problem 2 at 72¼. What would have been the total amount you received from the sale if the broker's commission was 3%?

4. Cooper Industries stock quotation for a particular day lists the approximate annual dividend paid per share to its shareholders at 1.32. If the closing price on the same day was 41½, what would be the amount listed in the "Yld." column of the quotation?

5. Chu-Ha Lee bought 50 shares of Rojon Enterprises common stock at 30⅜ per share and one year later sold them at 42½. If the broker charges a 2% commission on the market price, what is the capital gain on the transaction.

6. A bond quotation lists a $1,000 bond's closing price at 98 and its interest rate at 4¾. Determine the amount that will be listed in the "current yield" column of the bond quotation.

7. You own five $1,000 bonds with an interest rate of 8¼%. How much interest would you receive on July 1 from the issuing company, if you bought the bonds on March 1?

8. Michael Ganley owns fifteen 6½% bonds with a face value of $1,000 when purchased. On October 1, he sold all fifteen bonds to Joyce Barber at that date's bond quote of 107. If the comission for each bond is $5, find the amount she paid for the bonds.

9. Using the information provided in problem 5, find (a) the amount Michael Ganley received from the sale of the bonds, and (b) the amount of interest he received if he held the bonds for 5 years and received 10 interest payments before the sale.

10. The current market value of a mutual fund investments is $34,800,000. The fund's current liabilities are $2,950,000, and there are 4,280,000 fund shares outstanding. Determine the fund's net asset value based on these figures.

11. Harry Morgan bought 100 shares of a mutual fund with a NAV of $26.42. If the fund has a front-end load charge of 4%, find (a) the offer price, and (b) the amount Harry invested in the fund.

12. The mutual fund quotation lists Kent International Growth Fund's NAV at $16.12 and its offer price at $16.52. Determine the load charge (sales commission) percent.

13. Alex Rubenstein bought 50 shares of the Putnam Voyager Fund with an NAV of $21.03. The fund has a 3% front-end load charge. Two years later Alex sold all his fund shares at a net asset value price of $24.10. What was the amount of his profit or loss?

14. Case Industrial Tractor Inc. has 40,000 shares of stock outstanding, of which 10,000 shares are $50 per share, 6% preferred, and the remaining 30,000 shares are $20 per share common stock. The company declared $120,000 of its profits for dividends. Determine the amount of the dividend that will be paid per share for (a) preferred stock, and (b) common stock.

15. Fine Furniture Company was unable to declare a dividend for its stockholders last year. However, this year the company declared a profit of $95,000 to be paid to its stockholders. Fine Furniture has 3,000 shares of cumulative preferred $100 par value stock paying a yearly dividend of 5%, and 10,000 shares of common stock outstanding at the time the dividend was declared. Determine the amount of dividend per share that will be paid this year for (a) preferred stock and (b) common stock.

16. The law firm of Germain, Ballard, and Abbott started their practice 5 years ago and distributes profits based on the percent of original investment. If Germain invested $200,000, Ballard invested $160,000, and Abbott invested $140,000, how much of the firm's yearly profit of $80,000 will each partner receive?

17. Han's Department Store needs to distribute a $555 lighting bill between its departments according to the percent of employees in each department. The furniture department has 5 people, and the housewares department has 6 people.

18. The Goodfellow car dealership needs to distribute a $3,000 water bill among its departments according to the square footage used by each. The new car department had 4,000 square feet, the used car department has 2,800 square feet, and the service department has 6,200 square feet.

19. If the Goodfellow car dealership decided to distribute the $3,000 water bill among its departments according to the amount of sales reported by each, how much should each department pay of the water bill? The new car department sales were $400,000; used car department sales were $28,000 and the service department's sales were $16,200.

20. Tom Green rents a vacant storefront each weekend of the summer and sets up a flea market with five booths, each of equal size. His rental cost of the storefront is $700. The sales of each booth this last weekend were: booth 1, $520; booth 2, $670; booth 3, $750; booth 4, $330; and booth 5, $900. Distribute the rental fee by (a) the amount of space used by each booth, and (b) the amount of sales of each booth.

EXPRESS YOUR UNDERSTANDING

Compose one or two well-written sentences to express the requested information in your own words.

1. Why might an investor select an investment in bonds versus an investment in stock?

2. Explain why a business would issue stock instead of bonds as a means of increasing the availability of money for operations.

3. In a stock quotation, what does the "Yld" column represent? How is the figure presented in this column determined?

4. Describe the procedure you would use to determine the capital gains or capital loss on a sale of stock.

5. In a bond quotation, what does the column "Cur. yld" represent? How is the amount shown determined for a bond with a face value of $1,000?

6. Explain how accrued interest on a $1,000-bond is determined. Create an example that illustrates the procedure.

7. What does the phrase "8½ quoted at 98" mean when applied to a bond with a $10,000 face value?

8. What is the key distinction between a load and a no-load mutual fund.

9. Explain what a mutual fund's net asset value (NAV) is and describe how it is determined.

10. Describe the steps necessary to distribute $80,000 in profits among three partners if profits are divided according to each partner's investment and the investments were $20,000, $50,000, and $30,000, respectively.

11. Explain how to calculate the amount of dividend to be paid to a common stockholder who owns stock in a company that does not have any preferred stock outstanding.

12. Describe the procedure necessary to determine the total dividend to be paid to preferred stockholders.

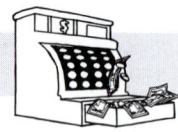

Mind Your Business

You and your partners have decided that each month you will look at each department in Media World and distribute the general operating expenses to the departments. You have all agreed to distribute those expenses most months by simply basing the distribution on floor space occupied by each. However, you have all agreed that during the months of March, June, September, and December, you will also distribute the expenses by sales and compare the two methods. The partners distribute income or losses of the business as defined by the terms of the partnership agreement.

Activity 1

Distribute the total operating expenses for the month of September to the three departments in the store (music, books, and video), basing the distribution on square footage of the store occupied by each department and proportion of net sales per department if the net sales were: music, $36,911.55; books, $14,196.75; and videos, $5,768.70.

Activity 2

Write out a comparison of the two methods of distribution of expenses in Activity 1 and make a recommendation of which method seems more appropriate to you, and why.

Activity 3

Based on the terms of the partnership agreement ("Mind Your Business," Chapter 4), determine the amount of September's net income that was distributed to each partner's capital balance, as reported in September's balance sheet.

SELF-TEST

A. Terminology

Complete the following items using the key terms presented at the beginning of the chapter. Check your responses against the answer key at the end of the test.

1. A _____ is a share or portion of a corporation.

2. _____ are a portion of the issuing company's profits, usually paid to the stockholder in the form of cash.

3. The life of a bond begins the day the bond is purchased and ends on the _____.

4. A _____ is the profit from the sale of assets such as securities or real estate.

5. A _____ is an interest-bearing certificate issued by a government or corporation, redeemable on a specified date.

6. A person who buys and sells securities is known as a _____.

7. Stocks and other securities are traded at a place called the stock exchange or sometimes called the _____.

8. The denomination of a bond is referred to as the _____ of a bond.

9. When the market price of a bond is lower than its face value, it is called a _____.

10. When the market price of a bond is higher than its face value, it is called a _____.

11. The value of one mutual fund share is referred to as the share's _____ _____.

12. A _____ is a type of mutual fund that requires the investor to pay a commission at the time the fund shares are either purchased or sold.

B. Calculation

The following concepts and short problems are designed to test your understanding of the objectives identified at the beginning of the chapter. Answers are provided at the end of the test.

13. If FGH Corporation wishes to distribute $400,000 in dividends among its 580,000 shareholders, how much will you receive in dividends if you own 200 shares of FGH?

Refer to the following stock quotation for problems 14–18.

52 Weeks		Stock	Div.	Yld %	P–E Ratio	Sales 100s	High	Low	Close	Net Change
High	Low									
61 ¼	49 ¼	ZERG	6.01	7.2	10	224	56	54	54 ½	+ ½

14. What was the high trading price for the day?
15. What was the high trading price for the year?
16. How many shares of stock were traded on this day?
17. What was yesterday's closing price?

18. Ahmad Dieg bought 200 shares of ZERG stock at yesterday's price and must sell some of it today. He has decided to sell 100 shares. The broker's fees were 3% in both transactions. What is Ahmad's capital gain or loss?

Refer to the following bond quotation for problems 19–21.

Cur. bonds	Net Yld.	Vol.	High	Low	Close	Change
Wills Cp	8⅜	910	86	85	85	− ⅝

19. How much would you pay for this bond if you bought it at the closing price?

20. How much interest would you receive from the issuing company if you held this bond for one year?

21. Calculate the current yield. Be sure to show your work.

22. Columbia Growth Fund has investments with a current total market value of $87,650,390, and the current liabilities of the fund are $6,485,370. If there are 5,362,150 fund shares outstanding, what is the net asset value (NAV) of the fund at this time?

23. Lisa Tassone bought 200 shares of Founders Balance Fund on January 4. The NAV on that date was $9.10. She sold all fund shares on November 25 of the same year. If the fund is a no-load fund and the NAV at the time she sold the shares was $12.06, what is the amount of her profit or loss?

24. Kevin Harvey owned five bonds whose face value was $1,000 each when he purchased them. He sold all five bonds on April 1 to Janet Green. The bond quotation for April 1 was 7½ quoted at 99, and the commission for each bond

is $10. Find (a) the amount that the buyer paid and (b) the amount the seller received.

25. Jackson's Floor and Tile Company has three departments. Divide an $875 utility bill among the departments according to the percent of employees in each department. The sales department has 13 employees, the billing and accounting department has 5 employees, and the manufacturing and shipping department has 30 employees.

26. The Bonwit Clothing Store wants to divide a $9,500 bill among its three departments based on the square footage used by each. The coats and outer apparel department has 230 square feet, the dress department occupies 520 square feet, and the intimate apparel department uses 250 square feet. How much should each department be charged?

27. The Thousand Islands Mini Mart has three shops within it. Distribute a one-month $2,500 utility bill by sales if each shop recorded the following sales: Big Mac's Gun Shop $12,000; Carm's Breakfast Nook $17,000; and Mario's Shoe Repair Shop $9,000.

Answers to Self-Test: **1.** stock **2.** Dividends **3.** maturity date **4.** capital gain **5.** bond **6.** stockbroker **7.** market **8.** face value **9.** discount **10.** premium **11.** net asset value **12.** load **13.** $137.93 **14.** 56 **15.** 61¼ **16.** 22,400 **17.** 54 **18.** −$275.50 **19.** $850 **20.** $83.75 **21.** 9.85% **22.** $15.14 **23.** $592 profit **24. a.** $5,093.75; **b.** $4,993.25 **25.** sales, $236.98; billing, $91.15; manufacturing, $546.88 **26.** apparel, $2,185; dress, $4,940; intimate, $2,375 **27.** gun shop, $789.47; breakfast nook, $1,118.42; shoe repair, $592.11

14

DEPRECIATION

Key Terms

depreciation
depreciation amount (expense)
accumulated depreciation
book value
useful life
tangible property
real property
straight-line method
declining-balance method
units-of-production method
sum-of-the-years'-digits method
depreciation schedule
salvage value
accumulated depreciation
accelerated method of depreciation
unadjusted basis
class life
recovery period
modified accelerated cost-recovery system (MACRS)

Learning Objectives

After completing the exercises in this chapter, you will be able to:

1. Explain why businesses depreciate assets.

2. Understand the conceptual difference between depreciation and cost recovery.

3. Prepare a depreciation schedule using the straight-line method.

4. Use the declining-balance method to develop a depreciation schedule.

5. Calculate depreciation using the sum-of-the-years'-digits method.

6. Calculate depreciation using the units-of-production method.

7. Recover an asset's cost using the modified accelerated cost recovery system (MACRS) guidelines.

8. Define key terms.

INTRODUCTION

Businesses purchase assets such as production machinery, vehicles, buildings, land, and equipment for use in their daily operations to produce goods and provide services. As we learned in Chapter 4, revenue generated from the sale of these goods and services is reduced by the cost of the goods sold and by the operating expenses reported in the period to determine the net income from operations. Because business assets are used to produce revenue, their cost is periodically deducted from revenue as an expense. **Depreciation** is the process of allocating the cost of an asset over the estimated useful life of the asset. The amount of the periodic deduction is reported as **depreciation expense** in the income statement and as an increase in the **accumulated depreciation** of the asset on the balance sheet. The **book value** of an asset is the net value reported on the balance sheet after deducting the accumulated depreciation from the total cost of the asset recorded at the time the asset was placed in service.

Learning objective
Explain why businesses depreciate assets.

Assets are depreciated to reflect the usage of the asset over time, and to match the cost of services provided by assets with related revenues in determining net income. Therefore, we must be able to predict the useful life of the asset. An asset's **useful life** is the estimated period of time the asset is used by a business. An automobile can be depreciated because its useful life can be determined by considering such factors as its mileage, repair costs, and model year. Land, on the other hand, is not depreciated because its useful life is indefinite. Property can be depreciated if it meets the following requirements:

1. It is used in a business or held for the production of income.

2. It has a determinable life and that life is longer than one year.

3. It is something that wears out, decays, is used up, becomes obsolete, or loses value from natural causes.

Depreciable assets are defined by categories, which serve in part to identify the method of depreciation used to determine the periodic depreciation expense. **Tangible property,** for example, is property that can be seen or touched, such as machinery, equipment, and vehicles. *Intangible assets* include copyrights, franchise fees, designs, patents, and customer lists. *Personal property* is property, such as equipment and machinery, that is not real estate. **Real property** is land and generally anything that is erected on, growing on, or attached to land.

Learning objective
Understand the conceptual difference between depreciation and cost recovery.

Depreciation as reported on a company's income statement (books) may not agree with the amount of depreciation claimed on the company's tax return. This discrepancy results from the use of traditional depreciation methods (straight-line, declining-balance, sum-of-the-year's-digits, and units-of-production) for book purposes, and IRS prescribed methods for tax purposes (Modified Accelerated Cost Recovery System, MACRS and Accelerated Cost Recovery System, ACRS).

The Internal Revenue Service (IRS) allows a number of different methods to determine depreciation. Businesses must select the appropriate method and apply the method consistently as required. In general, a business must use the depreciation guidelines reported in Publications 534 and 946, which are published by the IRS. Two tax methods, the Accelerated Cost Recovery System (ACRS) and the Modified Accelerated Cost Recovery System (MACRS), are required by the IRS. The ACRS method was enacted as part of the Economic Recovery Act of 1981 and allows businesses to recover the cost of most assets placed in service from 1981 through 1986. In 1986, the Tax Reform Act produced the MACRS method, which is used to recover the cost of most property placed in service after 1986.

In this chapter, we will examine the concepts and procedures of the following methods used to calculate depreciation and cost recovery.

1. *Traditional methods*   These methods are used by the accounting profession to compute depreciation on assets in accordance with *Generally Accepted Accounting Principles (GAAP)* guidelines. They include, straight-line, declining-balance, sum-of-the-years' digits, and units-of-production methods.

2. *Modified Accelerated Cost Recovery System (MACRS)*   This method is used to calculate cost recovery on all property placed in service after 1986 for federal tax purposes (see Section 14.5).

14.1   TRADITIONAL METHOD: STRAIGHT-LINE DEPRECIATION

Learning objective
Prepare a depreciation schedule using the straight-line method.

The traditional methods discussed in Sections 14.1, 14.2, 14.3, and 14.4 are presented because they are the accepted methods of calculating depreciation under GAAP guidelines. Businesses that either placed assets into service before 1981 or that are depreciating assets placed in service after 1981, which are not covered by the ACRS or MACRS must use these methods for tax purposes.

As you work through the examples presented, you will see that each method varies in focus and procedure. The **straight-line method,** for example, allows a business to evenly distribute the cost of an asset over its useful life. The **declining-balance method,** on the other hand, is considered an accelerated method of depreciating assets because the amount of depreciation will be greater in the earlier years and less in the latter years. The **sum-of-the-years'-digits method** is also an accelerated method, but is less accelerated than the declining-balance method. The **units-of-production method** is used to depreciate assets when the actual service rendered by the asset is a more appropriate base for depreciation than its years of service.

Now let's work through examples of each of these traditional methods so that we clearly understand how to calculate the periodic depreciation expense, and prepare a useful life-depreciation schedule. The *straight-line method* allows a business to depreciate an average amount of an asset's depreciable amount during each year of the asset's useful life. Before we analyze an example, let's familiarize ourselves with a few terms and formula symbols that apply to most of the depreciation methods discussed in this chapter.

- *Total cost*   The purchase cost plus freight and/or installation charges required to place the asset into service.
- *Salvage value*   The estimated value of an asset at the end of its useful life; sometimes referred to as residual value or scrap value.
- *Useful life*   The number of years the asset is expected to be in service. The useful life may be based on industry standards or on the experience of an individual company for book purposes. The useful life is mandated by the IRS for tax purposes.
- *Depreciable amount*   The amount of the asset to be depreciated. (Total cost − slavage value = depreciable amount.)
- *Accumulated depreciation*   The depreciable amount of the asset that has been depreciated prior to the current depreciation period.
- *Book value*   The net value of the asset after deducting the accumulated depreciation from the total cost of the asset.

The formula symbols and the variables that they represent are listed in Table 14.1. You should use the formulas to calculate the depreciation expense and book value for each of the traditional methods as applicable.

To calculate the straight-line periodic depreciation expense, we divide the depreciable amount by the useful life of the asset. This process is expressed in the following formulas:

$$\text{periodic depreciation expense} = \frac{\text{total cost} - \text{salvage value}}{\text{useful life (years)}}$$

Table 14.1

Depreciation formula symbols

Symbol	Variable representation
c	Asset's original total cost
s	Asset's salvage value
n	Estimated useful life
e	Periodic depreciation expense
b	Current year's book value
d	Accumulated depreciation amount
r	Depreciation rate (percent, fraction, or unit)
f	Depreciation factor

or

$$e = \frac{c - s}{n}$$

Example 1

Let us assume that Wilfred's Window and Glass Company has just purchased an asset for use in its business. The cost of the asset is $250,000; it cost $4,000 to have it installed; the estimated salvage value is $25,875; and the estimated useful life is 4 years. Find (a) the annual depreciation expense and (b) the book value at the end of the first year if the asset was purchased on January 5.

Solution

a.   Calculate the first year depreciation expense.

$$\text{Periodic depreciation expense} = \frac{\text{total cost} - \text{salvage value}}{\text{useful life}}$$

$$e = \frac{c - s}{n}$$

$$= \frac{\$254,000 - \$25,875}{4}$$

$$= \frac{\$228,125}{4}$$

$$= \$57,031.25$$

b.   The book value at the end of the first year is:

book value = total cost − accumulated depreciation

or

$$b = c - d$$
$$= \$254,000 - \$57,031.25$$
$$= \$196,968.75$$

Since the straight-line method of depreciation allocates an equal amount of depreciation expense to each period, the straight-line rate of depreciation can be determined as follows:

$$\text{straight-line rate} = \frac{100\%}{\text{estimated useful life}}$$

or

$$r = \frac{1}{n}$$

This approach can be used as an alternative method to compute the annual depreciation expense. Using the same information provided in Example 1, this method would be applied as follows:

$$r = \frac{1}{4}$$

$$= .25 \text{ or } 25\%$$

Periodic depreciation expense = depreciable amount × straight-line rate

or

$$e = (c - s) \times r$$

$$= (\$254,000 - \$25,875) \times .25$$

$$= \$228,125 \times .25$$

$$= \$57,031.25$$

A business will often summarize the depreciation of an asset in a table called a **depreciation schedule.** The table shows the annual depreciation expense, accumulated depreciation, and the book value of the asset each year of its useful life. Table 14.2 is a straight-line depreciation schedule for the asset listed in Example 1.

The **accumulated depreciation** increases as the current year's depreciation amount is added to the previous years' balance ($57,031.25 + 57,031.25 = $114,062.50). The current year-end book value is found by subtracting the accumulated depreciation from the total cost at the time the asset was placed in service ($254,000.00 − $114,062.50 = $139,937.50). The book value at the end of an asset's useful life should always be equal to its **salvage value.** Neither GAAP nor IRS guidelines allow an asset to be depreciated below its estimated salvage value.

If an asset is estimated to have no salvage value at the end of its useful life or if its salvage value is estimated to be less than 10% of its total cost, the total cost of the asset can be depreciated for tax purposes. To illustrate, if the equipment in Example 1 had been estimated to have a salvage value

Table 14.2

Straight-line depreciation schedule

Year	Depreciation expense	Accumulated depreciation	Book value
			$254,000.00
1	$57,031.25	$ 57,031.25	196,968.75
2	57,031.25	114,062.50	139,937.50
3	57,031.25	171,093.75	82,906.25
4	57,031.25	228,125.00	25,875.00

of $20,000, the annual amount of depreciation would have been $63,500 ($254,000 ÷ 4 = $63,500). We depreciate the total cost of the asset because the estimated salvage value ($20,000) is less than 10% of the asset's total cost ($254,000 × .10 = $25,400).

The problems encountered thus far in this section have all had purchase dates before January 15. It is unlikely that a business would purchase all of its tangible assets the first 15 days of the year. Therefore, it is necessary to calculate depreciation for part of a fiscal year.

When an asset is placed in service during any month other than January, the annual depreciation expense is prorated based on the number of months the asset was in service. If the asset is placed in service on or before the 15th of the month, that month is included in the calculation of the first year's depreciation expense. The procedure explained in Example 2 is applied only to those assets being depreciated under the straight-line, declining-balance, sum-of-the-years' digits, and units-of-production methods.

Example 2

On April 10, 1999, Brown Industries placed an injection molding machine in service. The company selected the straight-line method to depreciate the asset, which cost $15,800. Find the book value of the asset at the end of the first year if freight charges were $200, installation costs totaled $1,500, and the salvage value is estimated to be $2,700. The asset has an expected useful life of 4 years.

Solution

Step 1:   Determine the annual depreciation expense.

$$\text{depreciation expense} = \frac{\text{total cost} - \text{salvage value}}{\text{useful life (years)}}$$

$$= \frac{(\$15,800 + \$200 + \$1,500) - \$2,700}{4}$$

$$= \frac{\$14,800}{4}$$

$$= \$3,700$$

Step 2:   Calculate the prorated amount of depreciation for the 9 months the asset was in service during the first year.

$$\text{prorated depreciation expense} = \text{annual depreciation expense} \times \text{prorated time factor}$$

$$= \$3,700 \times 9/12$$

$$= \$2,775$$

Step 3: Determine the book value at the end of the first year.

$$\text{book value} = \text{total cost} - \text{accumulated depreciation}$$
$$= \$17,500 - \$2,775$$
$$= \$14,725$$

```
CALCULATOR  SOLUTION
Keyed entry                                              Display

AC  15800  +  200  +  1500  =                            17500
          Min  -  2700  =                                14800
                    ÷  4  =                              3700
              ×  9  ÷  12  =                             2775
          +/-  +  MR  =                                  14725
```

The useful life depreciation schedule for this asset would be extended into the fifth calendar year to completely depreciate the asset's total cost to its salvage value. The 3 months of depreciation not claimed during the first year would be claimed during the fifth year ($925 = $3,700 × 3/12) resulting in a book value equal to the salvage value as shown in Table 14.3.

Table 14.3

Prorated straight-line depreciation schedule

Year	Depreciation expense	Accumulated depreciation	Book value
			$17,500
1999	$2,775 (9/12 of annual)	$ 2,775	14,725
2000	3,700	6,475	11,025
2001	3,700	10,175	7,325
2002	3,700	13,875	3,625
2003	925 (3/12 of annual)	14,800	2,700

CHECK YOUR KNOWLEDGE

Straight-Line Depreciation

1. On January 12, John Brown, a farmer from Gary, Indiana, purchased a small airplane to be used for crop dusting. The original cost of the airplane was $85,000, the estimated salvage value is $12,250, and the useful life is 15 years. Calculate (a) annual depreciation expense, (b) the straight-line rate of depreciation, and (c) the book value at the end of the first year using the straight-line method.

2.   Find the book value at the end of the second year for an asset costing $28,000 and having a useful life of 4 years. The company paid $625 to have the asset shipped and an additional $1,800 to have it installed. The asset has an estimated salvage value of $6,425, was placed in service on January 4, and is depreciated under the straight-line method.

3.   If the asset in problem 2 is placed in service on August 8, what is the book value at the end of (a) year 1 and (b) year 2?

4.   A construction company purchased an asset costing $35,000 on January 10 with a salvage value of $2,800 and an estimated useful life of 7 years. Find (a) the first-year depreciation expense and (b) the book value at the end of the first year using the straight-line method.

5.   An asset was purchased by Rojon Enterprises on March 10 for $8,600. Freight and installation costs totaled $1,200. The asset's life expectancy is 5 years and its salvage value is estimated to be $2,300. Prepare a straight-line depreciation schedule using these headings:

Year	Depreciation expense	Accumulated depreciation	Book value

14.1   EXERCISES

Find the annual depreciation expense for the following assets using the straight-line method. Round answers to nearest cent.

	Cost	Freight charge	Installation cost	Salvage value	Estimated life	Depreciation expense
1.	$15,000	$0	$0	$3,000	4 years	_____
2.	$20,300	$0	$1,200	$1,800	10 years	_____
3.	$50,000	$840	$2,160	$11,000	7 years	_____
4.	$120,000	$1,800	$5,750	$10,500	20 years	_____
5.	$250,800	$0	$12,250	$33,000	12 years	_____

Complete the following straight-line depreciation schedule. The asset's salvage value is $4,800.

Year	Depreciation expense	Accumulated depreciation	Year-end book value
			$31,000
6. 1	$5,420	_____	_____
7. 2	_____	_____	_____

8.	3	_____	_____	_____
9.	4	_____	_____	_____
10.	5	_____	_____	_____

11. Tyson Zinc has just purchased equipment for his business. The original cost is $10,000, plus an installation cost of $300. The salvage value is $1,300 and the estimated useful life is 6 years. Find (a) the amount of depreciation expense and (b) the book value at the end of the first year.

12. Community Hospital uses the straight-line method of depreciation to depreciate an asset costing $30,500. If the asset is estimated to have a useful life of 10 years and a salvage value of $1,200, what is (a) the annual amount of depreciation expense, (b) the straight-line rate of depreciation, and (c) the book value at the end of the first year?

13. The Sterne Construction Company purchased a backhoe on July 8, 1998, that cost $375,000. The company estimated the asset would have a use-ful life of 15 years and a salvage value of $45,000. What was (a) the amount of depreciation expense claimed in 1998 and (b) the book value at the end of 1999?

14. An asset is purchased for $65,000 on January 10, 2000. The owner paid $425 for freight charges and $575 to have the asset installed. The asset has an estimated life of 8 years and a salvage value of $4,500. Prepare a depreciation schedule for the asset using the straight-line method of depreciation.

15. Pilgrim Packaging placed an asset in service on September 12, 1999. The company selected the straight-line method to depreciate the asset, which cost $6,300 and has a 5-year useful life with an estimated salvage value of $900. Prepare a depreciation schedule for the asset.

14.2 TRADITIONAL METHOD: DECLINING-BALANCE DEPRECIATION

Learning objective
Use the declining-balance method to develop a depreciation schedule.

The declining-balance method is an **accelerated method of depreciation** because it results in larger amounts of depreciation expense in the beginning years of the depreciation schedule compared to the straight-line method. A company might select the declining-balance method for a number of reasons. First, the book value of the asset may compare more favorably with the fair market value of the asset. Second, since the business reports a greater amount of depreciation expense, the taxable income will be lower in the earlier years, giving the company the use of those tax dollars. This method could also be attractive to companies that use high-technology machinery and equipment; such companies often depreciate their assets as rapidly as possible because of frequent technological change.

The **declining-balance method** allows a business to use a rate of depreciation equal to 125%, 150%, 175%, or 200% of the straight-line rate. The selection of which percentage to use depends on how rapidly the business wants to depreciate the asset. If a business chooses the 200% rate to depreciate an asset, it is using a method often referred to as the *double-declining-balance* method. When using this method, you must first determine the rate of depreciation. The rate of depreciation is determined by dividing the number 1 by the asset's useful life. For example, if the asset has a useful life of 5 years, its straight-line rate would be 20% ($1/5 = .20$). The straight-line rate is then multiplied by the selected percentage, say 150%, to determine the declining-balance rate ($.20 \times 1.50 = .30$). The rate of depreciation can also be found by dividing the 150% by the asset's useful life ($1.50/5 = .30$). To further accelerate the depreciation, we do not subtract the salvage value when calculating the yearly depreciation expense. We adjust the amount of depreciation expense in the last year of the depreciation schedule to equal the salvage value. Guidelines prohibit depreciating an asset below its salvage value.

In this chapter we will use various declining-balance percentages to determine depreciation, so read each problem carefully. The formulas used to calculate the declining-balance depreciation rate and periodic depreciation expense are

$$r = \frac{1}{n} \times \text{selected percentage}$$

$$e = b \times r$$

Note: In this case, b is the book value at the beginning of the period before the depreciation is calculated.

Example 3

On January 12, 1999, Fast-Roll Bearings purchased and placed in service an asset for $35,000. If the asset is depreciated under the double-declining-balance method using a 5-year useful life and a salvage value of $4,000, find the depreciation expense and the book value of the asset at the end of the first year.

Solution

Step 1:   Calculate the declining-balance rate.

$$\text{declining-balance rate} = \frac{1}{\text{useful life}} \times 2$$

$$= \frac{1}{5} \times 2$$

$$= .20 \times 2$$

$$= .40, \text{ or } 40\%$$

Step 2:   Calculate the depreciation expense using the asset's total cost and the declining-balance rate found in step 1. (Remember, we do not subtract the salvage value.)

$$\text{1st year's depreciation expense} = \text{book value} \times \text{declining-balance rate}$$

$$= 35,000 \times .40$$

$$= \$14,000$$

Step 3: Calculate the book value at the end of the first year.

1st year book value = total cost − accumulated depreciation

= $35,000 − $14,000

= $21,000

This procedure would be repeated for each of the remaining four years to fully depreciate the asset, as shown in Table 14.4. After the first year, the book value is found by adding the current year's depreciation expense to the previous year's accumulated depreciation ($8,400 + $14,000 = $22,400). The accumulated depreciation is then subtracted from the asset's total cost ($35,000 − 22,400 = $12,600) to give us the 2000 book value.

CALCULATOR SOLUTION

Keyed entry	Display
AC 35000 Min	35000
1 ÷ 5 × 1.5 =	.3
× MR =	10500
+/− M+ MR	24500
First year calculation	
× .3 =	7350
+/− M+ MR	17150
Second year calculation	
× .3 =	5145
+/− M+ MR	12005
Third year calculation	

Table 14.4

Declining-balance depreciation schedule

Year	Beginning book value	×	Rate	=	Depreciation expense	Accumulated depreciation	Book value
							$35,000
1999	$35,000		.40		$14,000	$14,000	21,000
2000	21,000		.40		8,400	22,400	12,600
2001	12,600		.40		5,040	27,440	7,560
2002	7,560		.40		3,024	30,464	4,536
2003	4,536		.40		536	31,000	4,000

In the final year (2003), we can claim a depreciation expense of only $536, since that is the amount that would reduce the previous year's book value to equal the salvage ($4,536 − $4,000 = $536). If we claimed the full 40% of the 2002 book value as the 2003 depreciation expense ($4,536.00 × .40 = $1,814.40), we would be depreciating the asset below the salvage value ($35,000.00 − $32,278.40 = $2,721.60), which is not in compliance with the rules.

The depreciation schedule shown in Table 14.4 for Example 3 is based on a purchase date of January 12, 1999. Consequently, a full year's depreciation was claimed the first year. Let us now analyze Example 4, which is the same asset used in Example 3, except the asset was placed in service on November 5, 1999, is being depreciated at 150%, and has a salvage value of $8,000. The depreciation expense must be prorated in the first year for the number of months the asset was in use. However, because larger amounts of depreciation are allowed in the early years and smaller or no depreciation is allowed in the later years, the unclaimed portion of the first year's depreciation is claimed in the second year. This procedure is continued until the asset is depreciated to the salvage value.

Example 4

On November 5, 1999, Fast-Roll Bearings purchased and placed in service an asset that cost $35,000. If the asset is depreciated under the 150% declining-balance method using a 5-year useful life and a salvage value of $8,000, find the depreciation expense and the book value for (a) 1999 and (b) 2000.

Solution

a.

Step 1: Calculate the declining balance rate.

$$r = \frac{1}{n} \times \text{Percentage}$$

$$.30 = 1/5 \times 1.5, \quad \text{or} \quad .30 = 1.5/5$$

Step 2: Calculate the first year's depreciation expense

$$e = b \times r$$

$$\$10,500 = \$35,000 \times .30$$

Note: The amounts calculated in the first two steps presented in Example 3 are the same for this example.

Step 3: Calculate the prorated amount of depreciation for the 2 months the asset was in service during the first year.

$$\text{prorated depreciation expense} = \text{(annual depreciation)} \times \text{(prorated time factor)}$$
$$= \$10,500 \times 2/12$$
$$= \$1,750$$

Step 4: Determine the book value at the end of the first year.

book value = total cost − accumulated depreciation
$$b = c - d$$
$$= \$35,000 - \$1,750$$
$$= \$33,250$$

b.

Step 1: Calculate the second year's annual depreciation expense using a book value based on a full year's depreciation the first year ($35,000 − 10,500 = \$24,500$). Refer to Table 14.4 for the book value amounts.

annual depreciation expense = book value × declining-balance rate
$$e = b \times r$$
$$= \$24,500 \times .30$$
$$= \$7,350$$

Step 2: Determine the prorated depreciation expense for the second year (2000).

$8,750.00 = 10,500 \times 10/12$ (10 months of 1999)

$1,225.00 = 7,350 \times 2/12$ (2 months of 2000)

$9,975.00 depreciation expense

Step 3: Determine the book value at the end of the second year.

book value = total cost − accumulated depreciation
$$b = c - d$$
$$= \$35,000 - (\$1,750 + \$9,975.00)$$
$$= \$35,000 - \$11,725$$
$$= \$23,275$$

Table 14.5 shows the calculations required to complete a depreciation schedule for this asset. Notice in year 5 (2003) the depreciation expense of the asset ($11,404.75 − \$3,001.25 = \$8,403.50$) is $8,403.50. We are not allowed to depreciate the asset below $8,000; therefore we must prorate the last year's depreciation expense ($8,403.50 − \$8,000 = \$403.50$) by claim-

Table 14.5

Prorated declining-balance depreciation schedule

Prorated declining-balance depreciation expense	Accumulated depreciation	Book value
		$35,000.00
1999 $35,000.00 × .30 = $10,500.00 × 2/12 = $1,750.00	$1,750.00	$33,250.00
2000 35,000 × .30 = 10,500.00 × 10/12 = 8,750.00		
24,500.00 × .30 = 7,350.00 × 2/12 = 1,225.00	$11,725.00	$23,275.00
$9,975.00		
2001 24,500.00 × .30 = 7,350.00 × 10/12 = 6,125.00		
17,150.00 × .30 = 5,145.00 × 2/12 = 857.50	$18,707.50	$16,292.50
$6,982.50		
2002 17,150.00 × .30 = 5,145.00 × 10/12 = 4,287.50		
12,005.00 × .30 = 3,601.50 × 2/12 = 600.25	$23,595.25	$11,404.75
$4,887.75		
2003 12,005.00 × .30 = 3,601.50 × 10/12 = 3,001.25		
8,403.50 × .30 = 403.50 × 2/12 = 67.25	$26,663.75	$8,336.25
$3,068.50		
2004 403.50 × 10/12 = $336.25	$27,000.00	$8,000.00

ing 2/12 of the $403.50 in 2003 ($403.50 × 2/12 = $67.25) and 10/12 of the $403.50 in 2004 ($403.50 × 10/12 = $336.25) to completely depreciate the asset.

CHECK YOUR KNOWLEDGE

Declining-balance Method

1. The original cost of an asset is $75,000, the estimated salvage value is $15,000, and the asset's useful life is 10 years. If the asset is depreciated at the double-declining balance rate, what is (a) the depreciation expense and (b) the book value for the first year?

2. Using the information provided in problem 1, determine the book value for (a) the second and (b) the third year.

3. A construction company purchased an asset on January 5 that cost $60,000 with a salvage value of $12,500 and an estimated useful life of 5 years. The asset is depreciated at the 150% declining-balance rate. Prepare a depreciation schedule using the form provided below.

Year	Depreciation amount	Accumulated depreciation	Book value
			$60,000.00
1	_____	_____	_____
2	_____	_____	_____
3	_____	_____	_____
4	_____	_____	_____
5	_____	_____	_____

4. Ultra-Tech placed in service on April 10 equipment that costs $9,600. It is estimated that the equipment will have a useful life of 4 years and a salvage value of $500. If the equipment is depreciated under the 150% declining-balance rate, what is the depreciation expense and book value at the end of (a) the first year and (b) the second year?

14.2 EXERCISES

Find the first-year depreciation expense for each of the following assets using the declining-balance method. Round answers to the nearest cent.

	Cost	Freight charge	Installa- tion charge	Salvage value	Declin- ing- balance rate	Estimated life	Depreciation expense
1.	$8,000	$500	$0	$3,200	150%	3 years	_____
2.	$17,500	$250	$1,350	$1,500	200%	5 years	_____
3.	$64,750	$0	$2,400	$8,000	125%	20 years	_____
4.	$47,800	$650	$0	$9,500	150%	12 years	_____
5.	$175,000	$1,200	$4,300	$15,000	125%	25 years	_____

Complete the following declining-balance depreciation schedule. The asset's salvage value is $2,000.

	Year	Beginning book value	Declin- ing rate	Depreciation expense	Accumulated depreciation	Year-end book value
6.	1	$12,000	.30	$_____	$_____	$_____
7.	2	_____	.30	_____	_____	_____

Year	Beginning book value	Declin-ing rate	Depreciation expense	Accumulated depreciation	Year-end book value
8. 3	_____	.30	_____	_____	_____
9. 4	_____	.30	_____	_____	_____
10. 5	_____	.30	_____	_____	_____

11. Marble Farms placed an asset in service on January 4, which can be depreciated under the double-declining-balance rate. The asset cost $12,000 and has a useful life of 5 years and an expected salvage value of $3,000. Find (a) the amount of depreciation expense and (b) the book value at the end of the first year.

12. Inland Supply erected a warehouse on its property at a total cost of $1,500,000. If the building has a useful life of 25 years and a salvage value of $250,000, what was the book value at the end of the second year? The asset was placed in service on January 7, 1978, and is depreciated at the 125% declining-balance rate.

13. On April 2, Healthcare purchased and placed in service an asset that cost $80,000. If the asset is depreciated under the 150% declining-balance method with a 10-year useful life and a salvage value of $5,000, find (a) the book value at the end of the first year and (b) the book value at the end of the second year.

14. Prepare a declining-balance depreciation schedule for exercise 11.

15. Niki's Body Shop purchased equipment for her business, which was placed in service on October 8. The equipment cost $8,000, has a useful life of 4 years, and a possible trade-in value of $2,000. Prepare a 150% declining-balance depreciation schedule.

14.3 TRADITIONAL METHOD: SUM-OF-THE-YEARS'-DIGITS DEPRECIATION

Learning objective
Calculate depreciation using the sum-of-the-years' digits method.

The sum-of-the-years'-digits method (SYD) is another accelerated method that results in higher depreciation expense in the earlier years and a lower depreciation expense in the later years of the asset's useful life. However, this method generates a depreciation expense that is less at the beginning and more at the end than the double-declining-balance method. The sum-of-the-years'-digits method allocates a fractional part of the periodic depreciation amount (original cost − salvage value) each year as the periodic depreciation expense. As the SYD fraction has both a numerator and a denominator, we must determine each. The numerator of the fraction changes each year because it is the number of periods remaining in the estimated useful life of the asset at the beginning of the period. The denominator of the fraction is the actual sum of the digits in the asset's useful life. Example 5 illustrates two different approaches you can use to calculate the SYD fraction.

Example 5

Determine the SYD fraction for an asset with an estimated useful life of 4 years.

Solution

Step 1:   Determine the denominator of the depreciation fraction.
Method 1:   Sum-the-digits

$$\text{denominator} = (1 + 2 + 3 + 4)$$
$$= 10$$

Method 2:   SYD formula

$$\text{denominator} = \frac{n(n + 1)}{2}$$
$$= \frac{4(4 + 1)}{2}$$
$$= \frac{20}{2}$$
$$= 10$$

Step 2:   Determine the numerator of the SYD fraction.

Year	SYD fraction
1	4/10
2	3/10
3	2/10
4	1/10

Notice that the fractions are never simplified. The denominators are all the same and the numerators are in the inverse order of the years. Each year a fractional part of the asset's depreciable amount becomes the depreciation expense (4/10 the first year, 3/10 the second year, etc.) until the entire amount (10/10, or 1) is expended.

Now that we know how to calculate the SYD fraction, we are ready to determine the periodic depreciation expense and the book value of an asset using the sum-of-the-year's-digits method. The formula for the depreciation expense is the same formula we used to determine straight-line depreciation, except that the rate (r) is expressed as a fraction instead of as a decimal.

Depreciation = (total cost − salvage value) × SYD fraction

or

$$e = (c - s) \times r$$

Example 6

Lakeside Graphics purchased a photocopier for $7,500, which was placed in service on January 8. The machine has an estimated useful life of 5 years and its salvage value is projected at $1,500. What is the first years' (a) depreciation expense, and (b) book value?

Solution

a.

Step 1:   Calculate the SYD fraction.

$$\text{denominator} = \frac{n(n + 1)}{2}$$

$$= \frac{5(5 + 1)}{2}$$

$$= \frac{30}{2}$$

$$= 15$$

Step 2:   Determine the periodic depreciation expense.

Depreciation = (total cost − salvage value) × SYD fraction expense

or

$$e = (c - s) \times r$$

$$= (\$7{,}500 - \$1{,}500) \times 5/15$$

$$= \$6{,}000 \times 5/15$$

$$= \$2{,}000$$

b.

Step 3:   Calculate the book value at the end of the first year.

Book value = total cost − accumulated depreciation

or

$$b = c - d$$

$$= \$7{,}500 - \$2{,}000$$

$$= \$5{,}500$$

Table 14.6 illustrates how you would develop a depreciation schedule for this problem. Notice that the column headings are similar to the declining-balance schedule except the depreciable amount and the SYD fractions are substituted for the beginning book value and the decimal rate.

When a business uses the sum-of-the-years'-digits method to depreciate an asset that has been placed in service after January, the annual depreciation expense is allocated between two accounting periods. This procedure was illustrated in Example 4, for the declining-balance method. However, the

Table 14.6

Sum-of-the-years' digits depreciation schedule

Year	Depreciable amount	Depreciation rate	Depreciation expense	Accumulated depreciation	Book value
					$7,500
1	$6,000	5/15	$2,000	$2,000	5,500
2	6,000	4/15	1,600	3,600	3,900
3	6,000	3/15	1,200	4,800	2,700
4	6,000	2/15	800	5,600	1,900
5	6,000	1/15	400	6,000	1,500

sum-of-the-years'-digits method poses a unique problem due to the use of a different fraction for allocating depreciation for each year of the asset's useful life. We will use the same asset we used in Example 6 to illustrate the allocation of depreciation if the asset was placed in service on April 5, instead of January 8.

Example 7

Lakeside Graphics purchased a photocopier for $7,500, which was placed in service on April 5. The machine has an estimated useful life of 5 years and its salvage value is projected at $1,500. What is the depreciation expense and book value for (a) year 1, and (b) year 2?

Solution

a.

Step 1: Calculate the SYD fraction.

$$denominator = \text{sum of the digits}$$
$$= (1 + 2 + 3 + 4 + 5)$$
$$= 15$$

Step 2: Determine the first year's annual depreciation expense.

$$e = (c - s) \times r$$
$$= (\$7,500 - \$1,500) \times 5/15$$
$$= \$6,000 \times 5/15$$
$$= \$2,000$$

Step 3: Calculate the prorated depreciation expense for the 9 months the asset was in service the first year.

$e = \$2,000 \times 9/12$

$\quad = \$1,500 \quad$ (partial year depreciation expense)

Step 4: Calculate the book value at the end of the first year.

$b = c - d$

$\quad = \$7,500 - \$1,500$

$\quad = \$6,000$

b.

Step 1: Calculate the second year's annual depreciation expense.

$e = (c - s) \times r$

$\quad = (\$7,500 - \$1,500) \times 4/15$

$\quad = \$6,000 \times 4/15$

$\quad = \$1,600$

Step 2: Calculate the prorated depreciation expense for year 2.

$\quad \$500 = \$2,000 \times 3/12 \quad$ (remaining 3 months of year 1)

$\quad \underline{1,200 \quad\;\; = \$1,600 \times 9/12} \quad$ (nine months of year 2)

$\quad \$1,700 \quad$ depreciation expense year 2

Step 3: Calculate the book value at the end of year 2.

$b = c - d$

$\quad = \$7,500 - (\$1,500 + \$1,700)$

$\quad = \$7,500 - \$3,200$

$\quad = \$4,300$

As the asset was placed in service on April 5, we must allocate 9/15 of the first year's annual depreciation to year one and 3/15 to year 2. The depreciation expense for year 2 is the 3 remaining months of the first year's depreciation and 9 months of the second year's annual depreciation. This procedure would be continued until the asset's depreciable amount has been allocated as depreciation expense. Table 14.7 shows the required calculations for this asset in the form of a depreciation schedule.

Note that the depreciation schedule indicates it takes 6 calendar years instead of 5 calendar years to depreciate the asset. This is because there are 3 months of depreciation that has not been allocated from the previous year's 12-month requirement ($\$6,000 \times 1/15 = \$400$) and we must depreciate the asset's entire depreciable amount ($\$6,000$). Figure 14.1 illustrates how depreciation is allocated to various calendar years and accounting periods for this example.

Table 14.7

Prorated sum-of-the-years'-digits depreciation schedule

Year	Depreciation calculations	Depreciation expense	Accumulated depreciation	Book value
				$7,500
1	$6,000 × 5/15 = $2,000 × 9/12 = $1,500		$1,500	6,000
2	$6,000 × 4/15 = $1,600 × 9/12 = $1,200 2,000 × 3/12 = 500 $1,700		3,200	4,300
3	$6,000 × 3/15 = $1,200 × 9/12 = $ 900 1,600 × 3/12 = 400 $1,300		4,500	3,000
4	$6,000 × 2/15 = $ 800 × 9/12 = $ 600 1,200 × 3/12 = 300 $ 900		5,400	2,100
5	$6,000 × 1/15 = $ 400 × 9/12 = $ 300 800 × 3/12 = 200 $ 500		5,900	1,600
6	$400 × 3/12 = $ 100		6,000	1,500

Figure 14.1

Allocation of depreciation by calendar years and accounting periods

Year	1999		2000		2001		2002		2003		2004
Fraction	9/12	3/12	9/12	3/12	9/12	3/12	9/12	3/12	9/12	3/12	
Amount	1,500	500	1,200	400	900	300	600	200	300	100	
Period	Year 1		Year 2		Year 3		Year 4		Year 5		

CHECK YOUR KNOWLEDGE

Sum-of-the-years'-Digits Method

1. Determine the sum-of-the-years'-digit fraction for the first year if the asset has an estimated useful life of (a) 3, (b) 12, (c) 20 years.

2. Medical Imaging purchased an x-ray machine for $120,000. The machine has an estimated useful life of 10 years and a salvage value of $10,000. If the sum-of-the-years'-digits method is used to depreciate the asset, what is the first year's (a) annual depreciation expense, and (b) book value.

3. Prepare a sum-of-the-years'-digits depreciation schedule for an asset that cost $5,400, has an estimated salvage value of $900, and is expected to be in service for 4 years. Use the following schedule format:

Year	Depreciation expense	Accumulated depreciation	Book value
			$5,400
1	_____	_____	_____
2	_____	_____	_____
3	_____	_____	_____
4	_____	_____	_____

4. Using the information in problem 2, determine (a) the depreciation expense the third year, (b) the accumulated depreciation at the end of the fifth year, and (c) the book value at the end of the eighth year.

5. Angelo's Bakery and Pizzeria bought a delivery truck that cost $32,000, which was placed in service on August 12. The truck has an estimated useful life of 8 years and a salvage value of $5,000. If the truck is depreciated under the sum-of-the-years' digits method, what is the book value at the end of (a) the first year, and (b) the second year?

14.3 EXERCISES

Determine the sum-of-the-years'-digits depreciation fractions for the years indicated, given the following estimated life for each.

Estimated life	Depreciation year	SYD fraction
1. 3	1	_____
2. 5	3	_____
3. 10	6	_____
4. 15	10	_____
5. 20	12	_____

Answers to CYK: **1. a.** 3/6; **b.** 12/78; **c.** 20/210   **2. a.** $20,000; **b.** $100,000   **3.** book value year 1, $3,600; book value year 2, $2,250; book value year 3, $1,350; book value year 4, $900 **4. a.** $16,000; **b.** $80,000; **c.** $16,000;   **5. a.** $29,500; **b.** $23,812.50

Find the first-year depreciation expense for each of the following assets using the sum-of-the-years' digits method.

	Cost	Freight charge	Installation charge	Salvage value	Estimated life	Depreciation expense
6.	$2,000	$150	$350	$ 500	3	_____
7.	$6,000	none	800	400	4	_____
8.	$18,550	450	none	2,500	5	_____
9.	$46,900	300	500	5,700	6	_____
10.	$65,000	none	none	3,200	8	_____

Complete the following sum-of-the-years'-digits depreciation schedule. The assets salvage value is $4,000.

	Year	Depreciable amount	Depreciation rate	Depreciation expense	Accumulated depreciation	Book value
						$12,000
11.	1	_____	4/10	_____	_____	_____
12.	2	_____	3/10	_____	_____	_____
13.	3	_____	2/10	_____	_____	_____
14.	4	_____	1/10	_____	_____	_____

15. Warner's Collision Services purchased a frame alignment machine for $28,000. The machine was put in service on January 5, has an estimated useful life of 8 years, and a salvage value of $4,200. Use the sum-of-the-years'-digits method to find the first year's (a) depreciation expense and (b) book value.

16. Using the information provided in exercise 15, determine (a) the fifth year's depreciation expense, (b) the accumulated depreciation expense at the end of the eighth year, and (c) the book value at the end of the eighth year.

17. Golf Warehouse purchased a computerized diagnostic sensor on January 8 to help its customers analyze their golf swings. The cost of the equipment was $3,250, and another $350 was charged for installation. Management estimates the useful life of the equipment will be 5 years and that it will have no salvage value due to technological change. Prepare a depreciation schedule for the asset using the sum-of-the-years'-digits method.

18. Brighton Chiropractic Health Center purchased five examination tables at a total cost of $17,500, which were placed in service on July 20. If the tables are depreciated under the sum-of-the-years'-digits method with a 10-year useful life and a salvage value of $2,500, determine (a) the book value at the end of the first year and (b) the book value at the end of the second year.

19. Northwestern Ready-Mix Inc. bought a new truck to deliver concrete on January 6. The truck cost $91,400, has a useful life of 8 years, and an estimated salvage value of $5,000. Use the sum-of-the-year's-digits method to find (a) the depreciation expense the second year, (b) the accumulated depreciation at the end of the fourth year, and (c) the book value at the end of the sixth year.

20. Commercial Photo Laboratory bought and placed in service a high-speed developing machine on November 10 that cost $30,000. The developer has an estimated useful life of 4 years and a salvage value of $6,000. Prepare a sum-of-the-years'-digits depreciation schedule.

14.4   TRADITIONAL METHOD: UNITS-OF-PRODUCTION DEPRECIATION

Learning objective
Calculate depreciation using the units-of-production method.

The units-of-production method is another traditional method used by businesses. An asset whose useful life is more accurately defined by the units produced, hours utilized, or miles driven could be depreciated by the *units-of-production method.* This method is also used to depreciate assets that are used seasonally. For example, construction, farm, and recreational equipment are often used quite heavily during some months of the year and not at all during other months of the year.

The amount of depreciation under the units-of-production method is based on a constant amount per unit of use as opposed to the time base of both the straight-line and declining-balance methods. To calculate the annual depreciation amount, we must first determine the per-unit of depreciation factor and then multiply the factor by the number of units produced, hours utilized, or miles driven. The following equation is used to calculate the depreciation amount:

Units-of-production depreciation method equation

annual depreciation expense

$$= \frac{(\text{total cost} - \text{salvage value})}{\text{units of life}} \times \text{units of production}$$

or

$$e = \frac{(c - s)}{n} \times x \text{ units of production}$$

$$e = f \times \text{units of production} \qquad (f = \text{depreciation factor})$$

Example 8

Santos Machine Works installed a stamping machine at a cost of $50,000 and expects the machine will produce 1,000,000 units of product before it must be replaced. If the salvage value of the machine is estimated to be $10,000 and the machine produced 80,000 units during the year, find (a) the first year's depreciation amount and (b) the book value at the end of the first year.

Solution

Step 1:   Determine the depreciation factor for the stamping machine using the equation,

$$\text{depreciation factor} = \frac{(\text{total cost} - \text{salvage value})}{\text{units of life}}$$

$$f = \frac{(\$50,000 - \$10,000)}{1,000,000}$$

$$f = \frac{(\$40,000)}{1,000,000} = \$.04 \text{ per unit}$$

Step 2: Calculate depreciation expense for first year.

$$e = \frac{(c - s)}{n} \times \text{units produced} = (\$.04)\, 80,000$$

$$e = \$3,200 \text{ annual depreciation expense}$$

Step 3: Calculate the book value at the end of the first year.

$$\text{book value} = \text{total cost} - \text{accumulated depreciation}$$

$$\begin{aligned} b &= c - d \\ &= \$50,000 - \$3,200 \\ &= \$46,800 \end{aligned}$$

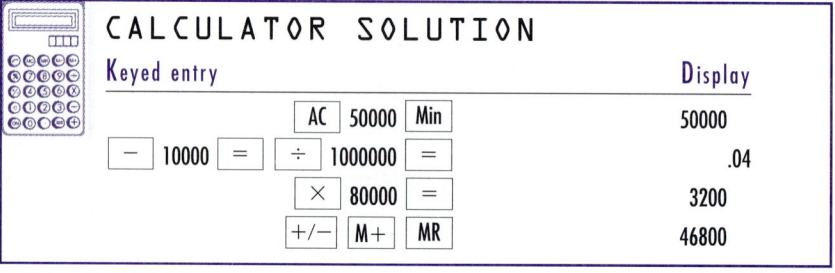

CALCULATOR SOLUTION

Keyed entry		Display
AC 50000 Min		50000
− 10000 = ÷ 1000000 =		.04
× 80000 =		3200
+/− M+ MR		46800

It should be noted that when an asset is depreciated under the units-of-production method, the date the asset is placed in service will not affect the amount of depreciation claimed during the year; only the usage of the asset will change its depreciation amount.

Let us now analyze an example that is based on the number of hours of operation and that requires the preparation of useful-life depreciation schedule.

Example 9

Silver Mountain Ski Center purchased a double chair lift for $60,000 and paid $20,000 to place the lift in service. The center estimates the lift will have a useful life of 25,000 hours and a salvage value of $15,000. Prepare a depreciation schedule using the units-of-production method if the asset was operated a total of 1,850 hours during the year.

Solution   **Step 1:**   Determine the depreciation factor (per-hour rate of depreciation) of the asset.

$$\text{unit depreciation factor} = \frac{\text{total cost} - \text{salvage value}}{\text{units of life}}$$

$$f = \frac{(c - s)}{n}$$

$$= \frac{60,000 + 20,000 - 15,000}{25,000 \text{ (hours)}}$$

$$= \$2.60 \text{ per hour unit}$$

Step 2:   Calculate the depreciation expense for the first year.

$$\text{depreciation expense} = \text{unit depreciation factor} \times \text{hours}$$

$$e = f \times \text{units produced}$$

$$e = \$2.60 \times 1,850$$

$$e = \$4,810$$

Step 3:   Calculate book value for the first year.

$$\text{book value} = \text{total cost} - \text{accumulated depreciation}$$

$$b = c - d$$

$$= \$80,000 - \$4,810$$

$$= \$75,190$$

Step 4:   Complete the depreciation schedule using the procedure shown in steps 2 and 3 for each year in the depreciation schedule shown in Table 14.8.

CALCULATOR SOLUTION

Keyed entry	Display
AC 80000 Min	80000
1850 × 2.6 =	4810
+/− M+ MR	75190

First year calculation

1975 × 2.6 =	5135
+/− M+ MR	70055

Second year calculation—use this same procedure to complete the depreciation schedule.

Table 14.8

Units-of-production depreciation schedule

Year	Hours utilized	Depreciation expense	Accumulated depreciation	Book value
0				$80,000
1	1,850	$4,810	$ 4,810	75,190
2	1,975	5,135	9,945	70,055
3	2,325	6,045	15,990	64,010
4	1,940	5,044	21,034	58,966
5	2,320	6,032	27,066	52,934
6	2,050	5,330	32,396	47,604
7	1,875	4,875	37,271	42,729
8	2,260	5,876	43,147	36,853
9	2,080	5,408	48,555	31,445
10	2,360	6,136	54,691	25,309
11	1,950	5,070	59,761	20,239
12	2,015	5,239	65,000	15,000

Notice in Table 14.8 that a column is designated for the unit measure of production (in this case, hours) and that the number of hours the equipment was utilized varies from year to year. When preparing a depreciation schedule using the units-of-production method, the schedule must contain information indicating the pattern of use for each year the asset is being depreciated.

Comparison of Traditional Methods

If the traditional methods of depreciation presented in this section were to be applied to the same asset, the total amount of depreciation expense claimed and the asset's book value at the end of its useful life would be the same. However, the annual depreciation expense will vary in any given year. Let's consider a $200,000 machine placed in service on January 1 that cost $30,000 to install, has a useful life of 5 years, and an estimated salvage value of $50,000. If the machine was estimated to produce 90,000 units and actually produced 20,000 units the first year, 50,000 units the second year, 2,500 units the third year, 12,500 units the fourth year, and 5,000 units the fifth year, the three methods of depreciation would result in the annual depreciation expenses shown in Table 14.9.

The depreciation expenses shown in Table 14.9 are based on the amount of the asset's value being depreciated ($180,000) over the 5-year period. (This procedure was explained earlier in this section.)

$$\text{amount to be depreciated} = \text{total cost} - \text{salvage value}$$
$$= (\$200,000 + \$30,000) - \$50,000$$
$$= \$180,000$$

Table 14.9

Comparison of
depreciation using
traditional methods

Year	Straight line	150% declining-balance	Sum-of-the years' digits	Units-of-produc-tion
1	$ 36,000	$ 69,000	$ 60,000	$ 40,000
2	36,000	48,300	48,000	100,000
3	36,000	33,810	36,000	5,000
4	36,000	14,445	24,000	25,000
5	36,000	14,445	12,000	10,000
Total	$180,000	$180,000	$180,000	$180,000

The annual depreciation expense in the first and subsequent years is determined by the three depreciation methods. The first year's depreciation expense for each of the three methods is calculated as follows:

Straight-line method:

$$\text{depreciation expense} = \frac{\text{total cost} - \text{salvage value}}{\text{useful life}}$$

$$= \frac{\$230,000 - \$50,000}{5}$$

$$= \$36,000$$

Declining-balance method:

$$\text{depreciation expense} = \text{book value} \times \text{declining-balance rate}$$

$$= \$230,000 \times .30 \qquad (1.50/5 = .30)$$

$$= \$69,000$$

Sum-of-the-years'-digits method:

$$\text{depreciation expense} = (\text{total cost} - \text{salvage value}) \times (\text{SYD rate})$$

$$= (\$230,000 - \$50,000) \times \left(\frac{5(6)}{2} = \frac{30}{2} = 15\right)$$

$$= \$180,000 \times 5/15$$

$$= \$60,000$$

Units-of-production method:

$$\text{depreciation expense} = \frac{(\text{total cost} - \text{salvage value})}{\text{units of life}} \times$$

$$\text{units produced}$$

$$= \frac{(\$230,000 - \$50,000)}{90,000} \times 20,000$$

$$= \$40,000$$

For Your Information

A COMPARISON OF THE TRADITIONAL METHODS

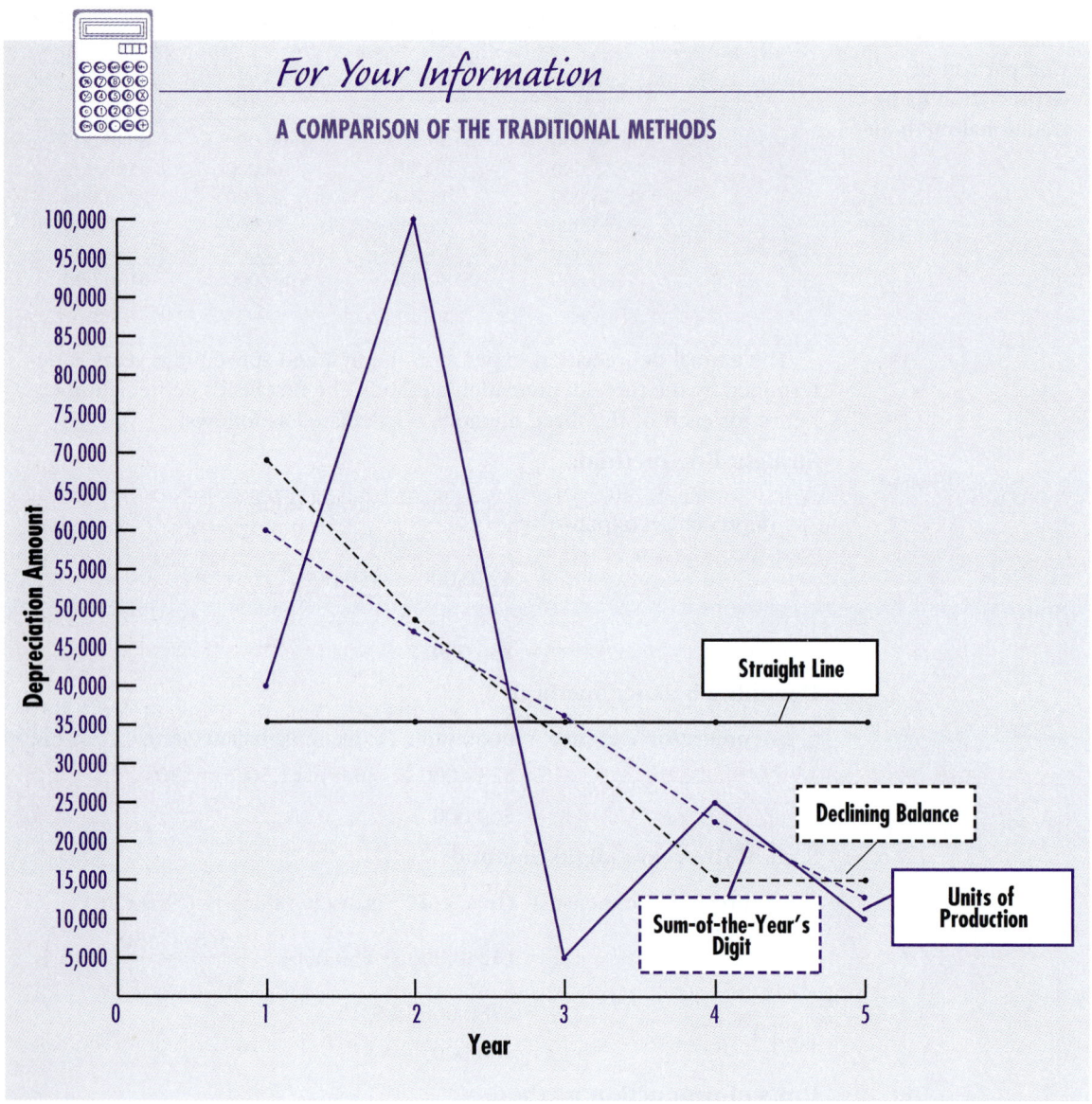

The method of depreciation chosen by a business for tax purposes is defined by the Internal Revenue Service. However, the method selected by a business to depreciate assets for "book" purposes within the business is determined by the firm's management. Therefore, managers must decide which of the traditional methods best achieves their financial objectives. The process of selecting which method of depreciation is most preferable is difficult, but, in most cases, the managers will consider the following factors in their search for the answer:

1. Obsolescence
2. Projected market values of the asset
3. Internal Revenue Service requirements
4. Impact on earnings per share

The four traditional methods of depreciation are compared visually on the accompanying graph. The graph is developed from the data presented in Table 14.9. Each depreciation method generates a different depreciation expense and book value each year. The slope of the curve defines the degree of acceleration for comparison purposes.

As you can see, each method of depreciation produces a different depreciation expense for the first year. The declining-balance and the sum-of-the-years'-digits methods accelerate the depreciation process, the units-of-production method depreciates the asset according to its use, while the straight-line method depreciates the asset in equal amounts each year. Managers must decide which of these methods best accommodates their financial strategies. This is not a simple decision, and the factors that affect their choice are complex and go beyond the scope of this text.

These traditional methods are used to record depreciation for bookkeeping purposes, regardless of the acquisition date, and are used for tax purposes on assets acquired prior to 1981.

CHECK YOUR KNOWLEDGE

Units-of-Production Method

Find the unit of depreciation factor and the annual depreciation amount for each of the following:

	Cost	Salvage value	Useful life	Units produced/ utilized	Unit depreciation factor	Annual depreciation amount
1.	$8,500	$1,000	50,000 units	3,500	$_____	$_____
2.	$32,000	$6,000	80,000 hours	6,250	_____	_____
3.	$75,000	$15,000	120,000 miles	18,720	_____	_____

4. A machine costs $89,500, has an estimated salvage value of $12,500, and has an estimated useful life of 275,000 units. If the machine produced 8,743 units the first year, find (a) the first year's depreciation amount and (b) the book value of the machine at the end of the first year.

5. National Transportation purchased a tractor at a cost of $95,000. The company estimates the tractor will be driven 200,000 miles and will have a salvage value of $35,000. Prepare a useful-life depreciation schedule using the units-of-production method and the mileage given in the schedule (.30 unit expense.)

Year	Annual mileage	Depreciation amount	Accumulated depreciation	Book value
				$95,000
1	20,540	$_____	$_____	_____
2	65,780	_____	_____	_____
3	72,165	_____	_____	_____
4	41,515	_____	_____	_____

14.4 EXERCISES

Find the units and yearly depreciation expense for each of the following assets using the units-of-production method. Round answers to the nearest cent.

	Cost	Installation charge	Salvage value	Useful life	Units produced/ utilized	Unit depreciation factor	Depreciation expense
1.	$10,000	$350	$1,400	100,000 units	3,250	_____	_____
2.	$27,500	$0	$2,000	50,000 miles	15,000	_____	_____
3.	$130,800	$4,200	$25,000	128,400 hours	6,420	_____	_____
4.	$3,400	$125	$500	75,000 units	12,500	_____	_____
5.	$80,450	$875	$9,200	250,000 units	16,840	_____	_____

Complete the following units-of-production depreciation schedule. The asset's salvage value is $5,400, and its useful life is estimated to be 50,000 units.

	Year	Miles driven	Depreciation factor	Depreciation expense	Accumulated depreciation	Book value
						$22,850
6.	1	4,340	_____	$_____	$_____	$_____
7.	2	13,965	_____	_____	_____	_____

Answers to CYK: 1. .15; $525   2. .325; $2,031.25   3. .50; $9,360.00   4. a. $2,448.04;
b. $87,051.96   5. $.30 depreciation rate/mile; $35,000 book value end of year 4

8.	3	15,250	_____	_____	_____	_____
9.	4	12,472	_____	_____	_____	_____
10.	5	3,973	_____	_____	_____	_____

Use each of the following depreciation methods to find the first year's depreciation expense and book value for an asset that cost $6,720, has a salvage value of $1,050, and a useful life of 6 years or 72,000 units. The asset produced 15,000 units the first year and was placed in service on January 10.

Depreciation method	Depreciation expense	Book value
11. Straight-line	_____	_____
12. Double-declining balance	_____	_____
13. Sum-of-the-years'-digits	_____	_____
14. Units-of-production	_____	_____

15. Rojon Enterprises installed a grinding machine in its production department. The company paid $52,500 for the machine and $3,250 to have it installed. Find the annual depreciation amount if the machine has a salvage value of $5,500, was operated 4,160 hours, and has an estimated life of 75,000 hours.

16. Using the information given in exercise 11, calculate the book value at the end of the third year the machine was operated when it was operated 6,280 hours the second year and 5,915 hours the third year.

17. A vehicle costs $78,500, has a trade-in value of $8,000, and has an estimated useful life of 150,000 miles. Find the book value at the end of the first year if the vehicle was driven 24,635 miles.

18. American Foods purchased a labeling machine to be used in its packing plant at a cost of $135,000. The machine is expected to produce 1,500,000 units during its useful life, at which time it will have a salvage value of $15,000. Find the book value of the machine at the end of the second year if the machine labeled 80,250 units the first year and 132,745 units the second year.

19. Johanson's Ice Cream Company purchased a delivery truck at a cost of $34,000. The company estimates the truck will have a trade-in value of $6,000, and company policy is to replace delivery vehicles every 100,000 miles. Prepare a depreciation schedule using the units-of-production method based on the following mileage:

first year	18,250 miles
second year	27,920 miles
third year	31,490 miles
fourth year	22,340 miles

20. Kleen-All Janitorial Services purchased a floor cleaning machine for $9,600, which was placed in service on November 8. The machine is estimated to have a useful life of 5 years, or 16,200 hours, and a $1,500 salvage value. Kleen-All wishes to depreciate the asset as rapidly as possible. Therefore, it has decided to compare the double-declining balance method with the units-of-production method, as the machine is used many hours each year. If the cleaner is used 680 hours the first year and 1,875 hours the second year, which depreciation method would you recommend Kleen-All use to achieve its objective? Base your decision on actual depreciation calculations.

14.5   MODIFIED ACCELERATED COST-RECOVERY SYSTEM (MACRS)

The **modified accelerated cost-recovery system (MACRS),** also known as the "general depreciation system," or "GDS," applies to all tangible property placed in service after 1986. As the name implies, MACRS is a modification of the accelerated cost-recovery system. The changes were enacted as part of the Tax Reform Act of 1986 and included the creation of two new classes of property, the reclassification of some property, and changes in the depreciation procedures and rates. Under MACRS, the depreciation deduction can be calculated in one of two ways. The depreciation can be computed using (1) a depreciation method and convention over the recovery period or (2) MACRS percentage tables. The depreciation deduction is approximately the same under both methods.

All property depreciated under MACRS is assigned to a property class. The class to which property is assigned is determined by its class life. The class life of an asset is the basis of its recovery period and the method of depreciation that can be used to recover the asset's cost. Table 14.10 shows the class life and recovery period for property depreciated under MACRS.

Under MACRS, we are allowed to recover the asset's total original cost. Therefore, we do not have to estimate its salvage value. Also, as mentioned earlier, we can elect to compute the depreciation deduction using a depreciation method (declining-balance or straight-line) and convention (midmonth or half-year) over the recovery period, or by using percentage tables. We will use the same example to illustrate both methods.

Declining-Balance Method

The declining-balance method can be used to depreciate all tangible assets. However, the class life of the asset determines the percentage method used to calculate the depreciation rate. Table 14.11 shows the applicable declining-balance method for each class of property and the corresponding depreciation rate.

The depreciation rates in Table 14.11 are determined by dividing the specified declining-balance percent (200% or 150%) by the recovery period. For example, for 5-year property, we divide 2.00 by 5 = .4, or 40%. For 20-year property, we divide 1.50 by 20 = .075, or 7.5%.

Under MACRS, the month of the year the asset is placed in service is not a factor in calculating the first year's depreciation expense. This is because the *half-year convention* treats all assets as placed in service on the midpoint of the first year. Consequently, the number of recovery years is always 1 year greater than the class life of the asset. If, however, more than 40% of a company's assets in any tax year are placed in service during the last 3 months of that year, then a *midquarter convention* must be used. To calculate the MACRS deduction for an asset subject to the midquarter convention, we first compute the depreciation for the full year and then multiply

Table 14.10

MACRS class lives
and recovery
periods

Recovery period/ (class life)	Asset description
3-year property (0–4 yrs)	Tractor units for use over-the-road, any race horse over 2 years old, and any horse that is over 12 years old.
5-year property (5–10 yrs)	Taxis, buses, heavy general-purpose trucks, computers and peripheral equipment, office machinery (typewriters, calculators), automobiles, and research equipment.
7-year property (10–16 yrs)	Office furniture and fixtures (desks, files) and any property that does not have a class life and that has not been designated by law as being in any other class.
10-year property (16–20 yrs)	Vessels, barges, tugs, similar water transportation equipment, any single-purpose agricultural or horticultural structure, and any tree or vine bearing fruits or nuts.
15-year property (20–25 yrs)	Roads, shrubbery, wharves, and any municipal wastewater treatment plant.
20-year property (25 yrs or more)	Farm buildings and municipal sewers.
27.5-year property	Residential rental real property such as rental houses, apartments, and mobile homes.
31.5-year property	Any nonresidential rental real property (office buildings, stores, warehouses, etc.) placed in service on or before May 12, 1993.
39-year property	Nonresidential property placed in service after May 12, 1993.

Table 14.11

Declining-balance
methods and rates

Recovery class	Declining-balance method	Depreciation rate
3	200%	66.67%
5	200	40.00
7	200	28.57
10	200	20.00
15	150	10.00
20	150	7.50

the appropriate percentages for the quarter of the tax year the asset is placed in service. Let's develop an example step-by-step using the declining-balance method and appropriate convention.

Example 10

Central Farm Supply purchased a full-color copier for $15,000. The copy machine was placed in service on April 20. Find the amount of depreciation and the book value at the end of the first year using the declining-balance method.

Solution

Step 1: Determine the class life of the property using Table 14.10. The asset is classified as a 5-year property and will be depreciated using the 200% declining-balance method (Table 14.11).

Step 2: Calculate the yearly depreciation expense.

$$\text{depreciation expense} = \text{original cost} \times \text{depreciation rate}$$
$$e = c \times r$$
$$= \$15,000 \times .40$$
$$= \$6,000$$

Step 3: Determine the depreciation deduction for the first year using the half-year convention.

$$\text{1st-year depreciation expense} = \text{annual depreciation} \times \text{half-year}$$
$$\text{depreciation rate}$$
$$= \$6,000 \times .5$$
$$= \$3,000$$

Step 4: Find the book value at the end of the first year.

$$\text{book value} = \text{original cost} - \text{accumulated depreciation}$$
$$b = c - d$$
$$= \$15,000 - \$3,000$$
$$= \$12,000$$

Notice in step 3 that the first year's depreciation amount ($3,000) is one-half the annual amount ($6,000). This is because the half-year convention allows us to claim only half of the annual depreciation amount the first year, regardless of when the asset is placed in service. The remaining half-year will be recovered during the sixth calendar year of the recovery schedule. The entire cost recovery calculations for this example are summarized in Table 14.12.

Table 14.12

200% declining-balance cost recovery schedule

Year	Depreciation calculations	Depreciation expense	Accumulated depreciation	Book value
0				$15,000.00
1	$15,000 × .40 × .5	$3,000.00	$ 3,000.00	12,000.00
2	12,000 × .40	4,800.00	7,800.00	7,200.00
3	7,200 × .40	2,880.00	10,680.00	4,320.00
4	4,320 × .40	1,728.00	12,408.00	2,592.00
5	2,592 × .6667	1,728.09	14,136.09	863.91
6		863.91	15,000.00	0.00

If we were to continue multiplying the adjusted basis of the property (book value) by the declining-balance rate, we would never be able to depreciate the asset's value to 0. Therefore, we apply the straight-line method for the last 2 years (4th and 5th) and claim the balance as the depreciation amount for the carry over year (6th). In the fourth year, the straight-line rate is 40% (1 divided by 2.5 remaining years), which in this case just happens to be the same as the declining-balance rate. In the fifth year, the straight-line rate is 66.67% (1 divided by 1.5 remaining years). For the sixth year, the depreciation expense is the fifth-year book value, since the remaining recovery period (a half-year) is less than one year and the asset's original value must be fully recovered.

Percentage Table Method

The Internal Revenue Service provides tables that can be used to figure depreciation under MACRS. The tables may be used for any property placed in service in a tax year and they are based on the appropriate depreciation method, recovery period, and convention. If the table method is elected, the percentage from the table must be used to figure depreciation deductions for the entire recovery period of the property. The rates shown in Table 14.13 are based on the 200% or 150% declining-balance method applying the half-year convention to each recovery period. To determine the rate of depreciation for any year, we find the recovery year in the left-hand column, and read across to the appropriate recovery period. For example, the depreciation rate for property with a 10-year recovery period in the sixth year of life is 7.37%.

Example 11

Let's assume the same facts as in Example 10, except that we will use the percentage table method to determine the depreciation deduction.

Table 14.13

MACRS percentage table: 200% or 150% declining-balance depreciation switching to straight-line for 3, 5, 7, 10, 15, and 20 years using half-year convention

Recovery year	Recovery period					
	3 years	5 years	7 years	10 years	15 years	20 years
1	33.33	20.00	14.29	10.00	5.00	3.750
2	44.45	32.00	24.49	18.00	9.50	7.219
3	14.81	19.20	17.49	14.40	8.55	6.677
4	7.41	11.52	12.49	11.52	7.70	6.177
5		11.52	8.93	9.22	6.93	5.713
6		5.76	8.92	7.37	6.23	5.285
7			8.93	6.55	5.90	4.888
8			4.46	6.55	5.90	4.522
9				6.56	5.91	4.462
10				6.55	5.90	4.461
11				3.28	5.91	4.462
12					5.90	4.461
13					5.91	4.462
14					5.90	4.461
15					5.91	4.462
16					2.95	4.461
17						4.462
18						4.461
19						4.462
20						4.461
21						2.231

Solution

Step 1: Determine the class life of the property using Table 14.10. The asset is classified as a 5-year property.

Step 2: Find the depreciation rate in Table 14.13 for a 5-year property in the first year of life (.20) and multiply this rate by the original cost of the property to determine the amount of depreciation.

depreciation expense = original cost $\times$ depreciation rate

$$e = c \times r$$
$$= \$15,000 \times .20$$
$$= \$3,000$$

Table 14.14

MACRS
depreciation
schedule using
percentage tables

Year	Depreciation calculations	Depreciation expense	Accumulated depreciation	Book value
				$15,000
1	($15,000 × .20)	$3,000	$3,000	12,000
2	($15,000 × .32)	4,800	7,800	7,200
3	($15,000 × .192)	2,880	10,680	4,320
4	($15,000 × .1152)	1,728	12,408	2,592
5	($15,000 × .1152)	1,728	14,136	864
6	($15,000 × .0576)	864	15,000	0

Step 3:   Find the book value at the end of the first year.

book value = original cost − accumulated depreciation

$$b = c - d$$
$$= \$15,000 - \$3,000$$
$$= \$12,000$$

Notice that the depreciation amount and the book value for the first year are the same for both methods. Table 14.14 illustrates how you would prepare a depreciation schedule for Example 11 using the percentage table method.

Straight-Line Method

A business can elect to depreciate tangible property using the straight-line method instead of the declining-balance method. To figure the MACRS deduction under the straight-line method, without percentage tables divide the number 1 by the years remaining in the recovery period at the beginning of the tax year. This procedure produces a different depreciation rate for each year of the recovery period. The rate is then applied to the previous year's book value to determine the current year's depreciation deduction. The half-year convention also applies to the straight-line method; therefore, a half-year of depreciation is deducted the first year the property is placed in service, and the unrecovered value of the property at the end of the recovery period is claimed in the carryover year.

When using the straight-line method, the periodic depreciation expense is calculated by both the declining-balance method and the straight-line method. The current depreciation expense will always be the larger of the two amounts. At some point in the calculation process the two methods will produce an equal amount of depreciation. From this point on, the straight-line method will provide a greater depreciation deduction and will be used as the current period's depreciation expense. Example 12 explains the steps required to determine the first year's depreciation expense and book value.

Example 12

Management Consulting Group purchased $40,000 of office furniture for its new facility in Chicago, which was placed in service on September 20. Prepare a cost recovery schedule using the MACRS straight-line method.

Solution

Step 1: Determine the class life and recovery period of the property. The property would be considered a 7-year property.

Step 2: Find the straight-line depreciation rate for a 7-year property in the first year of life (1/7 = .1429) and multiply this rate by the original cost of the property to determine the annual amount of depreciation.

annual depreciation amount = original cost × depreciation rate

$$e = c \times r$$
$$= \$40,000 \times .1429$$
$$= \$5,716$$

Step 3: Calculate the annual depreciation deduction using the declining-balance method. (Table 14.11 lists a 28.57% rate of depreciation for 7-year property.)

depreciation expense = original cost × depreciation rate

$$e = c \times r$$
$$= \$40,000 \times .2857$$
$$= \$11,428$$

Step 4: Determine the depreciation deduction using the half-year convention. Convention is applied to larger declining-balance deduction.

1st year depreciation amount = annual depreciation × half-year depreciation rate

$$= \$11,428 \times .5$$
$$= \$5,714$$

Step 5: Find the book value at the end of the first year.

book value = original cost − accumulated depreciation

$$b = c - d$$
$$= \$40,000 - \$5,714$$
$$= \$34,286$$

These steps would be completed for each year of the recovery period to prepare a cost recovery schedule. Because of the amount of work required, and the fact that the percentage table method generates almost the same

Table 14.15

Straight-line cost
recovery schedule

Year	Depreciation calculations	Depreciation expense	Accumulated depreciation	Book value
				$40,000
1	($40,000 × .1429)	$5,716	$5,716	34,284
2	($40,000 × .2449)	9,796	15,512	24,488
3	($40,000 × .1749)	6,996	22,508	17,492
4	($40,000 × .1249)	4,996	27,504	12,496
5	($40,000 × .0893)	3,572	31,076	8,924
6	($40,000 × .0892)	3,568	34,644	5,356
7	($40,000 × .0893)	3,572	38,216	1,784
8		1,784	40,000	0

amounts of depreciation expense, we will use the percentage table method for recovery schedules in this text. Table 14.15 illustrates how you would develop a straight-line cost recovery schedule for Example 12 using the percentage tables.

Real Property

Real property, or real estate, is land and generally anything that is erected on, growing on, or attached to land. The land itself, however, is not depreciable property.

For depreciation purposes, real property is classified as either *residential rental property* or *nonresidential rental property.* Table 14.10 gives a description of property that qualifies under each class in the last three recovery periods. The IRS tax guidelines require that businesses use the straight-line method to depreciate real property. The method is applied exactly as explained in the preceding section. There is, however, one modification involving the date the property is placed in service. For all depreciable real property, we use a *midmonth* convention. Under the midmonth convention, property is treated as placed in service on the midpoint of the month regardless of the day of purchase. Therefore, to determine the first year's depreciation deduction, we include one-half of the first month and each of the months that follow. Each year thereafter, we are allowed a full year's depreciation deduction based on the unrecovered value of the property and the depreciation rate for that year. Because of the number of years (27.5, 31.5, and 39) and since these calculations are repetitive, we will use the percentage table method discussed earlier to determine the depreciation deduction for real property. The depreciation rates can be found in Tables 14.16, 14.17, and 14.18.

Note that the depreciation rates found in Table 14.18 must be used to depreciate nonresidential property placed in service after May 12, 1993, and Table 14.17 rates are used to depreciate nonresidential property placed in service on or before May 12, 1993.

Table 14.16

MACRS percentage table: Straight-line depreciation over 27.5 years using midmonth convention

Recovery year	Month in the first recovery year the property is placed in service											
	1	2	3	4	5	6	7	8	9	10	11	12
1	3.485	3.182	2.879	2.576	2.273	1.970	1.667	1.364	1.061	0.758	0.455	0.152
2	3.636	3.636	3.636	3.636	3.636	3.636	3.636	3.636	3.636	3.636	3.636	3.636
3	3.636	3.636	3.636	3.636	3.636	3.636	3.636	3.636	3.636	3.636	3.636	3.636
4	3.636	3.636	3.636	3.636	3.636	3.636	3.636	3.636	3.636	3.636	3.636	3.636
5	3.636	3.636	3.636	3.636	3.636	3.636	3.636	3.636	3.636	3.636	3.636	3.636
6	3.636	3.636	3.636	3.636	3.636	3.636	3.636	3.636	3.636	3.636	3.636	3.636
7	3.636	3.636	3.636	3.636	3.636	3.636	3.636	3.636	3.636	3.636	3.636	3.636
8	3.636	3.636	3.636	3.636	3.636	3.636	3.636	3.636	3.636	3.636	3.636	3.636
9	3.636	3.636	3.636	3.636	3.636	3.636	3.636	3.636	3.636	3.636	3.636	3.636
10	3.637	3.637	3.637	3.637	3.637	3.637	3.636	3.636	3.636	3.636	3.636	3.636
11	3.636	3.636	3.636	3.636	3.636	3.636	3.637	3.637	3.637	3.637	3.637	3.637
12	3.637	3.637	3.637	3.637	3.637	3.637	3.636	3.636	3.636	3.636	3.636	3.636
13	3.636	3.636	3.636	3.636	3.636	3.636	3.637	3.637	3.637	3.637	3.637	3.637
14	3.637	3.637	3.637	3.637	3.637	3.637	3.636	3.636	3.636	3.636	3.636	3.636
15	3.636	3.636	3.636	3.636	3.636	3.636	3.637	3.637	3.637	3.637	3.637	3.637
16	3.637	3.637	3.637	3.637	3.637	3.637	3.636	3.636	3.636	3.636	3.636	3.636
17	3.636	3.636	3.636	3.636	3.636	3.636	3.637	3.637	3.637	3.637	3.637	3.637
18	3.637	3.637	3.637	3.637	3.637	3.637	3.636	3.636	3.636	3.636	3.636	3.636
19	3.636	3.636	3.636	3.636	3.636	3.636	3.637	3.637	3.637	3.637	3.637	3.637
20	3.637	3.637	3.637	3.637	3.637	3.637	3.636	3.636	3.636	3.636	3.636	3.636
21	3.636	3.636	3.636	3.636	3.636	3.636	3.637	3.637	3.637	3.637	3.637	3.637
22	3.637	3.637	3.637	3.637	3.637	3.637	3.636	3.636	3.636	3.636	3.636	3.636
23	3.636	3.636	3.636	3.636	3.636	3.636	3.637	3.637	3.637	3.637	3.637	3.637
24	3.637	3.637	3.637	3.637	3.637	3.637	3.636	3.636	3.636	3.636	3.636	3.636
25	3.636	3.636	3.636	3.636	3.636	3.636	3.637	3.637	3.637	3.637	3.637	3.637
26	3.637	3.637	3.637	3.637	3.637	3.637	3.636	3.636	3.636	3.636	3.636	3.636
27	3.636	3.636	3.636	3.636	3.636	3.636	3.637	3.637	3.637	3.637	3.637	3.637
28	1.970	2.273	2.576	2.879	3.182	3.485	3.636	3.636	3.636	3.636	3.636	3.636
29	0.000	0.000	0.000	0.000	0.000	0.000	0.152	0.455	0.758	1.061	1.364	1.667

Table 14.17

MACRS percentage table: Straight-line depreciation over 31.5 years using midmonth convention

Recovery year	\multicolumn{12}{c}{Month in the first recovery year the property is placed in service}											
	1	2	3	4	5	6	7	8	9	10	11	12
1	3.042	2.778	2.513	2.249	1.984	1.720	1.455	1.190	0.926	0.661	0.397	0.132
2	3.175	3.175	3.175	3.175	3.175	3.175	3.175	3.175	3.175	3.175	3.175	3.175
3	3.175	3.175	3.175	3.175	3.175	3.175	3.175	3.175	3.175	3.175	3.175	3.175
4	3.175	3.175	3.175	3.175	3.175	3.175	3.175	3.175	3.175	3.175	3.175	3.175
5	3.175	3.175	3.175	3.175	3.175	3.175	3.175	3.175	3.175	3.175	3.175	3.175
6	3.175	3.175	3.175	3.175	3.175	3.174	3.175	3.175	3.174	3.175	3.175	3.175
7	3.175	3.175	3.174	3.175	3.174	3.175	3.174	3.175	3.175	3.175	3.174	3.175
8	3.175	3.174	3.175	3.174	3.175	3.174	3.175	3.174	3.175	3.174	3.175	3.174
9	3.174	3.175	3.174	3.175	3.174	3.175	3.174	3.175	3.174	3.175	3.174	3.175
10	3.175	3.174	3.175	3.174	3.175	3.174	3.175	3.174	3.175	3.174	3.175	3.174
11	3.174	3.175	3.174	3.175	3.174	3.175	3.174	3.175	3.174	3.175	3.174	3.175
12	3.175	3.174	3.175	3.174	3.175	3.174	3.175	3.174	3.175	3.174	3.175	3.174
13	3.174	3.175	3.174	3.175	3.174	3.175	3.174	3.175	3.174	3.175	3.174	3.175
14	3.175	3.174	3.175	3.174	3.175	3.174	3.175	3.174	3.175	3.174	3.175	3.174
15	3.174	3.175	3.174	3.175	3.174	3.175	3.174	3.175	3.174	3.175	3.174	3.175
16	3.175	3.174	3.175	3.174	3.175	3.174	3.175	3.174	3.175	3.174	3.175	3.174
17	3.174	3.175	3.174	3.175	3.174	3.175	3.174	3.175	3.174	3.175	3.174	3.175
18	3.175	3.174	3.175	3.174	3.175	3.174	3.175	3.174	3.175	3.174	3.175	3.174
19	3.174	3.175	3.174	3.175	3.174	3.175	3.174	3.175	3.174	3.175	3.174	3.175
20	3.175	3.174	3.175	3.174	3.175	3.174	3.175	3.174	3.175	3.174	3.175	3.174
21	3.174	3.175	3.174	3.175	3.174	3.175	3.174	3.175	3.174	3.175	3.174	3.175
22	3.175	3.174	3.175	3.174	3.175	3.174	3.175	3.174	3.175	3.174	3.175	3.174
23	3.174	3.175	3.174	3.175	3.174	3.175	3.174	3.175	3.174	3.175	3.174	3.175
24	3.175	3.174	3.175	3.174	3.175	3.174	3.175	3.174	3.175	3.174	3.175	3.174
25	3.174	3.175	3.174	3.175	3.174	3.175	3.174	3.175	3.174	3.175	3.174	3.175
26	3.175	3.174	3.175	3.174	3.175	3.174	3.175	3.174	3.175	3.174	3.175	3.174
27	3.174	3.175	3.174	3.175	3.174	3.175	3.174	3.175	3.174	3.175	3.174	3.175
28	3.175	3.174	3.175	3.174	3.175	3.174	3.175	3.174	3.175	3.174	3.175	3.174
29	3.174	3.175	3.174	3.175	3.174	3.175	3.174	3.175	3.174	3.175	3.174	3.175
30	3.175	3.174	3.175	3.174	3.175	3.174	3.175	3.174	3.175	3.174	3.175	3.174
31	3.174	3.175	3.174	3.175	3.174	3.175	3.174	3.175	3.174	3.175	3.175	3.175
32	1.720	1.984	2.249	2.513	2.778	3.042	3.175	3.174	3.175	3.174	3.175	3.174
33	0.000	0.000	0.000	0.000	0.000	0.000	0.132	0.397	0.661	0.926	1.190	1.455

Table 14.18

MACRS percentage table: Straight-line depreciation over 39 years using midmonth convention

	Month in the first recovery year the property is placed in service											
Year	**1**	**2**	**3**	**4**	**5**	**6**	**7**	**8**	**9**	**10**	**11**	**12**
1	2.461	2.247	2.033	1.819	1.605	1.391	1.177	0.963	0.749	0.535	0.321	0.107
2–39	2.564	2.564	2.564	2.564	2.564	2.564	2.564	2.564	2.564	2.564	2.564	2.564
40	0.107	0.321	0.535	0.749	0.963	1.177	1.391	1.605	1.819	2.033	2.247	2.461

Example 13

Cedarvale Development Corporation purchased a 10-unit apartment complex for $800,000, which included land valued at $100,000. The property was placed in service on August 15. Prepare a depreciation schedule using the MACRS method of calculating depreciation.

Solution

Step 1: Determine the class life and the recovery period of the property. The property is considered residential rental real property and has a recovery period of 27.5 years.

Step 2: Find the depreciation rate for a 27.5-year real property using the midmonth convention placed in service in August (see Table 14.16). Multiply this rate times the unadjusted basis of the property ($800,000 − 100,000 = $700,000) to determine the annual amount of depreciation for the first year.

$$\text{annual depreciation} = \text{unadjusted basis} \times \text{depreciation rate}$$
$$= \$700,000 \times .01364$$
$$= \$9,548$$

Step 3: Calculate the book value at the end of the first year.

$$\text{book value} = \text{original cost} - \text{accumulated depreciation}$$
$$= \$700,000 - \$9,548$$
$$= \$690,452$$

Since the property was placed in service during the eighth month of the tax year, we must multiply the previous year's book value of the property each year by the percentages for the eighth month in Table 14.16 to determine the annual depreciation. The rates in the table reflect the midmonth convention and require no further calculations. However, if we wanted to determine the monthly depreciation expense for financial purposes, we would divide the first year's depreciation amount by 4.5 ($9,548 ÷ 4.5 =

$2,121.78). Under the midmonth convention, we are allowed to claim only one-half of the monthly depreciation amount the month the property is placed in service ($2,121.78 × .5 = $1,060.89), and a full month's depreciation for the remaining 4 months ($2,121.78 × 4 = $8,487.12) for a total of $9,548.01 ($8,487.12 + $1,060.89 = $9,548.01) for the first year. Each year thereafter, we would divide the annual depreciation amount by 12 to determine the monthly depreciation amount.

The calculations required to complete a depreciation schedule for Example 13 are shown in Table 14.19.

Table 14.19

Percentage method depreciation schedule

Year	Depreciation calculation	Depreciation amount	Accumulated depreciation	Book value
0				$700,000
1	$700,000 × .01364	$ 9,548	$ 9,548	690,452
2	700,000 × .03636	25,452	35,000	665,000
3	700,000 × .03636	25,452	60,452	639,548
4	700,000 × .03636	25,452	85,904	614,096
5	700,000 × .03636	25,452	111,356	588,644
6	700,000 × .03636	25,452	136,808	563,192
7	700,000 × .03636	25,452	162,260	537,740
8	700,000 × .03636	25,452	187,712	512,288
9	700,000 × .03636	25,452	213,164	486,836
10	700,000 × .03636	25,452	238,616	461,384
11	700,000 × .03637	25,459	264,075	435,925
12	700,000 × .03636	25,452	289,527	410,473
13	700,000 × .03637	25,459	314,986	385,014
14	700,000 × .03636	25,452	340,438	359,562
15	700,000 × .03637	25,459	365,897	334,103
16	700,000 × .03636	25,452	391,349	308,651
17	700,000 × .03637	25,459	416,808	283,192
18	700,000 × .03636	25,452	442,260	257,740
19	700,000 × .03637	25,459	467,719	232,281
20	700,000 × .03636	25,452	493,171	206,829
21	700,000 × .03637	25,459	518,630	181,370
22	700,000 × .03636	25,452	544,082	155,918
23	700,000 × .03637	25,459	569,541	130,459
24	700,000 × .03636	25,452	594,993	105,007
25	700,000 × .03637	25,459	620,452	79,548
26	700,000 × .03636	25,452	645,904	54,096
27	700,000 × .03637	25,459	671,363	28,637
28	700,000 × .03636	25,452	696,815	3,185
29	700,000 × .00455	3,185	700,000	0

Notice that it takes 29 calendar years to depreciate property with a 27.5-year recovery period. This is because we can only claim part of a year's depreciation in the first and twenty-ninth years, as indicated by the reduced percentages for these years shown in Table 14.16.

The MACRS method of calculating the recovery cost of property is less complicated compared to other methods presented in this chapter and provides for an accelerated rate of recovery that can be beneficial to a business and the economy. The tax considerations involving the depreciation methods presented in this chapter can be found in IRS Publication 946.

CHECK YOUR KNOWLEDGE

Modified Accelerated Cost-Recovery System (MACRS)

Use Tables 14.10 and 14.11 to determine the cost-recovery period and cost-recovery rates for the following property.

Type of recovery property	Class-life period	Recovery rate
1. delivery truck	_____	_____
2. winery grape vineyard	_____	_____
3. office furniture	_____	_____
4. dairy building	_____	_____

Use the appropriate tables in Section 14.5 to solve the following word problems.

5. Using the declining-balance method, what is the amount of depreciation and the book value for a $20,000 automobile at the end of the first year. Round to nearest dollar.

6. Southside Press purchased an offset printer for $7,500 that was placed in service on October 15. Prepare a depreciation schedule for the property using the MACRS percentage table method. Round to the nearest dollar.

7. Powers General Contractors purchased and placed in service a small backhoe that cost $50,000 on May 5. Prepare a cost-recovery schedule for the machine using the MACRS straight-line method. Round to the nearest dollar.

8. The law firm of Rivera & Ponto built an office building for its practice that was placed in service on June 29, 1992. The facility cost $295,000 including land, which was valued at $70,000. Prepare a partial cost-recovery schedule for the first 5 years using the MACRS percentage table method. Round to the nearest dollar.

9. Determine the book value at the end of the second year for a building classified as nonresidential property that was placed in service on March 20, 1997. The building cost $350,000, excluding the cost of the land, which was valued at $25,000. Round to the nearest dollar.

14.5 EXERCISES

Use Table 14.13 to complete the following.

Type of property	Recovery year	Class-life recovery period	Recovery rate
1. microcomputer	3	_____	_____
2. apple orchard	5	_____	_____
3. 3-year-old race horse	2	_____	_____
4. wastewater treatment plant	10	_____	_____
5. commercial bus	4	_____	_____
6. office desks/chairs	6	_____	_____
7. farm equipment storage building	15	_____	_____

Use Table 14.13 to compute the first year depreciation for the following. Round to nearest dollar.

Property cost	Recovery period	Recovery rate	Recovery amount	Book value
8. $5,000	3	_____	_____	_____
9. $18,500	5	_____	_____	_____
10. $32,700	10	_____	_____	_____
11. $85,250	20	_____	_____	_____
12. $48,600	7	_____	_____	_____
13. $125,000	15	_____	_____	_____

Use the appropriate tables presented in Section 14.5 to solve the following word problems.

14. Lockner Cards and Gifts purchased display fixtures for $12,800. Calculate the book value at the end of the first year using the MACRS declining-balance method. Round to the nearest dollar.

15. Prepare a cost recovery schedule using the information given in exercise 12. Round to the nearest dollar.

16. Allied Transport Systems purchased a tractor to haul freight trailers for $80,000. Prepare a depreciation schedule using the MACRS percentage table method (Table 14.13). Round to the nearest dollar.

Answers to CYK: **1.** 5-year; 40% **2.** 10-year; 20%   **3.** 7-year; 28.7%   **4.** 10-year; 20%   **5.** $4,000, $16,000   **6.** book value year 1, $6,000; book value year 3, $2,160; book value year 6, $0   **7.** book value year 1, $45,000; book value year 3, $25,000; book value year 6, $0   **8.** book value year 5, $192,555   **9.** book value year 2, $333,910

17. Georgio's Professional Cleaning Services purchased three automobiles for use by its sales personnel at $16,500 each. Prepare a depreciation schedule for the automobiles using the MACRS percentage table method (Table 14.13). Round to the nearest dollar.

18. Springwood Farms erected a horse stable for $175,000. Find the book value at the end of the tenth year using the MACRS percentage table method (Table 14.13). Round to the nearest dollar.

19. Metroplex Transportation purchased a bus for $120,000, which was placed in service on November 12. Find the book value at the end of the first year using the MACRS straight-line method. Round to the nearest dollar.

20. Prepare a cost recovery schedule using the information given in exercise 18 for the first 10 years. Round to the nearest dollar.

21. Marine Salvage purchased a tugboat for $350,000. The vessel was placed in service on July 20. Find the book value at the end of year 1 using the MACRS straight-line method. Round to the nearest dollar.

22. Marcia Perez purchased a three-family house for $145,000 that she intends to use as rental property. The property was placed in service on April 5. Find the book value at the end of the first year using the MACRS percentage table method. Round to the nearest dollar.

23. Prepare a cost recovery schedule using the percentage table method for exercise 21.

24. Realty Management purchased a six-unit apartment building for $375,000, which included land valued at $50,000. The property was placed in service on May 2, 1990. Find the book value of the property for each of the first 5 years using the MACRS percentage table method.

25. Metropolitan Development built a new retail store complex for $850,000. The rental property was placed in service on September 28, 1998. Find the book value of the property for each of the first 5 years.

CHAPTER REVIEW EXERCISES

The following problems review all mathematics and business concepts presented in the chapter. If you experience any difficulty solving a problem, go to the appropriate chapter section and review the examples provided. Express all answers as decimals rounded to tenths, or dollars rounded to cents, unless otherwise instructed.

1. On January 8, Ontario Plastics Inc. purchased a computerized molding machine for $145,000. The company paid $3,500 to have the system shipped to their location and $1,500 to be installed. Management estimates the machine will have a $30,000 salvage value at the end of its 5-year useful life. Determine the book value at the end of the first year using the straight-line depreciation method.

2. Use the information provided in problem 1 to determine the book value of the asset at the end of the first year if Ontario Plastics Inc.

uses the double-declining-balance depreciation method.

3. Determine the book value at the end of the first year using the sum-of-the-years' digits depreciation method for the moulding machine described in problem 1.

4. Find the book value at the end of the first year for the machine in problem 1. using the units-of-production depreciation method. The machine's estimated production is 3 million units of which 520,800 units were produced during the current year.

5. Top-Hat Limousine Service purchased a 12-passenger stretch limousine for $75,000 on May 20. It is estimated the vehicle will have a useful life of 4 years and a salvage value of $15,000. Determine the book value of the asset at the end of the second year using the straight-line depreciation method.

6. Determine the book value of the vehicle at the end of the first year if Top-Hat Limousine Service uses the 150% declining-balance method to depreciate the limousine described in exercise 5.

7. Find the book value of the limousine at the end of the first year if the company uses the sum-of-the-years'-digits method to depreciate the vehicle.

8. Determine the book value at the end of the second year using the units-of-production depreciation method if the limousine was driven 12,542 miles the first year and 28,637 miles the second year. The vehicle's estimated useful life is 120,000 miles.

9. Prepare a cost recovery schedule for the limousine described in exercise 5 using the MACRS percentage table method.

10. Charles Jefferson purchased a three-bay garage for his automobile repair business for $150,000, which included land valued at $25,000. The garage was placed in service for depreciation purposes on April 17, 1998. Prepare a partial cost recovery schedule for the first 5 years of the buildings depreciable life using the MACRS percentage table method. Round amounts to the nearest dollar.

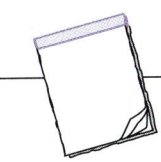

EXPRESS YOUR UNDERSTANDING

Compose one or two well-written sentences to express the requested information in your own words.

1. What is depreciation and why do businesses depreciate their assets?

2. How would you determine the monthly amount of depreciation expense of an asset using the traditional straight-line method, if the asset was purchased on January 10 and is expected to have no value once it is depreciated?

3. Explain how the traditional declining-balance method of depreciating assets differs from the traditional straight-line method.

4. What does the term book value mean and how is an asset's book value determined?

5. Describe how you would determine the annual depreciation expense for an asset using the traditional units-of-production method. How does this method of depreciation differ from the other traditional methods?

6. Explain how you would determine the sum-of-the-years'-digits fraction for an asset with a useful life of 5 years.

7. Under MACRS, how is the class life of an asset determined and categorized?

8. Describe the steps required to determine an asset's book value at the end of the first year using the MACRS straight-line method of cost recovery. Assume the asset is classified as personal property and is placed in service on October 3 of the first year.

9. Explain how you would determine the book value of an asset at the end of the first year using the MACRS percentage-table method for an asset with a 10-year recovery period.

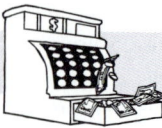

Mind Your Business

On January 10, 1997 when Media World opened its doors for business, many decisions were made involving the purchase of equipment and inventory necessary to get the business started on a solid financial foundation. At that time, you and your partners purchased various styles of shelving and storage cabinets to display merchandise. You also had to purchase computers, printers, computerized cash registers, and other types of equipment to properly equip the store and business office. As the business manager, you decided to group this equipment into two categories; store equipment and office equipment for depreciation purposes. The financial records of Media World include depreciation schedules for these assets, which were prepared the first month the business opened. These depreciation schedules are used each month by the bookkeeper, Kate Bradley, to develop financial data necessary to prepare the financial statements.

Activity 1

All equipment classified as store equipment was valued at $50,000 at the time of purchase. The equipment is estimated to have a salvage value of $2,000 after 10 years of service, and is being depreciated under the straight-line method. You have asked the bookkeeper to supply you with a copy of the schedule. What financial information should you expect to receive in the statement? (Present a complete depreciation schedule.)

Activity 2

The equipment considered to be in the office equipment category had a total purchase price of $12,000. This equipment is expected to have a useful life of 5 years, a salvage value of $2,500, and is being depreciated by the 150% declining-balance method. What financial information should you expect to receive in the statement? (Present a complete depreciation schedule.)

Activity 3

You have received the September financial statement drafts from the bookkeeper. (Refer to Chapter 4, "Mind Your Business.") Analyze each statement carefully, and determine if the amounts presented in each statement involving the depreciation of store equipment and office equipment are correct. Your analysis will include a quantitative assessment that validates the entries.

SELF-TEST

A. Terminology

Complete the following items using the key terms presented at the beginning of the chapter. Check your responses against the answer key at the end of the test.

1. The periodic deduction allowed by the Internal Revenue Service to recover the cost of a business asset that is used more than 1 year is called _____.

2. The amount of the periodic deduction is reported as _____ in the income statement and as an increase in the _____ of the asset in the balance sheet.

3. The _____ of an asset is the value reported in the balance sheet after deducting the accumulated depreciation from the total cost of the asset when it is placed in service.

4. The _____ method of depreciation allows a business to evenly distribute the cost of an asset over its useful life.

5. Traditional methods used to accelerate the depreciation of an asset's cost are called _____, and _____.

6. When the actual service rendered by an asset is a more appropriate base for depreciation than its years of service, a business may use the _____ method.

7. The _____ allows businesses to write off the cost of most tangible depreciable assets placed in service after 1980 and before 1987 more quickly than traditional methods.

8. Under MACRS, the type of recovery property being depreciated determines which _____ will be used to determine the recovery amount.

9. The _____ is also known as the General Depreciation System and applies to all tangible property placed in service after 1986.

10. Under MACRS, the _____ treats all property as being placed in service on the midpoint of the year regardless of when the property is actually placed in service.

B. Calculation

The following concepts and short problems are designed to test your understanding of the objectives identified at the beginning of the chapter. Answers are at the end of the test. Use the appropriate tables presented in this chapter to solve.

11. Jo-Ann's placed in service $8,000 of equipment for use in its business. The equipment will be depreciated using the straight-line method. If the equipment is estimated to have a useful life of 5 years and a salvage value of $1,200, find the book value at the end of the first year.

12. An asset costing $25,000 is depreciated under the 150% declining-balance method. The asset has a useful life of 5 years and a salvage value of $3,500. Find the book value at the end of the first year if the asset was placed in service on October 2.

13. Poland Casting Company placed in service a stamping machine that cost $80,650. The company paid an additional $2,500 to have the machine installed. Find the amount of depreciation claimed the first year if the machine has a salvage value of $3,200, it was operated 1,640 hours during the year, and has an estimated life of 65,000 hours. Round your answer to the nearest dollar.

14. Hastings Bottling Company purchased a bottle capping machine at a cost of $48,000. The machine is expected to cap 4,000,000 units during its useful life and is estimated to have a salvage value of $2,000. Find the book value of the machine at the end of the second year if the machine capped 60,250 units the first year and 580,470 units the second year. (Round your answer to the nearest dollar.)

15. Ever-Green Landscaping Company purchased a light-duty truck for $28,000 on March 12. If the asset has a salvage value of $4,000 and a useful life of 5 years, what is the book value at the end of the third year under the sum-of-the-years'-digits depreciation method?

16. Determine the book value at the end of the first year of a tractor used to haul trailers over the road using the MACRS declining-balance method. The tractor cost $200,000 and was placed in service on March 15. Round your answer to the nearest dollar.

17. Nugent Foods purchased a computer for $3,500 that was placed in service on December 6. What was the book value at the end of the second year if the property is being depreciated by the MACRS straight-line method? Round your answer to the nearest dollar.

18. Trim-N-Fit Conditioners purchased furniture for its office at a cost of $18,200. Find the book value at the end of the third year using the MACRS percentage table method.

19. James Anderson purchased a duplex for rental purposes for $140,000. The property was placed in service on April 8. Find the book value of the property at the end of the second year using the MACRS percentage tables. Round your answer to the nearest dollar.

20. Three-Star Investment purchased an eight-store mini mall for $750,000 that included land valued at $125,000. The property was placed in service on May 20. What was the book value of the property at the end of the first year?

Answers to Self-Test: *1.* depreciation   *2.* depreciation expense, accumulated depreciation   *3.* book value   *4.* straight-line   *5.* declining-balance and sum-of-the-years'-digits   *6.* units-of-production method   *7.* accelerated cost recovery system   *8.* class   *9.* modified accelerated cost recovery system   *10.* half-year convention   *11.* $6,640   *12.* $23,125   *13.* $2,017.20   *14.* $40,311   *15.* $8,800   *16.* $133,330   *17.* $2,450   *18.* $7,958.86   *19.* $131,304   *20.* $612,600

15

Key Terms

inventory
periodic inventory
  system
specific identification
  method
average cost method
lower-of-cost-or-market
  method
current market value
first-in, first-out (FIFO)
  method
last-in, first-out (LIFO)
  method
retail method
gross margin method

INVENTORY VALUATION

Learning Objectives

After completing the exercises in this chapter, you will be able to:

1. Calculate the value of ending inventory using the specific identification method.

2. Calculate the value of ending inventory using the average cost method.

3. Calculate the value of ending inventory using the lower-of-cost-or-market method.

4. Calculate the value of ending inventory using the FIFO method.

5. Calculate the value of ending inventory using the LIFO method.

6. Estimate the value of ending inventory using the retail method.

7. Estimate the value of ending inventory by using the gross margin method.

8. Define key terms.

INTRODUCTION

An **inventory** is a list of items that a company or business has in its possession at any specific time. For a manufacturer, this list may be made up of a variety of items ranging from raw materials to be used in a production process, to finished goods waiting to be shipped to wholesalers and retailers. In a small business, the list might consist primarily of goods, such as sneakers, bicycles, hammers, or rakes, owned by the business and held for resale to customers.

Inventory is recorded as a current asset on a business's balance sheet, and because it is very important for businesses to know the value of their assets, they must be able to determine the value of their remaining inventory at any time in the business cycle, such as at the end of a month, quarter, or year.

Initially, this might sound like a simple task, but when businesses have very large inventories of a variety of goods and materials, it is far from a simple task to maintain an accurate inventory. Ideally, a company would like to maintain a *continuous inventory* that would indicate at any instant in the business cycle exactly what items were in the business's possession. The use of computers and scanners is helping businesses to come closer to this ideal. In most cases, this ideal is far from being attained. Consequently, a form of *periodic inventory* is used:

1. The company or business periodically takes a physical count of the inventory to determine how *many* items are currently in the company's possession.

2. The company determines the *dollar value* of those items.

We refer to this as a **periodic inventory system** because the inventory is taken periodically rather than continuously.

In this chapter, we will present several different methods used to place a value on *beginning inventory* and *ending inventory,* which are phrases we will use to mean the value of inventory either at the beginning or at the end of some specified period of time. In order to better demonstrate these various methods, we will refer to the following illustration of the Lake Superior Oil Company.

The Lake Superior Oil Company is a retailer of home heating oil in a metropolitan area. It purchases fuel oil from companies that refine the oil and sell it to homeowners and businesses. Immediately following, is a partial income statement for Superior showing barrels of fuel oil purchased during the year.

Income Statement for Lake Superior Oil Company
(this is not a complete income statement)

Cost of goods sold:
   Beginning inventory (carried over
from last year)   1,000 barrels @ $34 = $ 34,000

   Purchases during the current year:
      January 14   5,000 barrels @ $30 = $150,000
      July 23   5,000 barrels @ $20 = $100,000
      October 18   8,000 barrels @ $17 = $136,000
      December 16   1,000 barrels @ $16 = $ 16,000
   Total goods available for sale   20,000 barrels $436,000

   Less value of ending inventory   ?
      units of physical inventory remaining = 2,000
   Unit value of inventory = $?

Cost of goods sold $?

    Superior must now place a dollar value on the remaining barrels of oil, but at what price? at $34? $30? $16? Since many barrels of oil may be placed in a single tank, Lake Superior Oil Company may not be able to distinguish between the oil purchased at $16 per barrel and that purchased at $34 per barrel. Sections 15.1 through 15.3 will demonstrate five different methods Lake Superior Oil (and other companies) can use to place a value on ending inventory.

15.1   SPECIFIC IDENTIFICATION, AVERAGE COST, AND LOWER-OF-COST-OR-MARKET METHODS

Specific Identification Method

Learning objective
Calculate the value of ending inventory using the specific identification method.

The specific identification method is a method of identifying and labeling each unit with the original cost of the unit. This makes the job of evaluating ending inventory very easy. As you physically count your remaining inventory, you also record the cost identified on the tag or label of each unit.

    If Lake Superior Oil Company were to pour all barrels of oil into a single tank, this method of specific identification would be impossible to use. On the other hand, if Superior kept each barrel of oil in its own container and labeled each container with the cost it paid for that barrel of oil, then the specific identification method could be used.

Example 1

Assume that Lake Superior Oil Company keeps the oil it receives in individual, 42-gallon barrels with labels that include a code indicating the price the company paid for that specific barrel of oil. Determine the value of the ending inventory if at the end of the period the company had 300 barrels that cost $34 each, 450 barrels that cost $30, 350 barrels that cost $20, and 900 barrels that cost $17.

Solution

Since the company stores the oil in well-marked barrels, it can specifically identify what it paid for each barrel and can easily determine the original cost of the oil in their ending inventory.

Ending inventory:

300 barrels @ $34 =	$10,200
450 barrels @ $30 =	13,500
350 barrels @ $20 =	7,000
900 barrels @ $17 =	15,300
2000	$46,000

So, the value of the ending inventory is $46,000.

Now that we have the value of the ending inventory, we can complete the cost of goods sold section of the income statement of the Lake Superior illustration.

$$\text{cost of goods sold} = \text{total goods available for sale} - \text{ending inventory}$$
$$= \$436,000 - \$46,000$$
$$= \$390,000$$

The ending inventory value is also shown as *merchandise inventory* in the *current assets* section of the company's balance sheet.

The specific identification method would be very time-consuming and impractical for companies that have a large number of units with low costs. However, this method is beneficial for companies with expensive units such as works of art or automobiles.

Average Cost Method

Learning objective
Calculate the value of ending inventory using the average cost method.

The **average cost method** identifies the average cost per unit of each item that remains in the inventory at the end of a period. To calculate the average cost of ending inventory, we use the *quantity times price equals cost* concept of the basic business transaction. In the Superior Oil illustration the *quantity* column is the number of barrels purchased, the *price* column is the dollar amounts for barrels purchased, and the *cost* column is obtained by multiplying the price times the quantity. Next, we divide the total of the cost column by the total of the quantity column (in this case, in barrels). This calculation will always result in the average cost per unit.

Example 2

Determine the value of Superior's ending inventory using the average cost method.

Solution

Step 1:   Determine the total value of the inventory for the period.

Quantity	×	Unit price	=	Cost
1,000 barrels		$34		$ 34,000
5,000		30		150,000
5,000		20		100,000
8,000		17		136,000
1,000		16		16,000
20,000 barrels				$436,000

Step 2:   Determine the average cost per unit (i.e., per barrel).

$436,000 ÷ 20,000 barrels = $21.80 average cost per barrel

Step 3:   Determine the value of the ending inventory by multiplying the number of units remaining in inventory by the average cost per unit.

$21.80 × 2,000 barrels = $43,600

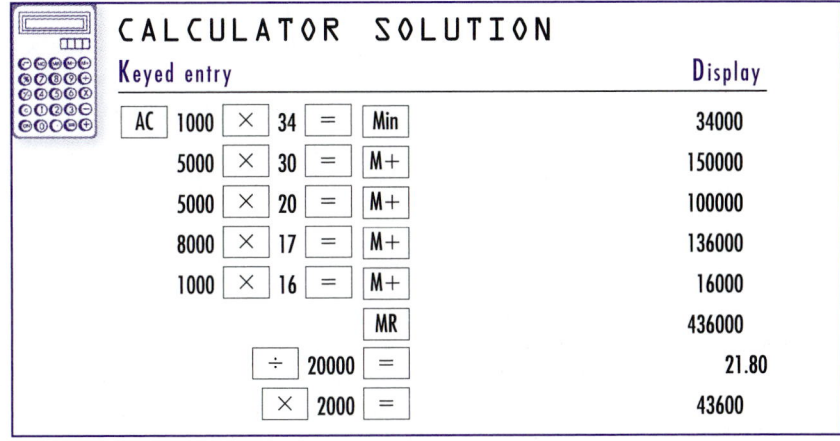

CALCULATOR SOLUTION	
Keyed entry	**Display**
AC 1000 × 34 = Min	34000
5000 × 30 = M+	150000
5000 × 20 = M+	100000
8000 × 17 = M+	136000
1000 × 16 = M+	16000
MR	436000
÷ 20000 =	21.80
× 2000 =	43600

This will allow Lake Superior Oil Company to complete the *less ending inventory* line of its income statement. Now that we have the value of the ending inventory using the average cost method, we can complete the cost of goods sold section of the income statement of the Lake Superior Oil illustration.

$$\text{cost of goods sold} = \text{total goods available for sale} - \text{ending inventory}$$
$$= \$436{,}000 - \$43{,}600$$
$$= \$392{,}400$$

The average cost method would be especially useful if Lake Superior Oil bought its oil in bulk, and it was delivered by the tank truck or railroad tank car and all of the oil was put into large oil tanks for storage and distribution. If shipments of fuel oil that had different costs were all loaded into one or more common storage tanks, it would be impossible to use the specific identification method. The average cost method would then be a very reasonable way to determine the value of ending inventory.

Lower-of-Cost-or-Market Method

Learning objective
Calculate the value of ending inventory using the lower-of-cost-or-market method.

A business may also elect to use the **lower-of-cost-or-market method,** when calculating the value of ending inventory under the specific identification method. This procedure allows individual units of the inventory to be valued at either the *actual cost at the time of purchase* or at the *current market value*, whichever is lower. The **current market value** is what it would cost to replace the unit—in this case a barrel of oil.

Example 3

In Example 1, the value of the ending inventory for the Lake Superior Oil Company using the specific identification method was $46,000. If the current market value is $16 per barrel, calculate the value of ending inventory using the lower-of-cost-or-market procedure.

Solution

The total number of barrels of oil in the inventory at the end of the period is 2,000 barrels. Therefore, the market value of the oil is $32,000 (2,000 @ $16/barrel, 2000 × $16 = $32,000). According to the procedure of the lower-of-cost-or-market method, the ending inventory would be valued at the lower price of $32,000, rather than the actual cost of $46,000.

Using the lower-of-cost-or-market method, the cost of goods sold section of the income statement would be

$$\text{cost of goods sold} = \text{total goods available for sale} - \text{ending inventory}$$
$$= \$436,000 - \$32,000$$
$$= \$404,000$$

A company might choose any one of these methods for determining the value of ending inventory, depending on which method was more convenient, more expedient, or more beneficial for accounting purposes.

CHECK YOUR KNOWLEDGE

Specific Identification, Average Cost, and Lower-of-Cost-or-Market Methods

Solve the following problems. Round dollar amounts to the nearest cent.

1. Let us suppose that the Impressionistic Art department at the National Gallery of Art in Washington, D.C., made three purchases during this past year:

June 5	a Monet painting for $75 million
October 7	a Renoir painting for $100 million
December 8	a Dégàs painting for $85 million

    If the Impressionistic Art department sold the Degas, and we make the assumption that the department had no beginning inventory, then what is the value of the ending inventory?

2. Using the information given in problem 1, determine the value of the Impressionistic Art department's ending inventory by the lower-of-cost-or-market procedure.

    Beginning inventory: none

    Purchases:

		Cost	**Market**
June 5	Monet	$ 75 million	$ 70 million
October 7	Renoir	$100 million	$100 million
December 8	Degas	$ 85 million	$ 90 million

    Ending inventory: Monet and Renoir

3. Shirley opened a shoe store during the month of June. Using the following information regarding the purchases for Shirley's Shoe Store, find the value of her ending inventory using the average cost method.

Purchases:	June 7	1,000 pair @ $10
	August 11	1,500 pair @ $11
	October 21	900 pair @ $17
	Ending inventory	1,100 pair

4. Use the following partial income statement for Higgin's Hardware Store to determine (a) the value of his ending inventory using the average cost method and (b) the cost of goods sold.

Beginning inventory (for 4″ carpenter nails)	7,000 @ $0.10
Purchases:	
February 1	15,000 @ $ .07
May 1	20,000 @ $ .04
December 1	5,000 @ $ .02
Ending inventory	10,000 nails

15.1 EXERCISES

1. Determine the value of the ending inventory by the specific identification method.

300 units @ $28 =
175 units @ $31 =
550 units @ $37 =
875 units @ $26 =

Value of ending inventory =

2. For the following information, find the value of the ending inventory by the average cost method.

Beginning inventory	200 units @ $15 = $_____
Purchases:	
March 10	300 units @ $20 = $_____
March 19	100 units @ $17 = $_____
March 24	400 units @ $24 = $_____
Total cost of goods available for sale	$_____
Calculation of value of ending inventory	
Units of physical inventory remaining = 350 units	
Average unit cost of inventory	$_____
Value of ending inventory	$_____

3. Determine the value of the ending inventory by the lower-of-cost-or-market method.

	Cost	Market
500 units @	$55	$60
230 units @	$53	$50
630 units @	$57	$54
725 units @	$54	$56

Value of ending inventory =

4. The Creepy Critter Pet Store had always had a large supply of insects and spiders, but Craig Croger, the owner of Creepy Critters, has decided to expand into the area of snakes. During this past month, Craig made the following purchases:

October 12	cobra for $100
October 13	python for $60
October 27	anaconda for $75

On December 31, Craig counted his inventory and noted that, of the three snakes purchased, he was able to sell only the cobra. Using the specific identification method, value Creepy Critter's ending inventory for snakes.

5. Adrian Kelsey inherited the Pirate's Cove Marina from his uncle—the late, great Captain Kelsey. Captain Kelsey had not only been in the business of docking yachts, but also of buying and selling sailing vessels. Adrian has decided to expand his business to include cabin cruisers.

Adrian made the following purchases:

April 22	36′ Carver	$100,000
June 29	35′ Criss Craft	$120,000
August 26	38′ Criss Craft	$169,000
August 31	40′ Criss Craft	$189,000
September 3	35′ Carver	$110,000

During the year, Adrian sold the 35′ and 36′ Carvers. Calculate the value of Adrian's ending inventory using the specific identification method.

6. Use the following information to determine the value of the cabin cruisers at the Pirate's Cove Marina. Apply the rule of lower-of-cost-or-market to this situation.

Beginning inventory   none

Purchases:		Cost	Market
April 22	36′ Carver	$100,000	$120,000
June 29	35′ Criss Craft	120,000	100,000
August 26	38′ Criss Craft	169,000	149,000
August 31	40′ Criss Craft	189,000	180,000
September 3	35′ Carver	110,000	130,000

Ending inventory   35′ Criss Craft
                    38′ Criss Craft
                    40′ Criss Craft

7. Rocco Kaminsky owns a TV repair shop and also sells a handful of new television sets. Use the following information to determine the value of Rocco's new television sets at the end of March. Apply the rule of lower-of-cost-or-market.

Beginning inventory

		Cost	Market
	13″ Zenith	$300	$200
	19″ RCA	$200	$325
Purchases			
February 13	25″ Sony	$700	$900
February 20	19″ Zenith	$450	$550
March 4	25″ RCA	$699	$899

Ending inventory   all but the two RCAs, which he sold

8. Dexter Ellsworth owns a sporting goods store in Bakersfield, California, where skateboards are an item in his inventory. Use the following information about skateboards to determine the value of Dexter's ending inventory using the average cost method.

Purchases:	
June 21	15 @ $45
August 23	30 @ $55
November 22	20 @ $65

Ending inventory   17 skateboards

9. Olivia Meredith owns a haircutting salon. Olivia has decided to purchase a new brand of shampoo for her salon. Using the following records,

calculate the value of ending inventory using the average cost method.

Purchases:

February 17	10 cases @ $36
April 13	20 cases @ $42
June 14	100 cases @ $40
Ending inventory	20 cases

10. Maria Vasquez is the accountant for the Peculiar Paint Store. It is her job to determine the value of the cans of Primrose Pink paint remaining in inventory at the year end. Determine the value of ending inventory using the average cost method.

Beginning inventory	
(Primrose Pink)	2 gal @ $8
Purchases:	
January 3	50 gal @ $10
April 3	20 gal @ $20
July 3	10 gal @ $25
Ending inventory	33 gal

11. An inventory of the room air conditioners in the appliance department of Hoffman's Department store at the end of March showed five 8,000-btu air conditioners at a cost of $350 each, four 6,000-btu air conditioners at a cost of $275 each, and seven 5,000-btu air conditioners at a cost of $125 each. In anticipation of summer, the manager of the department purchased ten 8,000-btu air conditioners at a cost of $400 each, six 6,000

btu at $300 each, and eight 5,000 btu at $200 each. What was the value of the inventory of room air conditioners by the specific identification method following the purchases?

12. At the end of September Hoffman's Department store still had in its inventory one 5,000-, two 6,000-, and three 8,000-btu room air conditioners from its end-of-March inventory on hand. In addition Hoffman's had two 5,000-, one 6,000- and one 8,000-btu room air conditioners from its April purchases. What was the value of the ending inventory in September using the specific identification method?

13. Using the average cost method for evaluating inventory and the information given in exercise 11, determine (a) the average cost per unit of the inventory of 5,000 btu air conditioners; (b) the average cost per unit of the inventory of 6,000 btu air conditioners; (c) the average cost per unit of the inventory of 8,000 btu air conditioners; (d) the total value of the inventory of room air conditioners.

14. What was the value of ending inventory of room air conditioners in September in the Hoffman's store (exercise 12) using the average cost method?

15. If the market value in September on the air conditioners was $250 for 8,000 btu, $225 for 6,000 btu and $125 for 5,000 btu, determine the value of Hoffman's ending inventory (exercise 13) in September using the lower of cost or market method.

15.2   THE FIRST-IN, FIRST-OUT (FIFO) AND LAST-IN, FIRST-OUT (LIFO) METHODS

The first-in, first-out and last-in, first-out methods of inventory valuation assume that the flow of items enter and leave the inventory in two different ways. Think of two different types of milk coolers in a store. The first is loaded from the back of the cooler and emptied from the front, so that the milk containers put in first are moved through the cooler and taken out of the front by customers first. The first containers put in the back of the cooler are the first to be taken out of the front (FIFO). The second type of cooler is loaded and emptied from the front of the cooler. The first milk containers loaded into the cooler are pushed to the back of the cooler, and the last containers loaded remain near the doors in the front. Now when customers take milk from the cooler, the last containers loaded are the first to be taken out of the cooler (LIFO).

Learning objective
Calculate the value of ending inventory using the FIFO method.

First-In, First-Out (FIFO) Method

The **first-in, first-out (FIFO) method** is an accounting procedure that describes the way a company might sell its goods. This method assumes that those purchases made first (earliest in the year) will be sold first. Hence, any goods remaining unsold at the end of the year will be valued at the price paid for the latest or most recent purchases, as shown in Figure 15.1.

When a company uses the FIFO method of inventory valuation, it does not mean that the company's actual physical flow of goods is to be sold in this way—this assumption is made for accounting purposes only.

Example 4

Assume that Lake Superior Oil receives its oil in barrels and they are unloaded into a warehouse at a loading dock at the back of the warehouse. The barrels are then moved through the warehouse on conveyor belts to the front of the warehouse where they are loaded onto trucks for delivery. The first barrels received at the warehouse move through the building and are the first to be sold and delivered from the other side of the warehouse. Calculate the value of the ending inventory for Lake Superior Oil Company using the FIFO method.

Solution

To calculate the value of the ending inventory using the FIFO method, we must refer to the purchase records of the Lake Superior Oil Company.

Figure 15.1

FIFO method of inventory valuation

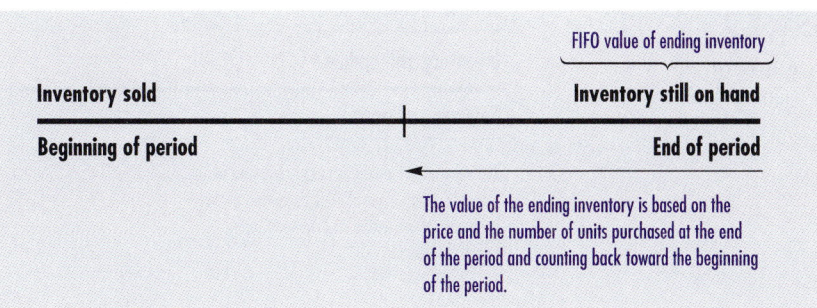

Beginning inventory (carried over
from last year)   1,000 barrels @ $34 = $ 34,000

Purchases during the current year:
January 14	5,000 barrels @ $30 = $150,000
July 23	5,000 barrels @ $20 = $100,000
October 18	8,000 barrels @ $17 = $136,000
December 16	1,000 barrels @ $16 = $ 16,000
Total goods available for sale	20,000 barrels $436,000

Starting with the most recent or latest purchases, we value the remaining
2,000 barrels as follows:

1,000 barrels @ $16 (from Dec. 16 purchases) = $16,000

+1,000 barrels @ $17 (from Oct 18 purchases) = $17,000

2,000 ending inventory $33,000

Learning objective
Calculate the value
of ending inventory
using the LIFO
method.

Last-In, First-Out (LIFO) Method

The **last-in, first-out (LIFO) method** for determining the value of ending
inventory more closely resembles the way in which a company actually
moves its physical inventory. That is, those purchases made during the
latter part of the year are sold first. Hence, any goods remaining unsold at
the end of the year will be valued at the price paid for the first or earliest
purchases, as shown in Figure 15.2.

When a company uses the LIFO method of inventory valuation, it does
not mean that the company's actual physical flow of goods occurs in this
way—this assumption is made for accounting purposes only.

Figure 15.2

LIFO value of end-
ing inventory

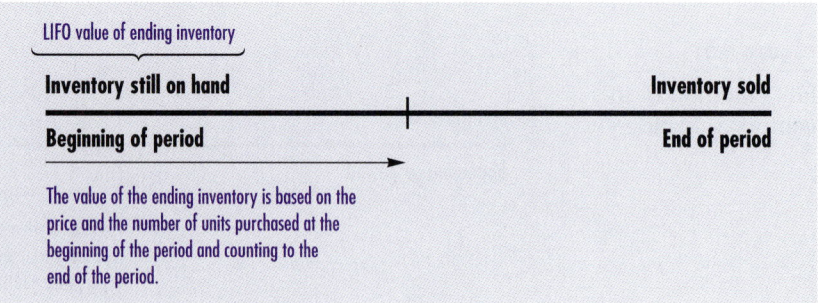

LIFO value of ending inventory

Inventory still on hand   Inventory sold

Beginning of period   End of period

The value of the ending inventory is based on the
price and the number of units purchased at the
beginning of the period and counting to the
end of the period.

Example 5

Assume the warehouse in which the barrels of oil are stored has only one door where the barrels are both entered into the warehouse and then taken out the same door. The first barrels in are moved to the back of the warehouse, and the last barrels remain nearer to the door and will be taken out first. Use the LIFO method to calculate the value of the ending inventory for the Lake Superior Oil Company.

Solution

To calculate the value of the ending inventory using the LIFO method, we must again refer to the purchase record of the Lake Superior Oil Company.

Beginning inventory (carried over from last year)	1,000 barrels @ $34 = $ 34,000

Purchases during the current year:

January 14	5,000 barrels @ $30 = $150,000
July 23	5,000 barrels @ $20 = $100,000
October 18	8,000 barrels @ $17 = $136,000
December 16	1,000 barrels @ $16 = $ 16,000

Ending inventory = 2,000 barrels

Starting with the earliest barrels on hand (in this case, beginning inventory), we value the ending inventory of 2,000 barrels as follows:

1,000 barrels @ $34 (from beginning inventory) = $34,000	
1,000 barrels @ $30 (from Jan. 24 purchases) = $30,000	
2,000 ending inventory $64,000	

Notice that the value of the ending inventory using the FIFO ($33,000) and LIFO ($64,000) methods is significantly different, even though the calculations are based on the same inventory information. Therefore, Lake Superior Oil might select the FIFO method if it were interested in reporting lower profits for tax purposes, or it might select the LIFO method if it were interested in reporting higher profits to its investors. Example 6 will illustrate the effect of the value of ending inventory on profits.

Example 6

During the current period, Lake Superior Oil Company reported revenues of $750,000 and operating expenses of $58,000. Calculate the profit before taxes based on the value of ending inventory determined by the FIFO method in Example 4 and the LIFO method in Example 5.

Solution

FIFO Method

Revenue		$750,000
Cost of goods sold:		
Beginning inventory (carried over from last year)	1,000 barrels @ $34 = $ 34,000	
Purchases during the current year:		
January 14	5,000 barrels @ $30 = $150,000	
July 23	5,000 barrels @ $20 = $100,000	
October 18	8,000 barrels @ $17 = $136,000	
December 16	1,000 barrels @ $16 = $ 16,000	
Total goods available for sale	20,000 barrels $435,000	
Less value of ending inventory	$ 33,000	
Cost of goods sold		$403,000
Gross profit		$347,000
Less operating expenses		$ 58,000
Profit before taxes		$289,000

LIFO Method

Revenue		$750,000
Cost of goods sold:		
Beginning inventory (carried over from last year)	1,000 barrels @ $34 = $ 34,000	
Purchases during the current year:		
January 14	5,000 barrels @ $30 = $150,000	
July 23	5,000 barrels @ $20 = $100,000	
October 18	8,000 barrels @ $17 = $136,000	
December 16	1,000 barrels @ $16 = $ 16,000	
Total goods available for sale	20,000 barrels $436,000	
Less value of ending inventory	$ 64,000	
Cost of goods sold		$372,000
Gross profit		$378,000
Less operating expenses		$ 58,000
Profit before taxes		$320,000

CHECK YOUR KNOWLEDGE

FIFO and LIFO

Solve the following problems.

1. Use the following information to evaluate the ending inventory of Shirley's Shoe Store by the FIFO method.

Beginning inventory	500 pair @ $12
Purchases:	
June 7	1,000 pair @ $10
August 11	1,500 pair @ $11
October 21	900 pair @ $17
Ending inventory	1,100 pair

2. Use the following partial income statement for Carmen's Hardware Store to determine the value of his ending inventory by the FIFO method.

Beginning inventory (for 4″ carpenter nails)	7,000 @ $.10
Purchases:	
February 1	15,000 @ $.07
May 1	20,000 @ $.04
December 1	5,000 @ $.02
Ending inventory	10,000 nails

3. Use the information given in problem 1 to determine the value the ending inventory of Shirley's Shoe Store using the LIFO method.

4. Use the information in problem 2 for Carmen's Hardware Store to determine the value of his ending inventory using the LIFO method.

15.2   EXERCISES

1. Find the value of the ending inventory by the first-in, first-out (FIFO) method for the following information.

Beginning inventory	150 units	@ $45	= $_____
Purchases:			
September 4	250 units	@ $55	= $_____
September 12	300 units	@ $40	= $_____
September 20	200 units	@ $47	= $_____
September 30	175 units	@ $50	= $_____

Units of physical inventory remaining at the end of the month = 560

_____units	@ $_____	= $_____
_____units	@ $_____	= $_____
_____units	@ $_____	= $_____
_____units	@ $_____	= $_____
Value of ending inventory		$_____

2. Find the value of the ending inventory by the last-in, first-out (LIFO) method for the following information.

Beginning inventory	150 units	@ $45	= $_____
Purchases:			
September 4	250 units	@ $55	= $_____
September 12	300 units	@ $40	= $_____
September 20	200 units	@ $47	= $_____
September 30	175 units	@ $50	= $_____

Units of physical inventory remaining at the end of the month = 560

_____units	@ $_____	= $_____
_____units	@ $_____	= $_____
_____units	@ $_____	= $_____
_____units	@ $_____	= $_____
Value of ending inventory		= $_____

3. Find the cost of goods sold using the following information.

Beginning inventory	3,000	@ $1.45	= $_____
Purchases:			
April 15	2,000	@ $1.50	= $_____
May 12	4,000	@ $1.65	= $_____
June 10	3,500	@ $1.75	= $_____
Total cost of goods available for sale			$_____
Less ending inventory	3,800	$1.61	= $_____
Cost of goods sold			$_____

4. Dexter Ellsworth owns a sporting goods store in Bakersfield, California, where skateboards are a new item in his inventory. Use the following information about skateboards to determine the value of Dexter's ending inventory using (a) the FIFO method, and (b) the LIFO method.

Purchases:	
June 21	15 @ $45
August 23	30 @ $55
November 22	20 @ $65
Ending inventory	17 skateboards

5. Olivia Meredith owns a haircutting salon. Olivia has decided to purchase a new brand of shampoo for her salon. Using the records below, calculate the value of ending inventory using (a) the FIFO method, and (b) the LIFO method.

Purchases:	
February 17	10 cases @ $36
April 13	20 cases @ $42
June 14	100 cases @ $40
Ending inventory	20 cases

6. Marìa Vasquez is the accountant for the Peculiar Paint Store. It is Marìa's job to determine the value of the cans of Primrose Pink paint remaining in inventory at the year end. Determine the value of ending inventory using (a) the FIFO method, and (b) the LIFO method.

Beginning inventory	
(Primrose Pink)	2 gal @ $ 8
Purchases:	
January 3	50 gal @ $10
April 3	20 gal @ $20
July 3	10 gal @ $25
Ending inventory	33 gal

7. Percy Longfellow is the proud owner of a bookstore in Worcester, Massachusetts, that sells only classic novels. Percy received word from the station manager at Channel 10 that, on December 8, *The Scarlet Pimpernel* would be shown on television. Percy knew from past experience that, after a classic film has been aired on TV, scores of people flock to his bookstore to buy

copies of the book on which the film was based. At the end of the year, Percy had five copies of *The Scarlet Pimpernel* remaining in inventory. Help Percy determine the value of these five remaining copies using (a) the average cost method, (b) the FIFO method, and (c) the LIFO method.

Beginning inventory *(The Scarlet Pimpernel)*	1 copy @ $1.00
Purchases:	
September 4	50 copies @ $2.00
October 4	60 copies @ $2.25
November 4	70 copies @ $2.95

8. Ophelia Jenkins owns a small nursery where 90% of the plants, trees, and shrubs that are for sale are grown on the premises. The other 10% are purchased from a local farmer. Use the partial balance sheet below to help Ophelia calculate the value of her ending inventory using (a) the average cost method, (b) the FIFO method, and (c) the LIFO method.

Beginning inventory (for geraniums)	10 plants @ $2
Purchases:	
April 1	70 plants @ $5
May 1	100 plants @ $4
June 1	50 plants @ $3
Ending inventory	40 plants

9. Mary Stewart is the owner, manager, and bookkeeper of the Downtown Diner. At the end of her first year in business, Mary took a physical count of all the items in the warehouse. Mary found that she had an inordinate amount of 32-ounce cans of tomato juice—800,000 cans to be exact. When Mary returned to the office she immediately looked at the records for tomato juice purchases (listed below). Calculate the value of these 800,000 cans using (a) the average cost method, (b) the FIFO method, and (c) the LIFO method.

Purchases:	
January	100,000 cans @ $ .38
March	600,000 cans @ $1.00
June	200,000 cans @ $1.75
December	500,000 cans @ $1.95

10. Mark Soffietti was the bookkeeper for Anthony Orsky's liquor distributorship. At year end, Anthony found that he had 2,000 cases of vodka remaining in inventory. Use the following information to determine the value of Anthony's vodka using method, (a) the FIFO method, and (b) the LIFO method.

Purchases:	
February 2	500 cases @ $ 75
April 2	500 cases @ $ 80
June 2	500 cases @ $ 85
August 2	500 cases @ $ 90
October 2	500 cases @ $ 95
December 2	500 cases @ $100

11. Penny Paulo manages the Picture Perfect Camera Store. She noted the following information about the store's inventory of Canon 35 mm automatic single-lens reflex cameras for the period April through June. Determine the value of ending inventory using (a) FIFO and (b) LIFO.

Beginning inventory Canon	4 cameras @ $350
Purchases:	
April 4	5 cameras @ $375
May 22	3 cameras @ $325
June 20	4 cameras @ $300
Ending inventory	10 cameras

12. Bismark Frankenstein, owner of Masquerades For Any Occasion costume store, discovered that he overestimated the demand for Zorro costumes this year and is therefore left with 75 such costumes. Determine the value of ending inventory of these costumes using (a) the FIFO method and (b) the LIFO method.

Beginning inventory	50 Zorro costumes @ $30
Purchases:	
August 30	70 Zorro costumes @ $50
September 30	30 Zorro costumes @ $70
Ending inventory	125 Zorro costumes

13. An inventory of the room air conditioners in the appliance department of Hoffman's department store at the end of March showed five 8,000-btu air conditioners at a cost of $350 each, four

6,000-btu air conditioners at a cost of $275 each, and seven 5,000-btu air conditioners at a cost of $125 each. In anticipation of summer, the manager of the department purchased ten 8,000-btu air conditioners at a cost of $400 each, six 6,000-btus at $300 each, and eight 5,000 btus at $200 each. At the end of September, Hoffman's department store still had in its inventory three 5,000-, seven 6,000-, and four 8,000-btu room air conditioners. (a) Use FIFO to put a value on the inventory at the end of September. (b) Use the LIFO method to evaluate ending inventory.

14. The Bright Spot lighting gallery is owned and operated by Isodora Bright (nicknamed Izzy). She has found that American Art Glass Orchid table lamps seem to be growing in popularity. Izzy Bright had 10 of these lamps on hand in June at a cost of $50 each. She purchased 15 lamps in July at a cost of $55 each, and 20 in August at $65 each. Unfortunately, fall sales of these lamps were not as good as Izzy had anticipated, and at the end of December she still had 30 in her inventory. What was the value of Izzy's ending inventory in December using (a) the FIFO method? (b) the LIFO method?

15. Branson's Tire City sells a Guardian all-season automobile tire with a 50,000-mile limited warranty for $30 each. It also sells a Touring 75,000-mile warranty tire for $40 each and a Trailblazer 80,000-mile warranty tire for $50 each. Calculate the value of Branson's ending inventory at the end of March based on the following information using (a) FIFO and (b) LIFO.

Beginning January inventory	20 Guardian @ $10 each
	10 Touring @ $20 each
	15 Trailblazers @ $25 each
Purchases:	
February	10 Guardian @ $12 each
	20 Touring @ $22 each
	15 Trailblazers @ $27 each
March	5 Guardian @ $11 each
	15 Touring @ $18 each
	9 Trailblazers @ $23 each
Ending inventory	25 Guardian
	10 Touring
	25 Trailblazers

15.3   INVENTORY ESTIMATION

Throughout this chapter, we have learned several methods used to value the units remaining in ending inventory. For each of these methods, the units in ending inventory were determined by a physical count. Since taking a physical count is very time-consuming and therefore very costly, it is usually done only once a year. During the remainder of the year, ending inventory must be estimated, because this information is necessary for the preparation of monthly and quarterly financial statements. This section illustrates two methods—the retail method and the gross margin method—used to estimate the cost of ending inventory.

Learning objective

Estimate the value of ending inventory using the retail method.

Retail Method

Retail merchandise outlets often use the **retail method** to estimate the value of their ending inventory. This method is based on the financial information reported in the income statement and does not require the business to calculate the individual cost of each item. The value of ending inventory is determined by first calculating the cost of goods available for sale at both cost and retail. Next, a *cost ratio* is formed by dividing the *cost of goods available for sale at cost* by the *cost of goods available for sale at retail.*

Formula for cost ratio

$$\text{cost ratio} = \frac{\text{cost of goods available for sale at cost}}{\text{cost of goods available for sale at retail}}$$

The cost ratio is then used to convert the value of ending inventory at retail to cost by multiplying the cost ratio by the ending inventory at retail. Example 7 illustrates this procedure step by step.

Example 7

Based on the following financial data, estimate the cost of the ending inventory for Solomon Software Company using the retail method.

	Cost	Retail
Beginning inventory for January	$ 58,500	$ 90,000
Purchases during January	$ 52,000	$ 80,000
Goods available for sale	$110,500	$170,000
Net sales during January		$ 70,000

Solution

Step 1: Calculate the value of ending inventory at retail:

Goods available for sale at retail	$170,000
Less net sales during the month	70,000
Ending inventory at retail	$100,000

Step 2: Compute the cost ratio

$$\text{cost ratio} = \frac{\text{cost of goods available for sale}}{\text{retail value of goods available for sale}}$$

$$= \frac{\$110,500}{\$170,000} = .65$$

For Your Information

INVENTORY CONTROL: ECONOMIC ORDER QUANTITY

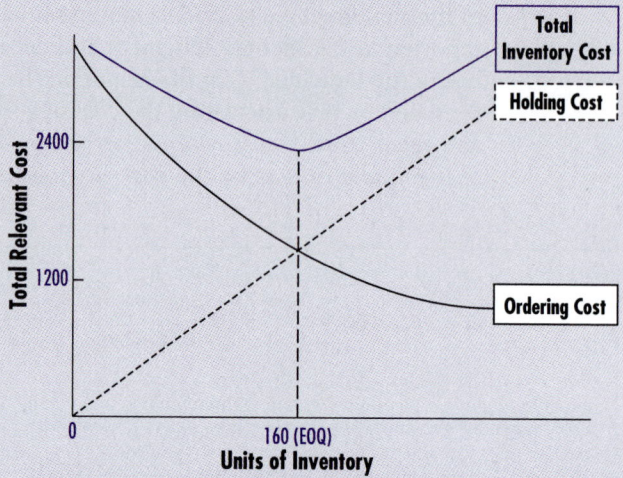

Inventory is an important consideration to business organizations because it represents cost. The aim of inventory control is to minimize the cost of ordering, holding, or running out of stock. Once a firm has determined the demand for its products, it often uses mathematical models to maintain inventories at optimum levels. The economic order quantity (EOQ) is a mathematical inventory control method that identifies the number of units to be ordered each time an order is placed. The equation includes annual demand (*D*), ordering costs (*O*), and holding costs (*H*). For example, assume a manufacturer estimates a total demand of 1,600 units of a part to be used in a manufacturing process, ordering costs amount to $120 per order, and holding costs are $15 per unit per year. Based on these values, the EOQ is 160 units.

$$\text{EOQ} = \sqrt{\frac{2DO}{H}} = \sqrt{\frac{2(1600)(120)}{15}} = \sqrt{25,600} = 160 \text{ units}$$

As shown in the accompanying graph, the EOQ (160) is the point where the ordering cost curve intersects the holding cost curve; the total inventory cost is minimized when these costs are equal. The graph also shows that ordering costs decrease as the number of units ordered in each lot increases because fewer orders are placed during the period, and that holding costs (storage) increase proportionally to the number of units ordered in each lot. The EOQ assumes that unit cost and demand are constant. If either of these variables changes, the EOQ must be adjusted accordingly for this method of inventory control to be effective.

Step 3: Determine the value of ending inventory at cost:

$$\text{ending inventory at cost} = \text{ending inventory at retail} \times \text{cost ratio}$$
$$= \$100,000 \times .65$$
$$= \$65,000$$

Learning objective
Estimate the ending inventory value using the gross margin method.

Gross Margin Method

The **gross margin method** is another method used to estimate inventory. In order to use this method, we must show gross margin as a percentage of sales revenues. This percentage is used to determine the cost of goods sold, which provides us with the necessary information to calculate the value of ending inventory.

Example 8

The Acme Company has a beginning inventory of $40,000 for March 1, purchases during March of $52,000, and March sales of $100,000. Gross margin as a percent of sales has been approximately 38% during the past year. Estimate inventory for the end of March assuming the gross margin percentage remains at 38%.

Solution

Step 1: Calculate the gross margin. We use the gross margin percentage for the past year (38%) to determine the part of sales ($100,000) that is the gross margin.

$$\text{gross margin} = \text{total sales for the period} \times \text{gross margin percent}$$
$$= \$100,000 \times .38$$
$$= \$38,000$$

Step 2: Recall that

$$\text{gross margin} = \text{sales} - \text{cost of goods sold}$$

Solving this equation for cost of goods sold, we have

$$\text{cost of goods sold} = \text{sales} - \text{gross margin}$$

Using the information in this example, this computation is

sales	$100,000
− gross margin	− 38,000
= cost of goods sold	$ 62,000

Step 3: Calculate the value of ending inventory. Recall that the procedure for calculation of the value of ending inventory is

beginning inventory
+ purchase
―――――――――――
= goods available for sale

and

goods available for sale
― cost of goods sold
―――――――――――
= ending inventory

The computation is as follows:

beginning inventory	$40,000
+ purchases	+ 52,000
= goods available for sale	$92,000

and

goods available for sale	$92,000
― cost of goods sold	― 62,000
= ending inventory	$30,000

CHECK YOUR KNOWLEDGE

Inventory Estimation

Solve the following problems. Round dollar amounts to the nearest cent and ratios to the nearest ten-thousandth.

1. The Spindler Corporation needs to estimate the value of its ending inventory in order to complete its financial statements for the month of July. (a) Calculate the cost ratio. (b) Use the retail method to estimate the value of ending inventory at the end of July.

	Cost	Retail
Beginning inventory for July	$20,000	$37,735.85
Purchases during July	30,000	66,603.77
Goods available for sale	50,000	104,339.62
Net sales during July		54,000.00

2.   The Peabody Corporation needs to estimate the cost of its ending inventory as of the end of June in order to complete its second-quarter financial statements. Sales for the month were $70,000, purchases during June were $30,000, beginning inventory for the month was $20,000, and it is assumed that gross margin is 40% of sales. (a) Find the dollar amount of cost of goods sold. (b) Use the gross margin method to estimate the value of ending inventory for the month of June.

15.3   EXERCISES

Solve the following problems. Round dollar amounts to the nearest cent and ratios to the nearest ten-thousandth.

1.   The following information refers to the Mercury Corporation. Calculate the cost ratio using the retail method.

	Cost	**Retail**
Beginning inventory	$ 58,000	$ 80,000
Purchases during the month	87,000	120,000
Goods available for sale	$145,000	$200,000
Net sales during the month		$140,000

2.   Use the retail method to calculate the cost ratio using the following information:

	Cost	**Retail**
Beginning inventory	$21,000	$ 30,000
Purchases during the month	59,500	85,000
Goods available for sale	$80,500	$115,000
Net sales during the month		$ 60,000

3.   The Konar Company needs to estimate the cost of its ending inventory in order to complete its financial statements for the month of April. Find the cost ratio using the retail method.

	Cost	**Retail**
Beginning inventory	$168,000	$280,000
Purchases during the month	348,000	580,000
Goods available for sale	$516,000	$860,000
Net sales during the month		$350,000

Answers to CYK: *1.*   **a.**   .4792;   **b.** $24,122.75   *2.*   **a.**   gross margin = $28,000; cost of goods sold = $42,000;   **b.**   ending inventory = $8,000

4. The MacKenzie Company needs to estimate the cost of its ending inventory in order to complete its financial statements for the month of May. Find the cost ratio using the retail method.

	Cost	Retail
Beginning inventory	$15,400	$22,000
Purchases during the month	12,600	18,000
Goods available for sale	$28,000	$40,000
Net sales during the month		$10,000

5. Calculate the estimated inventory from the information in exercise 1 using the retail method.

6. Calculate the estimated inventory from the information in exercise 2 using the retail method.

7. Calculate the estimated inventory from the information in exercise 3 using the retail method.

8. Calculate the estimated inventory from the information in exercise 4 using the retail method.

9. The Stonybrook Corporation needs to estimate the cost of its ending inventory as of the end of August. Sales for the month were $140,000, purchases during August were $60,000, beginning inventory for the month was $40,000, and it is assumed that gross margin is 35% of sales. Find the dollar amount of the cost of goods sold using the gross margin method.

10. The Birnbaum Company needs to estimate the cost of its ending inventory as of the end of October in order to complete its financial statements. Sales for the month were $300,000, purchases during October were $100,000, beginning inventory was $80,000, and it is assumed that gross margin is 45% of sales. Find the dollar amount of the cost of goods sold using the gross margin method.

11. The Harrison Company needs to estimate the cost of its ending inventory as of the end of November. Sales for the month were $200,000, purchases during November were $75,000, beginning inventory was $95,000, and gross margin was 30% of sales. Using the gross margin method find the dollar amount of the cost of goods sold.

12. The El Guapo Company needs to estimate its ending inventory as of the end of February in order to complete its monthly financial statements. Beginning inventory was $20,000, purchases during February were $24,000, sales for the month were $50,000, and gross margin is 36% of sales. Find the dollar amount of the cost of goods sold using the gross margin method.

13. Find the cost of ending inventory using the information from exercise 9.

14. Calculate the ending inventory as of the end of October using the information from exercise 10.

15. Calculate the cost of ending inventory using the information from exercise 11.

16. Find the ending inventory as of the end of February using the information from exercise 12.

CHAPTER REVIEW EXERCISES

Use the following inventory information from Yolanda's Yarn Shoppe to complete problems 1 through 5.

	Cost	Market
Beginning inventory	400 skeins @ $3.00	$2.00
Purchases		
January	1,000 skeins @ $1.00	$1.00
February	500 skeins @ $1.50	$1.25
March	300 skeins @ $2.00	$1.50

1. Use the specific identification method to determine the value of the ending inventory at the end of March if Yolanda's still has 200 skeins of its beginning inventory, 500 skeins of its January purchase, 100 skeins of its February purchase, and 150 skeins of its March purchase.

2. Calculate the total cost of the 2,200 skeins of yarn that were in the inventory at some time from January through March.

3. Calculate the average cost of a skein in the inventory from January through March (round your answer to the nearest cent).

4. Find the value of the ending inventory for Yolanda's at the end of March using the average cost method, knowing that there are 950 skeins of yarn left in the inventory.

5. Determine the market value of the ending inventory in March if Yolanda's still has 200 skeins of its beginning inventory, 500 skeins of its January purchase, 100 skeins of its February purchase, and 150 skeins of its March purchase.

6. Using the lower of cost or market value method, what is the ending value of Yolanda's inventory?

7. Determine the value of Yolanda's ending inventory at the end of March using the FIFO method if she still has 950 skeins of yarn left in the inventory.

8. Use the LIFO method to determine the ending value of Yolanda's inventory at the end of March if she still has 950 skeins of yarn.

9. If the total of the retail prices on the 2,200 skeins of yarn in Yolanda's inventory was $5,917, calculate the retail method cost ratio between cost of goods for sale at cost and the cost of goods for sale at retail. Use the actual cost of goods from your work in exercise 2, and round your answer to the nearest one-hundredth.

10. If Yolanda's total retail sales of yarn during the months of January through March were $3,362.50, use the retail method cost ratio to determine the cost of goods sold during the 3-month period.

11. If Yolanda's total purchase price for the yarn during this period was $3,550, what would you estimate the value of ending inventory to be by the retail method?

12. If Yolanda's gross margin percentage of sales is .40, or 40%, determine the part of her $3,362.50 sales from January through March that is gross margin.

13. Use the information of exercise 12 and your answer to 12 to determine the cost of goods sold by Yolanda's for the period of January through March.

14. If the beginning inventory value was $3,550, estimate the ending inventory value by the gross margin method.

15. Complete the following chart using the appropriate information that you determined about Yolanda's Yarn Shoppe.

Method	Value of ending inventory
Specific identification	_____
Average cost	_____
Lower of cost or market	_____
FIFO	_____
LIFO	_____
Retail	_____
Gross margin	_____

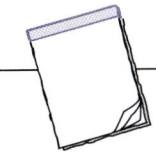

EXPRESS YOUR UNDERSTANDING

Compose one or two well-written sentences to express the requested information in your own words.

1. Give a description of the steps necessary to determine the average cost of a unit remaining in ending inventory.

2. How would you describe the difference between the FIFO and LIFO methods of inventory valuation?

3. How would you go about determining the value of ending inventory using the specific identification method?

4. Explain the basis of the "lower-of-cost-or-market" approach to determining the value of ending inventory.

5. Describe the procedure required to estimate the value of ending inventory for the retail method.

6. Explain how you would calculate the cost ratio to estimate the value of ending inventory under the retail method.

7. Assuming that a business has a gross margin of 40% of sales, explain how the margin can be used to estimate the value of ending inventory.

8. Why do businesses estimate the value of their ending inventories?

Mind Your Business

As one of her responsibilities, Kate Bradley, your bookkeeper, has to check the inventory at the end of each month. She usually has the clerks do a quick count of the number of music CDs, and cassettes, books, and videos in stock at the end of the last day of the month. At the end of September the clerks reported that there were 4,348 music CDs and cassettes, 2,300 books, and 881 videos in your inventory.

At the beginning of September you had 2,199 music CDs and cassettes, which cost $17,316; 1,336 books, which cost $12,120; and 742 videos, which cost $5,194. During the month, you purchased 4,396 music CDs for $25,550; 2,100 books for $13,650; and 334 videos for $3,415.

Kate has also estimated that the cost of goods for sale at retail is about $120,734 at the end of September.

Activity 1

Determine the value of your inventory by the average cost method.

Activity 2

Estimate the value of your inventory by the retail method.

Activity 3

Write a short paragraph indicating which of the methods has been used in the financial statements, and which of the methods should be used to be able to report the highest net profit for the month.

SELF-TEST

A. Terminology

Complete the following items using the key terms presented at the beginning of the chapter. Check your responses against the answer key at the end of the test.

1. Goods that are owned by the business and held for resale are called _____.

2. _____ is a method used to value inventory that assumes purchases made during the latter part of the year are sold first.

3. _____ is a method used to value inventory that assumes purchases made during the first part of the year will be sold first.

4. Inventory may be valued at either the cost of the unit or at the current market value—whichever is lower. The name of this rule is _____.

5. _____ is a method used to value inventory that assumes each item purchased is identified and labeled with the original cost of the unit.

6. Since continuous inventory systems are not common, most businesses use a _____ _____ system.

7. The _____ method uses the cost ratio to estimate the value of ending inventory.

8. The _____ _____ method employs the percentage of sales revenues to determine cost of goods sold and the estimated value of ending inventory.

B. Calculation

The following concepts and short problems are designed to test your understanding of the objectives identified at the beginning of the chapter. Answers are provided at the end of the test.

9. The Better Buy Clothing Store sells men's suits. Use the following information to determine the value of ending inventory using the average cost method.

Purchases:

January 31	30 @ $220
May 22	60 @ $170
September 4	40 @ $100
Ending inventory	20 suits

10. Value the ending inventory in exercise 9 using the FIFO method.

11. Value the ending inventory in exercise 9 using the LIFO method.

12. A department store has just started a stereo department. Determine the value of ending inventory using the specific identification method.

Purchases:		**Cost**	**Market**
January 2	Hitachi	$400	$450
February 15	Bose	$780	$750
April 30	Samsung	$600	$625
July 31	Sony	$850	$800
August 23	Pioneer	$930	$900
Ending inventory	Hitachi, Sony, and Pioneer		

13. Apply the rule of lower-of-cost-or-market to determine the value of ending inventory in exercise 12.

14. Using the retail method, estimate the value of ending inventory for the month of August based on the following financial information:

	Cost	**Retail**
Beginning inventory August 1	$ 40,000	$ 54,000
Purchases during August	$ 70,000	$ 94,500
Goods available for sale	$110,000	$148,500
Net sales for August		$ 90,500

15. A company wishes to estimate the cost of its ending inventory, as of the end of October, to prepare financial statements. If sales for the month are $80,000, purchases are $40,000, beginning inventory is $30,000, and the firm's gross margin is 40% of sales, what is the estimated value of ending inventory for the month based on the gross margin method?

Answers to Self-Test: **1.** inventory   **2.** LIFO   **3.** FIFO   **4.** lower-of-cost-or-market method   **5.** specific identification   **6.** periodic inventory   **7.** retail method   **8.** gross margin   **9.** $3,200   **10.** $2,000   **11.** $4,400   **12.** $2,180   **13.** $2,100   **14.** $42,960.60   **15.** $22,000

16

PAYROLL SYSTEMS

Learning Objectives

After completing the exercises in this chapter, you will be able to:

1. Calculate gross earnings based on hourly incentive, or salary payroll systems.

2. Identify and calculate the various types of employee payroll deductions.

3. Identify and calculate the various types of employer payroll taxes.

4. Complete payroll records and quarterly payroll tax returns.

5. Define key terms.

INTRODUCTION

The payroll is an extremely important subsystem of the accounting system because most organizations expend close to one-third of their revenues for labor costs and associated taxes. Expressed another way, generally, the largest single expense for many businesses is the payroll. Control of a cost as significant as payroll is vital to the achievement of the profit goals of an enterprise.

Employees of a firm provide services for which they expect and are guaranteed under legal statutes to receive compensation at established intervals. Therefore, the payroll system must be accurate and capable of processing data within the required time limitations. In addition, the system must provide internal audits of the procedures and technology to assure the proper use of funds.

Municipal agencies at the federal, state, and local levels have established laws that require employers to collect data in their payroll records concerning compensation, deductions, and taxes. This information is then used by the employer to prepare periodic reports for government agencies that identify the payments required for amounts withheld from employees and for employer payroll taxes.

Employees generally are compensated on the basis of time and work performed. For example, managerial or administrative personnel usually receive a compensation for their services expressed in terms of a week, month, or year. Most skilled and unskilled workers receive compensation for their labor based on hours of working time or piecework. Employees may receive other forms of compensation in addition to their base earnings such as overtime premiums, cost-of-living adjustments (COLA), profit-sharing plans, commissions, shift differentials, bonuses, and compensatory time. The payment of compensation to an employee is generally in the form of a check or cash. However, an employee may also be compensated in other forms such as merchandise, services, or property, based on the fair value of cash in payment for the employee's services. In this chapter, we will present the most common methods used to determine an employee's gross earnings, the procedures required to calculate deductions and net earnings, and the computations necessary to complete payroll tax records.

16.1   GROSS EARNINGS: HOURLY WAGES AND PIECEWORK

Learning objective
Calculate gross earnings based on hourly payroll systems.

When an employee receives compensation for his or her services by the hour, the employee's **wages** are determined by the number of hours worked during the pay period and the hourly rate paid by the employer for the job. These two variables, time and rate, are the basis of determining the employee's **gross earnings** under the hourly wage method. Gross earnings is the total compensation earned by an employee during a pay period before mandatory and elective deductions are withheld by the employer. The computation of gross earnings can be expressed by the following equation.

Figure 16.1

Sample payroll
time card

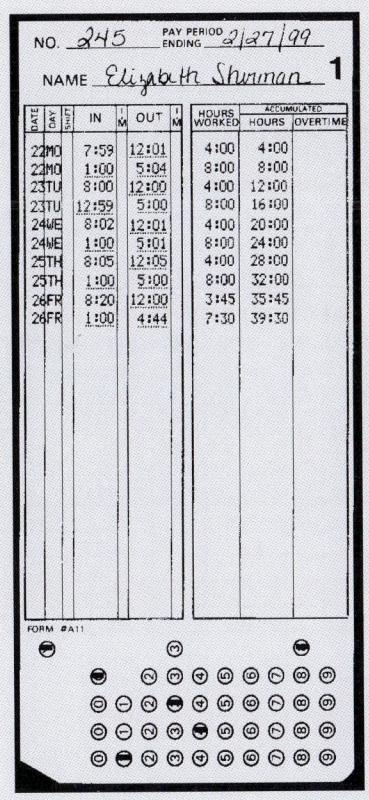

NO. 245	PAY PERIOD ENDING 2/27/99

NAME *Elizabeth Sherman*   **1**

DATE	DAY	SHIFT	IN	M	OUT	M	HOURS WORKED	ACCUMULATED HOURS	OVERTIME
22	MO		7:59		12:01		4:00	4:00	
22	MO		1:00		5:04		8:00	8:00	
23	TU		8:00		12:00		4:00	12:00	
23	TU		12:59		5:00		8:00	16:00	
24	WE		8:02		12:01		4:00	20:00	
24	WE		1:00		5:01		8:00	24:00	
25	TH		8:05		12:05		4:00	28:00	
25	TH		1:00		5:00		8:00	32:00	
26	FR		8:20		12:00		3:45	35:45	
26	FR		1:00		4:44		7:30	39:30	

FORM #A11

Source: Syracuse Time & Alarm Co., Inc.

**Gross earnings
for regular
time**

> gross earnings = hours worked × rate per hour

Hourly employees are usually required to maintain time cards similar to the one shown in Figure 16.1. The data on the time card is later recorded on an employee earnings record (Figure 16.2) and in the payroll register (Figure 16.3), which is a record of all employees' payroll information.

Example 1

Compute the gross earnings for Elizabeth Sherman based on the regular time and rate data on the time card shown in Figure 16.1.

Figure 16.2

Sample employee earnings record

EMPLOYEE'S INDIVIDUAL EARNINGS RECORD

2nd Quarter of 19 9X

Name   George Anderson   Social Security No.   364-56-8573   Phone No.   487-2406
Address   136 First Street   City   Detroit   State   Michigan   ZIP   48226
Position   QC Inspector   Pay Rate $10.50   Pay Period   Weekly
Male   X   Female ____   Married X Single ____   Number of Exemptions   3   Date Employed   9/12/86

| | EARNINGS | | | DEDUCTIONS | | | | | | | | | Check | Year-to- |
Wk	Regular	Overtime	Total	FICA	MEDCR	FIT	SIT	SDI	INS	RET	MISC	NET PAY	number	date
1	$420		$420.00	26.04	6.09	92.40	28.60	3.20	8.40	24.00		231.27	1549	5,482.35
2	420	85.00	505.00	31.31	7.32	99.80	34.25	3.20	8.40	28.50		292.22	1638	5,987.35
3	420	52.00	472.00	29.26	6.84	95.75	31.10	3.20	8.40	26.30		271.50	1789	6,459.35
4	420		420.00	26.04	6.09	92.40	28.60	3.20	8.40	24.00		231.27	1854	6,879.35
5	420	78.75	498.75	30.92	7.23	97.60	32.80	3.20	8.40	27.45		298.90	1931	7,378.10
6														
7														
8														
9														
10														
11														
12														
13														
QUARTER TOTALS														

Solution

Insert the total hours worked for the week and the regular rate of pay into the equation and multiply as shown.

$$\text{gross earnings} = \text{hours worked} \times \text{regular rate per hour}$$
$$= 39.5 \times \$6.00$$
$$= \$237.00$$

Overtime and Premium Wages

Salary and hourly wage rates are determined by a number of factors; one is the *Fair Labor Standards Act,* commonly known as the *wage and hour law.* This legislation, which covers the majority of full-time hourly wage employees, establishes the minimum hourly wage ($5.15 as of September 1, 1997) and sets the regular workweek at 40 hours. Although the law has been expanded in recent years to include state, county, and city hourly workers,

Figure 16.3

Sample payroll register

Week no. 7
Beginning 2/10 Ending 2/16 Record for 1st Quarter of 199X

Payroll Register

Name	Status	Exemptions	Rates Reg.	Rates O.T.	Hrs. worked Reg.	Hrs. worked O.T.	Earnings Reg.	Earnings O.T.	Gross earnings	FICA	MEDCR	FWT	SWT	RET	Net earnings
Adams, T	S	1	6.50	9.75	40	0	260.00		260.00	16.12	3.77	46.80	24.45	12.00	156.86
Buttler, G	M	3	9.40	14.10	40	10	376.00	141.00	517.00	32.05	7.50	98.40	42.55	24.00	312.50
Duncan, L	M	2	10.00	15.00	35	0	350.00		350.00	21.70	5.08	62.40	28.50	14.00	218.32
Ford, Z	S	1	8.60	12.90	40	5	344.00	64.50	408.50	25.33	5.92	73.44	34.50	18.20	251.11
Harter, P	M	4	12.25	18.38	40	12	490.00	220.56	710.56	44.05	10.30	134.90	58.30	30.50	432.51
Jordan, R	M	2	10.00	15.00	40	0	400.00		400.00	24.80	5.80	72.20	33.10	15.00	249.10
Kowalski, G	S	1	8.60	12.90	40	0	344.00		344.00	21.33	4.99	60.25	26.75	12.80	217.88
Martino, J	M	2	9.40	14.10	40	6	376.00	84.60	460.60	28.56	6.68	82.80	37.40	20.00	285.16
Nowak, J	S	2	8.60	12.90	40	0	344.00		344.00	21.33	4.99	52.40	24.50	12.80	227.98
Potter, S	M	3	12.25	18.38	40	0	490.00		490.00	30.38	7.11	88.20	39.60	22.40	302.31
Reddy, M	S	1	6.50	9.75	40	0	260.00		260.00	16.12	3.77	46.80	24.45	12.00	156.86
Sawyer, T	M	2	10.00	15.00	40	8	400.00	120.00	520.00	32.24	7.54	84.40	38.30	25.00	332.52
Williamson, F	M	3	9.40	14.10	40	0	376.00		376.00	23.31	5.45	58.20	25.60	16.00	247.44
Pay period total							4,810.00	630.66	5,440.66	337.32	78.90	961.19	438.00	234.70	3,390.55

it continues to exclude business executives, administrators, employees of small retail organizations, specified seasonal agricultural workers, and those in professional occupations like teachers, lawyers, and medical doctors. The 40-hour workweek requires the employer to pay an **overtime rate** (for hours worked in excess of 40 per week) of at least one and one-half the **regular rate.** This overtime rate is expressed as **time-and-a-half** (1 ½) and can be determined by one of the two methods presented in Example 2 using the following equation.

Formula for gross earnings with overtime

gross earnings = regular earnings + overtime earnings

$$(H \times R) + (H \times 1.5R)$$

Example 2

Todd Marshall's time card indicated that he worked 8 hours Monday; 10 hours Tuesday; 9 ½ hours Wednesday; 7 hours Thursday; and 10 ¾ hours Friday. Todd receives a regular rate of $8.50 per hour and time-and-a-half for overtime. Determine his gross earnings for the week.

Solution 1

Standard Overtime Method

Step 1:   Determine total number of hours worked and overtime hours.

$8 + 10 + 9.5 + 7 + 10.75 = \quad 45.25$ total hours

$\qquad\qquad\qquad\qquad\quad -40.00$ regular hours

$\qquad\qquad\qquad\qquad\qquad 5.25$ overtime hours

Step 2:   Determine the overtime rate.

$8.50 \times 1.5 = \$12.75$ per hour

Step 3:   Insert the data into the earnings equation and solve.

gross earnings = regular earnings + overtime earnings

$\qquad\quad = (40 \times 8.50) \qquad + (5.25 \times 12.75)$

$\qquad\quad = \$340.00 \qquad\qquad + \$66.9375$

$\qquad\quad = \$406.94$

Note: When computing gross earnings, round your answer to the nearest whole cent in the final calculation.

CALCULATOR SOLUTION

Keyed entry	Display
AC 40 × 8.5 =	340
Min	340
5.25 × 1.5 × 8.50 =	66.9375
M+	66.9375
MR	406.9375

Some companies prefer the *overtime premium method* to calculate gross wages with overtime. This method produces the same gross wage but is computed by adding the total hours worked at the regular rate to the overtime hours at the premium rate (one-half regular rate).

Solution 2

Overtime Premium Method

Step 1:   Determine the total hours worked and overtime hours.

$$8 + 10 + 9.5 + 7 + 10.75 = \quad 45.25 \text{ total hours}$$
$$-40.00 \text{ regular hours}$$
$$5.25 \text{ overtime hours}$$

Step 2:   Determine the overtime premium rate.

$$\$8.50 \times .5 = \$4.25$$

Step 3:   Insert the data into the earnings equation and solve.

gross earnings = regular earnings   + overtime earnings
           = (45.25 × 8.50)   + (5.25 × 4.25)
           = \$384.625   + \$22.3125
           = \$406.9375 or \$406.94

The overtime premium method provides a more distinct analysis of labor costs since it considers \$4.25 (one-half the regular rate) to be the actual additional cost of each overtime hour, compared to the \$12.75 (time-and-a-half) rate used with the standard overtime method.

Many companies pay the time-and-a-half rate for all time worked over eight hours in one day regardless of how many hours are worked during a week. This *daily overtime* exceeds the requirements of the Fair Labor Standards Act and is usually determined through collective bargaining and stated in the labor agreement. The next example shows how to calculate gross earnings using the daily overtime method to determine the premium wage.

Example 3

Lynette Elbert worked 8½ hours on Monday, 8 hours on Tuesday, 10 hours on Wednesday, 9 hours on Thursday, and was absent from work on Friday. Her regular hourly rate of pay is $10.20. Find her gross earnings for the week.

Solution

Step 1:   Determine the regular hours and overtime hours worked.

	M	T	W	T	F	Total hours
Regular	8	8	8	8	0	32.0
Overtime	.5	0	2	1	0	3.5

Step 2:   Determine the overtime rate.

$10.20 \times 1.5 = \$15.30$

Step 3:   Insert the rate into the earnings equation and solve.

gross earnings = regular earnings + overtime earnings

$$= (32 \times \$10.20) \quad + (3.5 \times \$15.30)$$
$$= \$326.40 \qquad\quad + \$53.55$$
$$= \$379.95$$

Elbert is paid time-and-a-half for the .5 hour of overtime Monday, 2 hours of overtime Wednesday, and 1 hour of overtime Thursday because she worked more than 8 hours each of these 3 days. For the week, she worked a total of 35.5 hours of which 3.5 hours were overtime and 32 were regular time.

The labor agreement is also the basis for other premium wage payments such as **double time** and **shift differentials.** Double time (twice the regular hourly rate) is paid for holidays and sometimes weekend work. A shift-differential is an additional amount per hour paid to employees who work other than the day shift, such as the second shift (4 P.M. to 12 A.M.) and the graveyard shift (12 A.M. to 8 A.M.). Example 4 illustrates how to calculate gross earnings when an employee receives both double time and a shift differential.

Example 4

Debbie Gonski works the 4 P.M.–12 A.M. shift at Rexford Handling. She is paid $7.58 per hour and a 20-cent shift differential for her work. Last week, she worked five 8-hour shifts. Find her gross earnings for the week if she is paid double-time for one of her 8-hour shifts because she worked a holiday.

Solution   *Step 1:*   Determine the regular hours and premium hours worked.

$$4 \times 8 = 32 \text{ regular hours}$$
$$1 \times 8 = \underline{8} \text{ premium hours}$$
$$40 \text{ total hours}$$

Step 2:   Determine the premium rate.

$$(7.58 + .20) \times 2 = \$15.56$$

Step 3:   Insert the data into the earnings equation and solve.

$$\text{gross earnings} = \text{regular earnings} + \text{premium earnings}$$
$$= (32 \times \$7.78) \quad + (8 \times \$15.56)$$
$$= \$248.96 \qquad\quad + \$124.48$$
$$= \$373.44$$

The $7.78 is the regular rate even though it includes the shift differential because this is the amount the employee regularly receives for each hour of work. Some employers will consider the extra 20-cent shift differential as premium earnings and record it separately on the employee's earnings record.

CHECK YOUR KNOWLEDGE

Calculating Hourly Wage Earnings and Overtime Earnings

1.   Robin Anderson is paid $4.50 an hour as a nurse's aid. Find her gross earnings for the current week if she worked a total of 38 hours.

2.   Last week, Ed Ramsey worked 48 hours. Find his (a) regular earnings, (b) overtime earnings, and (c) gross earnings for the week if he is paid $6.40 per hour and receives time-and-a-half for hours worked in excess of 40. (Use the standard overtime method.)

3.   Using the information in problem 2, recalculate parts a, b, and c based on the overtime premium method.

4.   Rick LaValle earns $5.80 per hour and is paid time-and-a-half for all time over 8 hours in any 1 day. His company also pays double time for holidays and Sundays. Find his gross earnings for a week in which he works the following hours: Monday, 10½; Tuesday, 8; Wednesday, 9; Thursday, 6; Friday, 9¾; and Sunday, 5 hours. (Use the standard overtime method.)

5.   Tad Nichols works the third shift (12:00 A.M.–8:00 A.M.) at Rapid Waters Chemical Company and is paid $9.62 per hour. Rapid Waters Chemical

Company pays its hourly workers an additional $.28 per hour for the third-shift hours and time-and-a-half for all overtime hours. Calculate Tad's gross earnings for the week if his time card shows that he worked 40 regular hours and 6 overtime hours.

Piecework

Learning objective
Calculate gross earnings based on incentive payroll systems.

Organizations in the manufacturing of goods, agriculture, and building construction may compensate some of their employees with **incentive rates** instead of *time rates*. When incentive rates are used, an employee's gross earnings are based on job performance instead of time worked. Piecework encourages employees to increase their productivity in order to increase their gross earnings. A **piece rate** is determined through a time and motion study of a job, and that rate is paid to the employee for each acceptable unit produced. Employees may also receive a premium rate for all units produced beyond established quotas. Employers are generally willing to pay higher wages for such increases in productivity as long as the units meet quality control requirements. Defective units often increase cost and decrease profits because of the loss of labor time and material. An employer, therefore, may require an employee to reimburse the company for part of the loss related to his or her defective units through a chargeback. *Chargebacks* reduce an employee's earnings; they are considered a penalty and are usually set at a rate lower than what was paid to produce the unit. There are a number of piecework plans in use in today's businesses. We will discuss the most commonly used plans in this section.

The basic *straight piecework* plan pays the employee a fixed amount for each unit of production. Under this plan the gross earnings are found using the following formula:

Gross earnings for straight piecework

gross earnings = (number of units × rate per unit)

Example 5

Jorge Swensen works in an electrical manufacturing firm and is paid $.75 for each switch he assembles that passes quality control inspections. His production card for the week indicated that he assembled 120 switches on

Answers to CYK: *1.* $171.00 *2.* **a.** $256.00; **b.** $76.80; **c.** $332.80
3. **a.** $307.20; **b.** $25.60; **c.** $332.80 *4.* $324.08 *5.* $485.10

Monday; 118 on Tuesday; 126 on Wednesday; 105 on Thursday; and 96 on Friday. Fifteen of those switches had been rejected by inspectors. If the company uses a chargeback of $.50 per defective unit, calculate Jorge's gross earnings for the week.

Solution   First, determine the total number of units produced.

Monday	120
Tuesday	118
Wednesday	126
Thursday	105
Friday	96
	565 units produced

Now, insert the appropriate variables into the equation and solve.

$$gross\ earnings = (number\ of\ units \times rate\ per\ unit)$$
$$- (defective\ units \times defective\ rate)$$
$$= (565 \times \$.75) - (15 \times \$.50)$$
$$= \$423.75 - \$7.50$$
$$= \$416.25$$

CALCULATOR   SOLUTION

Keyed entry						Display
AC	565	×	.75	=		423.75
				Min		423.75
	15	×	.5	=		7.5
				+/−		−7.5
				M+		−7.5
				MR		416.25

Many companies use the *differential piece-rate plan,* which incorporates a premium rate for units produced above established standards. Differential piece-rate schedules are prepared that identify the required level of production and the corresponding rate paid per unit. The schedule can be based on a 40-hour workweek, 8-hour workday, or hourly productivity. This plan provides the incentive to produce more because the employee's gross earnings increase more in proportion to units produced than with the straight piece-rate plan. The reason being, the employee receives higher rates of pay per unit for different levels of production. To illustrate this concept, let's compare the information in Example 5 to that in Example 6.

Example 6

The electrical manufacturing firm's differential piece-rate schedule is as follows:

Weekly net standard units	Rate per unit
0–300	$ .70
301–400	.85
401–500	.95
501–600	1.05
Over 600	1.15

Determine Jorge Swensen's gross earnings using this schedule.

Solution

The gross earnings for Jorge would be calculated in the following manner.

first	300 units at $ .70	$210.00
next	100 units at .85	85.00
next	100 units at .95	95.00
last	65 units at 1.05	68.25
	565 total units accepted	$458.25 piecework earnings

gross earnings = piecework earnings − chargeback
= $458.25   − (15 × $.50)
= $458.25   − $7.50
= $450.75

As you can see, using the differential piece-rate plan, Jorge receives $34.50 ($450.75 − 416.25) more in gross earnings for the 565 units he produced.

```
CALCULATOR  SOLUTION
Keyed entry                                    Display

AC    300   ×   .7    =                          210
                     Min                         210
      100   ×   .85   =                           85
                     M+                           85
      100   ×   .95   =                           95
                     M+                           95
       65   ×  1.05   =                        68.25
                     M+                        68.25
                     MR                       458.25
       15   ×   .5    =                          7.5
                     +/−                        −7.5
                     M+                         −7.5
                     MR                       450.75
```

Earlier we discussed the Fair Labor Standards Act, which sets the minimum wage for employees. To conform with this law, a company will provide an *hourly rate/piecework plan,* which includes an hourly rate guarantee if gross earnings from piecework is less than the required minimum wage rate. While the hourly rate cannot be below the minimum wage level, it can be as high as that determined by management and labor. Whatever the hourly rate, the employer must pay either the hourly rate or the piece rate, whichever is higher. This plan can be based on each hour of work, each job order, or each 8-hour workday. For example, if the hourly base is used, a worker who is paid $.55 per unit or $6.60 per hour would be paid $6.60 for any hour that 12 or fewer units were produced ($6.60 ÷ .55 = 12 units) and straight piece-rate for any hour that 13 or more units were produced (13 × $.55 = $7.15). Let's look at an example based on an 8-hour day.

Example 7

The Armstrong Manufacturing Company pays its production workers $8.40 an hour or $.30 per unit of production, whichever is higher. Find the weekly gross earnings for an employee, based on the following production record, if the employee worked 8 hours each day.

Day	Units of production
Monday	210
Tuesday	235
Wednesday	220
Thursday	240
Friday	215

Solution

We must calculate the pay for each day based on the piece rate and compare it to the hourly rate (8 × $8.40 = $67.20).

Day	Piece rate	Hourly rate	Gross earnings
Monday	210 × .30 = $63.00	$ 67.20	$ 67.20 hourly
Tuesday	235 × .30 = 70.50	67.20	70.50 piece
Wednesday	220 × .30 = 66.00	67.20	67.20 hourly
Thursday	240 × .30 = 72.00	67.20	72.00 piece
Friday	215 × .30 = 64.50	67.20	67.20 hourly
		$336.00	$344.10 total gross earnings

Another approach to solving Example 7 would be to divide the daily hourly wage by the piece rate, which will give you the daily task standard ($67.20 ÷ $.30 = 224 units). Any units produced above this daily task standard will result in higher daily wages that must be calculated, such as Tuesday's 235 and Thursday's 240 units of production.

The final piecework plan we will discuss in this section is called the **task and bonus plan.** Under this plan, the task (productivity per time period) is established and, if the worker produces less than task, the hourly rate is paid similar to the hourly rate/piecework plan. If, however, the worker performs equal to or greater than task, a *bonus* (additional compensation) is paid. The following equation can be used to calculate gross earnings when a bonus is to be paid.

Gross earnings when bonus is paid

gross earnings = (bonus factor) × (hourly rate) × (hours)
× (production task)

The *bonus factor* in the equation is a percentage agreed upon by management and labor as the additional monetary reward paid to the worker for equaling or exceeding the task requirement.

Example 8

Lisa Jarrett packages small appliances and has a task requirement of 120 units during an 8-hour period. Her job pays a bonus factor of 115%. Find her gross earnings for an 8-hour day in which she packaged 162 appliances. Lisa's hourly rate is $4.87 an hour.

Solution

Insert the required variables into the equation and solve. *Note:* Convert the bonus factor and the production ratio to decimals.

gross earnings = bonus factor × hourly rate × hours
$$\times \text{ production task}$$
$$= 115\% \times \$4.87 \times 8 \times 162/120$$
$$= 1.15 \ \ \times \$4.87 \times 8 \times 1.35$$
$$= \$60.49$$

If Lisa had packaged less than 120 appliances, her daily gross earnings would be $38.96 (8 × $4.87).

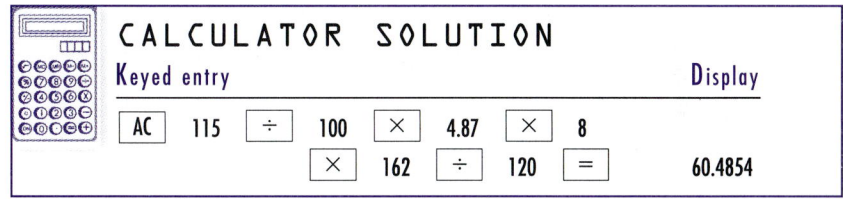

CALCULATOR SOLUTION

Keyed entry											Display
AC	115	÷	100	×	4.87	×	8				
		×	162	÷	120	=					60.4854

Employers must pay time-and-a-half for overtime to piecework employees if their earnings are regulated by the 40-hour workweek. The overtime rate is usually 1½ times the regular rate for each hour worked over 40 in the workweek. The regular rate of pay is computed by adding together earnings for the week from piecework, other hours worked, waiting time, and bonuses. This total is then divided by the total number of hours worked during the pay period to determine the regular rate for the week. The gross earnings for the pay period will be the total of the regular earnings and overtime earnings (gross earnings = regular earnings + overtime earnings). Example 9 illustrates how to calculate overtime based on the piece-rate method.

Example 9

Luis Santiago is paid $.62 per unit produced under his company's straight piece-rate plan and $5.00 per hour for waiting time. His production record for a recent week showed that he produced 390 units during a 50-hour work period with 6 hours of waiting time. What were Luis's gross earnings for the week if he receives time-and-a-half for overtime?

Solution

Step 1: Determine the regular earnings from piecework and waiting time.

$241.80   (390 × $.62) piecework
 30.00   (6 × $5.00) waiting time
$271.80   regular earnings

Step 2: Compute the regular hourly rate.

regular hourly rate = regular earnings ÷ total hours worked
$$= \$271.80/50$$
$$= \$5.436$$

Step 3: Determine the overtime rate.

overtime rate = regular rate × 1.5
$$= \$5.436 \times 1.5$$
$$= \$8.154$$

Step 4: Calculate the gross earnings based on the piece rate.

gross earnings = regular earnings + overtime earnings
$$= \$271.80 \qquad + (10 \times \$8.154)$$
$$= \$271.80 \qquad + \$81.54$$
$$= \$353.34$$

CHECK YOUR KNOWLEDGE

Calculating Piecework Earnings

1. John Michels is paid $.86 for each subassembly he produces and is charged back $.60 for defective units. Calculate his gross earnings for the week if he produced 478 subassemblies of which 12 were considered defective.

2. Rolling Hills Orchards pays its apple pickers daily based on a differential rate plan that provides $.50 per box for the first 20 boxes; $.75 for

the 21st through 35th box; and $1.00 per box above 35 per day. Find the gross earnings for a picker who picked 47 boxes on a given day.

3. Find the gross earnings for a worker who is paid $5.80 per hour or $.20 per unit, whichever is greater. The worker produced 238 units on Monday; 226 units on Tuesday; 218 units on Wednesday; 232 units on Thursday; and 247 units on Friday. In addition, calculate the daily task standard.

4. Robert Emerson is a machine operator at a local manufacturing facility and is paid by the terms of a task/bonus plan as follows. He receives a bonus of 130% for a task requirement of 150 units per 8-hour shift. If his hourly wage is $6.58, find his gross earnings for 2 days in which he produced 190 and 148 units, respectively.

5. Lucy Cooper is paid $.82 for each toy truck she assembles and $5.50 for each hour she has to wait to have her machine set up or repaired. During a recent week, she assembled 296 trucks in 42 hours including 3 hours of waiting time. What were her gross earnings for the week if she is paid time-and-a-half for overtime?

16.1 EXERCISES

Calculate the gross earnings for each of the following employees.

	Name	M	T	W	Th	F	Total hours	Hourly rate	Gross wage
1.	Adams, P.	8	7	8	6	8	_____	$10.42	_____
2.	Bedell, R.	8	0	8	8	8	_____	7.78	_____
3.	Kane, J.	3	8	8	8	8	_____	9.65	_____
4.	Peters, D.	8	8	8	8	8	_____	6.87	_____
5.	Watson, R.	0	0	8	8	8	_____	8.90	_____

Complete the following payroll record. Employees receive time-and-a-half for all hours over 40 hours per week. Use the standard overtime method to calculate overtime rates.

	Name	M	T	W	Th	F	S	Reg. hrs.	O.T. hrs.
6.	Abboud, C.	8	8	6	6	6		_____	_____
7.	Bacon, R.	8	8.5	8	10	8	4	_____	_____
8.	Choi, G.	8	8	8	8	8		_____	_____
9.	Healy, S.	8.25	12	7	8	9	3.75	_____	_____
10.	Montone, J.	9	3	8	8	8	6	_____	_____
11.	Regan, W.	8	9.5	8	10	8	3.25	_____	_____

Answers to CYK: **1.** $403.88 **2.** $33.25 **3.** $236.20; 232 units task standard **4.** $139.32
5. $277.74

Name	Rate Reg. rate	O.T. rate	Regular earnings	Overtime earnings	Total earnings
Abboud, C.	$ 6.40	_____	_____	_____	_____
Bacon, R.	8.10	_____	_____	_____	_____
Choi, G.	10.50	_____	_____	_____	_____
Healy, S.	6.80	_____	_____	_____	_____
Montone, J.	9.30	_____	_____	_____	_____
Regan, W.	7.60	_____	_____	_____	_____

Complete the following payroll record. Employee earnings are based on the straight piecework plan with time-and-a-half over 40 hours and $6.00 per hour waiting time.

	Employee	Units produced M	T	W	Th	F	Total units	Hours worked	Waiting time
12.	Ackley, F.	270	268	279	283	186	_____	36	4
13.	Didziak, F.	950	937	896	943	960	_____	44	0
14.	Eberle, D.	188	174	168	190	179	_____	47	3
15.	Jenkins, M.	112	85	102	98	87	_____	40	6
16.	Riccardo, T.	889	925	963	972	987	_____	40	0
17.	Shirtz, A.	250	237	261	278	290	_____	43	2

Employee	Rates Piece rate	Regular hourly rate	O.T. rate	Earnings Regular	Overtime	Total
Ackley, F.	$.248	_____	_____	_____	_____	_____
Didziak, F.	.0866	_____	_____	_____	_____	_____
Eberle, D.	.485	_____	_____	_____	_____	_____
Jenkins, M.	.7972	_____	_____	_____	_____	_____
Riccardo, T.	.0675	_____	_____	_____	_____	_____
Shirtz, A.	.375	_____	_____	_____	_____	_____

Calculate the gross earnings for each employee based on the following differential pay scale. Assume a charge-back of $.95 per unit.

Units	Rate
0–150	$ .85
151–300	.97
301–450	1.10
451–600	1.22
Over 600	1.30

	Name	Units	Rejects	Piecework earnings	Chargeback	Gross earnings
18.	Aiello, M.	520	5	_____	_____	_____
19.	Guckert, T.	348	12	_____	_____	_____
20.	Lutz, N.	625	21	_____	_____	_____
21.	Quilty, D.	290	0	_____	_____	_____
22.	Teachout, P.	487	8	_____	_____	_____

23. Robin Evans's time card shows she worked a 48½-hour week during the current pay period. She is paid a regular rate of $8.38 per hour and time-and-a-half for all hours over 40 per week. Calculate her gross earnings for the week based on the standard overtime method.

24. From the payroll information in problem 23, calculate the employee's gross earnings using the overtime premium method.

25. Philip Vannelli earns $9.20 per hour and is paid time-and-a-half for all time over 8 hours in any 1 day. His company also pays double time for Sundays and holidays. Calculate his gross earnings for the week based on the following hours: Monday, 8; Tuesday, 10½; Wednesday, 8; Thursday, 4; Friday, 8; Saturday, 6; and Sunday, 4 (use the standard overtime method).

26. An employee at Omni Engineering is paid $11.40 per hour and a $.10 shift differential for working the second shift. The employee is paid time-and-a-half for overtime, and he worked a total of 46 hours this pay period. Find the employee's gross earnings.

27. Find the gross earnings for a packer who is paid $.63 for each case she packs that passes in-spection. Calculate her gross earnings if her production for the week is 780 cases of which 58 were rejected. She is charged $.40 per rejection.

28. Betty Wright operates a drill press and is paid $6.45 per hour or $.15 per unit, whichever is greater. Her production record for the week shows that she produced 380 units Monday, 368 units Tuesday, 330 units Wednesday, 390 units Thursday, and 340 units Friday. Determine her gross earnings based on a daily task standard for the pay period.

29. Custom Fabricating pays its production worker by the terms of a task/bonus plan agreement. The plan provides a 125% bonus and a task requirement of 200 units per 8-hour workday. Find the gross earnings of an employee who produced 258 units during an 8-hour day and is paid $7.15 an hour.

30. An employee is paid $.75 per unit produced and $5.75 per waiting time hour. During the weekly pay period, the employee produced 450 units over a 45-hour period and had 5 hours of waiting time. Find the employee's gross earnings for the week if time-and-a-half is paid for overtime.

16.2   GROSS EARNINGS: SALARIES AND COMMISSIONS

Salaries

Learning objective
Calculate gross earnings based on salary payroll systems.

Employees who are engaged in administrative, managerial, and sales positions are most often paid a salary for their services. A **salary** is a fixed amount of compensation usually expressed in terms of a year or a month. For example, a manager may receive an annual salary of $30,000, where an office supervisor may be paid $1,800 per month. The actual payment of annual and monthly salaries to employees is prorated into *pay periods* that identify the time between paychecks. Most companies pay their employees weekly or biweekly; however, some still prefer semimonthly or monthly. Table 16.1 contains the prorated pay periods for an annual salary. Let's look at an example that explains how to prorate an annual salary into various pay periods.

Example 10

Elaine Smart is the Director of Patient Services at Community Hospital and receives an annual salary of $28,000. Determine her gross earnings if she is paid (a) weekly, (b) biweekly, or (c) semimonthly.

Solution

	annual salary		pay periods		weekly salary
a.	$28,000	÷	52	=	$538.46

	annual salary		pay periods		biweekly salary
b.	$28,000	÷	26	=	$1,076.92

	annual salary		pay periods		semimonthly salary
c.	$28,000	÷	24	=	$1,166.67

Many salaried employees are classified **exempt** and are compensated for their work, not the number of hours worked; hourly wage employees are **nonexempt.** The exempt status provides for payment of earnings under conditions that hourly employees may not be paid, such as leaving the work-

Table 16.1

Prorated annual pay periods

Period	Number of paychecks
Weekly	52
Biweekly	26; one every 2 weeks
Semimonthly	24; two each month
Monthly	12; one every month

place early or missing work due to illness. On the other hand, exempt salaried employees may not receive overtime compensation for additional hours worked beyond the required workday or workweek. There are, however, some salaried positions covered by the terms of the Fair Labor Standards Act that are paid overtime. When overtime is paid to salaried workers they usually receive time-and-a-half for all hours worked in excess of 40 hours per week. For example, if a salaried employee's regular workweek is 35 hours, he or she would receive an overtime rate equal to $1\frac{1}{2}$ times his or her prorated regular hourly rate for all hours worked over 40 in a given week plus his or her regular salary for all hours worked up to 40 hours per week. Examples 11 and 12 further illustrate this concept.

Example 11

Harold Jurkiewicz is an account clerk; he is paid a weekly salary of $280 and time-and-a-half for all hours over 40 worked in a week. Find Harold's gross earnings for the week if he worked 45 hours and his regular workweek is 35 hours.

Solution

Step 1: Prorate the weekly salary to an hourly rate based on the hours worked during the pay period.

$$\frac{\text{weekly salary}}{\text{regular workweek}} \quad \frac{\$280}{35} = \$8 \text{ prorated hourly rate}$$

Step 2: Determine the overtime rate, which is $12 (1.5 × $8) and the overtime hours (45 − 40 = 5).

Step 3: Then calculate the gross earnings for the week as follows:

$$
\begin{aligned}
\text{gross earnings} &= \text{regular earnings} + \text{overtime earnings} \\
&= (40 \times \$8) \quad + (5 \times \$12) \\
&= \$320 \quad\quad + \$60 \\
&= \$380
\end{aligned}
$$

Notice that the employee's regular earnings are based on the prorated regular hourly rate ($8.00) for the first 40 hours worked. In other words, any hours worked beyond the number of hours that the salary is intended to compensate for (40 − 35 = 5) are paid at the prorated regular hourly rate. The overtime rate ($12) is paid for each hour worked over 40 (45 − 40) during the pay period.

There are some salaried employees who receive a fixed salary for working hours that fluctuate week to week. The fixed salary is paid whatever the number of hours worked during the pay period; however, the employee must receive additional overtime compensation for each hour over 40 hours worked in the pay period at a rate not less than one-half the prorated regular rate.

Example 12

Janet Willis is a police sergeant and has working hours that are subject to change. Find her gross earnings for the week if she worked 48 hours, is paid an annual salary of $29,952, and is paid time-and-one-half for all hours over 40 worked in a week.

Solution

Step 1: Prorate the annual salary to a weekly salary ($29,952 ÷ 52 = $576 per week) and then to a prorated hourly rate by dividing the weekly salary by the hours worked.

$$\frac{\$576}{48} = \$12 \text{ prorated hourly rate}$$

Step 2: Determine the overtime rate (.5 × $12 = $6) and the overtime hours (48 − 40 = 8).

Step 3: Calculate the gross earnings.

$$
\begin{aligned}
\text{gross earnings} &= \text{regular earnings} + \text{overtime earnings} \\
&= \$576 \qquad\qquad + (8 \times \$6) \\
&= \$576 \qquad\qquad + \$48 \\
&= \$624
\end{aligned}
$$

Commissions

The ultimate success of any business can be measured in its ability to sell its goods and services. The sales force, therefore, is often compensated by **commission,** which is an incentive payment designed to generate a high volume of sales. The commission paid to the employee is always based on *net sales* and is usually expressed as a percent. Net sales is the result of deducting from gross sales all *sales returns* (returned merchandise) and *sales discounts* (reduction in the price of merchandise). This process was explained in detail in Chapters 4 and 5.

Formula for determining net sales

net sales = gross sales − sales returns − sales discounts

There are two basic commission plans, namely: *straight commission* and *variable,* or *graduated, commission.* In either case, the gross earnings based on commission can be calculated with the following equation:

Formula for determining gross earnings

gross earnings = (value of net sales) × (commission rate)

When a salesperson is paid a straight commission, he or she receives a fixed percent of net sales for a given pay period. As commission earnings are usually based on monthly sales figures, many companies allow their salespersons to draw against their commissions during the pay period. A *draw* is an advance on projected earnings, which is deducted from the employee's commission at the time the employee's earnings are normally determined. If the draw exceeds the employee's commission, the employee owes the company the difference, which is charged against future commissions.

Example 13

Matthew Higgins recorded gross sales of $25,750 for October. Determine his gross earnings for the period if his company offered a 5% discount to all customers. Matthew is paid a straight commission of $8\frac{1}{2}\%$ of net sales and received a draw of $800 in October.

Solution

Step 1: Calculate the discount ($25,750 × .05 = $1,287.50) and subtract the discount amount from gross sales to find the net sales amount.

$$net\ sales = gross\ sales - allowances\ and\ discounts$$
$$= \$25,750\ \ - \$1,287.50$$
$$= \$24,462.50$$

Step 2: Insert the required variables into the equation and find the gross earnings.

$$gross\ earnings = net\ sales\ \ \ \times commission\ rate$$
$$= \$24,462.50 \times .085$$
$$= \$2,079.31$$

Step 3: Subtract the draw from the gross earnings amount to determine the gross earnings that remain unpaid by the employer.

$$\text{unpaid gross earnings} = \text{gross earnings} - \text{draw}$$
$$= \$2,079.31 \quad - \$800$$
$$= \$1,279.31$$

Some businesses elect to pay a fixed amount as the commission, based on the number of units sold. If this is the case, the gross earnings would be calculated in a manner similar to an hourly employee's piecework wages. The equation would be modified as shown in the next example.

Example 14

Martha Deinhart works for a mailing service company and receives a commission of $.04 for each piece of mailing she prepares. Find her gross earnings if she prepared 5,625 mailings during a recent pay period.

Solution

$$\text{gross earnings} = \text{number of units} \times \text{commission rate per unit}$$
$$= 5,625 \qquad \times .04$$
$$= \$225.00$$

The variable, or graduated commission, plan is similar in motive to the differential-piecework plan for hourly workers as it pays higher commission rates for increased net sales levels. A *commission rate schedule* identifies the net sales range and the corresponding commission rate to be paid, as shown in Example 15.

Example 15

Salespersons employed by International Frozen Foods are paid a graduated commission on sales per month as follows:

Sales	Rate
$0–$8,000	8%
$8,001–$12,000	9%
$12,001–$15,000	10%
Over $15,000	12%

Find the gross earnings for Kara Andrews whose net sales amounted to $16,840 for the month.

Solution

Insert the variables into the equation and solve for gross earnings.

	Net sales	×	Commission rate	=	Gross earnings
gross earnings on first	$ 8,000	=	$8,000 × .08	=	$ 640.00
gross earnings on next	4,000	=	4,000 × .09	=	360.00
gross earnings on next	3,000	=	3,000 × .10	=	300.00
gross earnings on remaining	1,840	=	1,840 × .12	=	220.80
total net sales	$16,840		total gross earnings		$1,520.80

Many large retail stores and industrial firms pay their sales personnel a *salary plus commission.* This method of payment requires a fixed salary per pay period plus a commission on net sales. This plan eliminates the additional record keeping required by the draw system and provides the salesperson with a regular income as well as an incentive. The following equation is used to determine gross earnings with the salary-plus-commission plan.

Formula for gross earnings on salary-plus-commission plan

gross earnings = fixed salary + (net sales × commission rate)

Example 16

Lisa Shrank is paid a salary of $2,000 per month plus $4\frac{3}{4}$% commission on net sales over $10,000. Calculate her gross earnings for the month if her net sales were $18,000.

Solution

Insert the required variables into the equation and solve.

gross earnings = fixed salary + (commission sales × commission rate)

= $2,000 + ($8,000   × .0475)

= $2,000 + $380

= $2,380

The net sales figure ($8,000) in the example is the amount of commission sales over the required standard sales ($10,000) that the employee must sell before she can earn the commission ($18,000 − $10,000 = $8,000).

CHECK YOUR KNOWLEDGE

Calculating Salaries and Commissions

1. Eileen Hyde receives an annual salary of $17,680. What is her (a) bi-weekly salary? (b) semimonthly salary?

2. Ray Jewell is paid a weekly salary of $420. He is paid time-and-a-half for all hours over 40 hours a week. Find his gross earnings for a week in which he worked 45 hours.

3. Christie Halloran is compensated on a fluctuating workweek basis at an annual salary of $19,500. Find her gross earnings for two separate pay periods if she worked (a) 44 hours and (b) 35 hours during each pay period.

4. Last month Jason Novak had net sales of $40,250. What were his gross earnings for the period if he is paid a $5\frac{3}{8}\%$ straight commission rate and his drawing account totaled $1,250?

5. The student association of Northeast State College recently conducted a fund-raiser involving the sale of boxed candies. The club is paid a $1.25 commission on each box of candy sold. Calculate the club's gross earnings if they sold 2,687 boxes during the fund-raiser.

6. Maria Rodriguez is a sales representative for an industrial machinery manufacturer. Her earnings are based on a graduated commission as follows: 3% of the first $10,000 in sales; $4\frac{1}{4}\%$ of the next $25,000; 6% of the next $30,000; and 8% of all sales over $65,000. Find her gross earnings for the month if her sales for the period amounted to $72,500.

7. Midtown Automobile Emporium pays its salespersons a base salary of $500 biweekly plus a commission of $2\frac{1}{2}\%$ on all sales. Determine Mark Edsall's gross earnings for the period; he sold $48,750 in cars for the dealership.

16.2   EXERCISES

Determine the prorated earnings for each of the following salaries based on the given amounts.

	Weekly	**Biweekly**	**Semimonthly**	**Monthly**	**Annual**
1.	$200.00	_____	_____	_____	_____
2.	_____	_____	$850.00	_____	_____
3.	_____	$645.00	_____	_____	_____
4.	_____	_____	_____	$1,500.00	_____
5.	_____	_____	_____	_____	$40,800

Answers to CYK: **1. a.** $680; **b.** $736.67   **2.** $498.75   **3. a.** $392.05; **b.** $375
4. $913.44   **5.** $3,358.75   **6.** $3,762.50   **7.** $1,718.75

Using the graduated straight commission schedule below, calculate the monthly gross earnings for the following salespeople.

Net sales	Commission rate
$1 to $20,000	3%
$20,001 to $35,000	4.5%
$35,001 to $50,000	6%
Over $50,000	6.5%

	Employee	Gross sales	Total discounts/returns	Net sales	Gross earnings
6.	Carey, P.	$38,500	$1,260	_____	_____
7.	Melnick, T.	52,750	1,525	_____	_____
8.	Sansone, R.	21,250	980	_____	_____
9.	Valentine, H.	43,375	1,325	_____	_____

Hitchcock Interiors pays its salespeople a weekly salary of $275 plus a commission of $1\frac{5}{8}\%$ on net sales over $1,200 each pay period. Calculate the biweekly gross earnings for each salesperson using the payroll information provided in the table.

	Name	Gross sales	Returns	Net sales	Commission sales
10.	Abbott, D.	$ 8,500	$375	_____	_____
11.	Butler, G.	22,250	245	_____	_____
12.	North, R.	15,620	430	_____	_____
13.	Smith, J.	39,875	605	_____	_____
14.	Zeigler, H.	26,460	240	_____	_____

Name	Commission	Regular earnings	Gross earnings
Abbott, D.	_____	_____	_____
Butler, G.	_____	_____	_____
North, R.	_____	_____	_____
Smith, J.	_____	_____	_____
Zeigler, H.	_____	_____	_____

15. Carol Ryan, a radiologist, is paid an annual salary of $18,850. She is paid on a weekly basis with a regular workweek of 35 hours; she is paid overtime at the rate of time-and-a-half for hours worked over 40 per week. During recent pay periods, she worked 43 and 46 hours, respectively. Find her gross earnings for each of the pay periods.

16. Carl Costello's salary is $23,842 per year. His company pays him weekly and his hours vary as conditions require. Find his gross earnings for the week if he worked 54 hours during the current pay period and is paid overtime for each hour over 40 hours.

17. This past month, Alex Marshall had gross sales of $87,250, subject to a 2% trade discount. What were his gross earnings if he is paid a 5% straight commission rate on net sales and had draws totaling $1,350 for the period?

18. Clara Kassler sells cookware door to door. She receives no salary but earns $42.50 for each set of cookware sold. What is her gross earnings for the month if she sold 67 sets?

19. Luber Corporation pays its salespersons on a graduated commission basis: 2% of the first $25,000 net sales; 3½% on net sales above $25,000 to $40,000; and 4⅝% on net sales over $40,000. If Frank Walls had net sales of $68,450 for the month, what would be the amount of his commission based on his gross earnings?

20. Christine Dunn sells microcomputers for Smart Office Systems. She receives a semimonthly salary of $600 plus a commission of 3¼% on all net sales over $2,500 per pay period. If she had sales of $10,620 and $840 in merchandise returns, what are her gross earnings for the pay period?

16.3   EMPLOYEE PAYROLL DEDUCTIONS

Learning objective
Identify and calculate the various types of employee payroll deductions.

The preceding sections of this chapter are devoted to the various methods used by employers to determine gross earnings. Gross earnings, however, is not what the employee actually receives for his or her work. *Net earnings, net pay,* or **take-home pay** is generally less than gross earnings due to the *deductions* that are subtracted. This section of the chapter will focus on a discussion of the various mandatory and elective deductions that reduce an employee's gross earnings. The relationship between the principal variables can be expressed by the following equation.

Formula for net earnings

net earnings = gross earnings − total deductions

Mandatory Deductions

Mandatory deductions are those deductions usually required by federal, state, and/or local governments and include *Social Security tax (FICA), Medicare tax, income withholding taxes,* and contributions to state *unemployment compensation* programs in the form of **state disability insurance.**

Most employers are required by the **Federal Insurance Contributions Act (FICA)** to withhold from an employee's earnings a tax at a specified rate based on a taxable amount of accumulated gross earnings paid during the calendar year. This tax is sent by the employer to the *Internal Revenue Service* (*IRS*) to be used by the federal government to finance the **Social Security Fund.** The fund provides to qualified individuals payments for retirement, disability, survivor benefits, and health insurance for the elderly. The portion of the tax that is used to fund health insurance for the elderly (Medicare) is called the **Medicare tax.** The schedule of tax rates and wage bases subject to tax has been revised periodically by the federal government to reflect economic conditions. Prior to 1991, the Social Security and Medicare tax rates were combined into one rate with a single wage base. Since 1991, Social Security and Medicare taxes have been withheld separately and have different tax rates and wage bases. Table 16.2 shows the tax rates for employees, employers, and self-employed individuals as well as the wage bases for the years 1993–1998. Notice in Table 16.2 that self-employed persons are required to pay a self-employment FICA tax and a Medicare tax equal to the combined rate of the employee-employer FICA tax. The employer is required to pay a FICA tax equal to the employee's for each employee as part of its total payroll taxes. These employer taxes will be discussed in more detail in Section 16.4.

Using rates given in Table 16.2, an employer in 1998 would deduct from an employee's earnings 6.20% on the first $68,400 of wages earned in the calendar year for a maximum Social Security tax of $4,240.80 ($68,400 × .0620 = $4,240.80). If an employee earns more than $68,400 the additional earnings are exempt from the tax. Whatever the tax rate and wage base subject to the tax, the procedure used to calculate the tax has remained unchanged. We can use the same procedure to calculate the Medicare

Table 16.2

Social Security and Medicare tax rates

Year	Wage base	Employer tax rate	Employee tax rate	Self-employed tax rate
1998 Social Security	$68,400	6.20	6.20	12.4
1998 Medicare	All wages	1.45	1.45	2.9
1997 Social Security	$65,400	6.20	6.20	12.4
1997 Medicare	All wages	1.45	1.45	2.9
1996 Social Security	$62,700	6.20	6.20	12.4
1996 Medicare	All wages	1.45	1.45	2.9
1995 Social Security	$61,200	6.20	6.20	12.4
1995 Medicare	All wages	1.45	1.45	2.9
1994 Social Security	$59,600	6.20	6.20	12.4
1994 Medicare	All wages	1.45	1.45	2.9
1993 Social Security	$57,600	6.20	6.20	12.4
1993 Medicare	$135,000	1.45	1.45	2.9

Note: Prior to 1994, Medicare had a wage base that increased each year, resulting in a maximum Medicare tax.

For Your Information

SOCIAL SECURITY: GLORIOUS PAST, SHAKY FUTURE

How the Program Works

Today's workers are taxed to pay retirement benefits both for their parents' generation and to pay part of their own retirement benefits. When today's workers retire, their benefits will be covered by a combination of the taxes they paid when they were working and taxes on their children's generation. The principle is different from an investment, an insurance policy, or an annuity, in which individuals pay premiums and the income from those premiums is used to pay future benefits. Since 1983, recognizing that the ratio of beneficiaries to workers would decline in future years, the government has collected more money in Social Security revenues.

The Problem

While there continues to be a surplus, the future presents some serious challenges. The ratio of workers paying taxes to retirees drawing benefits has been shrinking for years and is projected to continue declining. The surplus has been placed in a trust fund. Unfortunately, the surplus has been used by the government to cover its operating expenses and to offset the budget deficit. In 1950, there were 16 workers for each Social Security beneficiary. In 1998 there were slightly more than three. As shown in the table, by the second and third decades of the next century, there will be only two people at work for each person receiving benefits.

Unless some changes are made to the program, the fund will run out of money by 2032. Economists and politicians agree that the sooner the problem is addressed the less painful and abrupt the solution will be. The government is considering proposals that will guarantee the survival of the popular Social Security program beyond the year 2032.

Source: Copyright 1999 by The New York Times Company. Reprinted by permission.

A portrait of Social Security

The beneficiaries
Population of the United States

	271 million

Number of people receiving a benefit

	44.2 million

62% Retired workers	7% Spouses and children of retired workers	16% Spouses, parents and children of deceased workers	11% Disabled people	4% Spouses and children of disabled people

MONTHLY BENEFIT (FIGURES AS OF DEC. 1998)	AVERAGE	MAXIMUM
Individual	$953	$1,373
Retired worker and spouse	$1,430	$2,060

The taxes

Workers covered	150 million
Social Security tax rate	12.4 percent equally divided between employer and employee
Medicare tax rate	2.9 percent equally divided
Total Social Security and Medicare tax rate for the self-employed	15.3 percent
Maximum annual earnings on which Social Security taxes are paid (1999)	$72,600
Annual earnings above which beneficiaries have their benefits cut (1999)	$15,500

Trust fund finances

IN BILLIONS	1998	2021	2031
Income	$484	$1,509	$2,050
Payments	$383	$1,528	$2,741
Total assets at year's end	$757	$3,758	$289

Source: Social Security Administration   N.Y. Times

Table 16.3

Maximum Social
Security
withholding tax

Year	Social Security	Medicare	Total	Percent change
1990	$3,180.60	$743.85	$3,924.45	0.0
1991	3,310.80	774.30	4,085.10	4.1
1992	3,441.00	804.75	4,245.75	3.9
1993	3,571.20	835.20	4,406.40	3.8
1994	3,695.20	864.20	4,559.40	3.5
1995	3,794.40	887.40	4,681.80	2.7
1996	3,887.40	909.15	4,796.55	2.5
1997	4,054.80	948.30	5,003.10	4.3
1998	4,240.80	991.80	5,232.60	4.6

Note: Since 1990 the maximum amount withheld from a person's annual wages for Social Security has increased 33.3 percent. The percent change is based on the combined rate of 7.65% (6.2 plus 1.45) of the annual FICA wage base.

tax that we used to determine the Social Security (FICA) tax. For example, in 1998, an employer had to withhold 1.45% of an employee's total earnings paid during the calendar year for Medicare tax. Therefore, there would be no maximum amount of Medicare tax. The amount of tax would be determined by the amount of earnings ($68,400 × .0145 = $991.80) Table 16.3 illustrates how the maximum withholding amounts for Social Security tax has increased since 1990.

The method used to determine these taxes is an application of the percentage equation ($P = BR$) presented in Chapter 2.

FICA tax = taxable earnings × tax rate

Medicare tax = taxable earnings × tax rate

Because the wage base amounts for these taxes change, we will use the 1998 tax rates in all examples and exercises in this chapter.

Example 17

The weekly gross earnings of Frank Taylor this pay period are $915.50, and his payroll record indicates he has year-to-date earnings of $63,524.75. How much of his current earnings must his employer withhold for FICA tax and for Medicare tax?

Solution

Step 1: Determine the taxable earnings.

$68,400.00	maximum taxable earnings
63,524.75	year-to-date earnings
$ 4,875.25	earnings subject to Social Security taxes

Step 2: Insert variables into the percentage equations and solve.

FICA tax = taxable earnings × tax rate

= $915.50 × .0620

= $56.76

Medicare tax = taxable earnings × tax rate

= $915.50 × .0145

= $13.27

The $915.50 earned is entirely taxable for both FICA and Medicare taxes since the current gross earnings are less than the earnings still subject to the taxes. If the current gross earnings are greater than the earnings subject to Social Security tax in step 1, the FICA tax is determined by multiplying the earnings still subject to Social Security tax by the FICA tax rate. Remember, all wages are subject to Medicare tax. Therefore, the Medicare tax is always determined by multiplying the current gross earnings by the Medicare tax rate.

The **federal withholding tax (FWT),** another mandatory deduction required by the federal government, is applied to the employee's federal income tax. The amount to be withheld by the employer from each employee is based on the following primary factors:

1. the type of pay period

2. the amount of the employee's gross earnings

3. the marital status of the employee

4. the number of exemptions claimed by the employee

The employees are required by their employer to fill out a W-4 form as shown in Figure 16.4. This form provides the employer with the information regarding each employee's marital status and the number of exemptions claimed so that the federal income withholding tax can be calculated. The employer can use one of two methods to determine the federal withholding tax to be deducted from the employee's gross earnings. The **wage bracket method** is a single-step operation using tables of amounts to be withheld. The **percentage method** uses tax rates and taxable earnings to calculate the amounts to be withheld. The tables that employers use to determine the FWT and the FICA tax are provided in the Internal Revenue Service's **Employer's Tax Guide Circular E.** The guide contains tables for both single and married employees paid daily, weekly, biweekly, semimonthly, monthly, and so forth. The taxes tables presented in this text (Figures 16.5, 16.6, 16.7, and 16.8) are excerpts from that publication. The rates and tables will change, but the procedures required to use these tables have remained unchanged.

Figure 16.4

Form W-4: Employee withholding allowance certificate

Form W-4 (1998)

Purpose. Complete Form W-4 so your employer can withhold the correct Federal income tax from your pay. Because your tax situation may change, you may want to refigure your withholding each year.

Exemption from withholding. If you are exempt, complete only lines 1, 2, 3, 4, and 7, and sign the form to validate it. Your exemption for 1998 expires February 16, 1999.

Note: You cannot claim exemption from withholding if (1) your income exceeds $700 and includes unearned income (e.g., interest and dividends) and (2) another person can claim you as a dependent on their tax return.

Basic instructions. If you are not exempt, complete the Personal Allowances Worksheet. The worksheets on page 2 adjust your

withholding allowances based on itemized deductions, adjustments to income, or two-earner/two-job situations. Complete all worksheets that apply. They will help you figure the number of withholding allowances you are entitled to claim. However, you may claim fewer allowances.

New—Child tax and higher education credits. For details on adjusting withholding for these and other credits, see Pub. 919, Is My Withholding Correct for 1998?

Head of household. Generally, you may claim head of household filing status on your tax return only if you are unmarried and pay more than 50% of the costs of keeping up a home for yourself and your dependent(s) or other qualifying individuals.

Nonwage income. If you have a large amount of nonwage income, such as interest or dividends, you should consider making estimated tax payments using Form 1040-ES. Otherwise, you may owe additional tax.

Two earners/two jobs. If you have a working spouse or more than one job, figure the total number of allowances you are entitled to claim on all jobs using worksheets from only one W-4. Your withholding will usually be most accurate when all allowances are claimed on the W-4 filed for the highest paying job and zero allowances are claimed for the others.

Check your withholding. After your W-4 takes effect, use Pub. 919 to see how the dollar amount you are having withheld compares to your estimated total annual tax. Get Pub. 919 especially if you used the Two-Earner/Two-Job Worksheet and your earnings exceed $150,000 (Single) or $200,000 (Married). To order Pub. 919, call 1-800-829-3676. Check your telephone directory for the IRS assistance number for further help.

Sign this form. Form W-4 is not valid unless you sign it.

Personal Allowances Worksheet

A Enter "1" for **yourself** if no one else can claim you as a dependent . . . . **A** _____

B Enter "1" if:
- You are single and have only one job; or
- You are married, have only one job, and your spouse does not work; or
- Your wages from a second job or your spouse's wages (or the total of both) are $1,000 or less.

B _____

C Enter "1" for your **spouse**. But, you may choose to enter -0- if you are married and have either a working spouse or more than one job. (This may help you avoid having too little tax withheld.) . . . . **C** _____

D Enter number of **dependents** (other than your spouse or yourself) you will claim on your tax return . . . . **D** _____

E Enter "1" if you will file as **head of household** on your tax return (see conditions under **Head of household** above) . **E** _____

F Enter "1" if you have at least $1,500 of **child or dependent care expenses** for which you plan to claim a credit . . **F** _____

G **New—Child Tax Credit:**
- If your total income will be between $16,500 and $47,000 ($21,000 and $60,000 if married), enter "1" for each eligible child.
- If your total income will be between $47,000 and $80,000 ($60,000 and $115,000 if married), enter "1" if you have two or three eligible children, or enter "2" if you have four or more . . . . **G** _____

H Add lines A through G and enter total here. **Note:** This amount may be different from the number of exemptions you claim on your return. ▶ **H** _____

For accuracy, complete all worksheets that apply.

- If you plan to **itemize or claim adjustments to income** and want to reduce your withholding, see the Deductions and Adjustments Worksheet on page 2.
- If you are **single**, have **more than one job**, and your combined earnings from all jobs exceed $32,000 OR if you are **married** and have a **working spouse or more than one job**, and the combined earnings from all jobs exceed $55,000, see the Two-Earner/Two-Job Worksheet on page 2 to avoid having too little tax withheld.
- If **neither** of the above situations applies, **stop here** and enter the number from line H on line 5 of Form W-4 below.

- - - - - - - - - - - - - - - **Cut here and give the certificate to your employer. Keep the top part for your records.** - - - - - - - - - - - - - - -

Form **W-4**
Department of the Treasury
Internal Revenue Service

Employee's Withholding Allowance Certificate

▶ **For Privacy Act and Paperwork Reduction Act Notice, see page 2.**

OMB No. 1545-0010

1998

| 1 Type or print your first name and middle initial | Last name | 2 Your social security number |

| Home address (number and street or rural route) | 3 ☐ Single ☐ Married ☐ Married, but withhold at higher Single rate. |

Note: If married, but legally separated, or spouse is a nonresident alien, check the Single box.

| City or town, state, and ZIP code | 4 If your last name differs from that on your social security card, check here and call 1-800-772-1213 for a new card . . . . . ▶ ☐ |

5 Total number of allowances you are claiming (from line H above or from the worksheets on page 2 if they apply) . **5** _____

6 Additional amount, if any, you want withheld from each paycheck . . . . **6** $ _____

7 I claim exemption from withholding for 1998, and I certify that I meet **BOTH** of the following conditions for exemption:
- Last year I had a right to a refund of **ALL** Federal income tax withheld because I had **NO** tax liability **AND**
- This year I expect a refund of **ALL** Federal income tax withheld because I expect to have **NO** tax liability.

If you meet both conditions, enter "EXEMPT" here . . . . ▶ **7** _____

Under penalties of perjury, I certify that I am entitled to the number of withholding allowances claimed on this certificate or entitled to claim exempt status.

Employee's signature ▶

Date ▶ _____ 19___

| 8 Employer's name and address (Employer: Complete 8 and 10 only if sending to the IRS) | 9 Office code (optional) | 10 Employer identification number |

Cat. No. 10220Q

Source: U.S. Dept. of the Treasury, Internal Revenue Service (U.S. Government Printing Office).

Figure 16.5

Employer's tax guide: Single person paid weekly

If the wages are–		And the number of withholding allowances claimed is—										
At least	But less than	0	1	2	3	4	5	6	7	8	9	10
		The amount of income tax to be withheld is—										
$0	$55	0	0	0	0	0	0	0	0	0	0	0
55	60	1	0	0	0	0	0	0	0	0	0	0
60	65	2	0	0	0	0	0	0	0	0	0	0
65	70	2	0	0	0	0	0	0	0	0	0	0
70	75	3	0	0	0	0	0	0	0	0	0	0
75	80	4	0	0	0	0	0	0	0	0	0	0
80	85	5	0	0	0	0	0	0	0	0	0	0
85	90	5	0	0	0	0	0	0	0	0	0	0
90	95	6	0	0	0	0	0	0	0	0	0	0
95	100	7	0	0	0	0	0	0	0	0	0	0
100	105	8	0	0	0	0	0	0	0	0	0	0
105	110	8	1	0	0	0	0	0	0	0	0	0
110	115	9	1	0	0	0	0	0	0	0	0	0
115	120	10	2	0	0	0	0	0	0	0	0	0
120	125	11	3	0	0	0	0	0	0	0	0	0
125	130	11	4	0	0	0	0	0	0	0	0	0
130	135	12	4	0	0	0	0	0	0	0	0	0
135	140	13	5	0	0	0	0	0	0	0	0	0
140	145	14	6	0	0	0	0	0	0	0	0	0
145	150	14	7	0	0	0	0	0	0	0	0	0
150	155	15	7	0	0	0	0	0	0	0	0	0
155	160	16	8	0	0	0	0	0	0	0	0	0
160	165	17	9	1	0	0	0	0	0	0	0	0
165	170	17	10	2	0	0	0	0	0	0	0	0
170	175	18	10	3	0	0	0	0	0	0	0	0
175	180	19	11	3	0	0	0	0	0	0	0	0
180	185	20	12	4	0	0	0	0	0	0	0	0
185	190	20	13	5	0	0	0	0	0	0	0	0
190	195	21	13	6	0	0	0	0	0	0	0	0
195	200	22	14	6	0	0	0	0	0	0	0	0
200	210	23	15	8	0	0	0	0	0	0	0	0
210	220	25	17	9	1	0	0	0	0	0	0	0
220	230	26	18	11	3	0	0	0	0	0	0	0
230	240	28	20	12	4	0	0	0	0	0	0	0
240	250	29	21	14	6	0	0	0	0	0	0	0
250	260	31	23	15	7	0	0	0	0	0	0	0
260	270	32	24	17	9	1	0	0	0	0	0	0
270	280	34	26	18	10	2	0	0	0	0	0	0
280	290	35	27	20	12	4	0	0	0	0	0	0
290	300	37	29	21	13	5	0	0	0	0	0	0
300	310	38	30	23	15	7	0	0	0	0	0	0
310	320	40	32	24	16	8	1	0	0	0	0	0
320	330	41	33	26	18	10	2	0	0	0	0	0
330	340	43	35	27	19	11	4	0	0	0	0	0
340	350	44	36	29	21	13	5	0	0	0	0	0
350	360	46	38	30	22	14	7	0	0	0	0	0
360	370	47	39	32	24	16	8	0	0	0	0	0
370	380	49	41	33	25	17	10	2	0	0	0	0
380	390	50	42	35	27	19	11	3	0	0	0	0
390	400	52	44	36	28	20	13	5	0	0	0	0
400	410	53	45	38	30	22	14	6	0	0	0	0
410	420	55	47	39	31	23	16	8	0	0	0	0
420	430	56	48	41	33	25	17	9	2	0	0	0
430	440	58	50	42	34	26	19	11	3	0	0	0
440	450	59	51	44	36	28	20	12	5	0	0	0
450	460	61	53	45	37	29	22	14	6	0	0	0
460	470	62	54	47	39	31	23	15	8	0	0	0
470	480	64	56	48	40	32	25	17	9	1	0	0
480	490	65	57	50	42	34	26	18	11	3	0	0
490	500	67	59	51	43	35	28	20	12	4	0	0
500	510	68	60	53	45	37	29	21	14	6	0	0
510	520	70	62	54	46	38	31	23	15	7	0	0
520	530	72	63	56	48	40	32	24	17	9	1	0
530	540	75	65	57	49	41	34	26	18	10	3	0
540	550	78	66	59	51	43	35	27	20	12	4	0
550	560	81	68	60	52	44	37	29	21	13	6	0
560	570	83	69	62	54	46	38	30	23	15	7	0
570	580	86	72	63	55	47	40	32	24	16	9	1
580	590	89	74	65	57	49	41	33	26	18	10	2
590	600	92	77	66	58	50	43	35	27	19	12	4

To illustrate both methods, we will continue to develop the calculation of Frank Taylor's deductions based on his earnings in Example 17.

Example 18

The gross earnings of Frank Taylor during this weekly pay period is $915.50. Using the information provided on his W-4 (Figure 16.4), determine the amount of federal withholding tax to be deducted from his gross earnings.

Figure 16.5

(Continued)

If the wages are—		And the number of withholding allowances claimed is—										
At least	But less than	0	1	2	3	4	5	6	7	8	9	10
		The amount of income tax to be withheld is—										
$600	$610	95	80	68	60	52	44	36	29	21	13	5
610	620	97	83	69	61	53	46	38	30	22	15	7
620	630	100	86	71	63	55	47	39	32	24	16	8
630	640	103	88	74	64	56	49	41	33	25	18	10
640	650	106	91	77	66	58	50	42	35	27	19	11
650	660	109	94	79	67	59	52	44	36	28	21	13
660	670	111	97	82	69	61	53	45	38	30	22	14
670	680	114	100	85	70	62	55	47	39	31	24	16
680	690	117	102	88	73	64	56	48	41	33	25	17
690	700	120	105	91	76	65	58	50	42	34	27	19
700	710	123	108	93	79	67	59	51	44	36	28	20
710	720	125	111	96	82	68	61	53	45	37	30	22
720	730	128	114	99	84	70	62	54	47	39	31	23
730	740	131	116	102	87	73	64	56	48	40	33	25
740	750	134	119	105	90	76	65	57	50	42	34	26
750	760	137	122	107	93	78	67	59	51	43	36	28
760	770	139	125	110	96	81	68	60	53	45	37	29
770	780	142	128	113	98	84	70	62	54	46	39	31
780	790	145	130	116	101	87	72	63	56	48	40	32
790	800	148	133	119	104	90	75	65	57	49	42	34
800	810	151	136	121	107	92	78	66	59	51	43	35
810	820	153	139	124	110	95	81	68	60	52	45	37
820	830	156	142	127	112	98	83	69	62	54	46	38
830	840	159	144	130	115	101	86	72	63	55	48	40
840	850	162	147	133	118	104	89	74	65	57	49	41
850	860	165	150	135	121	106	92	77	66	58	51	43
860	870	167	153	138	124	109	95	80	68	60	52	44
870	880	170	156	141	126	112	97	83	69	61	54	46
880	890	173	158	144	129	115	100	86	71	63	55	47
890	900	176	161	147	132	118	103	88	74	64	57	49
900	910	179	164	149	135	120	106	91	77	66	58	50
910	920	181	167	152	138	123	109	94	80	67	60	52
920	930	184	170	155	140	126	111	97	82	69	61	53
930	940	187	172	158	143	129	114	100	85	71	63	55
940	950	190	175	161	146	132	117	102	88	73	64	56
950	960	193	178	163	149	134	120	105	91	76	66	58
960	970	195	181	166	152	137	123	108	94	79	67	59
970	980	198	184	169	154	140	125	111	96	82	69	61
980	990	201	186	172	157	143	128	114	99	85	70	62
990	1,000	204	189	175	160	146	131	116	102	87	73	64
1,000	1,010	207	192	177	163	148	134	119	105	90	76	65
1,010	1,020	209	195	180	166	151	137	122	108	93	78	67
1,020	1,030	212	198	183	168	154	139	125	110	96	81	68
1,030	1,040	215	200	186	171	157	142	128	113	99	84	70
1,040	1,050	218	203	189	174	160	145	130	116	101	87	72
1,050	1,060	221	206	191	177	162	148	133	119	104	90	75
1,060	1,070	223	209	194	180	165	151	136	122	107	92	78
1,070	1,080	226	212	197	182	168	153	139	124	110	95	81
1,080	1,090	229	214	200	185	171	156	142	127	113	98	84
1,090	1,100	232	217	203	188	174	159	144	130	115	101	86
1,100	1,110	235	220	205	191	176	162	147	133	118	104	89
1,110	1,120	238	223	208	194	179	165	150	136	121	106	92
1,120	1,130	241	226	211	196	182	167	153	138	124	109	95
1,130	1,140	244	228	214	199	185	170	156	141	127	112	98
1,140	1,150	247	231	217	202	188	173	158	144	129	115	100
1,150	1,160	250	234	219	205	190	176	161	147	132	118	103
1,160	1,170	253	237	222	208	193	179	164	150	135	120	106
1,170	1,180	256	240	225	210	196	181	167	152	138	123	109
1,180	1,190	259	243	228	213	199	184	170	155	141	126	112
1,190	1,200	262	246	231	216	202	187	172	158	143	129	114
1,200	1,210	266	249	233	219	204	190	175	161	146	132	117
1,210	1,220	269	253	236	222	207	193	178	164	149	134	120
1,220	1,230	272	256	240	224	210	195	181	166	152	137	123
1,230	1,240	275	259	243	227	213	198	184	169	155	140	126
1,240	1,250	278	262	246	230	216	201	186	172	157	143	128

$1,250 and over	Use Table 1(a) for a **SINGLE** person on page 34. Also see the instructions on page 32.

Source: Circular E: *Employer's Tax Guide*, U.S. Dept. of the Treasury, Internal Revenue Service (U.S. Government Printing Office).

Solution 1

Wage Bracket Method

Since Frank's W-4 indicates that he is married and claims four exemptions, we can use the table in Figure 16.6 to determine his FWT deduction. We find his gross earning amount, $915.50, in the wage-bracket columns headed *at least* $910 *but less than* $920 and move to the right across the table to

Figure 16.6

Employer's tax guide: Married person paid weekly

If the wages are—		And the number of withholding allowances claimed is—										
At least	But less than	0	1	2	3	4	5	6	7	8	9	10
		The amount of income tax to be withheld is—										
$0	$125	0	0	0	0	0	0	0	0	0	0	0
125	130	1	0	0	0	0	0	0	0	0	0	0
130	135	1	0	0	0	0	0	0	0	0	0	0
135	140	2	0	0	0	0	0	0	0	0	0	0
140	145	3	0	0	0	0	0	0	0	0	0	0
145	150	4	0	0	0	0	0	0	0	0	0	0
150	155	4	0	0	0	0	0	0	0	0	0	0
155	160	5	0	0	0	0	0	0	0	0	0	0
160	165	6	0	0	0	0	0	0	0	0	0	0
165	170	7	0	0	0	0	0	0	0	0	0	0
170	175	7	0	0	0	0	0	0	0	0	0	0
175	180	8	0	0	0	0	0	0	0	0	0	0
180	185	9	1	0	0	0	0	0	0	0	0	0
185	190	10	2	0	0	0	0	0	0	0	0	0
190	195	10	2	0	0	0	0	0	0	0	0	0
195	200	11	3	0	0	0	0	0	0	0	0	0
200	210	12	4	0	0	0	0	0	0	0	0	0
210	220	14	6	0	0	0	0	0	0	0	0	0
220	230	15	7	0	0	0	0	0	0	0	0	0
230	240	17	9	1	0	0	0	0	0	0	0	0
240	250	18	10	3	0	0	0	0	0	0	0	0
250	260	20	12	4	0	0	0	0	0	0	0	0
260	270	21	13	6	0	0	0	0	0	0	0	0
270	280	23	15	7	0	0	0	0	0	0	0	0
280	290	24	16	9	1	0	0	0	0	0	0	0
290	300	26	18	10	2	0	0	0	0	0	0	0
300	310	27	19	12	4	0	0	0	0	0	0	0
310	320	29	21	13	5	0	0	0	0	0	0	0
320	330	30	22	15	7	0	0	0	0	0	0	0
330	340	32	24	16	8	0	0	0	0	0	0	0
340	350	33	25	18	10	2	0	0	0	0	0	0
350	360	35	27	19	11	3	0	0	0	0	0	0
360	370	36	28	21	13	5	0	0	0	0	0	0
370	380	38	30	22	14	6	0	0	0	0	0	0
380	390	39	31	24	16	8	0	0	0	0	0	0
390	400	41	33	25	17	9	2	0	0	0	0	0
400	410	42	34	27	19	11	3	0	0	0	0	0
410	420	44	36	28	20	12	5	0	0	0	0	0
420	430	45	37	30	22	14	6	0	0	0	0	0
430	440	47	39	31	23	15	8	0	0	0	0	0
440	450	48	40	33	25	17	9	1	0	0	0	0
450	460	50	42	34	26	18	11	3	0	0	0	0
460	470	51	43	36	28	20	12	4	0	0	0	0
470	480	53	45	37	29	21	14	6	0	0	0	0
480	490	54	46	39	31	23	15	7	0	0	0	0
490	500	56	48	40	32	24	17	9	1	0	0	0
500	510	57	49	42	34	26	18	10	3	0	0	0
510	520	59	51	43	35	27	20	12	4	0	0	0
520	530	60	52	45	37	29	21	13	6	0	0	0
530	540	62	54	46	38	30	23	15	7	0	0	0
540	550	63	55	48	40	32	24	16	9	1	0	0
550	560	65	57	49	41	33	26	18	10	2	0	0
560	570	66	58	51	43	35	27	19	12	4	0	0
570	580	68	60	52	44	36	29	21	13	5	0	0
580	590	69	61	54	46	38	30	22	15	7	0	0
590	600	71	63	55	47	39	32	24	16	8	1	0
600	610	72	64	57	49	41	33	25	18	10	2	0
610	620	74	66	58	50	42	35	27	19	11	4	0
620	630	75	67	60	52	44	36	28	21	13	5	0
630	640	77	69	61	53	45	38	30	22	14	7	0
640	650	78	70	63	55	47	39	31	24	16	8	0
650	660	80	72	64	56	48	41	33	25	17	10	2
660	670	81	73	66	58	50	42	34	27	19	11	3
670	680	83	75	67	59	51	44	36	28	20	13	5
680	690	84	76	69	61	53	45	37	30	22	14	6
690	700	86	78	70	62	54	47	39	31	23	16	8
700	710	87	79	72	64	56	48	40	33	25	17	9
710	720	89	81	73	65	57	50	42	34	26	19	11
720	730	90	82	75	67	59	51	43	36	28	20	12
730	740	92	84	76	68	60	53	45	37	29	22	14

the *withholding allowances claimed* column headed *4* to find the required FWT of 87 as follows:

At least	But less than	0	1	2	3	4	5	etc.
$910	$920	121	111	103	95	87	80	

Figure 16.6

(Continued)

(For Wages Paid in 1998)											
If the wages are—		And the number of withholding allowances claimed is—										
At least	But less than	0	1	2	3	4	5	6	7	8	9	10
		The amount of income tax to be withheld is—										
$740	$750	93	85	78	70	62	54	46	39	31	23	15
750	760	95	87	79	71	63	56	48	40	32	25	17
760	770	96	88	81	73	65	57	49	42	34	26	18
770	780	98	90	82	74	66	59	51	43	35	28	20
780	790	99	91	84	76	68	60	52	45	37	29	21
790	800	101	93	85	77	69	62	54	46	38	31	23
800	810	102	94	87	79	71	63	55	48	40	32	24
810	820	104	96	88	80	72	65	57	49	41	34	26
820	830	105	97	90	82	74	66	58	51	43	35	27
830	840	107	99	91	83	75	68	60	52	44	37	29
840	850	108	100	93	85	77	69	61	54	46	38	30
850	860	110	102	94	86	78	71	63	55	47	40	32
860	870	111	103	96	88	80	72	64	57	49	41	33
870	880	113	105	97	89	81	74	66	58	50	43	35
880	890	114	106	99	91	83	75	67	60	52	44	36
890	900	116	108	100	92	84	77	69	61	53	46	38
900	910	118	109	102	94	86	78	70	63	55	47	39
910	920	121	111	103	95	87	80	72	64	56	49	41
920	930	124	112	105	97	89	81	73	66	58	50	42
930	940	126	114	106	98	90	83	75	67	59	52	44
940	950	129	115	108	100	92	84	76	69	61	53	45
950	960	132	117	109	101	93	86	78	70	62	55	47
960	970	135	120	111	103	95	87	79	72	64	56	48
970	980	138	123	112	104	96	89	81	73	65	58	50
980	990	140	126	114	106	98	90	82	75	67	59	51
990	1,000	143	129	115	107	99	92	84	76	68	61	53
1,000	1,010	146	131	117	109	101	93	85	78	70	62	54
1,010	1,020	149	134	120	110	102	95	87	79	71	64	56
1,020	1,030	152	137	122	112	104	96	88	81	73	65	57
1,030	1,040	154	140	125	113	105	98	90	82	74	67	59
1,040	1,050	157	143	128	115	107	99	91	84	76	68	60
1,050	1,060	160	145	131	116	108	101	93	85	77	70	62
1,060	1,070	163	148	134	119	110	102	94	87	79	71	63
1,070	1,080	166	151	136	122	111	104	96	88	80	73	65
1,080	1,090	168	154	139	125	113	105	97	90	82	74	66
1,090	1,100	171	157	142	128	114	107	99	91	83	76	68
1,100	1,110	174	159	145	130	116	108	100	93	85	77	69
1,110	1,120	177	162	148	133	119	110	102	94	86	79	71
1,120	1,130	180	165	150	136	121	111	103	96	88	80	72
1,130	1,140	182	168	153	139	124	113	105	97	89	82	74
1,140	1,150	185	171	156	142	127	114	106	99	91	83	75
1,150	1,160	188	173	159	144	130	116	108	100	92	85	77
1,160	1,170	191	176	162	147	133	118	109	102	94	86	78
1,170	1,180	194	179	164	150	135	121	111	103	95	88	80
1,180	1,190	196	182	167	153	138	124	112	105	97	89	81
1,190	1,200	199	185	170	156	141	126	114	106	98	91	83
1,200	1,210	202	187	173	158	144	129	115	108	100	92	84
1,210	1,220	205	190	176	161	147	132	117	109	101	94	86
1,220	1,230	208	193	178	164	149	135	120	111	103	95	87
1,230	1,240	210	196	181	167	152	138	123	112	104	97	89
1,240	1,250	213	199	184	170	155	140	126	114	106	98	90
1,250	1,260	216	201	187	172	158	143	129	115	107	100	92
1,260	1,270	219	204	190	175	161	146	131	117	109	101	93
1,270	1,280	222	207	192	178	163	149	134	120	110	103	95
1,280	1,290	224	210	195	181	166	152	137	123	112	104	96
1,290	1,300	227	213	198	184	169	154	140	125	113	106	98
1,300	1,310	230	215	201	186	172	157	143	128	115	107	99
1,310	1,320	233	218	204	189	175	160	145	131	116	109	101
1,320	1,330	236	221	206	192	177	163	148	134	119	110	102
1,330	1,340	238	224	209	195	180	166	151	137	122	112	104
1,340	1,350	241	227	212	198	183	168	154	139	125	113	105
1,350	1,360	244	229	215	200	186	171	157	142	128	115	107
1,360	1,370	247	232	218	203	189	174	159	145	130	116	108
1,370	1,380	250	235	220	206	191	177	162	148	133	119	110
1,380	1,390	252	238	223	209	194	180	165	151	136	121	111
$1,390 and over		Use Table 1(b) for a **MARRIED** person on page 34. Also see the instructions on page 32.										

Source: Circular E: *Employer's Tax Guide*, U.S. Dept. of the Treasury, Internal Revenue Service (U.S. Government Printing Office).

Frank's employer should withhold $87 from his weekly gross earnings for federal withholding tax.

Solution 2

Percentage Method

Step 1: Determine the amount of the withholding allowance. Multiply the amount of one withholding allowance for the pay period from the table in

Figure 16.7

Employer's tax guide: Single person paid biweekly

If the wages are—		And the number of withholding allowances claimed is—										
At least	But less than	0	1	2	3	4	5	6	7	8	9	10
		The amount of income tax to be withheld is—										
$0	$105	0	0	0	0	0	0	0	0	0	0	0
105	110	1	0	0	0	0	0	0	0	0	0	0
110	115	2	0	0	0	0	0	0	0	0	0	0
115	120	2	0	0	0	0	0	0	0	0	0	0
120	125	3	0	0	0	0	0	0	0	0	0	0
125	130	4	0	0	0	0	0	0	0	0	0	0
130	135	5	0	0	0	0	0	0	0	0	0	0
135	140	5	0	0	0	0	0	0	0	0	0	0
140	145	6	0	0	0	0	0	0	0	0	0	0
145	150	7	0	0	0	0	0	0	0	0	0	0
150	155	8	0	0	0	0	0	0	0	0	0	0
155	160	8	0	0	0	0	0	0	0	0	0	0
160	165	9	0	0	0	0	0	0	0	0	0	0
165	170	10	0	0	0	0	0	0	0	0	0	0
170	175	11	0	0	0	0	0	0	0	0	0	0
175	180	11	0	0	0	0	0	0	0	0	0	0
180	185	12	0	0	0	0	0	0	0	0	0	0
185	190	13	0	0	0	0	0	0	0	0	0	0
190	195	14	0	0	0	0	0	0	0	0	0	0
195	200	14	0	0	0	0	0	0	0	0	0	0
200	205	15	0	0	0	0	0	0	0	0	0	0
205	210	16	0	0	0	0	0	0	0	0	0	0
210	215	17	1	0	0	0	0	0	0	0	0	0
215	220	17	2	0	0	0	0	0	0	0	0	0
220	225	18	3	0	0	0	0	0	0	0	0	0
225	230	19	3	0	0	0	0	0	0	0	0	0
230	235	20	4	0	0	0	0	0	0	0	0	0
235	240	20	5	0	0	0	0	0	0	0	0	0
240	245	21	6	0	0	0	0	0	0	0	0	0
245	250	22	6	0	0	0	0	0	0	0	0	0
250	260	23	7	0	0	0	0	0	0	0	0	0
260	270	24	9	0	0	0	0	0	0	0	0	0
270	280	26	10	0	0	0	0	0	0	0	0	0
280	290	27	12	0	0	0	0	0	0	0	0	0
290	300	29	13	0	0	0	0	0	0	0	0	0
300	310	30	15	0	0	0	0	0	0	0	0	0
310	320	32	16	1	0	0	0	0	0	0	0	0
320	330	33	18	2	0	0	0	0	0	0	0	0
330	340	35	19	4	0	0	0	0	0	0	0	0
340	350	36	21	5	0	0	0	0	0	0	0	0
350	360	38	22	7	0	0	0	0	0	0	0	0
360	370	39	24	8	0	0	0	0	0	0	0	0
370	380	41	25	10	0	0	0	0	0	0	0	0
380	390	42	27	11	0	0	0	0	0	0	0	0
390	400	44	28	13	0	0	0	0	0	0	0	0
400	410	45	30	14	0	0	0	0	0	0	0	0
410	420	47	31	16	0	0	0	0	0	0	0	0
420	430	48	33	17	2	0	0	0	0	0	0	0
430	440	50	34	19	3	0	0	0	0	0	0	0
440	450	51	36	20	5	0	0	0	0	0	0	0
450	460	53	37	22	6	0	0	0	0	0	0	0
460	470	54	39	23	8	0	0	0	0	0	0	0
470	480	56	40	25	9	0	0	0	0	0	0	0
480	490	57	42	26	11	0	0	0	0	0	0	0
490	500	59	43	28	12	0	0	0	0	0	0	0
500	520	61	46	30	14	0	0	0	0	0	0	0
520	540	64	49	33	17	2	0	0	0	0	0	0
540	560	67	52	36	20	5	0	0	0	0	0	0
560	580	70	55	39	23	8	0	0	0	0	0	0
580	600	73	58	42	26	11	0	0	0	0	0	0
600	620	76	61	45	29	14	0	0	0	0	0	0
620	640	79	64	48	32	17	1	0	0	0	0	0
640	660	82	67	51	35	20	4	0	0	0	0	0
660	680	85	70	54	38	23	7	0	0	0	0	0
680	700	88	73	57	41	26	10	0	0	0	0	0
700	720	91	76	60	44	29	13	0	0	0	0	0
720	740	94	79	63	47	32	16	1	0	0	0	0
740	760	97	82	66	50	35	19	4	0	0	0	0
760	780	100	85	69	53	38	22	7	0	0	0	0
780	800	103	88	72	56	41	25	10	0	0	0	0

Figure 16.9 by the number of allowances claimed by the employee on the W-4 form.

$$\text{withholding allowance} = 51.92 \times 4$$
$$= \$207.68$$

Step 2: Determine the earnings subject to withholding tax. Subtract the total withholding allowance found in step 1 from the total earnings to determine the amount of earnings subject to withholding tax.

Figure 16.7

(Continued)

If the wages are–		And the number of withholding allowances claimed is—										
At least	But less than	0	1	2	3	4	5	6	7	8	9	10
		The amount of income tax to be withheld is—										
$800	$820	106	91	75	59	44	28	13	0	0	0	0
820	840	109	94	78	62	47	31	16	0	0	0	0
840	860	112	97	81	65	50	34	19	3	0	0	0
860	880	115	100	84	68	53	37	22	6	0	0	0
880	900	118	103	87	71	56	40	25	9	0	0	0
900	920	121	106	90	74	59	43	28	12	0	0	0
920	940	124	109	93	77	62	46	31	15	0	0	0
940	960	127	112	96	80	65	49	34	18	3	0	0
960	980	130	115	99	83	68	52	37	21	6	0	0
980	1,000	133	118	102	86	71	55	40	24	9	0	0
1,000	1,020	136	121	105	89	74	58	43	27	12	0	0
1,020	1,040	139	124	108	92	77	61	46	30	15	0	0
1,040	1,060	144	127	111	95	80	64	49	33	18	2	0
1,060	1,080	150	130	114	98	83	67	52	36	21	5	0
1,080	1,100	155	133	117	101	86	70	55	39	24	8	0
1,100	1,120	161	136	120	104	89	73	58	42	27	11	0
1,120	1,140	167	139	123	107	92	76	61	45	30	14	0
1,140	1,160	172	143	126	110	95	79	64	48	33	17	1
1,160	1,180	178	149	129	113	98	82	67	51	36	20	4
1,180	1,200	183	154	132	116	101	85	70	54	39	23	7
1,200	1,220	189	160	135	119	104	88	73	57	42	26	10
1,220	1,240	195	166	138	122	107	91	76	60	45	29	13
1,240	1,260	200	171	142	125	110	94	79	63	48	32	16
1,260	1,280	206	177	148	128	113	97	82	66	51	35	19
1,280	1,300	211	182	153	131	116	100	85	69	54	38	22
1,300	1,320	217	188	159	134	119	103	88	72	57	41	25
1,320	1,340	223	194	164	137	122	106	91	75	60	44	28
1,340	1,360	228	199	170	141	125	109	94	78	63	47	31
1,360	1,380	234	205	176	147	128	112	97	81	66	50	34
1,380	1,400	239	210	181	152	131	115	100	84	69	53	37
1,400	1,420	245	216	187	158	134	118	103	87	72	56	40
1,420	1,440	251	222	192	163	137	121	106	90	75	59	43
1,440	1,460	256	227	198	169	140	124	109	93	78	62	46
1,460	1,480	262	233	204	175	146	127	112	96	81	65	49
1,480	1,500	267	238	209	180	151	130	115	99	84	68	52
1,500	1,520	273	244	215	186	157	133	118	102	87	71	55
1,520	1,540	279	250	220	191	162	136	121	105	90	74	58
1,540	1,560	284	255	226	197	168	139	124	108	93	77	61
1,560	1,580	290	261	232	203	174	144	127	111	96	80	64
1,580	1,600	295	266	237	208	179	150	130	114	99	83	67
1,600	1,620	301	272	243	214	185	156	133	117	102	86	70
1,620	1,640	307	278	248	219	190	161	136	120	105	89	73
1,640	1,660	312	283	254	225	196	167	139	123	108	92	76
1,660	1,680	318	289	260	231	202	172	143	126	111	95	79
1,680	1,700	323	294	265	236	207	178	149	129	114	98	82
1,700	1,720	329	300	271	242	213	184	155	132	117	101	85
1,720	1,740	335	306	276	247	218	189	160	135	120	104	88
1,740	1,760	340	311	282	253	224	195	166	138	123	107	91
1,760	1,780	346	317	288	259	230	200	171	142	126	110	94
1,780	1,800	351	322	293	264	235	206	177	148	129	113	97
1,800	1,820	357	328	299	270	241	212	183	153	132	116	100
1,820	1,840	363	334	304	275	246	217	188	159	135	119	103
1,840	1,860	368	339	310	281	252	223	194	165	138	122	106
1,860	1,880	374	345	316	287	258	228	199	170	141	125	109
1,880	1,900	379	350	321	292	263	234	205	176	147	128	112
1,900	1,920	385	356	327	298	269	240	211	181	152	131	115
1,920	1,940	391	362	332	303	274	245	216	187	158	134	118
1,940	1,960	396	367	338	309	280	251	222	193	164	137	121
1,960	1,980	402	373	344	315	286	256	227	198	169	140	124
1,980	2,000	407	378	349	320	291	262	233	204	175	146	127
2,000	2,020	413	384	355	326	297	268	239	209	180	151	130
2,020	2,040	419	390	360	331	302	273	244	215	186	157	133
2,040	2,060	424	395	366	337	308	279	250	221	192	163	136
2,060	2,080	430	401	372	343	314	284	255	226	197	168	139
2,080	2,100	435	406	377	348	319	290	261	232	203	174	145

$2,100 and over   Use Table 2(a) for a **SINGLE person** on page 34. Also see the instructions on page 32.

Source: Circular E: *Employer's Tax Guide,* U.S. Dept. of the Treasury, Internal Revenue Service (U.S. Government Printing Office).

$$\text{earnings subject to tax} = \$915.50 - 207.68$$
$$= \$707.82$$

Step 3: Calculate the FWT from the tables for the percentage method of withholding. Using the table in Figure 16.10, locate in part b, married person paid weekly, the wage amount *over* $124 *but not over* $899 and calculate

Figure 16.8

Employer's tax guide: Married person paid biweekly

If the wages are—		And the number of withholding allowances claimed is—										
At least	But less than	0	1	2	3	4	5	6	7	8	9	10
		The amount of income tax to be withheld is—										
$0	$250	0	0	0	0	0	0	0	0	0	0	0
250	260	1	0	0	0	0	0	0	0	0	0	0
260	270	3	0	0	0	0	0	0	0	0	0	0
270	280	4	0	0	0	0	0	0	0	0	0	0
280	290	6	0	0	0	0	0	0	0	0	0	0
290	300	7	0	0	0	0	0	0	0	0	0	0
300	310	9	0	0	0	0	0	0	0	0	0	0
310	320	10	0	0	0	0	0	0	0	0	0	0
320	330	12	0	0	0	0	0	0	0	0	0	0
330	340	13	0	0	0	0	0	0	0	0	0	0
340	350	15	0	0	0	0	0	0	0	0	0	0
350	360	16	0	0	0	0	0	0	0	0	0	0
360	370	18	2	0	0	0	0	0	0	0	0	0
370	380	19	3	0	0	0	0	0	0	0	0	0
380	390	21	5	0	0	0	0	0	0	0	0	0
390	400	22	6	0	0	0	0	0	0	0	0	0
400	410	24	8	0	0	0	0	0	0	0	0	0
410	420	25	9	0	0	0	0	0	0	0	0	0
420	430	27	11	0	0	0	0	0	0	0	0	0
430	440	28	12	0	0	0	0	0	0	0	0	0
440	450	30	14	0	0	0	0	0	0	0	0	0
450	460	31	15	0	0	0	0	0	0	0	0	0
460	470	33	17	1	0	0	0	0	0	0	0	0
470	480	34	18	3	0	0	0	0	0	0	0	0
480	490	36	20	4	0	0	0	0	0	0	0	0
490	500	37	21	6	0	0	0	0	0	0	0	0
500	520	39	24	8	0	0	0	0	0	0	0	0
520	540	42	27	11	0	0	0	0	0	0	0	0
540	560	45	30	14	0	0	0	0	0	0	0	0
560	580	48	33	17	2	0	0	0	0	0	0	0
580	600	51	36	20	5	0	0	0	0	0	0	0
600	620	54	39	23	8	0	0	0	0	0	0	0
620	640	57	42	26	11	0	0	0	0	0	0	0
640	660	60	45	29	14	0	0	0	0	0	0	0
660	680	63	48	32	17	1	0	0	0	0	0	0
680	700	66	51	35	20	4	0	0	0	0	0	0
700	720	69	54	38	23	7	0	0	0	0	0	0
720	740	72	57	41	26	10	0	0	0	0	0	0
740	760	75	60	44	29	13	0	0	0	0	0	0
760	780	78	63	47	32	16	0	0	0	0	0	0
780	800	81	66	50	35	19	3	0	0	0	0	0
800	820	84	69	53	38	22	6	0	0	0	0	0
820	840	87	72	56	41	25	9	0	0	0	0	0
840	860	90	75	59	44	28	12	0	0	0	0	0
860	880	93	78	62	47	31	15	0	0	0	0	0
880	900	96	81	65	50	34	18	3	0	0	0	0
900	920	99	84	68	53	37	21	6	0	0	0	0
920	940	102	87	71	56	40	24	9	0	0	0	0
940	960	105	90	74	59	43	27	12	0	0	0	0
960	980	108	93	77	62	46	30	15	0	0	0	0
980	1,000	111	96	80	65	49	33	18	2	0	0	0
1,000	1,020	114	99	83	68	52	36	21	5	0	0	0
1,020	1,040	117	102	86	71	55	39	24	8	0	0	0
1,040	1,060	120	105	89	74	58	42	27	11	0	0	0
1,060	1,080	123	108	92	77	61	45	30	14	0	0	0
1,080	1,100	126	111	95	80	64	48	33	17	2	0	0
1,100	1,120	129	114	98	83	67	51	36	20	5	0	0
1,120	1,140	132	117	101	86	70	54	39	23	8	0	0
1,140	1,160	135	120	104	89	73	57	42	26	11	0	0
1,160	1,180	138	123	107	92	76	60	45	29	14	0	0
1,180	1,200	141	126	110	95	79	63	48	32	17	1	0
1,200	1,220	144	129	113	98	82	66	51	35	20	4	0
1,220	1,240	147	132	116	101	85	69	54	38	23	7	0
1,240	1,260	150	135	119	104	88	72	57	41	26	10	0
1,260	1,280	153	138	122	107	91	75	60	44	29	13	0
1,280	1,300	156	141	125	110	94	78	63	47	32	16	1
1,300	1,320	159	144	128	113	97	81	66	50	35	19	4
1,320	1,340	162	147	131	116	100	84	69	53	38	22	7
1,340	1,360	165	150	134	119	103	87	72	56	41	25	10
1,360	1,380	168	153	137	122	106	90	75	59	44	28	13

the FWT in the following manner:

$$\text{withholding tax} = \text{taxable earnings} \times \text{tax rate}$$
$$= \$583.82 \times .15$$
$$= \$87.57$$

There is no tax withheld on the first $124 after subtracting withholding allowances. Therefore, the tax is calculated on $583.82 ($707.82 − $124 =

Figure 16.8

(Continued)

If the wages are—		And the number of withholding allowances claimed is—										
At least	But less than	0	1	2	3	4	5	6	7	8	9	10
		The amount of income tax to be withheld is—										
$1,380	$1,400	171	156	140	125	109	93	78	62	47	31	16
1,400	1,420	174	159	143	128	112	96	81	65	50	34	19
1,420	1,440	177	162	146	131	115	99	84	68	53	37	22
1,440	1,460	180	165	149	134	118	102	87	71	56	40	25
1,460	1,480	183	168	152	137	121	105	90	74	59	43	28
1,480	1,500	186	171	155	140	124	108	93	77	62	46	31
1,500	1,520	189	174	158	143	127	111	96	80	65	49	34
1,520	1,540	192	177	161	146	130	114	99	83	68	52	37
1,540	1,560	195	180	164	149	133	117	102	86	71	55	40
1,560	1,580	198	183	167	152	136	120	105	89	74	58	43
1,580	1,600	201	186	170	155	139	123	108	92	77	61	46
1,600	1,620	204	189	173	158	142	126	111	95	80	64	49
1,620	1,640	207	192	176	161	145	129	114	98	83	67	52
1,640	1,660	210	195	179	164	148	132	117	101	86	70	55
1,660	1,680	213	198	182	167	151	135	120	104	89	73	58
1,680	1,700	216	201	185	170	154	138	123	107	92	76	61
1,700	1,720	219	204	188	173	157	141	126	110	95	79	64
1,720	1,740	222	207	191	176	160	144	129	113	98	82	67
1,740	1,760	225	210	194	179	163	147	132	116	101	85	70
1,760	1,780	228	213	197	182	166	150	135	119	104	88	73
1,780	1,800	231	216	200	185	169	153	138	122	107	91	76
1,800	1,820	236	219	203	188	172	156	141	125	110	94	79
1,820	1,840	241	222	206	191	175	159	144	128	113	97	82
1,840	1,860	247	225	209	194	178	162	147	131	116	100	85
1,860	1,880	253	228	212	197	181	165	150	134	119	103	88
1,880	1,900	258	231	215	200	184	168	153	137	122	106	91
1,900	1,920	264	235	218	203	187	171	156	140	125	109	94
1,920	1,940	269	240	221	206	190	174	159	143	128	112	97
1,940	1,960	275	246	224	209	193	177	162	146	131	115	100
1,960	1,980	281	252	227	212	196	180	165	149	134	118	103
1,980	2,000	286	257	230	215	199	183	168	152	137	121	106
2,000	2,020	292	263	234	218	202	186	171	155	140	124	109
2,020	2,040	297	268	239	221	205	189	174	158	143	127	112
2,040	2,060	303	274	245	224	208	192	177	161	146	130	115
2,060	2,080	309	280	250	227	211	195	180	164	149	133	118
2,080	2,100	314	285	256	230	214	198	183	167	152	136	121
2,100	2,120	320	291	262	233	217	201	186	170	155	139	124
2,120	2,140	325	296	267	238	220	204	189	173	158	142	127
2,140	2,160	331	302	273	244	223	207	192	176	161	145	130
2,160	2,180	337	308	278	249	226	210	195	179	164	148	133
2,180	2,200	342	313	284	255	229	213	198	182	167	151	136
2,200	2,220	348	319	290	261	232	216	201	185	170	154	139
2,220	2,240	353	324	295	266	237	219	204	188	173	157	142
2,240	2,260	359	330	301	272	243	222	207	191	176	160	145
2,260	2,280	365	336	306	277	248	225	210	194	179	163	148
2,280	2,300	370	341	312	283	254	228	213	197	182	166	151
2,300	2,320	376	347	318	289	260	231	216	200	185	169	154
2,320	2,340	381	352	323	294	265	236	219	203	188	172	157
2,340	2,360	387	358	329	300	271	242	222	206	191	175	160
2,360	2,380	393	364	334	305	276	247	225	209	194	178	163
2,380	2,400	398	369	340	311	282	253	228	212	197	181	166
2,400	2,420	404	375	346	317	288	258	231	215	200	184	169
2,420	2,440	409	380	351	322	293	264	235	218	203	187	172
2,440	2,460	415	386	357	328	299	270	241	221	206	190	175
2,460	2,480	421	392	362	333	304	275	246	224	209	193	178
2,480	2,500	426	397	368	339	310	281	252	227	212	196	181
2,500	2,520	432	403	374	345	316	286	257	230	215	199	184
2,520	2,540	437	408	379	350	321	292	263	234	218	202	187
2,540	2,560	443	414	385	356	327	298	269	240	221	205	190
2,560	2,580	449	420	390	361	332	303	274	245	224	208	193
2,580	2,600	454	425	396	367	338	309	280	251	227	211	196
2,600	2,620	460	431	402	373	344	314	285	256	230	214	199
2,620	2,640	465	436	407	378	349	320	291	262	233	217	202
2,640	2,660	471	442	413	384	355	326	297	268	238	220	205
2,660	2,680	477	448	418	389	360	331	302	273	244	223	208

$2,680 and over	Use Table 2(b) for a **MARRIED** person on page 34. Also see the instructions on page 32.

Source: Circular E: *Employer's Tax Guide,* U.S. Dept. of the Treasury, Internal Revenue Service (U.S. Government Printing Office).

$583.82) at the 15% rate provided in the table. The percentage method produced a higher amount to be withheld ($87.57) compared to the wage-bracket method ($87). The difference is generally minimal (as in this example) and is adjusted when the employee prepares his or her individual income tax returns.

Figure 16.9

Percentage method income tax withholding table

Payroll period	One with-holding allowance
Weekly	$ 51.92
Biweekly	103.85
Semimonthly	112.50
Monthly	225.00
Quarterly	675.00
Semiannually	1,350.00
Annually	2,700.00
Daily or miscellaneous (each day of the payroll period)	10.38

Source: U.S. Dept. of the Treasury, Internal Revenue Service (U.S. Government Printing Office).

The employer must provide each employee with a W-2 (Figure 16.11) by January 31 of the following year, which shows the total wages paid, federal income tax withheld, state and local income taxes withheld, as well as FICA and Medicare taxes withheld. If the employee's tax liability on the tax return is less than the FWT withheld on the W-2, the employee will receive a refund. On the other hand, if the tax liability is greater than the FWT withheld, the employee will have to pay the additional amount when filing the tax return. As a rule, the amount of FWT withheld should closely approximate the employee's income tax liability for the year.

Another mandatory deduction withheld from some employees' gross earnings is **state withholding tax (SWT).** The methods used to collect state income tax when applicable are similar to the methods used to collect federal withholding tax. Many states use tables to determine the amount of withholding based on wage-brackets and withholding allowances claimed. Some states prefer to use a fixed percentage rate to determine the employee's SWT deduction. The *flat tax rate method* requires the employer to apply the same tax rate to all taxable income up to a specified amount of gross earnings paid during the calendar year. The calculation of this tax is similar to the method used in computing FICA tax, which we discussed in Section 16.3. Again, we will use Frank Taylor's example to further illustrate the calculation of his mandatory payroll deductions.

Example 19

Frank Taylor's gross earnings for the week are $915.50 and his year-to-date earnings amount to $63,524.75. Frank works in a state that has a flat tax rate of $3\frac{1}{2}\%$ on gross earnings with maximum taxable earnings of $64,900.

Figure 16.10

Tables for percentage method of withholding

(For Wages Paid in 1998)

TABLE 1—Weekly Payroll Period

(a) SINGLE person (including head of household)—

If the amount of wages (after subtracting withholding allowances) is:		The amount of income tax to withhold is:	
Not over $51		$0	
Over—	**But not over—**		**of excess over—**
$51	—$517	15%	—$51
$517	—$1,105	$69.90 plus 28%	—$517
$1,105	—$2,493	$234.54 plus 31%	—$1,105
$2,493	—$5,385	$664.82 plus 36%	—$2,493
$5,385	. . . . .	$1,705.94 plus 39.6%	—$5,385

(b) MARRIED person—

If the amount of wages (after subtracting withholding allowances) is:		The amount of income tax to withhold is:	
Not over $124		$0	
Over—	**But not over—**		**of excess over—**
$124	—$899	15%	—$124
$899	—$1,855	$116.25 plus 28%	—$899
$1,855	—$3,084	$383.93 plus 31%	—$1,855
$3,084	—$5,439	$764.92 plus 36%	—$3,084
$5,439	. . . . .	$1,612.72 plus 39.6%	—$5,439

TABLE 2—Biweekly Payroll Period

(a) SINGLE person (including head of household)—

If the amount of wages (after subtracting withholding allowances) is:		The amount of income tax to withhold is:	
Not over $102		$0	
Over—	**But not over—**		**of excess over—**
$102	—$1,035	15%	—$102
$1,035	—$2,210	$139.95 plus 28%	—$1,035
$2,210	—$4,987	$468.95 plus 31%	—$2,210
$4,987	—$10,769	$1,329.82 plus 36%	—$4,987
$10,769	. . . . .	$3,411.34 plus 39.6%	—$10,769

(b) MARRIED person—

If the amount of wages (after subtracting withholding allowances) is:		The amount of income tax to withhold is:	
Not over $248		$0	
Over—	**But not over—**		**of excess over—**
$248	—$1,798	15%	—$248
$1,798	—$3,710	$232.50 plus 28%	—$1,798
$3,710	—$6,167	$767.86 plus 31%	—$3,710
$6,167	—$10,879	$1,529.53 plus 36%	—$6,167
$10,879	. . . . .	$3,225.85 plus 39.6%	—$10,879

TABLE 3—Semimonthly Payroll Period

(a) SINGLE person (including head of household)—

If the amount of wages (after subtracting withholding allowances) is:		The amount of income tax to withhold is:	
Not over $110		$0	
Over—	**But not over—**		**of excess over—**
$110	—$1,121	15%	—$110
$1,121	—$2,394	$151.65 plus 28%	—$1,121
$2,394	—$5,402	$508.09 plus 31%	—$2,394
$5,402	—$11,667	$1,440.57 plus 36%	—$5,402
$11,667	. . . . .	$3,695.97 plus 39.6%	—$11,667

(b) MARRIED person—

If the amount of wages (after subtracting withholding allowances) is:		The amount of income tax to withhold is:	
Not over $269		$0	
Over—	**But not over—**		**of excess over—**
$269	—$1,948	15%	—$269
$1,948	—$4,019	$251.85 plus 28%	—$1,948
$4,019	—$6,681	$831.73 plus 31%	—$4,019
$6,681	—$11,785	$1,656.95 plus 36%	—$6,681
$11,785	. . . . .	$3,494.39 plus 39.6%	—$11,785

TABLE 4—Monthly Payroll Period

(a) SINGLE person (including head of household)—

If the amount of wages (after subtracting withholding allowances) is:		The amount of income tax to withhold is:	
Not over $221		$0	
Over—	**But not over—**		**of excess over—**
$221	—$2,242	15%	—$221
$2,242	—$4,788	$303.15 plus 28%	—$2,242
$4,788	—$10,804	$1,016.03 plus 31%	—$4,788
$10,804	—$23,333	$2,880.99 plus 36%	—$10,804
$23,333	. . . . .	$7,391.43 plus 39.6%	—$23,333

(b) MARRIED person—

If the amount of wages (after subtracting withholding allowances) is:		The amount of income tax to withhold is:	
Not over $538		$0	
Over—	**But not over—**		**of excess over—**
$538	—$3,896	15%	—$538
$3,896	—$8,038	$503.70 plus 28%	—$3,896
$8,038	—$13,363	$1,663.46 plus 31%	—$8,038
$13,363	—$23,571	$3,314.21 plus 36%	—$13,363
$23,571	. . . . .	$6,989.09 plus 39.6%	—$23,571

Source: U.S. Dept. of the Treasury, Internal Revenue Service (U.S. Government Printing Office).

How much of Frank's gross earnings will be deducted for state withholding tax?

Solution

Step 1: Determine the taxable earnings.

$ 64,900.00	maximum taxable earnings
− 63,524.75	year-to-date earnings
$ 1,375.25	earnings subject to tax

Figure 16.11

Form W-2: Wage and tax statement

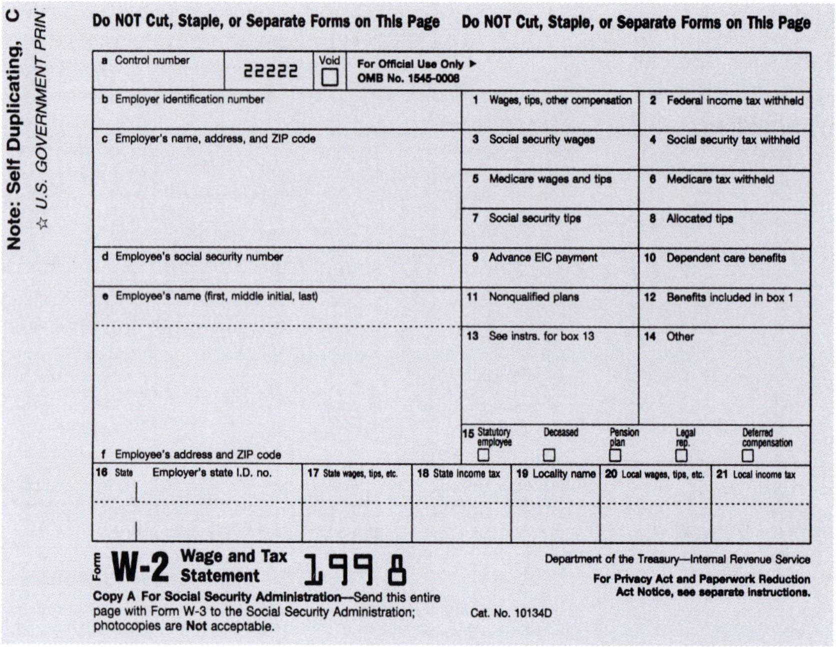
Note: Self Duplicating, C

☆ U.S. GOVERNMENT PRIN

Do NOT Cut, Staple, or Separate Forms on This Page Do NOT Cut, Staple, or Separate Forms on This Page

a Control number 22222	Void ☐	For Official Use Only ▶ OMB No. 1545-0008	
b Employer identification number		1 Wages, tips, other compensation	2 Federal income tax withheld
c Employer's name, address, and ZIP code		3 Social security wages	4 Social security tax withheld
		5 Medicare wages and tips	6 Medicare tax withheld
		7 Social security tips	8 Allocated tips
d Employee's social security number		9 Advance EIC payment	10 Dependent care benefits
e Employee's name (first, middle initial, last)		11 Nonqualified plans	12 Benefits included in box 1
		13 See instrs. for box 13	14 Other
f Employee's address and ZIP code		15 Statutory employee ☐ Deceased ☐ Pension plan ☐ Legal rep. ☐ Deferred compensation ☐	

16 State Employer's state I.D. no.	17 State wages, tips, etc.	18 State income tax	19 Locality name	20 Local wages, tips, etc.	21 Local income tax

Form **W-2** Wage and Tax Statement 1998

Copy A For Social Security Administration—Send this entire page with Form W-3 to the Social Security Administration; photocopies are Not acceptable.

Cat. No. 10134D

Department of the Treasury—Internal Revenue Service

For Privacy Act and Paperwork Reduction Act Notice, see separate instructions.

Source: U.S. Dept. of the Treasury, Internal Revenue Service (U.S. Government Printing Office).

Since the current earnings ($915.50) is less than the earnings still subject to tax ($1,375.25), the total amount of gross earnings is used to calculate the state withholding tax.

Step 2: Use the percentage equation to calculate the amount of tax withheld.

$$\text{SWT} = \text{taxable earnings} \times \text{tax rate}$$

$$\text{SWT} = \$915.50 \times .035$$

$$= \$32.04$$

Elective Deductions

Elective deductions are amounts withheld from an employee's gross earnings that have been authorized by the employee and agreed upon by the employer. Elective deductions may include payments for insurance premiums, union dues, retirement annuities, charitable contributions, U.S. savings bonds, personal savings or loan payments, and work uniform expenses. The elective as well as the mandatory deductions comprise the total deductions to be subtracted from gross earnings, which result in net pay or take-home pay.

Net pay formula

net pay = gross earnings − total deductions

To continue the analysis of the Frank Taylor example, let's look at the elective deductions he has authorized his employer to subtract from his gross earnings; then we can develop a summary of his total deductions for the pay period.

Example 20

Frank Taylor has requested his employer to deduct from his gross earnings of $915.50 the following amounts for company benefits and personal expenditures per pay period: group health insurance, $12.50; credit union savings, $25.00; dental insurance, $5.75; charitable contribution, $10.00; and retirement annuity, $40.75. His mandatory deductions for the period were: FICA tax, $56.76; Medicare tax, $13.27; federal withholding tax, $87.00; and state withholding tax, $32.04. Find his net pay for the pay period.

Solution

Gross earnings for period			$915.50
Mandatory deductions:			
FICA tax	$56.76		
Medicare tax	13.27		
Federal withholding tax	87.00		
State withholding tax	32.04	189.07	
Elective deductions:			
Health insurance	$12.50		
Credit union savings	25.00		
Dental insurance	5.75		
Charitable contribution	10.00		
Retirement annuity	40.75	94.00	
Total deductions			283.07
NET PAY			$632.43

or

net pay = gross earnings − total deductions

net pay = $915.50   − $283.07

= $632.43

```
┌──────────────────────────────────────────────────────────────────┐
│  ▭▭▭▭▭                                                            │
│  ⦿⦿⦿⦿⦿   CALCULATOR   SOLUTION                                   │
│  ⦿⦿⦿⦿⦿   Keyed entry                                  Display    │
│  ⦿⦿⦿⦿⦿  ─────────────────────────────────────────────────────── │
│  ⦿⦿⦿⦿⦿                                                            │
│  ⦿⦿⦿⦿⦿        AC  56.76  +  13.27  +  100                         │
│                          +  32.04  =              189.07          │
│                                       Min         189.07          │
│            AC  12.5  +  25  +  5.75  +  10                         │
│                          +  40.75  =               94.00          │
│                                       M+           94.00          │
│                   915.50  −  MR *  =              632.43          │
│                                                                    │
│         * MR  in last step before depressing                      │
│           =   will display total deductions                       │
│               figure (296.07)                                     │
└──────────────────────────────────────────────────────────────────┘
```

CHECK YOUR KNOWLEDGE

Employee Deductions

1. Sharon Reynolds, a physician's assistant, is single and claims three withholding allowances. She has year-to-date earnings of $19,265, and her current biweekly gross earnings amount to $865. How much of her gross earnings must be withheld for (a) FICA tax and (b) Medicare tax? Use Table 16.2 to calculate the taxes.

2. From the information provided in problem 1, calculate the federal withholding tax for Sharon. Use the (a) wage bracket method as well as the (b) percentage method for computing FWT.

3. If Sharon works in a state that has a flat tax rate of 6¾% on gross earnings up to maximum taxable earnings of $20,000, how much of her gross earnings will be deducted for state withholding tax (SWT)?

4. Sharon's employer, at her request, deducts $10.50 for health insurance, $18.75 for U.S. savings bonds, and $38.50 for her retirement fund. Determine Sharon's total deductions for the pay period.

5. Determine Sharon's net pay for the pay period using the results of your calculations from problems 1 through 4.

6. Vern Beckwith is married and claims five withholding allowances. His year-to-date earnings amount to $53,225 and his current semimonthly gross earnings are $2,800. Use the (a) percent method to calculate the FWT and the (b) FICA and Medicare taxes.

Answers to CYK: *1.* **a.** $53.63; **b.** $12.54   *2.* **a.** $68; **b.** $62.72   *3.* $49.61
4. $215.25   *5.* $613.75   *6.* **a.** $332.91; **b.** FICA $173.60, Medicare $40.60

16.3 EXERCISES

Calculate the FICA and Medicare tax deductions for each of the following employees using the rates from Table 16.2.

	Employee	Current gross earnings	Medicare tax	FICA tax
1.	Cooper, L.	$156.20	_____	_____
2.	Bevard, M.	375.80	_____	_____
3.	Hazeltine, M.	420.75	_____	_____
4.	Solomon, B.	694.90	_____	_____

Calculate the deduction(s) for each of the following employees based on 1998 tax rates in Table 16.2 (percentage method).

	Employee	Year-to-date earnings	Current gross earnings	FICA tax	Medicare tax
5.	Charles, T.	$18,680	$ 450	_____	_____
6.	Espinoza, C.	53,750	2,280	_____	_____
7.	Sith, P.	66,200	3,145	_____	_____
8.	Willis, R.	38,150	860	_____	_____

Use the wage-bracket method (Figures 16.5, 16.6, 16.7, and 16.8) to find the federal withholding tax (FWT) for each of the following employees.

	Employee	Current gross earnings	Pay period	Marital status	Exemp- tions	FWT tax
9.	Alhussein, R.	$ 489.70	Weekly	M	4	_____
10.	Hammond, P.	1,265.50	Biweekly	M	8	_____
11.	Keller, J.	620.45	Biweekly	S	1	_____
12.	Norton, B.	572.30	Weekly	M	6	_____
13.	Wong, D.	376.25	Weekly	S	2	_____
14.	Zimmerman, P.	847.80	Biweekly	M	2	_____

Using the percentage method, calculate the FWT tax for each of the following employees (Figures 16.9 and 16.10).

	Employee	Current gross earnings	Pay period	Marital status	Exemp- tions	FWT tax
15.	Anderson, D.	$ 520.65	Weekly	M	4	_____
16.	Dombowski, T.	748.20	Semimonthly	S	2	_____
17.	Loftus, P.	2,630.45	Monthly	M	2	_____
18.	Ramos, M.	987.30	Biweekly	M	5	_____
19.	Steel, J.	1,238.75	Monthly	S	1	_____

Complete the following biweekly payroll register as indicated. Use the appropriate wage bracket tables provided in the chapter for the FWT, FICA, and Medicare tax rates. All earnings are subject to FICA and Medicare taxes.

Employee	Marital status	Exemp- tions	Gross earnings	FWT tax
20. Bright, P.	S	2	$ 846.20	_____
21. Denny, M.	M	4	1,010.50	_____
22. Hartz, T.	M	2	910.80	_____
23. Lee, R.	S	1	1,240.75	_____
24. Udell, S.	M	6	1,418.30	_____

Employee	Medicare tax	FICA	Other deductions	Net pay
Bright, P.	_____	_____	$102.10	_____
Denny, M.	_____	_____	146.80	_____
Hartz, T.	_____	_____	98.75	_____
Lee, R.	_____	_____	162.40	_____
Udell, S.	_____	_____	187.25	_____

Complete the following weekly payroll register. Use the percentage method to calculate the FWT. Assume that all earnings paid during the period are subject to a 6.2% FICA tax and 1.45% Medicare tax rate.

Employee	Marital status	Exemptions	Gross earnings	FWT tax
25. Boco, P.	M	3	$397.60	_____
26. Deitz, T.	S	1	285.30	_____
27. Schmidt, P.	M	7	528.97	_____
28. Zello, E.	S	2	465.40	_____

Employee	Medicare tax	FICA	Other deductions	Net pay
Boco, P	_____	_____	$28.40	_____
Deitz, T.	_____	_____	10.90	_____
Schmidt, P.	_____	_____	65.50	_____
Zello, E.	_____	_____	42.20	_____

29. Judy Terrell's earnings this pay period are $1,520.75 and her year-to-date earnings are $42,520.80. How much of her current earnings must her employer withhold for (a) FICA tax and (b) Medicare tax based on the tax rates given in Table 16.2.

30. If an employee is paid $6.50 per hour for 46 hours of working time, how much must be deducted for FICA and medicare taxes if the employer uses the percentage method? All wages are taxable and the employee receives time-and-a-half for all hours worked over 40 in a workweek.

31. Martin Prosser is married and claims three exemptions. He receives an annual salary of $21,970. If he is paid on a biweekly basis, what is the amount of FWT deducted from his earnings each pay period? (Use the wage bracket method.)

32. Art Rice, a licensed practical nurse (LPN), is single and claims 2 exemptions. He is paid weekly at an hourly rate of $7.80, with time-and-a-half paid for hours in excess of 40 hours a week. He worked 42 hours last week. If his regular workweek is 35 hours, what are his (a) FICA, (b) Medicare, and (c) FWT taxes based on the percentage method, assuming all earnings are taxable? (Use 1998 tax rates.)

33. Sarah Fillingham is an electrical engineer and receives a semimonthly salary of $3,250. She is married and claims four withholding allowances. If her year-to-date earnings are $52,875, how much

will her employer deduct from her current earnings for (a) FWT, (b) FICA, and (c) Medicare taxes.

34. Bob Bigsbee's gross earnings for the week are $523.85, and his year-to-date earnings amount to $29,785.20. He works in a state that has a flat state withholding tax rate (SWT) of $2\frac{1}{2}\%$ on gross earnings up to $30,000. How much of Bob's gross earnings will be deducted for state withholding tax?

35. Alice Reschke works for the local power authority. She is paid a weekly salary of $420 and is married with 2 exemptions. What is her net pay if her FICA, Medicare, and FWT taxes are based on the wage bracket method and her employer deducts the following: group health insurance, $6.50; dental insurance, $2.50; credit union, $25.00; and retirement contribution, $22.00?

16.4 EMPLOYER PAYROLL TAXES AND REPORTS

Learning objective
Identify and calculate the various types of employer payroll taxes.

Employers are required by law to maintain a *payroll register* that indicates how gross earnings are determined, the amount of mandatory and elective deductions withheld, and the net pay of each employee. The payroll register (Figure 16.3) contains the information necessary to calculate the employer's quarterly payroll taxes.

At the end of March, June, September, and December, employers must file *Form 941,* the *Employer's Quarterly Federal Tax Return* (Figure 16.12). Form 941 is used to report the (FWT) federal withholding tax, the Social Security (FICA), and Medicare taxes collected from employees, and the Social Security and Medicare taxes to be paid by the employer.

FICA Tax and Medicare Tax

Employers must pay a FICA tax and Medicare tax equal to the FICA Medicare taxes withheld from their employees. The amount of these taxes is reported on line 8 of form 941 and is determined by multiplying the total taxable Social Security and Medicare wages of employees by the combined rate of each tax. The combined FICA rate (12.4%) is found in Table 16.2 by adding the employee rate (6.2%) to the employer's rate (6.2%) for the same period. The combined Medicare rate (2.9%) is determined in the same manner (1.45% + 1.45% = 2.9%).

The employer pays the taxes reported in Form 941 each quarter by depositing the required tax with an authorized financial institution or a federal reserve bank in accordance with the deposit rules shown in Figure 16.13.

Source: U.S. Dept. of the Treasury, Internal Revenue Service (U.S. Government Printing Office).

Notice that if the undeposited taxes at the end of any quarter are less than $500, the employer can pay the taxes directly to the IRS with Form 941. If a deposit is required, the employer must file Form 8109 (Figure 16.14) along with the proper amount of money at an authorized financial institution within the allotted time periods. Form 941 must be filed by the end of the month

Figure 16.13

Chart B—Summary of deposit rules for Social Security and Medicare taxes and withheld income tax

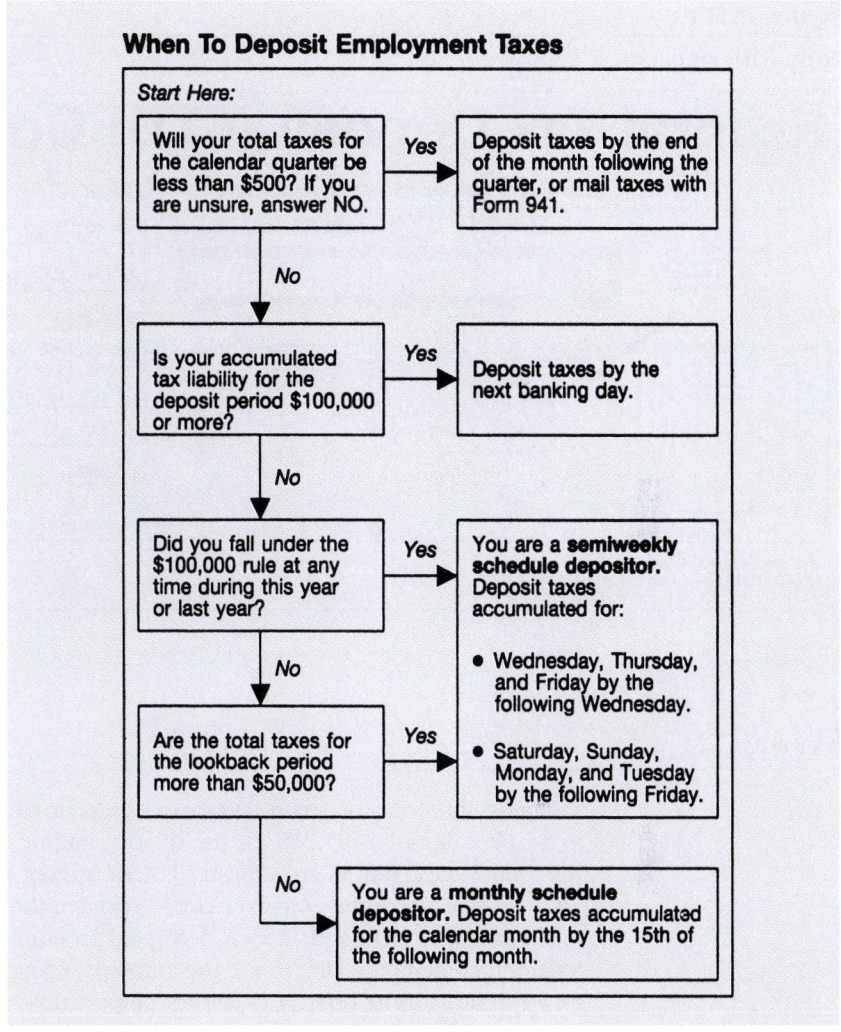

When To Deposit Employment Taxes

Start Here:

Will your total taxes for the calendar quarter be less than $500? If you are unsure, answer NO. — **Yes** → Deposit taxes by the end of the month following the quarter, or mail taxes with Form 941.

No

Is your accumulated tax liability for the deposit period $100,000 or more? — **Yes** → Deposit taxes by the next banking day.

No

Did you fall under the $100,000 rule at any time during this year or last year? — **Yes** → You are a **semiweekly schedule depositor.** Deposit taxes accumulated for:

- Wednesday, Thursday, and Friday by the following Wednesday.

No

Are the total taxes for the lookback period more than $50,000? — **Yes** →

- Saturday, Sunday, Monday, and Tuesday by the following Friday.

No

You are a **monthly schedule depositor.** Deposit taxes accumulated for the calendar month by the 15th of the following month.

Source: U.S. Dept. of the Treasury, Internal Revenue Service (U.S. Government Printing Office).

following each quarter. Therefore, if an employer makes deposits within the deposit period, the balance due on line 15 of Form 941 will be 0.

Completing Form 941

Learning objective
Complete payroll records and quarterly payroll tax returns.

To help you better understand this employer payroll tax, Example 21 will illustrate the calculation procedures necessary to complete selected lines of Form 941 (Figure 16.12) that involve the reporting of employer payroll taxes. If will not be possible to complete the entire form as this report must be prepared from a complete set of payroll records for each quarter.

Figure 16.14

Form 8109: Federal tax deposit coupon

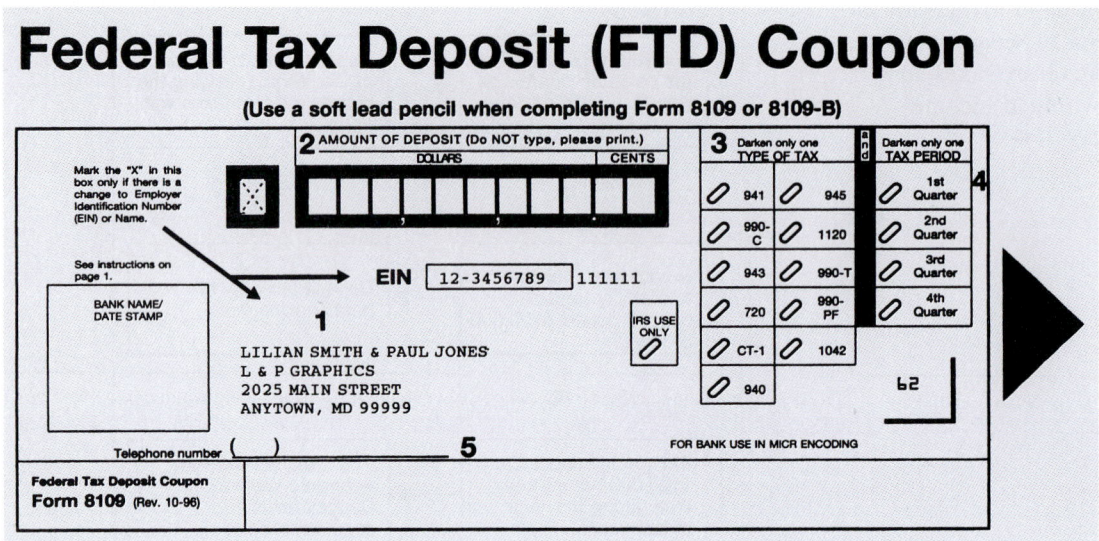

Source: U.S. Dept. of the Treasury, Internal Revenue Service (U.S. Government Printing Office).

Example 21

The payroll records of Rojon Enterprises indicate total earnings paid subject to withholding of $485,235 for the quarter ending September 30, 1998. Of the total wages paid, all are subject to Medicare tax; $12,650 is exempt from FICA tax because some employees had exceeded the maximum taxable earnings base ($68,400); and $155,275.20 paid to employees was withheld for federal income taxes. Determine the amount of payroll taxes to be reported on Form 941 for the quarter. (Refer to the partially completed Form 941, Figure 16.12.)

Solution

First, enter the total earnings ($485,235) paid during the quarter on line 2. Then, enter the federal income tax withheld ($155,275.20) on line 3. Next, determine the amount of wages reported for the quarter that are subject to FICA tax and Medicare tax and enter those on line 6a and line 7a in the calculation columns ($_____ × 12.4%, or .124), and ($_____ × 2.9%, or .029).

$$\text{taxable FICA wages} = \text{total wages reported} - \text{exempt wages}$$
$$= \$485,235 \qquad - \$12,650$$
$$= \$472,585$$

As all earnings paid during the period are subject to the Medicare tax, $485,235 would be entered on line 7a in the Medicare tax calculation column. We can now multiply the taxable FICA wages and the Medicare wages by the combined rates (the employer must match the employees' contribution) and enter on lines 6a and 7.

$$\text{FICA tax} = \text{taxable FICA wages} \times \text{combined rate}$$
$$= \$472,585 \qquad \times .124 \text{ (1998 rate)}$$
$$= \$58,600.54$$

$$\text{Medicare tax} = \text{taxable Medicare wages} \times \text{combined rate}$$
$$= \$485,235 \qquad \times .029$$
$$= \$14,071.82$$

At this point, we have explained how the entries on lines 2, 3, 6a, and 7a were determined. The other entries follow the instructions provided for each line on the form.

Line 5:   Enter the amount from line 3 on line 5, as there are no adjustments for preceding quarters.

Lines 8, 9, and 10:   Enter the amount from lines 6b, 6d, and 7b on lines 8 and 10, as Rojon Enterprises did not have to make any adjustments for errors in Social Security taxes reported on line 9.

Line 11:   Total taxes. Add lines 5 and 10.

Line 12:   No adjustments necessary, enter 0.

Line 13:   Net taxes. Subtract line 12 from line 11.

Line 14:   Enter same amount listed on line 13 (net taxes).

Line 15:   Balance due is 0. The amount of tax owed by Rojon Enterprises for the quarter and the amount deposited are the same. Consequently, no payment is required with the return (line 15), nor is there a refund due for overpayment (line 16).

Line 17:   The amounts on lines 17a, 17b, and 17c should equal the amount on line 17d and the amount deposited on line 14.

Unemployment Taxes

Every employer is required by the **Federal Unemployment Tax Act (FUTA)** to pay a tax on each employee's earnings for unemployment insurance. An employer may also be required by a **State Unemployment Tax Act (SUTA)** to pay a similar state tax. The federal and state systems support the unemployment insurance program, which provides payments of unemployment compensation to workers who have lost their jobs. The tax is paid entirely by the employer. The tax rate and wage limits can change. All changes are published annually in the *Employer's Tax Guide,* which can be obtained from the IRS. The problems in this text will be based on the 1998 federal unemployment tax (FUTA) rate of 6.2% of the first $7,000 in earnings paid to employees during the year. The tax guidelines also state that an employer is allowed a credit of up to 5.4% for the state unemployment tax (SUTA) they pay. Therefore, the federal rate can be reduced to .8% by the maximum state experience rate of 5.4% (6.2% − 5.4% = .8%). If the state experience rate is less than 5.4%, the employer is allowed the full 5.4% credit.

Figure 16.15

Form 940: Employer's annual federal unemployment (FUTA) tax returns

Figure 16.15

(Continued)

Form 940 (1997)   Page **2**

Part II	**Tax Due or Refund**

1	Gross FUTA tax. Multiply the wages in Part I, line 5, by .062	1	
2	Maximum credit. Multiply the wages in Part I, line 5, by .054 . . .	2			
3	**Computation of tentative credit** (Note: *All taxpayers must complete the applicable columns.*)

(a) Name of state	(b) State reporting number(s) as shown on employer's state contribution returns	(c) Taxable payroll (as defined in state act)	(d) State experience rate period From	(d) State experience rate period To	(e) State experience rate	(f) Contributions if rate had been 5.4% (col. (c) x .054)	(g) Contributions payable at experience rate (col. (c) x col. (e))	(h) Additional credit (col. (f) minus col.(g)) If 0 or less, enter -0-.	(i) Contributions actually paid to state

3a	Totals . . . ▶								
3b	Total tentative credit (add line 3a, columns (h) and (i) only—see instructions for limitations on late payments) ▶		
4	
5	
6	**Credit:** Enter the smaller of the amount in Part II, line 2 or line 3b	6	
7	**Total FUTA tax** (subtract line 6 from line 1)	7	
8	Total FUTA tax deposited for the year, including any overpayment applied from a prior year . .	8	
9	**Balance due** (subtract line 8 from line 7). This should be $100 or less. Pay to the Internal Revenue Service. See page 4 of the **Instructions for Form 940** for details . . . . ▶	9	
10	**Overpayment** (subtract line 7 from line 8). Check if it is to be: ☐ **Applied to next return** or ☐ **Refunded** . . . . ▶	10	

Part III	**Record of Quarterly Federal Unemployment Tax Liability** (*Do not include state liability.*) Complete only if line 7 is over $100.

Quarter	First (Jan. 1–Mar. 31)	Second (Apr. 1–June 30)	Third (July 1–Sept. 30)	Fourth (Oct. 1–Dec. 31)	Total for year
Liability for quarter					

Under penalties of perjury, I declare that I have examined this return, including accompanying schedules and statements, and to the best of my knowledge and belief, it is true, correct, and complete, and that no part of any payment made to a state unemployment fund claimed as a credit was, or is to be, deducted from the payments to employees.

Signature ▶   Title (Owner, etc.) ▶   Date ▶

Source: U.S. Dept. of the Treasury, Internal Revenue Service (U.S. Government Printing Office).

However, if the employer is exempt from state unemployment tax, it must pay the full 6.2% FUTA rate.

Employers are required to file Form 940 (Figure 16.15) annually and make any deposits that may be required quarterly. A deposit is required if an employer has more than a $100 unemployment tax liability at the end of a calendar quarter.

Example 22

Gancy Chemical Distributors has a state unemployment experience rating of 3.6% for the current year. Using the following payroll information, determine the FUTA tax liability and the SUTA tax liability on earnings paid during the quarter ending June 30.

Employee	First-quarter earnings	Second-quarter earnings
J. Marko	$ 4,270	$ 4,350
B. Netti	6,580	5,785
D. Hills	7,240	6,800
R. Hsu	2,800	4,600
T. Parker	3,200	3,900
L. VanAuken	2,600	4,250
Totals	$26,690	$29,685

Solution

Employee	First-quarter earnings	Unemployment tax wage limit	Second-quarter taxable earnings
J. Marko	$ 4,270	7,000	$ 2,730
B. Netti	6,580	7,000	420
D. Hills	7,240	7,000	0
R. Hsu	2,800	7,000	4,200
T. Parker	3,200	7,000	3,800
L. VanAuken	2,600	7,000	4,250
Totals	$26,690		$15,400

The solution table shows how much of each employee's total earnings for the second quarter are subject to federal unemployment (FUTA) tax. Of the total wages paid during the second quarter ($29,685), only $15,400 is subject to the tax. This is because most of the employees reached the yearly limit ($7,000) during the quarter and only the difference between their first-quarter earnings and the yearly limit can be taxed. One employee (D. Hills) had already reached the taxable limit during the first quarter; therefore, the employer pays no FUTA tax on his second-quarter earnings. Another employee (L. VanAuken) has not yet been paid more than the taxable limit for the year ($2,600 + $4,250 = $6,850). The employer must pay FUTA tax on his total second-quarter earnings. Now that we know the amount of second-quarter earnings subject to unemployment tax, we calculate the FUTA and SUTA tax as follows:

$$\text{FUTA tax} = \text{taxable earnings} \times \text{tax rate}$$
$$= \$15,400 \qquad \times .008$$
$$= \$123.20$$

$$\text{SUTA tax} = \text{taxable earnings} \times \text{tax rate}$$
$$= \$15,400 \qquad \times .036$$
$$= \$554.40$$

The FUTA tax rate is .8% because Gancy Chemical Distributors has been assigned a state experience tax rate of 3.6%, which is less than the 5.4% maximum that allows them to take full credit (6.2% − 5.4% = .8%).

Gancy Chemical Distributors must deposit the $123.20 in FUTA tax because the quarterly tax liability exceeds the $100 limit allowed in each quarter. The FUTA tax will be deposited along with any FICA tax or federal withholding tax that is required on June 30 using Form 8109 (Figure 16.14). At the end of the year, the company will file Form 940 (Figure 16.15), the Employer's Annual Federal Unemployment (FUTA) Tax Return.

If the company's state experiential tax rate was higher (4.5%), its SUTA tax would also be higher ($15,400 × .045 = $693), but its FUTA would be the same ($123.20), resulting in a greater total unemployment tax ($123.20 + $693 = $816.20) for the quarter. Therefore, many businesses will keep this concept in mind when they make decisions related to workforce reductions.

Most states use a procedure similar to the federal system presented to report and deposit the state unemployment tax liability (SUTA). Due to a lack of conformity among states, this text will not go into any more detail regarding SUTA than has already been presented. If you are interested in knowing more about your state's employer tax system, guidelines are available from your state tax agency.

CHECK YOUR KNOWLEDGE

Employer Payroll Taxes and Reports

1. Julia Cornish is paid $10.50 an hour and is paid time-and-a-half for all time over 40 hours a week. During the current pay period, Julia worked 46 hours. Her accumulated gross earnings are $6,775.00. Using the following rates: FICA, 6.2%; Medicare, 1.45%; SUTA, 2.7%, calculate the following. (You must determine the FUTA rate using the data provided.)
   a. gross earnings
   b. employer's FICA tax
   c. employer's Medicare tax
   d. FUTA tax
   e. SUTA tax
2. Using the following payroll information of McDowell Company, calculate:
   a. the entries required to complete Form 941, the Employer's Quarterly Federal Tax Return, lines 2, 3, 6b, 7b, and 8 as required

b. the third-quarter FUTA taxes

c. the third-quarter SUTA taxes

Employee	Accumulated gross earnings	Third-quarter gross earnings	Third-quarter FWT
G. Berson	$22,570	$11,310	$3,166.80
F. Castro	15,420	7,735	1,933.75
S. Devall	4,980	6,200	1,488.00
W. LaGrange	26,150	13,050	3,654.20
M. Thayer	2,980	3,650	789.50
P. Wheeler	13,725	6,825	1,638.00

McDowell has a state experiential rate of 4.2%. Its records indicate no adjustments will be required on their employer tax reports for the third quarter. Use other rates and wage limits presented in Section 16.4 as they pertain to the required calculations.

16.4   EXERCISES

Given the first-quarter payroll information for Kleen-All Janitorial Services, calculate the required employer payroll taxes on each employee's earnings for the period. Use the following rates: FICA 6.20%; Medicare 1.45%; FUTA .8%; and SUTA 4.2%.

Employee	First-quarter gross earnings	FICA	Medicare	FUTA	SUTA
1. Ferry, M.	$6,687.40	_____	_____	_____	_____
2. Lawson, B.	1,275.00	_____	_____	_____	_____
3. Orcutt, R.	3,865.20	_____	_____	_____	_____
4. Romano, F.	5,189.65	_____	_____	_____	_____

Answers to CYK: **1. a.** $514.50; **b.** $31.90; **c.** $7.46; **d.** $1.80; **e.** $6.08   **2. a.** line 2 $48,770; line 3 $12,670.25; line 6b $6,047.48; line 7b $1,414.33; line 8 $7,461.81; **b.** $45.36; **c.** $238.14

Using the first-quarter payroll summary and tax rates provided above, calculate the second-quarter payroll taxes on each employee's earnings.

Employee	Second-quarter gross earnings	FICA	Medicare	FUTA	SUTA
5. Ferry, M.	$5,247.20	_____	_____	_____	_____
6. Lawson, B.	4,865.80	_____	_____	_____	_____
7. Orcutt, R.	3,990.10	_____	_____	_____	_____
8. Romano, F.	6,420.00	_____	_____	_____	_____

9. Ted Farnholtz is the sole proprietor of Farnholtz Plumbing. Ted's employees' taxable earnings for the first quarter of 1998 totaled $50,000. Assuming tax rates of 5% SUTA and .8% FUTA, what quarterly payroll tax did Farnholtz Plumbing have to pay for (a) FICA, (b) Medicare, (c) FUTA, and (d) SUTA? (Earnings are subject to all taxes.)

10. The Reverend David Griffith is considered to be self-employed as a minister of St. Charles's Church. He was paid $45,000 in 1998 for his services to the church. Assuming a FICA rate of 12.4%, a Medicare tax rate of 2.9%, a 3.8% SUTA tax rate, and a .8% FUTA tax rate, how much must the reverend pay in taxes for (a) FICA, (b) Medicare, (c) SUTA, and (d) FUTA?

11. Sara Jaynes earned $500 last week, to bring her cumulative earnings to $7,300. Sara works in a state that does not collect state unemployment taxes. How much must Sara's employer pay in payroll taxes on her current earnings based on 1998 tax rates?

12. Marie Penafeather earns a salary of $65,000 per year from Jenkins Modular Homes, a company that has a state unemployment experience rating of 5.2%. Find the amount of payroll taxes paid by Jenkins Modular Homes on Marie's salary for the 1998 fiscal year.

13. Jim Nowicki works on a salary-plus-commission basis. He is paid a monthly base salary of $1,200 and a 2.25% commission on all net sales. If his monthly net sales amounted to $45,000 and the employer tax rates are 6.20% FICA; 1.45% medicare; 4.3% SUTA, and .8% FUTA, calculate the following: (a) gross earnings, (b) employer's Medicare tax, (c) employer's FICA tax, (d) FUTA tax, and (e) SUTA.

14. The payroll records of Hartson Enterprises at the end of the quarter indicated that the company had taxable Social Security wages of $12,689.20; taxable Medicare wages of $452,457.80; and that $142,178.50 had been withheld in FWT from employees' earnings. Determine the total amount of taxes Hartson Enterprises reported for the quarter on Form 941, the Employer's Quarterly Federal Tax Return.

15. Given the following payroll information:

Employee	Accumulated gross earnings	Second-quarter gross earnings	Second-quarter federal withholding tax
Butler, D.	$6,420.00	$5,980.75	$1,245.60
Franz, T.	3,175.20	3,250.50	680.25
Hernandez, F.	8,345.10	9,175.60	2,487.90
Taylor, P.	5,680.75	4,245.20	972.50

a. Calculate the employer's FICA tax for the second quarter (assume FICA rate of 6.20% on $55,500).

b. Calculate the employer's Medicare tax. (Assume a tax rate of 1.45% on $130,200.)

c. Calculate the FUTA tax for the second quarter.

d. Calculate SUTA tax for the second quarter (assume a 3.5% SUTA experience rate on the first $7,000).

e. Determine the total amount of taxes to be deposited for the second quarter indicated on Form 941, the employer's quarterly tax return.

CHAPTER REVIEW EXERCISES

The following problems review all mathematics and business concepts presented in the chapter. If you experience any difficulty solving a problem, go to the appropriate chapter section and review the examples provided. Express all answers as decimals rounded to tenths, or dollars rounded to cents, unless otherwise instructed.

1. Hilda Andrew earns $6.50 as an assembler at Sutton Electronics and receives time-and-a-half for overtime. If Hilda worked 45 hours this pay period, what are her gross earnings? (Use standard overtime method.)

2. Calculate the gross earnings for Hilda (exercise 1) using the overtime premium method.

3. Steve Francis is paid $5.85 per hour as a checkout clerk at Wegman's Super Market. He seldom works over 25 hours a week, but is paid $1 more per hour on Sundays and $2 more per hour on holidays. His current pay period time card indicates he worked 4 hours Monday; 4 hours Wednesday, 6 hours Thursday, and 5 hours Sunday. If Thursday was a holiday, determine Steve's gross earnings for the pay period.

4. A manufacturing company pays its small machine operators $.65 for each piece of product produced and $.35 chargeback for each piece that fails inspection standards. If an employee produced 210 units Monday; 198 units Tuesday; 175 units Wednesday; 0 units Thursday; 186 units Friday; and had a total of 25 units rejected for the week, what were the employee's gross earnings for the pay period?

5. Joy Phillips receives $.37 for each piece she finishes on her grinding machine and $8.40 for each hour she waits to have her machine repaired or set-up for jobs. This pay period she finished 1,218 pieces in 47 hours and had 5 hours of waiting time. If she receives time-and-a-half

for overtime, what are her gross earnings for the period?

6. Southern Packaging Corporation pays its employees by the task and bonus plan. The plan has a task requirement for one group of workers of 30 units per hour and a bonus factor of 120%. Find the gross earnings of an employee who packaged 1,500 units during a 40 hour week and is paid $8.72 an hour.

7. Vincent Massena, Marketing Manager for Largent Technologies Inc., receives an annual salary of $55,900. Determine his biweekly gross earnings.

8. Matthew Ryan is paid an annual salary of $26,000 and time-and-a-half for all hours over 40 per week. As a merchandiser, his hours fluctuate on a weekly basis. Find his gross earnings for the current pay period if he worked a total of 50 hours.

9. Northwest Forest Products pays its sales people a fixed salary plus commission on net sales over a specified quota. Andrea Christopher is paid $1,200 per month plus a 2% commission on net sales over $20,000. If she reported gross sales of $47,200, received a draw of $650, and the company gives all customers a 5% trade discount, what is her current month's gross earnings?

10. Georgia Copley's earnings to date are $67,550. Her current biweekly gross earnings are $970. Determine the amount of (a) Social Security tax, and (b) the amount of Medicare tax withheld for the pay period.

11. Jeff Allen, a self-employed landscaper, receives a gross weekly salary of $875. If his gross earnings to date are $42,000, what are his (a) Social Security tax and (b) Medicare tax?

12. Ben Goldman operates heavy equipment for a construction company and is paid $22.50 per hour with time-and-a-half for all hours over 8 hours a day. This pay period he worked 36 hours, of which 4 hours were overtime. Ben is married and claims four withholding allowances. How much of his gross earnings will be withheld for federal withholding tax? (Use the percentage method.)

13. Use the information provided in exercise 12 to determine Ben Goldman's net pay for the week. Ben lives in a state that has a flat state withholding tax rate of 2%, and he had additional deductions of $27.50 for health insurance, $50 for

credit union savings, and $33.08 for retirement. (All earnings are taxable.)

14. Bob Hunter is paid a weekly salary of $450 as a salesperson for Michelson Beverage Distributors, and his cumulative earnings prior to the current pay period was $6,600. Find the amount of (a) FUTA tax, and (b) SUTA tax the distributor must pay on Bob's current earnings if their state unemployment experience rating is 4.2%.

15. The third quarter payroll records of Niland's Protective Services show that the company will report taxable Social Security wages of $82,678.40; taxable Medicare wages of $127,430.90; and federal withholding taxes of $23,195.75. Determine the amount of taxes that Niland Protective Services will report on their third-quarter federal tax return, Form 941.

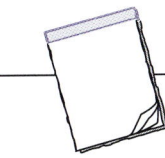 **EXPRESS YOUR UNDERSTANDING**

Compose one or two well-written sentences to express the requested information.

1. Explain how gross earnings are calculated for an employee paid by the hour if the employee worked 45 hours in one pay period.

2. Compare the standard overtime method and the premium overtime method of calculating overtime earnings.

3. Describe how gross earnings are determined using the differential piece-rate plan and a chargeback for defective units produced.

4. Identify and describe the methods used to prorate an annual salary into specific pay periods.

5. Explain how you would determine gross earnings for an employee who is paid a straight commission on net sales and is allowed to draw on projected earnings.

6. Describe how you would calculate an employee's net earnings.

7. What factors determine the amount to be withheld by employers from an employee's earnings for federal income taxes?

8. Give a step-by-step description of how to calculate federal withholding tax using the percentage method.

9. Discuss the purpose and content of Form 941, the Employer's Quarterly Federal Tax Return.

10. Illustrate how an employer's state unemployment experience tax rate of 3.5% affects the amount of federal unemployment tax paid by the employer on an employee's earnings.

Mind Your Business

Media World currently employs five people full-time and two people part-time. A work schedule has been developed that accommodates all job descriptions, daily traffic patterns, and store hours. Employees work the same hours each week and receive their weekly paychecks every Friday. Managers are paid a salary while all other employees are paid an hourly rate based on their position and length of service. Being a small company, Media World has few employee benefits. However, full-time employees are provided an opportunity to participate in a group health insurance and 401(K) retirement plan. Employees are required to contribute $15 per pay period of their gross earnings for health insurance and 3% of their earnings for retirement. The managers must also contribute 3% of their earnings for their 401(K) plan, but receive health insurance as a company paid benefit. Part-time employees cannot participate in any benefit program at this time. Kate Bradley, the bookkeeper, prepares the weekly payroll and submits the checks to you each Friday for your signature.

Activity 1

Prepare the payroll register created by Kate Bradley for the week of September 10–16 using the payroll information just provided and the information that follows. (Base all tax calculations on the appropriate tax schedules provided in the text.)

Name	Status	Exemptions	Rates	Hours worked
Bradley, Kate	M	2	Salary $500	40
Kane, Diane	S	1	Salary $475	40
Welch, Thomas	S	1	Salary $625	40
Stephens, Tyrone	M	3	Hourly $11.25	40
Mendez, Luis	S	1	Hourly $6.00 P/T	20
Evans, Susan	M	2	Hourly $10.50	40
McCaffery, Sheila	S	1	Hourly $6.00 P/T	25

Activity 2

It is your responsibility as business manager to prepare all federal and state tax reports and to assure the payment of all taxes on required dates. Determine the amount to be reported on the federal tax deposit coupon Form 8109 for the taxes deposited for the month of July. Taxes for the month are based on four payroll periods, all of which were exactly the same as the payroll presented in Activity 1.

Activity 3

Prepare Form 941, Employers Quarterly Federal Tax Return, for the third quarter ending September 30. The quarter consisted of 12 payroll periods, all of which were exactly the same as the payroll provided in Activity 1. Use Figure 16.12 as a guide and determine the amounts to be reported on lines 2, 3, 5, 6a, 6b, 7a, 7b, 8, 10, 11, 13, 14, 15, and 17a–d. There are no adjustments on earned income credit (EIC).

SELF-TEST

A. Terminology

Complete the following items using the key terms presented at the beginning of the chapter. Check your responses against the answer key at the end of the test.

1. A _____ is a form of earnings based on an hourly payment for a time period expressed in hours or weeks.

2. The Fair Labor Standards Act requires an employer to pay an _____ rate of _____ for hours worked over 40 per week.

3. A _____ is an additional amount per hour paid to employees who work other than day shifts.

4. _____ rates are paid when employees' gross earnings are based on job performance instead of time worked.

5. Under the _____ plan, a productivity level is established; when the employee produces less than that level, an hourly rate is paid; when the employee exceeds the production level a _____ is paid.

6. A _____ is a fixed amount of compensation usually expressed in terms of a year or a month.

7. A _____ is determined through a time and motion study of a job, and the amount is paid to the employee for all acceptable units produced.

8. Employees who are classified _____ are compensated for their work and not the numbers of hours worked.

9. An incentive payment designed to generate a high volume of net sales is called a _____.

10. Deductions required by federal, state, and local governments are referred to as _____.

11. Mandatory deductions include _____, _____, _____, and _____ insurance.

12. _____ tax is paid equally by both employee and employer based on the employee's gross earnings for the period.

13. The _____ method uses tax rates and taxable earnings to calculate the amount of FICA tax and FWT to be withheld from employee earnings.

14. Employer's payroll taxes include _____, _____, _____, and _____; they are based on total taxable earnings and rates reported in the _____ provided by the IRS.

15. The Internal Revenue Service publishes the _____, which contains tables used by employers to determine the amount of withholding tax based on employee wages.

B. Calculation

The following concepts and short problems are designed to test your understanding of the objectives identified at the beginning of the chapter. Answers are provided at the end of the test.

16. Jodi Larson is a front-desk clerk at the Bayside Motel and receives $6.48 per hour. She is also

paid time-and-a-half for hours worked in excess of 40 per week and double time for holidays. Find her gross wage for the current pay period if she worked 57 ½ hours of which 6 hours were worked on a holiday.

17. An assembler at the Ace Manufacturing Company produced 42 units per hour on Monday, 56 units per hour on Tuesday, and 30 units per hour on Wednesday. She worked 8 hours each of these days, but did not report to work due to illness the remainder of the week. Find her gross earnings if she is paid $8.50 per hour or $.25 per unit, whichever is greater, based on an hourly task standard.

18. Bill Ford is a machinist at Jenson's Machine Shop and has a task requirement of 200 units per 8-hour shift with a 130% bonus factor. Find his gross earnings if he is paid $12.80 per hour and produced 280 units during an 8-hour period.

19. Louis Pisegna is paid an annual salary of $19,760 on a weekly basis. He also receives time-and-a-half for all hours worked over 40 hours a week. If his regular workweek is 35 hours, find his gross earnings for the week if he worked a total of 46 hours.

20. Ahmad Ashkar is employed as a sales representative for Executive Systems. He is paid a base salary of $300 semimonthly plus a graduated commission of 2% on the first $50,000 net sales, and 1 ½% on net sales over $50,000. What were his gross earnings for the period if he had gross sales of $68,500 and discounts amounting to 3% of total sales?

21. Kelly Hauser is single and claims two exemptions. She has year-to-date earnings of $18,650 and her current weekly earnings amount to $482.70. What is her net pay if her deductions include FICA tax, Medicare tax, FWT, and SWT of 3% on gross earnings up to maximum taxable earnings of $18,500? Use the wage bracket method to calculate the appropriate payroll taxes.

22. Byron Stone is paid weekly at an hourly rate of $19.40 with time-and-a-half pay over 40 hours. He is married and claims three exemptions. What is his net pay if he worked 48 hours this pay period and his deductions include $75.90 for health insurance; $75.00 for savings; $12.50 for disability insurance; and $132.50 for retirement in addition to his FICA, Medicare, and FWT taxes? Use the percentage method to calculate employee payroll taxes based on 1998 FICA and Medicare rates. Assume all earnings are taxable.

23. Jack Winfield is employed at Eastern Community College as the Director of Financial Services. He receives an annual salary of $52,650 and is paid biweekly. His year-to-date earnings are $6,075. Using the 1998 FICA and Medicare tax rates and a SUTA tax rate of 3.2% with a tax limit of $7,000, calculate the following:

a. gross earnings
b. employer FICA tax
c. SUTA tax
d. FUTA tax
e. Medicare tax.

24. Given the following payroll summary, calculate the employer's payroll taxes to be reported for the second quarter. Assume a SUTA tax experience rate of 4.2% on the first $7,000. (Use applicable 1998 tax rates.)

	Payroll summary	
Employee	First-quarter accumulated earnings	Second-quarter accumulated earnings
Bedell, T.	$4,780	$5,960
Hartman, R.	7,200	7,840
Selmon, M.	2,115	4,875

25. The payroll records of Lakeside Marina at the end of the third quarter indicated the company had paid total wages of $132,680.90 of which $124,580.75 were taxable Social Security wages; the total wages paid during the quarter were also taxable Medicare wages. The company had deducted $7,843.20 in FWT tax from employees' earnings. Determine the total amount of employer taxes Lakeside Marina reported for the quarter on Form 941, the Employer's Quarterly Federal Tax Return. (Use 1998 tax rates.)

Answers to Self-Test: *1.* wage *2.* overtime, time-and-a-half *3.* shift differential *4.* incentive rates *5.* task and bonus, plan *6.* salary *7.* piece rate *8.* exempt *9.* commission *10.* mandatory deductions *11.* FICA taxes, Medicare taxes, FWT, disability (SDI) *12.* FICA (Social Security) *13.* percentage *14.* FICA, Medicare, FUTA, SUTA, Circular E Tax Guide *15.* Circular E Tax Guide *16.* $448.74 *17.* $264.00 *18.* $186.37 *19.* $477.74 *20.* $1,546.68 *21.* $389.77 *22.* $488.73 *23.* **a.** $2,025; **b.** $125.55; **c.** $29.60; **d.** $7.40; **e.** $29.36 *24.* **a.** FICA = $1,157.85; **b.** Medicare = $270.79; **c.** SUTA = $297.99; **d.** FUTA = $56.76 *25.* $27,138.96

17

Key Terms

sales tax
excise tax
real property
personal property
property assessor
fair market value
assessment rate
assessment value
property tax rate
mills
filing status
exemptions
total income
W-2 form
Form 1099
adjusted gross income
  (AGI)
taxable income
standard deduction
personal deductions
tax due
refund

TAXES

Learning Objectives

After completing the exercises in this chapter, you will be able to:

1. Compute the sales tax and the total price on goods sold when discounts and other charges are involved.

2. Determine the marked price of an item when the sales tax rate and total price is known.

3. Find the total price of goods subject to both an excise and sales tax.

4. Determine the assessed value of real property for property tax.

5. Calculate property tax rates with various bases.

6. Find the property taxes of real property using the property tax formula.

7. Use IRS procedures to determine individual income tax or refund due.

8. Determine individual income tax using IRS schedule 1040.

9. Prepare U.S. Individual Income Tax Return, Form 1040EZ.

10. Define key terms.

INTRODUCTION

Article I of the United States Constitution empowers Congress to "lay and collect taxes" as one of its provisions. As a result, the federal government derives its revenues through taxes on individuals and businesses to pay for the cost of government goods and services. Public reaction to federal, state, and local tax policies has seldom been supportive. In general, the public believes taxes are to high, unfair, and never used efficiently. Economic theory states that the public interest is best served by a tax system that is universal, simple, fair, and that enhances growth and quality of life. However, there is evidence to suggest that government tax systems promote the opposite outcomes. The fact that a high percentage of personal income is exempt from taxes, different types of income are taxed differently, and that work, savings, and investment are penalized support the commonly held belief that tax laws are complex, inefficient, and unfair. Regardless of one's opinion on taxes, governments will continue to use taxes as the primary source of revenue to pay for national defense, health care, education, law enforcement, public assistance, highways, and park and recreation facilities.

Governments generate income from many different types of taxes, such as individual income, payroll, corporate income, excise, sales, estate, gift, and property. This chapter presents information concerning and the calculations necessary to determine sales, excise, property, and individual income taxes. Payroll taxes were discussed in the previous chapter.

17.1   SALES AND EXCISE TAXES

Many sellers (manufacturers, wholesalers, and retailers) are required to collect sales and/or excise taxes on the sale of merchandise. In this section, we will discuss these two taxes and the procedures used in their calculations.

Sales Tax

A **sales tax** is a tax that is charged on the sale of particular goods and services sold by retail merchants. The tax is expressed as a percent of the net sales price and is collected by the seller. As required by regulation, the sales taxes are then remitted to the state, county, or city government in which the goods are sold.

Learning objective
Compute the sales tax and total price on goods sold when discounts and other charges are involved.

Sales taxes are an important source of revenue for governmental agencies. Consequently, many states and cities are increasing their sales tax rates as well as expanding the list of taxable goods and services. In addition, some states and municipalities are beginning to collect sales taxes on goods sold outside the jurisdiction of the taxing agency. This departure from the practice of collecting taxes only on goods sold within the tax district is designed to recover the amount of sales tax that is lost on tax-exempt sales that result from mail-order and telemarketing sales.

Sales tax rates vary from state to state and city to city. State tax rates range between 2% and 8%; city and county taxes are usually lower, 1% to 4%. To assist the sales clerk in situations where the cash register does not automatically calculate the sales tax, charts are available. Sales tax charts, such as shown in Figure 17.1, identify the tax to be collected for each consecutive price interval. The price intervals in sales tax charts can be based on a fixed amount or as a percentage determined by the tax rate.

When a merchant reduces the marked price or list price of merchandise by applying a markdown or a trade discount, the sales tax is computed on the marked-down selling price or the net price, respectively. If a cash discount is applicable, the discount would be calculated on the net price. The sales tax is also calculated on the net price but added to the net price before the cash discount is deducted to determine the amount to be paid. Other charges such as freight, handling, and late charges are exempt from sales tax and would not be included in the taxable amount of a sale. The examples that follow will explain how to calculate sales tax on regular sales and on sales when discounts and other charges are involved. To calculate the sales tax without a tax chart, we multiply the total amount of the sale that is taxable by the tax rate. The sales tax is then added to the marked price to determine the total price.

Sales tax equations

> taxable sales × sales tax rate = sales tax
>
> marked price + sales tax = total price

Example 1

Lisa Reynolds, a sales clerk for the Cycle Emporium, sold a bicycle for $249.99, the sales slip for which is shown in Figure 17.2. The store must collect a 5% state sales tax and a 2% city tax. Determine (a) how much sales tax she charged and (b) how much she charged the customer for the bicycle.

Solution

a.   marked price × combined sales tax rate = sales tax

  $249.99   ×   (.05 + .02)   = $17.499, or $17.50

b.   marked price + sales tax = total price

  $249.99   + $17.50 = $267.49

Figure 17.1

Sales and use tax bracket schedule for state and local tax purposes

4% Sales and use tax collection chart

Amount of sale	Tax to be collected	Amount of sale	Tax to be collected	Amount of sale	Tax to be collected	Amount of sale	Tax to be collected
$0.01 to $0.12	$.00	2.63 to 2.87	.11	5.13 to 5.37	.21	7.63 to 7.87	.31
.13 to .33	.01	2.88 to 3.12	.12	5.38 to 5.62	.22	7.88 to 8.12	.32
.34 to .58	.02	3.13 to 3.37	.13	5.63 to 5.87	.23	8.13 to 8.37	.33
.59 to .83	.03	3.38 to 3.62	.14	5.88 to 6.12	.24	8.38 to 8.62	.34
.84 to 1.12	.04	3.63 to 3.87	.15	6.13 to 6.37	.25	8.63 to 8.87	.35
1.13 to 1.37	.05	3.88 to 4.12	.16	6.38 to 6.62	.26	8.88 to 9.12	.36
1.38 to 1.62	.06	4.13 to 4.37	.17	6.63 to 6.87	.27	9.13 to 9.37	.37
1.63 to 1.87	.07	4.38 to 4.62	.18	6.88 to 7.12	.28	9.38 to 9.62	.38
1.88 to 2.12	.08	4.63 to 4.87	.19	7.13 to 7.37	.29	9.63 to 9.87	.39
2.13 to 2.37	.09	4.88 to 5.12	.20	7.38 to 7.62	.30	9.88 to 10.00	.40
2.38 to 2.62	.10						

5% Combined sales and use tax collection chart

Amount of sale	Tax to be collected	Amount of sale	Tax to be collected	Amount of sale	Tax to be collected	Amount of sale	Tax to be collected
$0.01 to $0.10	$.00	3.10 to 3.29	.16	6.10 to 6.29	.31	9.10 to 9.29	.46
.11 to .27	.01	3.30 to 3.49	.17	6.30 to 6.49	.32	9.30 to 9.49	.47
.28 to .47	.02	3.50 to 3.69	.18	6.50 to 6.69	.33	9.50 to 9.69	.48
.48 to .67	.03	3.70 to 3.89	.19	6.70 to 6.89	.34	9.70 to 9.89	.49
.68 to .87	.04	3.90 to 4.09	.20	6.90 to 7.09	.35	9.90 to 10.00	.50
.88 to 1.09	.05	4.10 to 4.29	.21	7.10 to 7.29	.36		
1.10 to 1.29	.06	4.30 to 4.49	.22	7.30 to 7.49	.37		
1.30 to 1.49	.07	4.50 to 4.69	.23	7.50 to 7.69	.38		
1.50 to 1.69	.08	4.70 to 4.89	.24	7.70 to 7.89	.39		
1.70 to 1.89	.09	4.90 to 5.09	.25	7.90 to 8.09	.40		
1.90 to 2.09	.10	5.10 to 5.29	.26	8.10 to 8.29	.41		
2.10 to 2.29	.11	5.30 to 5.49	.27	8.30 to 8.49	.42		
2.30 to 2.49	.12	5.50 to 5.69	.28	8.50 to 8.69	.43		
2.50 to 2.69	.13	5.70 to 5.89	.29	8.70 to 8.89	.44		
2.70 to 2.89	.14	5.90 to 6.09	.30	8.90 to 9.09	.45		
2.90 to 3.09	.15						

(Continued)

Figure 17.1

(Continued)

6% Combined sales and use tax collection chart

Amount of sale	Tax to be collected	Amount of sale	Tax to be collected	Amount of sale	Tax to be collected	Amount of sale	Tax to be collected
$0.01 to $0.10	$.00	2.59 to 2.74	.16	5.09 to 5.24	.31	7.59 to 7.74	.46
.11 to .22	.01	2.75 to 2.91	.17	5.25 to 5.41	.32	7.75 to 7.91	.47
.23 to .38	.02	2.92 to 3.08	.18	5.42 to 5.58	.33	7.92 to 8.08	.48
.39 to .56	.03	3.09 to 3.24	.19	5.59 to 5.74	.34	8.09 to 8.24	.49
.57 to .72	.04	3.25 to 3.41	.20	5.75 to 5.91	.35	8.25 to 8.41	.50
.73 to .88	.05	3.42 to 3.58	.21	5.92 to 6.08	.36	8.42 to 8.58	.51
.89 to 1.08	.06	3.59 to 3.74	.22	6.09 to 6.24	.37	8.59 to 8.74	.52
1.09 to 1.24	.07	3.75 to 3.91	.23	6.25 to 6.41	.38	8.75 to 8.91	.53
1.25 to 1.41	.08	3.92 to 4.08	.24	6.42 to 6.58	.39	8.92 to 9.08	.54
1.42 to 1.58	.09	4.09 to 4.24	.25	6.59 to 6.74	.40	9.09 to 9.24	.55
1.59 to 1.74	.10	4.25 to 4.41	.26	6.75 to 6.91	.41	9.25 to 9.41	.56
1.75 to 1.91	.11	4.42 to 4.58	.27	6.92 to 7.08	.42	9.42 to 9.58	.57
1.92 to 2.08	.12	4.59 to 4.74	.28	7.09 to 7.24	.43	9.59 to 9.74	.58
2.09 to 2.24	.13	4.75 to 4.91	.29	7.25 to 7.41	.44	9.75 to 9.91	.59
2.25 to 2.41	.14	4.92 to 5.08	.30	7.42 to 7.58	.45	9.92 to 10.00	.60
2.42 to 2.58	.15						

7% Combined sales and use tax collection chart

Amount of sale	Tax to be collected	Amount of sale	Tax to be collected	Amount of sale	Tax to be collected	Amount of sale	Tax to be collected
$0.01 to $0.10	$.00	2.93 to 3.07	.21	5.79 to 5.92	.41	8.65 to 8.78	.61
.11 to .20	.01	3.08 to 3.21	.22	5.93 to 6.07	.42	8.79 to 8.92	.62
.21 to .33	.02	3.22 to 3.35	.23	6.08 to 6.21	.43	8.93 to 9.07	.63
.34 to .47	.03	3.36 to 3.49	.24	6.22 to 6.35	.44	9.08 to 9.21	.64
.48 to .62	.04	3.50 to 3.64	.25	6.36 to 6.49	.45	9.22 to 9.35	.65
.63 to .76	.05	3.65 to 3.78	.26	6.50 to 6.64	.46	9.36 to 9.49	.66
.77 to .91	.06	3.79 to 3.92	.27	6.65 to 6.78	.47	9.50 to 9.64	.67
.92 to 1.07	.07	3.93 to 4.07	.28	6.79 to 6.92	.48	9.65 to 9.78	.68
1.08 to 1.21	.08	4.08 to 4.21	.29	6.93 to 7.07	.49	9.79 to 9.92	.69
1.22 to 1.35	.09	4.22 to 4.35	.30	7.08 to 7.21	.50	9.93 to 10.00	.70
1.36 to 1.49	.10	4.36 to 4.49	.31	7.22 to 7.35	.51		
1.50 to 1.64	.11	4.50 to 4.64	.32	7.36 to 7.49	.52		
1.65 to 1.78	.12	4.65 to 4.78	.33	7.50 to 7.64	.53		
1.79 to 1.92	.13	4.79 to 4.92	.34	7.65 to 7.78	.54		
1.93 to 2.07	.14	4.93 to 5.07	.35	7.79 to 7.92	.55		
2.08 to 2.21	.15	5.08 to 5.21	.36	7.93 to 8.07	.56		
2.22 to 2.35	.16	5.22 to 5.35	.37	8.08 to 8.21	.57		
2.36 to 2.49	.17	5.36 to 5.49	.38	8.22 to 8.35	.58		
2.50 to 2.64	.18	5.50 to 5.64	.39	8.36 to 8.49	.59		
2.65 to 2.78	.19	5.65 to 5.78	.40	8.50 to 8.64	.60		
2.79 to 2.92	.20						

(Continued)

Figure 17.1

(Continued)

8% Combined sales and use tax collection chart

Amount of sale	Tax to be collected	Amount of sale	Tax to be collected	Amount of sale	Tax to be collected	Amount of sale	Tax to be collected
$.01 to $0.10	$.00	2.57 to 2.68	.21	5.07 to 5.18	.41	7.57 to 7.68	.61
.11 to .17	.01	2.69 to 2.81	.22	5.19 to 5.31	.42	7.69 to 7.81	.62
.18 to .29	.02	2.82 to 2.93	.23	5.32 to 5.43	.43	7.82 to 7.93	.63
.30 to .42	.03	2.94 to 3.06	.24	5.44 to 5.56	.44	7.94 to 8.06	.64
.43 to .54	.04	3.07 to 3.18	.25	5.57 to 5.68	.45	8.07 to 8.18	.65
.55 to .67	.05	3.19 to 3.31	.26	5.69 to 5.81	.46	8.19 to 8.31	.66
.68 to .79	.06	3.32 to 3.43	.27	5.82 to 5.93	.47	8.32 to 8.43	.67
.80 to .92	.07	3.44 to 3.56	.28	5.94 to 6.06	.48	8.44 to 8.56	.68
.93 to 1.06	.08	3.57 to 3.68	.29	6.07 to 6.18	.49	8.57 to 8.68	.69
1.07 to 1.18	.09	3.69 to 3.81	.30	6.19 to 6.31	.50	8.69 to 8.81	.70
1.19 to 1.31	.10	3.82 to 3.93	.31	6.32 to 6.43	.51	8.82 to 8.93	.71
1.32 to 1.43	.11	3.94 to 4.06	.32	6.44 to 6.56	.52	8.94 to 9.06	.72
1.44 to 1.56	.12	4.07 to 4.18	.33	6.57 to 6.68	.53	9.07 to 9.18	.73
1.57 to 1.68	.13	4.19 to 4.31	.34	6.69 to 6.81	.54	9.19 to 9.31	.74
1.69 to 1.81	.14	4.32 to 4.43	.35	6.82 to 6.93	.55	9.32 to 9.43	.75
1.82 to 1.93	.15	4.44 to 4.56	.36	6.94 to 7.06	.56	9.44 to 9.56	.76
1.94 to 2.06	.16	4.57 to 4.68	.37	7.07 to 7.18	.57	9.57 to 9.68	.77
2.07 to 2.18	.17	4.69 to 4.81	.38	7.19 to 7.31	.58	9.69 to 9.81	.78
2.19 to 2.31	.18	4.82 to 4.93	.39	7.32 to 7.43	.59	9.82 to 9.93	.79
2.32 to 2.43	.19	4.94 to 5.06	.40	7.44 to 7.56	.60	9.94 to 10.00	.80
2.44 to 2.56	.20						

Example 2

George Pakova, owner of the Deli-Box luncheonette, purchased a meat slicer for $650. He was granted a trade discount of 15% and terms of 2%, 10 days. If the sales tax was 7% and he remitted payment during the discount period, how much did he pay for the meat slicer?

Solution

$650.00	list price
− 97.50	less trade discount (650 × .15)
$552.50	net price (taxable amount)
+ 38.68	plus sales tax ($552.50 × .07)
$591.18	total price
− 11.05	less cash discount (552.50 × .02)
$580.13	amount paid

Figure 17.2

Sales slip

<table>
<tr><td colspan="8" align="center">**SALES SLIP**</td></tr>
<tr><td colspan="2">Customer's
Order No. _____</td><td colspan="3">Phone
No. _____</td><td colspan="3">Date _4/12_ 19 _91_</td></tr>
<tr><td colspan="8">Sold To __Mrs. Elizabeth Ryan_____</td></tr>
<tr><td colspan="8">Address __127 Bryant Ave._____</td></tr>
<tr><td>SOLD BY</td><td>CASH</td><td>C.O.D.</td><td>CHARGE</td><td>ON ACCT.</td><td>MDSE. RETD.</td><td colspan="2">PAID OUT</td></tr>
<tr><td>QUANTITY</td><td colspan="3" align="center">DESCRIPTION</td><td></td><td>PRICE</td><td colspan="2" align="center">AMOUNT</td></tr>
<tr><td></td><td colspan="4"></td><td></td><td></td><td></td></tr>
<tr><td>1</td><td colspan="4">Model # 127X</td><td></td><td>249</td><td>99</td></tr>
<tr><td></td><td colspan="4">BMX Bicycle</td><td></td><td></td><td></td></tr>
<tr><td></td><td colspan="4"></td><td></td><td></td><td></td></tr>
<tr><td></td><td colspan="4"></td><td></td><td></td><td></td></tr>
<tr><td></td><td colspan="4"></td><td></td><td></td><td></td></tr>
<tr><td></td><td colspan="4"></td><td></td><td></td><td></td></tr>
<tr><td></td><td colspan="4" align="center">ck# 1263</td><td></td><td></td><td></td></tr>
<tr><td></td><td colspan="4"></td><td></td><td></td><td></td></tr>
<tr><td></td><td colspan="4"></td><td></td><td></td><td></td></tr>
<tr><td></td><td colspan="4"></td><td></td><td></td><td></td></tr>
<tr><td></td><td colspan="4"></td><td>7%</td><td>TAX</td><td>17</td><td>50</td></tr>
<tr><td></td><td colspan="4"></td><td></td><td>TOTAL</td><td>267</td><td>49</td></tr>
<tr><td colspan="8">All claims and returned goods MUST be accompanied by this bill.
12299
Rec'd by</td></tr>
</table>

Learning objective
Determine the marked price of an item when the sales tax and total price is known.

Note: The sales tax and the cash discount are both calculated on the net price. However, the sales tax is added to the net price where the cash discount is subtracted from the net price.

Example 3

A sales clerk in a drug store charged a customer a total of $16.12 for an item that included 4% sales tax. Find the marked price of the item.

Solution

Algebra:

marked price + sales tax = total price

$$P + .04P = \$16.12$$
$$1.04P = \$16.12$$
$$P = \$15.50 \text{ marked price}$$

Formula:

$$\text{marked price} = \frac{\text{total price}}{1 + \text{sales tax rate}}$$
$$= \frac{\$16.12}{1 + .04}$$
$$= \frac{\$16.12}{1.04}$$
$$= \$15.50$$

Excise Tax

Learning objective
Find the total price
of goods subject to
both an excise and
sales tax.

Another tax levied on the sale or manufacture of specific goods and services by governmental agencies is the **excise tax.** Governmental agencies may charge an excise tax on the sale of gasoline, tobacco, alcoholic beverages, jewelry, recreational and sporting goods, firearms, entertainment, motor vehicles, luggage, fur, cosmetics, telephone service, tires, and licenses. Excise taxes are paid in addition to sales taxes, and like sales taxes, excise taxes are usually expressed as a percent. However, in some cases, excise taxes are expressed as a fixed amount based on the quantity sold. For example, notice in Table 17.1 the federal excise tax on the sale of gasoline is 18.4 cents per gallon. Excise taxes collected on the sale of goods by sellers are often included in the markup percentage and are passed on to the buyer through the unit selling price. When the excise tax is not included in the price of an item, it must be added to the selling price to determine the taxable amount of the sale before the sales tax can be computed. Examine Examples 4 and 5 carefully so that you thoroughly understand this distinction.

Example 4

On September 15, Alice Littlejohn purchased a set of steelbelted radial tires at $89.95 each from Goodman's Tire and Service Center. How much did the

Table 17.1

Federal excise tax schedule

Tax/Item	Rate/Amount
Air transportation	8% of sales price
International travel	$12 for arrival and departures
Air freight	6.25% of freight charges
Ship passenger tax	$3 per person
Communication service	3% of charges
Firearms	10% of mfg. price
Fishing equipment	10% of mfg. price
Archery bows	11% of price
Arrows	12.4% of price
Coal (underground mined)	$1.10 per ton or 4.4% of sales price, whichever is lower
Coal (surface mined)	$0.55 per ton or 4.4% of sales price, whichever is lower
Alcohol: distilled spirits	$13.50 per gallon
Beer	$18 per barrel
Gasoline	$0.184 per gallon
Diesel fuel	$0.244 per gallon
Aviation fuel	$0.214 per gallon
Tires	
40 pounds to 70 pounds	$0.15 per pound
70 pounds to 90 pounds	$4.50 plus 30 cents a pound in excess of 70 pounds
over 90 pounds	$10.50 plus 50 cents a pound in excess of 90 pounds
Tractors and trailers over 26,000 pounds in gross weight	12% of sales price
Luxury tax	
passenger cars	7% of sales price amount over $36,000
Vaccines	$0.75 on each vaccine contained in the named vaccine

tire center charge Alice for the tires if the excise tax on each tire was $4.29 and the sales tax rate was 6%?

Solution

$359.80	sales price ($89.95 × 4)
+ 17.16	plus excise tax ($4.29 × 4)
$376.96	total sales price (taxable amount)
+ 22.62	plus sales tax ($376.96 × .06)
$399.58	total price of sale

Note: The amount of the sale subject to sales tax includes the excise tax when the excise tax is not expressed as part of the unit price.

Example 5

Millie's Service Station sold 10,675 gallons of unleaded gasoline during the month of October at $1.159 a gallon. If the price per gallon included a federal excise tax of 18.4 cents and a state excise tax of 12 cents, find (a) the total revenue from sales of unleaded gasoline during the month, (b) the federal excise tax to be paid for the month, and (c) the state excise tax to be paid for the month.

Solution

a. $10,675 \times \$1.159 = \$12,372.33$ total revenue from sales

b. $10,675 \times \$.184 = \$1,964.20$ federal excise tax

c. $10,675 \times \$.12 = \$1,281.00$ state excise tax

CHECK YOUR KNOWLEDGE

Sales and Excise Taxes

1. Calculate the sales tax and the total amount on each of the following transactions using a sales tax rate of 3%.

  a. $1.49

  b. $125.99

  c. $4.750

2. Find the sales tax on each of these transactions using the chart for 7% in Figure 17.1

  a. 19 cents

  b. 68 cents

  c. $1.49

  d. $6.05

  e. $9.95

3. Before leaving the grocery store, Marvin Emmons checked his receipt to make sure he was charged properly for his purchases. If all food items are exempt from sales tax and the sales tax rate is 7%, is the amount shown on the receipt correct? Recalculate the receipt if necessary.

```
        1/11  Store # 77
        Reg. 10  OPR 32

Italian bread        .89
cookies             1.79
dish detergent      1.49 tax
2% milk             2.15
peanut butter       1.89
6 pack Cola         2.59 tax
grapes               .99
nacho chips         2.39 tax
margarine            .59
deodorant           2.39 tax
sub total          $17.16

tax paid             .62

total              $17.78
```

4. Louis Cirillo purchased a sportcoat at the Gentlemen's Store that had been marked down 30%. If the coat's marked price was $95 and he was charged a sales tax of 6%, how much did he pay for the sportcoat?

5. Mel's Customobile Service Center purchased a brake rotor from Upstate Distributors for $35. How much did Mel pay Upstate Distributors for the rotor if he received a cash discount of 3%, was charged $2.50 to have the part delivered, and paid a 5% sales tax?

6. The records of the Hot Rock Music Company indicate sales of $13,330.60 for the month, including a 7% sales tax. What was the amount of (a) taxable sales for the month and (b) sales tax.

7. Mark Keller purchased a diamond necklace from Wilmont Jewelers for $15,000 plus taxes. How much did the jewelry store receive for the necklace if the excise tax was 10% and the sales tax was 5%?

8. General Petroleum sells 20,000 gallons of diesel fuel to All-State Trucking at $.985 per gallon. If the price includes a federal excise tax of 24.4 cents and a state tax of 6.5 cents, find (a) the amount of federal excise tax collected and (b) the amount of state tax collected on the sale of diesel fuel.

Answers to CYK: **1.** **a.** $.04; **b.** $3.78; **c.** $142.50 **2.** **a.** $.01; **b.** $.05; **c.** $.10; **d.** $.42; **e.** $.70 **3.** total incorrect; $17.61 is correct total **4.** $70.49 **5.** $38.20 **6.** **a.** $12,458.50; **b.** $872.10 **7.** $17,325.00 **8.** **a.** $4,880; **b.** $1,300

17.1   EXERCISES

Fill in the blanks with the correct amounts. Round amounts to the nearest whole cent.

	Marked price	Mark-down percent	Sale price	Sales tax rate	Sales tax amount	Total sale amount
1.	$34	10%	_____	3%	_____	_____
2.	$282.50	25%	_____	5%	_____	_____
3.	$7.99	8%	_____	7%	_____	_____
4.	$1,450	30%	_____	4%	_____	_____
5.	$168	42%	_____	6%	_____	_____

Find the missing amounts. Use the complement method to calculate your answers.

	List price	Trade discount	Net price	Excise tax	Total sales price	Sales tax	Total amount of sale
6.	$25	5%	_____	2%	_____	4%	_____
7.	$380	10%	_____	1%	_____	2%	_____
8.	$659	25%	_____	3%	_____	3%	_____
9.	$1,850	37%	_____	1½%	_____	5%	_____
10.	$74.60	12%	_____	4%	_____	6%	_____

Solve each of the following problems. Round dollar amounts to the nearest cent, and rates to the nearest tenth of a percent.

11. An automobile stereo cassette system costs the dealer $159. If the system is sold for $230, and the dealer collects a state sales tax of 5% and a city sales tax of 3%, find the amount the customer paid for the system.

12. Jack Karpinski purchased a sportcoat that sold for $139 plus a sales tax of 6%. How much sales tax did Jack pay on the coat?

13. A clerk for Hanson's Department Store sells an expandable watch band to a customer for $8.99. How much should she charge the customer for the band if she uses the 7% sales tax schedule in Figure 7.1

14. All-Sports Distributors sold merchandise to Jenson's Sport Shop listed at $4,250 with terms of 2/10, *n*/30, and a trade discount of 10%. A sales tax of 5% is required. If Jenson's pays for the merchandise within the discount period, how much must be remitted?

15. The R. D. Jones Company made a purchase of $12,750. The company received a trade discount of 20% and paid a sales tax of 5%. What was the total amount paid for the purchase?

16. Julie Mendoza bought a pair of shoes for $69.54 that included a 7% sales tax. Find (a) the marked price and (b) the sales tax paid.

17. A manufacturer of electric outboard motors sold a distributor 150 motors at $75 each. The manufacturer is required to collect a 3% federal excise tax. How much will the manufacturer charge the distributor for the motors?

18. Amanda Jefferson pays a commercial airline $278 for a round-trip flight from Boston to Orlando, excluding taxes. How much will Amanda's ticket cost if she has to pay an 8% federal excise tax and a 4% state sales tax?

19. Panther Lake Marina sold 7,500 gallons of fuel for motor boats during the month of June at $1.259 a gallon. If the price per gallon included a federal excise tax of 18.4 cents and a state tax of 12.5 cents, find (a) the federal excise tax col-

lected during the month, (b) state tax collected during the month, and (c) the amount of sales from fuel excluding taxes.

20.   The billing clerk for Upstate Telephone Co. is preparing a customer's bill for monthly service. Local charges are $15.75, and toll charges total $23.50. If the telephone company must collect a federal excise tax of 3% and a local tax of 4%, find (a) the amount of each tax and (b) the total amount the customer must pay for the month.

17.2   PROPERTY TAXES

State and local governments also obtain revenues to support the cost of goods and services provided by levying a tax on the owners of **real property.** Generally, property is something tangible or intangible to which its owner has legal title. *Real property* consists of land and any buildings on the land. **Personal property** consists of all other possessions that an individual or company might hold. Once a value has been established for property, then a tax can be levied on that property, either real or personal. We will consider only real property taxes.

Assessing the Value of Real Property

To determine the total value of the taxable property in a city, village, county, township, school district, and so on, a **property assessor** is employed to determine the value of each parcel of taxable land and its building (eliminating nontaxable government-owned property, religious property, etc.). The assessor first establishes the **fair market value** for the property—the amount of money that the property could be sold for on the real estate market. The assessor may use a variety of methods to accomplish this task, including the comparison of the property with similar property that has recently sold in the area. Using the selling prices of those other properties, the assessor arrives at a fair market value for the property being assessed. Many communities use an **assessment rate** that establishes the **asssement value** of the property as a percentage of the fair market value of the property. For example, if the assessment rate is 25% of the fair market value, then a parcel of property with a fair market value of $80,000 will have an assessment value of $80,000 $\times$ .25 = $20,000 for tax purposes. Other communities use an assessment rate that is 100% of the fair market value, or a full market value assessment of the property.

Calculating Tax Rate

Once each parcel of taxable property has been assessed, the total assessed value of all of the taxable property in the governmental unit is computed. The amount of revenue that must be obtained through property taxes is usually determined as part of the tax district's budgeting process. At the time that revenues are needed, property taxes are levied on the owners of property in the governmental unit by determining a **property tax rate**, which is expressed as a certain amount of cents per dollar, or an amount per $100 or per $1,000. The tax rate is determined by dividing the amount of revenues needed by the total assessed value

$$\text{tax rate} = \frac{\text{total revenues to be realized from property taxes}}{\text{total assessed value of all taxable property}}$$

The tax rate is also expressed in terms of **mills**, in which one mill is a monetary unit of 1/1000 of one U.S. dollar, or 1/10 of one U.S. cent. The amount of property tax revenue generated by a specific parcel of land and the buildings on it is determined by multiplying the assessed value of that property by the tax rate.

Example 6

The Stuyvesant City School District needs $2,500,000 from property taxes from all taxable property in the school district, which has a total fair market value of $100,000,000. Calculate the tax rate if the assessed value of the land in the school district is (a) full value (100%) of fair market value, (b) 60% of fair market value, (c) 10% of fair market value.

Solution

a.   The total assessed value = (assessment rate) × (total fair market value)

total assessed value = 100% of $100,000,000

total assessed value = 1.00 × $100,000,000 = $100,000,000

$$\text{tax rate} = \frac{\text{total revenues to be raised from property taxes}}{\text{total assessed value}}$$

$$\text{tax rate} = \frac{\$2,500,000}{\$100,000,000} = .025$$

which can be expressed as a 2.5% tax rate, or as 2.5 cents per dollar of taxable assessed value.

b.   total assessed value = (assessment rate) × (total fair market value)

total assessed value = 60% of $100,000,000

total assessed value = .60 × $100,000,000 = $60,000,000

$$\text{tax rate} = \frac{\text{total revenues to be raised from property taxes}}{\text{total assessed value}}$$

$$\text{tax rate} = \frac{\$2,500,000}{\$60,000,000} = .041666$$

This would most likely be rounded up to a 4.17% tax rate, or as 4.17 cents per dollar of taxable assessed value.

c. The total assessed value = (assessment rate) × (total fair market value)

total assessed value = 10% of $100,000,000

total assessed value = .10 × $100,000,000 = $10,000,000

$$\text{tax rate} = \frac{\text{total revenues to be raised from property taxes}}{\text{total assessed value}}$$

$$\text{tax rate} = \frac{\$2,500,000}{\$10,000,000} = .25$$

The tax rate may be expressed in four different ways. If part a of Example 6, we noted that the tax rate could be expressed as 2.5 cents on each dollar of assessed value, or as 2.5% of the assessed value of the property. It could also be expressed as $25 per $100 of assessed value, or as $250 per $1,000 of assessed value, or as 25 mills per dollar of assessed value, which is 25/1000 of a dollar (i.e., .025 of one dollar). Figure 17.3 is an illustration of a tax district that determines school tax based on dollars of assessed value.

Calculating the Amount of Property Tax

To determine the specific amount of tax on a piece of property, we multiply the assessed value of the property by the tax rate.

Example 7

Donovan Christie's home in the Stuyvesant Central School District has a fair market value of $125,000 and is assessed at full market value. The school tax rate for this year on the properties in the school district is .025. Determine the amount of Donovan's school tax this year.

Solution

tax = (assessed value) × (tax rate)

tax = $125,000 × .025 = $3,125.00

Example 8

The Stuyvesant School District tax rate could also be expressed as $2.50 per $100 of assessed value, or as $25 per $1,000 of assessed value. Calculate Donovan Christie's school tax if the tax rate is expressed as (a) $2.50 per $100 of assessed value, and (b) $25 per $1,000 of assessed value.

Solution

a. First determine the number of $100 units that are in Donovan's assessment by dividing the assessment by 100.

$$\text{number of \$100 units} = \frac{\$125,000}{\$100} = 1,250$$

Figure 17.3

School tax bill
based on property
tax value

1998 - 1999 SCHOOL TAX
TOWN OF MELVILLE - CHARTER COUNTY, NEW YORK

Fiscal Year: 07/01/98   Warrant Date: 08/19/98   Estimated State Aid: $17020,340

TAX MAP NUMBER	BANK	NYS TAX & FINANCE SCHOOL CODE	BILL NO.
587359 087. -09-34-8	3FDE056	29576	28634

MAKE CHECK
PAYABLE TO   Receiver of Taxes

PROPERTY INFORMATION:
Dimension: 75.00 x 150.00
RS 1   210 Single Family
Address: 45 West Street

In Person
Payment   Municipal Building

FULL MARKET VALUE   89,000
% OF VALUE   100.00
ASSESSMENT   JULY 1   89,000

Exemptions:

OWNER:

Robbins, George E.
Eleanor P
210 West Street
Melville, N.Y. 13596

Levy Description	Tax Levy	% Change	Rate Per 1000	Taxable Value	Amount Due
School Tax	20.629.730	1.1	17.803794	89,000	1,584.54
Library Tax	47.500		.041008	89,000	3.65

PARTIAL PAYMENTS ARE ALLOWED BY THIS DISTRICT
1. First partial payment must be 50% of the   3. No delinquent taxes may be due on this property.
   total bill by 10/08/98.   4. Entire balance due to Finance Department on 11/02/98
2. Second and third partial payment must be at
   least 50% of the balance plus penalities.

PENALTY SCHEDULE:

TOTAL   DUE $1,588.19
DATE   DUE   10/08/98

Sept 09 - Oct 08 $31.76
Oct 09 - Nov 02   79.41

Next, multiply the number of $100 units by the tax rate expressed as number of dollars per $100.

$$\text{tax} = 1{,}250 \times \$2.50 = \$3{,}125$$

b. Determine the number of $1,000 units that are in Donovan's assessment by dividing the assessment by 1,000.

$$\text{number of \$1,000 units} = \frac{\$125{,}000}{\$1000} = 125$$

Next, multiply the number of $1,000 units by the tax rate expressed as number of dollars per $1,000.

$$\text{tax} = 125 \times \$25 = \$3{,}125$$

Still another way to express tax rate is in *mills*, where one mill is defined to be either 1/000 of a dollar or 1/10 of a cent.

Example 9

Solution

Express the Stuyvesant Central School District tax rate in terms of mills.

The tax rate is computed the same way as previously in Example 6.

$$\text{tax rate} = \frac{\$2,500,000}{\$100,000,000} = .025$$

which can easily be seen as 25/1000 of one dollar. This is stated as 25 mills per dollar.

We also stated in Example 6 that the tax rate could be stated as 2.5 cents per dollar. Since there are 25 tenths of a cent in 2.5 cents, the tax rate in terms of the definition of a mill as a tenth of a cent can also be stated as 25 mills.

In either of these cases, we must convert the number of mills from 25 to a factor of .025 to determine the amount of tax that Donovan will have to pay.

$$\text{tax} = (\text{assessed value}) \times (\text{mills}/1,000) = \$125,000 \times .025 = \$3,125$$

Property taxes as a revenue resource tend to be used by smaller governmental units such as local school districts, public libraries, recreational districts, townships, cities, and villages, whereas sales taxes and personal income taxes tend to be used by counties, states, and the federal government. As people demand more services, however, a mix of these taxes is being used in many communities to pay for those services.

CHECK YOUR KNOWLEDGE

Property Taxes

Complete each of the following and express answers to the nearest cent or to the nearest one-thousandth of a percent.

1. The village of Steuben has established the fair market value of its taxable property to be $90,000,000. The village uses an assessment rate of 40% of fair market value. The village needs to realize revenues of $1,800,000 next year from property taxes to maintain village services to its residents. (a) What is the total assessed value of the property in Steuben? (b) What tax rate will be necessary to realize the $1,800,000?

2. Express the tax rate in part (b) of problem 1 as (a) cents per dollar of assessed value, (b) dollars per hundred dollars of assessed value, (c) dollars per thousand dollars of assessed value, and (d) mills.

3. Sara Phong owns a townhouse in the village of Steuben with a fair market value of $50,000. (a) Determine the assessed value of her home. (b) Determine the amount of tax she will be asked to pay the village this year.

17.2   EXERCISES

Complete each of the following and express dollar answers to the nearest cent and tax rates to the nearest thousandth.

1. For each of the following fair market values, determine the assessed values using the assessment rate given.

Fair market value	Assessment rate	Assessment value
$30,000	10%	_____
$45,000	25%	_____
$150,000	40%	_____
$73,000	full value	_____

2. Determine the tax rate for each of the following municipalities, given the total assessed value of the property and the revenue needed by the municipality.

Municipality	Revenue	Assessed value	Tax rate
Newtown	$3,000,000	$300,000,000	_____
Old York	$700,000	$3,500,000	_____
South Town	$950,000	$72,000,000	_____
Central City	$1,500,000	$91,450,000	_____

Answers to CYK:

1. a. $36,000,000   b. .05 or 5% of assessed value   2. a. 5 cents per dollar
b. 100 × .05 = 5 dollars per hundred   c. 1,000 × 50 dollars per thousand
d. mills = number of tenth of a cent = $\dfrac{\text{number of cents}}{1/10}$ = $\dfrac{5 \text{ cents}}{1/10}$ = 50 mills

mills = number of thousandths of a dollar = $\dfrac{\text{number of dollars}}{1/1000}$ = $\dfrac{.05 \text{ dollars}}{1/1000}$ = 50 mills

3. a. assessed value = $20,000   b. tax = tax rate × assessed value = .05 × $20,000 = $1,000

3. Determine the number of $100 units and the number of $1,000 units there are in each of the following property values.

Property value	Number of $100 units	Number of $1,000 units
$2,500	_____	_____
$4,500	_____	_____
$9,000	_____	_____
$15,000	_____	_____
$38,000	_____	_____
$364,000	_____	_____
$128,000	_____	_____
$1,300,000	_____	_____

4. Calculate the tax bill for each of the following properties given the assessed value of the property and the tax rate.

Assessed value	Tax rate	Tax
$65,000	.050	_____
$54,000	.023	_____
$125,000	.015	_____

5. Calculate the tax bill for each of the following properties given the assessed value of the property and the tax rate.

Assessed value	Tax rate	Tax
$85,000	$5 per hundred	_____
$47,000	$3.45 per hundred	_____
$69,000	$2.50 per hundred	_____
$64,000	$60 per thousand	_____
$130,000	$35 per thousand	_____
$29,500	$15 per thousand	_____

6. Calculate the tax bill for each of the following properties given the assessed value of the property and the tax rate.

Assessed value	Tax rate	Tax
$53,000	40 mills	_____
$250,000	35 mills	_____
$72,000	15 mills	_____

7. Bernie Holtzworth owns a home on a one-acre piece of land in Portland County that has a fair market value of $115,000. Portland County's full value of taxable properties in the county is $750,000,000, and it uses a full-value assessment rate. The county Library and Parks District must realize revenue of $1,500,000 from property taxes. (a) What is the total assessed value of the property in Portland County? (b) Calculate the tax rate for the Library and Parks District. (c) What is the property tax assessment of Bernie's property in Portland County? (d) Calculate how much Bernie's tax will be for Library and Parks.

8. Charysse Abate holds title to her mother's home in Alameda County. It has a fair market value of $98,000. Alameda County's assessment rate is 25% of fair market value, and the assessors have determined that the total fair market value of property in their county is $375,000,000. The county budget for next year calls for $4,575,390 to be realized through property taxes. (a) What is the total assessed value of the property in Alameda County? (b) Calculate the tax rate for the county. (c) What is the property tax assessment on Charysse's property in Alameda County? (d) Calculate how much Charysse's tax will be for Alameda County.

9. Kalun Chen's Electronics store has an assessed value of $17,500 and is located in the city of Otisco. The total assessed value of Otisco's taxable properties is $10,800,000, and the county has a 10% assessment rate. The city school district needs $1,500,000 this year from property taxes. (a) What is the total fair market value of the property in Otisco? (b) Calculate the tax rate for the school district. (c) What is the fair market value on Kalun's property? (d) Calculate how much Kalun's tax will be for schools.

10. The total assessed value of Sennet Township's taxable properties is $35,200,000 and the township has a 25% assessment rate. Jesse Mark's home has an assessed value of $21,000 and is located in the township. The Department of Public Works needs $950,000 this year from property taxes. (a) What is the total fair market value of the property in Sennet? (b) Calculate the tax rate for the Public Works Dept. (c) What is the fair market value on Jesse's property? (d) Calculate how much Jesse's tax will be for Sennet Township.

17.3   PERSONAL INCOME TAX

Learning objective
Use IRS procedures to determine individual income tax or refund due.

In this section we will examine the concepts and procedures involved in filing federal individual income tax using forms 1040 and 1040EZ. The procedures can appear complicated due to an individual's financial situation, so some individuals contract with professional tax preparers to prepare their tax returns. However, many people prepare their own returns using the instructions provided by the Internal Revenue Service with the tax forms or with tax software that will complete all calculations and print reports. Under certain conditions, the IRS will compute your personal income tax or allow you to file your taxes electronically from your home computer. With all these options, you might ask, "Why is it necessary to know how to prepare an individual income tax return"? As a taxpayer, you should be aware of the annual changes in tax regulations so that you can provide the required data necessary to complete your tax return, no matter which option you select.

Filing Requirements

Not everyone has to file a federal income tax return. Tax regulations define who must file a tax return based on a person's filing status and gross income. Table 17.2 lists the filing requirements for most individuals. Any person who has had federal income tax withheld on earned income amounts less than those identified in Table 17.2 should file a tax return to get those taxes refunded. Tax guidelines also suggest that individuals should file a tax return if they are eligible for the earned income credit. The earned income credit reduces the tax liability and therefore may increase the refund amount. Federal tax guidelines require taxpayers to use one of three forms to file their tax return—Form 1040, Form 1040A, or Form 1040EZ. The form required is determined primarily by the taxpayer's filing status, age, number of dependents, sources of income, and amount of income. In this section, we will learn how to file an individual income tax return using either Form 1040 or Form 1040 EZ.

The process of preparing a federal income tax return is systematic and generally not difficult, provided there are no complicated tax issues. To determine an individual's tax liability, the following steps must be completed.

Step 1: Determine filing and exemption status of taxpayer.

Table 17.2

Federal income tax filing requirements

Filing status	Age	Gross Income
Single	dependent under 65 with unearned income of $1 or more	$ 700
Single	under 65	6,950
Single	65 or older	8,000
Married filing jointly	under 65 (both spouses)	$12,500
	65 or older (one spouse)	13,350
	65 or older (both spouses)	14,200
Married filing separately	any age	2,700
Head of household	under 65	8,950
	65 or older	10,000
Qualifying widow(er) with dependent child	under 65	9,800
	65 or older	10,650

Gross income: All income received in the form of money, goods, property, and services that is not exempt from tax.

Unearned income: Investment-type income such as interest, dividends, and capital gains. Also includes unemployment compensation, taxable Social Security, pensions, annuities, and income from trusts.

Step 2:   Determine total income.

Step 3:   Determine adjusted gross income (AGI).

Step 4:   Calculate the tax amount based on taxable income.

Step 5.   Determine the tax owed or the refund due.

Preparing U.S. Individual Income Tax Return Form 1040

Learning objective
Determine individual
income tax using IRS
Schedule 1040.

Form 1040 can be used by anyone required to file an individual income tax return. However, the procedure to complete this form is more complicated than the procedures required to complete the other forms. For this reason, individuals generally do not use this form unless they are required to do so. Taxpayers must use Form 1040 if they meet any of the following basic conditions:

1.   Taxable income is $50,000 or more.
2.   They claim dependents.
3.   They itemize deductions.
4.   They have earned income that cannot be reported on Form 1040A or Form 1040EZ.
5.   They pay tax on self-employment income.

There are other conditions under which this form must be used, so the tax regulations should be read carefully to determine the most appropriate form based on the taxpayer's specific circumstances.

Examples 10 through 15 contain information that is necessary to apply the steps outlined earlier to complete Form 1040. The solutions will be presented as part of each step and illustrated in Figure 17.5.

Step 1: Determine Filing and Exemption Status of Taxpayer.   An individual's **filing status** is tied to his or her marital status. People who are single pay a higher tax rate per dollar on earnings than those who are married, widowed, or are filing as a head of household with a qualifying person (total care dependent). **Exemptions** are the number of individuals in the household and include the head of household, spouse, and dependents. We will explain later how exemptions reduce income taxes.

Example 10

Jack and Mary Reynolds are married and have two dependent children, Jessica and Matthew. Jessica is 15 and has a part-time job as a cashier. Matthew is 10. The Reynolds reside at 327 Elm Street, Auburn, New York, 13945, and file a joint return. Determine the Reynolds's (a) filing status and (b) number of exemptions.

Solution

a. Determine filing status
Filing status: Check box 2, married filing jointly

b. Calculate the number of exemptions
Exemptions: Check box 6a and 6b for Jack and Mary's exemptions. Then, complete 6c listing dependents name, relationship and age. Total number of exemptions claimed (line d) is 4

Step 2: Determine Total Income. **Total income** includes wages, salaries, tips, interest, dividends, alimony, business income, taxable pensions and annuities, unemployment benefits, and Social Security payments. Income (wages, tips, and selected other compensation) is reported on a **W-2 form** by employers (Figure 17.4). The W-2 is also used to report total federal and state income taxes withheld, as well as Social Security and Medicare taxes withheld. Interest, dividends, and some types of wages are reported by

Figure 17.4

Wage and interest income statements

a Employer's name, address, and ZIP code
Jack Reynolds
327 Elm Street
Auburn, N. Y. 13945

1 Wages, tips, other compensation $48,200	2 Federal income tax withheld $7,840
3 Social security wages $48,200	4 Social security tax withheld $2,988.40
5 Medicare wages and tips	6 Medicare tax withheld $2,024.40
7 Social security tips	8 Allocated tips
9 Advance EIC payment	10 Dependent care benefits
11 Nonqualified plans	12 Benefits included in Box 1
13 See Instrs. For Box 13	14 Other

Form W-2 Wage and Tax Statement

b Employer's identification number
145-47653547-8676

c Employee's social security number
235-57-9842

d Employee's name, address and ZIP code
Natural Springs Water Co.
2849 Center Rd.
Geneva, N. Y. 14038

15 Statutory employee ☐ Deceased ☐ Pension plan ☒ Legal rep ☐ Hshld. emp. ☐ Sub-total ☐ Deferred compensation ☐ Void ☐

This information is being furnished to the Internal Revenue Service. If you are required to file a tax return, a negligence penalty or other sanction may be imposed on you if this income is taxable and you fail to report it.

16 State	Employer's state I.D. number	17 State wages, tips, etc.	18 State income tax	19 Locality name	20 Local wages, tips, etc.	21 Local income tax
NY	2764-8576	$48,200	$3,856			

COPY C
For Employee's Records
(See Notice on back)

☐ CORRECTED (if checked)

PAYER'S name, street address, city, state, ZIP code, and telephone no.
Solvay Bank
247 Milton Avenue
Solvay, N. Y. 13289

1 Rents $	OMB No. 1545-0115
2 Royalties $	
3 Other income $ 642	Form 1099-MISC

Miscellaneous Income

PAYER'S Federal identification number	RECIPIENT'S identification number
3874-8576-48	235-57-9842

RECIPIENT'S name: Jack Reynolds
Street address (including apt. no.): 327 Elm St.
City, state, and ZIP code: Auburn, N. Y. 13945
Account number (optional)

4 Federal income tax withheld $	5 Fishing boat proceeds $
6 Medical and health care payments $	7 Nonemployee compensation $
8 Substitute payments in lieu of dividends or interest $	9 Payer made direct sales of $5,000 or more of consumer products to a buyer (recipient) for resale ▶ ☐
10 Crop insurance proceeds $	11 State income tax withheld $
12 State/Payer's state number	13 $

Copy B
For Recipient
This is important tax information and is being furnished to the Internal Revenue Service. If you are required to file a return, a negligence penalty or other sanction may be imposed on you if this income is taxable and the IRS determines that it has not been reported.

Form 1099-MISC (Keep for your records.) Department of the Treasury - Internal Revenue Service business on **Form 1099** (Figure 17.4). Copies of the W-2 and 1099 forms are also sent to the IRS to verify such payments to individuals for tax purposes. Therefore, taxpayers do not have to include copies of these forms when they file their income tax.

Example 11

Jack Reynolds's salary this past year was $48,200 as the controller for Natural Springs Water Co. Mary Reynolds earned $34,870 during the same period from the school district as a science teacher. The Reynolds received a number of 1099 forms from various businesses, indicating that they had been paid a total of $642 in interest and $1,813 in dividends. What total income amount should be reported on Form 1040, line 22?

Solution

$$\text{Total income} = \text{total wages} + \text{total interest} + \text{total dividends}$$
$$= (\$48,200 + \$34,870) + \$642 + \$1,813$$
$$= \$85,525$$

Because the Reynolds elected to file a joint return, all sources of income are added together and reported as the total. Also, for tax purposes, the amounts entered on Form 1040 are rounded to the nearest whole dollar. For example, total income of $28,562.30 would be entered as $28,562.

Step 3: Determine Adjusted Gross Income (AGI). Total income reported on line 22 is reduced by adjustments such as contributions to an Individual Retirement Account (IRA), medical savings accounts, moving expenses, self-employment tax, and alimony payments. Most of these adjustments require additional information, which is provided by attaching appropriate forms and documentation to the tax return. The **adjusted gross income (AGI)** is the amount reported on line 32 and is found with the following formula derived from the Form 1040 tax instructions:

$$\text{Adjusted gross} = \text{total income} - \text{total adjustments income}$$

Example 12

The Reynolds financial records verify that Jack had $1,600 in IRA contributions and Mary had $1,500 in IRA contributions, and that an $800 Student Loan Interest deduction was made during the year. Determine the Reynolds adjusted gross income.

Solution

$$\text{Adjusted gross} = \text{total income} - \text{total adjustments}$$
$$\text{income} = \$85,525 - (\$1,600 + \$1,500 + \$800)$$
$$= \$81,625$$

Figure 17.5

Individual income tax Form 1040

Form **1040** Department of the Treasury—Internal Revenue Service
U.S. Individual Income Tax Return (P) **1998** IRS Use Only—Do not write or staple in this space.

For the year Jan. 1–Dec. 31, 1998, or other tax year beginning , 1998, ending , 19 OMB No. 1545-0074

Label (See instructions on page 18.) **Use the IRS label.** Otherwise, please print or type.
LABEL HERE

Your first name and initial: JACK A.	Last name: REYNOLDS		Your social security number: 123 : 45 : 6789
If a joint return, spouse's first name and initial: MARY P.	Last name: REYNOLDS		Spouse's social security number: 987 : 65 : 4321
Home address (number and street). If you have a P.O. box, see page 18.: 327 Elm Street		Apt. no.	▲ **IMPORTANT!** ▲ You must enter your SSN(s) above.
City, town or post office, state, and ZIP code. If you have a foreign address, see page 18.: Auburn, N.Y. 13509			

Presidential Election Campaign (See page 18.)

	Yes	No	Note: Checking "Yes" will not change your tax or reduce your refund.
Do you want $3 to go to this fund?		X	
If a joint return, does your spouse want $3 to go to this fund?		X	

Filing Status

Check only one box.

1 ☐ Single
2 ☒ Married filing joint return (even if only one had income)
3 ☐ Married filing separate return. Enter spouse's social security no. above and full name here. ▶
4 ☐ Head of household (with qualifying person). (See page 18.) If the qualifying person is a child but not your dependent, enter this child's name here. ▶
5 ☐ Qualifying widow(er) with dependent child (year spouse died ▶ 19). (See page 18.)

Exemptions

6a ☒ **Yourself.** If your parent (or someone else) can claim you as a dependent on his or her tax return, **do not** check box 6a.
b ☒ **Spouse**

No. of boxes checked on 6a and 6b: 2

c **Dependents:**

(1) First name Last name	(2) Dependent's social security number	(3) Dependent's relationship to you	(4) ✓ if qualifying child for child tax credit (see page 19)
Jessica Reynolds	111 : 22 : 3333	daughter	☒
Matthew Reynolds	555 : 44 : 6666	Son	☒
			☐
			☐
			☐
			☐

If more than six dependents, see page 19.

No. of your children on 6c who: • lived with you: 2
• did not live with you due to divorce or separation (see page 19)
Dependents on 6c not entered above
Add numbers entered on lines above ▶ 4

d Total number of exemptions claimed

Income

Attach **Copy B of your Forms W-2, W-2G, and 1099-R here.**

If you did not get a W-2, see page 20.

Enclose, but do not staple, any payment. Also, please use **Form 1040-V.**

Line	Description		Amount
7	Wages, salaries, tips, etc. Attach Form(s) W-2	7	83,070
8a	**Taxable** interest. Attach Schedule B if required	8a	642
b	**Tax-exempt** interest. DO NOT include on line 8a — 8b		
9	Ordinary dividends. Attach Schedule B if required	9	1,813
10	Taxable refunds, credits, or offsets of state and local income taxes (see page 21)	10	
11	Alimony received	11	
12	Business income or (loss). Attach Schedule C or C-EZ	12	
13	Capital gain or (loss). Attach Schedule D	13	
14	Other gains or (losses). Attach Form 4797	14	
15a	Total IRA distributions 15a — b Taxable amount (see page 22)	15b	
16a	Total pensions and annuities 16a — b Taxable amount (see page 22)	16b	
17	Rental real estate, royalties, partnerships, S corporations, trusts, etc. Attach Schedule E	17	
18	Farm income or (loss). Attach Schedule F	18	
19	Unemployment compensation	19	
20a	Social security benefits 20a — b Taxable amount (see page 24)	20b	
21	Other income. List type and amount—see page 24	21	
22	Add the amounts in the far right column for lines 7 through 21. This is your **total income** ▶	22	85,525

Adjusted Gross Income

If line 33 is under $30,095 (under $10,030 if a child did not live with you), see EIC inst. on page 36.

Line	Description		Amount		Amount
23	IRA deduction (see page 25)	23	3,100		
24	Student loan interest deduction (see page 27)	24	800		
25	Medical savings account deduction. Attach Form 8853	25			
26	Moving expenses. Attach Form 3903	26			
27	One-half of self-employment tax. Attach Schedule SE	27			
28	Self-employed health insurance deduction (see page 28)	28			
29	Keogh and self-employed SEP and SIMPLE plans	29			
30	Penalty on early withdrawal of savings	30			
31a	Alimony paid b Recipient's SSN ▶	31a			
32	Add lines 23 through 31a			32	3,900
33	Subtract line 32 from line 22. This is your **adjusted gross income** ▶			33	81,625

For Disclosure, Privacy Act, and Paperwork Reduction Act Notice, see page 51. Cat. No. 12599G Form **1040** (1998)

(Continued)

Note: The total income ($85,525) was determined in step 2, Example 11.

Step 4: Calculate the Income Tax Amount Based on Taxable Income. The amount of federal income tax paid by a taxpayer is based on taxable income. **Taxable income** is the amount of adjusted gross income

Figure 17.5

(Continued)

Form 1040 (1998)   Page **2**

Tax and Credits	34	Amount from line 33 (adjusted gross income)		34	81,625

35a Check if: ☐ You were 65 or older, ☐ Blind; ☐ Spouse was 65 or older, ☐ Blind.
Add the number of boxes checked above and enter the total here . . . . ▶ 35a

b If you are married filing separately and your spouse itemizes deductions or you were a dual-status alien, see page 29 and check here . . . . . ▶ 35b ☐

Standard Deduction for Most People

Single: $4,250

Head of household: $6,250

Married filing jointly or Qualifying widow(er): $7,100

Married filing separately: $3,550

	36	Enter the **larger** of your **itemized deductions** from Schedule A, line 28, **OR standard deduction** shown on the left. **But see page 30** to find your standard deduction if you checked any box on line 35a or 35b **or** if someone can claim you as a dependent . .		36	7,380
	37	Subtract line 36 from line 34		37	74,245
	38	If line 34 is $93,400 or less, multiply $2,700 by the total number of exemptions claimed on line 6d. If line 34 is over $93,400, see the worksheet on page 30 for the amount to enter		38	10,800
	39	**Taxable income.** Subtract line 38 from line 37. If line 38 is more than line 37, enter -0-		39	63,445
	40	**Tax.** See page 30. Check if any tax from a ☐ Form(s) 8814 b ☐ Form 4972 ▶		40	12,245
	41	Credit for child and dependent care expenses. Attach Form 2441	41		
	42	Credit for the elderly or the disabled. Attach Schedule R .	42		
	43	Child tax credit (see page 31)	43		
	44	Education credits. Attach Form 8863	44		
	45	Adoption credit. Attach Form 8839	45		
	46	Foreign tax credit. Attach Form 1116 if required	46		
	47	Other. Check if from a ☐ Form 3800 b ☐ Form 8396 c ☐ Form 8801 d ☐ Form (specify)	47		
	48	Add lines 41 through 47. These are your **total credits** . . .		48	0
	49	Subtract line 48 from line 40. If line 48 is more than line 40, enter -0- . . . . ▶		49	12,245

Other Taxes	50	Self-employment tax. Attach Schedule SE		50	
	51	Alternative minimum tax. Attach Form 6251		51	
	52	Social security and Medicare tax on tip income not reported to employer. Attach Form 4137		52	
	53	Tax on IRAs, other retirement plans, and MSAs. Attach Form 5329 if required . .		53	
	54	Advance earned income credit payments from Form(s) W-2 . . .		54	
	55	Household employment taxes. Attach Schedule H . . .		55	
	56	Add lines 49 through 55. This is your **total tax** . . . . . . ▶		56	12,245

Payments	57	Federal income tax withheld from Forms W-2 and 1099 .	57	13,116	

Attach Forms W-2 and W-2G on the front. Also attach Form 1099-R if tax was withheld.

	58	1998 estimated tax payments and amount applied from 1997 return .	58		
	59a	**Earned income credit.** Attach Schedule EIC if you have a qualifying child b Nontaxable earned income: amount ▶ ____ and type ▶	59a		
	60	Additional child tax credit. Attach Form 8812 . . .	60		
	61	Amount paid with Form 4868 (request for extension) . . .	61		
	62	Excess social security and RRTA tax withheld (see page 43) .	62		
	63	Other payments. Check if from a ☐ Form 2439 b ☐ Form 4136	63		
	64	Add lines 57, 58, 59a, and 60 through 63. These are your **total payments** ▶		64	13,116

Refund	65	If line 64 is more than line 56, subtract line 56 from line 64. This is the amount you **OVERPAID**		65	862
	66a	Amount of line 65 you want **REFUNDED TO YOU** . . . . . . ▶		66a	862

Have it directly deposited! See page 44 and fill in 66b, 66c, and 66d.

▶ b Routing number [_____]   ▶ c Type: ☐ Checking ☐ Savings
▶ d Account number [_____]

67 Amount of line 65 you want **APPLIED TO YOUR 1999 ESTIMATED TAX** ▶ | 67 |

Amount You Owe	68	If line 56 is more than line 64, subtract line 64 from line 56. This is the **AMOUNT YOU OWE.** For details on how to pay, see page 44 . . . . ▶		68	
	69	Estimated tax penalty. Also include on line 68 . . .	69		

Sign Here

Joint return? See page 18. Keep a copy for your records.

Under penalties of perjury, I declare that I have examined this return and accompanying schedules and statements, and to the best of my knowledge and belief, they are true, correct, and complete. Declaration of preparer (other than taxpayer) is based on all information of which preparer has any knowledge.

Your signature *Jack Reynolds*   Date 3/20   Your occupation Controller   Daytime telephone number (optional)

Spouse's signature. If a joint return, BOTH must sign. *Mary Reynolds.*   Date 3/20   Spouse's occupation Teacher   ( )

Paid Preparer's Use Only

Preparer's signature ▶   Date   Check if self-employed ☐   Preparer's social security no.

Firm's name (or yours if self-employed) and address ▶   EIN   ZIP code

♻ Printed on recycled paper   *U.S. Government Printing Office: 1998 — 435-538

(determined in step 3) that remains after deductions and total exemptions are subtracted. The formula is again derived from the instructions provided in the tax computation section of Form 1040, lines 34 through 39.

$$\text{Taxable income} = \text{adjusted gross income} - (\text{total deductions} + \text{total exemptions})$$

Personal deductions are allowable expenditures made by the taxpayer during the tax period. Deductions include medical and dental payments, state and local taxes, property taxes, home mortgage interest payments, gifts to charities, casualty and theft losses, job expenses, and other type of miscellaneous deductions. The taxpayer can elect to itemize deductions (Figure 17.6,

Figure 17.6

Schedule A - Itemized deductions

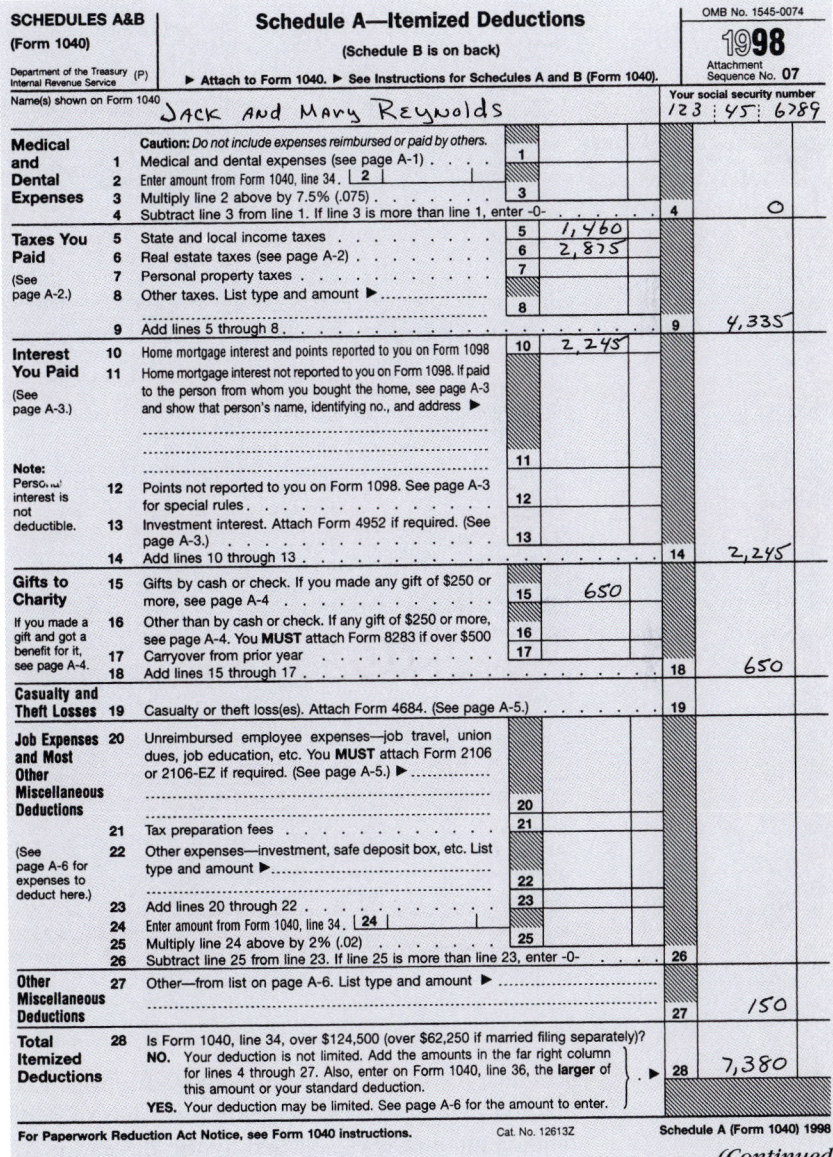

Schedule A (Form 1040) 1998

(Continued)

Figure 17.6

(Continued)

Schedules A&B (Form 1040) 1998

OMB No. 1545-0074   Page **2**

Name(s) shown on Form 1040. Do not enter name and social security number if shown on other side.

Jack and Mary Reynolds

Your social security number

123 : 45 : 6789

Schedule B—Interest and Ordinary Dividends

Attachment Sequence No. **08**

Note: *If you had over $400 in taxable interest income, you must also complete Part III.*

Part I
Interest

(See pages 20 and B-1.)

Note: If you received a Form 1099-INT, Form 1099-OID, or substitute statement from a brokerage firm, list the firm's name as the payer and enter the total interest shown on that form.

1 List name of payer. If any interest is from a seller-financed mortgage and the buyer used the property as a personal residence, see page B-1 and list this interest first. Also, show that buyer's social security number and address ▶

	Amount
Auburn Savings and Loan	*394*
Fleet Bank	*248*

2 Add the amounts on line 1 . . . .

| **2** | *642* |

3 Excludable interest on series EE U.S. savings bonds issued after 1989 from Form 8815, line 14. You MUST attach Form 8815 to Form 1040

| **3** | |

4 Subtract line 3 from line 2. Enter the result here and on Form 1040, line 8a ▶

| **4** | |

Note: *If you had over $400 in ordinary dividends, you must also complete Part III.*

Part II
Ordinary Dividends

(See pages 21 and B-1.)

Note: If you received a Form 1099-DIV or substitute statement from a brokerage firm, list the firm's name as the payer and enter the ordinary dividends shown on that form.

5 List name of payer. Include only ordinary dividends. Report any capital gain distributions on Schedule D, line 13 ▶

	Amount
Bell Atlantic Co.	*530*
Carrier Corporation	*784*
Prudential Insurance Co	*190*
Bristol-Myers-Squibb	*309*
	1,813

6 Add the amounts on line 5. Enter the total here and on Form 1040, line 9 . ▶

| **6** | *1,813* |

Part III
Foreign Accounts and Trusts

(See page B-2.)

You must complete this part if you **(a)** had over $400 of interest or ordinary dividends; **(b)** had a foreign account; or **(c)** received a distribution from, or were a grantor of, or a transferor to, a foreign trust.

		Yes	No

7a At any time during 1998, did you have an interest in or a signature or other authority over a financial account in a foreign country, such as a bank account, securities account, or other financial account? See page B-2 for exceptions and filing requirements for Form TD F 90-22.1 . . . .

b If "Yes," enter the name of the foreign country ▶

8 During 1998, did you receive a distribution from, or were you the grantor of, or transferor to, a foreign trust? If "Yes," you may have to file Form 3520. See page B-2 . . . . . .

For Paperwork Reduction Act Notice, see Form 1040 instructions.   Printed on recycled paper   **Schedule B (Form 1040) 1998**

*U.S. Government Printing Office: 1998 — 435-538

Schedule A) or take the **standard deduction.** Taxpayers usually select the method that produces the greater amount of deductions. The standard deduction amount is established by the IRS and is based on the taxpayer's filing status (determined in step 1). Table 17.3 lists the standard deductions used for problems in this text.

Table 17.3

Standard deduction amounts

Filing status	Standard deduction
Single	$4,250
Married, filing jointly	7,100
Married, filing separately	3,550
Head of household	6,250
Widow(er)	7,100

Exemptions, like deductions, reduce adjusted gross income to determine taxable income. However, exemptions are allowances based on the filing status and number of dependents claimed by the taxpayer. This information is determined in step 1 and is now used in this step to calculate the total exemption amount. If the adjusted gross income (line 33) is $90,900 or less, the total number of exemptions claimed on line 6d is multiplied by $2,700 to determine the total exemption amount. Example 13 illustrates how to calculate taxable income for the Reynolds and uses financial information from the examples presented in steps 1 thru 3.

Example 13

Jack and Mary Reynolds's adjusted gross income is $81,625 and they are claiming four personal exemptions on their joint income tax return. The Schedule A—Itemized Deductions lists the following: state and local income taxes, $1,460; real estate taxes, $2,875; home mortgage interest, $2,245; gifts to charity, $650; and other miscellaneous deductions, $150. Determine the Reynolds's taxable income.

Solution

a. Determine which type of deduction (itemized or standard) should be used to find taxable income.

total itemized = $1,460 + $2,875 + $2,245 + $650 + $150

= $7,380

standard deduction = $7,100 (married filing jointly, Table 17.3)

b. Determine total exemption amount.

total exemption = $2,700 × 4

= 10,800

c. Determine taxable income.

$$\text{taxable income} = \text{adjusted gross} - \underset{\text{(deductions}}{\text{total}} + \underset{\text{exemptions)}}{\text{total}}$$

= $81,625 − ($7,380 + $10,800)

= $81,625 − $18,180

= $63,445

Now that we know the taxable income amount, we can calculate the tax. The tax can be determined using tax tables that are based on taxable income amounts and filing status. When taxable income is less than $100,000, tax tables such as Table 17.4 can be used. If taxable income is greater than $100,000, tax rate schedules (Table 17.5) are required to calculate the tax. The tax rates shown in these tables are the rates that were in effect at the time this text was published. Although the rates and amounts shown are subject to change, the procedures to calculate the tax have remained generally unchanged.

Example 14

Determine Jack and Mary Reynolds's income tax based on the taxable income determined in Example 13 of $63,445, using the appropriate tax rate table.

Solution

Taxable income is $63,445 and the Reynolds are filing "married filing jointly." The tax is calculated using Table 17.4 because taxable income is less than $100,000. The tax is found in "taxable income heading $63,445" at the intersection of "married filing jointly" and taxable income of "at least" $63,400 "but less than" $63,450. The Reynold's tax is $12,254, and is reported on line 40 of Form 1040.

If the tax were calculated using Table 17.5, Schedule Y-1, the tax would be determined as follows:

tax on $42,350 ($42,350 × .15)	$6,352.50
plus .28 (63,445 − 42,350)	5,906.60
total tax	$12,259.10

Notice that the amount of the tax is not the same for this example because the taxable income $63,445 is above the midpoint in the taxable income interval ($63,400–63,450). If the taxable income were above or below $63,445, the tax using Table 17.4 would be the same, but the tax using Table 17.5 would be higher or lower. Therefore, you should use the correct table to calculate the tax. Also, the tax regulations require taxpayers to use Table 17.5 if taxable income is $100,000 or more.

Step 5: Determine the Tax Due or the Refund Due to the Taxpayer. The tax amount determined in step 4 can be increased by "other taxes" and decreased by "credits" as defined by tax regulations. However, most individuals filing a Form 1040 report the same amount of tax on line

Table 17.4

Federal income tax table

Tax Table—Continued

If line 39 (taxable income) is— At least	But less than	Single	Married filing jointly *	Married filing separately	Head of a household
59,000					
59,000	59,050	13,232	11,022	14,011	12,114
59,050	59,100	13,246	11,036	14,026	12,128
59,100	59,150	13,260	11,050	14,042	12,142
59,150	59,200	13,274	11,064	14,057	12,156
59,200	59,250	13,288	11,078	14,073	12,170
59,250	59,300	13,302	11,092	14,088	12,184
59,300	59,350	13,316	11,106	14,104	12,198
59,350	59,400	13,330	11,120	14,119	12,212
59,400	59,450	13,344	11,134	14,135	12,226
59,450	59,500	13,358	11,148	14,150	12,240
59,500	59,550	13,372	11,162	14,166	12,254
59,550	59,600	13,386	11,176	14,181	12,268
59,600	59,650	13,400	11,190	14,197	12,282
59,650	59,700	13,414	11,204	14,212	12,296
59,700	59,750	13,428	11,218	14,228	12,310
59,750	59,800	13,442	11,232	14,243	12,324
59,800	59,850	13,456	11,246	14,259	12,338
59,850	59,900	13,470	11,260	14,274	12,352
59,900	59,950	13,484	11,274	14,290	12,366
59,950	60,000	13,498	11,288	14,305	12,380
60,000					
60,000	60,050	13,512	11,302	14,321	12,394
60,050	60,100	13,526	11,316	14,336	12,408
60,100	60,150	13,540	11,330	14,352	12,422
60,150	60,200	13,554	11,344	14,367	12,436
60,200	60,250	13,568	11,358	14,383	12,450
60,250	60,300	13,582	11,372	14,398	12,464
60,300	60,350	13,596	11,386	14,414	12,478
60,350	60,400	13,610	11,400	14,429	12,492
60,400	60,450	13,624	11,414	14,445	12,506
60,450	60,500	13,638	11,428	14,460	12,520
60,500	60,550	13,652	11,442	14,476	12,534
60,550	60,600	13,666	11,456	14,491	12,548
60,600	60,650	13,680	11,470	14,507	12,562
60,650	60,700	13,694	11,484	14,522	12,576
60,700	60,750	13,708	11,498	14,538	12,590
60,750	60,800	13,722	11,512	14,553	12,604
60,800	60,850	13,736	11,526	14,569	12,618
60,850	60,900	13,750	11,540	14,584	12,632
60,900	60,950	13,764	11,554	14,600	12,646
60,950	61,000	13,778	11,568	14,615	12,660
61,000					
61,000	61,050	13,792	11,582	14,631	12,674
61,050	61,100	13,806	11,596	14,646	12,688
61,100	61,150	13,820	11,610	14,662	12,702
61,150	61,200	13,834	11,624	14,677	12,716
61,200	61,250	13,848	11,638	14,693	12,730
61,250	61,300	13,862	11,652	14,708	12,744
61,300	61,350	13,876	11,666	14,724	12,758
61,350	61,400	13,890	11,680	14,739	12,772
61,400	61,450	13,904	11,694	14,755	12,786
61,450	61,500	13,920	11,708	14,770	12,800
61,500	61,550	13,935	11,722	14,786	12,814
61,550	61,600	13,951	11,736	14,801	12,828
61,600	61,650	13,966	11,750	14,817	12,842
61,650	61,700	13,982	11,764	14,832	12,856
61,700	61,750	13,997	11,778	14,848	12,870
61,750	61,800	14,013	11,792	14,863	12,884
61,800	61,850	14,028	11,806	14,879	12,898
61,850	61,900	14,044	11,820	14,894	12,912
61,900	61,950	14,059	11,834	14,910	12,926
61,950	62,000	14,075	11,848	14,925	12,940

If line 39 (taxable income) is— At least	But less than	Single	Married filing jointly *	Married filing separately	Head of a household
62,000					
62,000	62,050	14,090	11,862	14,941	12,954
62,050	62,100	14,106	11,876	14,956	12,968
62,100	62,150	14,121	11,890	14,972	12,982
62,150	62,200	14,137	11,904	14,987	12,996
62,200	62,250	14,152	11,918	15,003	13,010
62,250	62,300	14,168	11,932	15,018	13,024
62,300	62,350	14,183	11,946	15,034	13,038
62,350	62,400	14,199	11,960	15,049	13,052
62,400	62,450	14,214	11,974	15,065	13,066
62,450	62,500	14,230	11,988	15,080	13,080
62,500	62,550	14,245	12,002	15,096	13,094
62,550	62,600	14,261	12,016	15,111	13,108
62,600	62,650	14,276	12,030	15,127	13,122
62,650	62,700	14,292	12,044	15,142	13,136
62,700	62,750	14,307	12,058	15,158	13,150
62,750	62,800	14,323	12,072	15,173	13,164
62,800	62,850	14,338	12,086	15,189	13,178
62,850	62,900	14,354	12,100	15,204	13,192
62,900	62,950	14,369	12,114	15,220	13,206
62,950	63,000	14,385	12,128	15,235	13,220
63,000					
63,000	63,050	14,400	12,142	15,251	13,234
63,050	63,100	14,416	12,156	15,266	13,248
63,100	63,150	14,431	12,170	15,282	13,262
63,150	63,200	14,447	12,184	15,297	13,276
63,200	63,250	14,462	12,198	15,313	13,290
63,250	63,300	14,478	12,212	15,328	13,304
63,300	63,350	14,493	12,226	15,344	13,318
63,350	63,400	14,509	12,240	15,359	13,332
63,400	63,450	14,524	12,254	15,375	13,346
63,450	63,500	14,540	12,268	15,390	13,360
63,500	63,550	14,555	12,282	15,406	13,374
63,550	63,600	14,571	12,296	15,421	13,388
63,600	63,650	14,586	12,310	15,437	13,402
63,650	63,700	14,602	12,324	15,452	13,416
63,700	63,750	14,617	12,338	15,468	13,430
63,750	63,800	14,633	12,352	15,483	13,444
63,800	63,850	14,648	12,366	15,499	13,458
63,850	63,900	14,664	12,380	15,514	13,472
63,900	63,950	14,679	12,394	15,530	13,486
63,950	64,000	14,695	12,408	15,545	13,500
64,000					
64,000	64,050	14,710	12,422	15,561	13,514
64,050	64,100	14,726	12,436	15,576	13,528
64,100	64,150	14,741	12,450	15,592	13,542
64,150	64,200	14,757	12,464	15,607	13,556
64,200	64,250	14,772	12,478	15,623	13,570
64,250	64,300	14,788	12,492	15,638	13,584
64,300	64,350	14,803	12,506	15,654	13,598
64,350	64,400	14,819	12,520	15,669	13,612
64,400	64,450	14,834	12,534	15,685	13,626
64,450	64,500	14,850	12,548	15,700	13,640
64,500	64,550	14,865	12,562	15,716	13,654
64,550	64,600	14,881	12,576	15,731	13,668
64,600	64,650	14,896	12,590	15,747	13,682
64,650	64,700	14,912	12,604	15,762	13,696
64,700	64,750	14,927	12,618	15,778	13,710
64,750	64,800	14,943	12,632	15,793	13,724
64,800	64,850	14,958	12,646	15,809	13,738
64,850	64,900	14,974	12,660	15,824	13,752
64,900	64,950	14,989	12,674	15,840	13,766
64,950	65,000	15,005	12,688	15,855	13,780

If line 39 (taxable income) is— At least	But less than	Single	Married filing jointly *	Married filing separately	Head of a household
65,000					
65,000	65,050	15,020	12,702	15,871	13,794
65,050	65,100	15,036	12,716	15,886	13,808
65,100	65,150	15,051	12,730	15,902	13,822
65,150	65,200	15,067	12,744	15,917	13,836
65,200	65,250	15,082	12,758	15,933	13,850
65,250	65,300	15,098	12,772	15,948	13,864
65,300	65,350	15,113	12,786	15,964	13,878
65,350	65,400	15,129	12,800	15,979	13,892
65,400	65,450	15,144	12,814	15,995	13,906
65,450	65,500	15,160	12,828	16,010	13,920
65,500	65,550	15,175	12,842	16,026	13,934
65,550	65,600	15,191	12,856	16,041	13,948
65,600	65,650	15,206	12,870	16,057	13,962
65,650	65,700	15,222	12,884	16,072	13,976
65,700	65,750	15,237	12,898	16,088	13,990
65,750	65,800	15,253	12,912	16,103	14,004
65,800	65,850	15,268	12,926	16,119	14,018
65,850	65,900	15,284	12,940	16,134	14,032
65,900	65,950	15,299	12,954	16,150	14,046
65,950	66,000	15,315	12,968	16,165	14,060
66,000					
66,000	66,050	15,330	12,982	16,181	14,074
66,050	66,100	15,346	12,996	16,196	14,088
66,100	66,150	15,361	13,010	16,212	14,102
66,150	66,200	15,377	13,024	16,227	14,116
66,200	66,250	15,392	13,038	16,243	14,130
66,250	66,300	15,408	13,052	16,258	14,144
66,300	66,350	15,423	13,066	16,274	14,158
66,350	66,400	15,439	13,080	16,289	14,172
66,400	66,450	15,454	13,094	16,305	14,186
66,450	66,500	15,470	13,108	16,320	14,200
66,500	66,550	15,485	13,122	16,336	14,214
66,550	66,600	15,501	13,136	16,351	14,228
66,600	66,650	15,516	13,150	16,367	14,242
66,650	66,700	15,532	13,164	16,382	14,256
66,700	66,750	15,547	13,178	16,398	14,270
66,750	66,800	15,563	13,192	16,413	14,284
66,800	66,850	15,578	13,206	16,429	14,298
66,850	66,900	15,594	13,220	16,444	14,312
66,900	66,950	15,609	13,234	16,460	14,326
66,950	67,000	15,625	13,248	16,475	14,340
67,000					
67,000	67,050	15,640	13,262	16,491	14,354
67,050	67,100	15,656	13,276	16,506	14,368
67,100	67,150	15,671	13,290	16,522	14,382
67,150	67,200	15,687	13,304	16,537	14,396
67,200	67,250	15,702	13,318	16,553	14,410
67,250	67,300	15,718	13,332	16,568	14,424
67,300	67,350	15,733	13,346	16,584	14,438
67,350	67,400	15,749	13,360	16,599	14,452
67,400	67,450	15,764	13,374	16,615	14,466
67,450	67,500	15,780	13,388	16,630	14,480
67,500	67,550	15,795	13,402	16,646	14,494
67,550	67,600	15,811	13,416	16,661	14,508
67,600	67,650	15,826	13,430	16,677	14,522
67,650	67,700	15,842	13,444	16,692	14,536
67,700	67,750	15,857	13,458	16,708	14,550
67,750	67,800	15,873	13,472	16,723	14,564
67,800	67,850	15,888	13,486	16,739	14,578
67,850	67,900	15,904	13,500	16,754	14,592
67,900	67,950	15,919	13,514	16,770	14,606
67,950	68,000	15,935	13,528	16,785	14,620

* This column must also be used by a qualifying widow(er).

Continued on next page

56, "total tax" that they reported on line 40. To determine if the taxpayer owes the government any tax, or if the government must refund prepaid taxes, the taxpayer's W-2 forms must be analyzed to find the total tax withheld by employers. This procedure was discussed in detail in Chapter 16.

Table 17.5

Federal income tax schedules

Schedule X—Use if your filing status is Single

If the amount on Form 1040, line 39, is: Over—	But not over—	Enter on Form 1040, line 40	of the amount over—
$0	$25,350	.... 15%	$0
25,350	61,400	$3,802.50 + 28%	25,350
61,400	128,100	13,896.50 + 31%	61,400
128,100	278,450	34,573.50 + 36%	128,100
278,450	....	88,699.50 + 39.6%	278,450

Schedule Y-1—Use if your filing status is Married filing jointly or Qualifying widow(er)

If the amount on Form 1040, line 39, is: Over—	But not over—	Enter on Form 1040, line 40	of the amount over—
$0	$42,350	.... 15%	$0
42,350	102,300	$6,352.50 + 28%	42,350
102,300	155,950	23,138.50 + 31%	102,300
155,950	278,450	39,770.00 + 36%	155,950
278,450	....	83,870.00 + 39.6%	278,450

Schedule Y-2—Use if your filing status is Married filing separately

If the amount on Form 1040, line 39, is: Over—	But not over—	Enter on Form 1040, line 40	of the amount over—
$0	$21,175	.... 15%	$0
21,175	51,150	$3,176.25 + 28%	21,175
51,150	77,975	11,569.25 + 31%	51,150
77,975	139,225	19,885.00 + 36%	77,975
139,225	....	41,935.00 + 39.6%	139,225

Schedule Z—Use if your filing status is Head of household

If the amount on Form 1040, line 39, is: Over—	But not over—	Enter on Form 1040, line 40	of the amount over—
$0	$33,950	.... 15%	$0
33,950	87,700	$5,092.50 + 28%	33,950
87,700	142,000	20,142.50 + 31%	87,700
142,000	278,450	36,975.50 + 36%	142,000
278,450	....	86,097.50 + 39.6%	278,450

When the total amount of Federal Withholding Tax withheld is greater than the total tax, the government will refund the difference to the taxpayer. However, if the amount of the withholding tax withheld is less than the total tax, the taxpayer must pay the government the difference between the two amounts. Example 15 illustrates the last step in completing the Reynold's individual income tax return, Form 1040.

For Your Information

FEDERAL TAXES: SOURCES AND OUTLAYS

Approximately 33 cents out of every $1.00 spent by American families is for some type of federal tax. This is nearly twice the amount spent on the next biggest item, housing, at 16 cents. In 1997, the federal government collected $1,453 billion in federal taxes and spent $1,560 billion, decreasing the deficit to $107 billion. The budget deficit is financed largely by government borrowing from the public. The government borrows from the public by selling bonds and other debt securities to private citizens, banks, businesses, and other governments.

The pie charts show how base, rate, and part are used by the federal government to illustrate the relative sizes of the major categories of federal tax sources and outlays.

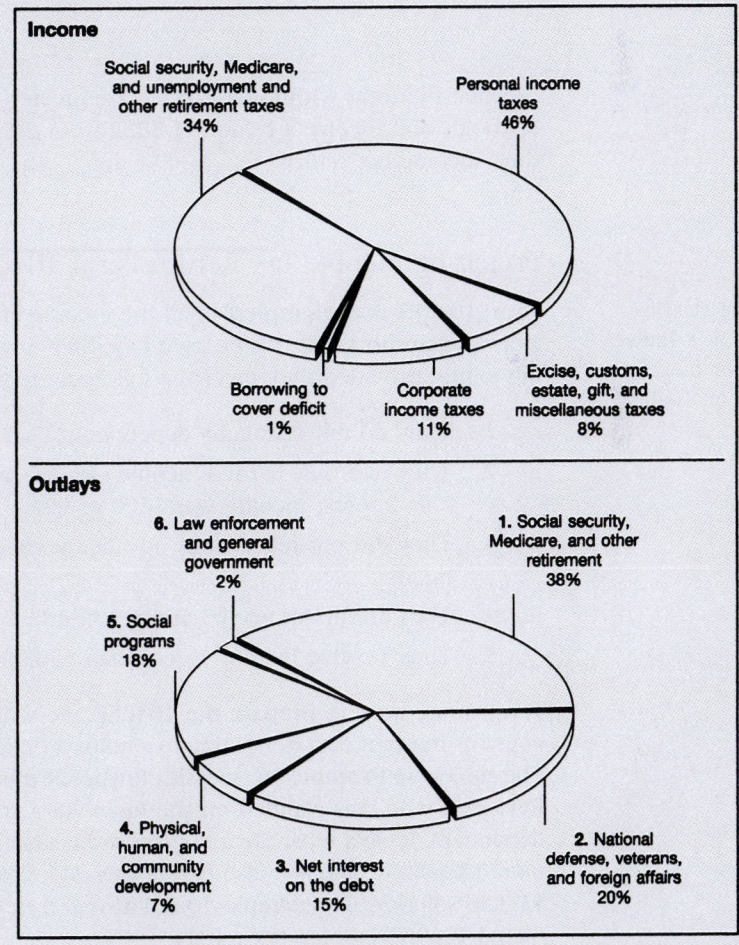

Income

Social security, Medicare, and unemployment and other retirement taxes
34%

Personal income taxes
46%

Borrowing to cover deficit
1%

Corporate income taxes
11%

Excise, customs, estate, gift, and miscellaneous taxes
8%

Outlays

6. Law enforcement and general government
2%

1. Social security, Medicare, and other retirement
38%

5. Social programs
18%

4. Physical, human, and community development
7%

3. Net interest on the debt
15%

2. National defense, veterans, and foreign affairs
20%

Source: Tax Foundation, I.R.S. Form 1040 Instruction booklet (U.S. Government Printing Office).

725

Example 15

Jack Reynolds's W-2 shows that his employer withheld $7,840 in federal with-holding taxes from his salary. Mary Reynolds's employer reported with-holding $5,276 from her salary during the year. Determine the amount of income tax or the amount of refund the Reynolds will pay or receive for the tax period using the information provided in Examples 10 through 14.

Solution

Determine the difference between the tax due and total taxes withheld as reported on the W-2s.

$13,116	($7,840 + $5,276) total taxes withheld, less
12,245	total tax due (Example 14)
$862	refund due

Since the total withholding taxes are greater than the total tax due, the Reynolds will receive a refund of $862 from the government after they file their income tax return.

Preparing Income Tax Return Form 1040EZ

Learning objective
Prepare U.S. Individual Income Tax Return Form 1040EZ.

Form 1040EZ is the simplest of all the income tax returns to complete. It is a one-page report that can be used by either single or married taxpayers filing jointly, provided they meet the following requirements:

1. They do not claim any dependents.
2. They had only earned income (wages, salaries, tips, etc.) and taxable interest income was $400 or less.
3. They did not receive any advance earned income credit payments.
4. They are under age 65 and not blind.
5. Their taxable income is less than $50,000.

To illustrate how to prepare the 1040EZ, we will develop an example that contains relevant data pertaining to a taxpayer eligible to use this tax report. The subject in Example 16 is Jessica Reynolds, the daughter of Jack and Mary Reynolds, who was claimed on the Reynolds's tax return (Form 1040) as a dependent. Jessica must file a tax return because she has earned income of more than $650, and she also had unearned income (interest) of more than $1. The solution for Example 16 is illustrated in Figure 17.7, Jessica's completed 1040EZ income tax report.

Figure 17.7

Individual income
tax Form 1040EZ

Department of the Treasury—Internal Revenue Service

Form 1040EZ **Income Tax Return for Single and Joint Filers With No Dependents** (P) **1998** OMB No. 1545-0675

Use the IRS label here

Your first name and initial: JESSICA A. Last name: REYNOLDS

Your social security number: 111 22 3333

If a joint return, spouse's first name and initial / Last name

Spouse's social security number

Home address (number and street). If you have a P.O. box, see page 7. Apt. no.: 327 ELM STREET

City, town or post office, state, and ZIP code. If you have a foreign address, see page 7.: AUBURN, N.Y. 13509

Presidential Election Campaign (See page 7.)

Note: Checking "Yes" will not change your tax or reduce your refund.
Do you want $3 to go to this fund? ▶ Yes ☐ No ☒
If a joint return, does your spouse want $3 to go to this fund? ▶ Yes ☐ No ☐

▲ **IMPORTANT!** ▲ You must enter your SSN(s) above.

		Dollars	Cents

Income

Attach Copy B of Form(s) W-2 here. Enclose, but do not staple, any payment.

1 Total wages, salaries, and tips. This should be shown in box 1 of your W-2 form(s). Attach your W-2 form(s). — 1 — 2,486

2 Taxable interest income. If the total is over $400, you cannot use Form 1040EZ. — 2 — 105

3 Unemployment compensation (see page 8). — 3

4 Add lines 1, 2, and 3. This is your **adjusted gross income**. If under $10,030, see page 9 to find out if you can claim the earned income credit on line 8a. — 4 — 2,591

Note: You must check Yes or No.

5 Can your parents (or someone else) claim you on their return?
Yes. ☒ Enter amount from worksheet on back.
No. ☐ If single, enter 6,950.00. If married, enter 12,500.00. See back for explanation. — 5 — 4,000

6 Subtract line 5 from line 4. If line 5 is larger than line 4, enter 0. This is your **taxable income**. ▶ 6 — 0

Payments and tax

7 Enter your Federal income tax withheld from box 2 of your W-2 form(s). — 7 — 286

8a **Earned income credit** (see page 9).
b Nontaxable earned income: enter type and amount below.
Type: ____ $ ____ — 8a

9 Add lines 7 and 8a. These are your **total payments**. — 9 — 286

10 **Tax.** Use the amount on **line 6** above to find your tax in the tax table on pages 20–24 of the booklet. Then, enter the tax from the table on this line. — 10 — 0

Refund

Have it directly deposited! See page 12 and fill in 11b, 11c, and 11d.

11a If line 9 is larger than line 10, subtract line 10 from line 9. This is your **refund**. — 11a — 286

b Routing number
c Type: ☐ Checking ☐ Savings
d Account number

Amount you owe

12 If line 10 is larger than line 9, subtract line 9 from line 10. This is the **amount you owe**. See page 14 for details on how to pay. — 12

I have read this return. Under penalties of perjury, I declare that to the best of my knowledge and belief, the return is true, correct, and accurately lists all amounts and sources of income I received during the tax year.

Sign here ▶
Keep copy for your records.

Your signature: Jessica Reynolds
Date: 3/10
Your occupation: Student
Spouse's signature if joint return. See page 7.
Date
Spouse's occupation

For Official Use Only

For Disclosure, Privacy Act, and Paperwork Reduction Act Notice, see page 18. Cat. No. 12616G 1998 Form 1040EZ

Example 16

Jessica Reynolds's W-2 report indicates that she received $2,486 in wages as a cashier from the Recordbook. Her W-2 also shows that $286 had been withheld in federal withholding taxes. First Federal Savings Bank sent Jessica a

Form 1099, as she had earned $105 in interest on funds deposited in her savings account. Complete Jessica's income tax return, Form 1040EZ.

Solution

Step 1:   Determine adjusted gross income (lines 1–4) and enter amount on line 4.

$$\text{adjusted gross income} = \text{total wages} + \text{taxable interest}$$
$$= \$2{,}486 + \$105$$
$$= \$2{,}591$$

Step 2:   Calculate taxable income (line 5 and 6) and enter amount on line 6.

$$\text{taxable income} = \text{adjusted gross income} - \text{standard deduction}$$
$$= \$2{,}591 - \$4{,}000$$
$$= \$0$$

Because Jessica's parents claimed her as a dependent on their tax return, Jessica must check the "yes" box on line 5 and take the standard deduction of $4,000. She is not entitled to an exemption, as her parents claimed her exemption. Her taxable income is $0 because her adjusted gross income is less than the standard deduction. If her adjusted gross income was greater than the standard deduction, she would have to pay taxes on the difference.

Step 3:   Calculate the tax amount based on taxable income and enter on line 10.

Since the taxable income (line 6) is $0, there is no tax liability. If the amount on line 6 is greater than $0, the tax tables would be used to determine the tax based on taxable income and filing status.

Step 4:   Determine the refund due (lines 7–10) and enter amount on line 11a.

$$\text{refund due or} = \text{total federal tax withheld} - \text{total tax}$$
$$\text{tax owed} = \$286 - \$0$$
$$= \$286 \text{ refund due}$$

Jessica would receive the entire amount of federal withholding tax reported on her W-2 as a refund, as she had no tax liability. In other words, her total payments $286 (line 7) are greater than her tax, $0 (line 10). Therefore, she will enter on line 11a the difference ($286). See Figure 17.7 for the entries described in this solution.

CHECK YOUR KNOWLEDGE

Personal Income Tax

1. Gary and Kristen Bradley had a combined income of $87,530. They received $1,200 in dividends from stock investments and $372 in interest from certificates of deposit. If the Bradleys deposited $3,000 in their IRAs, what is the adjusted gross income reported on their tax return?

2. The Bradleys (problem 1) file a joint return and claim 5 exemptions. Calculate the Bradleys taxable income, if their total itemized deductions are $10,640.

3. Determine the Bradleys income tax based on information provided in problems 1 and 2 using the appropriate tax table.

4. The Bradleys W-2 forms show that Gary had $10,647 and Kristen had $4,360 withheld from their wages for federal withholding taxes. Determine the amount of refund or the amount of tax owed for the Bradleys.

5. Mike Grabowski attends college and worked part-time this past year. His W-2 form shows that he had total wages of $4,380, and that $573 had been withheld from his wages for federal withholding tax. Mike is single, has no itemized deductions, and is claimed as a dependent by his parents. Calculate the tax liability or the refund due on Mike's income tax return, Form 1040EZ.

17.3 **EXERCISES**

Calculate the adjusted gross income.

	Total wages	Dividends earned	Interest earned	Adjustments to income	Adjusted gross income
1.	$10,250	$ 0	$ 147	$ 0	$_____
2.	35,600	590	240	845	_____
3.	65,932	1,264	785	2,000	_____
4.	120,475	3,870	1,090	6,349	_____
5.	82,358	1,036	674	4,152	_____

Answers to CYK: *1.* $86,102 refund due   *2.* $62,212   *3.* $12,067   *4.* $2,940 refund due   *5.* $515

Calculate the taxable income and the income tax for each of the following. Use Tables in Appendix D. Filing status abbreviations are as follows: S, single; MJ, married filing jointly; MS, married filing separately; HH, head of household.

	Name	Filing status	Exemptions	Adjusted gr. income	Deductions	Taxable income
6.	Falso	S	1	$12,500	$ 5,150	$_____
7.	Bryant	MJ	6	64,365	10,346	_____
8.	Lawson	HH	2	48,872	8,026	_____
9.	Pilcher	MS	1	24,398	4,250	_____
10.	Squadrito	MJ	3	92,134	14,649	_____

Calculate the missing amounts using Tables in Appendix D. Filing status abbreviations are the same as those given for exercises 6–10.

	Name	Filing status	Taxable income	Income tax	Tax withheld	Refund due	Tax owed
11.	Lyons	S	$32,640	_____	$ 5,274	$_____	$_____
12.	Kleins	MJ	78,293	_____	17,367	_____	_____
13.	Abbott	MS	18,765	_____	2,492	_____	_____
14.	McCabe	HH	46,387	_____	9,126	_____	_____
15.	Welch	S	53,460	_____	10,810	_____	_____

16. Peggy Walton's total wages are $27,490. She received $268 in dividends and $124 in interest from investments during the year. If she deposited $1,000 in an IRA, what is her adjusted gross income?

17. The Sheldons filed a joint tax return on combined income of $74,630, and claimed four exemptions. If they take the standard deduction and have adjustments to income of $2,250, what is the amount of their taxable income?

18. Using the information in exercise 17, determine the amount of tax owed or refund due if the Sheldons had $11,216 withheld from their wages in federal withholding tax.

19. Calculate the refund due or the tax owed on taxable income of $105,240 if the taxpayer's filing status is "married filing jointly," and total taxes withheld is $23,469.

20. Bill and Kathy Matthew's adjusted gross income is $64,930. They file a joint tax return and claim five exemptions. If their itemized deductions are state and local taxes, $9,215; mortgage interest,

$4,875; gifts to charity, $925; and miscellaneous deductions, $270, what is the amount of their income tax liability?

21. Judy Dunn is single, lives separate from her parents, and earned $32,170 as a medical records technician this year. She received interest on investments of $380 and had taxes withheld of $4,762 from her earnings. Determine the amount of refund due or tax owed for Judy if she files her income tax on Form 1040 EZ.

22. Tom Wilkenson has an adjusted gross income of $48,295, files as a head of household, and claims two exemptions. His W-2 form reports $5,184 was withheld from his wages for income taxes. If he takes the standard deduction, what is the amount of refund due or tax owed?

23. Kim Clawson is single and has an adjusted gross income of $69,250. She has deductions of $12,140 for taxes paid, $3,609 in mortgage interest, and $1,450 in charitable contributions. If she claims one exemption on her income tax report Form 1040, what is the amount of her tax?

24. Hong Nguyen is a full-time college student, is single, and lives with her parents. Hong works part time and had adjusted gross income last year of $3,692. Find the tax amount to be reported on Form 1040EZ, if Hong's parents claim him as an exemption.

25. Jeff and Ellen Parks had combined earnings of $42,305. They file a joint return claiming two exemptions and the standard deduction. If they had adjustments to income of $1,000, what is their income tax?

CHAPTER REVIEW EXERCISES

The following problems review all mathematics and business concepts presented in the chapter. If you experience any difficulty solving a problem, go to the appropriate chapter section and review the examples provided. Express all answers as decimals rounded to tenths, or dollars rounded to cents, unless otherwise instructed.

1. Merit Electrical Supply marked down all overhead lighting fixtures 25%. If a customer purchases a fixture with a marked price of $249.99 and the sales tax on the sale is 7%, how much must the sales clerk charge the customer?

2. A retailer purchased supplies from a wholesaler for $1,520. The wholesaler offers the retailer a trade discount of 10/5/10 and terms of 3/10, $n/30$. If the wholesaler is required to collect a sales tax on the sale of 4% and the payment was remitted during the discount period, how much did the retailer pay for the supplies?

3. What was the marked price of a television set that was purchased by a customer for a total price of $418.95, which included a sales tax of 5%?

4. Dean Farley received his monthly telephone bill, which stated that he owed the telephone company $28.72. Is the amount of the bill correct if his standard monthly charge is $10.50, his long distance charges were $15.85, and he is charged an excise tax of 3% and a sales tax of 6%?

5. Cody Mitchell purchased a firearm (revolver) to be used in competition as a member of the Rod and Gun Club for $295. How much will the firearm cost if he has to pay a 10% federal excise tax and a 7% sales tax?

6. Nelson's Travel Center collected $4,260 in state excise taxes from gasoline sales in July. If gasoline prices for the entire month remained at $1.259 per gallon and the excise tax was $.08 per gallon, what were (a) the number of gallons sold during the month, and (b) the total revenue from sales of gasoline for the period?

7. Geraldo County's taxable property has a fair market value of $600,000,000 and needs to realize revenues of $9,000,000 from property taxes. The assessment rate for Geraldo County is 20% of fair market value. Determine (a) the assessed value of the taxable property in Geraldo county, (b) the tax rate of Geraldo County expressed in percentage, (c) the tax rate expressed in dollars per hundred dollars of assessed value, (d) the tax rate expressed in dollars per thousand dollars of assessed value, and (e) the tax rate expressed in mills.

8. Sydney Corcoran owns a home in Geraldo county which has a fair market value of $80,000. Using the information in exercise 7, calculate (a) the property tax assessment on Sydney's property, and (b) the amount of Sydney's property tax.

9. Lucy and Bill Lewbowski filed a joint tax return on combined income of $84,965 and claim six exemptions. If their itemized deductions are $15,408 and they list income adjustments of $3,800, what is the amount of their taxable income?

10. Using the information in exercise 9, determine the amount of tax owed or refund due if the Lewbowskis had $12,114 withheld from their wages in federal withholding tax.

11. Jennifer Andrews has adjusted gross income of $28,500 and files as a single person claiming one exemption. If she had $3,749 withheld in withholding taxes and takes the standard deduction, what is the amount of her tax owed or refund due?

12. Phil Moss is attending college full time and lives in a dormitory on campus. This past year he had adjusted gross income of $4,950, and $428 was withheld from his earnings for taxes. Find the amount of tax owed or refund due to be reported on Form 1040EZ if Phil's parents claim him as an exemption.

13. The Wilsons have taxable income of $162,840. They file "married filing jointly" and had $38,632 withheld in withholding taxes from their wages during the year. Determine their refund due or tax owed.

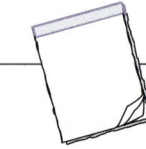 **EXPRESS YOUR UNDERSTANDING**

Compose one or two well-written sentences to express the requested information in your own words.

1. Explain how sales tax is applied to a purchase of merchandise when both a trade discount and cash discount are offered to the buyer.

2. Describe how to determine the total price of a sale when the seller is required to charge both a federal excise tax and a state sales tax.

3. How would you determine the marked price of an item if the total price of the sale ($15.50) included a 6% sales tax?

4. Explain what fair market value is for a parcel of property, and how the assessed value of the property is determined using the fair market value.

5. Describe how the property tax rate is calculated.

6. Explain how the actual property tax for a specific parcel of property is determined.

7. How is adjusted gross income determined for income tax purposes?

8. Explain how deductions and exemptions are used to determine income tax.

9. Describe the procedure (steps) to determine taxable income.

10. Explain how a taxpayer would determine the "refund due" or "tax owed" on taxable income.

 *Mind Your Business*

As a business, Media World has tax obligations that include payroll taxes and sales taxes. At this time, the business is not levied a property tax because it does not own the building within which the business is located. The owner of the building is responsible for property taxes and has prorated the taxes to all tenants as part of their monthly rent. Media World's profits are also not taxable as the business has been organized as a partnership. Consequently, each partner is responsible for the payment of individual income taxes based on their personal financial circumstances. For this reason, the company files reports that identify the distribution of profits to the partners. Media World's largest tax obligation is sales tax. The company must report sales tax information to the state and county and submit payment in accordance with each government level's requirements. Kate Bradley prepares the tax reports and submits them to you for your evaluation and authorization of payment.

Activity 1

On the 12th of each month, Kate Bradley submits the county sales tax report and check in full payment of the previous month's tax liability to the County Sales Tax Bureau. The county sales tax rate is 2.75% of net monthly sales. Kate has requested your signature on check No. 1579 for $1,750 as payment of August's county sales tax. What amount of August's gross sales did she use to determine the monthly sales tax expense?

Activity 2

During the month of August, Media World's records indicate that 2.6% of net sales were returned by customers for credit and that customers had been given sales discounts amounting to $5,092.50. What amount of gross sales was reported on the August income statement?

Activity 3

State sales taxes are reported and paid quarterly. The payment must be received by the State Tax Bureau by the 15th day of the month following the last day of the month of the quarter. To comply with this requirement, Kate Bradley has given you the State Quarterly Sales Tax Report listing a sales tax of 5.5%. Net sales for July were $66,145. Using the August and September net sales figures previously calculated in Mind Your Business activities, determine the amount of state sales tax reported on the quarterly report ending September 30.

Activity 4

Explain why the amount of sales tax payable reported on the September 30 Balance Sheet ("Mind Your Business," chapter 4) does not equal the amount of state sales tax reported on the quarterly sales tax report.

SELF-TEST

A. Terminology

Complete the following items using the key terms presented at the beginning of the chapter. Check your responses against the answer key at the end of the test.

1. A _____ is charged on the sale of goods and services sold by retail merchants in most states to cover the cost of governmental services.

2. An _____ is levied on the sale or manufacture of specific goods and services by the federal government.

3. Generally, _____ property is land or any buildings on the land.

4. In a district in which property taxes are to be levied, the person who determines the assessed value of individual parcels of land is called an _____.

5. The _____ value of real property is anywhere from 0% to 100% of fair market value of the property.

6. The _____ rate for a tax district is determined by dividing the revenue needed from property taxes by the total _____ value of the property in the community.

7. The actual tax to be paid by a property owner is determined by multiplying the _____ rate by the _____ value of the property.

8. When calculating individual income tax, the _____ is the difference between total income and total adjustments.

9. Medical expenses, state and local taxes, and mortgage interest are examples of _____ that reduce a taxpayer's _____.

10. _____ are allowances based on the filing status and number of dependents claimed by a taxpayer.

B. Calculation

The following concepts and short problems are designed to test your understanding of the objectives identified at the beginning of the chapter. Answers are provided at the end of the test.

11. A sales clerk sold merchandise with a marked price of $420. If the clerk must mark down the merchandise 25% and charge a sales tax of 4%, how much should the customer be charged?

12. Parsons Farms Convenient Mart collected $3,275 in state sales taxes this week. If the stores' state sales tax rate is 5%, what was the amount of taxable sales for the week?

13. Karen Elliott purchased a car for $18,250. She lives in a community where businesses must collect a state sales tax of 4%, a county sales tax of 3%, and a city tax of 1%. Find (a) the amount of the county tax and (b) the amount Karen paid for the car, including taxes.

14. Bradford Office Interiors purchased six conference tables with eight matching chairs at $1,250 per set. Bradford was allowed a trade discount of 12% and terms of 3/10, $n/30$. The company is required to pay a 5% sales tax. How much must be remitted if Bradford makes payment during the discount period?

15. Jed Clapton purchased a hunting rifle from Northwest Sport Equipment for $425. What was the total amount of the purchase if the store collected a 10% excise tax and a 7% sales tax on the sale?

16. Kristen Holden owns the land and building which houses her beauty salon in Jacksonville. The fair market value of her shop is $20,000. The city of Jacksonville's assessment rate is 40% of fair market value, and they have determined that the total fair market value of property in their small city is $150,000,000. The city budget for next year includes $2,100,000 from property taxes. Determine (a) the total assessed value of the property in Jacksonville, (b) the tax rate for the city, (c) the property tax assessment on Kristen's property, and (d) Kristen's property tax amount.

17. Harry Dickenson earned $18,630 last year. If he received dividends of $380 and deposited $500 in an IRA, what was his adjusted gross income?

18. Shannon Kelly's adjusted gross income is $36,425 and she claims one exemption as a single person. If she takes the standard deduction, what is the amount of her taxable income?

19. Use the information in exercise 12 to determine the amount of refund due or the tax owed if

Shannon had $4,582 in withholding tax deducted from her wages.

20. Find the refund due or the tax owed on taxable income of $124,637 if the taxpayers filing status is "married filing jointly" and total taxes withheld is $31,898.

18

Key Terms

risk
risk management
risk manager
actuary
insurance
life insurance
face value
beneficiary
premium
term life insurance
whole life insurance
limited pay life
   insurance
endowment life
   insurance
annuity life insurance
universal life insurance
cash surrender value
decreasing term life
   insurance
property insurance
fire insurance
coinsurance clause
automobile insurance
health insurance
business insurance

RISK MANAGEMENT

Learning Objectives

After completing the exercises in this chapter, you will be able to:

1. Calculate a premium for term, ordinary life, decreasing term, limited pay life, and endowment life insurance.

2. Calculate the cash surrender value of an insurance policy.

3. Calculate the premium for a fire insurance policy.

4. Calculate the amount of claim covered using the coinsurance clause.

5. Calculate liability insurance coverage of an automobile.

6. Define the key terms.

INTRODUCTION

We live in a world of *uncertainty*. As much as we might like to think that we have control of the world around us, that is only true in a very limited sense.

The mathematics of business does provide a mathematical model of *certainty*. Most business transactions have been conceived using pure mathematics. The concepts of simple interest, simple discounts, compound interest, and annuities were all conceived by applying algebra to business concepts in a very logical manner.

We know that if we invest $1,000 at 10% interest compounded quarterly for 8 years, it will have a value of $2,203.76 at the end of the 8-year term. This is one example of many considered in this text that clearly involve certainty.

While it is true that business, unlike the natural sciences, does provide us with a true example of the mathematics of certainty, many aspects of the business field are fraught with uncertainty. How do we deal with the uncertainties that surround us? Is there any way to add some order or control to these uncertainties? Is there an area of mathematics that helps us to deal with uncertainty? How do we confront, control, and cope with the risks that are ever present as a result of that uncertainty?

Risk

Personal Risk.   Many decisions we make every day involve **risk** due to uncertainty. The route we choose to travel in an automobile on any given day can result in an accident, or in delays that make us late for an appointment. Decisions we elect can result in damage to our self-image or the images others have of us.

Some decisions we make, however, directly affect our financial condition, both present and future. In this chapter, we will confine our concern for risk management to financial risks.

Business Risk.   Whenever we are involved in a business venture, the success or failure of that venture is measured in terms of financial gain or loss. Although, in business, we do encounter some degree of personal risk, we clearly regard risk in the business world to mean *risks of financial loss incurred by uncertainty*.

Definition of Risk.   We shall regard the definition of risk for both personal risk and business risk as *the uncertainty of financial loss*.* When we apply this definition to our personal lives, we can consider risks to our personal earning ability due to death, illness, accident, injury, or age. Other such risks might involve the loss or damage to property we own, such as our

---

*David Bicklehaupt, *General Insurance* (Homewood, IL: Richard D. Irwin, 1983).

homes or automobiles. We might also be held accountable (liable) for any damage or harm we cause to other people and their property.

When we apply the definition to our business lives, financial risks to our business can be caused by uncertainties such as loss of property or product due to accident or theft; loss of business income due to illness of employees; loss due to employee dishonesty; liability loss due to civil or criminal wrongs to the public or to individuals; or, simply, losses due to unexpected economic changes in our society.

Risk Management

Risk management can be defined as the attempt to minimize loss by (a) controlling risk, and (b) using insurance to compensate for any losses incurred due to accidents. Risk control in our personal lives might include monitoring our health to minimize loss of work time due to illness, injury, or death. We might drive defensively to decrease the chance of an auto accident. We also might monitor our property to decrease the risk of loss or damage due to fire, theft, or vandalism. Such measures might decrease the probability of the risks mentioned above, but they won't eliminate the risks entirely.

Our risk control efforts might be backed up by insurance to cover lost wages due to lost work time caused by illness or injury (disability insurance), or death (life insurance). We also protect our financial losses with health insurance to offset the costs of regaining our health. Other insurance includes auto (liability and collision) and property (fire, theft, and comprehensive).

Similar efforts at control and financial back-up are exercised by businesses. Large companies employ **risk managers** whose primary function is to investigate all aspects of a business with the goal of determining the existence of risk, recommending and implementing plans of action to control risk, and exploring and obtaining programs of economic back-up (such as insurance) to help maintain the fiscal stability to the business should any adversity occur due to uncertainty.

18.1   **LIFE INSURANCE**

The insurance industry is a business that provides financial protection to offset economic losses due to uncertainty. Just as probability is the area of mathematics that deals with uncertainty, insurance is the area of business that deals with uncertainty.

Recall from Chapter 3 that probability is the mathematics of uncertainty, and that probability is simply a ratio.

probability that an event will occur in an experiment =

$$\frac{\text{number of ways the event can occur}}{\text{total number of ways the experiment can occur}}$$

Since risk management is clearly involved with uncertainty, probability plays an important role in this component of doing business. We will demonstrate how probability is used with an example of how the life insurance business might predict the number of people it insures would die and therefore the amount of money it would have to pay out in benefits. By estimating the amount it would have to pay out in benefits, it can determine how much it must charge each insured person in premiums so that the company can cover its costs for those benefits.

Example 1

Assume that an insurance company sold life insurance policies with a face value of $1,000 to 100,000 twenty-year-olds in the USA during 1987. (a) How much would the insurance company have had to pay to the beneficiaries by the end of the year? (b) How much would the insurance company have had to collect from each of the insured in order to have collected enough money in premiums to cover the cost in benefits paid out?

Solution

The U.S. Department of Health and Human Services National Health Center for Statistics publishes *Vital Statistics of the United States* yearly to provide information on various aspects of the country's population.

a. Among those statistics are mortality tables, which indicate the number of people of a given age who survived (i.e., did not die) during that year. Those tables indicate that for every 98,122 people who lived to be 20 years old in 1987, 104 did not live to be 21 years old. So the proportion of 20-year-old persons who did not survive to be 21 was

proportion who died $= 104/98,122 = .00106$

Multiplying this by 100,000 tells us that the insurance company expected 106 of every 100,000 twenty-year-olds to die in 1987. Therefore, the amount of benefits that it would have expected to pay out would have been

Benefits paid per 100,000 policies $= 106 \times \$1,000 = \$106,000$

b. Each of the 100,000 people insured then would have had to pay $1.06 in premiums in order for the company to have been able to meet its obligations to the beneficiaries of those 106 insured people who died.

Premium per each $1,000 policy $= \$106,000/100,000$ policies $=$
$$\$1.06 \text{ per policy}$$

For every 100,000 twenty-year-olds who purchased a $1,000 life insurance policy in 1987, each would have had to pay $1.06 for the policy to cover the cost of benefits (claims) paid out to the families of the 106 who died in 1987.

Now projecting into the future, the company could use the same proportion as a *probability* to plan for the future costs to the company for claims on twenty-year-olds.

Probability that a twenty-year-old will die this year = 104/98,122 = .00106, and therefore the minimum premium the company must charge each insured for each $1,000 of life insurance is $1.06.

Now the company is using probability to predict what it expects will happen in the future (inferring what will happen), and making decisions accordingly. Insurance companies employ **actuaries,** people who perform the mathematics to make these inferences, to help the company make these decisions for all age groups whose life they will insure, and for all types of insurance that they offer.

The insurance company would also have to add additional charges to the premiums mentioned to help cover the cost of running its business, and to provide profit for the company.

total premium = costs to cover claims +
expenses of doing business + profit

As mentioned earlier, mathematicians known as actuaries work with this type of data and even more sophisticated mathematical principles to help insurance companies cope with *uncertainty* and make the best possible decisions to minimize risk and maximize profits while staying competitive in the open market.

Probability provides us with a means of assessing the likelihood that a person will die in a given year. Expressed differently *probability gives us a means of estimating how many people we expect will die in a group of people in a given year.* **Life insurance** provides us with a means of assessing and offsetting the economic losses due to the death of a person. Insurance in a broader sense provides economic protection in situations that cause economic loss such as loss of life; loss of health; loss of employment; loss of, or damage to, property due to fire, theft, or vandalism, and so on.

Probability combined with statistical data helps us to know what to expect in many situations of uncertainty. What is the likelihood of economic loss due to loss of health from an accident or illness? transportation losses, such as automobile or marine accidents? property losses, such as damage to our home, business, or other property?

Life insurance provides compensation for economic loss due to the death of the insured. The basic types of life insurance include term, decreasing term, ordinary (whole or straight) life, limited pay, endowment, and universal life. Each of these types of insurance has a **face value,** which is the amount of money paid to a *beneficiary* upon the death of the insured person. The **beneficiary** is the person designated to receive the insurance award (money). The **premium** is the amount of money paid to the insurance company to purchase the insurance.

- **Term life insurance** pays the face value of the policy to the beneficiary if the insured dies within the specified time period (term) of coverage. Term life insurance generally provides a higher amount of coverage per premium dollar than other forms of life insurance. On the other hand, term life insurance usually requires that the insured requalify for coverage at the end of each term.

- **Whole life insurance** provides protection for the insured's entire life, so long as the premiums are paid.

- **Limited pay life** provides protection for the insured's entire life, but is paid up with a limited number of premium payments.

- **Endowment life insurance** provides life insurance coverage that pays the face value of the policy upon the death of the insured, or pays an amount of money to the insured at the end of a specified period of time if the insured is still living at that time.

- **Annuity life insurance** provides life insurance that pays a specified amount upon the death of the insured, or if the insured lives past a specified age, pays a regular periodic payment to the insured for the remainder of his/her life (or for a predetermined time period).

- **Universal life insurance** is a flexible form of insurance that incorporates many of the features of the above policies and tailors them to the needs of the insurance customer. This form of insurance provides protection, savings, and annuity benefits that can be adapted to the changing needs of the insured.

Learning objective
Calculate a premium for term, ordinary life, decreasing term, limited pay life, and endowment life insurance.

Table 18.1 lists some typical premiums for life insurance with a $1,000 face value.

Table 18.1

Typical annual premiums for $1,000 face value life insurance

Age	5-year renewable	Whole life	20-year limited pay	20-year endowment
20	$1.10	$17.50	$27.30	$48.30
25	1.10	19.30	29.80	48.70
30	1.16	21.90	33.10	49.60
35	1.40	25.10	36.60	50.10
40	1.94	29.60	40.20	51.80
45	2.90	35.10	46.10	52.30
50	5.00	43.10	54.30	56.50

Example 2

Arnie Johnson is 25 years old and wishes to purchase $10,000 of life insurance. Determine the annual premium payments he would be required to pay for (a) 5-year renewable term life, (b) whole life, (c) 20-year limited pay life, and (d) 20-year endowment life.

Solution

Step 1:   First, we determine the number of $1,000 units of face value of life insurance Arnie wants to purchase:

$$\text{number of } \$1{,}000 \text{ units} = \frac{\text{face value}}{\$1{,}000} = \frac{\$10{,}000}{\$1{,}000} = 10$$

Step 2:   Next, we multiply the number of $1,000 units by the appropriate premium per $1,000 indicated in Table 18.1 for Arnie's age (25 years) and the type of insurance desired.

a.   5-year term premium = number of units × table value
 = 10 × table value of 5-year
   term and age 25
 = 10 × $1.10
 = $11.00

b.   whole life premium = number of units × table value
 = 10 × table value of
   whole life at age 25
 = 10 × $19.30 = $193.00

c.   20-year limited pay premium = number units
   × table value for age 25
 = 10 × $29.80
 = $298.00

d.   20-year endowment = number of units
   × table value for age 25
 = 10 × $48.70
 = $487.00

Learning objective
Calculate the cash surrender value of an insurance policy.

All forms of insurance just mentioned, except term insurance, have a **cash surrender value.** Cash surrender value is similar to a savings account. The insured's premium payments earn savings while providing insurance protection. If the insured cancels (surrenders) the policy after a certain time period (usually at least 3 years) money earned is paid to her or him. The amount earned and paid to the insured upon surrendering the policy is called the *cash surrender value.* Table 18.2 shows typical 20-year limited payment cash surrender values.

Table 18.2

Cash surrender
value for 20-year
limited payment
life

| Age | \multicolumn{7}{c}{End of year} |
	1	2	3	5	10	15	20
20		12	32	73	190	320	480
25		16	36	85	213	355	530
30		20	43	97	238	395	582
35		26	52	110	263	435	636
40		32	60	123	288	475	690
45		38	69	135	313	515	741

Example 3

Assuming Arnie Johnson purchased a $10,000 twenty-year limited pay life policy at age 25, what will be the cash surrender value of the policy at the end of (a) 5 years? (b) 10 years? (c) 20 years?

Solution

We calculate the cash surrender value in much the same way we calculated the premium cost.

Step 1: First we determine the number of $1,000 units of face value of insurance Arnie bought.

$$\text{number of units} = \frac{\text{face value of policy}}{\$1,000} = \frac{\$10,000}{\$1,000} = 10$$

Step 2: Next, we determine cash surrender value by multiplying the number of units by the table value corresponding to Arnie's age at the time he purchased the policy and the number of years for which Arnie has paid the premiums on the policy.

$$\text{cash surrender value} = \text{number of units} \times \frac{\text{table value for age}}{\text{and years paid}}$$

a. $\text{cash surrender value at end of 5 years} = 10 \times \$85 = \$850$

b. $\begin{aligned}\text{cash surrender value at end of 10 years} &= \frac{10 \times \text{table value for}}{25 \text{ years and } 10 \text{ years}} \\ &= 10 \times \$213 \\ &= \$2,130\end{aligned}$

c. $\begin{aligned}\text{cash surrender at end of 20 years} &= \frac{10 \times \text{table value for}}{25 \text{ years and } 20 \text{ years}} \\ &= 10 \times \$530 \\ &= 10 \times \$5,300\end{aligned}$

Table 18.3

20-year decreasing term life premium per $1,000 face value

Age	Premium
20	$ .90
25	1.00
30	1.15
35	1.50
40	2.25
45	3.60
50	6.00

Table 18.4

20-year decreasing term face value for $100,000

Year	Face value
1	$100,000
5	91,000
10	75,000
15	50,000
20	36,000

Notice that if Arnie pays his fixed premium of $298 each year for 20 years, he will have $10,000 worth of life insurance protection for each of those years. If he dies before the 20-year period is completed, his beneficiaries will receive $10,000. If he survives the 20-year period, he will have accumulated a savings of $5,300 in the policy.

Term life insurance provides protection *only*, without the savings benefit, but the premiums are much lower, which provides the same or greater protection at lower cost. Although term life insurance may cost less for the same amount of face value protection, the premiums do increase when the policy is renewed at the end of the specified term. **Decreasing term life insurance** is an alternate form of term insurance for which the premium remains the same throughout the life of the policy, but the face value decreases at the end of each specified term (see Tables 18.3 and 18.4).

Example 4

A 35-year-old male purchased $100,000 of decreasing term insurance for 20 years. Determine (a) his premium, (b) the face value of the policy at age 40, and (c) the face value at age 50.

Solution

a. The amount he would pay is a premium of 100 units × $1.50 per unit = $150 per year for each of the 20 years in the life of the policy.

For Your Information

HOW INFLATION AFFECTS INSURANCE VALUES

The graph shows the purchasing power of a $100,000 life insurance policy at successive 5-year intervals, assuming a 5% per year rate of inflation. The original $100,000 of purchasing power is reduced to $78,350 5 years later, $61,390 10

Policy Amount: $100,000
Assumed Annual inflation Rate: 5%

Purchasing Power After:

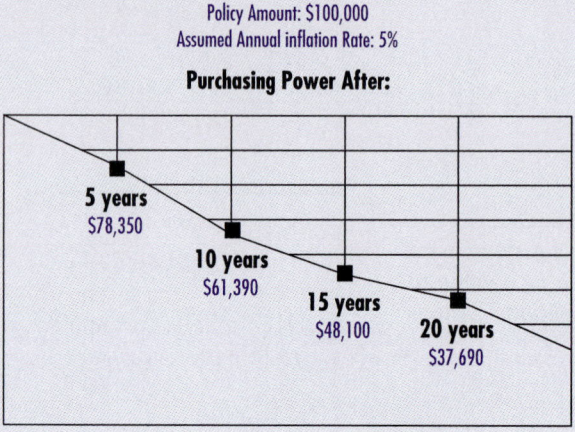

years later, and to less than half the original purchasing power after 15 years.

Insurance companies recommend that the face value of a life insurance policy be 6 to 10 times the insured's yearly income. This factor is designed to help offset inflation's effect on purchasing power.

Source: Teacher's Insurance and Annuity Association—College Retirement Equities Fund. Used with permission.

b.   At age 40 the face value of his coverage would be $91,000 (determined from Table 18.4 for year 5 since he would have had the policy for $40 - 35 = 5$ years).

c.   At age 50, 15 years after the original purchase of the policy, the face value would be $50,000.

Decreasing term life insurance provides greater protection early in the life of the policy, and the benefit of a fixed premium throughout the life of the policy. A family with young children might prefer such a policy, since it will provide high coverage when the children are young and lower coverage as the children mature and become less dependent on their parents.

Underwriters (persons or companies in the insurance business) are providing an increasing number of variations on the basic types of life insurance we noted. Actuaries working for insurance companies use computers to "customize" policies to accommodate customers' needs for protection, savings, investment, and retirement.

CHECK YOUR KNOWLEDGE

Life Insurance

Solve the following problems. Round dollar amounts to the nearest cent.

1. Jake Malone is 35 years old and wants to purchase $20,000 face value of life insurance. Use Table 18.1 to determine the premium Jake will have to pay for each type of insurance: (a) 5-year renewable term life, (b) whole life, (c) 20-year limited pay life, and (d) 20-year endowment life.

2. Marsha Sorensen purchased $50,000 face value of 20-year limited pay life insurance when she was 25 years old. Now, at age 30, she intends to surrender the policy. Use Table 18.2 to determine the cash surrender value.

3. Frank Zajac purchased a $100,000 20-year decreasing term life insurance policy 15 years ago at age 25. (a) Use Table 18.3 to determine Frank's yearly premium for the policy. (b) Use Table 18.4 to determine the current face value of Frank's policy.

18.1 EXERCISES

Solve exercises 1–9 using Tables A–D below. Round dollar amounts to the nearest cent.

Table A

Annual premiums for $1,000 face value of life insurance

Age	5-year renewable	Whole life	20-year limited pay	20-year endow- ment
20	$1.21	$19.25	$30.00	$53.13
25	1.21	21.23	32.78	53.57
30	1.28	24.10	36.40	54.56
35	1.54	27.60	40.26	55.10
40	2.13	32.56	44.22	56.98
45	3.19	38.60	50.71	57.53
50	5.50	47.40	59.75	62.15

Answers to CYK: *1.* **a.** $28; **b.** $502; **c.** $732; **d.** $1,002   *2.* $4,250.00   *3.* **a.** $100; **b.** $50,000

Table B

Cash surrender value for 20-year limited pay life insurance

Age		End of year					
	1	2	3	5	10	15	20
20		19	39	82	199	329	489
25		25	45	94	222	364	539
30		29	52	106	247	404	591
35		35	61	119	272	444	645
40		41	69	132	297	484	699
45		47	78	144	322	524	750

Table C

Decreasing term life insurance premium per $1,000 face value

Age	Premium
20	$1.05
25	1.15
30	1.30
35	1.65
40	2.40
45	3.75
50	6.15

Table D

20-year decreasing term life insurance face value for $100,000

Year	Face value
1	$100,000
5	92,000
10	77,000
15	52,000
20	38,000

1. Jim Smith is 30 years old and plans to purchase $40,000 face value of life insurance. Determine the premium Jim will have to pay for each type of insurance: (a) 5-year renewable term life, (b) whole life, (c) 20-year limited pay life, and (d) 20-year endowment life.

2. Sam Harris is 45 years old and plans to purchase $28,000 face value of life insurance. Determine the premium Sam will have to pay for each type of insurance: (a) 5-year renewable term life, (b) whole life, (c) 20-year limited pay life, and (d) 20-year endowment life.

3. Diran Abdul is 25 years old and plans to purchase $39,000 face value of life insurance. Determine the premium Diran will have to pay for each type of insurance: (a) 5-year renewable term life, (b) whole life, (c) 20-year limited pay life, and (d) 20-year endowment life.

4. Sara Hedges purchased $20,000 face value of 20-year limited pay life insurance when she was 20 years old. Now, at age 40, she intends to surrender the policy. Determine the cash surrender value.

5. My-Lin Sing purchased $40,000 face value of 20-year limited pay life insurance when she was 30 years old. Now, at age 45, she intends to surrender the policy. Determine the cash surrender value.

6. Jack Marmin purchased $100,000 face value of 20-year limited pay life insurance when he was 25 years old. Now, at age 40, he intends to surrender the policy. Determine the cash surrender value.

7. Arthur Levinson purchased a $100,000 twenty-year decreasing term life insurance policy at age 30. Determine (a) his yearly premium for the policy and (b) what the cash value of the policy is now that his age is 50.

8. Ann Michels purchased a $100,000 twenty-year decreasing term life insurance policy at age 20. Determine (a) her yearly premium for the policy and (b) what the cash value of the policy is now that she is 35 years old.

9. Cesar Escobar purchased a $100,000 twenty-year decreasing term life insurance policy at age 25. Determine (a) his yearly premium for the policy and (b) what the cash value of the policy is now that his age is 45.

18.2   PROPERTY INSURANCE

Economic loss due to damage to property we own or are purchasing is another form of risk that must be considered. Damage to real estate by fire, theft, vandalism, or other dangers could be economically devastating in terms of repair or replacement costs.

Fire Insurance

Learning objective
Calculate the premium for a fire insurance policy.

Fire and lightning pose very definite risks to the loss of a building. Whether we consider the protection of our home or our business structures, physical loss of either could result in a corresponding risk of economic loss to us.

The premium paid for fire insurance is based on many factors regarding the structure to be insured. A structure built with fire resistant materials such as concrete or steel would cost less to insure than a structure made of wood and less fire-resistant materials. Structures with sprinkler systems or that are near fire hydrants or in close proximity to fire departments would also have a lower premium than structures without such features.

Fire insurance premiums are usually based on each $100 of property valuation. A rate per $100 is determined based on an inspection of the property to be insured (for example, $.46 per $100 valuation) and the premium calculated.

Example 5

Determine the annual premium for a $200,000 building at a rate of $.46 per $100.

Solution

$$\text{premium} = \frac{\$200,000}{\$100} \times \$.46 = 2,000 \times \$.46 = \$920$$

Insurance companies provide incentives to purchase insurance for long terms such as 2, 3, or 4 years by providing long-term rate discounts (Table 18.5).

Table 18.5

Rate discount percentages

Number of years	1-year premium
1	100%
2	185%
3	270%
4	355%

Consequently, a 2-year premium in Example 5 would be

$$\text{2-year premium} = \text{1-year rate} \times \text{percent for 2-year term}$$
$$= \$920 \times 1.85$$
$$= \$1,702$$

rather than $1,840 ($920 × 2). For a 3-year policy, the premium would be

$$\text{3-year premium} = \text{1-year rate} \times \text{percent for 3-year term}$$
$$= \$920 \times 2.70$$
$$= \$2,484$$

rather than $2,760 ($920 × 3 = $2,760).

Example 6

Determine the fire insurance premium on a $120,000 building with a rate of $.54 per $100 valuation for (a) 1 year, (b) 2 years, and (c) 3 years.

Solution

a. $\text{premium for 1 year} = \dfrac{\text{value of property}}{\$100} \times \text{rate per } \$100$

$$= \dfrac{\$120,000}{\$100} \times \$.54$$
$$= 1,200 \times \$.54$$
$$= \$648$$

b. premium for 2 years = premium for 1 year
$\qquad\qquad\qquad\quad \times$ multiple year premium for 2 years
$$= \$648 \times \$1.85$$
$$= \$1,198.80$$

c. premium for 3 years = premium for 1 year
$\qquad\qquad\qquad\quad \times$ percent for 3-year term
$$= \$648 \times \$2.70$$
$$= \$1,749.60$$

Coinsurance Clause

Learning objective
Calculate the amount of claim covered using the coinsurance clause.

A fire insurance policy usually includes a **coinsurance clause** that states the minimum amount the building must be insured for so that a claim will be fully paid by the insurance company. Although a building may be valued at $120,000, it is unlikely that it will be totally destroyed. Most likely, the foundation, at least, will be left intact. The coinsurance clause requires the owners to insure a certain percentage, often 80%, of the value of the structure.

Example 7

Determine the required amount of fire insurance for a $120,000 building if the coinsurance clause requires 80% coverage.

Solution

An 80% coinsurance clause would require the owners of the $120,000 building to carry $120,000 × .80 = $96,000 of fire insurance in order to receive 100% coverage of a claim filed.

The amount of a claim that would be covered by the insurance company if the owners insured the building for *less* than the required 80% is determined by the following formula:

Coinsurance clause formula

$$\frac{\text{amount of claim}}{\text{to be paid}} = \frac{\text{amount of insurance carried}}{\text{amount of insurance required}} \times \frac{\text{amount}}{\text{of claim}}$$
$$\text{by coinsurance clause}$$

Example 8

Bob McClosky owns a building valued at $200,000. He purchased $120,000 of fire insurance with an 80% coinsurance clause. Three months later, Bob suffered $80,000 worth of fire damage to the building. (a) How much insurance should Bob have purchased in order to have been fully covered? (b) How much of the $80,000 claim will the insurance company pay?

Solution

a. To determine the amount of insurance Bob should have purchased, we multiply the percentage of the coinsurance clause by the value of Bob's building:

amount of insurance should carry = .80 × $200,000 = $160,000

b. To determine the amount of Bob's claim that the insurance company will pay.

Step 1:   We determine the ratio:

$$\frac{\text{amount of insurance carried}}{\text{amount of insurance that should be carried}} = \frac{\$120,000}{\$160,000} = .75$$

Step 2:   We then multiply this factor by the claim of $80,000.

amount of claim insurance company will pay = .75 × $80,000
$$= \$60,000$$

So, Bob will have to pay the $20,000 difference.

Notice that if Bob had carried the required $160,000 of insurance, the amount of the claim paid by the insurance company would have been:

$$\frac{\$160,000}{\$160,000} \times \$80,000 = 1 \times \$80,000 = \$80,000$$

and Bob's loss would have been fully covered.

CHECK YOUR KNOWLEDGE

Property Insurance

Solve the following problems. Round dollar amounts to the nearest cent.

1. E.J.'s Electrical Supply company owns a $300,000 warehouse. The rate for fire insurance on the structure is $.53 per $100. (a) How much will a 1-year premium for the insurance cost? (b) Use Table 18.5 to determine the cost of a 2-year premium and a 4-year premium.

2. Marney Hill owns a building valued at $350,000. She purchased $250,000 of fire insurance with an 80% coinsurance clause on the structure, and had a fire that caused $200,000 of damage. (a) How much should Marney have insured the building for according to the coinsurance clause? (b) How much of the $200,000 claim will be covered by the insurance company?

18.2   EXERCISES

Solve the following problems. Round dollar amounts to the nearest cent.

1. Abe's Plumbing supply owns a retail/wholesale complex valued at $600,000. Ann Jones's Independent Insurance Agency has quoted a rate of $.49 per $100 for fire insurance to cover the complex. How much will a 1-year premium for the insurance cost? Use Table 18.5 to determine the cost of (a) a 2-year premium, (b) a 3-year premium, and (c) a 4-year premium.

2. Alexandra's Greenery Service decorates commercial malls with decorative plantings. She has a $250,000 combination greenhouse/warehouse where she stores her supplies and plantings. Alex's fire insurance rate on the building is $.45 per $100. How much will a 1-year premium for the insurance cost? Use Table 18.5 to determine the cost of (a) a 2-year premium, (b) a 3-year premium, and (c) a 4-year premium.

3. Amy Felton has purchased a single-family home in her town of Belleville and has converted the downstairs into the Yarn Shoppe and small office, and the upstairs into a two-bedroom apartment. The value of her commercial enterprise is $125,000. She has been quoted a rate of $.42 per $100 for fire insurance to cover the complex.

Answers to CYK: **1. a.** $1,590; **b.** $2,941.50; $5,644.50   **2. a.** $280,000; **b.** $178,571.42

How much will a 1-year premium for the insurance cost? Use Table 18.5 to determine the cost of (a) a 2-year premium, (b) a 3-year premium, and (c) a 4-year premium.

4. Sally Feldman owns a building valued at $400,000. She purchased $300,000 of fire insurance with an 80% coinsurance clause on the structure. She recently had a fire that totally destroyed the building. How much should Sally have insured the building for, according to the coinsurance clause? How much of the $400,000 claim will be covered by the insurance company?

5. Harbinger Real Estate owns a building valued at $350,000. The firm purchased $250,000 of fire insurance with an 80% coinsurance clause on the structure. The building had a fire that caused $180,000 damage to the building. How much should Harbinger have insured the building for according to the coinsurance clause? How much of the $180,000 claim will be covered by the insurance company?

6. The B'ville Ambulance Corps houses its emergency equipment in a building valued at $275,000. The Corps purchased $200,000 of fire insurance with an 80% coinsurance clause on the structure. They recently had a fire that totally destroyed the building. How much should the

Corps have insured the building for according to the coinsurance clause? How much of the $275,000 claim will be covered by the insurance company?

7. Abe's Plumbing Supply, in exercise 1, purchased $400,000 of fire insurance with an 80% coinsurance clause on its building. A fire totally destroyed the building causing $600,000 damage. How much should Abe's have insured the building for, and how much of the $600,000 claim will the insurance company be responsible to cover.

8. Alexandra's Greenery Service, in exercise 2, purchased $200,000 of fire insurance with an 80% coinsurance clause on her building. A fire damaged the Greenery Service building causing $175,000 damage. How much should Alex have insured the building for, and how much of the $175,000 claim will the insurance company cover?

9. Amy Felton, in exercise 3, purchased $80,000 of fire insurance with an 80% coinsurance clause on her building. A fire destroyed part of the building causing $50,000 damage. How much should Amy have insured the building for, and how much of the $50,000 claim will the insurance company be responsible to cover?

18.3 AUTOMOBILE INSURANCE

Automobile insurance provides protection to offset financial losses due to liability and collision claims involving the insured's vehicle when used for either business or personal transport. Vehicles used for commercial (business) purposes pay a different premium rate than vehicles used strictly for personal purposes.

Liability automobile insurance provides benefits to offset costs due to accidents that cause bodily harm to others, or physical damage to others' property. Liability insurance on an automobile is often stated in a three-number sequence, such as, 25/50/10. The first number indicates the maximum level of medical coverage to be paid by the insurance company for medical expenses incurred as a result of an accident to a passenger in the insured's vehicle (in this example, $25,000); the second number indicates the maximum total medical costs the insurance company will pay for a single accident (here, $50,000); the third number indicates the maximum amount to be paid for physical damage to the property of others caused by the insured's vehicle (here, $10,000).

Automobile insurance quotation

maximum medical /maximum total medical /maximum total property
payments per person/ payments per accident / payments per accident
25/50/10
$25,000/$50,000/$10,000

Example 9

Wendy Willis had an automobile accident for which she was responsible. She had two passengers, Antoine and Cassie, who were injured. Antoine's medical costs totaled $23,000, and Cassie's were $8,000. Wendy's auto caused $3,000 worth of damage to the parked auto with which she collided. Wendy's auto was insured with 20/40/10 liability coverage. (a) How much of the total costs will be covered by insurance? (b) How much of the total costs will not be covered by insurance?

Solution

a. Insurance will cover $20,000 of the $23,000 medical claim of the first person injured. Insurance will cover all of the $8,000 claim of the second person injured. Since the total costs of the medical injury claim is $28,000 for medical costs and Wendy's coverage for medical costs is $40,000, the insurance company will pay $28,000 for medical costs. Since Wendy is covered for a maximum of $10,000 property damage, the $3,000 claim would be totally covered by insurance.

total insurance coverage of costs = $20,000 + $8,000 + $3,000
= $31,000

b. The total costs for the accident were

total costs = $23,000 + $8,000 + $3,000 = $34,000

The insurance company will not be responsible for paying the $3,000 beyond the $31,000 limit.

uninsured costs = $34,000 − $31,000 = $3,000

In addition to liability insurance, automobile owners can purchase extended protection in the form of *collision* and *comprehensive* insurance.

Automobile *collision* insurance provides benefits to cover all or part of the cost of damage to the *insured's vehicle. Comprehensive* insurance provides additional coverage for other physical damage to the insured's vehicle due to theft, vandalism, and so on. The cost of premiums for collision and comprehensive is based on where the vehicle is registered, its value, and the purpose for which it is used (personal or commercial). Premiums for these forms of insurance are determined in much the same way as other types of insurance already covered in this chapter.

CHECK YOUR KNOWLEDGE

Automobile Insurance

Solve the following problem. Round dollar amounts to the nearest cent.

1. Sally Mack's automobile liability policy is for 15/30/10. She had an accident and subsequent medical liability claims of $9,000, $8,500, $7,000, $5,500, and $1,500 for other people hurt in the accident; and $13,000 for damage done to other people's property. How much will the insurance company pay? How much will insurance not have to pay of the costs of the accident?

18.3 EXERCISES

Solve the following problems. Round dollar amounts to the nearest cent.

1. Tanya Magoon's auto liability policy is for 10/20/5. She had an accident and subsequent medical liability claims of $3,000, $4,500, and $1,500 for other people hurt in the accident; and $4,000 for damage done to other people's property. How much will the insurance company pay of the costs of the accident?

2. Marty Gale's auto liability policy is for 10/20/10. He had an accident and subsequent medical liability claims of $5,000, $4,500, and $12,000 for other people hurt in the accident; and $4,000 for damage to other people's property. How much will the insurance company pay of the costs of the accident?

3. Jan Sargent's auto liability policy is 10/20/5. She had an accident and subsequent medical liability claims of $500, $1,500, and $1,200 for other people hurt in the accident, and $7,000 for damage to other people's property. How much will the insurance company pay of the costs of the accident?

4. Bob Jackson's auto liability policy is for 25/500/20. He had a one-car accident in which his car left the road and destroyed several road signs, shrubbery, and a small storage building. Four passengers in Bob's car were injured. Bob is faced with the possible loss of his license for driving while intoxicated; he has medical liability claims of $2,500, $6,500, $9,000, and $7,500 for other people hurt in the accident; and $30,000 for damage to other people's property. How much will the insurance company pay of the costs of the accident?

Answers to CYK: Insurance will cover $30,000 of the medical costs and $10,000 of the property loss; insurance will not cover $1,500 of the medical costs and $3,000 of the property loss.

18.4   HEALTH INSURANCE

Health insurance provides protection to offset economic losses due to injury or illness. Health insurance provides payment of benefits for

a.   health care expenses incurred from illness or injury (medical insurance)

b.   loss of income from inability to work caused by injury or illness (disability insurance)

Health insurance is provided by two types of insurers, government insurers and private insurers. The social, or government, type of insurance includes "Old Age, Survivors, Disability, and Health Insurance (OASDHI)" often referred to as Social Security insurance, and worker's compensation insurance.

Private health insurance companies provide a broad spectrum of coverages including

a.   medical insurance to cover costs of health professionals, hospitals, medications, and medical laboratory work

b.   disability income insurance to cover loss of earnings while a wage earner is unable to work due to illness or injury

Health insurance, medical and disability income, is provided by both individual and group plans. Usually, group insurance plans cost less than individual plans. The larger number of people paying premiums into the company help to offset the costs of claims paid out by the insurance company, much the same as our earlier example concerning life insurance (Example 1). It is also less expensive for the insurance company to write a single policy to cover a large group than it is to write a separate policy for each person in the group. Other factors, such as the type of work or life-style of the group, may affect premium costs as well. A group of workers engaged in a hazardous line of work would be more likely to have higher premium costs than a group of people whose work is less dangerous and less likely to lead to illness or injury.

Actuaries determine premium costs based on a variety of statistical information pertaining to the expected payout of claims for a particular group of insureds. People in different age groups, occupations, or living circumstances have a different likelihood of making claims on their policies, and the amount of those claims will vary. Insurance actuaries gather statistical data describing these different characteristics and determine premium costs accordingly, similar to the way we noted for determining life insurance premiums.

18.5 BUSINESS INSURANCE

The basic concepts discussed earlier in this chapter apply to commercial (business) insurance. The *risk manager* in a business is responsible for establishing and maintaining a plan of action of (1) controlling risk and (2) risk financing. The risk manager wants to protect the assets and income of the business by providing protection against the risks of accidental loss. The risk manager of a company is not only responsible for determining ways to reduce risk in the company, but is also responsible for the company's insurance program of risk protection.

Business insurance includes a wide variety of protection plans.

- *Employee protection plans* including medical, disability, life, retirement, and social security insurance.

- *Transportation insurance,* including ocean marine and inland marine insurance, covers sea perils of ships and cargoes, and inland transportation by rail, motor truck, and other methods of transport.

- *Liability insurance,* including bodily injury, property damage liability, worker's compensation, and criminal and civil wrongs liability insurance, protects the company in the event it violates contract or tort law (e.g., intentional acts, omissions, or negligence).

- *Crime insurance* protects against nonemployee acts such as theft or vandalism, and *suretyship* protects against an employee failing to fulfill an obligation.

Other forms of insurance to protect a business, its employees, and its customers from the many risks due to uncertainty are available as well.

CHAPTER REVIEW EXERCISES

The following problems review all mathematics and business concepts presented in the chapter. If you experience any difficulty solving a problem, go to the appropriate chapter section and review the examples provided. Express all answers as decimals rounded to the nearest one hundredth, or dollars rounded to the nearest cent, unless otherwise instructed.

1. Al Borel is 30 years old and is considering purchasing $150,000 in life insurance. Use Table 18.1 to determine the premium he would have to pay for a 5-year renewable life policy.

2. Ed Simplson is 35 years old and is considering purchasing $25,000 in life insurance. Use Table 18.1 to determine the premium he would have to pay for a whole life policy.

3. Cindi Scarborough is 40 years old and is considering purchasing $50,000 in life insurance. Use Table 18.1 to determine the premium she would have to pay for a 20-year limited pay life policy.

4. Suki Robinson is 30 years old and is considering purchasing $20,000 in life insurance. Use Table

18.1 to determine the premium she would have to pay for a 20-year endowment life policy.

5. Assuming Cindi purchases the 20-year endowment policy mentioned in exercise 3, what will the policy's cash surrender value be in 10 years? (Use Table 18.2 for this problem.)

6. Bob Einstein is 45 years old and is considering purchasing $100,000 in life insurance. Use Table 18.3 to determine the premium he would have to pay for a 20-year decreasing term life policy.

7. Use Table 18.4 to determine the face value of Bob Einstein's decreasing term policy 15 years after he purchases it.

8. Otsego Industrial Corporation owns a $400,000 building that All Purpose Insurance has offered to insure against fire at a premium of 63 cents per $100 of coverage with an 80% coinsurance clause. (a) How much coverage minimally would the coinsurance clause require? (b) Calculate the premium for the fire insurance if they insure for the minimum required in part (a).

9. If Otsego Industries insured the building in exercise 8 for $280,000 and had a fire and filed a claim for $240,000, how much would the coinsurance clause stipulate that All Purpose Insurance would have to pay?

10. Sally Markes had automobile insurance coverage of 20/100/15 when she had an accident in which her friend Marsha received $18,000 in injuries, Steve received injuries requiring $22,000 in medical care, Mark received $24,000 in medical care, and the damage done to others' property totaled $13,000. How much of the costs were not covered by Sally's policy?

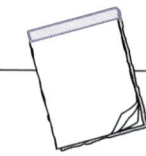 **EXPRESS YOUR UNDERSTANDING**

Compose one or two well-written sentences to express the requested information in your own words.

1. Define risk, in the business sense.

2. What is risk management?

3. Describe the difference between term life insurance and endowment life insurance.

4. Describe the difference between five-year term life insurance and decreasing term life insurance.

5. What is cash surrender value?

6. Describe how you would calculate the amount of claim to be paid on a $10,000 fire insurance claim if a coinsurance clause required $100,000 of insurance, but the owner purchased only $75,000 of insurance.

7. Explain the coverage of a 30/60/15 automobile liability policy.

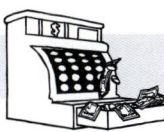

Mind Your Business

Media World's fiscal year will close out on December 31, and among your duties you must review the company's current insurance and recommend the types and amounts of insurance coverage that the business should have for the next year.

Activity 1

Contact an insurance company in your community and arrange to meet with one of its agents. Describe Media World to the agent, and ask for recommendations on the kinds of insurance that a business like Media World might want to have to reduce its financial risk in your community. Write an executive summary that would include the following:

a. All of the different kinds of coverages for the business, its partners, and its employees that the business could purchase

b. A list of any and all insurance that the business must have to do business in your community

Recommend minimal insurance coverage you feel the business should have. Also, provide a cost analysis identifying the annual cost of the minimum insurance coverage you will recommend to your partners, and explain briefly how these costs would be reflected in the financial statements of the business.

SELF-TEST

A. Terminology

Complete the following items using the key terms presented at the beginning of the chapter. Check your responses against the answer key at the end of the test.

1. The uncertainty of financial loss was our definition of _____.

2. _____ is the attempt to minimize loss by controlling risk and using insurance to compensate for accidental losses.

3. The _____ _____ of a policy is the amount of money that would be paid to the beneficiary in the event of the death of the insured.

4. _____ life insurance provides protection for the insured's entire life, whereas _____ life insurance covers the insured for a specific time period.

5. Decreasing term life insurance policies have the feature that the _____ amount remains the same as time passes, but the _____ _____ of the policy decreases.

6. _____ life insurance policies have the feature that, if the insured survives through the life of the policy, the insurance pays a regular periodic payment to the insured for the remainder of his or her life, or for a specified period of time.

7. _____ life insurance pays the insured a fixed sum of money if he/she survives the life of the policy.

8. If a person cancels a policy, the policy may have a _____ _____ value, which is a cash savings that the policy earned while the premiums were paid.

9. A _____ clause for a property insurance policy specifies the amount of insurance that should be purchased in order for losses to be fully covered by the insurance.

B. Calculation

The following concepts and short problems are designed to test your understanding of the objectives identified at the beginning of the chapter. Answers are provided at the end of the test.

10. Sam Harris is 45 years old and plans to purchase $28,000 face value of life insurance. Use Table 18.1 to determine the premium Sam will have to pay for each type of insurance: (a) 5-year renewable term life, (b) whole life, (c) 20-year limited pay life, and (d) 20-year endowment life.

11. Miguel Espinoza purchased $20,000 face value of 20-year limited pay life insurance when he was 30 years old. Now at age 45, he intends to surrender the policy. Use Table 18.2 to determine the cash surrender value.

12. Jake Nichols purchased a $100,000 20-year decreasing term life insurance policy at age 30. Determine (a) what his yearly premium for the policy is and (b) what the cash value of the policy is now that his age is 50. (Use Tables 18.3 and 18.4.)

13. Sally Feldman owns a building valued at $400,000. She purchased $300,000 of fire insurance with an 80% coinsurance clause on the structure. A fire totally destroyed the building. How much should Sally have insured the building for according to the coinsurance clause? How much of the $400,000 claim will be covered by the insurance company?

14. Tanya Magoon's auto liability policy is for 10/20/5. She had an accident and subsequent medical liability claims of $3,000, $4,500, and $1,500 for other people hurt in the accident, and $4,000 for damage done to other people's property. How much will the insurance company pay of the costs of the accident?

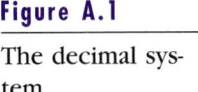

Appendix A
Review of Arithmetic

The focus of this appendix is to provide a review of the concepts of basic arithmetic that are essential to your successful completion of this course and other courses in a business curriculum, such as accounting, marketing, and management. The concepts presented should not be new to you. You should already possess the required proficiency in all of these skills. The use of new technology increases the importance of understanding the fundamentals of arithmetic, since the most crucial part of solving a mathematical problem is to set it up properly and to determine the correct quantitative procedures required to analyze the problem data.

DECIMAL NUMBERS AND ROUNDING

The system that we use for writing numbers is called the *Hindu-Arabic numeration system*. This system is a *decimal system* because it is based on the ten digits 0, 1, 2, 3, 4, 5, 6, 7, 8, 9, which means any number can be written using the proper selection and positioning of these digits. The decimal system is a *place value system* because the value of each digit depends on its position relative to the decimal point. As Figure A.1 shows, each position has a name that indicates its value. The decimal point in a decimal number separates whole numbers from their fractional parts. Whole numbers are written to the left of the decimal point and increase by factors of ten from ones to tens to hundreds to thousands, and so on. Commas are inserted every three places to the left of the decimal point to facilitate reading the number. Numbers that are not whole numbers are written to the right of the decimal point and decrease by factors of one-tenth from tenths to hundredths to thousandths, and so on. Commas are not used for positions to the right of the decimal.

Let us analyze a decimal number using some of the numbers provided in Figure A.1.

Figure A.1

The decimal system

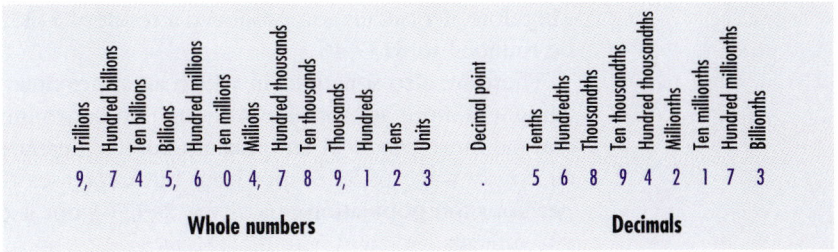

Example 1

Write 9,123.568 in extended place value form.

Solution

$$9,123.568 = (9 \times 1,000) + (1 \times 100) + (2 \times 10) + (3 \times 1) +$$
$$(5 \times .1) + (6 \times .01) + (8 \times .001)$$

We are often required to write numbers out in word form. For example, legal documents such as negotiable instruments, checks, notes, and bills of sale all require amounts of money to be written in both number and word form. In fact, if there is an error in the presentation of amounts, the written amount is always considered the legal figure.

Decimal numbers are expressed in word form exactly as they are read. To read a decimal number, the whole numbers are read as their place value indicates, followed by the name of the group. The decimal point in the number is replaced by the word *and.* The fractional parts to the right of the decimal are read as whole numbers, followed by the name of the smallest place value position.

Example 2

Write 47,682.52463 in word form.

Solution

Forty-seven thousand, six hundred eighty-two and fifty-two thousand four hundred sixty-three hundred thousandths.

ROUNDING AND ESTIMATING DECIMAL NUMBERS

When working with decimal numbers, it is necessary to express the results of a calculation to the place value most appropriate to the situation. For example, the U.S. monetary system is expressed in dollars and cents, which requires an answer to be expressed to the nearest whole cent (hundredths). Therefore, if a calculation produced a result of $137.4642, the answer would be rounded to $137.46.

There are also situations in which an approximation rather than an exact amount is more appropriate. Although an approximation is less accurate, it may be more practical and meaningful. To illustrate, an employee receives an hourly wage of $3.87 per hour but expresses the wage as about $4.00 per hour, the population of a city is 249,761 but is expressed as 250,000 by city officials.

The rounding of a number is nothing more than the determination of its significant place value. Therefore, any number may be rounded and, quite often, in several ways (thousands, tens, integers, tenths, etc.).

To round a whole number or a decimal number to any place value, complete the following steps:

ROUNDING A NUMBER

Step 1:   Identify the place value of accuracy desired in the number to be rounded.

Step 2:   Evaluate the numerical value of the first digit to the right of the selected place value to be rounded.

Step 3:   If the first digit to the right is 5 or greater, increase the selected place value by 1 and change all digits to the right of the place value rounded to 0s.

Step 4:   If the first digit to the right of the selected place value to be rounded is 4 or less, simply change all digits to the right of the place value selected for rounding to 0s.

Step 5:   If the selected place value for rounding is a decimal, follow steps 1 through 4 and drop all digits to the right of the place value selected for rounding.

The following examples provide a thorough application of the rounding procedure just described. Analyze these examples carefully, as you will be required to round answers in all chapters of this text.

Example 3

Round $5,847,389 to the nearest million.

Solution

Select the place value:

millions position

↓

$5,847,389

↑

digit to the right of selected place value is 8

The selected place value of the accuracy desired is millions. Since the digit to the right of the millions position is 5 or greater (8), we round up, adding 1 to the 5 and changing all digits to the right of the 5 to 0s.

$5,847,389 is rounded to the nearest million as $6,000,000.

Example 4

Round 43,275 to the nearest ten thousand.

Solution   Select the place value:

ten thousands position

↓

43,275

↑

digit to the right of selected place value is 3

The selected place value of the accuracy desired is ten thousand. Since the digit to the right of the ten thousands position is 4 or less (3), we round down, the ten thousand position is not changed, and all digits to the right of the 4 are changed to 0s.

43,275 is rounded to the nearest ten thousand as 40,000.

Example 5

Round $945.2682 to the nearest cent.

Solution   Select the place value:

hundredths position

↓

$945.2682

↑

digit to the right of selected place value is 8

The selected place value of the accuracy desired is hundredths. Since the digit to the right of the hundredths position is 5 or greater (8), we round up, adding 1 to the 6 in the cents (or hundredths) position and dropping the digits to the right of it.

$945.2682 is rounded to the nearest cent to be $945.27.

Example 6

Round 12.4465 to the nearest tenth.

Solution   Select the place value:

tenths position

↓

12.4465

↑

digit to the right of selected place value position is 4

The selected place value of the accuracy desired is tenths. Since the digit to the right of the tenths position is 4 or less (4), we round down: the tenths position is not changed, and all digits to the right of the 4 are dropped.

12.4465 is rounded to the nearest tenth to be 12.4.

ESTIMATING THE RESULT OF AN ARITHMETIC CALCULATION

A practical use of rounding is the process of estimating answers. We should mentally estimate answers when we wish to check the accuracy of our calculations or when we want an approximation instead of an exact amount. To estimate sums, differences, products, or quotients, we round values to the appropriate place value (10s, 100s, 1,000s, etc.) indicated by the degree of accuracy required. For example, if you purchase items that cost $21.30, $5.69, and $12.15, an estimate of your total cost would be $39.00 ($21 + $6 + $12 = $39). Here we rounded each cost to the nearest whole dollar and then added the rounded values to get an estimate of the actual total cost. We rounded to whole number values because it is *mentally* easier to sum whole numbers and because the nearest whole dollar is a reasonable degree of accuracy.

To further illustrate estimation, let's say you wish to buy 48 videotapes costing $9.95 each. You can estimate your total cost by rounding $9.95 to $10.00 and 48 to 50 and multiplying ($10 × 50 = $500). When estimating products, multiply the integers alone (1 × 5) and then add back in the total number of 0s in the multiplier (50) and multiplicand ($10)—in this case, 2—to the product: $500.

When we estimate a quotient, we round the numbers and then drop the common 0s and divide the remaining numbers. For instance, if we wanted to estimate the average amount of each sale based on total sales of $4,350 from 769 separate sales, we would round the total sales to $4,000 and the number of separate sales to 800 and mentally divide, as follows:

$$\frac{\$4{,}000}{8} \frac{\$40}{8} = \$5 \text{ per sale}$$

Mental estimation of calculations being done on a calculator or computer is a very important factor in determining if the results shown by those devices are *reasonable* results.

CHECK YOUR KNOWLEDGE

Decimal Numbers and Rounding

Write the following numbers in extended place value form.

1. 2,460   *2.* 78,258.931   *3.* 3.00472

Write the following numbers in word form.

4. 6,275   *5.* 265,892,326   *6.* 57.03680

In the number 619,587.325:

7. The 6 is in the _____ position.

8. The 2 is in the _____ position.

9. The 8 is in the _____ position.

Round each of the following values as indicated.

10. $1,279.57 to the nearest dollar

11. 4,627,385 to the nearest hundred thousand

12. 9.78% to the nearest tenth of a percent

13. 3.0005185 to the nearest hundred thousandth

14. $75.0543 to the nearest cent

Estimate each of the following:

15. $1.20 + $.95 + $6.10 + $.65

16. 31×45

17. 82.50×24

18. $64 \div 17$

19. 38 + 71 − 52

20. $2,130 \div 270$

Answers to CYK: *1.* 2,460 = (2 × 1,000) + (4 × 100) + 6 × 10) + (0 × 1)   *2.* 78,258.931 = (7 × 10,000) + (8 × 1,000) + (2 × 100) + (5 × 10) + (8 × 1) + (9 × .1) + (3 × .01) + (1 × .001)   *3.* 3.00472 = (3 × 1) + (0 × .1) + (0 × .01) + (4 × .001) + (7 × .0001) + (2 × .00001)   *4.* 6,275 = six thousand, two hundred seventy-five   *5.* 265,892,326 = two hundred sixty-five million, eight hundred ninety two thousand, three hundred twenty-six   *6.* 57.03680 = fifty-seven and three thousand six hundred eighty hundred-thousandths   *7.* hundred thousands   *8.* hundredths.   *9.* tens   *10.* $ 1,280   *11.* 4,600,000   *12.* 9.8%   *13.* 3.00052   *14.* $75.05   *15.* $9   *16.* 1,500   *17.* $1,600   *18.* 3   *19.* 60   *20.* 7

Figure A.2

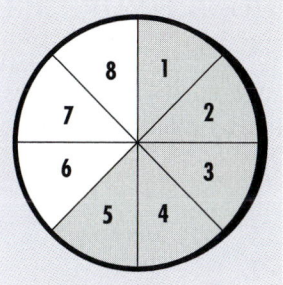

FRACTIONS

A *fraction* is a numerical expression used to indicate a value that is a part of a whole unit. The fraction ⅝ indicates that the whole consists of eight equal parts and we are interested in five of those eight parts. In Figures A.2 and A.3, we have divided graphic illustrations (the whole) each into eight equal parts, of which five are shaded. The unshaded portion represents three-eighths of the whole, or ⅜.

A fraction consists of two numbers separated by a line called a *bar* that indicates a division or a *ratio*. The number above the bar is called the *numerator* and represents the part of the whole under consideration. The number below the bar is called the *denominator* and represents the number of equal parts into which the whole has been divided.

$$\frac{5}{8} = \text{⅝} \qquad \frac{\text{numerator}}{\text{denominator}}$$

The basic types of fractions are proper fractions, improper fractions, mixed numbers, and complex fractions. A *proper fraction* has a numerator that is smaller than the denominator, and its value is always less than 1. For example, ¾, ⁶⁄₈, and ¹²⁄₂₄ are proper fractions. An *improper fraction* has a numerator that is equal to or greater than the denominator. Therefore, ⁶⁄₆, ¹²⁄₄, and ²⁵⁄₈ are examples of improper fractions. A *mixed number* is a combination of a whole number and a proper fraction. The following are mixed numbers: 2½, 12¾, and 25⅖. A *complex fraction* has a numerator and/or a denominator containing a fraction or a mixed number. For example (½)/(¾). (3⅝)/100, and (10⅖)/(50¾) are complex fractions.

Figure A.3

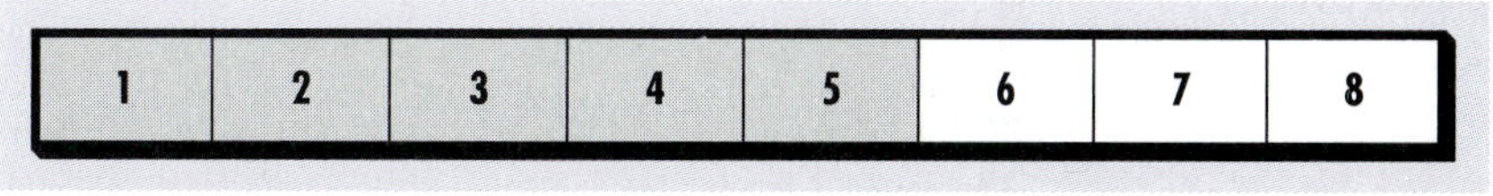

1	2	3	4	5	6	7	8

CONVERTING FRACTIONS TO HIGHER OR LOWER TERMS

A fraction may be converted to higher terms or lower terms without changing the value of the fraction by multiplying or dividing the terms by the same factor.

Example 7

Raise a fraction to higher terms by converting $\frac{3}{5}$ to a proper fraction with a denominator of 35.

Solution

$$\frac{3}{5} = \frac{3 \times 7}{5 \times 7} = \frac{21}{35}$$ Both numerator and denominator are multiplied by the factor 7 since $5 \times 7 = 35$.

Whenever we multiply by 1, the value of the number (whole or fraction) remains the same:

$$\frac{3}{5} \times 1 = \frac{3}{5}$$

The "1" we've used in Example 7 is written as a fraction with the same number in the numerator and denominator, which always equals 1:

$$\frac{3}{5} \times 1 = \frac{3}{5} \times \frac{7}{7}$$

When we perform the multiplication:

$$\frac{3}{5} \times \frac{7}{7} = \frac{21}{35},$$

So $\frac{21}{35}$ has the same value as $\frac{3}{5}$.

A proper fraction is considered reduced (converted) to its *lowest terms* when there is no factor greater than 1 by which both the numerator and the denominator can be divided evenly.

Example 8

Reduce a fraction to lowest terms by converting $\frac{27}{45}$ to lowest terms.

Solution
$$\frac{27}{45} = \frac{27 \div 9}{45 \div 9} = \frac{3}{5}$$
Both numerator and denominator are divided by the factor 9 since 9 produces the lowest terms of $^{27}/_{45}$.

We are reducing the fraction by factoring a common factor, 9, out of the numerator and denominator, and reversing the multiplication.

$$\frac{27}{45} = \frac{3 \times 9}{5 \times 9} = \frac{3}{5} \times \frac{9}{9} = \frac{3}{5} \times 1 = \frac{3}{5}$$

Reducing fractions can be challenging because it is not always easy to identify which numbers will divide into another number. The following rules of divisibility will help you with the task of reducing fractions.

Rules of divisibility

A number can be divided evenly by:
2 if the last digit is an even number
3 if the sum of the digits is divisible by 3
4 if the last two digits are divisible by 4
5 if the last digit is 0 or 5
6 if the number is even and the sum of the digits is divisible by 3
8 if the last three digits are divisible by 8
9 if the sum of the digits is divisible by 9
10 if the last digit is 0

Example 9

Show that 15,384 is divisible by 3.

Solution

$1 + 5 + 3 + 8 + 4 = 21$; 21 is divisible by 3.

Example 10

Show that 3,620,560 is divisible by 8.

Solution

The last 3 digits of 3,620,560 (560) form a number divisible by 8.

CONVERTING IMPROPER FRACTIONS AND MIXED NUMBERS

Improper fractions can be converted to mixed numbers by dividing the numerator by the denominator, as the following example illustrates.

Example 11

Convert the improper fraction $^{25}/_4$ to a mixed number.

Solution

$$\frac{25}{4} = 6\frac{1}{4}$$

The denominator 4 is divided into the numerator 25, producing a quotient of 6 and a remainder of 1.

If an improper fraction can be converted to a mixed number, then a mixed number can be converted to an improper fraction. The procedure is to multiply the denominator by the whole numbers and add the result to the numerator. This figure becomes the numerator of the improper fraction and is placed over the original denominator.

Example 12

Convert the mixed number $7^5/_8$ to an improper fraction.

Solution

$$7\frac{5}{8} = \frac{(7 \times 8) + 5}{8}$$

The original denominator is retained as the denominator in the improper fraction.

$$= \frac{61}{8}$$

A mixed number is really two numbers being added together. $7^5/_8 = 7 + {}^5/_8$. The 7 can be written as a fraction with 1 as its denominator: $^7/_1$. When we add fractions, though, we need a common denominator, so we raise the fraction to $^7/_1 \times {}^8/_8 = {}^{56}/_8$

$$\frac{7}{1} + \frac{5}{8} = \left(\frac{7}{1} \times \frac{8}{8}\right) + \frac{5}{8} = \frac{56}{8} + \frac{5}{8} = \frac{61}{8}$$

CHECK YOUR KNOWLEDGE

Fractions

Identify each as a proper fraction, improper fraction, or a mixed number.

1. $8/12$ *2.* $15/9$ *3.* $12/6$ *4.* $15\frac{1}{2}$

Convert each fraction to a new fraction with the indicated denominator.

5. $1/3, ?/12$ *6.* $6/12, ?/60$ *7.* $17/20, ?/100$ *8.* $3/30, ?/240$

Reduce each fraction to its lowest terms.

9. $9/15$ *10.* $26/98$ *11.* $98/182$ *12.* $102/969$

Convert each improper fraction to a mixed number.

13. $19/8$ *14.* $33/12$ *15.* $97/14$ *16.* $300/9$

Convert each mixed number to an improper fraction.

17. $2\frac{1}{8}$ *18.* $18\frac{4}{9}$ *19.* $29\frac{11}{15}$ *20.* $4\frac{103}{229}$

Answers to CYK: *1.* proper fraction   *2.* improper fraction   *3.* improper fraction   *4.* mixed number   *5.* $4/12$ *6.* $30/60$ *7.* $85/100$ *8.* $24/240$ *9.* $3/5$ *10.* $13/49$ *11.* $7/13$ *12.* $2/19$ *13.* $2\frac{3}{8}$ *14.* $2\frac{3}{4}$ *15.* $6\frac{13}{14}$ *16.* $33\frac{1}{3}$ *17.* $17/8$ *18.* $166/9$ *19.* $446/15$ *20.* $1,109/215$

Compound Interest and Annuity Tables

Rate = .5%	A	B	C	D	E	F
Period	Compound Int.	Compound Int.	Ordinary	Sinking	Present Value	Amortization
n	Future Value	Present Value	Annuity	Fund	Annuity	
1	1.005	0.995024876	1	1	0.995024876	1.005
2	1.010025	0.990074503	2.005	0.498753117	1.985099379	0.50375312
3	1.015075125	0.985148759	3.015025	0.331672208	2.970248138	0.33667221
4	1.020150501	0.980247522	4.030100125	0.248132793	3.95049566	0.25313279
5	1.025251253	0.975370668	5.050250626	0.198009975	4.925866328	0.20300997
6	1.030377509	0.970518078	6.075501879	0.164595456	5.896384406	0.16959546
7	1.035529397	0.96568963	7.105879388	0.140728536	6.862074036	0.14572854
8	1.040707044	0.960885204	8.141408785	0.122828865	7.82295924	0.12782886
9	1.045910579	0.95610468	9.182115829	0.108907361	8.77906392	0.11390736
10	1.051140132	0.951347941	10.22802641	0.097770573	9.730411861	0.10277057
11	1.056395833	0.946614866	11.27916654	0.088659033	10.67702673	0.09365903
12	1.061677812	0.94190534	12.33556237	0.08106643	11.61893207	0.08606643
13	1.066986201	0.937219243	13.39724018	0.074642239	12.55615131	0.07964224
14	1.072321132	0.932556461	14.46422639	0.069136086	13.48870777	0.07413609
15	1.077682738	0.927916877	15.53654752	0.064364364	14.41662465	0.06936436
16	1.083071151	0.923300375	16.61423026	0.060189367	15.33992502	0.06518937
17	1.088486507	0.918706841	17.69730141	0.05650579	16.25863186	0.06150579
18	1.09392894	0.91413616	18.78578791	0.053231731	17.17276802	0.05823173
19	1.099398584	0.909588219	19.87971685	0.050302527	18.08235624	0.05530253
20	1.104895577	0.905062904	20.97911544	0.047666452	18.98741915	0.05266645
21	1.110420055	0.900560104	22.08401101	0.045281629	19.88797925	0.05028163
22	1.115972155	0.896079705	23.19443107	0.043113797	20.78405896	0.0481138
23	1.121552016	0.891621597	24.31040322	0.041134653	21.67568055	0.04613465
24	1.127159776	0.887185669	25.43195524	0.03932061	22.56286622	0.04432061
25	1.132795575	0.88277181	26.55911502	0.037651857	23.44563803	0.04265186
26	1.138459553	0.87837991	27.69191059	0.036111629	24.32401794	0.04111163
27	1.144151851	0.874009861	28.83037015	0.034685646	25.1980278	0.03968565
28	1.14987261	0.869661553	29.974522	0.033361666	26.06768936	0.03836167
29	1.155621973	0.865334879	31.12439461	0.032129139	26.93302423	0.03712914
30	1.161400083	0.86102973	32.28001658	0.030978918	27.79405397	0.03597892
31	1.167207083	0.856746	33.44141666	0.029903039	28.65079997	0.03490304
32	1.173043119	0.852483582	34.60862375	0.028894532	29.50328355	0.03389453
33	1.178908334	0.84824237	35.78166686	0.027947273	30.35152592	0.03294727
34	1.184802876	0.844022259	36.9605752	0.027055856	31.19554818	0.03205586
35	1.19072689	0.839823143	38.14537807	0.026215496	32.03537132	0.0312155
36	1.196680525	0.835644919	39.33610496	0.025421937	32.87101624	0.03042194
37	1.202663927	0.831487481	40.53278549	0.024671386	33.70250372	0.02967139
38	1.208677247	0.827350728	41.73544942	0.023960446	34.52985445	0.02896045
39	1.214720633	0.823234555	42.94412666	0.023286071	35.353089	0.02828607
40	1.220794236	0.819138861	44.1588473	0.022645519	36.17222786	0.02764552
41	1.226898208	0.815063543	45.37964153	0.022036313	36.98729141	0.02703631
42	1.233032699	0.8110085	46.60653974	0.021456216	37.79829991	0.02645622
43	1.239197862	0.806973632	47.83957244	0.020903197	38.60527354	0.0259032
44	1.245393852	0.802958838	49.0787703	0.020375409	39.40823238	0.02537541
48	1.270489161	0.787098411	54.09783222	0.018485029	42.58031778	0.02348503
60	1.348850153	0.741372196	69.77003051	0.014332802	51.72556075	0.0193328
72	1.432044278	0.69830243	86.4088557	0.011572888	60.33951394	0.01657289
84	1.520369636	0.657734788	104.0739272	0.009608554	68.45304244	0.01460855
96	1.614142708	0.619523909	122.8285417	0.00814143	76.09521825	0.01314143
120	1.819396734	0.549632733	163.8793468	0.00610205	90.07345333	0.01110205
180	2.454093562	0.407482427	290.8187124	0.003438568	118.5035147	0.00843857
240	3.310204476	0.302096142	462.0408952	0.002164311	139.5807717	0.00716431
360	6.022575212	0.166041928	1004.515042	0.000995505	166.7916144	0.00599551

Rate =1%	A	B	C	D	E	F
Period	Compound Int.	Compound Int.	Ordinary	Sinking	Present Value	Amortization
n	Future Value	Present Value	Annuity	Fund	Annuity	
1	1.01	0.99009901	1	1	0.99009901	1.01
2	1.0201	0.980296049	2.01	0.497512438	1.970395059	0.50751244
3	1.030301	0.970590148	3.0301	0.330022111	2.940985207	0.34002211
4	1.04060401	0.960980344	4.060401	0.246281094	3.901965552	0.25628109
5	1.05101005	0.951465688	5.10100501	0.1960398	4.853431239	0.2060398
6	1.061520151	0.942045235	6.15201506	0.162548367	5.795476475	0.17254837
7	1.072135352	0.932718055	7.213535211	0.138628283	6.728194529	0.14862828
8	1.082856706	0.923483222	8.285670563	0.120690292	7.651677752	0.13069029
9	1.093685273	0.914339824	9.368527268	0.106740363	8.566017576	0.11674036
10	1.104622125	0.905286955	10.46221254	0.095582077	9.471304531	0.10558208
11	1.115668347	0.896323718	11.56683467	0.086454076	10.36762825	0.09645408
12	1.12682503	0.887449225	12.68250301	0.078848789	11.25507747	0.08884879
13	1.13809328	0.878662599	13.80932804	0.07241482	12.13374007	0.08241482
14	1.149474213	0.86996297	14.94742132	0.066901172	13.00370304	0.07690117
15	1.160968955	0.861349475	16.09689554	0.06212378	13.86505252	0.07212378
16	1.172578645	0.852821262	17.25786449	0.057944597	14.71787378	0.0679446
17	1.184304431	0.844377487	18.43044314	0.054258055	15.56225127	0.06425806
18	1.196147476	0.836017314	19.61474757	0.050982048	16.39826858	0.06098205
19	1.20810895	0.827739915	20.81089504	0.048051754	17.2260085	0.05805175
20	1.22019004	0.81954447	22.01900399	0.045415315	18.04555297	0.05541531
21	1.23239194	0.811430169	23.23919403	0.043030752	18.85698313	0.05303075
22	1.24471586	0.803396207	24.47158598	0.040863718	19.66037934	0.05086372
23	1.257163018	0.795441789	25.71630183	0.03888584	20.45582113	0.04888584
24	1.269734649	0.787566127	26.97346485	0.037073472	21.24338726	0.04707347
25	1.282431995	0.779768443	28.2431995	0.035406753	22.0231557	0.04540675
26	1.295256315	0.772047963	29.5256315	0.033868878	22.79520366	0.04386888
27	1.308208878	0.764403924	30.82088781	0.032445529	23.55960759	0.04244553
28	1.321290967	0.756835568	32.12909669	0.031124436	24.31644316	0.04112444
29	1.334503877	0.749342147	33.45038766	0.02989502	25.0657853	0.03989502
30	1.347848915	0.741922918	34.78489153	0.028748113	25.80770822	0.03874811
31	1.361327404	0.734577146	36.13274045	0.027675731	26.54228537	0.03767573
32	1.374940679	0.727304105	37.49406785	0.026670886	27.26958947	0.03667089
33	1.388690085	0.720103075	38.86900853	0.025727438	27.98969255	0.03572744
34	1.402576986	0.712973341	40.25769862	0.024839969	28.70266589	0.03483997
35	1.416602756	0.705914199	41.6602756	0.024003682	29.40858009	0.03400368
36	1.430768784	0.69892495	43.07687836	0.02321431	30.10750504	0.03321431
37	1.445076471	0.692004901	44.50764714	0.022468049	30.79950994	0.03246805
38	1.459527236	0.685153367	45.95272361	0.021761496	31.4846633	0.0317615
39	1.474122509	0.67836967	47.41225085	0.021091595	32.16303298	0.0310916
40	1.488863734	0.671653139	48.88637336	0.020455598	32.83468611	0.0304556
41	1.503752371	0.665003108	50.37523709	0.019851023	33.49968922	0.02985102
42	1.518789895	0.658418919	51.87898946	0.019275626	34.15810814	0.02927563
43	1.533977794	0.651899919	53.39777936	0.018727371	34.81000806	0.02872737
44	1.549317572	0.645445465	54.93175715	0.018204406	35.45545352	0.02820441
48	1.612226078	0.620260405	61.22260777	0.016333835	37.97395949	0.02633384
60	1.816696699	0.550449616	81.66966986	0.012244448	44.95503841	0.02224445
72	2.047099312	0.488496085	104.7099312	0.009550193	51.15039148	0.01955019
84	2.306722744	0.433515472	130.6722744	0.007652733	56.64845276	0.01765273
96	2.599272926	0.38472297	159.9272926	0.006252841	61.52770299	0.01625284
120	3.300386895	0.30299478	230.0386895	0.004347095	69.70052203	0.01434709
180	5.995801975	0.16678336	499.5801975	0.002001681	83.32166399	0.01200168
240	10.89255365	0.091805837	989.2553654	0.001010861	90.81941635	0.01101086
360	35.94964133	0.027816689	3494.964133	0.000286126	97.21833108	0.01028613

Rate =1.5% Period	A Compound Int. Future Value	B Compound Int. Present Value	C Ordinary Annuity	D Sinking Fund	E Present Value Annuity	F Amortization
n						
1	1.015	0.985221675	1	1	0.985221675	1.015
2	1.030225	0.970661749	2.015	0.496278	1.955883424	0.51127792
3	1.045678375	0.956316994	3.045225	0.328383	2.912200417	0.34338296
4	1.061363551	0.94218423	4.090903	0.244445	3.854384648	0.25944479
5	1.077284004	0.928260325	5.152267	0.194089	4.782644973	0.20908932
6	1.093443264	0.914542193	6.229551	0.160525	5.697187165	0.17552521
7	1.109844913	0.901026791	7.322994	0.136556	6.598213956	0.15155616
8	1.126492587	0.887711124	8.432839	0.118584	7.48592508	0.13358402
9	1.143389975	0.87459224	9.559332	0.10461	8.36051732	0.11960982
10	1.160540825	0.861667232	10.70272	0.093434	9.222184552	0.10843418
11	1.177948937	0.848933233	11.86326	0.084294	10.07111779	0.09929384
12	1.195618171	0.836387422	13.04121	0.07668	10.90750521	0.09167999
13	1.213552444	0.824027017	14.23683	0.07024	11.73153222	0.08524036
14	1.231755731	0.811849277	15.45038	0.064723	12.5433815	0.07972332
15	1.250232067	0.799851505	16.68214	0.059944	13.34323301	0.07494436
16	1.268985548	0.788031039	17.93237	0.055765	14.13126405	0.07076508
17	1.288020331	0.77638526	19.20136	0.05208	14.90764931	0.06707966
18	1.307340636	0.764911587	20.48938	0.048806	15.67256089	0.06380578
19	1.326950745	0.753607474	21.79672	0.045878	16.42616837	0.06087847
20	1.346855007	0.742470418	23.12367	0.043246	17.16863879	0.05824574
21	1.367057832	0.731497949	24.47052	0.040865	17.90013673	0.0558655
22	1.387563699	0.720687634	25.83758	0.038703	18.62082437	0.05370332
23	1.408377155	0.710037078	27.22514	0.036731	19.33086145	0.05173075
24	1.429502812	0.69954392	28.63352	0.034924	20.03040537	0.0499241
25	1.450945354	0.689205832	30.06302	0.033263	20.7196112	0.04826345
26	1.472709534	0.679020524	31.51397	0.031732	21.39863172	0.04673196
27	1.494800177	0.668985738	32.98668	0.030315	22.06761746	0.04531527
28	1.51722218	0.659099249	34.48148	0.029001	22.72671671	0.04400108
29	1.539980513	0.649358866	35.9987	0.027779	23.37607558	0.04277878
30	1.56308022	0.63976243	37.53868	0.026639	24.01583801	0.04163919
31	1.586526424	0.630307813	39.10176	0.025574	24.64614582	0.0405743
32	1.61032432	0.620992919	40.68829	0.024577	25.26713874	0.0395771
33	1.634479185	0.611815684	42.29861	0.023641	25.87895442	0.03864144
34	1.658996373	0.602774073	43.93309	0.022762	26.48172849	0.03776189
35	1.683881318	0.593866081	45.59209	0.021934	27.07559458	0.03693363
36	1.709139538	0.585089735	47.27597	0.021152	27.66068431	0.0361524
37	1.734776631	0.576443089	48.98511	0.020414	28.2371274	0.03541437
38	1.760798281	0.567924226	50.71989	0.019716	28.80505163	0.03471613
39	1.787210255	0.559531257	52.48068	0.019055	29.36458288	0.03405463
40	1.814018409	0.551262322	54.26789	0.018427	29.9158452	0.0334271
41	1.841228685	0.543115588	56.08191	0.017831	30.45896079	0.03283106
42	1.868847115	0.535089249	57.92314	0.017264	30.99405004	0.03226426
43	1.896879822	0.527181526	59.79199	0.016725	31.52123157	0.03172465
44	1.925333019	0.519390666	61.68887	0.01621	32.04062223	0.03121038
48	2.043478289	0.489361695	69.56522	0.014375	34.04255365	0.029375
60	2.443219776	0.409295967	96.21465	0.010393	39.38026889	0.02539343
72	2.921157961	0.342329998	128.0772	0.007808	43.84466677	0.02280779
84	3.49258954	0.286320505	166.1726	0.006018	47.57863301	0.02101784
96	4.175803519	0.239474869	211.7202	0.004723	50.70167541	0.01972321
120	5.969322872	0.167523188	331.2882	0.003019	55.49845411	0.01801852
180	14.58436769	0.068566565	905.6245	0.001104	62.09556231	0.01610421
240	35.63281555	0.028064019	2308.854	0.000433	64.79573209	0.01543312
360	212.7037809	0.004701374	14113.59	7.09E-05	66.35324174	0.01507085

Rate = 2%	A	B	C	D	E	F
Period	Compound Int.	Compound Int.	Ordinary	Sinking	Present Value	Amortization
n	Future Value	Present Value	Annuity	Fund	Annuity	
1	1.02	0.980392157	1	1	0.980392157	1.02
2	1.0404	0.961168781	2.02	0.49505	1.941560938	0.5150495
3	1.061208	0.942322335	3.0604	0.326755	2.883883273	0.34675467
4	1.08243216	0.923845426	4.121608	0.242624	3.807728699	0.26262375
5	1.104080803	0.90573081	5.20404	0.192158	4.713459509	0.21215839
6	1.126162419	0.887971382	6.308121	0.158526	5.601430891	0.17852581
7	1.148685668	0.870560179	7.434283	0.134512	6.471991069	0.15451196
8	1.171659381	0.853490371	8.582969	0.11651	7.32548144	0.1365098
9	1.195092569	0.836755266	9.754628	0.102515	8.162236706	0.12251544
10	1.21899442	0.8203483	10.94972	0.091327	8.982585006	0.11132653
11	1.243374308	0.804263039	12.16872	0.082178	9.786848045	0.10217794
12	1.268241795	0.788493176	13.41209	0.07456	10.57534122	0.0945596
13	1.29360663	0.773032525	14.68033	0.068118	11.34837375	0.08811835
14	1.319478763	0.757875025	15.97394	0.062602	12.10624877	0.08260197
15	1.345868338	0.74301473	17.29342	0.057825	12.8492635	0.07782547
16	1.372785705	0.728445814	18.63929	0.05365	13.57770931	0.07365013
17	1.400241419	0.714162562	20.01207	0.04997	14.29187188	0.06996984
18	1.428246248	0.700159375	21.41231	0.046702	14.99203125	0.0667021
19	1.456811173	0.68643076	22.84056	0.043782	15.67846201	0.06378177
20	1.485947396	0.672971333	24.29737	0.041157	16.35143334	0.06115672
21	1.515666344	0.659775817	25.78332	0.038785	17.01120916	0.05878477
22	1.545979671	0.646839036	27.29898	0.036631	17.6580482	0.0566314
23	1.576899264	0.634155918	28.84496	0.034668	18.29220412	0.0546681
24	1.608437249	0.621721488	30.42186	0.032871	18.9139256	0.0528711
25	1.640605994	0.609530871	32.0303	0.03122	19.52345647	0.05122044
26	1.673418114	0.597579285	33.67091	0.029699	20.12103576	0.04969923
27	1.706886477	0.585862044	35.34432	0.028293	20.7068978	0.04829309
28	1.741024206	0.574374553	37.05121	0.02699	21.28127234	0.04698967
29	1.77584469	0.563112307	38.79223	0.025778	21.84438466	0.04577836
30	1.811361584	0.552070889	40.56808	0.02465	22.39645555	0.04464992
31	1.847588816	0.54124597	42.37944	0.023596	22.93770152	0.04359635
32	1.884540592	0.530633304	44.22703	0.022611	23.46833482	0.04261061
33	1.922231404	0.520228729	46.11157	0.021687	23.98856355	0.04168653
34	1.960676032	0.510028166	48.0338	0.020819	24.49859172	0.04081867
35	1.999889553	0.500027613	49.99448	0.020002	24.99861933	0.04000221
36	2.039887344	0.49022315	51.99437	0.019233	25.48884248	0.03923285
37	2.080685091	0.480610932	54.03425	0.018507	25.96945341	0.03850678
38	2.122298792	0.471187188	56.11494	0.017821	26.4406406	0.03782057
39	2.164744768	0.461948223	58.23724	0.017171	26.90258883	0.03717114
40	2.208039664	0.452890415	60.40198	0.016556	27.35547924	0.03655575
41	2.252200457	0.444010211	62.61002	0.015972	27.79948945	0.03597188
42	2.297244466	0.435304128	64.86222	0.015417	28.23479358	0.03541729
43	2.343189355	0.426768753	67.15947	0.01489	28.66156233	0.03488993
44	2.390053142	0.418400739	69.50266	0.014388	29.07996307	0.03438794
48	2.587070385	0.386537609	79.35352	0.012602	30.67311957	0.03260184
60	3.281030788	0.304782266	114.0515	0.008768	34.76088668	0.02876797
72	4.161140375	0.240318737	158.057	0.006327	37.98406314	0.02632683
84	5.277332137	0.189489684	213.8666	0.004676	40.52551579	0.02467581
96	6.69293318	0.149411323	284.6467	0.003513	42.52943386	0.02351313
120	10.76516303	0.09289223	488.2582	0.002048	45.3553885	0.0220481
180	35.32083136	0.028311904	1716.042	0.000583	48.58440478	0.02058274
240	115.8887352	0.008628966	5744.437	0.000174	49.56855168	0.02017408
360	1247.561128	0.000801564	62328.06	1.6E-05	49.9599218	0.02001604

Rate = 2.5%	A	B	C	D	E	F
Period	Compound Int.	Compound Int.	Ordinary	Sinking	Present Value	Amortization
r	Future Value	Present Value	Annuity	Fund	Annuity	
1	1.025	0.975609756	1	1	0.975609756	1.025
2	1.050625	0.951814396	2.025	0.493827	1.927424152	0.518827
3	1.076890625	0.928599411	3.075625	0.325137	2.856023563	0.350137
4	1.103812891	0.905950645	4.152516	0.240818	3.761974208	0.265818
5	1.131408213	0.883854288	5.256329	0.190247	4.645828496	0.215247
6	1.159693418	0.862296866	6.387737	0.15655	5.508125362	0.18155
7	1.188685754	0.841265235	7.54743	0.132495	6.349390597	0.157495
8	1.218402898	0.820746571	8.736116	0.114467	7.170137167	0.139467
9	1.24886297	0.800728362	9.954519	0.100457	7.970865529	0.125457
10	1.280084544	0.781198402	11.20338	0.089259	8.752063931	0.114259
11	1.312086658	0.762144782	12.48347	0.080106	9.514208713	0.105106
12	1.344888824	0.743555885	13.79555	0.072487	10.2577646	0.097487
13	1.378511045	0.725420376	15.14044	0.066048	10.98318497	0.091048
14	1.412973821	0.707727196	16.51895	0.060537	11.69091217	0.085537
15	1.448298166	0.690465557	17.93193	0.055766	12.38137773	0.080766
16	1.484505621	0.673624934	19.38022	0.051599	13.05500266	0.076599
17	1.521618261	0.657195057	20.86473	0.047928	13.71219772	0.072928
18	1.559658718	0.641165909	22.38635	0.04467	14.35336363	0.06967
19	1.598650186	0.625527716	23.94601	0.041761	14.97889134	0.066761
20	1.63861644	0.610270943	25.54466	0.039147	15.58916229	0.064147
21	1.679581851	0.595386286	27.18327	0.036787	16.18454857	0.061787
22	1.721571398	0.580864669	28.86286	0.034647	16.76541324	0.059647
23	1.764610683	0.566697238	30.58443	0.032696	17.33211048	0.057696
24	1.80872595	0.552875354	32.34904	0.030913	17.88498583	0.055913
25	1.853944098	0.539390589	34.15776	0.029276	18.42437642	0.054276
26	1.900292701	0.526234721	36.01171	0.027769	18.95061114	0.052769
27	1.947800018	0.513399728	37.912	0.026377	19.46401087	0.051377
28	1.996495019	0.500877784	39.8598	0.025088	19.96488866	0.050088
29	2.046407394	0.488661252	41.8563	0.023891	20.45354991	0.048891
30	2.097567579	0.476742685	43.9027	0.022778	20.93029259	0.047778
31	2.150006769	0.465114815	46.00027	0.021739	21.39540741	0.046739
32	2.203756938	0.453770551	48.15028	0.020768	21.84917796	0.045768
33	2.258850861	0.442702977	50.35403	0.019859	22.29188094	0.044859
34	2.315322133	0.431905343	52.61289	0.019007	22.72378628	0.044007
35	2.373205186	0.421371066	54.92821	0.018206	23.14515734	0.043206
36	2.432535316	0.411093723	57.30141	0.017452	23.55625107	0.042452
37	2.493348699	0.401067047	59.73395	0.016741	23.95731812	0.041741
38	2.555682416	0.391284924	62.2273	0.01607	24.34860304	0.04107
39	2.619574476	0.381741389	64.78298	0.015436	24.73034443	0.040436
40	2.685063838	0.372430624	67.40255	0.014836	25.10277505	0.039836
41	2.752190434	0.36334695	70.08762	0.014268	25.466122	0.039268
42	2.820995195	0.354484829	72.83981	0.013729	25.82060683	0.038729
43	2.891520075	0.345838858	75.6608	0.013217	26.16644569	0.038217
44	2.963808077	0.337403764	78.55232	0.01273	26.50384945	0.03773
48	3.271489561	0.305671157	90.85958	0.011006	27.77315371	0.036006
60	4.399789749	0.227283588	135.9916	0.007353	30.90865649	0.032353
72	5.917228062	0.168998049	196.6891	0.005084	33.24007803	0.030084
84	7.958013891	0.125659494	278.3206	0.003593	34.97362023	0.028593
96	10.70264395	0.093434856	388.1058	0.002577	36.26260574	0.027577
120	19.35814983	0.051657829	734.326	0.001362	37.93368683	0.026362
180	85.17178919	0.011740977	3366.872	0.000297	39.53036093	0.025297
240	374.737965	0.002668531	14949.52	6.69E-05	39.89325875	0.025067
360	7254.233675	0.000137851	290129.3	3.45E-06	39.99448598	0.025003

Rate =3% Period n	A Compound Int. Future Value	B Compound Int. Present Value	C Ordinary Annuity	D Sinking Fund	E Present Value Annuity	F Amortization
1	1.03	0.970873786	1	1	0.970873786	1.03
2	1.0609	0.942595909	2.03	0.492611	1.913469696	0.52261084
3	1.092727	0.915141659	3.0909	0.32353	2.828611355	0.35353036
4	1.12550881	0.888487048	4.183627	0.239027	3.717098403	0.26902705
5	1.159274074	0.862608784	5.309136	0.188355	4.579707187	0.21835457
6	1.194052297	0.837484257	6.46841	0.154598	5.417191444	0.1845975
7	1.229873865	0.813091511	7.662462	0.130506	6.230282955	0.16050635
8	1.266770081	0.789409234	8.892336	0.112456	7.01969219	0.14245639
9	1.304773184	0.766416732	10.15911	0.098434	7.786108922	0.12843386
10	1.343916379	0.744093915	11.46388	0.087231	8.530202837	0.11723051
11	1.384233871	0.722421277	12.8078	0.078077	9.252624113	0.10807745
12	1.425760887	0.70137988	14.19203	0.070462	9.954003994	0.10046209
13	1.468533713	0.68095134	15.61779	0.06403	10.63495533	0.09402954
14	1.512589725	0.661117806	17.08632	0.058526	11.29607314	0.08852634
15	1.557967417	0.641861947	18.59891	0.053767	11.93793509	0.08376658
16	1.604706439	0.623166939	20.15688	0.049611	12.56110203	0.07961085
17	1.652847632	0.605016446	21.76159	0.045953	13.16611847	0.07595253
18	1.702433061	0.587394608	23.41444	0.042709	13.75351308	0.0727087
19	1.753506053	0.570286027	25.11687	0.039814	14.32379911	0.06981388
20	1.806111235	0.553675754	26.87037	0.037216	14.87747486	0.06721571
21	1.860294572	0.537549276	28.67649	0.034872	15.41502414	0.06487178
22	1.916103409	0.521892501	30.53678	0.032747	15.93691664	0.06274739
23	1.973586511	0.506691748	32.45288	0.030814	16.44360839	0.0608139
24	2.032794106	0.491933736	34.42647	0.029047	16.93554212	0.05904742
25	2.09377793	0.477605569	36.45926	0.027428	17.41314769	0.05742787
26	2.156591268	0.463694727	38.55304	0.025938	17.87684242	0.05593829
27	2.221289006	0.450189056	40.70963	0.024564	18.32703147	0.05456421
28	2.287927676	0.437076753	42.93092	0.023293	18.76410823	0.05329323
29	2.356565506	0.424346362	45.21885	0.022115	19.18845459	0.05211467
30	2.427262471	0.41198676	47.57542	0.021019	19.60044135	0.05101926
31	2.500080345	0.399987145	50.00268	0.019999	20.00042849	0.04999893
32	2.575082756	0.388337034	52.50276	0.019047	20.38876553	0.04904662
33	2.652335238	0.377026247	55.07784	0.018156	20.76579178	0.04815612
34	2.731905296	0.3660449	57.73018	0.017322	21.13183668	0.04732196
35	2.813862454	0.355383398	60.46208	0.016539	21.48722007	0.04653929
36	2.898278328	0.345032425	63.27594	0.015804	21.8322525	0.04580379
37	2.985226678	0.334982937	66.17422	0.015112	22.16723544	0.04511162
38	3.074783478	0.325226152	69.15945	0.014459	22.49246159	0.04445934
39	3.167026983	0.315753546	72.23423	0.013844	22.80821513	0.04384385
40	3.262037792	0.306556841	75.40126	0.013262	23.11477197	0.04326238
41	3.359898926	0.297628001	78.6633	0.012712	23.41239997	0.04271241
42	3.460695894	0.288959224	82.0232	0.012192	23.7013592	0.04219167
43	3.56451677	0.280542936	85.48389	0.011698	23.98190213	0.04169811
44	3.671452273	0.272371782	89.04841	0.01123	24.25427392	0.04122985
48	4.132251879	0.241998801	104.4084	0.009578	25.26670664	0.03957777
60	5.891603104	0.16973309	163.0534	0.006133	27.67556367	0.03613296
72	8.400017267	0.119047374	246.6672	0.004054	29.36508752	0.03405404
84	11.97641607	0.083497433	365.8805	0.002733	30.55008556	0.03273313
96	17.07550559	0.05856342	535.8502	0.001866	31.38121934	0.03186619
120	34.71098714	0.028809322	1123.7	0.00089	32.37302261	0.03088992
180	204.5033596	0.004889895	6783.445	0.000147	33.17033683	0.03014742
240	1204.852628	0.000829977	40128.42	2.49E-05	33.30566743	0.03002492
360	41821.62407	2.39111E-05	1394021	7.17E-07	33.3325363	0.03000072

Rate =3.5%	A	B	C	D	E	F
Period	Compound Int.	Compound Int.	Ordinary	Sinking	Present Value	Amortization
n	Future Value	Present Value	Annuity	Fund	Annuity	
1	1.035	0.966183575	1	1	0.966183575	1.035
2	1.071225	0.942595909	2.035	0.4914	1.899694275	0.52640049
3	1.108717875	0.915141659	3.106225	0.321934	2.801636981	0.35693418
4	1.147523001	0.888487048	4.214943	0.237251	3.673079209	0.27225114
5	1.187686306	0.862608784	5.362466	0.186481	4.515052375	0.22148137
6	1.229255326	0.837484257	6.550152	0.152668	5.32855302	0.18766821
7	1.272279263	0.813091511	7.779408	0.128544	6.11454398	0.16354449
8	1.316809037	0.789409234	9.051687	0.110477	6.873955537	0.14547665
9	1.362897353	0.766416732	10.3685	0.096446	7.607686509	0.13144601
10	1.410598761	0.744093915	11.73139	0.085241	8.316605323	0.12024137
11	1.459969717	0.722421277	13.14199	0.076092	9.001551036	0.11109197
12	1.511068657	0.70137988	14.60196	0.068484	9.663334335	0.10348395
13	1.56395606	0.68095134	16.11303	0.062062	10.30273849	0.09706157
14	1.618694522	0.661117806	17.67699	0.056571	10.92052028	0.09157073
15	1.675348831	0.641861947	19.29568	0.051825	11.5174109	0.08682507
16	1.73398604	0.623166939	20.97103	0.047685	12.09411681	0.08268483
17	1.794675551	0.605016446	22.70502	0.044043	12.65132059	0.07904313
18	1.857489196	0.587394608	24.49969	0.040817	13.18968173	0.07581684
19	1.922501317	0.570286027	26.35718	0.03794	13.70983742	0.07294033
20	1.989788863	0.553675754	28.27968	0.035361	14.2124033	0.07036108
21	2.059431474	0.537549276	30.26947	0.033037	14.6979742	0.06803659
22	2.131511575	0.521892501	32.3289	0.030932	15.16712484	0.06593207
23	2.20611448	0.506691748	34.46041	0.029019	15.62041047	0.0640188
24	2.283328487	0.491933736	36.66653	0.027273	16.0583676	0.06227283
25	2.363244984	0.477605569	38.94986	0.025674	16.48151459	0.06067404
26	2.445958559	0.463694727	41.3131	0.024205	16.89035226	0.0592054
27	2.531567108	0.450189056	43.75906	0.022852	17.28536451	0.05785241
28	2.620171957	0.437076753	46.29063	0.021603	17.66701885	0.05660265
29	2.711877976	0.424346362	48.9108	0.020445	18.035767	0.05544538
30	2.806793705	0.41198676	51.62268	0.019371	18.39204541	0.05437133
31	2.905031484	0.399987145	54.42947	0.018372	18.73627576	0.0533724
32	3.006707586	0.388337034	57.3345	0.017442	19.06886547	0.0524415
33	3.111942352	0.377026247	60.34121	0.016572	19.39020818	0.05157242
34	3.220860334	0.3660449	63.45315	0.01576	19.70068423	0.05075966
35	3.333590446	0.355383398	66.67401	0.014998	20.0006611	0.04999835
36	3.450266111	0.345032425	70.0076	0.014284	20.29049381	0.04928416
37	3.571025425	0.334982937	73.45787	0.013613	20.57052542	0.04861325
38	3.696011315	0.325226152	77.02889	0.012982	20.84108736	0.04798214
39	3.825371711	0.315753546	80.72491	0.012388	21.10249987	0.04738775
40	3.959259721	0.306556841	84.55028	0.011827	21.35507234	0.04682728
41	4.097833811	0.297628001	88.50954	0.011298	21.59910371	0.04629822
42	4.241257995	0.288959224	92.60737	0.010798	21.83488281	0.04579828
43	4.389702025	0.280542936	96.84863	0.010325	22.0626887	0.04532539
44	4.543341595	0.272371782	101.2383	0.009878	22.28279102	0.04487768
48	5.213588981	0.241998801	120.3883	0.008306	23.09124425	0.04330646
60	7.878090901	0.16973309	196.5169	0.005089	24.94473412	0.04008862
72	11.90433624	0.119047374	311.5525	0.00321	26.17134275	0.03820973
84	17.98826938	0.083497433	485.3791	0.00206	26.98309186	0.03706025
96	27.18151006	0.05856342	748.0431	0.001337	27.52029387	0.03633682
120	62.06431624	0.028809322	1744.695	0.000573	28.11107663	0.03557317
180	488.948325	0.004889895	13941.38	7.17E-05	28.51299412	0.03507173
240	3851.97935	0.000829977	110028	9.09E-06	28.56401123	0.03500909
360	239070.4646	2.39111E-05	6830556	1.46E-07	28.57130906	0.03500015

Rate =4%	A	B	C	D	E	F
Period	Compound Int.	Compound Int.	Ordinary	Sinking	Present Value	Amortization
n	Future Value	Present Value	Annuity	Fund	Annuity	
1	1.04	0.961538462	1	1	0.961538462	1.04
2	1.0816	0.924556213	2.04	0.490196	1.886094675	0.53019608
3	1.124864	0.888996359	3.1216	0.320349	2.775091033	0.36034854
4	1.16985856	0.854804191	4.246464	0.23549	3.629895224	0.27549005
5	1.216652902	0.821927107	5.416323	0.184627	4.451822331	0.22462711
6	1.265319018	0.790314526	6.632975	0.150762	5.242136857	0.1907619
7	1.315931779	0.759917813	7.898294	0.12661	6.00205467	0.16660961
8	1.36856905	0.730690205	9.214226	0.108528	6.732744875	0.14852783
9	1.423311812	0.702586736	10.5828	0.094493	7.435331611	0.13449299
10	1.480244285	0.675564169	12.00611	0.083291	8.110895779	0.12329094
11	1.539454056	0.649580932	13.48635	0.074149	8.760476711	0.11414904
12	1.601032219	0.62459705	15.02581	0.066552	9.38507376	0.10655217
13	1.665073507	0.600574086	16.62684	0.060144	9.985647847	0.10014373
14	1.731676448	0.577475083	18.29191	0.054669	10.56312293	0.09466897
15	1.800943506	0.555264503	20.02359	0.049941	11.11838743	0.0899411
16	1.872981246	0.533908176	21.82453	0.04582	11.65229561	0.08582
17	1.947900496	0.513373246	23.69751	0.042199	12.16566885	0.08219852
18	2.025816515	0.493628121	25.64541	0.038993	12.65929697	0.07899333
19	2.106849176	0.474642424	27.67123	0.036139	13.1339394	0.07613862
20	2.191123143	0.456386946	29.77808	0.033582	13.59032634	0.07358175
21	2.278768069	0.438833602	31.9692	0.03128	14.02915995	0.07128011
22	2.369918792	0.421955387	34.24797	0.029199	14.45111533	0.06919881
23	2.464715543	0.405726333	36.61789	0.027309	14.85684167	0.06730906
24	2.563304165	0.390121474	39.0826	0.025587	15.24696314	0.06558683
25	2.665836331	0.375116802	41.64591	0.024012	15.62207994	0.06401196
26	2.772469785	0.360689233	44.31174	0.022567	15.98276918	0.06256738
27	2.883368576	0.34681657	47.08421	0.021239	16.32958575	0.06123854
28	2.998703319	0.333477471	49.96758	0.020013	16.66306322	0.06001298
29	3.118651452	0.320651415	52.96629	0.01888	16.98371463	0.05887993
30	3.24339751	0.308318668	56.08494	0.01783	17.2920333	0.0578301
31	3.37313341	0.296460258	59.32834	0.016855	17.58849356	0.05685535
32	3.508058747	0.28505794	62.70147	0.015949	17.8735515	0.05594859
33	3.648381097	0.274094173	66.20953	0.015104	18.14764567	0.05510357
34	3.794316341	0.26355209	69.85791	0.014315	18.41119776	0.05431477
35	3.946088994	0.253415471	73.65222	0.013577	18.66461323	0.05357732
36	4.103932554	0.243668722	77.59831	0.012887	18.90828195	0.05288688
37	4.268089856	0.234296848	81.70225	0.01224	19.1425788	0.05223957
38	4.43881345	0.225285431	85.97034	0.011632	19.36786423	0.05163192
39	4.616365988	0.216620606	90.40915	0.011061	19.58448484	0.05106083
40	4.801020628	0.208289045	95.02552	0.010523	19.79277388	0.05052349
41	4.993061453	0.200277928	99.82654	0.010017	19.99305181	0.05001738
42	5.192783911	0.19257493	104.8196	0.00954	20.18562674	0.0495402
43	5.400495268	0.185168202	110.0124	0.00909	20.37079494	0.04908989
44	5.616515078	0.178046348	115.4129	0.008665	20.54884129	0.04866454
48	6.570528242	0.152194765	139.2632	0.007181	21.19513088	0.04718065
60	10.51962741	0.095060401	237.9907	0.004202	22.62348997	0.04420185
72	16.84226241	0.059374446	396.0566	0.002525	23.51563885	0.04252489
84	26.96500475	0.037085104	649.1251	0.001541	24.07287241	0.04154054
96	43.17184138	0.023163246	1054.296	0.000949	24.42091884	0.0409485
120	110.6625608	0.00903648	2741.564	0.000365	24.774088	0.04036476
180	1164.128908	0.000859011	29078.22	3.44E-05	24.97852472	0.04003439
240	12246.20236	8.1658E-05	306130.1	3.27E-06	24.99795855	0.04000327
360	1355196.114	7.37901E-07	33879878	2.95E-08	24.99998155	0.04000003

Note: 7.38E-07 means move the decimal point 7 places to the left

Rate =4.5%	A	B	C	D	E	F
Period	Compound Int.	Compound Int.	Ordinary	Sinking	Present Value	Amortization
n	Future Value	Present Value	Annuity	Fund	Annuity	
1	1.045	0.956937799	1	1	0.956937799	1.045
2	1.092025	0.915729951	2.045	0.488997555	1.87266775	0.533997555
3	1.141166125	0.876296604	3.137025	0.31877336	2.748964354	0.36377336
4	1.192518601	0.838561344	4.278191125	0.233743648	3.587525698	0.278743648
5	1.246181938	0.802451047	5.470709726	0.18279164	4.389976744	0.22779164
6	1.302260125	0.767895738	6.716891663	0.148878388	5.157872483	0.193878388
7	1.36086183	0.734828458	8.019151788	0.124701468	5.89270094	0.169701468
8	1.422100613	0.703185127	9.380013619	0.106609653	6.595886067	0.151609653
9	1.48609514	0.672904428	10.80211423	0.09257447	7.268790495	0.13757447
10	1.552969422	0.643927682	12.28820937	0.081378822	7.912718177	0.126378822
11	1.622853046	0.616198739	13.84117879	0.072248182	8.528916916	0.117248182
12	1.695881433	0.589663865	15.46403184	0.064666189	9.118580781	0.109666189
13	1.772196097	0.564271641	17.15991327	0.058275353	9.682852422	0.103275353
14	1.851944922	0.539972862	18.93210937	0.052820316	10.22282528	0.097820316
15	1.935282443	0.516720442	20.78405429	0.048113808	10.73954573	0.093113808
16	2.022370153	0.494469323	22.71933673	0.044015369	11.23401505	0.089015369
17	2.11337681	0.473176385	24.74170689	0.040417583	11.70719143	0.085417583
18	2.208478766	0.452800369	26.8550837	0.037236898	12.1599918	0.082236898
19	2.307860311	0.433301788	29.06356246	0.034407344	12.59329359	0.079407344
20	2.411714025	0.41464286	31.37142277	0.031876144	13.00793645	0.076876144
21	2.520241156	0.396787426	33.7831368	0.029600567	13.40472388	0.074600567
22	2.633652008	0.379700886	36.30337795	0.027545646	13.78442476	0.072545646
23	2.752166348	0.36335013	38.93702996	0.025682493	14.14777489	0.070682493
24	2.876013834	0.347703474	41.68919631	0.02398703	14.49547837	0.06898703
25	3.005434457	0.332730597	44.56521015	0.022439028	14.82820896	0.067439028
26	3.140679007	0.318402485	47.5706446	0.021021367	15.14661145	0.066021367
27	3.282009562	0.304691373	50.71132361	0.019719462	15.45130282	0.064719462
28	3.429699993	0.291570692	53.99333317	0.018520805	15.74287351	0.063520805
29	3.584036492	0.279015016	57.42303316	0.017414615	16.02188853	0.062414615
30	3.745318135	0.267000016	61.00706966	0.016391543	16.28888854	0.061391543
31	3.913857451	0.255502407	64.75238779	0.015443446	16.54439095	0.060443446
32	4.089981036	0.244499911	68.66624524	0.014563196	16.78889086	0.059563196
33	4.274030182	0.233971207	72.75622628	0.013744528	17.02286207	0.058744528
34	4.466361541	0.223895892	77.03025646	0.012981912	17.24675796	0.057981912
35	4.66734781	0.214254442	81.496618	0.012270448	17.4610124	0.057270448
36	4.877378461	0.205028174	86.16396581	0.01160578	17.66604058	0.05660578
37	5.096860492	0.19619921	91.04134427	0.010984021	17.86223979	0.055984021
38	5.326219214	0.18775044	96.13820476	0.010401692	18.04999023	0.055401692
39	5.565899079	0.179665493	101.464424	0.009855671	18.22965572	0.054855671
40	5.816364538	0.171928701	107.0303231	0.009343147	18.40158442	0.054343147
41	6.078100942	0.164525073	112.8466876	0.00886158	18.56610949	0.05386158
42	6.351615484	0.157440261	118.9247885	0.008408676	18.72354975	0.053408676
43	6.637438181	0.150660537	125.276404	0.007982349	18.87421029	0.052982349
44	6.936122899	0.144172763	131.9138422	0.007580706	19.01838305	0.052580706
48	8.271455573	0.120897706	161.5879016	0.006188582	19.53560654	0.051188582
60	14.02740793	0.071289008	289.497954	0.003454256	20.63802204	0.048454256
72	23.78882066	0.042036552	506.4182368	0.001974652	21.28807662	0.046974652
84	40.34301926	0.024787436	874.2893169	0.001143786	21.67139032	0.046143786
96	68.4169773	0.014616255	1498.155051	0.000667488	21.89741655	0.045667488
120	196.7681732	0.005082123	4350.403849	0.000229864	22.10928616	0.045229864

Rate =5%	A	B	C	D	E	F
Period	Compound Int.	Compound Int.	Ordinary	Sinking	Present Value	Amortization
n	Future Value	Present Value	Annuity	Fund	Annuity	
1	1.05	0.952380952	1	1	0.952380952	1.05
2	1.1025	0.907029478	2.05	0.488998	1.859410431	0.53780488
3	1.157625	0.863837599	3.1525	0.318773	2.723248029	0.36720856
4	1.21550625	0.822702475	4.310125	0.233744	3.545950504	0.28201183
5	1.276281563	0.783526166	5.525631	0.182792	4.329476671	0.2309748
6	1.340095641	0.746215397	6.801913	0.148878	5.075692067	0.19701747
7	1.407100423	0.71068133	8.142008	0.124701	5.786373397	0.17281982
8	1.477455444	0.676839362	9.549109	0.10661	6.463212759	0.15472181
9	1.551328216	0.644608916	11.02656	0.092574	7.107821676	0.14069008
10	1.628894627	0.613913254	12.57789	0.081379	7.721734929	0.12950457
11	1.710339358	0.584679289	14.20679	0.072248	8.306414218	0.12038889
12	1.795856326	0.556837418	15.91713	0.064666	8.863251636	0.11282541
13	1.885649142	0.530321351	17.71298	0.058275	9.393572987	0.10645577
14	1.979931599	0.505067953	19.59863	0.05282	9.89864094	0.10102397
15	2.078928179	0.481017098	21.57856	0.048114	10.37965804	0.09634229
16	2.182874588	0.458111522	23.65749	0.044015	10.83776956	0.09226991
17	2.292018318	0.436296688	25.84037	0.040418	11.27406625	0.08869914
18	2.406619234	0.415520655	28.13238	0.037237	11.6895869	0.08554622
19	2.526950195	0.395733957	30.539	0.034407	12.08532086	0.08274501
20	2.653297705	0.376889483	33.06595	0.031876	12.46221034	0.08024259
21	2.78596259	0.358942365	35.71925	0.029601	12.82115271	0.07799611
22	2.92526072	0.341849871	38.50521	0.027546	13.16300258	0.07597051
23	3.071523756	0.325571306	41.43048	0.025682	13.48857388	0.07413682
24	3.225099944	0.31006791	44.502	0.023987	13.79864179	0.0724709
25	3.386354941	0.295302772	47.7271	0.022439	14.09394457	0.07095246
26	3.555672688	0.281240735	51.11345	0.021021	14.3751853	0.06956432
27	3.733456322	0.267848319	54.66913	0.019719	14.64303362	0.06829186
28	3.920129138	0.255093637	58.40258	0.018521	14.89812726	0.06712253
29	4.116135595	0.242946321	62.32271	0.017415	15.14107358	0.06604551
30	4.321942375	0.231377449	66.43885	0.016392	15.37245103	0.06505144
31	4.538039494	0.220359475	70.76079	0.015443	15.5928105	0.06413212
32	4.764941469	0.209866167	75.29883	0.014563	15.80267667	0.06328042
33	5.003188542	0.19987254	80.06377	0.013745	16.00254921	0.06249004
34	5.253347969	0.1903548	85.06696	0.012982	16.19290401	0.06175545
35	5.516015368	0.181290285	90.32031	0.01227	16.37419429	0.06107171
36	5.791816136	0.172657415	95.83632	0.011606	16.54685171	0.06043446
37	6.081406943	0.164435633	101.6281	0.010984	16.71128734	0.05983979
38	6.38547729	0.156605365	107.7095	0.010402	16.86789271	0.05928423
39	6.704751154	0.149147966	114.095	0.009856	17.01704067	0.05876462
40	7.039988712	0.142045682	120.7998	0.009343	17.15908635	0.05827816
41	7.391988148	0.135281602	127.8398	0.008862	17.29436796	0.05782229
42	7.761587555	0.128839621	135.2318	0.008409	17.42320758	0.05739471
43	8.149666933	0.122704401	142.9933	0.007982	17.54591198	0.05699333
44	8.55715028	0.116861334	151.143	0.007581	17.66277333	0.05661625
48	10.40126965	0.096142109	188.0254	0.006189	18.07715782	0.05531843
60	18.67918589	0.053535524	353.5837	0.003454	18.92928953	0.05282818
72	33.54513415	0.029810583	650.9027	0.001975	19.40378834	0.05153633
84	60.24224138	0.016599648	1184.845	0.001144	19.66800704	0.05084399
96	108.1864103	0.009243305	2143.728	0.000667	19.8151339	0.05046648
120	348.9119857	0.002866052	6958.24	0.00023	19.94267895	0.05014371

Rate =5.5%	A	B	C	D	E	F
Period	Compound Int.	Compound Int.	Ordinary	Sinking	Present Value	Amortization
n	Future Value	Present Value	Annuity	Fund	Annuity	
1	1.055	0.947867299	1	1	0.947867299	1.055
2	1.113025	0.898452416	2.055	0.486618	1.846319714	0.541618
3	1.174241375	0.851613664	3.168025	0.315654	2.697933378	0.37065407
4	1.238824651	0.807216743	4.342266	0.230294	3.505150122	0.28529449
5	1.306960006	0.765134354	5.581091	0.179176	4.270284476	0.23417644
6	1.378842807	0.725245833	6.888051	0.145179	4.995530309	0.20017895
7	1.454679161	0.687436809	8.266894	0.120964	5.682967117	0.17596442
8	1.534686515	0.651598871	9.721573	0.102864	6.334565988	0.15786401
9	1.619094273	0.617629261	11.25626	0.088839	6.952195249	0.14383946
10	1.708144458	0.585430579	12.87535	0.077668	7.537625829	0.13266777
11	1.802092404	0.554910502	14.5835	0.068571	8.09253633	0.12357065
12	1.901207486	0.525981518	16.38559	0.061029	8.618517849	0.11602923
13	2.005773897	0.498560681	18.2868	0.054684	9.11707853	0.10968426
14	2.116091462	0.472569366	20.29257	0.049279	9.589647895	0.10427912
15	2.232476492	0.447933048	22.40866	0.044626	10.03758094	0.0996256
16	2.355262699	0.424581088	24.64114	0.040583	10.46216203	0.09558254
17	2.484802148	0.402446529	26.9964	0.037042	10.86460856	0.09204197
18	2.621466266	0.381465904	29.4812	0.03392	11.24607447	0.08891992
19	2.765646911	0.361579056	32.10267	0.03115	11.60765352	0.08615006
20	2.917757491	0.342728963	34.86832	0.028679	11.95038248	0.08367933
21	3.078234153	0.324861577	37.78608	0.026465	12.27524406	0.08146478
22	3.247537031	0.307925665	40.86431	0.024471	12.58316973	0.07947123
23	3.426151568	0.291872668	44.11185	0.02267	12.87504239	0.07766965
24	3.614589904	0.276656558	47.538	0.021036	13.15169895	0.0760358
25	3.813392349	0.262233704	51.15259	0.019549	13.41393266	0.07454935
26	4.023128928	0.248562753	54.96598	0.018193	13.66249541	0.07319307
27	4.244401019	0.235604505	58.98911	0.016952	13.89809991	0.07195228
28	4.477843075	0.223321805	63.23351	0.015814	14.12142172	0.0708144
29	4.724124444	0.211679436	67.71135	0.014769	14.33310116	0.06976857
30	4.983951288	0.200644016	72.43548	0.013805	14.53374517	0.06880539
31	5.258068609	0.190183901	77.41943	0.012917	14.72392907	0.06791665
32	5.547262383	0.180269101	82.6775	0.012095	14.90419817	0.06709519
33	5.852361814	0.170871185	88.22476	0.011335	15.07506936	0.06633469
34	6.174241714	0.161963209	94.07712	0.01063	15.23703257	0.06562958
35	6.513825008	0.153519629	100.2514	0.009975	15.3905522	0.06497493
36	6.872085383	0.145516236	106.7652	0.009366	15.53606843	0.06436635
37	7.250050079	0.137930082	113.6373	0.0088	15.67399851	0.06379993
38	7.648802834	0.130739414	120.8873	0.008272	15.80473793	0.06327217
39	8.06948699	0.123923615	128.5361	0.00778	15.92866154	0.06277991
40	8.513308774	0.117463142	136.6056	0.00732	16.04612469	0.06232034
41	8.981540757	0.111339471	145.1189	0.006891	16.15746416	0.0618909
42	9.475525498	0.105535044	154.1005	0.006489	16.2629992	0.06148927
43	9.996679401	0.100033217	163.576	0.006113	16.36303242	0.06111337
44	10.54649677	0.094818215	173.5727	0.005761	16.45785063	0.06076128
48	13.06526017	0.076538851	219.3684	0.004559	16.79020271	0.05955854
60	24.83977045	0.040258021	433.4504	0.002307	17.44985416	0.05730707
72	47.22555751	0.021174975	840.4647	0.00119	17.79681864	0.05618982
84	89.78558347	0.011137646	1614.283	0.000619	17.97931554	0.05561947

Rate =6%	A	B	C	D	E	F
Period	Compound Int.	Compound Int.	Ordinary	Sinking	Present Value	Amortization
n	Future Value	Present Value	Annuity	Fund	Annuity	
1	1.06	0.943396226	1	1	0.943396226	1.06
2	1.1236	0.88999644	2.06	0.485437	1.833392666	0.54543689
3	1.191016	0.839619283	3.1836	0.31411	2.673011949	0.37410981
4	1.26247696	0.792093663	4.374616	0.228591	3.465105613	0.28859149
5	1.338225578	0.747258173	5.637093	0.177396	4.212363786	0.2373964
6	1.418519112	0.70496054	6.975319	0.143363	4.917324326	0.20336263
7	1.503630259	0.665057114	8.393838	0.119135	5.58238144	0.17913502
8	1.593848075	0.627412371	9.897468	0.101036	6.209793811	0.16103594
9	1.689478959	0.591898464	11.49132	0.087022	6.801692274	0.14702224
10	1.790847697	0.558394777	13.18079	0.075868	7.360087051	0.13586796
11	1.898298558	0.526787525	14.97164	0.066793	7.886874577	0.12679294
12	2.012196472	0.496969364	16.86994	0.059277	8.38384394	0.11927703
13	2.13292826	0.468839022	18.88214	0.05296	8.852682963	0.11296011
14	2.260903956	0.442300964	21.01507	0.047585	9.294983927	0.10758491
15	2.396558193	0.417265061	23.27597	0.042963	9.712248988	0.10296276
16	2.540351685	0.393646284	25.67253	0.038952	10.10589527	0.09895214
17	2.692772786	0.371364419	28.21288	0.035445	10.47725969	0.0954448
18	2.854339153	0.350343791	30.90565	0.032357	10.82760348	0.09235654
19	3.025599502	0.33051301	33.75999	0.029621	11.15811649	0.08962086
20	3.207135472	0.311804727	36.78559	0.027185	11.46992122	0.08718456
21	3.399563601	0.294155403	39.99273	0.025005	11.76407662	0.08500455
22	3.603537417	0.277505097	43.39229	0.023046	12.04158172	0.08304557
23	3.819749662	0.261797261	46.99583	0.021278	12.30337898	0.08127848
24	4.048934641	0.246978548	50.81558	0.019679	12.55035753	0.079679
25	4.29187072	0.232998631	54.86451	0.018227	12.78335616	0.07822672
26	4.549382963	0.219810029	59.15638	0.016904	13.00316619	0.07690435
27	4.822345941	0.207367952	63.70577	0.015697	13.21053414	0.07569717
28	5.111686697	0.195630143	68.52811	0.014593	13.40616428	0.07459255
29	5.418387899	0.184556739	73.6398	0.01358	13.59072102	0.07357961
30	5.743491173	0.174110131	79.05819	0.012649	13.76483115	0.07264891
31	6.088100643	0.16425484	84.80168	0.011792	13.92908599	0.07179222
32	6.453386682	0.154957397	90.88978	0.011002	14.08404339	0.07100234
33	6.840589883	0.146186223	97.34316	0.010273	14.23022961	0.07027293
34	7.251025276	0.137911531	104.1838	0.009598	14.36814114	0.06959843
35	7.686086792	0.130105218	111.4348	0.008974	14.49824636	0.06897386
36	8.147252	0.122740772	119.1209	0.008395	14.62098713	0.06839483
37	8.63608712	0.115793181	127.2681	0.007857	14.73678031	0.06785743
38	9.154252347	0.10923885	135.9042	0.007358	14.84601916	0.06735812
39	9.703507488	0.103055519	145.0585	0.006894	14.94907468	0.06689377
40	10.28571794	0.097222188	154.762	0.006462	15.04629687	0.06646154
41	10.90286101	0.091719045	165.0477	0.006059	15.13801592	0.06605886
42	11.55703267	0.086527401	175.9505	0.005683	15.22454332	0.06568342
43	12.25045463	0.081629624	187.5076	0.005333	15.30617294	0.06533312
44	12.98548191	0.077009079	199.758	0.005006	15.38318202	0.06500606
48	16.39387173	0.060998403	256.5645	0.003898	15.65002661	0.06389765
60	32.98769085	0.030314338	533.1282	0.001876	16.16142771	0.06187572
72	66.37771515	0.015065297	1089.629	0.000918	16.41557838	0.06091774
84	133.5650042	0.007486991	2209.417	0.000453	16.54188348	0.06045261

Rate =6.5%	A	B	C	D	E	F
Period	Compound Int.	Compound Int.	Ordinary	Sinking	Present Value	Amortization
n	Future Value	Present Value	Annuity	Fund	Annuity	
1	1.065	0.938967136	1	1	0.938967136	1.065
2	1.134225	0.881659283	2.065	0.484262	1.820626419	0.5492615
3	1.207949625	0.827849092	3.199225	0.312576	2.648475511	0.3775757
4	1.286466351	0.777323091	4.407175	0.226903	3.425798602	0.29190274
5	1.370086663	0.729880837	5.693641	0.175635	4.155679438	0.24063454
6	1.459142297	0.685334119	7.063728	0.141568	4.841013557	0.20656831
7	1.553986546	0.643506215	8.52287	0.117331	5.484519772	0.18233137
8	1.654995671	0.604231188	10.07686	0.099237	6.088750959	0.1642373
9	1.76257039	0.567353228	11.73185	0.085238	6.656104187	0.15023803
10	1.877137465	0.532726036	13.49442	0.074105	7.188830223	0.13910469
11	1.999151401	0.50021224	15.37156	0.065055	7.689042463	0.13005521
12	2.129096242	0.469682854	17.37071	0.057568	8.158725317	0.12256817
13	2.267487497	0.441016765	19.49981	0.051283	8.599742082	0.11628256
14	2.414874185	0.414100249	21.7673	0.04594	9.01384233	0.11094048
15	2.571841007	0.388826524	24.18217	0.041353	9.402668855	0.10635278
16	2.739010672	0.365095328	26.75401	0.037378	9.767764183	0.10237757
17	2.917046366	0.342812515	29.49302	0.033906	10.1105767	0.09890633
18	3.106654379	0.321889685	32.41007	0.030855	10.43246638	0.09585461
19	3.308586914	0.302243836	35.51672	0.028156	10.73471022	0.09315575
20	3.523645064	0.283797029	38.82531	0.025756	11.01850725	0.0907564
21	3.752681993	0.266476083	42.34895	0.023613	11.28498333	0.08861333
22	3.996606322	0.250212285	46.10164	0.021691	11.53519562	0.0866912
23	4.256385733	0.234941113	50.09824	0.019961	11.77013673	0.08496078
24	4.533050806	0.220601984	54.35463	0.018398	11.99073871	0.0833977
25	4.827699108	0.207138013	58.88768	0.016981	12.19787673	0.08198148
26	5.14149955	0.194495787	63.71538	0.015695	12.39237251	0.0806948
27	5.475697021	0.182625152	68.85688	0.014523	12.57499766	0.07952288
28	5.831617327	0.171479016	74.33257	0.013453	12.74647668	0.07845305
29	6.210672454	0.16101316	80.16419	0.012474	12.90748984	0.0774744
30	6.614366163	0.151186066	86.37486	0.011577	13.05867591	0.07657744
31	7.044299964	0.141958748	92.98923	0.010754	13.20063465	0.07575393
32	7.502179461	0.133294599	100.0335	0.009997	13.33392925	0.07499665
33	7.989821126	0.125159248	107.5357	0.009299	13.4590885	0.07429924
34	8.509159499	0.11752042	115.5255	0.008656	13.57660892	0.0736561
35	9.062254867	0.110347812	124.0347	0.008062	13.68695673	0.07306226
36	9.651301433	0.103612969	133.0969	0.007513	13.7905697	0.07251332
37	10.27863603	0.097289173	142.7482	0.007005	13.88785887	0.07200534
38	10.94674737	0.091351336	153.0269	0.006535	13.97921021	0.0715348
39	11.65828595	0.085775903	163.9736	0.006099	14.06498611	0.07109854
40	12.41607453	0.080540754	175.6319	0.005694	14.14552687	0.07069373
41	13.22311938	0.075625121	188.048	0.005318	14.22115199	0.07031779
42	14.08262214	0.071009503	201.2711	0.004968	14.29216149	0.06996842
43	14.99799258	0.06667559	215.3537	0.004644	14.35883708	0.06964352
44	15.97286209	0.062606188	230.3517	0.004341	14.42144327	0.06934119
48	20.54854961	0.048665235	300.7469	0.003325	14.63591946	0.06832505
60	43.74983974	0.022857227	657.6898	0.00152	15.03296574	0.06652047
72	93.14761936	0.010735647	1417.656	0.000705	15.21945158	0.06570539
84	198.3202463	0.00504235	3035.696	0.000329	15.30704078	0.06532941

Rate =7% Period n	A Compound Int. Future Value	B Compound Int. Present Value	C Ordinary Annuity	D Sinking Fund	E Present Value Annuity	F Amortization
1	1.07	0.934579439	1	1	0.934579439	1.07
2	1.1449	0.873438728	2.07	0.483092	1.808018168	0.55309179
3	1.225043	0.816297877	3.2149	0.311052	2.624316044	0.38105167
4	1.31079601	0.762895212	4.439943	0.225228	3.387211256	0.29522812
5	1.402551731	0.712986179	5.750739	0.173891	4.100197436	0.24389069
6	1.500730352	0.666342224	7.153291	0.139796	4.76653966	0.2097958
7	1.605781476	0.622749742	8.654021	0.115553	5.389289402	0.18555322
8	1.71818618	0.582009105	10.2598	0.097468	5.971298506	0.16746776
9	1.838459212	0.543933743	11.97799	0.083486	6.515232249	0.15348647
10	1.967151357	0.508349292	13.81645	0.072378	7.023581541	0.1423775
11	2.104851952	0.475092796	15.7836	0.063357	7.498674337	0.1333569
12	2.252191589	0.444011959	17.88845	0.055902	7.942686297	0.12590199
13	2.409845	0.414964448	20.14064	0.049651	8.357650744	0.11965085
14	2.57853415	0.387817241	22.55049	0.044345	8.745467985	0.11434494
15	2.759031541	0.36244602	25.12902	0.039795	9.107914005	0.10979462
16	2.952163749	0.338734598	27.88805	0.035858	9.446648603	0.10585765
17	3.158815211	0.31657439	30.84022	0.032425	9.763222993	0.10242519
18	3.379932276	0.295863916	33.99903	0.029413	10.05908691	0.0994126
19	3.616527535	0.276508333	37.37896	0.026753	10.33559524	0.09675301
20	3.869684462	0.258419003	40.99549	0.024393	10.59401425	0.09439293
21	4.140562375	0.241513087	44.86518	0.022289	10.83552733	0.092289
22	4.430401741	0.225713165	49.00574	0.020406	11.0612405	0.09040577
23	4.740529863	0.210946883	53.43614	0.018714	11.27218738	0.08871393
24	5.072366953	0.19714662	58.17667	0.017189	11.469334	0.08718902
25	5.42743264	0.184249178	63.24904	0.015811	11.65358318	0.08581052
26	5.807352925	0.172195493	68.67647	0.014561	11.82577867	0.08456103
27	6.21386763	0.160930367	74.48382	0.013426	11.98670904	0.08342573
28	6.648838364	0.150402212	80.69769	0.012392	12.13711125	0.08239193
29	7.114257049	0.140562815	87.34653	0.011449	12.27767407	0.08144865
30	7.612255043	0.131367117	94.46079	0.010586	12.40904118	0.0805864
31	8.145112896	0.122773007	102.073	0.009797	12.53181419	0.07979691
32	8.715270798	0.114741128	110.2182	0.009073	12.64655532	0.07907292
33	9.325339754	0.107234699	118.9334	0.008408	12.75379002	0.07840807
34	9.978113537	0.100219345	128.2588	0.007797	12.85400936	0.07779674
35	10.67658148	0.093662939	138.2369	0.007234	12.9476723	0.07723396
36	11.42394219	0.087535457	148.9135	0.006715	13.03520776	0.07671531
37	12.22361814	0.081808838	160.3374	0.006237	13.1170166	0.07623685
38	13.07927141	0.076456858	172.561	0.005795	13.19347345	0.07579505
39	13.99482041	0.071455008	185.6403	0.005387	13.26492846	0.07538676
40	14.97445784	0.066780381	199.6351	0.005009	13.33170884	0.07500914
41	16.02266989	0.062411571	214.6096	0.00466	13.39412041	0.07465962
42	17.14425678	0.058328571	230.6322	0.004336	13.45244898	0.07433591
43	18.34435475	0.054512683	247.7765	0.004036	13.50696167	0.0740359
44	19.62845959	0.050946433	266.1209	0.003758	13.5579081	0.07375769
48	25.72890651	0.03886679	353.2701	0.002831	13.73047443	0.0728307
60	57.94642683	0.017257319	813.5204	0.001229	14.03918115	0.07122923
72	130.5064551	0.007662456	1850.092	0.000541	14.17625063	0.07054051
84	293.9255405	0.003402222	4184.651	0.000239	14.23711111	0.07023897

Rate =7.5%	A	B	C	D	E	F
Period	Compound Int.	Compound Int.	Ordinary	Sinking	Present Value	Amortization
n	Future Value	Present Value	Annuity	Fund	Annuity	
1	1.075	0.930232558	1	1	0.930232558	1.075
2	1.155625	0.865332612	2.075	0.481928	1.79556517	0.55692771
3	1.242296875	0.80496057	3.230625	0.309538	2.60052574	0.38453763
4	1.335469141	0.74880053	4.472922	0.223568	3.34932627	0.29856751
5	1.435629326	0.696558632	5.808391	0.172165	4.045884902	0.24716472
6	1.543301526	0.647961518	7.24402	0.138045	4.69384642	0.21304489
7	1.65904914	0.602754901	8.787322	0.1138	5.296601321	0.18880032
8	1.783477826	0.560702233	10.44637	0.095727	5.857303555	0.17072702
9	1.917238662	0.521583473	12.22985	0.081767	6.378887028	0.15676716
10	2.061031562	0.485193928	14.14709	0.070686	6.864080956	0.14568593
11	2.215608929	0.451343189	16.20812	0.061697	7.315424145	0.13669747
12	2.381779599	0.419854129	18.42373	0.054278	7.735278275	0.12927783
13	2.560413069	0.390561981	20.80551	0.048064	8.125840255	0.1230642
14	2.752444049	0.363313471	23.36592	0.042797	8.489153726	0.11779737
15	2.958877353	0.337966019	26.11836	0.038287	8.827119745	0.11328724
16	3.180793154	0.314386995	29.07724	0.034391	9.14150674	0.10939116
17	3.419352641	0.292453018	32.25804	0.031	9.433959758	0.10600003
18	3.675804089	0.272049319	35.67739	0.028029	9.706009077	0.10302896
19	3.951489396	0.253069134	39.35319	0.025411	9.959078211	0.1004109
20	4.2478511	0.235413148	43.30468	0.023092	10.19449136	0.09809219
21	4.566439933	0.218988975	47.55253	0.021029	10.41348033	0.09602937
22	4.908922928	0.203710674	52.11897	0.019187	10.61719101	0.09418687
23	5.277092147	0.189498302	57.0279	0.017535	10.80668931	0.09253528
24	5.672874058	0.17627749	62.30499	0.01605	10.9829668	0.09105008
25	6.098339613	0.16397906	67.97786	0.014711	11.14694586	0.08971067
26	6.555715084	0.152538661	74.0762	0.0135	11.29948452	0.08849961
27	7.047393715	0.141896429	80.63192	0.012402	11.44138095	0.08740204
28	7.575948244	0.131996678	87.67931	0.011405	11.57337763	0.0864052
29	8.144144362	0.122787607	95.25526	0.010498	11.69616524	0.08549811
30	8.754955189	0.11422103	103.3994	0.009671	11.81038627	0.08467124
31	9.411576828	0.106252121	112.1544	0.008916	11.91663839	0.08391628
32	10.11744509	0.098839182	121.5659	0.008226	12.01547757	0.08322599
33	10.87625347	0.091943425	131.6834	0.007594	12.10742099	0.08259397
34	11.69197248	0.085528768	142.5596	0.007015	12.19294976	0.08201461
35	12.56887042	0.079561644	154.2516	0.006483	12.27251141	0.08148291
36	13.5115357	0.074010832	166.8205	0.005994	12.34652224	0.08099447
37	14.52490088	0.068847286	180.332	0.005545	12.41536952	0.08054533
38	15.61426844	0.064043987	194.8569	0.005132	12.47941351	0.08013197
39	16.78533858	0.059575802	210.4712	0.004751	12.53898931	0.07975124
40	18.04423897	0.05541935	227.2565	0.0044	12.59440866	0.07940031
41	19.39755689	0.051552884	245.3008	0.004077	12.64596155	0.07907663
42	20.85237366	0.047956171	264.6983	0.003778	12.69391772	0.07877789
43	22.41630168	0.044610392	285.5507	0.003502	12.73852811	0.07850201
44	24.09752431	0.041498039	307.967	0.003247	12.78002615	0.0782471
48	32.18150008	0.031073753	415.7533	0.002405	12.91901662	0.07740527
60	76.64924036	0.013046444	1008.657	0.000991	13.15938075	0.07599142
72	182.561597	0.005477603	2420.821	0.000413	13.26029862	0.07541308
84	434.8214872	0.002299794	5784.286	0.000173	13.30266941	0.07517288

Rate =8%	A	B	C	D	E	F
Period	Compound Int.	Compound Int.	Ordinary	Sinking	Present Value	Amortization
n	Future Value	Present Value	Annuity	Fund	Annuity	
1	1.08	0.925925926	1	1	0.925925926	1.08
2	1.1664	0.85733882	2.08	0.480769	1.783264746	0.56076923
3	1.259712	0.793832241	3.2464	0.308034	2.577096987	0.38803351
4	1.36048896	0.735029853	4.506112	0.221921	3.31212684	0.3019208
5	1.469328077	0.680583197	5.866601	0.170456	3.992710037	0.25045645
6	1.586874323	0.630169627	7.335929	0.136315	4.622879664	0.21631539
7	1.713824269	0.583490395	8.922803	0.112072	5.206370059	0.1920724
8	1.85093021	0.540268885	10.63663	0.094015	5.746638944	0.17401476
9	1.999004627	0.500248967	12.48756	0.08008	6.246887911	0.16007971
10	2.158924997	0.463193488	14.48656	0.069029	6.710081399	0.14902949
11	2.331638997	0.428882859	16.64549	0.060076	7.138964258	0.14007634
12	2.518170117	0.397113759	18.97713	0.052695	7.536078017	0.13269502
13	2.719623726	0.367697925	21.4953	0.046522	7.903775942	0.12652181
14	2.937193624	0.340461041	24.21492	0.041297	8.244236983	0.12129685
15	3.172169114	0.315241705	27.15211	0.03683	8.559478688	0.11682954
16	3.425942643	0.291890468	30.32428	0.032977	8.851369155	0.11297687
17	3.700018055	0.270268951	33.75023	0.029629	9.121638107	0.10962943
18	3.996019499	0.250249029	37.45024	0.026702	9.371887136	0.1067021
19	4.315701059	0.231712064	41.44626	0.024128	9.6035992	0.10412763
20	4.660957144	0.214548207	45.76196	0.021852	9.818147407	0.10185221
21	5.033833715	0.198655748	50.42292	0.019832	10.01680316	0.09983225
22	5.436540413	0.183940507	55.45676	0.018032	10.20074366	0.09803207
23	5.871463646	0.170315284	60.8933	0.016422	10.37105895	0.09642217
24	6.341180737	0.157699337	66.76476	0.014978	10.52875828	0.09497796
25	6.848475196	0.146017905	73.10594	0.013679	10.67477619	0.09367878
26	7.396353212	0.135201764	79.95442	0.012507	10.80997795	0.09250713
27	7.988061469	0.125186818	87.35077	0.011448	10.93516477	0.0914481
28	8.627106386	0.115913721	95.33883	0.010489	11.05107849	0.09048891
29	9.317274897	0.107327519	103.9659	0.009619	11.15840601	0.08961854
30	10.06265689	0.099377333	113.2832	0.008827	11.25778334	0.08882743
31	10.86766944	0.092016049	123.3459	0.008107	11.34979939	0.08810728
32	11.737083	0.085200045	134.2135	0.007451	11.43499944	0.08745081
33	12.67604964	0.078888931	145.9506	0.006852	11.51388837	0.08685163
34	13.69013361	0.073045306	158.6267	0.006304	11.58693367	0.08630411
35	14.78534429	0.067634543	172.3168	0.005803	11.65456822	0.08580326
36	15.96817184	0.062624577	187.1021	0.005345	11.71719279	0.08534467
37	17.24562558	0.057985719	203.0703	0.004924	11.77517851	0.0849244
38	18.62527563	0.053690481	220.3159	0.004539	11.82886899	0.08453894
39	20.11529768	0.049713408	238.9412	0.004185	11.8785824	0.08418513
40	21.7245215	0.046030933	259.0565	0.00386	11.92461333	0.08386016
41	23.46248322	0.042621235	280.781	0.003561	11.96723457	0.08356149
42	25.33948187	0.039464106	304.2435	0.003287	12.00669867	0.08328684
43	27.36664042	0.036540839	329.583	0.003034	12.04323951	0.08303414
44	29.55597166	0.03383411	356.9496	0.002802	12.07707362	0.08280152
48	40.21057314	0.024869081	490.1322	0.00204	12.18913649	0.08204027
60	101.2570637	0.009875854	1253.213	0.000798	12.37655182	0.08079795
72	254.9825118	0.003921838	3174.781	0.000315	12.45097703	0.08031498
84	642.0893416	0.001557416	8013.617	0.000125	12.4805323	0.08012479

Rate =8.5%	A	B	C	D	E	F
Period	Compound Int.	Compound Int.	Ordinary	Sinking	Present Value	Amortization
n	Future Value	Present Value	Annuity	Fund	Annuity	
1	1.085	0.921658986	1	1	0.921658986	1.085
2	1.177225	0.849455287	2.085	0.479616	1.771114273	0.564616
3	1.277289125	0.782908098	3.262225	0.306539	2.554022371	0.391539
4	1.385858701	0.721574284	4.539514	0.220288	3.275596656	0.305288
5	1.50365669	0.665045423	5.925373	0.168766	3.940642079	0.253766
6	1.631467509	0.612945091	7.42903	0.134607	4.55358717	0.219607
7	1.770142247	0.564926351	9.060497	0.110369	5.11851352	0.195369
8	1.920604338	0.520669448	10.83064	0.092331	5.639182968	0.177331
9	2.083855707	0.479879675	12.75124	0.078424	6.119062643	0.163424
10	2.260983442	0.442285415	14.8351	0.067408	6.561348058	0.152408
11	2.453167034	0.407636327	17.09608	0.058493	6.968984386	0.143493
12	2.661686232	0.375701684	19.54925	0.051153	7.34468607	0.136153
13	2.887929562	0.346268833	22.21094	0.045023	7.690954903	0.130023
14	3.133403575	0.319141782	25.09887	0.039842	8.010096685	0.124842
15	3.399742879	0.294139891	28.23227	0.03542	8.304236576	0.12042
16	3.688721024	0.271096674	31.63201	0.031614	8.57533325	0.116614
17	4.002262311	0.249858686	35.32073	0.028312	8.825191935	0.113312
18	4.342454607	0.230284503	39.323	0.02543	9.055476438	0.11043
19	4.711563249	0.212243781	43.66545	0.022901	9.267720219	0.107901
20	5.112046125	0.195616388	48.37701	0.020671	9.463336608	0.105671
21	5.546570045	0.180291602	53.48906	0.018695	9.64362821	0.103695
22	6.018028499	0.166167375	59.03563	0.016939	9.809795585	0.101939
23	6.529560921	0.153149655	65.05366	0.015372	9.96294524	0.100372
24	7.0845736	0.141151755	71.58322	0.01397	10.104097	0.09897
25	7.686762356	0.130093784	78.66779	0.012712	10.23419078	0.097712
26	8.340137156	0.119902105	86.35455	0.01158	10.35409288	0.09658
27	9.049048814	0.110508852	94.69469	0.01056	10.46460174	0.09556
28	9.818217964	0.101851477	103.7437	0.009639	10.56645321	0.094639
29	10.65276649	0.093872329	113.562	0.008806	10.66032554	0.093806
30	11.55825164	0.086518276	124.2147	0.008051	10.74684382	0.093051
31	12.54070303	0.079740346	135.773	0.007365	10.82658416	0.092365
32	13.60666279	0.073493407	148.3137	0.006742	10.90007757	0.091742
33	14.76322913	0.067735859	161.9203	0.006176	10.96781343	0.091176
34	16.0181036	0.062429363	176.6836	0.00566	11.03024279	0.09066
35	17.37964241	0.057538583	192.7017	0.005189	11.08778137	0.090189
36	18.85691201	0.053030952	210.0813	0.00476	11.14081233	0.08976
37	20.45974953	0.048876454	228.9382	0.004368	11.18968878	0.089368
38	22.19882824	0.045047423	249.398	0.00401	11.2347362	0.08901
39	24.08572865	0.041518362	271.5968	0.003682	11.27625457	0.088682
40	26.13301558	0.038265771	295.6825	0.003382	11.31452034	0.088382
41	28.3543219	0.035267992	321.8156	0.003107	11.34978833	0.088107
42	30.76443927	0.032505062	350.1699	0.002856	11.38229339	0.087856
43	33.3794166	0.029958582	380.9343	0.002625	11.41225197	0.087625
44	36.21666702	0.027611597	414.3137	0.002414	11.43986357	0.087414
48	50.19118309	0.019923818	578.7198	0.001728	11.53030802	0.086728
60	133.593181	0.007485412	1559.92	0.000641	11.67664221	0.085641
72	355.5831307	0.002812282	4171.566	0.00024	11.73162021	0.08524
84	946.4507234	0.001056579	11122.95	8.99E-05	11.75227554	0.08509

Rate =9%	A	B	C	D	E	F
Period	Compound Int.	Compound Int.	Ordinary	Sinking	Present Value	Amortization
r	Future Value	Present Value	Annuity	Fund	Annuity	
1	1.09	0.917431193	1	1	0.917431193	1.09
2	1.1881	0.841679993	2.09	0.478469	1.759111186	0.5684689
3	1.295029	0.77218348	3.2781	0.305055	2.531294666	0.39505476
4	1.41158161	0.708425211	4.573129	0.218669	3.239719877	0.30866866
5	1.538623955	0.649931386	5.984711	0.167092	3.889651263	0.25709246
6	1.677100111	0.596267327	7.523335	0.13292	4.48591859	0.22291978
7	1.828039121	0.547034245	9.200435	0.108691	5.032952835	0.19869052
8	1.992562642	0.50186628	11.02847	0.090674	5.534819115	0.18067438
9	2.171893279	0.46042778	13.02104	0.076799	5.995246894	0.1667988
10	2.367363675	0.422410807	15.19293	0.06582	6.417657701	0.15582009
11	2.580426405	0.38753285	17.56029	0.056947	6.805190552	0.14694666
12	2.812664782	0.355534725	20.14072	0.049651	7.160725277	0.13965066
13	3.065804612	0.326178647	22.95338	0.043567	7.486903924	0.13356656
14	3.341727027	0.299246465	26.01919	0.038433	7.786150389	0.12843317
15	3.64248246	0.274538041	29.36092	0.034059	8.06068843	0.12405888
16	3.970305881	0.251869763	33.0034	0.0303	8.312558193	0.12029991
17	4.32763341	0.231073177	36.9737	0.027046	8.543631369	0.11704625
18	4.717120417	0.21199374	41.30134	0.024212	8.755625109	0.11421229
19	5.141661255	0.19448967	46.01846	0.02173	8.950114779	0.11173041
20	5.604410768	0.17843089	51.16012	0.019546	9.128545669	0.10954648
21	6.108807737	0.163698064	56.76453	0.017617	9.292243733	0.10761663
22	6.658600433	0.15018171	62.87334	0.015905	9.442425443	0.10590499
23	7.257874472	0.137781385	69.53194	0.014382	9.580206829	0.10438188
24	7.911083175	0.126404941	76.78981	0.013023	9.706611769	0.10302256
25	8.62308066	0.115967836	84.7009	0.011806	9.822579605	0.10180625
26	9.39915792	0.10639251	93.32398	0.010715	9.928972115	0.10071536
27	10.24508213	0.097607807	102.7231	0.009735	10.02657992	0.09973491
28	11.16713952	0.089548447	112.9682	0.008852	10.11612837	0.09885205
29	12.17218208	0.082154538	124.1354	0.008056	10.19828291	0.09805572
30	13.26767847	0.075371136	136.3075	0.007336	10.27365404	0.09733635
31	14.46176953	0.069147831	149.5752	0.006686	10.34280187	0.0966856
32	15.76332879	0.063438377	164.037	0.006096	10.40624025	0.09609619
33	17.18202838	0.058200346	179.8003	0.005562	10.4644406	0.09556173
34	18.72841093	0.053394813	196.9823	0.005077	10.51783541	0.0950766
35	20.41396792	0.048986067	215.7108	0.004636	10.56682148	0.09463584
36	22.25122503	0.044941346	236.1247	0.004235	10.61176282	0.09423505
37	24.25383528	0.041230593	258.3759	0.00387	10.65299342	0.09387033
38	26.43668046	0.037826232	282.6298	0.003538	10.69081965	0.0935382
39	28.8159817	0.034702965	309.0665	0.003236	10.72552261	0.09323555
40	31.40942005	0.031837582	337.8824	0.00296	10.7573602	0.09295961
41	34.23626786	0.029208791	369.2919	0.002708	10.78656899	0.09270789
42	37.31753197	0.026797056	403.5281	0.002478	10.81336604	0.09247814
43	40.67610984	0.024584455	440.8457	0.002268	10.8379505	0.09226837
44	44.33695973	0.022554546	481.5218	0.002077	10.86050504	0.09207675
48	62.585237	0.015978209	684.2804	0.001461	10.93357546	0.09146139
60	176.031292	0.005680808	1944.792	0.000514	11.04799102	0.09051419
72	495.1170154	0.002019725	5490.189	0.000182	11.08866973	0.09018214
84	1392.598192	0.000718082	15462.2	6.47E-05	11.10313242	0.09006467

Rate =10%	A	B	C	D	E	F
Period	Compound Int.	Compound Int.	Ordinary	Sinking	Present Value	Amortization
n	Future Value	Present Value	Annuity	Fund	Annuity	
1	1.1	0.909090909	1	1	0.909090909	1.1
2	1.21	0.826446281	2.1	0.4761905	1.73553719	0.576190476
3	1.331	0.751314801	3.31	0.3021148	2.486851991	0.402114804
4	1.4641	0.683013455	4.641	0.2154708	3.169865446	0.315470804
5	1.61051	0.620921323	6.1051	0.1637975	3.790786769	0.263797481
6	1.771561	0.56447393	7.71561	0.1296074	4.355260699	0.22960738
7	1.9487171	0.513158118	9.487171	0.1054055	4.868418818	0.2054055
8	2.14358881	0.46650738	11.4358881	0.087444	5.334926198	0.187444018
9	2.357947691	0.424097618	13.57947691	0.0736405	5.759023816	0.173640539
10	2.59374246	0.385543289	15.9374246	0.0627454	6.144567106	0.162745395
11	2.853116706	0.350493899	18.53116706	0.0539631	6.495061005	0.153963142
12	3.138428377	0.318630818	21.38428377	0.0467633	6.813691823	0.146763315
13	3.452271214	0.28966438	24.52271214	0.0407785	7.103356203	0.140778524
14	3.797498336	0.263331254	27.97498336	0.0357462	7.366687457	0.135746223
15	4.177248169	0.239392049	31.77248169	0.0314738	7.606079506	0.131473777
16	4.594972986	0.217629136	35.94972986	0.0278166	7.823708642	0.127816621
17	5.054470285	0.197844669	40.54470285	0.0246641	8.021553311	0.124664134
18	5.559917313	0.17985879	45.59917313	0.0219302	8.201412101	0.121930222
19	6.115909045	0.163507991	51.15909045	0.0195469	8.364920092	0.119546868
20	6.727499949	0.148643628	57.27499949＊	0.0174596	8.51356372	0.117459625
21	7.400249944	0.135130571	64.00249944	0.0156244	8.648694291	0.11562439
22	8.140274939	0.122845974	71.40274939	0.0140051	8.771540264	0.114005063
23	8.954302433	0.111678158	79.54302433	0.0125718	8.883218422	0.112571813
24	9.849732676	0.101525598	88.49732676	0.0112998	8.98474402	0.111299776
25	10.83470594	0.092295998	98.34705943	0.0101681	9.077040018	0.110168072
26	11.91817654	0.083905453	109.1817654	0.009159	9.160945471	0.109159039
27	13.10999419	0.076277684	121.0999419	0.0082576	9.237223156	0.108257642
28	14.42099361	0.069343349	134.2099361	0.007451	9.306566505	0.107451013
29	15.86309297	0.063039409	148.6309297	0.0067281	9.369605914	0.106728075
30	17.44940227	0.057308553	164.4940227	0.0060792	9.426914467	0.106079248
31	19.1943425	0.052098685	181.943425	0.0054962	9.479013152	0.105496214
32	21.11377675	0.047362441	201.1377675	0.0049717	9.526375593	0.104971717
33	23.22515442	0.043056764	222.2515442	0.0044994	9.569432357	0.104499406
34	25.54766986	0.039142513	245.4766986	0.0040737	9.60857487	0.104073706
35	28.10243685	0.035584103	271.0243685	0.0036897	9.644158973	0.103689705
36	30.91268053	0.032349184	299.1268053	0.0033431	9.676508157	0.103343064
37	34.00394859	0.029408349	330.0394859	0.0030299	9.705916506	0.10302994
38	37.40434344	0.026734863	364.0434344	0.0027469	9.732651369	0.102746925
39	41.14477779	0.024304421	401.4477779	0.002491	9.75695579	0.102490984
40	45.25925557	0.022094928	442.5925557	0.0022594	9.779050718	0.102259414
41	49.78518112	0.020086298	487.8518112	0.0020498	9.799137017	0.102049803
42	54.76369924	0.018260271	537.6369924	0.00186	9.817397288	0.101859991
43	60.24006916	0.016600247	592.4006916	0.001688	9.833997535	0.101688047
44	66.26407608	0.015091133	652.6407608	0.0015322	9.849088668	0.101532237
48	97.01723378	0.010307447	960.1723378	0.0010415	9.89692553	0.10104148
60	304.4816395	0.00328427	3034.816395	0.0003295	9.967157297	0.100329509
72	955.5938177	0.00104647	9545.938177	0.0001048	9.989535303	0.100104757
84	2999.062754	0.000333438	29980.62754	3.335E-05	9.996665625	0.100033355

Annual Percentage Rate Table for Monthly Payment Plans

Number of Payments	ANNUAL PERCENTAGE RATE															
	10.00%	10.25%	10.50%	10.75%	11.00%	11.25%	11.50%	11.75%	12.00%	12.25%	12.50%	12.75%	13.00%	13.25%	13.50%	13.75%
	Finance charge per $100 of amount financed															
1	0.83	0.85	0.87	0.90	0.92	0.94	0.96	0.98	1.00	1.02	1.04	1.06	1.08	1.10	1.12	1.15
2	1.25	1.28	1.31	1.35	1.38	1.41	1.44	1.47	1.50	1.53	1.57	1.60	1.63	1.66	1.69	1.73
3	1.67	1.71	1.76	1.80	1.84	1.88	1.92	1.96	2.01	2.05	2.09	2.13	2.17	2.22	2.26	2.30
4	2.09	2.14	2.20	2.25	2.30	2.35	2.41	2.46	2.51	2.57	2.62	2.67	2.72	2.78	2.83	2.88
5	2.51	2.58	2.64	2.70	2.77	2.83	2.89	2.96	3.02	3.08	3.15	3.21	3.27	3.34	3.40	3.46
6	2.94	3.01	3.08	3.16	3.23	3.31	3.38	3.45	3.53	3.60	3.68	3.75	3.83	3.90	3.97	4.05
7	3.36	3.45	3.53	3.62	3.70	3.78	3.87	3.95	4.04	4.12	4.21	4.29	4.38	4.47	4.55	4.64
8	3.79	3.88	3.98	4.07	4.17	4.26	4.36	4.46	4.55	4.65	4.74	4.84	4.94	5.03	5.13	5.22
9	4.21	4.32	4.43	4.53	4.64	4.75	4.85	4.96	5.07	5.17	5.28	5.39	5.49	5.60	5.71	5.82
10	4.64	4.76	4.88	4.99	5.11	5.23	5.35	5.46	5.58	5.70	5.82	5.94	6.05	6.17	6.29	6.41
11	5.07	5.20	5.33	5.45	5.58	5.71	5.84	5.97	6.10	6.23	6.36	6.49	6.62	6.75	6.88	7.01
12	5.50	5.64	5.78	5.92	6.06	6.20	6.34	6.48	6.62	6.76	6.90	7.04	7.18	7.32	7.46	7.60
13	5.93	6.08	6.23	6.38	6.53	6.68	6.84	6.99	7.14	7.29	7.44	7.59	7.75	7.90	8.05	8.20
14	6.36	6.52	6.69	6.85	7.01	7.17	7.34	7.50	7.66	7.82	7.99	8.15	8.31	8.48	8.64	8.81
15	6.80	6.97	7.14	7.32	7.49	7.66	7.84	8.01	8.19	8.36	8.53	8.71	8.88	9.06	9.23	9.41
16	7.23	7.41	7.60	7.78	7.97	8.15	8.34	8.53	8.71	8.90	9.08	9.27	9.46	9.64	9.83	10.02
17	7.67	7.86	8.06	8.25	8.45	8.65	8.84	9.04	9.24	9.44	9.63	9.83	10.03	10.23	10.43	10.63
18	8.10	8.31	8.52	8.73	8.93	9.14	9.35	9.56	9.77	9.98	10.19	10.40	10.61	10.82	11.03	11.24
19	8.54	8.76	8.98	9.20	9.42	9.64	9.86	10.08	10.30	10.52	10.74	10.96	11.18	11.41	11.63	11.85
20	8.98	9.21	9.44	9.67	9.90	10.13	10.37	10.60	10.83	11.06	11.30	11.53	11.76	12.00	12.23	12.46
21	9.42	9.66	9.90	10.15	10.39	10.63	10.88	11.12	11.36	11.61	11.85	12.10	12.34	12.59	12.84	13.08
22	9.86	10.12	10.37	10.62	10.88	11.13	11.39	11.64	11.90	12.16	12.41	12.67	12.93	13.19	13.44	13.70
23	10.30	10.57	10.84	11.10	11.37	11.63	11.90	12.17	12.44	12.71	12.97	13.24	13.51	13.78	14.05	14.32
24	10.75	11.02	11.30	11.58	11.86	12.14	12.42	12.70	12.98	13.26	13.54	13.82	14.10	14.38	14.66	14.95
25	11.19	11.48	11.77	12.06	12.35	12.64	12.93	13.22	13.52	13.81	14.10	14.40	14.69	14.98	15.28	15.57
26	11.64	11.94	12.24	12.54	12.85	13.15	13.45	13.75	14.06	14.36	14.67	14.97	15.28	15.59	15.89	16.20
27	12.09	12.40	12.71	13.03	13.34	13.66	13.97	14.29	14.60	14.92	15.24	15.56	15.87	16.19	16.51	16.83
28	12.53	12.86	13.18	13.51	13.84	14.16	14.49	14.82	15.15	15.48	15.81	16.14	16.47	16.80	17.13	17.46
29	12.98	13.32	13.66	14.00	14.33	14.67	15.01	15.35	15.70	16.04	16.38	16.72	17.07	17.41	17.75	18.10
30	13.43	13.78	14.13	14.48	14.83	15.19	15.54	15.89	16.24	16.60	16.95	17.31	17.66	18.02	18.38	18.74
31	13.89	14.25	14.61	14.97	15.33	15.70	16.06	16.43	16.79	17.16	17.53	17.90	18.27	18.63	19.00	19.38
32	14.34	14.71	15.09	15.46	15.84	16.21	16.59	16.97	17.35	17.73	18.11	18.49	18.87	19.25	19.63	20.02
33	14.79	15.18	15.57	15.95	16.34	16.73	17.12	17.51	17.90	18.29	18.65	19.08	19.47	19.87	20.26	20.66
34	15.25	15.65	16.05	16.44	16.85	17.25	17.65	18.05	18.46	18.86	19.27	19.67	20.08	20.49	20.90	21.31
35	15.70	16.11	16.53	16.94	17.35	17.77	18.18	18.60	19.01	19.43	19.85	20.27	20.69	21.11	21.53	21.95
36	16.16	16.58	17.01	17.43	17.86	18.29	18.71	19.14	19.57	20.00	20.43	20.87	21.30	21.73	22.17	22.60
37	16.62	17.06	17.49	17.93	18.37	18.81	19.25	19.69	20.13	20.58	21.02	21.46	21.91	22.36	22.81	23.25
38	17.08	17.53	17.98	18.43	18.88	19.33	19.78	20.24	20.69	21.15	21.61	22.07	22.52	22.99	23.45	23.91
39	17.54	18.00	18.46	18.93	19.39	19.86	20.32	20.79	21.26	21.73	22.20	22.67	23.14	23.61	24.09	24.56
40	18.00	18.48	18.95	19.43	19.90	20.38	20.86	21.34	21.82	22.30	22.79	23.27	23.76	24.25	24.73	25.22
41	18.47	18.95	19.44	19.93	20.42	20.91	21.40	21.89	22.39	22.88	23.38	23.88	24.38	24.88	25.38	25.88
42	18.93	19.43	19.93	20.43	20.93	21.44	21.94	22.45	22.96	23.47	23.98	24.49	25.00	25.51	26.03	26.55
43	19.40	19.91	20.42	20.94	21.45	21.97	22.49	23.01	23.53	24.05	24.57	25.10	25.62	26.15	26.68	27.21
44	19.86	20.39	20.91	21.44	21.97	22.50	23.03	23.57	24.10	24.64	25.17	25.71	26.25	26.79	27.33	27.88
45	20.33	20.87	21.41	21.95	22.49	23.03	23.58	24.12	24.67	25.22	25.77	26.32	26.88	27.43	27.99	28.55
46	20.80	21.35	21.90	22.46	23.01	23.57	24.13	24.69	25.25	25.81	26.37	26.94	27.51	28.08	28.65	29.22
47	21.27	21.83	22.40	22.97	23.53	24.10	24.68	25.25	25.82	26.40	26.98	27.56	28.14	28.72	29.31	29.89
48	21.74	22.32	22.90	23.48	24.06	24.64	25.23	25.81	26.40	26.99	27.58	28.18	28.77	29.37	29.97	30.57
49	22.21	22.80	23.39	23.99	24.58	25.18	25.78	26.38	26.98	27.59	28.19	28.80	29.41	30.02	30.63	31.34
50	22.69	23.29	23.89	24.50	25.11	25.72	26.33	26.95	27.56	28.18	28.80	29.42	30.04	30.67	31.29	31.92
51	23.16	23.78	24.40	25.02	25.64	26.26	26.89	27.52	28.15	28.78	29.41	30.05	30.68	31.32	31.96	32.60
52	23.64	24.27	24.90	25.53	26.17	26.81	27.45	28.09	28.73	29.38	30.02	30.67	31.32	31.98	32.63	33.29
53	24.11	24.76	25.40	26.05	26.70	27.35	28.00	28.66	29.32	29.98	30.64	31.30	31.97	32.63	33.30	33.97
54	24.59	25.25	25.91	26.57	27.23	27.90	28.56	29.23	29.91	30.58	31.25	31.93	32.61	33.29	33.98	34.66
55	25.07	25.74	26.41	27.09	27.77	28.44	29.13	29.81	30.50	31.18	31.87	32.56	33.26	33.95	34.65	35.35
56	25.55	26.23	26.92	27.61	28.30	28.99	29.69	30.39	31.09	31.79	32.49	33.20	33.91	34.62	35.33	36.04
57	26.03	26.73	27.43	28.13	28.84	29.54	30.25	30.97	31.68	32.39	33.11	33.83	34.56	35.28	36.01	36.74
58	26.51	27.23	27.94	28.66	29.37	30.10	30.82	31.55	32.27	33.00	33.74	34.47	35.21	35.95	36.69	37.43
59	27.00	27.72	28.45	29.18	29.91	30.65	31.39	32.13	32.87	33.61	34.36	35.11	35.86	36.62	37.37	38.13
60	27.48	28.22	28.96	29.71	30.45	31.20	31.96	32.71	33.47	34.23	34.99	35.75	36.52	37.29	38.06	38.83

Source: Roueche & Graves, *Business Mathematics*, 7th ed., Upper Saddle River, NJ: Prentice Hall, 1997.

Number of Payments	ANNUAL PERCENTAGE RATE															
	14.00%	14.25%	14.50%	14.75%	15.00%	15.25%	15.50%	15.75%	16.00%	16.25%	16.50%	16.75%	17.00%	17.25%	17.50%	17.75%
	Finance charge per $100 of amount financed															
1	1.17	1.19	1.21	1.23	1.25	1.27	1.29	1.31	1.33	1.35	1.37	1.40	1.42	1.44	1.46	1.48
2	1.75	1.78	1.82	1.85	1.88	1.91	1.94	1.97	2.00	2.04	2.07	2.10	2.13	2.16	2.19	2.22
3	2.34	2.38	2.43	2.47	2.51	2.55	2.59	2.64	2.68	2.72	2.76	2.80	2.85	2.89	2.93	2.97
4	2.93	2.99	3.04	3.09	3.14	3.20	3.25	3.30	3.36	3.41	3.46	3.51	3.57	3.62	3.67	3.73
5	3.53	3.59	3.65	3.72	3.78	3.84	3.91	3.97	4.04	4.10	4.16	4.23	4.29	4.35	4.42	4.48
6	4.12	4.20	4.27	4.35	4.42	4.49	4.57	4.64	4.72	4.79	4.87	4.94	5.02	5.09	5.17	5.24
7	4.72	4.81	4.89	4.98	5.06	5.15	5.23	5.32	5.40	5.49	5.58	5.66	5.75	5.83	5.92	6.00
8	5.32	5.42	5.51	5.61	5.71	5.80	5.90	6.00	6.09	6.19	6.29	6.38	6.48	6.58	6.67	6.77
9	5.92	6.03	6.14	6.25	6.35	6.46	6.57	6.68	6.78	6.89	7.00	7.11	7.22	7.32	7.43	7.54
10	6.53	6.65	6.77	6.88	7.00	7.12	7.24	7.36	7.48	7.60	7.72	7.84	7.96	8.08	8.19	8.31
11	7.14	7.27	7.40	7.53	7.66	7.79	7.92	8.05	8.18	8.31	8.44	8.57	8.70	8.83	8.96	9.09
12	7.74	7.89	8.03	8.17	8.31	8.45	8.59	8.74	8.88	9.02	9.16	9.30	9.45	9.59	9.73	9.87
13	8.36	8.51	8.66	8.81	8.97	9.12	9.27	9.43	9.58	9.73	9.89	10.04	10.20	10.35	10.50	10.66
14	8.97	9.13	9.30	9.46	9.63	9.79	9.96	10.12	10.29	10.45	10.62	10.78	10.95	11.11	11.28	11.45
15	9.59	9.76	9.94	10.11	10.29	10.47	10.64	10.82	11.00	11.17	11.35	11.51	11.71	11.88	12.06	12.24
16	10.20	10.39	10.58	10.77	10.95	11.14	11.33	11.52	11.71	11.90	12.09	12.28	12.46	12.65	12.84	13.03
17	10.82	11.02	11.22	11.42	11.62	11.82	12.02	12.22	12.42	12.62	12.83	13.03	13.23	13.43	13.63	13.83
18	11.45	11.66	11.87	12.08	12.29	12.50	12.72	12.93	13.14	13.35	13.57	13.78	13.99	14.21	14.42	14.64
19	12.07	12.30	12.52	12.74	12.97	13.19	13.41	13.64	13.86	14.09	14.31	14.54	14.76	14.99	15.22	15.44
20	12.70	12.93	13.17	13.41	13.64	13.88	14.11	14.35	14.59	14.82	15.06	15.30	15.54	15.77	16.01	16.25
21	13.33	13.58	13.82	14.07	14.32	14.57	14.82	15.06	15.31	15.56	15.81	16.05	16.31	16.56	16.81	17.07
22	13.96	14.22	14.48	14.74	15.00	15.26	15.52	15.78	16.04	16.30	16.57	16.83	17.09	17.36	17.62	17.88
23	14.59	14.87	15.14	15.41	15.68	15.96	16.23	16.50	16.78	17.05	17.32	17.60	17.88	18.15	18.43	18.70
24	15.23	15.51	15.80	16.08	16.37	16.65	16.94	17.22	17.51	17.80	18.09	18.37	18.65	18.95	19.24	19.53
25	15.87	16.17	16.46	16.76	17.06	17.35	17.65	17.95	18.25	18.55	18.85	19.15	19.45	19.75	20.05	20.36
26	16.51	16.82	17.13	17.44	17.75	18.06	18.37	18.68	18.99	19.30	19.62	19.93	20.24	20.56	20.87	21.19
27	17.15	17.47	17.80	18.12	18.44	18.76	19.09	19.41	19.74	20.06	20.39	20.71	21.04	21.37	21.69	22.02
28	17.80	18.13	18.47	18.80	19.14	19.47	19.81	20.15	20.48	20.82	21.16	21.50	21.84	22.18	22.52	22.86
29	18.45	18.79	19.14	19.49	19.83	20.18	20.53	20.88	21.23	21.58	21.94	22.29	22.64	22.99	23.35	23.70
30	19.10	19.45	19.81	20.17	20.54	20.90	21.26	21.62	21.99	22.35	22.72	23.08	23.45	23.81	24.18	24.55
31	19.75	20.12	20.49	20.87	21.24	21.61	21.99	22.37	22.74	23.12	23.50	23.88	24.26	24.64	25.02	25.40
32	20.40	20.79	21.17	21.56	21.95	22.33	22.72	23.11	23.50	23.89	23.28	24.68	25.07	25.46	25.86	26.25
33	21.06	21.46	21.85	22.25	22.65	23.06	23.46	23.86	24.26	24.67	25.07	25.48	25.88	26.29	26.70	27.11
34	21.72	22.13	22.54	22.95	23.37	23.78	24.19	24.61	25.03	25.44	25.86	26.28	26.70	27.12	27.54	27.97
35	22.38	22.80	23.23	23.65	24.08	24.51	24.94	25.36	25.79	26.23	26.66	27.09	27.52	27.96	28.39	28.83
36	23.04	23.48	23.92	24.35	24.80	25.24	25.68	26.12	26.57	27.01	27.46	27.90	28.35	28.80	29.25	29.70
37	23.70	24.16	24.61	25.06	25.51	25.97	26.42	26.88	27.34	27.80	28.26	28.72	29.18	29.64	30.10	30.57
38	24.37	24.84	25.30	25.77	26.24	26.70	27.17	27.64	28.11	28.59	29.06	29.53	30.01	30.49	30.96	31.44
39	25.04	25.52	26.00	26.48	26.96	27.44	27.92	28.41	28.89	29.38	29.87	30.36	30.85	31.34	31.83	32.32
40	25.71	26.20	26.70	27.19	27.69	28.18	28.68	29.18	29.68	30.18	30.68	31.18	31.68	32.19	32.69	33.20
41	26.39	26.89	27.40	27.91	28.41	28.92	29.44	29.95	30.46	30.97	31.49	32.01	32.52	33.04	33.56	34.08
42	27.06	27.58	28.10	28.62	29.15	29.67	30.19	30.72	31.25	31.78	32.31	32.84	33.37	33.90	34.44	34.97
43	27.74	28.27	28.81	29.34	29.88	30.42	30.96	31.50	32.04	32.58	33.13	33.67	34.22	34.76	35.31	35.86
44	28.42	28.97	29.52	30.07	30.62	31.17	31.72	32.28	32.83	33.39	33.95	34.51	35.07	35.63	36.19	36.76
45	29.11	29.67	30.23	30.79	31.36	31.92	32.49	33.06	33.63	34.20	34.77	35.35	35.92	36.50	37.08	37.66
46	29.79	30.36	30.94	31.52	32.10	32.68	33.26	33.84	34.43	35.01	35.60	36.19	36.78	37.37	37.96	38.56
47	30.48	31.07	31.66	32.25	32.84	33.44	34.03	34.63	35.23	35.83	36.43	37.04	37.64	38.25	38.86	39.46
48	31.17	31.77	32.37	32.98	33.59	34.20	34.81	35.42	36.03	36.65	37.27	37.88	38.50	39.13	39.75	40.37
49	31.86	32.48	33.09	33.71	34.34	34.96	35.59	36.21	36.84	37.47	38.10	38.74	39.37	40.01	40.65	41.29
50	32.55	33.18	33.82	34.45	35.09	35.73	36.37	37.01	37.65	38.30	38.94	39.59	40.24	40.89	41.55	42.20
51	33.25	33.89	34.54	35.19	35.84	36.49	37.15	37.81	38.46	39.12	39.79	40.45	41.11	41.78	42.45	43.12
52	33.95	34.61	35.27	35.93	36.60	37.27	37.94	38.61	39.28	39.96	40.63	41.31	41.99	42.67	43.36	44.04
53	34.65	35.32	36.00	36.68	37.36	38.04	38.72	39.41	40.10	40.79	41.48	42.17	42.87	43.57	44.27	44.97
54	35.35	36.04	36.73	37.42	38.12	38.82	39.52	40.22	40.92	41.63	42.33	43.04	43.75	44.47	45.18	45.90
55	36.05	36.76	37.46	38.17	38.88	39.60	40.31	41.03	41.74	42.47	43.19	43.91	44.64	45.37	46.10	46.83
56	36.76	37.48	38.20	38.92	39.65	40.38	41.11	41.84	42.57	43.31	44.05	44.79	45.53	46.27	47.02	47.77
57	37.47	38.20	38.94	39.68	40.42	41.16	41.91	42.65	53.40	44.15	44.91	45.66	46.42	47.18	47.94	48.71
58	38.18	38.93	39.68	40.43	41.19	41.95	42.71	43.47	44.23	45.00	45.77	46.54	47.32	48.09	48.87	49.65
59	38.89	39.66	40.42	41.19	41.96	42.74	43.51	44.29	45.07	45.85	46.64	47.42	48.41	49.01	49.80	50.60
60	39.61	40.39	41.17	41.95	42.74	43.53	44.32	45.11	45.91	46.71	47.51	48.31	49.12	49.92	50.73	51.55

Source: Roueche & Graves, *Business Mathematics*, 7th ed., Upper Saddle River, NJ: Prentice Hall, 1997.

Number of Payments	\multicolumn{16}{c}{ANNUAL PERCENTAGE RATE}
	18.00%	18.25%	18.50%	18.75%	19.00%	19.25%	19.50%	19.75%	20.00%	20.25%	20.50%	20.75%	21.00%	21.25%	21.50%	21.75%
	\multicolumn{16}{c}{Finance charge per \$100 of amount financed}
1	1.50	1.52	1.54	1.56	1.58	1.60	1.62	1.65	1.67	1.69	1.71	1.73	1.75	1.77	1.79	1.81
2	2.26	2.29	2.32	2.35	2.38	2.41	2.44	2.48	2.51	2.54	2.57	2.60	2.63	2.66	2.70	2.73
3	3.01	3.06	3.10	3.14	3.18	3.23	3.27	3.31	3.35	3.39	3.44	3.48	3.52	3.56	3.60	3.65
4	3.78	3.83	3.88	3.94	3.99	4.04	4.10	4.15	4.20	4.25	4.31	4.36	4.41	4.47	4.52	4.57
5	4.54	4.61	4.67	4.74	4.80	4.86	4.93	4.99	5.06	5.12	5.18	5.25	5.31	5.37	5.44	5.50
6	5.32	5.39	5.46	5.54	5.61	5.69	5.76	5.84	5.91	5.99	6.06	6.14	6.21	6.29	6.36	6.44
7	6.09	6.18	6.26	6.35	6.43	6.52	6.60	6.69	6.78	6.86	6.95	7.04	7.12	7.21	7.29	7.38
8	6.87	6.96	7.06	7.16	7.26	7.35	7.45	7.55	7.64	7.74	7.84	7.94	8.03	8.13	8.23	8.33
9	7.65	7.76	7.87	7.97	8.08	8.19	8.30	8.41	8.52	8.63	8.73	8.84	8.95	9.06	9.17	9.28
10	8.43	8.55	8.67	8.79	8.91	9.03	9.15	9.27	9.39	9.51	9.63	9.75	9.88	10.00	10.12	10.24
11	9.22	9.35	9.49	9.62	9.75	9.88	10.01	10.14	10.28	10.41	10.54	10.67	10.80	10.94	11.07	11.20
12	10.02	10.16	10.30	10.44	10.59	10.73	10.87	11.02	11.16	11.31	11.45	11.59	11.74	11.88	12.02	12.17
13	10.81	10.97	11.12	11.28	11.43	11.59	11.74	11.90	12.05	12.21	12.36	12.52	12.67	12.83	12.99	13.14
14	11.61	11.78	11.95	12.11	12.28	12.45	12.61	12.78	12.95	13.11	13.28	13.45	13.62	13.79	13.95	14.12
15	12.42	12.59	12.77	12.95	13.13	13.31	13.49	13.67	13.85	14.03	14.21	14.39	14.57	14.75	14.93	15.11
16	13.22	13.41	13.60	13.80	13.99	14.18	14.37	14.56	14.75	14.94	15.13	15.33	15.52	15.71	15.90	16.10
17	14.04	14.24	14.44	14.64	14.85	15.05	15.25	15.46	15.66	15.86	16.07	16.27	16.48	16.68	16.89	17.09
18	14.85	15.07	15.28	15.49	15.71	15.93	16.14	16.36	16.57	16.79	17.01	17.22	17.44	17.66	17.88	18.09
19	15.67	15.90	16.12	16.35	16.58	16.81	17.03	17.26	17.49	17.72	17.95	18.18	18.41	18.64	18.87	19.10
20	16.49	16.73	16.97	17.21	17.45	17.69	17.93	18.17	18.41	18.66	18.90	19.14	19.38	19.63	19.87	20.11
21	17.32	17.57	17.82	18.07	18.33	18.58	18.83	19.09	19.34	19.60	19.85	20.11	20.36	20.62	20.87	21.13
22	18.15	18.41	18.68	18.94	19.21	19.47	19.74	20.01	20.27	20.54	20.81	21.08	21.34	21.61	21.88	22.15
23	18.98	19.26	19.54	19.81	20.09	20.37	20.65	20.93	21.21	21.49	21.77	22.05	22.33	22.61	22.90	23.18
24	19.82	20.11	20.40	20.69	20.98	21.27	21.56	21.86	22.15	22.44	22.74	23.03	23.33	23.62	23.92	24.21
25	20.66	20.96	21.27	21.57	21.87	22.18	22.48	22.79	23.10	23.40	23.71	24.02	24.32	24.63	24.94	25.25
26	21.50	21.82	22.14	22.45	22.77	23.09	23.41	23.73	24.04	24.36	24.68	25.01	25.33	25.65	25.97	26.29
27	22.35	22.68	23.01	23.34	23.67	24.00	24.33	24.67	25.00	25.33	25.67	26.00	26.34	26.67	27.01	27.34
28	23.20	23.55	23.89	24.23	24.58	24.92	25.27	25.61	25.96	26.30	26.65	27.00	27.35	27.70	28.05	28.40
29	24.06	24.41	24.77	25.13	25.49	25.84	26.20	26.56	26.92	27.28	27.64	28.00	28.37	28.73	29.09	29.46
30	24.92	25.29	25.66	26.03	26.40	26.77	27.14	27.52	27.89	28.26	28.64	29.01	29.39	29.77	30.14	30.52
31	25.78	26.16	26.55	26.93	27.32	27.70	28.09	28.47	28.86	29.25	29.64	30.03	30.42	30.81	31.20	31.59
32	26.65	27.04	27.44	27.84	28.24	28.64	29.04	29.44	29.84	30.24	30.64	31.05	31.45	31.85	32.26	32.67
33	27.52	27.93	28.34	28.75	29.16	29.57	29.99	30.40	30.82	31.23	31.65	32.07	32.49	32.91	33.33	33.75
34	28.39	28.81	29.24	29.66	30.09	30.52	30.95	31.37	31.80	32.23	32.67	33.10	33.53	33.96	34.40	34.83
35	29.27	29.71	30.14	30.58	31.02	31.47	31.91	32.35	32.79	33.24	33.68	34.13	34.58	35.03	35.47	35.92
36	30.15	30.60	31.05	31.51	31.96	32.42	32.87	33.33	33.79	34.25	34.71	35.17	35.63	36.09	36.56	37.02
37	31.03	31.50	31.97	32.43	32.90	33.37	33.84	34.32	34.79	35.26	35.74	36.21	36.69	37.16	37.64	38.12
38	31.92	32.40	32.88	33.37	33.85	34.33	34.82	35.30	35.79	36.28	36.77	37.26	37.75	38.24	38.73	39.23
39	32.81	33.31	33.80	34.30	34.80	35.30	35.80	36.30	36.80	37.30	37.81	38.31	38.82	39.32	39.83	40.34
40	33.71	34.22	34.73	35.24	35.75	36.26	36.78	37.29	37.81	38.33	38.85	39.37	39.89	40.41	40.93	41.46
41	34.61	35.13	35.66	36.18	36.71	37.24	37.77	38.30	38.83	39.36	39.89	40.43	40.96	41.50	42.04	42.58
42	35.51	36.05	36.59	37.13	37.67	38.21	38.76	39.30	39.85	40.40	40.95	41.50	42.05	42.60	43.15	43.71
43	36.42	36.97	37.52	38.08	38.63	39.19	39.75	40.31	40.87	41.44	42.00	42.57	43.13	43.70	44.27	44.84
44	37.33	37.89	38.46	39.03	39.60	40.18	40.75	41.33	41.90	42.48	43.06	43.64	44.22	44.81	45.39	45.98
45	38.24	38.82	39.41	39.99	40.58	41.17	41.75	42.35	42.94	43.53	44.13	44.72	45.32	45.92	46.52	47.12
46	39.16	39.75	40.35	40.95	41.55	42.16	42.76	43.37	43.98	44.58	45.20	45.81	46.42	47.03	47.65	48.27
47	40.08	40.69	41.30	41.92	42.54	43.15	43.77	44.40	45.02	45.64	46.27	46.90	47.53	48.16	48.79	49.42
48	41.00	41.63	42.26	42.89	43.52	44.15	44.79	45.43	46.07	46.71	47.35	47.99	48.64	49.28	49.93	50.58
49	41.93	42.57	43.22	43.86	44.51	45.16	45.81	46.46	47.12	47.77	48.43	49.09	49.75	50.41	51.08	51.74
50	42.86	43.52	44.18	44.84	45.50	46.17	46.83	47.50	48.17	48.84	49.52	50.19	50.87	51.55	52.23	52.91
51	43.79	44.47	45.14	45.82	46.50	47.18	47.86	48.55	49.23	49.92	50.61	51.30	51.99	52.69	53.38	54.08
52	44.73	45.42	46.11	46.80	47.50	48.20	48.89	49.59	50.30	51.00	51.71	52.41	53.12	53.83	54.55	54.26
53	45.67	46.38	47.08	47.79	48.50	49.22	49.93	50.65	51.37	52.09	52.81	53.53	54.26	54.98	55.71	56.44
54	46.62	47.34	48.06	48.79	49.51	50.24	50.97	51.70	52.44	53.17	53.91	54.65	55.39	56.14	56.88	57.63
55	47.57	48.30	49.04	49.78	50.52	51.27	52.02	52.76	53.52	54.27	55.02	55.78	56.54	57.30	58.06	58.82
56	48.52	49.27	50.03	50.78	51.54	52.30	53.06	53.83	54.60	55.37	56.14	56.91	57.68	58.46	59.24	60.02
57	49.47	50.24	51.01	51.79	52.56	53.34	54.12	54.90	55.68	56.47	57.25	58.04	58.84	59.63	60.43	61.22
58	50.43	51.22	52.00	52.79	53.58	54.38	55.17	55.97	56.77	57.57	58.38	59.18	59.99	60.80	61.62	62.43
59	51.39	52.20	53.00	53.80	54.61	55.42	56.23	57.05	57.87	58.68	59.51	60.33	61.15	61.98	62.81	63.64
60	52.36	53.18	54.00	54.82	55.64	56.47	57.30	58.13	58.96	59.80	60.64	61.48	62.32	63.17	64.01	64.86

Source: Roueche & Graves, *Business Mathematics*, 7th ed., Upper Saddle River, NJ: Prentice Hall, 1997.

Number of Payments	ANNUAL PERCENTAGE RATE															
	26.00%	26.25%	26.50%	26.75%	27.00%	27.25%	27.50%	27.75%	28.00%	28.25%	28.50%	28.75%	29.00%	29.25%	29.50%	29.75%
	Finance charge per $100 of amount financed															
1	2.17	2.19	2.21	2.23	2.25	2.27	2.29	2.31	2.33	2.35	2.37	2.40	2.42	2.44	2.46	2.48
2	3.26	3.29	3.32	3.36	3.39	3.42	3.45	3.48	3.51	3.54	3.58	3.61	3.64	3.67	3.70	3.73
3	4.36	4.41	4.45	4.49	4.53	4.58	4.62	4.66	4.70	4.74	4.79	4.83	4.87	4.91	4.96	5.00
4	5.47	5.53	5.58	5.63	5.69	5.74	5.79	5.85	5.90	5.95	6.01	6.06	6.11	6.17	6.22	6.27
5	6.59	6.66	6.72	6.79	6.85	6.91	6.98	7.04	7.11	7.17	7.24	7.30	7.37	7.43	7.49	7.56
6	7.72	7.79	7.87	7.95	8.02	8.10	8.17	8.25	8.32	8.40	8.48	8.55	8.63	8.70	8.78	8.85
7	8.85	8.94	9.03	9.11	9.20	9.29	9.37	9.46	9.55	9.64	9.72	9.81	9.90	9.98	10.07	10.16
8	9.99	10.09	10.19	10.29	10.39	10.49	10.58	10.68	10.78	10.88	10.98	11.08	11.18	11.28	11.38	11.47
9	11.14	11.25	11.36	11.47	11.58	11.69	11.80	11.91	12.03	12.14	12.25	12.36	12.47	12.58	12.69	12.80
10	12.30	12.42	12.54	12.67	12.79	12.91	13.03	13.15	13.28	13.40	13.52	13.64	13.77	13.89	14.01	14.14
11	13.46	13.60	13.73	13.87	14.00	14.13	14.27	14.40	14.54	14.67	14.81	14.94	15.08	15.21	15.35	15.48
12	14.64	14.78	14.93	15.07	15.22	15.37	15.51	15.66	15.81	15.95	16.10	16.25	16.40	16.54	16.69	16.84
13	15.82	15.97	16.13	16.29	16.45	16.61	16.77	16.93	17.09	17.24	17.40	17.56	17.72	17.88	18.04	18.20
14	17.00	17.17	17.35	17.52	17.69	17.86	18.03	18.20	18.37	18.54	18.72	18.89	19.06	19.23	19.41	19.58
15	18.20	18.38	18.57	18.75	18.93	19.12	19.30	19.48	19.67	19.85	20.04	20.22	20.41	20.59	20.78	20.96
16	19.40	19.60	19.79	19.99	20.19	20.38	20.58	20.78	20.97	21.17	21.37	21.57	21.76	21.96	22.16	22.36
17	20.61	20.82	21.03	21.24	21.45	21.66	21.87	22.08	22.29	22.50	22.71	22.92	23.13	23.34	23.55	23.77
18	21.83	22.05	22.27	22.50	22.72	22.94	23.16	23.39	23.61	23.83	24.06	24.28	24.51	24.73	24.96	25.18
19	23.06	23.29	23.53	23.76	24.00	24.23	24.47	24.71	24.94	25.18	25.42	25.65	25.89	26.13	26.37	26.61
20	24.29	24.54	24.79	25.04	25.28	25.53	25.78	26.03	26.28	26.53	26.78	27.04	27.29	27.54	27.79	28.04
21	25.53	25.79	26.05	26.32	26.58	26.84	27.11	27.37	27.63	27.90	28.16	28.43	28.69	28.96	29.22	29.49
22	26.78	27.05	27.33	27.61	27.88	28.16	28.44	28.71	28.99	29.27	29.55	29.82	30.10	30.38	30.66	30.94
23	28.04	28.32	28.61	28.90	29.19	29.48	29.77	30.07	30.36	30.65	30.94	31.23	31.53	31.82	32.11	32.41
24	29.30	29.60	29.90	30.21	30.51	30.82	31.12	31.43	31.73	32.04	32.34	32.65	32.96	33.27	33.57	33.88
25	30.57	30.89	31.20	31.52	31.84	32.16	32.48	32.80	33.12	33.44	33.76	34.08	34.40	34.72	35.04	35.37
26	31.85	32.18	32.51	32.84	33.18	33.51	33.84	34.18	34.51	34.84	35.18	35.51	35.85	36.19	36.52	36.86
27	33.14	33.48	33.83	34.17	34.52	34.87	35.21	35.56	35.91	36.26	36.61	36.96	37.31	37.66	38.01	38.36
28	34.43	34.79	35.15	35.51	35.87	36.23	36.59	36.96	37.32	37.68	38.05	38.41	38.78	39.15	39.51	39.88
29	35.73	36.10	36.48	36.85	37.23	37.61	37.98	38.36	38.74	39.12	39.50	39.88	40.26	40.64	41.02	41.40
30	37.04	37.43	37.82	38.21	38.60	38.99	39.38	39.77	40.17	40.56	40.95	41.35	41.75	42.14	42.54	42.94
31	38.35	38.76	39.16	39.57	39.97	40.38	40.79	41.19	41.60	42.01	42.42	42.83	43.24	43.65	44.06	44.48
32	39.68	40.10	40.52	40.94	41.36	41.78	42.20	42.62	43.05	43.47	43.90	44.32	44.75	45.17	45.60	46.03
33	41.01	41.44	41.88	42.31	42.75	43.19	43.62	44.06	44.50	44.94	45.38	45.82	46.26	46.70	47.15	47.59
34	42.35	42.80	43.25	43.70	44.15	44.60	45.05	45.51	45.96	46.42	46.87	47.33	47.79	48.24	48.70	49.16
35	43.69	44.16	44.62	45.09	45.56	46.02	46.49	46.96	47.43	47.90	48.37	48.85	49.32	49.79	50.27	50.74
36	45.05	45.53	46.01	46.49	46.97	47.45	47.94	48.42	48.91	49.40	49.88	50.37	50.86	51.35	51.84	52.33
37	46.41	46.90	47.40	47.90	48.39	48.89	49.39	49.89	50.40	50.90	51.40	51.91	52.41	52.92	53.42	53.93
38	47.77	48.29	48.80	49.31	49.82	50.34	50.86	51.37	51.89	52.41	52.93	53.45	53.97	54.49	55.02	55.54
39	49.15	49.68	50.20	50.73	51.26	51.79	52.33	52.86	53.39	53.93	54.46	55.00	55.54	56.08	56.62	57.16
40	50.53	51.07	51.62	52.16	52.71	53.26	53.81	54.35	54.90	55.46	56.01	56.56	57.12	57.67	58.23	58.79
41	51.92	52.48	53.04	53.60	54.16	54.73	55.29	55.86	56.42	56.99	57.56	58.13	58.70	59.28	59.85	60.42
42	53.32	53.89	54.47	55.05	55.63	56.21	56.79	57.37	57.95	58.54	59.12	59.71	60.30	60.89	61.48	62.07
43	54.72	55.31	55.90	56.50	57.09	57.69	58.29	58.89	59.49	60.09	60.69	61.30	61.90	62.51	63.11	63.72
44	56.13	56.74	57.35	57.96	58.57	59.19	59.80	60.42	61.03	61.65	62.27	62.89	63.51	64.14	64.76	65.39
45	57.55	58.17	58.80	59.43	60.06	60.69	61.32	61.95	62.59	63.22	63.86	64.50	65.13	65.77	66.42	67.06
46	58.97	59.61	60.26	60.90	61.55	62.20	62.84	63.49	64.15	64.80	65.45	66.11	66.76	67.42	68.08	68.74
47	60.40	61.06	61.72	62.38	63.05	63.71	64.38	65.05	65.71	66.38	67.06	67.73	68.40	69.08	69.75	70.43
48	61.84	62.52	63.20	63.87	64.56	65.24	65.92	66.60	67.29	67.98	68.67	69.36	70.05	70.74	71.44	72.13
49	63.29	63.98	64.68	65.37	66.07	66.77	67.47	68.17	68.87	69.58	70.29	70.99	71.70	72.41	73.13	73.84
50	64.74	65.45	66.16	66.88	67.59	68.31	69.03	69.75	70.47	71.19	71.91	72.64	73.37	74.10	74.83	75.56
51	66.20	66.93	67.66	68.39	69.12	69.86	70.59	71.33	72.07	72.81	73.55	74.29	75.04	75.78	76.53	77.28
52	67.67	68.41	69.16	69.91	70.66	71.41	72.16	72.92	73.67	74.43	75.19	75.95	76.72	77.48	78.25	79.02
53	69.14	69.90	70.67	71.43	72.20	72.97	73.74	74.52	75.29	76.07	76.85	77.62	78.41	79.19	79.97	80.76
54	70.62	71.40	72.18	72.97	73.75	74.54	75.33	76.12	76.91	77.71	78.50	79.30	80.10	80.90	81.71	82.51
55	72.11	72.91	73.71	74.51	75.31	76.12	76.92	77.73	78.55	79.36	80.17	80.99	81.81	82.63	83.45	84.27
56	73.60	74.42	75.24	76.06	76.88	77.70	78.53	79.35	80.18	81.02	81.85	82.68	83.52	84.36	85.20	86.04
57	75.10	75.94	76.77	77.61	78.45	79.29	80.14	80.98	81.83	82.68	83.53	84.39	85.24	86.10	86.96	87.82
58	76.61	77.46	78.32	79.17	80.03	80.89	81.75	82.62	83.48	84.35	85.22	86.10	86.97	87.85	88.72	89.60
59	78.12	78.99	79.87	80.74	81.62	82.50	83.38	84.26	85.15	86.03	86.92	87.81	88.71	89.60	90.50	91.40
60	79.64	80.53	81.42	82.32	83.21	84.11	85.01	85.91	86.81	87.72	88.63	89.54	90.45	91.37	92.28	93.20

Source: Roueche & Graves, *Business Mathematics*, 7th ed., Upper Saddle River, NJ: Prentice Hall, 1997.

Number of Payments	ANNUAL PERCENTAGE RATE															
	30.00%	30.25%	30.50%	30.75%	31.00%	31.25%	31.50%	31.75%	32.00%	32.25%	32.50%	32.75%	33.00%	33.25%	33.50%	33.75%
	Finance charge per $100 of amount financed															
1	2.50	2.52	2.54	2.56	2.58	2.60	2.62	2.65	2.67	2.69	2.71	2.73	2.75	2.77	2.79	2.81
2	3.77	3.80	3.83	3.86	3.89	3.92	3.95	3.99	4.02	4.05	4.08	4.11	4.14	4.18	4.21	4.24
3	5.04	5.08	5.13	5.17	5.21	5.25	5.30	5.34	5.38	5.42	5.46	5.51	5.55	5.59	5.63	5.68
4	6.33	6.38	6.43	6.49	6.54	6.59	6.65	6.70	6.75	6.81	6.86	6.91	6.97	7.02	7.08	7.13
5	7.62	7.69	7.75	7.82	7.88	7.95	8.01	8.08	8.14	8.20	8.27	8.33	8.40	8.46	8.53	8.59
6	8.93	9.01	9.08	9.16	9.23	9.31	9.39	9.46	9.54	9.61	9.69	9.77	9.84	9.92	9.99	10.07
7	10.25	10.33	10.42	10.51	10.60	10.68	10.77	10.86	10.95	11.03	11.12	11.21	11.30	11.39	11.47	11.56
8	11.57	11.67	11.77	11.87	11.97	12.07	12.17	12.27	12.37	12.47	12.57	12.67	12.77	12.87	12.97	13.07
9	12.91	13.02	13.13	13.24	13.36	13.47	13.58	13.69	13.80	13.91	14.02	14.14	14.25	14.36	14.47	14.58
10	14.26	14.38	14.50	14.63	14.75	14.87	15.00	15.12	15.24	15.37	15.49	15.62	15.74	15.86	15.99	16.11
11	15.62	15.75	15.89	16.02	16.16	16.29	16.43	16.56	16.70	16.84	16.97	17.11	17.24	17.38	17.52	17.65
12	16.98	17.13	17.28	17.43	17.58	17.72	17.87	18.02	18.17	18.32	18.47	18.61	18.76	18.91	19.06	19.21
13	18.36	18.52	18.68	18.84	19.00	19.16	19.33	19.49	19.65	19.81	19.97	20.13	20.29	20.45	20.62	20.78
14	19.75	19.92	20.10	20.27	20.44	20.62	20.79	20.96	21.14	21.13	21.49	21.66	21.83	22.01	22.18	22.36
15	21.15	21.34	21.52	21.71	21.89	22.08	22.27	22.45	22.64	22.83	23.01	23.20	23.39	23.58	23.76	23.95
16	22.56	22.76	22.96	23.16	23.35	23.55	23.75	23.95	24.15	24.35	24.55	24.75	24.96	25.16	25.36	25.56
17	23.98	24.19	24.40	24.61	24.83	25.04	25.25	25.47	25.68	25.89	26.11	26.32	26.53	26.75	26.96	27.18
18	25.41	25.63	25.86	26.08	26.31	26.54	26.76	26.99	27.22	27.44	27.67	27.90	28.13	28.35	28.58	28.81
19	26.85	27.08	27.32	27.56	27.80	28.04	28.28	28.52	28.76	29.00	29.25	29.49	29.73	29.97	30.21	30.45
20	28.29	28.55	28.80	29.05	29.31	29.56	29.81	30.07	30.32	30.58	30.83	31.09	31.34	31.60	31.86	32.11
21	29.75	30.02	30.29	30.55	30.82	31.09	31.36	31.62	31.89	32.16	32.43	32.70	32.97	33.24	33.51	33.78
22	31.22	31.50	31.78	32.06	32.35	32.63	32.91	33.19	33.48	33.76	34.04	34.33	34.61	34.89	35.18	35.46
23	32.70	33.00	33.29	33.59	33.88	34.18	34.48	34.77	35.07	35.37	35.66	35.96	36.26	36.56	36.86	37.16
24	34.19	34.50	34.81	35.12	35.43	35.74	36.05	36.26	36.67	36.99	37.30	37.61	37.92	38.24	38.55	38.87
25	35.69	36.01	36.34	36.66	36.99	37.31	37.64	37.96	38.29	38.62	38.94	39.27	39.60	39.93	40.26	40.59
26	37.20	37.54	37.88	38.21	38.55	38.89	39.23	39.58	39.92	40.26	40.60	40.94	41.29	41.63	41.97	42.32
27	38.72	39.07	39.42	39.78	40.13	40.49	40.84	41.20	41.56	41.91	42.27	42.63	42.99	43.34	43.70	44.06
28	40.25	40.61	40.98	41.35	41.72	42.09	42.46	42.83	43.20	43.58	43.95	44.32	44.70	45.07	45.45	45.82
29	41.78	42.17	42.55	42.94	43.32	43.71	44.09	44.48	44.87	45.25	45.64	46.03	46.42	46.81	47.20	47.59
30	43.33	43.73	44.13	44.53	44.93	45.33	45.73	46.13	46.54	46.94	47.34	47.75	48.15	48.56	48.96	49.37
31	44.89	45.30	45.72	46.13	46.55	46.97	47.38	47.80	48.22	48.64	49.06	49.48	49.90	50.32	50.74	51.17
32	46.46	46.89	47.32	47.75	48.18	48.61	49.05	49.48	49.91	50.35	50.78	51.22	51.66	52.09	52.53	52.97
33	48.04	48.48	48.93	49.37	49.82	50.27	50.72	51.17	51.62	52.07	52.52	52.97	53.43	53.88	54.33	54.79
34	49.62	50.08	50.55	51.01	51.47	51.94	52.40	52.87	53.33	53.80	54.27	54.74	55.21	55.68	56.15	56.62
35	51.22	51.70	52.17	52.65	53.13	53.61	54.09	54.58	55.06	55.54	56.03	56.51	57.00	57.48	57.97	58.46
36	52.83	53.32	53.81	54.31	54.80	55.30	55.80	56.30	56.80	57.30	57.80	58.30	58.80	59.30	59.81	60.31
37	54.44	54.95	55.46	55.97	56.49	57.00	57.51	58.03	58.54	59.06	59.58	60.10	60.62	61.14	61.66	62.18
38	56.07	56.59	57.12	57.65	58.18	58.71	59.24	59.77	60.30	60.84	61.37	61.90	62.44	62.98	63.52	64.06
39	57.70	58.24	58.79	59.33	59.88	60.42	60.97	61.52	62.07	62.62	63.17	63.72	64.28	64.83	65.39	65.94
40	59.34	59.90	60.47	61.03	61.59	62.15	62.72	63.28	63.85	64.42	64.99	65.56	66.13	66.70	67.27	67.84
41	61.00	61.57	62.15	62.73	63.31	63.89	64.47	65.06	65.64	66.22	66.81	67.40	67.99	68.57	69.16	69.76
42	62.66	63.25	63.85	64.44	65.04	65.64	66.24	66.84	67.44	68.04	68.65	69.25	69.86	70.46	71.07	71.68
43	64.33	64.94	65.56	66.17	66.78	67.40	68.01	68.63	69.25	69.87	70.49	71.11	71.74	72.36	72.99	73.61
44	66.01	66.64	67.27	67.90	68.53	69.17	69.80	70.43	71.07	71.71	72.35	72.99	73.63	74.27	74.91	75.56
45	67.70	68.35	69.00	69.64	70.29	70.94	71.60	72.25	72.90	73.56	74.21	74.87	75.53	76.19	76.85	77.52
46	69.40	70.07	70.73	71.40	72.06	72.73	73.40	74.07	74.74	75.42	76.09	76.77	77.44	78.12	78.80	79.48
47	71.11	71.79	72.47	73.16	73.84	74.53	75.22	75.90	76.60	77.29	77.98	78.67	79.37	80.07	80.76	81.46
48	72.83	73.53	74.23	74.93	75.63	76.34	77.04	77.75	78.46	79.17	79.88	80.59	81.30	82.02	82.74	83.45
49	74.55	75.27	75.99	76.71	77.43	78.15	78.88	79.60	80.33	81.06	81.79	82.52	83.25	83.98	84.72	84.45
50	76.29	77.02	77.76	78.50	79.24	79.98	80.72	81.46	82.21	82.96	83.70	84.45	85.20	85.96	86.71	87.47
51	78.03	78.79	79.54	80.30	81.06	81.81	82.58	83.34	84.10	84.87	85.63	86.40	87.17	87.94	88.71	89.49
52	79.79	80.56	81.33	82.11	82.88	83.66	84.44	85.22	86.00	86.78	87.57	88.36	89.15	89.94	90.73	91.52
53	81.55	82.34	83.13	83.92	84.72	85.51	86.31	87.11	87.91	88.72	89.52	90.33	91.13	91.94	92.75	93.57
54	83.32	84.13	84.94	85.75	86.56	87.38	88.19	89.01	89.83	90.66	91.48	92.30	92.13	93.96	94.79	95.62
55	85.10	85.93	86.75	87.58	88.42	89.25	90.09	90.92	91.76	92.60	93.45	94.29	95.14	95.99	96.83	97.69
56	86.89	87.73	88.58	89.43	90.28	91.13	91.99	92.84	93.70	94.56	95.43	96.29	97.15	98.02	98.89	99.76
57	88.68	89.55	90.41	91.28	92.15	93.02	93.90	94.77	95.65	96.53	97.41	98.30	99.18	100.07	100.96	101.85
58	90.49	91.37	92.26	93.14	94.03	94.92	95.82	96.71	97.61	98.51	99.41	100.31	101.22	102.12	103.03	103.94
59	92.30	93.20	94.11	95.01	95.92	96.83	97.75	98.66	99.58	100.50	101.42	102.34	103.26	104.19	105.12	106.05
60	94.12	95.04	95.97	96.89	97.82	98.75	99.68	100.62	101.56	102.49	103.43	104.38	105.32	106.27	107.21	108.16

Source: Roueche & Graves, *Business Mathematics*, 7th ed., Upper Saddle River, NJ: Prentice Hall, 1997.

Number of Payments	ANNUAL PERCENTAGE RATE															
	34.00%	34.25%	34.50%	34.75%	35.00%	35.25%	35.50%	35.75%	36.00%	36.25%	36.50%	36.75%	37.00%	37.25%	37.50%	37.75%
	Finance charge per $100 of amount financed															
1	2.83	2.85	2.87	2.90	2.92	2.94	2.96	2.98	3.00	3.02	3.04	3.06	3.08	3.10	3.12	3.15
2	4.27	4.30	4.33	4.36	4.40	4.43	4.46	4.49	4.52	4.55	4.59	4.62	4.65	4.68	4.71	4.74
3	5.72	5.76	5.80	5.85	5.89	5.93	5.97	6.02	6.06	6.10	6.14	6.19	6.23	6.27	6.31	6.36
4	7.18	7.24	7.29	7.34	7.40	7.45	7.50	7.56	7.61	7.66	7.72	7.77	7.83	7.88	7.93	7.99
5	8.66	8.72	8.79	8.85	8.92	8.98	9.05	9.11	9.18	9.24	9.31	9.37	9.44	9.50	9.57	9.63
6	10.15	10.22	10.30	10.38	10.45	10.53	10.61	10.68	10.76	10.83	10.91	10.99	11.06	11.14	11.22	11.29
7	11.65	11.74	11.83	11.91	12.00	12.09	12.18	12.27	12.35	12.44	12.53	12.62	12.71	12.80	12.88	12.97
8	13.17	13.27	13.36	13.46	13.56	13.66	13.76	13.86	13.97	14.07	14.17	14.27	14.37	14.47	14.57	14.67
9	14.69	14.81	14.92	15.03	15.14	15.25	15.37	15.48	15.59	15.70	15.82	15.93	16.04	16.15	16.27	16.38
10	16.24	16.36	16.48	16.61	16.73	16.86	16.98	17.11	17.23	17.36	17.48	17.60	17.73	17.85	17.98	18.10
11	17.79	17.93	18.06	18.20	18.34	18.47	18.61	18.75	18.89	19.02	19.16	19.30	19.43	19.57	19.71	19.85
12	19.36	19.51	19.66	19.81	19.96	20.11	20.25	20.40	20.55	20.70	20.85	21.00	21.15	21.31	21.46	21.61
13	20.94	21.10	21.26	21.43	21.59	21.75	21.91	22.08	22.24	22.40	22.56	22.73	22.89	23.05	23.22	23.38
14	22.53	22.71	22.88	23.06	23.23	23.41	23.59	23.76	23.94	24.11	24.29	24.47	24.64	24.82	25.00	25.17
15	24.14	24.33	24.52	24.71	24.89	25.08	25.27	25.46	25.65	25.84	26.03	26.22	26.41	26.60	26.79	26.98
16	25.76	25.96	26.16	26.37	26.57	26.77	26.97	27.17	27.38	27.58	27.78	27.99	28.19	28.39	28.60	28.80
17	27.39	27.61	27.82	28.04	28.25	28.47	28.69	28.90	29.12	29.34	29.55	29.77	29.99	30.20	30.42	30.64
18	29.04	29.27	29.50	29.73	29.96	30.19	30.42	30.65	30.88	31.11	31.34	31.57	31.80	32.03	32.26	32.49
19	30.70	30.94	31.18	31.43	31.67	31.91	32.16	32.40	32.65	32.89	33.14	33.38	33.63	33.87	34.12	34.36
20	32.37	32.63	32.88	33.14	33.40	33.66	33.91	34.17	34.43	34.69	34.95	35.21	35.47	35.73	35.99	36.25
21	34.05	34.32	34.60	34.87	35.14	35.41	35.68	35.96	36.23	36.50	36.78	37.05	37.33	37.60	37.88	38.15
22	35.75	36.04	36.32	36.61	36.89	37.18	37.47	37.76	38.04	38.33	38.62	38.91	39.20	39.49	39.78	40.07
23	37.46	37.76	38.06	38.36	38.66	38.96	39.27	39.57	39.87	40.18	40.48	40.78	41.09	41.39	41.70	42.00
24	39.18	39.50	39.81	40.13	40.44	40.76	41.08	41.40	41.71	42.03	42.35	42.67	42.99	43.31	43.63	43.95
25	40.92	41.25	41.58	41.91	42.24	42.57	42.90	43.24	43.57	43.90	44.24	44.57	44.91	45.24	45.58	45.91
26	42.66	43.01	43.36	43.70	44.05	44.40	44.74	45.09	45.44	45.79	46.14	46.49	46.84	47.19	47.54	47.89
27	44.42	44.78	45.15	45.51	45.87	46.23	46.60	46.96	47.32	47.69	48.05	48.42	48.78	49.15	49.52	49.88
28	46.20	46.57	46.95	47.33	47.70	48.08	48.46	48.84	49.22	49.60	49.98	50.36	50.75	51.13	51.51	51.89
29	47.98	48.37	48.77	49.16	49.55	49.95	50.34	50.74	51.13	51.53	51.93	52.32	52.72	53.12	53.52	53.92
30	49.78	50.19	50.60	51.00	51.41	51.82	52.23	52.65	53.06	53.47	53.88	54.30	54.71	55.13	55.54	55.96
31	51.59	52.01	52.44	52.86	53.29	53.71	54.14	54.57	55.00	55.43	55.85	56.28	56.72	57.15	57.58	58.01
32	53.41	53.85	54.29	54.73	55.17	55.62	56.06	56.50	56.95	57.39	57.84	58.29	58.73	59.18	59.63	60.08
33	55.24	55.70	56.16	56.62	57.07	57.53	57.99	58.45	58.92	59.38	59.84	60.30	60.77	61.23	61.70	62.16
34	57.09	57.56	58.04	58.51	58.99	59.46	59.94	60.42	60.89	61.37	61.85	62.33	62.81	63.30	63.78	64.26
35	58.95	59.44	59.93	60.42	60.91	61.40	61.90	62.39	62.89	63.38	63.88	64.38	64.88	65.37	65.87	66.37
36	60.82	61.33	61.83	62.34	62.85	63.36	63.87	64.38	64.89	65.41	65.92	66.43	66.95	67.47	67.98	68.50
37	62.70	63.22	63.75	64.27	64.80	65.33	65.85	66.38	66.91	67.44	67.97	68.51	69.04	69.57	70.11	70.64
38	64.59	65.14	65.68	66.22	66.76	67.31	67.85	68.40	68.95	69.49	70.04	70.59	71.14	71.69	72.25	72.80
39	66.50	67.06	67.62	68.18	68.74	69.30	69.86	70.43	70.99	71.56	72.12	72.69	73.26	73.83	74.40	74.97
40	68.42	68.99	69.57	70.15	70.13	71.31	71.89	72.47	73.05	73.63	74.22	74.80	75.39	75.98	76.56	77.15
41	70.35	70.94	71.53	72.13	72.73	73.32	73.92	74.52	75.12	75.72	76.32	76.93	77.53	78.14	78.74	79.35
42	72.29	72.90	73.51	74.12	74.74	75.35	75.97	76.59	77.20	77.82	78.44	79.07	79.69	80.31	80.94	81.56
43	74.24	74.87	75.50	76.13	76.76	77.40	78.03	78.67	79.30	79.94	80.58	81.22	81.86	82.50	83.14	83.79
44	76.20	76.85	77.50	78.15	78.80	79.45	80.10	80.76	81.41	82.07	82.72	83.38	84.04	84.70	85.36	86.03
45	78.18	78.84	79.51	80.18	80.85	81.52	82.19	82.86	83.53	84.21	84.88	85.56	86.24	86.92	87.60	88.28
46	80.17	80.85	81.53	82.22	82.91	83.60	84.28	84.98	85.67	86.36	87.06	87.75	88.45	89.15	89.85	90.55
47	82.16	82.87	83.57	84.27	84.98	85.69	86.39	87.10	87.81	88.53	89.24	89.95	90.67	91.39	92.11	92.83
48	84.17	84.89	85.61	86.34	87.06	87.79	88.52	89.24	89.97	90.70	91.44	92.17	92.91	93.64	94.38	95.12
49	86.19	86.93	87.67	88.41	89.16	89.90	90.65	91.40	92.14	92.89	93.65	94.40	95.15	95.91	96.67	97.42
50	88.22	88.98	89.74	90.50	91.26	92.03	92.79	93.56	94.33	95.10	95.87	96.64	97.41	98.19	98.96	99.74
51	90.26	91.04	91.82	92.60	93.38	94.16	94.95	95.74	96.52	97.31	98.10	98.89	99.69	100.48	101.28	102.07
52	92.32	93.11	93.91	94.71	95.51	96.31	97.12	97.92	98.73	99.54	100.35	101.16	101.97	102.79	103.60	104.42
53	94.38	95.20	96.01	96.83	97.65	98.47	99.30	100.12	100.95	101.78	102.61	103.44	104.27	105.10	105.94	106.78
54	96.45	97.29	98.13	98.96	99.80	100.64	101.49	102.33	103.18	104.03	104.87	105.73	106.58	107.43	108.29	109.14
55	98.54	99.39	100.25	101.11	101.97	102.83	103.69	104.55	105.42	106.29	107.16	108.03	108.90	109.77	110.65	111.53
56	100.63	101.51	102.38	103.26	104.14	105.02	105.90	106.79	107.67	108.56	109.45	110.34	111.23	112.13	113.02	113.92
57	102.74	103.63	104.53	105.43	106.32	107.22	108.13	109.03	109.94	110.85	111.75	112.67	113.58	114.49	115.41	116.33
58	104.85	105.77	106.68	107.60	108.52	109.44	110.36	111.29	112.21	113.14	114.07	115.00	115.93	116.87	117.81	118.74
59	106.98	107.91	108.85	109.79	110.73	111.67	112.61	113.55	114.50	115.45	116.40	117.35	118.30	119.26	120.22	121.17
60	109.12	110.07	111.02	111.98	112.94	113.90	114.87	115.83	116.80	117.77	118.74	119.71	120.68	121.66	122.64	123.62

Source: Roueche & Graves, *Business Mathematics*, 7th ed., Upper Saddle River, NJ: Prentice Hall, 1997.

Number of Payments	ANNUAL PERCENTAGE RATE															
	38.00%	38.25%	38.50%	38.75%	39.00%	39.25%	39.50%	39.75%	40.00%	40.25%	40.50%	40.75%	41.00%	41.25%	41.50%	41.75%
	Finance charge per $100 of amount financed															
1	3.17	3.19	3.21	3.23	3.25	3.27	3.29	3.31	3.33	3.35	3.37	3.40	3.42	3.44	3.46	3.48
2	4.77	4.81	4.84	4.87	4.90	4.93	4.96	5.00	5.03	5.06	5.09	5.12	5.15	5.19	5.22	5.25
3	6.40	6.44	6.48	6.53	6.57	6.61	6.65	6.70	6.74	6.78	6.82	6.87	6.91	6.95	7.00	7.04
4	8.04	8.09	8.15	8.20	8.25	8.31	8.36	8.42	8.47	8.52	8.58	8.63	8.69	8.74	8.79	8.85
5	9.70	9.76	9.83	9.89	9.96	10.02	10.09	10.15	10.22	10.28	10.35	10.41	10.48	10.54	10.61	10.68
6	11.37	11.45	11.52	11.60	11.68	11.75	11.83	11.91	11.99	12.06	12.14	12.22	12.29	12.37	12.45	12.52
7	13.06	13.15	13.24	13.33	13.42	13.50	13.59	13.68	13.77	13.86	13.95	14.04	14.13	14.21	14.30	14.39
8	14.77	14.87	14.97	15.07	15.17	15.27	15.37	15.47	15.57	15.67	15.77	15.88	15.98	16.08	16.18	16.28
9	16.49	16.60	16.72	16.83	16.94	17.05	17.17	17.28	17.39	17.51	17.62	17.73	17.85	17.96	18.07	18.19
10	18.23	18.35	18.48	18.61	18.73	18.86	18.98	19.11	19.23	19.36	19.48	19.61	19.74	19.86	19.99	20.12
11	19.99	20.12	20.26	20.40	20.54	20.68	20.81	20.95	21.09	21.23	21.37	21.51	21.65	21.78	21.92	22.06
12	21.76	21.91	22.06	22.21	22.36	22.51	22.66	22.81	22.97	23.12	23.27	23.42	23.57	23.72	23.88	24.03
13	23.54	23.71	23.87	24.04	24.20	24.37	24.53	24.69	24.86	25.02	25.19	25.35	25.52	25.68	25.85	26.01
14	25.35	25.53	25.70	25.88	26.06	26.24	26.41	26.59	26.77	26.95	27.13	27.30	27.48	27.66	27.84	28.02
15	27.17	27.36	27.55	27.74	27.93	28.12	28.32	28.51	28.70	28.89	29.08	29.27	29.47	29.66	29.85	30.04
16	29.01	29.21	29.41	29.62	29.82	30.03	30.23	30.44	30.65	30.85	31.06	31.26	31.47	31.68	31.88	32.09
17	30.86	31.08	31.29	31.51	31.73	31.95	32.17	32.39	32.61	32.83	33.05	33.27	33.49	33.71	33.93	34.15
18	32.73	32.96	33.19	33.42	33.66	33.89	34.12	34.36	34.59	34.83	35.06	35.29	35.53	35.76	36.00	36.23
19	34.61	34.86	35.10	35.35	35.60	35.85	36.09	36.34	36.59	36.84	37.09	37.34	37.59	37.84	38.09	38.34
20	36.51	36.77	37.03	37.30	37.56	37.82	38.08	38.35	38.61	38.87	39.14	39.40	39.66	39.93	40.19	40.46
21	38.43	38.70	38.98	39.26	39.53	39.81	40.09	40.36	40.64	40.92	41.20	41.48	41.76	42.04	42.32	42.60
22	40.36	40.65	40.94	41.23	41.52	41.82	42.11	42.40	42.69	42.99	43.28	43.58	43.87	44.16	44.46	44.75
23	42.31	42.61	42.92	43.23	43.53	43.84	44.15	44.46	44.76	45.07	45.38	45.69	46.00	46.31	46.62	46.93
24	44.27	44.59	44.91	45.23	45.56	45.88	46.20	46.53	46.85	47.17	47.50	47.82	48.15	48.48	48.80	49.13
25	46.25	46.59	46.92	47.26	47.60	47.94	48.28	48.61	48.95	49.29	49.63	49.98	50.32	50.66	51.00	51.34
26	48.24	48.60	48.95	49.30	49.66	50.01	50.36	50.72	51.07	51.43	51.79	52.14	52.50	52.86	53.22	53.58
27	50.25	50.62	50.99	51.36	51.73	52.10	52.47	52.84	53.21	53.58	53.96	54.33	54.70	55.08	55.45	55.83
28	52.28	52.66	53.05	53.43	53.82	54.20	54.59	54.98	55.37	55.76	56.14	56.53	56.92	57.31	57.70	58.10
29	54.32	54.72	55.12	55.52	55.92	56.33	56.73	57.13	57.54	57.94	58.35	58.75	59.16	59.57	59.97	60.38
30	56.37	56.79	57.21	57.63	58.05	58.46	58.88	59.30	59.73	60.15	60.57	60.99	61.42	61.84	62.26	62.69
31	58.44	58.88	59.31	59.75	60.18	60.62	61.05	61.49	61.93	62.37	62.81	63.25	63.69	64.13	64.57	65.01
32	60.53	60.98	61.43	61.88	62.34	62.79	63.24	63.70	64.15	64.61	65.06	65.52	65.98	66.43	66.89	67.35
33	62.63	63.10	63.57	64.03	64.50	64.97	65.44	65.92	66.39	66.86	67.33	67.81	68.28	68.76	69.23	69.71
34	64.75	65.23	65.72	66.20	66.69	67.18	67.66	68.15	68.64	69.13	69.62	70.11	70.61	71.10	71.59	72.09
35	66.88	67.38	67.88	68.38	68.89	69.39	69.90	70.40	70.91	71.42	71.93	72.44	72.95	73.46	73.97	74.48
36	69.02	69.54	70.06	70.58	71.10	71.62	72.15	72.67	73.20	73.72	74.25	74.78	75.30	75.83	76.36	76.89
37	71.18	71.72	72.25	72.79	73.33	73.87	74.41	74.96	75.50	76.04	76.59	77.13	77.68	78.22	78.77	79.32
38	73.35	73.91	74.46	75.02	75.58	76.14	76.69	77.25	77.81	78.38	78.94	79.50	80.07	80.63	81.20	81.76
39	75.54	76.11	76.69	77.26	77.84	78.41	78.99	79.57	80.15	80.73	81.31	81.89	82.47	83.06	83.64	84.23
40	77.74	78.33	78.93	79.52	80.11	80.71	81.30	81.90	82.50	83.09	83.69	84.29	84.90	85.50	86.10	86.70
41	79.96	80.57	81.18	81.79	82.40	83.01	83.63	84.24	84.86	85.48	86.09	86.71	87.33	87.95	88.58	89.20
42	82.19	82.82	83.45	84.07	84.71	85.34	85.97	86.60	87.24	87.87	88.51	89.15	89.79	90.43	91.07	91.71
43	84.43	85.08	85.73	86.37	87.02	87.67	88.33	88.98	89.63	90.29	90.94	91.60	92.26	92.92	93.58	94.24
44	86.69	87.36	88.02	88.69	89.36	90.03	90.70	91.37	92.04	92.72	93.39	94.07	94.74	95.42	96.10	96.78
45	88.96	89.65	90.33	91.02	91.70	92.39	93.08	93.77	94.47	95.16	95.85	96.55	97.24	97.94	98.64	99.34
46	91.25	91.95	92.65	93.36	94.07	94.77	95.48	96.19	96.90	97.62	98.33	99.04	99.76	100.48	101.20	101.91
47	93.55	94.27	94.99	95.72	96.44	97.17	97.90	98.63	99.36	100.09	100.82	101.56	102.29	103.03	103.77	104.50
48	95.86	96.60	97.34	98.09	98.83	99.58	100.33	101.07	101.82	102.58	103.33	104.08	104.84	105.59	106.35	107.11
49	98.18	98.94	99.71	100.47	101.23	102.00	102.77	103.54	104.31	105.08	105.85	106.62	107.40	108.18	108.95	109.73
50	100.52	101.30	102.08	102.87	103.65	104.44	105.22	106.01	106.80	107.59	108.39	109.18	109.98	110.77	111.57	112.37
51	102.87	103.67	104.47	105.28	106.08	106.89	107.69	108.50	109.31	110.12	110.94	111.75	112.57	113.38	114.20	115.02
52	105.24	106.06	106.88	107.70	108.53	109.35	110.18	111.01	111.84	112.67	113.50	114.33	115.17	116.01	116.85	117.69
53	107.61	108.45	109.29	110.14	110.98	111.83	112.68	113.52	114.37	115.23	116.08	116.93	117.79	118.65	119.51	120.37
54	110.00	110.86	111.72	112.59	113.45	114.32	115.19	116.05	116.93	117.80	118.67	119.55	120.42	121.30	122.18	123.06
55	112.40	113.28	114.17	115.05	115.94	116.82	117.71	118.60	119.49	120.38	121.28	122.17	123.07	123.97	124.87	125.77
56	114.82	115.72	116.62	117.53	118.43	119.34	120.25	121.16	122.07	122.98	123.90	124.81	125.73	126.65	127.57	128.49
57	117.25	118.17	119.09	120.01	120.94	121.87	122.80	123.73	124.66	125.59	126.53	127.47	128.40	129.34	130.29	131.23
58	119.68	120.63	121.57	122.51	123.46	124.41	125.36	126.31	127.26	128.22	129.17	130.13	131.09	132.05	133.02	133.98
59	122.13	123.10	124.06	125.03	125.99	126.96	127.93	128.91	129.88	130.86	131.83	132.81	133.79	134.78	135.76	136.75
60	124.60	125.58	126.56	127.55	128.54	129.53	130.52	131.51	132.51	133.51	134.51	135.51	136.51	137.51	138.52	139.52

Source: Roueche & Graves, *Business Mathematics*, 7th ed., Upper Saddle River, NJ: Prentice Hall, 1997.

Number of Payments	ANNUAL PERCENTAGE RATE															
	42.00%	42.25%	42.50%	42.75%	43.00%	43.25%	43.50%	43.75%	44.00%	44.25%	44.50%	44.75%	45.00%	45.25%	45.50%	45.75%
	Finance charge per $100 of amount financed															
1	3.50	3.52	3.54	3.56	3.58	3.60	3.62	3.65	3.67	3.69	3.71	3.73	3.75	3.77	3.79	3.81
2	5.28	5.31	5.34	5.37	5.41	5.44	5.47	5.50	5.53	5.56	5.60	5.63	5.66	5.69	5.72	5.75
3	7.08	7.12	7.17	7.21	7.25	7.29	7.34	7.38	7.42	7.46	7.51	7.55	7.59	7.63	7.68	7.72
4	8.90	8.95	9.01	9.06	9.12	9.17	9.22	9.28	9.33	9.39	9.44	9.49	9.55	9.60	9.66	9.71
5	10.74	10.81	10.87	10.94	11.00	11.07	11.13	11.20	11.26	11.33	11.39	11.46	11.53	11.59	11.66	11.72
6	12.60	12.68	12.76	12.83	12.91	12.99	13.06	13.14	13.22	13.30	13.37	13.45	13.53	13.60	13.68	13.76
7	14.48	14.57	14.66	14.75	14.84	14.93	15.02	15.10	15.19	15.28	15.37	15.46	15.55	15.64	15.73	15.82
8	16.38	16.48	16.58	16.69	16.79	16.89	16.99	17.09	17.19	17.29	17.40	17.50	17.60	17.70	17.80	17.90
9	18.30	18.42	18.53	18.64	18.76	18.87	18.98	19.10	19.21	19.33	19.44	19.55	19.67	19.78	19.90	20.01
10	20.24	20.37	20.49	20.62	20.75	20.87	21.00	21.13	21.25	21.38	21.51	21.63	21.76	21.89	22.02	22.14
11	22.20	22.34	22.48	22.62	22.76	22.90	23.04	23.18	23.32	23.46	23.60	23.74	23.88	24.02	24.16	24.30
12	24.18	24.33	24.49	24.64	24.79	24.94	25.10	25.25	25.40	25.55	25.71	25.86	26.01	26.17	26.32	26.48
13	26.18	26.35	26.51	26.68	26.84	27.01	27.18	27.34	27.51	27.67	27.84	28.01	28.18	28.34	28.51	28.68
14	28.20	28.38	28.56	28.74	28.92	29.10	29.28	29.46	29.64	29.82	30.00	30.18	30.36	30.54	30.72	30.90
15	30.24	30.43	30.62	30.82	31.01	31.20	31.40	31.59	31.79	31.98	32.17	32.37	32.56	32.76	32.95	33.15
16	32.30	32.50	32.71	32.92	33.12	33.33	33.54	33.75	33.96	34.17	34.37	34.58	34.79	35.00	35.21	35.42
17	34.37	34.59	34.82	35.04	35.26	35.48	35.70	35.93	36.15	36.37	36.60	36.82	37.04	37.27	37.49	37.71
18	36.47	36.71	36.94	37.18	37.41	37.65	37.89	38.13	38.36	38.60	38.84	39.08	39.31	39.55	39.79	40.03
19	38.59	38.84	39.09	39.34	39.59	39.84	40.09	40.34	40.60	40.85	41.10	41.35	41.61	41.86	42.11	42.37
20	40.72	40.99	41.25	41.52	41.79	42.05	42.32	42.59	42.85	43.12	43.39	43.66	43.92	44.19	44.46	44.73
21	42.88	43.16	43.44	43.72	44.00	44.28	44.56	44.85	45.13	45.41	45.69	45.98	46.26	46.55	46.83	47.11
22	45.05	45.35	45.64	45.94	46.24	46.53	46.83	47.13	47.43	47.72	48.02	48.32	48.62	48.92	49.22	49.52
23	47.24	47.55	47.87	48.18	48.49	48.80	49.12	49.43	49.74	50.06	50.37	50.69	51.00	51.32	51.63	51.95
24	49.45	49.78	50.11	50.44	50.77	51.09	51.42	51.75	52.08	52.41	52.74	53.07	53.41	53.74	54.07	54.40
25	51.69	52.03	52.37	52.72	53.06	53.40	53.75	54.10	54.44	54.79	55.13	55.48	55.83	56.18	56.53	56.87
26	53.93	54.29	54.65	55.01	55.37	55.73	56.10	56.46	56.82	57.18	57.55	57.91	58.27	58.64	59.00	59.37
27	56.20	56.58	56.95	57.33	57.71	58.08	58.46	58.84	59.22	59.60	59.98	60.36	60.74	61.12	61.50	61.89
28	58.49	58.88	59.27	59.67	60.06	60.45	60.85	61.24	61.64	62.04	62.43	62.83	63.23	63.63	64.02	64.42
29	60.79	61.20	61.61	62.02	62.43	62.84	63.25	63.67	64.08	64.49	64.91	65.32	65.73	66.15	66.57	66.98
30	63.11	63.54	63.97	64.39	64.82	65.25	65.68	66.11	66.54	66.97	67.40	67.83	68.26	68.70	69.13	69.56
31	65.45	65.90	66.34	66.79	67.23	67.68	68.12	68.57	69.02	69.46	69.91	70.36	70.81	71.26	71.71	72.16
32	67.81	68.27	68.73	69.20	69.66	70.12	70.59	71.05	71.51	71.98	72.45	72.91	73.38	73.85	74.32	74.79
33	70.19	70.67	71.15	71.63	72.11	72.59	73.07	73.55	74.03	74.52	75.00	75.48	75.97	76.45	76.94	77.43
34	72.58	73.08	73.58	74.07	74.57	75.07	75.57	76.07	76.57	77.07	77.57	78.07	78.58	79.08	79.59	80.09
35	74.99	75.51	76.02	76.54	77.05	77.57	78.09	78.61	79.12	79.64	80.16	80.68	81.21	81.73	82.25	82.78
36	77.42	77.95	78.49	79.02	79.55	80.09	80.62	81.16	81.70	82.24	82.77	83.31	83.85	84.39	84.94	85.48
37	79.87	80.42	80.97	81.52	82.07	82.63	83.18	83.74	84.29	84.85	85.40	85.96	86.52	87.08	87.64	88.20
38	82.33	82.90	83.47	84.04	84.61	85.18	85.76	86.33	86.90	87.48	88.05	88.63	89.21	89.79	90.37	90.95
39	84.81	85.40	85.99	86.58	87.17	87.76	88.35	88.94	89.53	90.13	90.72	91.32	91.91	92.51	93.11	93.71
40	87.31	87.91	88.52	89.13	89.74	90.35	90.96	91.57	92.18	92.79	93.41	94.02	94.64	95.25	95.87	96.49
41	89.82	90.45	91.07	91.70	92.33	92.96	93.58	94.22	94.85	95.48	96.11	96.75	97.38	98.02	98.65	99.29
42	92.35	93.00	93.64	94.29	94.93	95.58	96.23	96.88	97.53	98.18	98.83	99.49	100.14	100.80	101.45	102.11
43	94.90	95.56	96.23	96.89	97.56	98.22	98.89	99.56	100.23	100.90	101.57	102.25	102.92	103.60	104.27	104.95
44	97.46	98.14	98.83	99.51	100.20	100.88	101.57	102.26	102.95	103.64	104.33	105.03	105.72	106.41	107.11	107.81
45	100.04	100.74	101.45	102.15	102.85	103.56	104.27	104.98	105.69	106.40	107.11	107.82	108.53	109.25	109.96	110.68
46	102.63	103.36	104.08	104.80	105.53	106.25	106.98	107.71	108.44	109.17	109.90	110.63	111.37	112.10	112.84	113.58
47	105.25	105.99	106.73	107.47	108.22	108.96	109.71	110.46	111.21	111.96	112.71	113.46	114.22	114.97	115.73	116.49
48	107.87	108.63	109.39	110.16	110.92	111.69	112.46	113.23	113.99	114.77	115.54	116.31	117.09	117.86	118.64	119.42
49	110.51	111.29	112.08	112.86	113.64	114.43	115.22	116.01	116.80	117.59	118.38	119.17	119.97	120.77	121.56	122.36
50	113.17	113.97	114.77	115.58	116.38	117.19	118.00	118.81	119.62	120.43	121.24	122.06	122.87	123.69	124.51	125.32
51	115.84	116.66	117.48	118.31	119.14	119.96	120.79	121.62	122.45	123.28	124.12	124.95	125.79	126.63	127.46	128.30
52	118.53	119.37	120.21	121.06	121.90	122.75	123.60	124.45	125.30	126.16	127.01	127.87	128.72	129.58	130.44	131.30
53	121.23	122.09	122.95	123.82	124.69	125.56	126.43	127.30	128.17	129.04	129.92	130.80	131.67	132.55	133.43	134.32
54	123.94	124.83	125.71	126.60	127.49	128.38	129.27	130.16	131.05	131.95	132.84	133.74	134.64	135.54	136.44	137.35
55	126.67	127.58	128.48	129.39	130.30	131.21	132.12	133.03	133.95	134.87	135.78	136.70	137.62	138.54	139.47	140.39
56	129.42	130.34	131.27	132.20	133.13	134.06	134.99	135.93	136.86	137.80	138.74	139.68	140.62	141.56	142.51	143.45
57	132.17	133.12	134.07	135.02	135.97	136.92	137.88	138.83	139.79	140.75	141.71	142.67	143.63	144.60	145.56	146.53
58	134.95	135.91	136.88	137.85	138.83	139.80	140.78	141.75	142.73	143.71	144.69	145.68	146.66	147.65	148.63	149.62
59	137.73	138.72	139.71	140.70	141.70	142.69	143.69	144.69	145.69	146.69	147.69	148.70	149.70	150.71	151.72	152.73
60	140.53	141.54	142.55	143.57	144.58	145.60	146.62	147.64	148.66	149.68	150.71	151.73	152.76	153.79	154.82	155.85

Source: Roueche & Graves, *Business Mathematics*, 7th ed., Upper Saddle River, NJ: Prentice Hall, 1997.

1998 Tax Table

Use if your taxable income is less than $100,000.
If $100,000 or more, use the Tax Rate Schedules.

Example. Mr. and Mrs. Brown are filing a joint return. Their taxable income on line 39 of Form 1040 is $25,300. First, they find the $25,300–25,350 income line. Next, they find the column for married filing jointly and read down the column. The amount shown where the income line and filing status column meet is $3,799. This is the tax amount they should enter on line 40 of their Form 1040.

Sample Table

At least	But less than	Single	Married filing jointly *	Married filing separately	Head of a household
			Your tax is—		
25,200	25,250	3,784	3,784	4,310	3,784
25,250	25,300	3,791	3,791	4,324	3,791
25,300	25,350	3,799	(3,799)	4,338	3,799
25,350	25,400	3,810	3,806	4,352	3,806

If line 39 (taxable income) is— At least	But less than	Single	Married filing jointly *	Married filing separately	Head of a house-hold
			Your tax is—		
0	5	0	0	0	0
5	15	2	2	2	2
15	25	3	3	3	3
25	50	6	6	6	6
50	75	9	9	9	9
75	100	13	13	13	13
100	125	17	17	17	17
125	150	21	21	21	21
150	175	24	24	24	24
175	200	28	28	28	28
200	225	32	32	32	32
225	250	36	36	36	36
250	275	39	39	39	39
275	300	43	43	43	43
300	325	47	47	47	47
325	350	51	51	51	51
350	375	54	54	54	54
375	400	58	58	58	58
400	425	62	62	62	62
425	450	66	66	66	66
450	475	69	69	69	69
475	500	73	73	73	73
500	525	77	77	77	77
525	550	81	81	81	81
550	575	84	84	84	84
575	600	88	88	88	88
600	625	92	92	92	92
625	650	96	96	96	96
650	675	99	99	99	99
675	700	103	103	103	103
700	725	107	107	107	107
725	750	111	111	111	111
750	775	114	114	114	114
775	800	118	118	118	118
800	825	122	122	122	122
825	850	126	126	126	126
850	875	129	129	129	129
875	900	133	133	133	133
900	925	137	137	137	137
925	950	141	141	141	141
950	975	144	144	144	144
975	1,000	148	148	148	148
1,000					
1,000	1,025	152	152	152	152
1,025	1,050	156	156	156	156
1,050	1,075	159	159	159	159
1,075	1,100	163	163	163	163
1,100	1,125	167	167	167	167
1,125	1,150	171	171	171	171
1,150	1,175	174	174	174	174
1,175	1,200	178	178	178	178
1,200	1,225	182	182	182	182
1,225	1,250	186	186	186	186
1,250	1,275	189	189	189	189
1,275	1,300	193	193	193	193

If line 39 (taxable income) is— At least	But less than	Single	Married filing jointly *	Married filing separately	Head of a house-hold
			Your tax is—		
1,300	1,325	197	197	197	197
1,325	1,350	201	201	201	201
1,350	1,375	204	204	204	204
1,375	1,400	208	208	208	208
1,400	1,425	212	212	212	212
1,425	1,450	216	216	216	216
1,450	1,475	219	219	219	219
1,475	1,500	223	223	223	223
1,500	1,525	227	227	227	227
1,525	1,550	231	231	231	231
1,550	1,575	234	234	234	234
1,575	1,600	238	238	238	238
1,600	1,625	242	242	242	242
1,625	1,650	246	246	246	246
1,650	1,675	249	249	249	249
1,675	1,700	253	253	253	253
1,700	1,725	257	257	257	257
1,725	1,750	261	261	261	261
1,750	1,775	264	264	264	264
1,775	1,800	268	268	268	268
1,800	1,825	272	272	272	272
1,825	1,850	276	276	276	276
1,850	1,875	279	279	279	279
1,875	1,900	283	283	283	283
1,900	1,925	287	287	287	287
1,925	1,950	291	291	291	291
1,950	1,975	294	294	294	294
1,975	2,000	298	298	298	298
2,000					
2,000	2,025	302	302	302	302
2,025	2,050	306	306	306	306
2,050	2,075	309	309	309	309
2,075	2,100	313	313	313	313
2,100	2,125	317	317	317	317
2,125	2,150	321	321	321	321
2,150	2,175	324	324	324	324
2,175	2,200	328	328	328	328
2,200	2,225	332	332	332	332
2,225	2,250	336	336	336	336
2,250	2,275	339	339	339	339
2,275	2,300	343	343	343	343
2,300	2,325	347	347	347	347
2,325	2,350	351	351	351	351
2,350	2,375	354	354	354	354
2,375	2,400	358	358	358	358
2,400	2,425	362	362	362	362
2,425	2,450	366	366	366	366
2,450	2,475	369	369	369	369
2,475	2,500	373	373	373	373
2,500	2,525	377	377	377	377
2,525	2,550	381	381	381	381
2,550	2,575	384	384	384	384
2,575	2,600	388	388	388	388
2,600	2,625	392	392	392	392
2,625	2,650	396	396	396	396
2,650	2,675	399	399	399	399
2,675	2,700	403	403	403	403

If line 39 (taxable income) is— At least	But less than	Single	Married filing jointly *	Married filing separately	Head of a house-hold
			Your tax is—		
2,700	2,725	407	407	407	407
2,725	2,750	411	411	411	411
2,750	2,775	414	414	414	414
2,775	2,800	418	418	418	418
2,800	2,825	422	422	422	422
2,825	2,850	426	426	426	426
2,850	2,875	429	429	429	429
2,875	2,900	433	433	433	433
2,900	2,925	437	437	437	437
2,925	2,950	441	441	441	441
2,950	2,975	444	444	444	444
2,975	3,000	448	448	448	448
3,000					
3,000	3,050	454	454	454	454
3,050	3,100	461	461	461	461
3,100	3,150	469	469	469	469
3,150	3,200	476	476	476	476
3,200	3,250	484	484	484	484
3,250	3,300	491	491	491	491
3,300	3,350	499	499	499	499
3,350	3,400	506	506	506	506
3,400	3,450	514	514	514	514
3,450	3,500	521	521	521	521
3,500	3,550	529	529	529	529
3,550	3,600	536	536	536	536
3,600	3,650	544	544	544	544
3,650	3,700	551	551	551	551
3,700	3,750	559	559	559	559
3,750	3,800	566	566	566	566
3,800	3,850	574	574	574	574
3,850	3,900	581	581	581	581
3,900	3,950	589	589	589	589
3,950	4,000	596	596	596	596
4,000					
4,000	4,050	604	604	604	604
4,050	4,100	611	611	611	611
4,100	4,150	619	619	619	619
4,150	4,200	626	626	626	626
4,200	4,250	634	634	634	634
4,250	4,300	641	641	641	641
4,300	4,350	649	649	649	649
4,350	4,400	656	656	656	656
4,400	4,450	664	664	664	664
4,450	4,500	671	671	671	671
4,500	4,550	679	679	679	679
4,550	4,600	686	686	686	686
4,600	4,650	694	694	694	694
4,650	4,700	701	701	701	701
4,700	4,750	709	709	709	709
4,750	4,800	716	716	716	716
4,800	4,850	724	724	724	724
4,850	4,900	731	731	731	731
4,900	4,950	739	739	739	739
4,950	5,000	746	746	746	746

Continued on next page

* This column must also be used by a qualifying widow(er).

1998 Tax Table—Continued

If line 39 (taxable income) is— At least	But less than	Single	Married filing jointly *	Married filing separately	Head of a household
5,000					
5,000	5,050	754	754	754	754
5,050	5,100	761	761	761	761
5,100	5,150	769	769	769	769
5,150	5,200	776	776	776	776
5,200	5,250	784	784	784	784
5,250	5,300	791	791	791	791
5,300	5,350	799	799	799	799
5,350	5,400	806	806	806	806
5,400	5,450	814	814	814	814
5,450	5,500	821	821	821	821
5,500	5,550	829	829	829	829
5,550	5,600	836	836	836	836
5,600	5,650	844	844	844	844
5,650	5,700	851	851	851	851
5,700	5,750	859	859	859	859
5,750	5,800	866	866	866	866
5,800	5,850	874	874	874	874
5,850	5,900	881	881	881	881
5,900	5,950	889	889	889	889
5,950	6,000	896	896	896	896
6,000					
6,000	6,050	904	904	904	904
6,050	6,100	911	911	911	911
6,100	6,150	919	919	919	919
6,150	6,200	926	926	926	926
6,200	6,250	934	934	934	934
6,250	6,300	941	941	941	941
6,300	6,350	949	949	949	949
6,350	6,400	956	956	956	956
6,400	6,450	964	964	964	964
6,450	6,500	971	971	971	971
6,500	6,550	979	979	979	979
6,550	6,600	986	986	986	986
6,600	6,650	994	994	994	994
6,650	6,700	1,001	1,001	1,001	1,001
6,700	6,750	1,009	1,009	1,009	1,009
6,750	6,800	1,016	1,016	1,016	1,016
6,800	6,850	1,024	1,024	1,024	1,024
6,850	6,900	1,031	1,031	1,031	1,031
6,900	6,950	1,039	1,039	1,039	1,039
6,950	7,000	1,046	1,046	1,046	1,046
7,000					
7,000	7,050	1,054	1,054	1,054	1,054
7,050	7,100	1,061	1,061	1,061	1,061
7,100	7,150	1,069	1,069	1,069	1,069
7,150	7,200	1,076	1,076	1,076	1,076
7,200	7,250	1,084	1,084	1,084	1,084
7,250	7,300	1,091	1,091	1,091	1,091
7,300	7,350	1,099	1,099	1,099	1,099
7,350	7,400	1,106	1,106	1,106	1,106
7,400	7,450	1,114	1,114	1,114	1,114
7,450	7,500	1,121	1,121	1,121	1,121
7,500	7,550	1,129	1,129	1,129	1,129
7,550	7,600	1,136	1,136	1,136	1,136
7,600	7,650	1,144	1,144	1,144	1,144
7,650	7,700	1,151	1,151	1,151	1,151
7,700	7,750	1,159	1,159	1,159	1,159
7,750	7,800	1,166	1,166	1,166	1,166
7,800	7,850	1,174	1,174	1,174	1,174
7,850	7,900	1,181	1,181	1,181	1,181
7,900	7,950	1,189	1,189	1,189	1,189
7,950	8,000	1,196	1,196	1,196	1,196

If line 39 (taxable income) is— At least	But less than	Single	Married filing jointly *	Married filing separately	Head of a household
8,000					
8,000	8,050	1,204	1,204	1,204	1,204
8,050	8,100	1,211	1,211	1,211	1,211
8,100	8,150	1,219	1,219	1,219	1,219
8,150	8,200	1,226	1,226	1,226	1,226
8,200	8,250	1,234	1,234	1,234	1,234
8,250	8,300	1,241	1,241	1,241	1,241
8,300	8,350	1,249	1,249	1,249	1,249
8,350	8,400	1,256	1,256	1,256	1,256
8,400	8,450	1,264	1,264	1,264	1,264
8,450	8,500	1,271	1,271	1,271	1,271
8,500	8,550	1,279	1,279	1,279	1,279
8,550	8,600	1,286	1,286	1,286	1,286
8,600	8,650	1,294	1,294	1,294	1,294
8,650	8,700	1,301	1,301	1,301	1,301
8,700	8,750	1,309	1,309	1,309	1,309
8,750	8,800	1,316	1,316	1,316	1,316
8,800	8,850	1,324	1,324	1,324	1,324
8,850	8,900	1,331	1,331	1,331	1,331
8,900	8,950	1,339	1,339	1,339	1,339
8,950	9,000	1,346	1,346	1,346	1,346
9,000					
9,000	9,050	1,354	1,354	1,354	1,354
9,050	9,100	1,361	1,361	1,361	1,361
9,100	9,150	1,369	1,369	1,369	1,369
9,150	9,200	1,376	1,376	1,376	1,376
9,200	9,250	1,384	1,384	1,384	1,384
9,250	9,300	1,391	1,391	1,391	1,391
9,300	9,350	1,399	1,399	1,399	1,399
9,350	9,400	1,406	1,406	1,406	1,406
9,400	9,450	1,414	1,414	1,414	1,414
9,450	9,500	1,421	1,421	1,421	1,421
9,500	9,550	1,429	1,429	1,429	1,429
9,550	9,600	1,436	1,436	1,436	1,436
9,600	9,650	1,444	1,444	1,444	1,444
9,650	9,700	1,451	1,451	1,451	1,451
9,700	9,750	1,459	1,459	1,459	1,459
9,750	9,800	1,466	1,466	1,466	1,466
9,800	9,850	1,474	1,474	1,474	1,474
9,850	9,900	1,481	1,481	1,481	1,481
9,900	9,950	1,489	1,489	1,489	1,489
9,950	10,000	1,496	1,496	1,496	1,496
10,000					
10,000	10,050	1,504	1,504	1,504	1,504
10,050	10,100	1,511	1,511	1,511	1,511
10,100	10,150	1,519	1,519	1,519	1,519
10,150	10,200	1,526	1,526	1,526	1,526
10,200	10,250	1,534	1,534	1,534	1,534
10,250	10,300	1,541	1,541	1,541	1,541
10,300	10,350	1,549	1,549	1,549	1,549
10,350	10,400	1,556	1,556	1,556	1,556
10,400	10,450	1,564	1,564	1,564	1,564
10,450	10,500	1,571	1,571	1,571	1,571
10,500	10,550	1,579	1,579	1,579	1,579
10,550	10,600	1,586	1,586	1,586	1,586
10,600	10,650	1,594	1,594	1,594	1,594
10,650	10,700	1,601	1,601	1,601	1,601
10,700	10,750	1,609	1,609	1,609	1,609
10,750	10,800	1,616	1,616	1,616	1,616
10,800	10,850	1,624	1,624	1,624	1,624
10,850	10,900	1,631	1,631	1,631	1,631
10,900	10,950	1,639	1,639	1,639	1,639
10,950	11,000	1,646	1,646	1,646	1,646

If line 39 (taxable income) is— At least	But less than	Single	Married filing jointly *	Married filing separately	Head of a household
11,000					
11,000	11,050	1,654	1,654	1,654	1,654
11,050	11,100	1,661	1,661	1,661	1,661
11,100	11,150	1,669	1,669	1,669	1,669
11,150	11,200	1,676	1,676	1,676	1,676
11,200	11,250	1,684	1,684	1,684	1,684
11,250	11,300	1,691	1,691	1,691	1,691
11,300	11,350	1,699	1,699	1,699	1,699
11,350	11,400	1,706	1,706	1,706	1,706
11,400	11,450	1,714	1,714	1,714	1,714
11,450	11,500	1,721	1,721	1,721	1,721
11,500	11,550	1,729	1,729	1,729	1,729
11,550	11,600	1,736	1,736	1,736	1,736
11,600	11,650	1,744	1,744	1,744	1,744
11,650	11,700	1,751	1,751	1,751	1,751
11,700	11,750	1,759	1,759	1,759	1,759
11,750	11,800	1,766	1,766	1,766	1,766
11,800	11,850	1,774	1,774	1,774	1,774
11,850	11,900	1,781	1,781	1,781	1,781
11,900	11,950	1,789	1,789	1,789	1,789
11,950	12,000	1,796	1,796	1,796	1,796
12,000					
12,000	12,050	1,804	1,804	1,804	1,804
12,050	12,100	1,811	1,811	1,811	1,811
12,100	12,150	1,819	1,819	1,819	1,819
12,150	12,200	1,826	1,826	1,826	1,826
12,200	12,250	1,834	1,834	1,834	1,834
12,250	12,300	1,841	1,841	1,841	1,841
12,300	12,350	1,849	1,849	1,849	1,849
12,350	12,400	1,856	1,856	1,856	1,856
12,400	12,450	1,864	1,864	1,864	1,864
12,450	12,500	1,871	1,871	1,871	1,871
12,500	12,550	1,879	1,879	1,879	1,879
12,550	12,600	1,886	1,886	1,886	1,886
12,600	12,650	1,894	1,894	1,894	1,894
12,650	12,700	1,901	1,901	1,901	1,901
12,700	12,750	1,909	1,909	1,909	1,909
12,750	12,800	1,916	1,916	1,916	1,916
12,800	12,850	1,924	1,924	1,924	1,924
12,850	12,900	1,931	1,931	1,931	1,931
12,900	12,950	1,939	1,939	1,939	1,939
12,950	13,000	1,946	1,946	1,946	1,946
13,000					
13,000	13,050	1,954	1,954	1,954	1,954
13,050	13,100	1,961	1,961	1,961	1,961
13,100	13,150	1,969	1,969	1,969	1,969
13,150	13,200	1,976	1,976	1,976	1,976
13,200	13,250	1,984	1,984	1,984	1,984
13,250	13,300	1,991	1,991	1,991	1,991
13,300	13,350	1,999	1,999	1,999	1,999
13,350	13,400	2,006	2,006	2,006	2,006
13,400	13,450	2,014	2,014	2,014	2,014
13,450	13,500	2,021	2,021	2,021	2,021
13,500	13,550	2,029	2,029	2,029	2,029
13,550	13,600	2,036	2,036	2,036	2,036
13,600	13,650	2,044	2,044	2,044	2,044
13,650	13,700	2,051	2,051	2,051	2,051
13,700	13,750	2,059	2,059	2,059	2,059
13,750	13,800	2,066	2,066	2,066	2,066
13,800	13,850	2,074	2,074	2,074	2,074
13,850	13,900	2,081	2,081	2,081	2,081
13,900	13,950	2,089	2,089	2,089	2,089
13,950	14,000	2,096	2,096	2,096	2,096

* This column must also be used by a qualifying widow(er).

Continued on next page

1998 Tax Table—Continued

If line 39 (taxable income) is— At least	But less than	Single	Married filing jointly *	Married filing separately	Head of a household
14,000					
14,000	14,050	2,104	2,104	2,104	2,104
14,050	14,100	2,111	2,111	2,111	2,111
14,100	14,150	2,119	2,119	2,119	2,119
14,150	14,200	2,126	2,126	2,126	2,126
14,200	14,250	2,134	2,134	2,134	2,134
14,250	14,300	2,141	2,141	2,141	2,141
14,300	14,350	2,149	2,149	2,149	2,149
14,350	14,400	2,156	2,156	2,156	2,156
14,400	14,450	2,164	2,164	2,164	2,164
14,450	14,500	2,171	2,171	2,171	2,171
14,500	14,550	2,179	2,179	2,179	2,179
14,550	14,600	2,186	2,186	2,186	2,186
14,600	14,650	2,194	2,194	2,194	2,194
14,650	14,700	2,201	2,201	2,201	2,201
14,700	14,750	2,209	2,209	2,209	2,209
14,750	14,800	2,216	2,216	2,216	2,216
14,800	14,850	2,224	2,224	2,224	2,224
14,850	14,900	2,231	2,231	2,231	2,231
14,900	14,950	2,239	2,239	2,239	2,239
14,950	15,000	2,246	2,246	2,246	2,246
15,000					
15,000	15,050	2,254	2,254	2,254	2,254
15,050	15,100	2,261	2,261	2,261	2,261
15,100	15,150	2,269	2,269	2,269	2,269
15,150	15,200	2,276	2,276	2,276	2,276
15,200	15,250	2,284	2,284	2,284	2,284
15,250	15,300	2,291	2,291	2,291	2,291
15,300	15,350	2,299	2,299	2,299	2,299
15,350	15,400	2,306	2,306	2,306	2,306
15,400	15,450	2,314	2,314	2,314	2,314
15,450	15,500	2,321	2,321	2,321	2,321
15,500	15,550	2,329	2,329	2,329	2,329
15,550	15,600	2,336	2,336	2,336	2,336
15,600	15,650	2,344	2,344	2,344	2,344
15,650	15,700	2,351	2,351	2,351	2,351
15,700	15,750	2,359	2,359	2,359	2,359
15,750	15,800	2,366	2,366	2,366	2,366
15,800	15,850	2,374	2,374	2,374	2,374
15,850	15,900	2,381	2,381	2,381	2,381
15,900	15,950	2,389	2,389	2,389	2,389
15,950	16,000	2,396	2,396	2,396	2,396
16,000					
16,000	16,050	2,404	2,404	2,404	2,404
16,050	16,100	2,411	2,411	2,411	2,411
16,100	16,150	2,419	2,419	2,419	2,419
16,150	16,200	2,426	2,426	2,426	2,426
16,200	16,250	2,434	2,434	2,434	2,434
16,250	16,300	2,441	2,441	2,441	2,441
16,300	16,350	2,449	2,449	2,449	2,449
16,350	16,400	2,456	2,456	2,456	2,456
16,400	16,450	2,464	2,464	2,464	2,464
16,450	16,500	2,471	2,471	2,471	2,471
16,500	16,550	2,479	2,479	2,479	2,479
16,550	16,600	2,486	2,486	2,486	2,486
16,600	16,650	2,494	2,494	2,494	2,494
16,650	16,700	2,501	2,501	2,501	2,501
16,700	16,750	2,509	2,509	2,509	2,509
16,750	16,800	2,516	2,516	2,516	2,516
16,800	16,850	2,524	2,524	2,524	2,524
16,850	16,900	2,531	2,531	2,531	2,531
16,900	16,950	2,539	2,539	2,539	2,539
16,950	17,000	2,546	2,546	2,546	2,546

If line 39 (taxable income) is— At least	But less than	Single	Married filing jointly *	Married filing separately	Head of a household
17,000					
17,000	17,050	2,554	2,554	2,554	2,554
17,050	17,100	2,561	2,561	2,561	2,561
17,100	17,150	2,569	2,569	2,569	2,569
17,150	17,200	2,576	2,576	2,576	2,576
17,200	17,250	2,584	2,584	2,584	2,584
17,250	17,300	2,591	2,591	2,591	2,591
17,300	17,350	2,599	2,599	2,599	2,599
17,350	17,400	2,606	2,606	2,606	2,606
17,400	17,450	2,614	2,614	2,614	2,614
17,450	17,500	2,621	2,621	2,621	2,621
17,500	17,550	2,629	2,629	2,629	2,629
17,550	17,600	2,636	2,636	2,636	2,636
17,600	17,650	2,644	2,644	2,644	2,644
17,650	17,700	2,651	2,651	2,651	2,651
17,700	17,750	2,659	2,659	2,659	2,659
17,750	17,800	2,666	2,666	2,666	2,666
17,800	17,850	2,674	2,674	2,674	2,674
17,850	17,900	2,681	2,681	2,681	2,681
17,900	17,950	2,689	2,689	2,689	2,689
17,950	18,000	2,696	2,696	2,696	2,696
18,000					
18,000	18,050	2,704	2,704	2,704	2,704
18,050	18,100	2,711	2,711	2,711	2,711
18,100	18,150	2,719	2,719	2,719	2,719
18,150	18,200	2,726	2,726	2,726	2,726
18,200	18,250	2,734	2,734	2,734	2,734
18,250	18,300	2,741	2,741	2,741	2,741
18,300	18,350	2,749	2,749	2,749	2,749
18,350	18,400	2,756	2,756	2,756	2,756
18,400	18,450	2,764	2,764	2,764	2,764
18,450	18,500	2,771	2,771	2,771	2,771
18,500	18,550	2,779	2,779	2,779	2,779
18,550	18,600	2,786	2,786	2,786	2,786
18,600	18,650	2,794	2,794	2,794	2,794
18,650	18,700	2,801	2,801	2,801	2,801
18,700	18,750	2,809	2,809	2,809	2,809
18,750	18,800	2,816	2,816	2,816	2,816
18,800	18,850	2,824	2,824	2,824	2,824
18,850	18,900	2,831	2,831	2,831	2,831
18,900	18,950	2,839	2,839	2,839	2,839
18,950	19,000	2,846	2,846	2,846	2,846
19,000					
19,000	19,050	2,854	2,854	2,854	2,854
19,050	19,100	2,861	2,861	2,861	2,861
19,100	19,150	2,869	2,869	2,869	2,869
19,150	19,200	2,876	2,876	2,876	2,876
19,200	19,250	2,884	2,884	2,884	2,884
19,250	19,300	2,891	2,891	2,891	2,891
19,300	19,350	2,899	2,899	2,899	2,899
19,350	19,400	2,906	2,906	2,906	2,906
19,400	19,450	2,914	2,914	2,914	2,914
19,450	19,500	2,921	2,921	2,921	2,921
19,500	19,550	2,929	2,929	2,929	2,929
19,550	19,600	2,936	2,936	2,936	2,936
19,600	19,650	2,944	2,944	2,944	2,944
19,650	19,700	2,951	2,951	2,951	2,951
19,700	19,750	2,959	2,959	2,959	2,959
19,750	19,800	2,966	2,966	2,966	2,966
19,800	19,850	2,974	2,974	2,974	2,974
19,850	19,900	2,981	2,981	2,981	2,981
19,900	19,950	2,989	2,989	2,989	2,989
19,950	20,000	2,996	2,996	2,996	2,996

If line 39 (taxable income) is— At least	But less than	Single	Married filing jointly *	Married filing separately	Head of a household
20,000					
20,000	20,050	3,004	3,004	3,004	3,004
20,050	20,100	3,011	3,011	3,011	3,011
20,100	20,150	3,019	3,019	3,019	3,019
20,150	20,200	3,026	3,026	3,026	3,026
20,200	20,250	3,034	3,034	3,034	3,034
20,250	20,300	3,041	3,041	3,041	3,041
20,300	20,350	3,049	3,049	3,049	3,049
20,350	20,400	3,056	3,056	3,056	3,056
20,400	20,450	3,064	3,064	3,064	3,064
20,450	20,500	3,071	3,071	3,071	3,071
20,500	20,550	3,079	3,079	3,079	3,079
20,550	20,600	3,086	3,086	3,086	3,086
20,600	20,650	3,094	3,094	3,094	3,094
20,650	20,700	3,101	3,101	3,101	3,101
20,700	20,750	3,109	3,109	3,109	3,109
20,750	20,800	3,116	3,116	3,116	3,116
20,800	20,850	3,124	3,124	3,124	3,124
20,850	20,900	3,131	3,131	3,131	3,131
20,900	20,950	3,139	3,139	3,139	3,139
20,950	21,000	3,146	3,146	3,146	3,146
21,000					
21,000	21,050	3,154	3,154	3,154	3,154
21,050	21,100	3,161	3,161	3,161	3,161
21,100	21,150	3,169	3,169	3,169	3,169
21,150	21,200	3,176	3,176	3,176	3,176
21,200	21,250	3,184	3,184	3,190	3,184
21,250	21,300	3,191	3,191	3,204	3,191
21,300	21,350	3,199	3,199	3,218	3,199
21,350	21,400	3,206	3,206	3,232	3,206
21,400	21,450	3,214	3,214	3,246	3,214
21,450	21,500	3,221	3,221	3,260	3,221
21,500	21,550	3,229	3,229	3,274	3,229
21,550	21,600	3,236	3,236	3,288	3,236
21,600	21,650	3,244	3,244	3,302	3,244
21,650	21,700	3,251	3,251	3,316	3,251
21,700	21,750	3,259	3,259	3,330	3,259
21,750	21,800	3,266	3,266	3,344	3,266
21,800	21,850	3,274	3,274	3,358	3,274
21,850	21,900	3,281	3,281	3,372	3,281
21,900	21,950	3,289	3,289	3,386	3,289
21,950	22,000	3,296	3,296	3,400	3,296
22,000					
22,000	22,050	3,304	3,304	3,414	3,304
22,050	22,100	3,311	3,311	3,428	3,311
22,100	22,150	3,319	3,319	3,442	3,319
22,150	22,200	3,326	3,326	3,456	3,326
22,200	22,250	3,334	3,334	3,470	3,334
22,250	22,300	3,341	3,341	3,484	3,341
22,300	22,350	3,349	3,349	3,498	3,349
22,350	22,400	3,356	3,356	3,512	3,356
22,400	22,450	3,364	3,364	3,526	3,364
22,450	22,500	3,371	3,371	3,540	3,371
22,500	22,550	3,379	3,379	3,554	3,379
22,550	22,600	3,386	3,386	3,568	3,386
22,600	22,650	3,394	3,394	3,582	3,394
22,650	22,700	3,401	3,401	3,596	3,401
22,700	22,750	3,409	3,409	3,610	3,409
22,750	22,800	3,416	3,416	3,624	3,416
22,800	22,850	3,424	3,424	3,638	3,424
22,850	22,900	3,431	3,431	3,652	3,431
22,900	22,950	3,439	3,439	3,666	3,439
22,950	23,000	3,446	3,446	3,680	3,446

* This column must also be used by a qualifying widow(er).

Continued on next page

1998 Tax Table—*Continued*

At least	But less than	Single	Married filing jointly *	Married filing separately	Head of a household
23,000					
23,000	23,050	3,454	3,454	3,694	3,454
23,050	23,100	3,461	3,461	3,708	3,461
23,100	23,150	3,469	3,469	3,722	3,469
23,150	23,200	3,476	3,476	3,736	3,476
23,200	23,250	3,484	3,484	3,750	3,484
23,250	23,300	3,491	3,491	3,764	3,491
23,300	23,350	3,499	3,499	3,778	3,499
23,350	23,400	3,506	3,506	3,792	3,506
23,400	23,450	3,514	3,514	3,806	3,514
23,450	23,500	3,521	3,521	3,820	3,521
23,500	23,550	3,529	3,529	3,834	3,529
23,550	23,600	3,536	3,536	3,848	3,536
23,600	23,650	3,544	3,544	3,862	3,544
23,650	23,700	3,551	3,551	3,876	3,551
23,700	23,750	3,559	3,559	3,890	3,559
23,750	23,800	3,566	3,566	3,904	3,566
23,800	23,850	3,574	3,574	3,918	3,574
23,850	23,900	3,581	3,581	3,932	3,581
23,900	23,950	3,589	3,589	3,946	3,589
23,950	24,000	3,596	3,596	3,960	3,596
24,000					
24,000	24,050	3,604	3,604	3,974	3,604
24,050	24,100	3,611	3,611	3,988	3,611
24,100	24,150	3,619	3,619	4,002	3,619
24,150	24,200	3,626	3,626	4,016	3,626
24,200	24,250	3,634	3,634	4,030	3,634
24,250	24,300	3,641	3,641	4,044	3,641
24,300	24,350	3,649	3,649	4,058	3,649
24,350	24,400	3,656	3,656	4,072	3,656
24,400	24,450	3,664	3,664	4,086	3,664
24,450	24,500	3,671	3,671	4,100	3,671
24,500	24,550	3,679	3,679	4,114	3,679
24,550	24,600	3,686	3,686	4,128	3,686
24,600	24,650	3,694	3,694	4,142	3,694
24,650	24,700	3,701	3,701	4,156	3,701
24,700	24,750	3,709	3,709	4,170	3,709
24,750	24,800	3,716	3,716	4,184	3,716
24,800	24,850	3,724	3,724	4,198	3,724
24,850	24,900	3,731	3,731	4,212	3,731
24,900	24,950	3,739	3,739	4,226	3,739
24,950	25,000	3,746	3,746	4,240	3,746
25,000					
25,000	25,050	3,754	3,754	4,254	3,754
25,050	25,100	3,761	3,761	4,268	3,761
25,100	25,150	3,769	3,769	4,282	3,769
25,150	25,200	3,776	3,776	4,296	3,776
25,200	25,250	3,784	3,784	4,310	3,784
25,250	25,300	3,791	3,791	4,324	3,791
25,300	25,350	3,799	3,799	4,338	3,799
25,350	25,400	3,810	3,806	4,352	3,806
25,400	25,450	3,824	3,814	4,366	3,814
25,450	25,500	3,838	3,821	4,380	3,821
25,500	25,550	3,852	3,829	4,394	3,829
25,550	25,600	3,866	3,836	4,408	3,836
25,600	25,650	3,880	3,844	4,422	3,844
25,650	25,700	3,894	3,851	4,436	3,851
25,700	25,750	3,908	3,859	4,450	3,859
25,750	25,800	3,922	3,866	4,464	3,866
25,800	25,850	3,936	3,874	4,478	3,874
25,850	25,900	3,950	3,881	4,492	3,881
25,900	25,950	3,964	3,889	4,506	3,889
25,950	26,000	3,978	3,896	4,520	3,896

At least	But less than	Single	Married filing jointly *	Married filing separately	Head of a household
26,000					
26,000	26,050	3,992	3,904	4,534	3,904
26,050	26,100	4,006	3,911	4,548	3,911
26,100	26,150	4,020	3,919	4,562	3,919
26,150	26,200	4,034	3,926	4,576	3,926
26,200	26,250	4,048	3,934	4,590	3,934
26,250	26,300	4,062	3,941	4,604	3,941
26,300	26,350	4,076	3,949	4,618	3,949
26,350	26,400	4,090	3,956	4,632	3,956
26,400	26,450	4,104	3,964	4,646	3,964
26,450	26,500	4,118	3,971	4,660	3,971
26,500	26,550	4,132	3,979	4,674	3,979
26,550	26,600	4,146	3,986	4,688	3,986
26,600	26,650	4,160	3,994	4,702	3,994
26,650	26,700	4,174	4,001	4,716	4,001
26,700	26,750	4,188	4,009	4,730	4,009
26,750	26,800	4,202	4,016	4,744	4,016
26,800	26,850	4,216	4,024	4,758	4,024
26,850	26,900	4,230	4,031	4,772	4,031
26,900	26,950	4,244	4,039	4,786	4,039
26,950	27,000	4,258	4,046	4,800	4,046
27,000					
27,000	27,050	4,272	4,054	4,814	4,054
27,050	27,100	4,286	4,061	4,828	4,061
27,100	27,150	4,300	4,069	4,842	4,069
27,150	27,200	4,314	4,076	4,856	4,076
27,200	27,250	4,328	4,084	4,870	4,084
27,250	27,300	4,342	4,091	4,884	4,091
27,300	27,350	4,356	4,099	4,898	4,099
27,350	27,400	4,370	4,106	4,912	4,106
27,400	27,450	4,384	4,114	4,926	4,114
27,450	27,500	4,398	4,121	4,940	4,121
27,500	27,550	4,412	4,129	4,954	4,129
27,550	27,600	4,426	4,136	4,968	4,136
27,600	27,650	4,440	4,144	4,982	4,144
27,650	27,700	4,454	4,151	4,996	4,151
27,700	27,750	4,468	4,159	5,010	4,159
27,750	27,800	4,482	4,166	5,024	4,166
27,800	27,850	4,496	4,174	5,038	4,174
27,850	27,900	4,510	4,181	5,052	4,181
27,900	27,950	4,524	4,189	5,066	4,189
27,950	28,000	4,538	4,196	5,080	4,196
28,000					
28,000	28,050	4,552	4,204	5,094	4,204
28,050	28,100	4,566	4,211	5,108	4,211
28,100	28,150	4,580	4,219	5,122	4,219
28,150	28,200	4,594	4,226	5,136	4,226
28,200	28,250	4,608	4,234	5,150	4,234
28,250	28,300	4,622	4,241	5,164	4,241
28,300	28,350	4,636	4,249	5,178	4,249
28,350	28,400	4,650	4,256	5,192	4,256
28,400	28,450	4,664	4,264	5,206	4,264
28,450	28,500	4,678	4,271	5,220	4,271
28,500	28,550	4,692	4,279	5,234	4,279
28,550	28,600	4,706	4,286	5,248	4,286
28,600	28,650	4,720	4,294	5,262	4,294
28,650	28,700	4,734	4,301	5,276	4,301
28,700	28,750	4,748	4,309	5,290	4,309
28,750	28,800	4,762	4,316	5,304	4,316
28,800	28,850	4,776	4,324	5,318	4,324
28,850	28,900	4,790	4,331	5,332	4,331
28,900	28,950	4,804	4,339	5,346	4,339
28,950	29,000	4,818	4,346	5,360	4,346

At least	But less than	Single	Married filing jointly *	Married filing separately	Head of a household
29,000					
29,000	29,050	4,832	4,354	5,374	4,354
29,050	29,100	4,846	4,361	5,388	4,361
29,100	29,150	4,860	4,369	5,402	4,369
29,150	29,200	4,874	4,376	5,416	4,376
29,200	29,250	4,888	4,384	5,430	4,384
29,250	29,300	4,902	4,391	5,444	4,391
29,300	29,350	4,916	4,399	5,458	4,399
29,350	29,400	4,930	4,406	5,472	4,406
29,400	29,450	4,944	4,414	5,486	4,414
29,450	29,500	4,958	4,421	5,500	4,421
29,500	29,550	4,972	4,429	5,514	4,429
29,550	29,600	4,986	4,436	5,528	4,436
29,600	29,650	5,000	4,444	5,542	4,444
29,650	29,700	5,014	4,451	5,556	4,451
29,700	29,750	5,028	4,459	5,570	4,459
29,750	29,800	5,042	4,466	5,584	4,466
29,800	29,850	5,056	4,474	5,598	4,474
29,850	29,900	5,070	4,481	5,612	4,481
29,900	29,950	5,084	4,489	5,626	4,489
29,950	30,000	5,098	4,496	5,640	4,496
30,000					
30,000	30,050	5,112	4,504	5,654	4,504
30,050	30,100	5,126	4,511	5,668	4,511
30,100	30,150	5,140	4,519	5,682	4,519
30,150	30,200	5,154	4,526	5,696	4,526
30,200	30,250	5,168	4,534	5,710	4,534
30,250	30,300	5,182	4,541	5,724	4,541
30,300	30,350	5,196	4,549	5,738	4,549
30,350	30,400	5,210	4,556	5,752	4,556
30,400	30,450	5,224	4,564	5,766	4,564
30,450	30,500	5,238	4,571	5,780	4,571
30,500	30,550	5,252	4,579	5,794	4,579
30,550	30,600	5,266	4,586	5,808	4,586
30,600	30,650	5,280	4,594	5,822	4,594
30,650	30,700	5,294	4,601	5,836	4,601
30,700	30,750	5,308	4,609	5,850	4,609
30,750	30,800	5,322	4,616	5,864	4,616
30,800	30,850	5,336	4,624	5,878	4,624
30,850	30,900	5,350	4,631	5,892	4,631
30,900	30,950	5,364	4,639	5,906	4,639
30,950	31,000	5,378	4,646	5,920	4,646
31,000					
31,000	31,050	5,392	4,654	5,934	4,654
31,050	31,100	5,406	4,661	5,948	4,661
31,100	31,150	5,420	4,669	5,962	4,669
31,150	31,200	5,434	4,676	5,976	4,676
31,200	31,250	5,448	4,684	5,990	4,684
31,250	31,300	5,462	4,691	6,004	4,691
31,300	31,350	5,476	4,699	6,018	4,699
31,350	31,400	5,490	4,706	6,032	4,706
31,400	31,450	5,504	4,714	6,046	4,714
31,450	31,500	5,518	4,721	6,060	4,721
31,500	31,550	5,532	4,729	6,074	4,729
31,550	31,600	5,546	4,736	6,088	4,736
31,600	31,650	5,560	4,744	6,102	4,744
31,650	31,700	5,574	4,751	6,116	4,751
31,700	31,750	5,588	4,759	6,130	4,759
31,750	31,800	5,602	4,766	6,144	4,766
31,800	31,850	5,616	4,774	6,158	4,774
31,850	31,900	5,630	4,781	6,172	4,781
31,900	31,950	5,644	4,789	6,186	4,789
31,950	32,000	5,658	4,796	6,200	4,796

* This column must also be used by a qualifying widow(er).

Continued on next page

1998 Tax Table—*Continued*

Columns for each group: At least | But less than | Single | Married filing jointly * | Married filing separately | Head of a household — Your tax is—

* This column must also be used by a qualifying widow(er).

32,000 – 34,999

At least	But less than	Single	Married filing jointly	Married filing separately	Head of a household
32,000	32,050	5,672	4,804	6,214	4,804
32,050	32,100	5,686	4,811	6,228	4,811
32,100	32,150	5,700	4,819	6,242	4,819
32,150	32,200	5,714	4,826	6,256	4,826
32,200	32,250	5,728	4,834	6,270	4,834
32,250	32,300	5,742	4,841	6,284	4,841
32,300	32,350	5,756	4,849	6,298	4,849
32,350	32,400	5,770	4,856	6,312	4,856
32,400	32,450	5,784	4,864	6,326	4,864
32,450	32,500	5,798	4,871	6,340	4,871
32,500	32,550	5,812	4,879	6,354	4,879
32,550	32,600	5,826	4,886	6,368	4,886
32,600	32,650	5,840	4,894	6,382	4,894
32,650	32,700	5,854	4,901	6,396	4,901
32,700	32,750	5,868	4,909	6,410	4,909
32,750	32,800	5,882	4,916	6,424	4,916
32,800	32,850	5,896	4,924	6,438	4,924
32,850	32,900	5,910	4,931	6,452	4,931
32,900	32,950	5,924	4,939	6,466	4,939
32,950	33,000	5,938	4,946	6,480	4,946
33,000	33,050	5,952	4,954	6,494	4,954
33,050	33,100	5,966	4,961	6,508	4,961
33,100	33,150	5,980	4,969	6,522	4,969
33,150	33,200	5,994	4,976	6,536	4,976
33,200	33,250	6,008	4,984	6,550	4,984
33,250	33,300	6,022	4,991	6,564	4,991
33,300	33,350	6,036	4,999	6,578	4,999
33,350	33,400	6,050	5,006	6,592	5,006
33,400	33,450	6,064	5,014	6,606	5,014
33,450	33,500	6,078	5,021	6,620	5,021
33,500	33,550	6,092	5,029	6,634	5,029
33,550	33,600	6,106	5,036	6,648	5,036
33,600	33,650	6,120	5,044	6,662	5,044
33,650	33,700	6,134	5,051	6,676	5,051
33,700	33,750	6,148	5,059	6,690	5,059
33,750	33,800	6,162	5,066	6,704	5,066
33,800	33,850	6,176	5,074	6,718	5,074
33,850	33,900	6,190	5,081	6,732	5,081
33,900	33,950	6,204	5,089	6,746	5,089
33,950	34,000	6,218	5,096	6,760	5,100
34,000	34,050	6,232	5,104	6,774	5,114
34,050	34,100	6,246	5,111	6,788	5,128
34,100	34,150	6,260	5,119	6,802	5,142
34,150	34,200	6,274	5,126	6,816	5,156
34,200	34,250	6,288	5,134	6,830	5,170
34,250	34,300	6,302	5,141	6,844	5,184
34,300	34,350	6,316	5,149	6,858	5,198
34,350	34,400	6,330	5,156	6,872	5,212
34,400	34,450	6,344	5,164	6,886	5,226
34,450	34,500	6,358	5,171	6,900	5,240
34,500	34,550	6,372	5,179	6,914	5,254
34,550	34,600	6,386	5,186	6,928	5,268
34,600	34,650	6,400	5,194	6,942	5,282
34,650	34,700	6,414	5,201	6,956	5,296
34,700	34,750	6,428	5,209	6,970	5,310
34,750	34,800	6,442	5,216	6,984	5,324
34,800	34,850	6,456	5,224	6,998	5,338
34,850	34,900	6,470	5,231	7,012	5,352
34,900	34,950	6,484	5,239	7,026	5,366
34,950	35,000	6,498	5,246	7,040	5,380

35,000 – 37,999

At least	But less than	Single	Married filing jointly	Married filing separately	Head of a household
35,000	35,050	6,512	5,254	7,054	5,394
35,050	35,100	6,526	5,261	7,068	5,408
35,100	35,150	6,540	5,269	7,082	5,422
35,150	35,200	6,554	5,276	7,096	5,436
35,200	35,250	6,568	5,284	7,110	5,450
35,250	35,300	6,582	5,291	7,124	5,464
35,300	35,350	6,596	5,299	7,138	5,478
35,350	35,400	6,610	5,306	7,152	5,492
35,400	35,450	6,624	5,314	7,166	5,506
35,450	35,500	6,638	5,321	7,180	5,520
35,500	35,550	6,652	5,329	7,194	5,534
35,550	35,600	6,666	5,336	7,208	5,548
35,600	35,650	6,680	5,344	7,222	5,562
35,650	35,700	6,694	5,351	7,236	5,576
35,700	35,750	6,708	5,359	7,250	5,590
35,750	35,800	6,722	5,366	7,264	5,604
35,800	35,850	6,736	5,374	7,278	5,618
35,850	35,900	6,750	5,381	7,292	5,632
35,900	35,950	6,764	5,389	7,306	5,646
35,950	36,000	6,778	5,396	7,320	5,660
36,000	36,050	6,792	5,404	7,334	5,674
36,050	36,100	6,806	5,411	7,348	5,688
36,100	36,150	6,820	5,419	7,362	5,702
36,150	36,200	6,834	5,426	7,376	5,716
36,200	36,250	6,848	5,434	7,390	5,730
36,250	36,300	6,862	5,441	7,404	5,744
36,300	36,350	6,876	5,449	7,418	5,758
36,350	36,400	6,890	5,456	7,432	5,772
36,400	36,450	6,904	5,464	7,446	5,786
36,450	36,500	6,918	5,471	7,460	5,800
36,500	36,550	6,932	5,479	7,474	5,814
36,550	36,600	6,946	5,486	7,488	5,828
36,600	36,650	6,960	5,494	7,502	5,842
36,650	36,700	6,974	5,501	7,516	5,856
36,700	36,750	6,988	5,509	7,530	5,870
36,750	36,800	7,002	5,516	7,544	5,884
36,800	36,850	7,016	5,524	7,558	5,898
36,850	36,900	7,030	5,531	7,572	5,912
36,900	36,950	7,044	5,539	7,586	5,926
36,950	37,000	7,058	5,546	7,600	5,940
37,000	37,050	7,072	5,554	7,614	5,954
37,050	37,100	7,086	5,561	7,628	5,968
37,100	37,150	7,100	5,569	7,642	5,982
37,150	37,200	7,114	5,576	7,656	5,996
37,200	37,250	7,128	5,584	7,670	6,010
37,250	37,300	7,142	5,591	7,684	6,024
37,300	37,350	7,156	5,599	7,698	6,038
37,350	37,400	7,170	5,606	7,712	6,052
37,400	37,450	7,184	5,614	7,726	6,066
37,450	37,500	7,198	5,621	7,740	6,080
37,500	37,550	7,212	5,629	7,754	6,094
37,550	37,600	7,226	5,636	7,768	6,108
37,600	37,650	7,240	5,644	7,782	6,122
37,650	37,700	7,254	5,651	7,796	6,136
37,700	37,750	7,268	5,659	7,810	6,150
37,750	37,800	7,282	5,666	7,824	6,164
37,800	37,850	7,296	5,674	7,838	6,178
37,850	37,900	7,310	5,681	7,852	6,192
37,900	37,950	7,324	5,689	7,866	6,206
37,950	38,000	7,338	5,696	7,880	6,220

38,000 – 40,999

At least	But less than	Single	Married filing jointly	Married filing separately	Head of a household
38,000	38,050	7,352	5,704	7,894	6,234
38,050	38,100	7,366	5,711	7,908	6,248
38,100	38,150	7,380	5,719	7,922	6,262
38,150	38,200	7,394	5,726	7,936	6,276
38,200	38,250	7,408	5,734	7,950	6,290
38,250	38,300	7,422	5,741	7,964	6,304
38,300	38,350	7,436	5,749	7,978	6,318
38,350	38,400	7,450	5,756	7,992	6,332
38,400	38,450	7,464	5,764	8,006	6,346
38,450	38,500	7,478	5,771	8,020	6,360
38,500	38,550	7,492	5,779	8,034	6,374
38,550	38,600	7,506	5,786	8,048	6,388
38,600	38,650	7,520	5,794	8,062	6,402
38,650	38,700	7,534	5,801	8,076	6,416
38,700	38,750	7,548	5,809	8,090	6,430
38,750	38,800	7,562	5,816	8,104	6,444
38,800	38,850	7,576	5,824	8,118	6,458
38,850	38,900	7,590	5,831	8,132	6,472
38,900	38,950	7,604	5,839	8,146	6,486
38,950	39,000	7,618	5,846	8,160	6,500
39,000	39,050	7,632	5,854	8,174	6,514
39,050	39,100	7,646	5,861	8,188	6,528
39,100	39,150	7,660	5,869	8,202	6,542
39,150	39,200	7,674	5,876	8,216	6,556
39,200	39,250	7,688	5,884	8,230	6,570
39,250	39,300	7,702	5,891	8,244	6,584
39,300	39,350	7,716	5,899	8,258	6,598
39,350	39,400	7,730	5,906	8,272	6,612
39,400	39,450	7,744	5,914	8,286	6,626
39,450	39,500	7,758	5,921	8,300	6,640
39,500	39,550	7,772	5,929	8,314	6,654
39,550	39,600	7,786	5,936	8,328	6,668
39,600	39,650	7,800	5,944	8,342	6,682
39,650	39,700	7,814	5,951	8,356	6,696
39,700	39,750	7,828	5,959	8,370	6,710
39,750	39,800	7,842	5,966	8,384	6,724
39,800	39,850	7,856	5,974	8,398	6,738
39,850	39,900	7,870	5,981	8,412	6,752
39,900	39,950	7,884	5,989	8,426	6,766
39,950	40,000	7,898	5,996	8,440	6,780
40,000	40,050	7,912	6,004	8,454	6,794
40,050	40,100	7,926	6,011	8,468	6,808
40,100	40,150	7,940	6,019	8,482	6,822
40,150	40,200	7,954	6,026	8,496	6,836
40,200	40,250	7,968	6,034	8,510	6,850
40,250	40,300	7,982	6,041	8,524	6,864
40,300	40,350	7,996	6,049	8,538	6,878
40,350	40,400	8,010	6,056	8,552	6,892
40,400	40,450	8,024	6,064	8,566	6,906
40,450	40,500	8,038	6,071	8,580	6,920
40,500	40,550	8,052	6,079	8,594	6,934
40,550	40,600	8,066	6,086	8,608	6,948
40,600	40,650	8,080	6,094	8,622	6,962
40,650	40,700	8,094	6,101	8,636	6,976
40,700	40,750	8,108	6,109	8,650	6,990
40,750	40,800	8,122	6,116	8,664	7,004
40,800	40,850	8,136	6,124	8,678	7,018
40,850	40,900	8,150	6,131	8,692	7,032
40,900	40,950	8,164	6,139	8,706	7,046
40,950	41,000	8,178	6,146	8,720	7,060

* This column must also be used by a qualifying widow(er).

Continued on next page

1998 Tax Table—Continued

If line 39 (taxable income) is— At least	But less than	Single	Married filing jointly *	Married filing separately	Head of a household
41,000					
41,000	41,050	8,192	6,154	8,734	7,074
41,050	41,100	8,206	6,161	8,748	7,088
41,100	41,150	8,220	6,169	8,762	7,102
41,150	41,200	8,234	6,176	8,776	7,116
41,200	41,250	8,248	6,184	8,790	7,130
41,250	41,300	8,262	6,191	8,804	7,144
41,300	41,350	8,276	6,199	8,818	7,158
41,350	41,400	8,290	6,206	8,832	7,172
41,400	41,450	8,304	6,214	8,846	7,186
41,450	41,500	8,318	6,221	8,860	7,200
41,500	41,550	8,332	6,229	8,874	7,214
41,550	41,600	8,346	6,236	8,888	7,228
41,600	41,650	8,360	6,244	8,902	7,242
41,650	41,700	8,374	6,251	8,916	7,256
41,700	41,750	8,388	6,259	8,930	7,270
41,750	41,800	8,402	6,266	8,944	7,284
41,800	41,850	8,416	6,274	8,958	7,298
41,850	41,900	8,430	6,281	8,972	7,312
41,900	41,950	8,444	6,289	8,986	7,326
41,950	42,000	8,458	6,296	9,000	7,340
42,000					
42,000	42,050	8,472	6,304	9,014	7,354
42,050	42,100	8,486	6,311	9,028	7,368
42,100	42,150	8,500	6,319	9,042	7,382
42,150	42,200	8,514	6,326	9,056	7,396
42,200	42,250	8,528	6,334	9,070	7,410
42,250	42,300	8,542	6,341	9,084	7,424
42,300	42,350	8,556	6,349	9,098	7,438
42,350	42,400	8,570	6,360	9,112	7,452
42,400	42,450	8,584	6,374	9,126	7,466
42,450	42,500	8,598	6,388	9,140	7,480
42,500	42,550	8,612	6,402	9,154	7,494
42,550	42,600	8,626	6,416	9,168	7,508
42,600	42,650	8,640	6,430	9,182	7,522
42,650	42,700	8,654	6,444	9,196	7,536
42,700	42,750	8,668	6,458	9,210	7,550
42,750	42,800	8,682	6,472	9,224	7,564
42,800	42,850	8,696	6,486	9,238	7,578
42,850	42,900	8,710	6,500	9,252	7,592
42,900	42,950	8,724	6,514	9,266	7,606
42,950	43,000	8,738	6,528	9,280	7,620
43,000					
43,000	43,050	8,752	6,542	9,294	7,634
43,050	43,100	8,766	6,556	9,308	7,648
43,100	43,150	8,780	6,570	9,322	7,662
43,150	43,200	8,794	6,584	9,336	7,676
43,200	43,250	8,808	6,598	9,350	7,690
43,250	43,300	8,822	6,612	9,364	7,704
43,300	43,350	8,836	6,626	9,378	7,718
43,350	43,400	8,850	6,640	9,392	7,732
43,400	43,450	8,864	6,654	9,406	7,746
43,450	43,500	8,878	6,668	9,420	7,760
43,500	43,550	8,892	6,682	9,434	7,774
43,550	43,600	8,906	6,696	9,448	7,788
43,600	43,650	8,920	6,710	9,462	7,802
43,650	43,700	8,934	6,724	9,476	7,816
43,700	43,750	8,948	6,738	9,490	7,830
43,750	43,800	8,962	6,752	9,504	7,844
43,800	43,850	8,976	6,766	9,518	7,858
43,850	43,900	8,990	6,780	9,532	7,872
43,900	43,950	9,004	6,794	9,546	7,886
43,950	44,000	9,018	6,808	9,560	7,900

If line 39 (taxable income) is— At least	But less than	Single	Married filing jointly *	Married filing separately	Head of a household
44,000					
44,000	44,050	9,032	6,822	9,574	7,914
44,050	44,100	9,046	6,836	9,588	7,928
44,100	44,150	9,060	6,850	9,602	7,942
44,150	44,200	9,074	6,864	9,616	7,956
44,200	44,250	9,088	6,878	9,630	7,970
44,250	44,300	9,102	6,892	9,644	7,984
44,300	44,350	9,116	6,906	9,658	7,998
44,350	44,400	9,130	6,920	9,672	8,012
44,400	44,450	9,144	6,934	9,686	8,026
44,450	44,500	9,158	6,948	9,700	8,040
44,500	44,550	9,172	6,962	9,714	8,054
44,550	44,600	9,186	6,976	9,728	8,068
44,600	44,650	9,200	6,990	9,742	8,082
44,650	44,700	9,214	7,004	9,756	8,096
44,700	44,750	9,228	7,018	9,770	8,110
44,750	44,800	9,242	7,032	9,784	8,124
44,800	44,850	9,256	7,046	9,798	8,138
44,850	44,900	9,270	7,060	9,812	8,152
44,900	44,950	9,284	7,074	9,826	8,166
44,950	45,000	9,298	7,088	9,840	8,180
45,000					
45,000	45,050	9,312	7,102	9,854	8,194
45,050	45,100	9,326	7,116	9,868	8,208
45,100	45,150	9,340	7,130	9,882	8,222
45,150	45,200	9,354	7,144	9,896	8,236
45,200	45,250	9,368	7,158	9,910	8,250
45,250	45,300	9,382	7,172	9,924	8,264
45,300	45,350	9,396	7,186	9,938	8,278
45,350	45,400	9,410	7,200	9,952	8,292
45,400	45,450	9,424	7,214	9,966	8,306
45,450	45,500	9,438	7,228	9,980	8,320
45,500	45,550	9,452	7,242	9,994	8,334
45,550	45,600	9,466	7,256	10,008	8,348
45,600	45,650	9,480	7,270	10,022	8,362
45,650	45,700	9,494	7,284	10,036	8,376
45,700	45,750	9,508	7,298	10,050	8,390
45,750	45,800	9,522	7,312	10,064	8,404
45,800	45,850	9,536	7,326	10,078	8,418
45,850	45,900	9,550	7,340	10,092	8,432
45,900	45,950	9,564	7,354	10,106	8,446
45,950	46,000	9,578	7,368	10,120	8,460
46,000					
46,000	46,050	9,592	7,382	10,134	8,474
46,050	46,100	9,606	7,396	10,148	8,488
46,100	46,150	9,620	7,410	10,162	8,502
46,150	46,200	9,634	7,424	10,176	8,516
46,200	46,250	9,648	7,438	10,190	8,530
46,250	46,300	9,662	7,452	10,204	8,544
46,300	46,350	9,676	7,466	10,218	8,558
46,350	46,400	9,690	7,480	10,232	8,572
46,400	46,450	9,704	7,494	10,246	8,586
46,450	46,500	9,718	7,508	10,260	8,600
46,500	46,550	9,732	7,522	10,274	8,614
46,550	46,600	9,746	7,536	10,288	8,628
46,600	46,650	9,760	7,550	10,302	8,642
46,650	46,700	9,774	7,564	10,316	8,656
46,700	46,750	9,788	7,578	10,330	8,670
46,750	46,800	9,802	7,592	10,344	8,684
46,800	46,850	9,816	7,606	10,358	8,698
46,850	46,900	9,830	7,620	10,372	8,712
46,900	46,950	9,844	7,634	10,386	8,726
46,950	47,000	9,858	7,648	10,400	8,740

If line 39 (taxable income) is— At least	But less than	Single	Married filing jointly *	Married filing separately	Head of a household
47,000					
47,000	47,050	9,872	7,662	10,414	8,754
47,050	47,100	9,886	7,676	10,428	8,768
47,100	47,150	9,900	7,690	10,442	8,782
47,150	47,200	9,914	7,704	10,456	8,796
47,200	47,250	9,928	7,718	10,470	8,810
47,250	47,300	9,942	7,732	10,484	8,824
47,300	47,350	9,956	7,746	10,498	8,838
47,350	47,400	9,970	7,760	10,512	8,852
47,400	47,450	9,984	7,774	10,526	8,866
47,450	47,500	9,998	7,788	10,540	8,880
47,500	47,550	10,012	7,802	10,554	8,894
47,550	47,600	10,026	7,816	10,568	8,908
47,600	47,650	10,040	7,830	10,582	8,922
47,650	47,700	10,054	7,844	10,596	8,936
47,700	47,750	10,068	7,858	10,610	8,950
47,750	47,800	10,082	7,872	10,624	8,964
47,800	47,850	10,096	7,886	10,638	8,978
47,850	47,900	10,110	7,900	10,652	8,992
47,900	47,950	10,124	7,914	10,666	9,006
47,950	48,000	10,138	7,928	10,680	9,020
48,000					
48,000	48,050	10,152	7,942	10,694	9,034
48,050	48,100	10,166	7,956	10,708	9,048
48,100	48,150	10,180	7,970	10,722	9,062
48,150	48,200	10,194	7,984	10,736	9,076
48,200	48,250	10,208	7,998	10,750	9,090
48,250	48,300	10,222	8,012	10,764	9,104
48,300	48,350	10,236	8,026	10,778	9,118
48,350	48,400	10,250	8,040	10,792	9,132
48,400	48,450	10,264	8,054	10,806	9,146
48,450	48,500	10,278	8,068	10,820	9,160
48,500	48,550	10,292	8,082	10,834	9,174
48,550	48,600	10,306	8,096	10,848	9,188
48,600	48,650	10,320	8,110	10,862	9,202
48,650	48,700	10,334	8,124	10,876	9,216
48,700	48,750	10,348	8,138	10,890	9,230
48,750	48,800	10,362	8,152	10,904	9,244
48,800	48,850	10,376	8,166	10,918	9,258
48,850	48,900	10,390	8,180	10,932	9,272
48,900	48,950	10,404	8,194	10,946	9,286
48,950	49,000	10,418	8,208	10,960	9,300
49,000					
49,000	49,050	10,432	8,222	10,974	9,314
49,050	49,100	10,446	8,236	10,988	9,328
49,100	49,150	10,460	8,250	11,002	9,342
49,150	49,200	10,474	8,264	11,016	9,356
49,200	49,250	10,488	8,278	11,030	9,370
49,250	49,300	10,502	8,292	11,044	9,384
49,300	49,350	10,516	8,306	11,058	9,398
49,350	49,400	10,530	8,320	11,072	9,412
49,400	49,450	10,544	8,334	11,086	9,426
49,450	49,500	10,558	8,348	11,100	9,440
49,500	49,550	10,572	8,362	11,114	9,454
49,550	49,600	10,586	8,376	11,128	9,468
49,600	49,650	10,600	8,390	11,142	9,482
49,650	49,700	10,614	8,404	11,156	9,496
49,700	49,750	10,628	8,418	11,170	9,510
49,750	49,800	10,642	8,432	11,184	9,524
49,800	49,850	10,656	8,446	11,198	9,538
49,850	49,900	10,670	8,460	11,212	9,552
49,900	49,950	10,684	8,474	11,226	9,566
49,950	50,000	10,698	8,488	11,240	9,580

* This column must also be used by a qualifying widow(er).

Continued on next page

1998 Tax Table—Continued

50,000

If line 39 (taxable income) is— At least	But less than	Single	Married filing jointly *	Married filing separately	Head of a household
50,000	50,050	10,712	8,502	11,254	9,594
50,050	50,100	10,726	8,516	11,268	9,608
50,100	50,150	10,740	8,530	11,282	9,622
50,150	50,200	10,754	8,544	11,296	9,636
50,200	50,250	10,768	8,558	11,310	9,650
50,250	50,300	10,782	8,572	11,324	9,664
50,300	50,350	10,796	8,586	11,338	9,678
50,350	50,400	10,810	8,600	11,352	9,692
50,400	50,450	10,824	8,614	11,366	9,706
50,450	50,500	10,838	8,628	11,380	9,720
50,500	50,550	10,852	8,642	11,394	9,734
50,550	50,600	10,866	8,656	11,408	9,748
50,600	50,650	10,880	8,670	11,422	9,762
50,650	50,700	10,894	8,684	11,436	9,776
50,700	50,750	10,908	8,698	11,450	9,790
50,750	50,800	10,922	8,712	11,464	9,804
50,800	50,850	10,936	8,726	11,478	9,818
50,850	50,900	10,950	8,740	11,492	9,832
50,900	50,950	10,964	8,754	11,506	9,846
50,950	51,000	10,978	8,768	11,520	9,860

51,000

At least	But less than	Single	Married filing jointly *	Married filing separately	Head of a household
51,000	51,050	10,992	8,782	11,534	9,874
51,050	51,100	11,006	8,796	11,548	9,888
51,100	51,150	11,020	8,810	11,562	9,902
51,150	51,200	11,034	8,824	11,577	9,916
51,200	51,250	11,048	8,838	11,593	9,930
51,250	51,300	11,062	8,852	11,608	9,944
51,300	51,350	11,076	8,866	11,624	9,958
51,350	51,400	11,090	8,880	11,639	9,972
51,400	51,450	11,104	8,894	11,655	9,986
51,450	51,500	11,118	8,908	11,670	10,000
51,500	51,550	11,132	8,922	11,686	10,014
51,550	51,600	11,146	8,936	11,701	10,028
51,600	51,650	11,160	8,950	11,717	10,042
51,650	51,700	11,174	8,964	11,732	10,056
51,700	51,750	11,188	8,978	11,748	10,070
51,750	51,800	11,202	8,992	11,763	10,084
51,800	51,850	11,216	9,006	11,779	10,098
51,850	51,900	11,230	9,020	11,794	10,112
51,900	51,950	11,244	9,034	11,810	10,126
51,950	52,000	11,258	9,048	11,825	10,140

52,000

At least	But less than	Single	Married filing jointly *	Married filing separately	Head of a household
52,000	52,050	11,272	9,062	11,841	10,154
52,050	52,100	11,286	9,076	11,856	10,168
52,100	52,150	11,300	9,090	11,872	10,182
52,150	52,200	11,314	9,104	11,887	10,196
52,200	52,250	11,328	9,118	11,903	10,210
52,250	52,300	11,342	9,132	11,918	10,224
52,300	52,350	11,356	9,146	11,934	10,238
52,350	52,400	11,370	9,160	11,949	10,252
52,400	52,450	11,384	9,174	11,965	10,266
52,450	52,500	11,398	9,188	11,980	10,280
52,500	52,550	11,412	9,202	11,996	10,294
52,550	52,600	11,426	9,216	12,011	10,308
52,600	52,650	11,440	9,230	12,027	10,322
52,650	52,700	11,454	9,244	12,042	10,336
52,700	52,750	11,468	9,258	12,058	10,350
52,750	52,800	11,482	9,272	12,073	10,364
52,800	52,850	11,496	9,286	12,089	10,378
52,850	52,900	11,510	9,300	12,104	10,392
52,900	52,950	11,524	9,314	12,120	10,406
52,950	53,000	11,538	9,328	12,135	10,420

53,000

At least	But less than	Single	Married filing jointly *	Married filing separately	Head of a household
53,000	53,050	11,552	9,342	12,151	10,434
53,050	53,100	11,566	9,356	12,166	10,448
53,100	53,150	11,580	9,370	12,182	10,462
53,150	53,200	11,594	9,384	12,197	10,476
53,200	53,250	11,608	9,398	12,213	10,490
53,250	53,300	11,622	9,412	12,228	10,504
53,300	53,350	11,636	9,426	12,244	10,518
53,350	53,400	11,650	9,440	12,259	10,532
53,400	53,450	11,664	9,454	12,275	10,546
53,450	53,500	11,678	9,468	12,290	10,560
53,500	53,550	11,692	9,482	12,306	10,574
53,550	53,600	11,706	9,496	12,321	10,588
53,600	53,650	11,720	9,510	12,337	10,602
53,650	53,700	11,734	9,524	12,352	10,616
53,700	53,750	11,748	9,538	12,368	10,630
53,750	53,800	11,762	9,552	12,383	10,644
53,800	53,850	11,776	9,566	12,399	10,658
53,850	53,900	11,790	9,580	12,414	10,672
53,900	53,950	11,804	9,594	12,430	10,686
53,950	54,000	11,818	9,608	12,445	10,700

54,000

At least	But less than	Single	Married filing jointly *	Married filing separately	Head of a household
54,000	54,050	11,832	9,622	12,461	10,714
54,050	54,100	11,846	9,636	12,476	10,728
54,100	54,150	11,860	9,650	12,492	10,742
54,150	54,200	11,874	9,664	12,507	10,756
54,200	54,250	11,888	9,678	12,523	10,770
54,250	54,300	11,902	9,692	12,538	10,784
54,300	54,350	11,916	9,706	12,554	10,798
54,350	54,400	11,930	9,720	12,569	10,812
54,400	54,450	11,944	9,734	12,585	10,826
54,450	54,500	11,958	9,748	12,600	10,840
54,500	54,550	11,972	9,762	12,616	10,854
54,550	54,600	11,986	9,776	12,631	10,868
54,600	54,650	12,000	9,790	12,647	10,882
54,650	54,700	12,014	9,804	12,662	10,896
54,700	54,750	12,028	9,818	12,678	10,910
54,750	54,800	12,042	9,832	12,693	10,924
54,800	54,850	12,056	9,846	12,709	10,938
54,850	54,900	12,070	9,860	12,724	10,952
54,900	54,950	12,084	9,874	12,740	10,966
54,950	55,000	12,098	9,888	12,755	10,980

55,000

At least	But less than	Single	Married filing jointly *	Married filing separately	Head of a household
55,000	55,050	12,112	9,902	12,771	10,994
55,050	55,100	12,126	9,916	12,786	11,008
55,100	55,150	12,140	9,930	12,802	11,022
55,150	55,200	12,154	9,944	12,817	11,036
55,200	55,250	12,168	9,958	12,833	11,050
55,250	55,300	12,182	9,972	12,848	11,064
55,300	55,350	12,196	9,986	12,864	11,078
55,350	55,400	12,210	10,000	12,879	11,092
55,400	55,450	12,224	10,014	12,895	11,106
55,450	55,500	12,238	10,028	12,910	11,120
55,500	55,550	12,252	10,042	12,926	11,134
55,550	55,600	12,266	10,056	12,941	11,148
55,600	55,650	12,280	10,070	12,957	11,162
55,650	55,700	12,294	10,084	12,972	11,176
55,700	55,750	12,308	10,098	12,988	11,190
55,750	55,800	12,322	10,112	13,003	11,204
55,800	55,850	12,336	10,126	13,019	11,218
55,850	55,900	12,350	10,140	13,034	11,232
55,900	55,950	12,364	10,154	13,050	11,246
55,950	56,000	12,378	10,168	13,065	11,260

56,000

At least	But less than	Single	Married filing jointly *	Married filing separately	Head of a household
56,000	56,050	12,392	10,182	13,081	11,274
56,050	56,100	12,406	10,196	13,096	11,288
56,100	56,150	12,420	10,210	13,112	11,302
56,150	56,200	12,434	10,224	13,127	11,316
56,200	56,250	12,448	10,238	13,143	11,330
56,250	56,300	12,462	10,252	13,158	11,344
56,300	56,350	12,476	10,266	13,174	11,358
56,350	56,400	12,490	10,280	13,189	11,372
56,400	56,450	12,504	10,294	13,205	11,386
56,450	56,500	12,518	10,308	13,220	11,400
56,500	56,550	12,532	10,322	13,236	11,414
56,550	56,600	12,546	10,336	13,251	11,428
56,600	56,650	12,560	10,350	13,267	11,442
56,650	56,700	12,574	10,364	13,282	11,456
56,700	56,750	12,588	10,378	13,298	11,470
56,750	56,800	12,602	10,392	13,313	11,484
56,800	56,850	12,616	10,406	13,329	11,498
56,850	56,900	12,630	10,420	13,344	11,512
56,900	56,950	12,644	10,434	13,360	11,526
56,950	57,000	12,658	10,448	13,375	11,540

57,000

At least	But less than	Single	Married filing jointly *	Married filing separately	Head of a household
57,000	57,050	12,672	10,462	13,391	11,554
57,050	57,100	12,686	10,476	13,406	11,568
57,100	57,150	12,700	10,490	13,422	11,582
57,150	57,200	12,714	10,504	13,437	11,596
57,200	57,250	12,728	10,518	13,453	11,610
57,250	57,300	12,742	10,532	13,468	11,624
57,300	57,350	12,756	10,546	13,484	11,638
57,350	57,400	12,770	10,560	13,499	11,652
57,400	57,450	12,784	10,574	13,515	11,666
57,450	57,500	12,798	10,588	13,530	11,680
57,500	57,550	12,812	10,602	13,546	11,694
57,550	57,600	12,826	10,616	13,561	11,708
57,600	57,650	12,840	10,630	13,577	11,722
57,650	57,700	12,854	10,644	13,592	11,736
57,700	57,750	12,868	10,658	13,608	11,750
57,750	57,800	12,882	10,672	13,623	11,764
57,800	57,850	12,896	10,686	13,639	11,778
57,850	57,900	12,910	10,700	13,654	11,792
57,900	57,950	12,924	10,714	13,670	11,806
57,950	58,000	12,938	10,728	13,685	11,820

58,000

At least	But less than	Single	Married filing jointly *	Married filing separately	Head of a household
58,000	58,050	12,952	10,742	13,701	11,834
58,050	58,100	12,966	10,756	13,716	11,848
58,100	58,150	12,980	10,770	13,732	11,862
58,150	58,200	12,994	10,784	13,747	11,876
58,200	58,250	13,008	10,798	13,763	11,890
58,250	58,300	13,022	10,812	13,778	11,904
58,300	58,350	13,036	10,826	13,794	11,918
58,350	58,400	13,050	10,840	13,809	11,932
58,400	58,450	13,064	10,854	13,825	11,946
58,450	58,500	13,078	10,868	13,840	11,960
58,500	58,550	13,092	10,882	13,856	11,974
58,550	58,600	13,106	10,896	13,871	11,988
58,600	58,650	13,120	10,910	13,887	12,002
58,650	58,700	13,134	10,924	13,902	12,016
58,700	58,750	13,148	10,938	13,918	12,030
58,750	58,800	13,162	10,952	13,933	12,044
58,800	58,850	13,176	10,966	13,949	12,058
58,850	58,900	13,190	10,980	13,964	12,072
58,900	58,950	13,204	10,994	13,980	12,086
58,950	59,000	13,218	11,008	13,995	12,100

* This column must also be used by a qualifying widow(er).

Continued on next page

1998 Tax Table—*Continued*

If line 39 (taxable income) is—		And you are—				If line 39 (taxable income) is—		And you are—				If line 39 (taxable income) is—		And you are—			
At least	But less than	Single	Married filing jointly *	Married filing sepa-rately	Head of a house-hold	At least	But less than	Single	Married filing jointly *	Married filing sepa-rately	Head of a house-hold	At least	But less than	Single	Married filing jointly *	Married filing sepa-rately	Head of a house-hold
		Your tax is—						Your tax is—						Your tax is—			
59,000						**62,000**						**65,000**					
59,000	**59,050**	13,232	11,022	14,011	12,114	**62,000**	**62,050**	14,090	11,862	14,941	12,954	**65,000**	**65,050**	15,020	12,702	15,871	13,794
59,050	**59,100**	13,246	11,036	14,026	12,128	**62,050**	**62,100**	14,106	11,876	14,956	12,968	**65,050**	**65,100**	15,036	12,716	15,886	13,808
59,100	**59,150**	13,260	11,050	14,042	12,142	**62,100**	**62,150**	14,121	11,890	14,972	12,982	**65,100**	**65,150**	15,051	12,730	15,902	13,822
59,150	**59,200**	13,274	11,064	14,057	12,156	**62,150**	**62,200**	14,137	11,904	14,987	12,996	**65,150**	**65,200**	15,067	12,744	15,917	13,836
59,200	**59,250**	13,288	11,078	14,073	12,170	**62,200**	**62,250**	14,152	11,918	15,003	13,010	**65,200**	**65,250**	15,082	12,758	15,933	13,850
59,250	**59,300**	13,302	11,092	14,088	12,184	**62,250**	**62,300**	14,168	11,932	15,018	13,024	**65,250**	**65,300**	15,098	12,772	15,948	13,864
59,300	**59,350**	13,316	11,106	14,104	12,198	**62,300**	**62,350**	14,183	11,946	15,034	13,038	**65,300**	**65,350**	15,113	12,786	15,964	13,878
59,350	**59,400**	13,330	11,120	14,119	12,212	**62,350**	**62,400**	14,199	11,960	15,049	13,052	**65,350**	**65,400**	15,129	12,800	15,979	13,892
59,400	**59,450**	13,344	11,134	14,135	12,226	**62,400**	**62,450**	14,214	11,974	15,065	13,066	**65,400**	**65,450**	15,144	12,814	15,995	13,906
59,450	**59,500**	13,358	11,148	14,150	12,240	**62,450**	**62,500**	14,230	11,988	15,080	13,080	**65,450**	**65,500**	15,160	12,828	16,010	13,920
59,500	**59,550**	13,372	11,162	14,166	12,254	**62,500**	**62,550**	14,245	12,002	15,096	13,094	**65,500**	**65,550**	15,175	12,842	16,026	13,934
59,550	**59,600**	13,386	11,176	14,181	12,268	**62,550**	**62,600**	14,261	12,016	15,111	13,108	**65,550**	**65,600**	15,191	12,856	16,041	13,948
59,600	**59,650**	13,400	11,190	14,197	12,282	**62,600**	**62,650**	14,276	12,030	15,127	13,122	**65,600**	**65,650**	15,206	12,870	16,057	13,962
59,650	**59,700**	13,414	11,204	14,212	12,296	**62,650**	**62,700**	14,292	12,044	15,142	13,136	**65,650**	**65,700**	15,222	12,884	16,072	13,976
59,700	**59,750**	13,428	11,218	14,228	12,310	**62,700**	**62,750**	14,307	12,058	15,158	13,150	**65,700**	**65,750**	15,237	12,898	16,088	13,990
59,750	**59,800**	13,442	11,232	14,243	12,324	**62,750**	**62,800**	14,323	12,072	15,173	13,164	**65,750**	**65,800**	15,253	12,912	16,103	14,004
59,800	**59,850**	13,456	11,246	14,259	12,338	**62,800**	**62,850**	14,338	12,086	15,189	13,178	**65,800**	**65,850**	15,268	12,926	16,119	14,018
59,850	**59,900**	13,470	11,260	14,274	12,352	**62,850**	**62,900**	14,354	12,100	15,204	13,192	**65,850**	**65,900**	15,284	12,940	16,134	14,032
59,900	**59,950**	13,484	11,274	14,290	12,366	**62,900**	**62,950**	14,369	12,114	15,220	13,206	**65,900**	**65,950**	15,299	12,954	16,150	14,046
59,950	**60,000**	13,498	11,288	14,305	12,380	**62,950**	**63,000**	14,385	12,128	15,235	13,220	**65,950**	**66,000**	15,315	12,968	16,165	14,060
60,000						**63,000**						**66,000**					
60,000	**60,050**	13,512	11,302	14,321	12,394	**63,000**	**63,050**	14,400	12,142	15,251	13,234	**66,000**	**66,050**	15,330	12,982	16,181	14,074
60,050	**60,100**	13,526	11,316	14,336	12,408	**63,050**	**63,100**	14,416	12,156	15,266	13,248	**66,050**	**66,100**	15,346	12,996	16,196	14,088
60,100	**60,150**	13,540	11,330	14,352	12,422	**63,100**	**63,150**	14,431	12,170	15,282	13,262	**66,100**	**66,150**	15,361	13,010	16,212	14,102
60,150	**60,200**	13,554	11,344	14,367	12,436	**63,150**	**63,200**	14,447	12,184	15,297	13,276	**66,150**	**66,200**	15,377	13,024	16,227	14,116
60,200	**60,250**	13,568	11,358	14,383	12,450	**63,200**	**63,250**	14,462	12,198	15,313	13,290	**66,200**	**66,250**	15,392	13,038	16,243	14,130
60,250	**60,300**	13,582	11,372	14,398	12,464	**63,250**	**63,300**	14,478	12,212	15,328	13,304	**66,250**	**66,300**	15,408	13,052	16,258	14,144
60,300	**60,350**	13,596	11,386	14,414	12,478	**63,300**	**63,350**	14,493	12,226	15,344	13,318	**66,300**	**66,350**	15,423	13,066	16,274	14,158
60,350	**60,400**	13,610	11,400	14,429	12,492	**63,350**	**63,400**	14,509	12,240	15,359	13,332	**66,350**	**66,400**	15,439	13,080	16,289	14,172
60,400	**60,450**	13,624	11,414	14,445	12,506	**63,400**	**63,450**	14,524	12,254	15,375	13,346	**66,400**	**66,450**	15,454	13,094	16,305	14,186
60,450	**60,500**	13,638	11,428	14,460	12,520	**63,450**	**63,500**	14,540	12,268	15,390	13,360	**66,450**	**66,500**	15,470	13,108	16,320	14,200
60,500	**60,550**	13,652	11,442	14,476	12,534	**63,500**	**63,550**	14,555	12,282	15,406	13,374	**66,500**	**66,550**	15,485	13,122	16,336	14,214
60,550	**60,600**	13,666	11,456	14,491	12,548	**63,550**	**63,600**	14,571	12,296	15,421	13,388	**66,550**	**66,600**	15,501	13,136	16,351	14,228
60,600	**60,650**	13,680	11,470	14,507	12,562	**63,600**	**63,650**	14,586	12,310	15,437	13,402	**66,600**	**66,650**	15,516	13,150	16,367	14,242
60,650	**60,700**	13,694	11,484	14,522	12,576	**63,650**	**63,700**	14,602	12,324	15,452	13,416	**66,650**	**66,700**	15,532	13,164	16,382	14,256
60,700	**60,750**	13,708	11,498	14,538	12,590	**63,700**	**63,750**	14,617	12,338	15,468	13,430	**66,700**	**66,750**	15,547	13,178	16,398	14,270
60,750	**60,800**	13,722	11,512	14,553	12,604	**63,750**	**63,800**	14,633	12,352	15,483	13,444	**66,750**	**66,800**	15,563	13,192	16,413	14,284
60,800	**60,850**	13,736	11,526	14,569	12,618	**63,800**	**63,850**	14,648	12,366	15,499	13,458	**66,800**	**66,850**	15,578	13,206	16,429	14,298
60,850	**60,900**	13,750	11,540	14,584	12,632	**63,850**	**63,900**	14,664	12,380	15,514	13,472	**66,850**	**66,900**	15,594	13,220	16,444	14,312
60,900	**60,950**	13,764	11,554	14,600	12,646	**63,900**	**63,950**	14,679	12,394	15,530	13,486	**66,900**	**66,950**	15,609	13,234	16,460	14,326
60,950	**61,000**	13,778	11,568	14,615	12,660	**63,950**	**64,000**	14,695	12,408	15,545	13,500	**66,950**	**67,000**	15,625	13,248	16,475	14,340
61,000						**64,000**						**67,000**					
61,000	**61,050**	13,792	11,582	14,631	12,674	**64,000**	**64,050**	14,710	12,422	15,561	13,514	**67,000**	**67,050**	15,640	13,262	16,491	14,354
61,050	**61,100**	13,806	11,596	14,646	12,688	**64,050**	**64,100**	14,726	12,436	15,576	13,528	**67,050**	**67,100**	15,656	13,276	16,506	14,368
61,100	**61,150**	13,820	11,610	14,662	12,702	**64,100**	**64,150**	14,741	12,450	15,592	13,542	**67,100**	**67,150**	15,671	13,290	16,522	14,382
61,150	**61,200**	13,834	11,624	14,677	12,716	**64,150**	**64,200**	14,757	12,464	15,607	13,556	**67,150**	**67,200**	15,687	13,304	16,537	14,396
61,200	**61,250**	13,848	11,638	14,693	12,730	**64,200**	**64,250**	14,772	12,478	15,623	13,570	**67,200**	**67,250**	15,702	13,318	16,553	14,410
61,250	**61,300**	13,862	11,652	14,708	12,744	**64,250**	**64,300**	14,788	12,492	15,638	13,584	**67,250**	**67,300**	15,718	13,332	16,568	14,424
61,300	**61,350**	13,876	11,666	14,724	12,758	**64,300**	**64,350**	14,803	12,506	15,654	13,598	**67,300**	**67,350**	15,733	13,346	16,584	14,438
61,350	**61,400**	13,890	11,680	14,739	12,772	**64,350**	**64,400**	14,819	12,520	15,669	13,612	**67,350**	**67,400**	15,749	13,360	16,599	14,452
61,400	**61,450**	13,904	11,694	14,755	12,786	**64,400**	**64,450**	14,834	12,534	15,685	13,626	**67,400**	**67,450**	15,764	13,374	16,615	14,466
61,450	**61,500**	13,920	11,708	14,770	12,800	**64,450**	**64,500**	14,850	12,548	15,700	13,640	**67,450**	**67,500**	15,780	13,388	16,630	14,480
61,500	**61,550**	13,935	11,722	14,786	12,814	**64,500**	**64,550**	14,865	12,562	15,716	13,654	**67,500**	**67,550**	15,795	13,402	16,646	14,494
61,550	**61,600**	13,951	11,736	14,801	12,828	**64,550**	**64,600**	14,881	12,576	15,731	13,668	**67,550**	**67,600**	15,811	13,416	16,661	14,508
61,600	**61,650**	13,966	11,750	14,817	12,842	**64,600**	**64,650**	14,896	12,590	15,747	13,682	**67,600**	**67,650**	15,826	13,430	16,677	14,522
61,650	**61,700**	13,982	11,764	14,832	12,856	**64,650**	**64,700**	14,912	12,604	15,762	13,696	**67,650**	**67,700**	15,842	13,444	16,692	14,536
61,700	**61,750**	13,997	11,778	14,848	12,870	**64,700**	**64,750**	14,927	12,618	15,778	13,710	**67,700**	**67,750**	15,857	13,458	16,708	14,550
61,750	**61,800**	14,013	11,792	14,863	12,884	**64,750**	**64,800**	14,943	12,632	15,793	13,724	**67,750**	**67,800**	15,873	13,472	16,723	14,564
61,800	**61,850**	14,028	11,806	14,879	12,898	**64,800**	**64,850**	14,958	12,646	15,809	13,738	**67,800**	**67,850**	15,888	13,486	16,739	14,578
61,850	**61,900**	14,044	11,820	14,894	12,912	**64,850**	**64,900**	14,974	12,660	15,824	13,752	**67,850**	**67,900**	15,904	13,500	16,754	14,592
61,900	**61,950**	14,059	11,834	14,910	12,926	**64,900**	**64,950**	14,989	12,674	15,840	13,766	**67,900**	**67,950**	15,919	13,514	16,770	14,606
61,950	**62,000**	14,075	11,848	14,925	12,940	**64,950**	**65,000**	15,005	12,688	15,855	13,780	**67,950**	**68,000**	15,935	13,528	16,785	14,620

* This column must also be used by a qualifying widow(er).

Continued on next page

1998 Tax Table—*Continued*

If line 39 (taxable income) is— At least	But less than	Single	Married filing jointly *	Married filing separately	Head of a household
68,000					
68,000	68,050	15,950	13,542	16,801	14,634
68,050	68,100	15,966	13,556	16,816	14,648
68,100	68,150	15,981	13,570	16,832	14,662
68,150	68,200	15,997	13,584	16,847	14,676
68,200	68,250	16,012	13,598	16,863	14,690
68,250	68,300	16,028	13,612	16,878	14,704
68,300	68,350	16,043	13,626	16,894	14,718
68,350	68,400	16,059	13,640	16,909	14,732
68,400	68,450	16,074	13,654	16,925	14,746
68,450	68,500	16,090	13,668	16,940	14,760
68,500	68,550	16,105	13,682	16,956	14,774
68,550	68,600	16,121	13,696	16,971	14,788
68,600	68,650	16,136	13,710	16,987	14,802
68,650	68,700	16,152	13,724	17,002	14,816
68,700	68,750	16,167	13,738	17,018	14,830
68,750	68,800	16,183	13,752	17,033	14,844
68,800	68,850	16,198	13,766	17,049	14,858
68,850	68,900	16,214	13,780	17,064	14,872
68,900	68,950	16,229	13,794	17,080	14,886
68,950	69,000	16,245	13,808	17,095	14,900
69,000					
69,000	69,050	16,260	13,822	17,111	14,914
69,050	69,100	16,276	13,836	17,126	14,928
69,100	69,150	16,291	13,850	17,142	14,942
69,150	69,200	16,307	13,864	17,157	14,956
69,200	69,250	16,322	13,878	17,173	14,970
69,250	69,300	16,338	13,892	17,188	14,984
69,300	69,350	16,353	13,906	17,204	14,998
69,350	69,400	16,369	13,920	17,219	15,012
69,400	69,450	16,384	13,934	17,235	15,026
69,450	69,500	16,400	13,948	17,250	15,040
69,500	69,550	16,415	13,962	17,266	15,054
69,550	69,600	16,431	13,976	17,281	15,068
69,600	69,650	16,446	13,990	17,297	15,082
69,650	69,700	16,462	14,004	17,312	15,096
69,700	69,750	16,477	14,018	17,328	15,110
69,750	69,800	16,493	14,032	17,343	15,124
69,800	69,850	16,508	14,046	17,359	15,138
69,850	69,900	16,524	14,060	17,374	15,152
69,900	69,950	16,539	14,074	17,390	15,166
69,950	70,000	16,555	14,088	17,405	15,180
70,000					
70,000	70,050	16,570	14,102	17,421	15,194
70,050	70,100	16,586	14,116	17,436	15,208
70,100	70,150	16,601	14,130	17,452	15,222
70,150	70,200	16,617	14,144	17,467	15,236
70,200	70,250	16,632	14,158	17,483	15,250
70,250	70,300	16,648	14,172	17,498	15,264
70,300	70,350	16,663	14,186	17,514	15,278
70,350	70,400	16,679	14,200	17,529	15,292
70,400	70,450	16,694	14,214	17,545	15,306
70,450	70,500	16,710	14,228	17,560	15,320
70,500	70,550	16,725	14,242	17,576	15,334
70,550	70,600	16,741	14,256	17,591	15,348
70,600	70,650	16,756	14,270	17,607	15,362
70,650	70,700	16,772	14,284	17,622	15,376
70,700	70,750	16,787	14,298	17,638	15,390
70,750	70,800	16,803	14,312	17,653	15,404
70,800	70,850	16,818	14,326	17,669	15,418
70,850	70,900	16,834	14,340	17,684	15,432
70,900	70,950	16,849	14,354	17,700	15,446
70,950	71,000	16,865	14,368	17,715	15,460

If line 39 (taxable income) is— At least	But less than	Single	Married filing jointly *	Married filing separately	Head of a household
71,000					
71,000	71,050	16,880	14,382	17,731	15,474
71,050	71,100	16,896	14,396	17,746	15,488
71,100	71,150	16,911	14,410	17,762	15,502
71,150	71,200	16,927	14,424	17,777	15,516
71,200	71,250	16,942	14,438	17,793	15,530
71,250	71,300	16,958	14,452	17,808	15,544
71,300	71,350	16,973	14,466	17,824	15,558
71,350	71,400	16,989	14,480	17,839	15,572
71,400	71,450	17,004	14,494	17,855	15,586
71,450	71,500	17,020	14,508	17,870	15,600
71,500	71,550	17,035	14,522	17,886	15,614
71,550	71,600	17,051	14,536	17,901	15,628
71,600	71,650	17,066	14,550	17,917	15,642
71,650	71,700	17,082	14,564	17,932	15,656
71,700	71,750	17,097	14,578	17,948	15,670
71,750	71,800	17,113	14,592	17,963	15,684
71,800	71,850	17,128	14,606	17,979	15,698
71,850	71,900	17,144	14,620	17,994	15,712
71,900	71,950	17,159	14,634	18,010	15,726
71,950	72,000	17,175	14,648	18,025	15,740
72,000					
72,000	72,050	17,190	14,662	18,041	15,754
72,050	72,100	17,206	14,676	18,056	15,768
72,100	72,150	17,221	14,690	18,072	15,782
72,150	72,200	17,237	14,704	18,087	15,796
72,200	72,250	17,252	14,718	18,103	15,810
72,250	72,300	17,268	14,732	18,118	15,824
72,300	72,350	17,283	14,746	18,134	15,838
72,350	72,400	17,299	14,760	18,149	15,852
72,400	72,450	17,314	14,774	18,165	15,866
72,450	72,500	17,330	14,788	18,180	15,880
72,500	72,550	17,345	14,802	18,196	15,894
72,550	72,600	17,361	14,816	18,211	15,908
72,600	72,650	17,376	14,830	18,227	15,922
72,650	72,700	17,392	14,844	18,242	15,936
72,700	72,750	17,407	14,858	18,258	15,950
72,750	72,800	17,423	14,872	18,273	15,964
72,800	72,850	17,438	14,886	18,289	15,978
72,850	72,900	17,454	14,900	18,304	15,992
72,900	72,950	17,469	14,914	18,320	16,006
72,950	73,000	17,485	14,928	18,335	16,020
73,000					
73,000	73,050	17,500	14,942	18,351	16,034
73,050	73,100	17,516	14,956	18,366	16,048
73,100	73,150	17,531	14,970	18,382	16,062
73,150	73,200	17,547	14,984	18,397	16,076
73,200	73,250	17,562	14,998	18,413	16,090
73,250	73,300	17,578	15,012	18,428	16,104
73,300	73,350	17,593	15,026	18,444	16,118
73,350	73,400	17,609	15,040	18,459	16,132
73,400	73,450	17,624	15,054	18,475	16,146
73,450	73,500	17,640	15,068	18,490	16,160
73,500	73,550	17,655	15,082	18,506	16,174
73,550	73,600	17,671	15,096	18,521	16,188
73,600	73,650	17,686	15,110	18,537	16,202
73,650	73,700	17,702	15,124	18,552	16,216
73,700	73,750	17,717	15,138	18,568	16,230
73,750	73,800	17,733	15,152	18,583	16,244
73,800	73,850	17,748	15,166	18,599	16,258
73,850	73,900	17,764	15,180	18,614	16,272
73,900	73,950	17,779	15,194	18,630	16,286
73,950	74,000	17,795	15,208	18,645	16,300

If line 39 (taxable income) is— At least	But less than	Single	Married filing jointly *	Married filing separately	Head of a household
74,000					
74,000	74,050	17,810	15,222	18,661	16,314
74,050	74,100	17,826	15,236	18,676	16,328
74,100	74,150	17,841	15,250	18,692	16,342
74,150	74,200	17,857	15,264	18,707	16,356
74,200	74,250	17,872	15,278	18,723	16,370
74,250	74,300	17,888	15,292	18,738	16,384
74,300	74,350	17,903	15,306	18,754	16,398
74,350	74,400	17,919	15,320	18,769	16,412
74,400	74,450	17,934	15,334	18,785	16,426
74,450	74,500	17,950	15,348	18,800	16,440
74,500	74,550	17,965	15,362	18,816	16,454
74,550	74,600	17,981	15,376	18,831	16,468
74,600	74,650	17,996	15,390	18,847	16,482
74,650	74,700	18,012	15,404	18,862	16,496
74,700	74,750	18,027	15,418	18,878	16,510
74,750	74,800	18,043	15,432	18,893	16,524
74,800	74,850	18,058	15,446	18,909	16,538
74,850	74,900	18,074	15,460	18,924	16,552
74,900	74,950	18,089	15,474	18,940	16,566
74,950	75,000	18,105	15,488	18,955	16,580
75,000					
75,000	75,050	18,120	15,502	18,971	16,594
75,050	75,100	18,136	15,516	18,986	16,608
75,100	75,150	18,151	15,530	19,002	16,622
75,150	75,200	18,167	15,544	19,017	16,636
75,200	75,250	18,182	15,558	19,033	16,650
75,250	75,300	18,198	15,572	19,048	16,664
75,300	75,350	18,213	15,586	19,064	16,678
75,350	75,400	18,229	15,600	19,079	16,692
75,400	75,450	18,244	15,614	19,095	16,706
75,450	75,500	18,260	15,628	19,110	16,720
75,500	75,550	18,275	15,642	19,126	16,734
75,550	75,600	18,291	15,656	19,141	16,748
75,600	75,650	18,306	15,670	19,157	16,762
75,650	75,700	18,322	15,684	19,172	16,776
75,700	75,750	18,337	15,698	19,188	16,790
75,750	75,800	18,353	15,712	19,203	16,804
75,800	75,850	18,368	15,726	19,219	16,818
75,850	75,900	18,384	15,740	19,234	16,832
75,900	75,950	18,399	15,754	19,250	16,846
75,950	76,000	18,415	15,768	19,265	16,860
76,000					
76,000	76,050	18,430	15,782	19,281	16,874
76,050	76,100	18,446	15,796	19,296	16,888
76,100	76,150	18,461	15,810	19,312	16,902
76,150	76,200	18,477	15,824	19,327	16,916
76,200	76,250	18,492	15,838	19,343	16,930
76,250	76,300	18,508	15,852	19,358	16,944
76,300	76,350	18,523	15,866	19,374	16,958
76,350	76,400	18,539	15,880	19,389	16,972
76,400	76,450	18,554	15,894	19,405	16,986
76,450	76,500	18,570	15,908	19,420	17,000
76,500	76,550	18,585	15,922	19,436	17,014
76,550	76,600	18,601	15,936	19,451	17,028
76,600	76,650	18,616	15,950	19,467	17,042
76,650	76,700	18,632	15,964	19,482	17,056
76,700	76,750	18,647	15,978	19,498	17,070
76,750	76,800	18,663	15,992	19,513	17,084
76,800	76,850	18,678	16,006	19,529	17,098
76,850	76,900	18,694	16,020	19,544	17,112
76,900	76,950	18,709	16,034	19,560	17,126
76,950	77,000	18,725	16,048	19,575	17,140

* This column must also be used by a qualifying widow(er).

Continued on next page

1998 Tax Table—*Continued*

If line 39 (taxable income) is—		And you are—			
At least	But less than	Single	Married filing jointly *	Married filing separately	Head of a household
		Your tax is—			
77,000					
77,000	77,050	18,740	16,062	19,591	17,154
77,050	77,100	18,756	16,076	19,606	17,168
77,100	77,150	18,771	16,090	19,622	17,182
77,150	77,200	18,787	16,104	19,637	17,196
77,200	77,250	18,802	16,118	19,653	17,210
77,250	77,300	18,818	16,132	19,668	17,224
77,300	77,350	18,833	16,146	19,684	17,238
77,350	77,400	18,849	16,160	19,699	17,252
77,400	77,450	18,864	16,174	19,715	17,266
77,450	77,500	18,880	16,188	19,730	17,280
77,500	77,550	18,895	16,202	19,746	17,294
77,550	77,600	18,911	16,216	19,761	17,308
77,600	77,650	18,926	16,230	19,777	17,322
77,650	77,700	18,942	16,244	19,792	17,336
77,700	77,750	18,957	16,258	19,808	17,350
77,750	77,800	18,973	16,272	19,823	17,364
77,800	77,850	18,988	16,286	19,839	17,378
77,850	77,900	19,004	16,300	19,854	17,392
77,900	77,950	19,019	16,314	19,870	17,406
77,950	78,000	19,035	16,328	19,885	17,420
78,000					
78,000	78,050	19,050	16,342	19,903	17,434
78,050	78,100	19,066	16,356	19,921	17,448
78,100	78,150	19,081	16,370	19,939	17,462
78,150	78,200	19,097	16,384	19,957	17,476
78,200	78,250	19,112	16,398	19,975	17,490
78,250	78,300	19,128	16,412	19,993	17,504
78,300	78,350	19,143	16,426	20,011	17,518
78,350	78,400	19,159	16,440	20,029	17,532
78,400	78,450	19,174	16,454	20,047	17,546
78,450	78,500	19,190	16,468	20,065	17,560
78,500	78,550	19,205	16,482	20,083	17,574
78,550	78,600	19,221	16,496	20,101	17,588
78,600	78,650	19,236	16,510	20,119	17,602
78,650	78,700	19,252	16,524	20,137	17,616
78,700	78,750	19,267	16,538	20,155	17,630
78,750	78,800	19,283	16,552	20,173	17,644
78,800	78,850	19,298	16,566	20,191	17,658
78,850	78,900	19,314	16,580	20,209	17,672
78,900	78,950	19,329	16,594	20,227	17,686
78,950	79,000	19,345	16,608	20,245	17,700
79,000					
79,000	79,050	19,360	16,622	20,263	17,714
79,050	79,100	19,376	16,636	20,281	17,728
79,100	79,150	19,391	16,650	20,299	17,742
79,150	79,200	19,407	16,664	20,317	17,756
79,200	79,250	19,422	16,678	20,335	17,770
79,250	79,300	19,438	16,692	20,353	17,784
79,300	79,350	19,453	16,706	20,371	17,798
79,350	79,400	19,469	16,720	20,389	17,812
79,400	79,450	19,484	16,734	20,407	17,826
79,450	79,500	19,500	16,748	20,425	17,840
79,500	79,550	19,515	16,762	20,443	17,854
79,550	79,600	19,531	16,776	20,461	17,868
79,600	79,650	19,546	16,790	20,479	17,882
79,650	79,700	19,562	16,804	20,497	17,896
79,700	79,750	19,577	16,818	20,515	17,910
79,750	79,800	19,593	16,832	20,533	17,924
79,800	79,850	19,608	16,846	20,551	17,938
79,850	79,900	19,624	16,860	20,569	17,952
79,900	79,950	19,639	16,874	20,587	17,966
79,950	80,000	19,655	16,888	20,605	17,980

If line 39 (taxable income) is—		And you are—			
At least	But less than	Single	Married filing jointly *	Married filing separately	Head of a household
		Your tax is—			
80,000					
80,000	80,050	19,670	16,902	20,623	17,994
80,050	80,100	19,686	16,916	20,641	18,008
80,100	80,150	19,701	16,930	20,659	18,022
80,150	80,200	19,717	16,944	20,677	18,036
80,200	80,250	19,732	16,958	20,695	18,050
80,250	80,300	19,748	16,972	20,713	18,064
80,300	80,350	19,763	16,986	20,731	18,078
80,350	80,400	19,779	17,000	20,749	18,092
80,400	80,450	19,794	17,014	20,767	18,106
80,450	80,500	19,810	17,028	20,785	18,120
80,500	80,550	19,825	17,042	20,803	18,134
80,550	80,600	19,841	17,056	20,821	18,148
80,600	80,650	19,856	17,070	20,839	18,162
80,650	80,700	19,872	17,084	20,857	18,176
80,700	80,750	19,887	17,098	20,875	18,190
80,750	80,800	19,903	17,112	20,893	18,204
80,800	80,850	19,918	17,126	20,911	18,218
80,850	80,900	19,934	17,140	20,929	18,232
80,900	80,950	19,949	17,154	20,947	18,246
80,950	81,000	19,965	17,168	20,965	18,260
81,000					
81,000	81,050	19,980	17,182	20,983	18,274
81,050	81,100	19,996	17,196	21,001	18,288
81,100	81,150	20,011	17,210	21,019	18,302
81,150	81,200	20,027	17,224	21,037	18,316
81,200	81,250	20,042	17,238	21,055	18,330
81,250	81,300	20,058	17,252	21,073	18,344
81,300	81,350	20,073	17,266	21,091	18,358
81,350	81,400	20,089	17,280	21,109	18,372
81,400	81,450	20,104	17,294	21,127	18,386
81,450	81,500	20,120	17,308	21,145	18,400
81,500	81,550	20,135	17,322	21,163	18,414
81,550	81,600	20,151	17,336	21,181	18,428
81,600	81,650	20,166	17,350	21,199	18,442
81,650	81,700	20,182	17,364	21,217	18,456
81,700	81,750	20,197	17,378	21,235	18,470
81,750	81,800	20,213	17,392	21,253	18,484
81,800	81,850	20,228	17,406	21,271	18,498
81,850	81,900	20,244	17,420	21,289	18,512
81,900	81,950	20,259	17,434	21,307	18,526
81,950	82,000	20,275	17,448	21,325	18,540
82,000					
82,000	82,050	20,290	17,462	21,343	18,554
82,050	82,100	20,306	17,476	21,361	18,568
82,100	82,150	20,321	17,490	21,379	18,582
82,150	82,200	20,337	17,504	21,397	18,596
82,200	82,250	20,352	17,518	21,415	18,610
82,250	82,300	20,368	17,532	21,433	18,624
82,300	82,350	20,383	17,546	21,451	18,638
82,350	82,400	20,399	17,560	21,469	18,652
82,400	82,450	20,414	17,574	21,487	18,666
82,450	82,500	20,430	17,588	21,505	18,680
82,500	82,550	20,445	17,602	21,523	18,694
82,550	82,600	20,461	17,616	21,541	18,708
82,600	82,650	20,476	17,630	21,559	18,722
82,650	82,700	20,492	17,644	21,577	18,736
82,700	82,750	20,507	17,658	21,595	18,750
82,750	82,800	20,523	17,672	21,613	18,764
82,800	82,850	20,538	17,686	21,631	18,778
82,850	82,900	20,554	17,700	21,649	18,792
82,900	82,950	20,569	17,714	21,667	18,806
82,950	83,000	20,585	17,728	21,685	18,820

If line 39 (taxable income) is—		And you are—			
At least	But less than	Single	Married filing jointly *	Married filing separately	Head of a household
		Your tax is—			
83,000					
83,000	83,050	20,600	17,742	21,703	18,834
83,050	83,100	20,616	17,756	21,721	18,848
83,100	83,150	20,631	17,770	21,739	18,862
83,150	83,200	20,647	17,784	21,757	18,876
83,200	83,250	20,662	17,798	21,775	18,890
83,250	83,300	20,678	17,812	21,793	18,904
83,300	83,350	20,693	17,826	21,811	18,918
83,350	83,400	20,709	17,840	21,829	18,932
83,400	83,450	20,724	17,854	21,847	18,946
83,450	83,500	20,740	17,868	21,865	18,960
83,500	83,550	20,755	17,882	21,883	18,974
83,550	83,600	20,771	17,896	21,901	18,988
83,600	83,650	20,786	17,910	21,919	19,002
83,650	83,700	20,802	17,924	21,937	19,016
83,700	83,750	20,817	17,938	21,955	19,030
83,750	83,800	20,833	17,952	21,973	19,044
83,800	83,850	20,848	17,966	21,991	19,058
83,850	83,900	20,864	17,980	22,009	19,072
83,900	83,950	20,879	17,994	22,027	19,086
83,950	84,000	20,895	18,008	22,045	19,100
84,000					
84,000	84,050	20,910	18,022	22,063	19,114
84,050	84,100	20,926	18,036	22,081	19,128
84,100	84,150	20,941	18,050	22,099	19,142
84,150	84,200	20,957	18,064	22,117	19,156
84,200	84,250	20,972	18,078	22,135	19,170
84,250	84,300	20,988	18,092	22,153	19,184
84,300	84,350	21,003	18,106	22,171	19,198
84,350	84,400	21,019	18,120	22,189	19,212
84,400	84,450	21,034	18,134	22,207	19,226
84,450	84,500	21,050	18,148	22,225	19,240
84,500	84,550	21,065	18,162	22,243	19,254
84,550	84,600	21,081	18,176	22,261	19,268
84,600	84,650	21,096	18,190	22,279	19,282
84,650	84,700	21,112	18,204	22,297	19,296
84,700	84,750	21,127	18,218	22,315	19,310
84,750	84,800	21,143	18,232	22,333	19,324
84,800	84,850	21,158	18,246	22,351	19,338
84,850	84,900	21,174	18,260	22,369	19,352
84,900	84,950	21,189	18,274	22,387	19,366
84,950	85,000	21,205	18,288	22,405	19,380
85,000					
85,000	85,050	21,220	18,302	22,423	19,394
85,050	85,100	21,236	18,316	22,441	19,408
85,100	85,150	21,251	18,330	22,459	19,422
85,150	85,200	21,267	18,344	22,477	19,436
85,200	85,250	21,282	18,358	22,495	19,450
85,250	85,300	21,298	18,372	22,513	19,464
85,300	85,350	21,313	18,386	22,531	19,478
85,350	85,400	21,329	18,400	22,549	19,492
85,400	85,450	21,344	18,414	22,567	19,506
85,450	85,500	21,360	18,428	22,585	19,520
85,500	85,550	21,375	18,442	22,603	19,534
85,550	85,600	21,391	18,456	22,621	19,548
85,600	85,650	21,406	18,470	22,639	19,562
85,650	85,700	21,422	18,484	22,657	19,576
85,700	85,750	21,437	18,498	22,675	19,590
85,750	85,800	21,453	18,512	22,693	19,604
85,800	85,850	21,468	18,526	22,711	19,618
85,850	85,900	21,484	18,540	22,729	19,632
85,900	85,950	21,499	18,554	22,747	19,646
85,950	86,000	21,515	18,568	22,765	19,660

* This column must also be used by a qualifying widow(er).

Continued on next page

1998 Tax Table—Continued

86,000 / 87,000 / 88,000

If line 39 (taxable income) is— At least	But less than	Single	Married filing jointly *	Married filing separately	Head of a household
86,000					
86,000	86,050	21,530	18,582	22,783	19,674
86,050	86,100	21,546	18,596	22,801	19,688
86,100	86,150	21,561	18,610	22,819	19,702
86,150	86,200	21,577	18,624	22,837	19,716
86,200	86,250	21,592	18,638	22,855	19,730
86,250	86,300	21,608	18,652	22,873	19,744
86,300	86,350	21,623	18,666	22,891	19,758
86,350	86,400	21,639	18,680	22,909	19,772
86,400	86,450	21,654	18,694	22,927	19,786
86,450	86,500	21,670	18,708	22,945	19,800
86,500	86,550	21,685	18,722	22,963	19,814
86,550	86,600	21,701	18,736	22,981	19,828
86,600	86,650	21,716	18,750	22,999	19,842
86,650	86,700	21,732	18,764	23,017	19,856
86,700	86,750	21,747	18,778	23,035	19,870
86,750	86,800	21,763	18,792	23,053	19,884
86,800	86,850	21,778	18,806	23,071	19,898
86,850	86,900	21,794	18,820	23,089	19,912
86,900	86,950	21,809	18,834	23,107	19,926
86,950	87,000	21,825	18,848	23,125	19,940
87,000					
87,000	87,050	21,840	18,862	23,143	19,954
87,050	87,100	21,856	18,876	23,161	19,968
87,100	87,150	21,871	18,890	23,179	19,982
87,150	87,200	21,887	18,904	23,197	19,996
87,200	87,250	21,902	18,918	23,215	20,010
87,250	87,300	21,918	18,932	23,233	20,024
87,300	87,350	21,933	18,946	23,251	20,038
87,350	87,400	21,949	18,960	23,269	20,052
87,400	87,450	21,964	18,974	23,287	20,066
87,450	87,500	21,980	18,988	23,305	20,080
87,500	87,550	21,995	19,002	23,323	20,094
87,550	87,600	22,011	19,016	23,341	20,108
87,600	87,650	22,026	19,030	23,359	20,122
87,650	87,700	22,042	19,044	23,377	20,136
87,700	87,750	22,057	19,058	23,395	20,150
87,750	87,800	22,073	19,072	23,413	20,166
87,800	87,850	22,088	19,086	23,431	20,181
87,850	87,900	22,104	19,100	23,449	20,197
87,900	87,950	22,119	19,114	23,467	20,212
87,950	88,000	22,135	19,128	23,485	20,228
88,000					
88,000	88,050	22,150	19,142	23,503	20,243
88,050	88,100	22,166	19,156	23,521	20,259
88,100	88,150	22,181	19,170	23,539	20,274
88,150	88,200	22,197	19,184	23,557	20,290
88,200	88,250	22,212	19,198	23,575	20,305
88,250	88,300	22,228	19,212	23,593	20,321
88,300	88,350	22,243	19,226	23,611	20,336
88,350	88,400	22,259	19,240	23,629	20,352
88,400	88,450	22,274	19,254	23,647	20,367
88,450	88,500	22,290	19,268	23,665	20,383
88,500	88,550	22,305	19,282	23,683	20,398
88,550	88,600	22,321	19,296	23,701	20,414
88,600	88,650	22,336	19,310	23,719	20,429
88,650	88,700	22,352	19,324	23,737	20,445
88,700	88,750	22,367	19,338	23,755	20,460
88,750	88,800	22,383	19,352	23,773	20,476
88,800	88,850	22,398	19,366	23,791	20,491
88,850	88,900	22,414	19,380	23,809	20,507
88,900	88,950	22,429	19,394	23,827	20,522
88,950	89,000	22,445	19,408	23,845	20,538

89,000 / 90,000 / 91,000

If line 39 (taxable income) is— At least	But less than	Single	Married filing jointly *	Married filing separately	Head of a household
89,000					
89,000	89,050	22,460	19,422	23,863	20,553
89,050	89,100	22,476	19,436	23,881	20,569
89,100	89,150	22,491	19,450	23,899	20,584
89,150	89,200	22,507	19,464	23,917	20,600
89,200	89,250	22,522	19,478	23,935	20,615
89,250	89,300	22,538	19,492	23,953	20,631
89,300	89,350	22,553	19,506	23,971	20,646
89,350	89,400	22,569	19,520	23,989	20,662
89,400	89,450	22,584	19,534	24,007	20,677
89,450	89,500	22,600	19,548	24,025	20,693
89,500	89,550	22,615	19,562	24,043	20,708
89,550	89,600	22,631	19,576	24,061	20,724
89,600	89,650	22,646	19,590	24,079	20,739
89,650	89,700	22,662	19,604	24,097	20,755
89,700	89,750	22,677	19,618	24,115	20,770
89,750	89,800	22,693	19,632	24,133	20,786
89,800	89,850	22,708	19,646	24,151	20,801
89,850	89,900	22,724	19,660	24,169	20,817
89,900	89,950	22,739	19,674	24,187	20,832
89,950	90,000	22,755	19,688	24,205	20,848
90,000					
90,000	90,050	22,770	19,702	24,223	20,863
90,050	90,100	22,786	19,716	24,241	20,879
90,100	90,150	22,801	19,730	24,259	20,894
90,150	90,200	22,817	19,744	24,277	20,910
90,200	90,250	22,832	19,758	24,295	20,925
90,250	90,300	22,848	19,772	24,313	20,941
90,300	90,350	22,863	19,786	24,331	20,956
90,350	90,400	22,879	19,800	24,349	20,972
90,400	90,450	22,894	19,814	24,367	20,987
90,450	90,500	22,910	19,828	24,385	21,003
90,500	90,550	22,925	19,842	24,403	21,018
90,550	90,600	22,941	19,856	24,421	21,034
90,600	90,650	22,956	19,870	24,439	21,049
90,650	90,700	22,972	19,884	24,457	21,065
90,700	90,750	22,987	19,898	24,475	21,080
90,750	90,800	23,003	19,912	24,493	21,096
90,800	90,850	23,018	19,926	24,511	21,111
90,850	90,900	23,034	19,940	24,529	21,127
90,900	90,950	23,049	19,954	24,547	21,142
90,950	91,000	23,065	19,968	24,565	21,158
91,000					
91,000	91,050	23,080	19,982	24,583	21,173
91,050	91,100	23,096	19,996	24,601	21,189
91,100	91,150	23,111	20,010	24,619	21,204
91,150	91,200	23,127	20,024	24,637	21,220
91,200	91,250	23,142	20,038	24,655	21,235
91,250	91,300	23,158	20,052	24,673	21,251
91,300	91,350	23,173	20,066	24,691	21,266
91,350	91,400	23,189	20,080	24,709	21,282
91,400	91,450	23,204	20,094	24,727	21,297
91,450	91,500	23,220	20,108	24,745	21,313
91,500	91,550	23,235	20,122	24,763	21,328
91,550	91,600	23,251	20,136	24,781	21,344
91,600	91,650	23,266	20,150	24,799	21,359
91,650	91,700	23,282	20,164	24,817	21,375
91,700	91,750	23,297	20,178	24,835	21,390
91,750	91,800	23,313	20,192	24,853	21,406
91,800	91,850	23,328	20,206	24,871	21,421
91,850	91,900	23,344	20,220	24,889	21,437
91,900	91,950	23,359	20,234	24,907	21,452
91,950	92,000	23,375	20,248	24,925	21,468

92,000 / 93,000 / 94,000

If line 39 (taxable income) is— At least	But less than	Single	Married filing jointly *	Married filing separately	Head of a household
92,000					
92,000	92,050	23,390	20,262	24,943	21,483
92,050	92,100	23,406	20,276	24,961	21,499
92,100	92,150	23,421	20,290	24,979	21,514
92,150	92,200	23,437	20,304	24,997	21,530
92,200	92,250	23,452	20,318	25,015	21,545
92,250	92,300	23,468	20,332	25,033	21,561
92,300	92,350	23,483	20,346	25,051	21,576
92,350	92,400	23,499	20,360	25,069	21,592
92,400	92,450	23,514	20,374	25,087	21,607
92,450	92,500	23,530	20,388	25,105	21,623
92,500	92,550	23,545	20,402	25,123	21,638
92,550	92,600	23,561	20,416	25,141	21,654
92,600	92,650	23,576	20,430	25,159	21,669
92,650	92,700	23,592	20,444	25,177	21,685
92,700	92,750	23,607	20,458	25,195	21,700
92,750	92,800	23,623	20,472	25,213	21,716
92,800	92,850	23,638	20,486	25,231	21,731
92,850	92,900	23,654	20,500	25,249	21,747
92,900	92,950	23,669	20,514	25,267	21,762
92,950	93,000	23,685	20,528	25,285	21,778
93,000					
93,000	93,050	23,700	20,542	25,303	21,793
93,050	93,100	23,716	20,556	25,321	21,809
93,100	93,150	23,731	20,570	25,339	21,824
93,150	93,200	23,747	20,584	25,357	21,840
93,200	93,250	23,762	20,598	25,375	21,855
93,250	93,300	23,778	20,612	25,393	21,871
93,300	93,350	23,793	20,626	25,411	21,886
93,350	93,400	23,809	20,640	25,429	21,902
93,400	93,450	23,824	20,654	25,447	21,917
93,450	93,500	23,840	20,668	25,465	21,933
93,500	93,550	23,855	20,682	25,483	21,948
93,550	93,600	23,871	20,696	25,501	21,964
93,600	93,650	23,886	20,710	25,519	21,979
93,650	93,700	23,902	20,724	25,537	21,995
93,700	93,750	23,917	20,738	25,555	22,010
93,750	93,800	23,933	20,752	25,573	22,026
93,800	93,850	23,948	20,766	25,591	22,041
93,850	93,900	23,964	20,780	25,609	22,057
93,900	93,950	23,979	20,794	25,627	22,072
93,950	94,000	23,995	20,808	25,645	22,088
94,000					
94,000	94,050	24,010	20,822	25,663	22,103
94,050	94,100	24,026	20,836	25,681	22,119
94,100	94,150	24,041	20,850	25,699	22,134
94,150	94,200	24,057	20,864	25,717	22,150
94,200	94,250	24,072	20,878	25,735	22,165
94,250	94,300	24,088	20,892	25,753	22,181
94,300	94,350	24,103	20,906	25,771	22,196
94,350	94,400	24,119	20,920	25,789	22,212
94,400	94,450	24,134	20,934	25,807	22,227
94,450	94,500	24,150	20,948	25,825	22,243
94,500	94,550	24,165	20,962	25,843	22,258
94,550	94,600	24,181	20,976	25,861	22,274
94,600	94,650	24,196	20,990	25,879	22,289
94,650	94,700	24,212	21,004	25,897	22,305
94,700	94,750	24,227	21,018	25,915	22,320
94,750	94,800	24,243	21,032	25,933	22,336
94,800	94,850	24,258	21,046	25,951	22,351
94,850	94,900	24,274	21,060	25,969	22,367
94,900	94,950	24,289	21,074	25,987	22,382
94,950	95,000	24,305	21,088	26,005	22,398

* This column must also be used by a qualifying widow(er).

Continued on next page

1998 Tax Table—*Continued*

If line 39 (taxable income) is—		And you are—				If line 39 (taxable income) is—		And you are—			
At least	But less than	Single	Married filing jointly *	Married filing sepa-rately *	Head of a house-hold	At least	But less than	Single	Married filing jointly *	Married filing sepa-rately *	Head of a house-hold
		Your tax is—						Your tax is—			
95,000						**98,000**					
95,000	95,050	24,320	21,102	26,023	22,413	98,000	98,050	25,250	21,942	27,103	23,343
95,050	95,100	24,336	21,116	26,041	22,429	98,050	98,100	25,266	21,956	27,121	23,359
95,100	95,150	24,351	21,130	26,059	22,444	98,100	98,150	25,281	21,970	27,139	23,374
95,150	95,200	24,367	21,144	26,077	22,460	98,150	98,200	25,297	21,984	27,157	23,390
95,200	95,250	24,382	21,158	26,095	22,475	98,200	98,250	25,312	21,998	27,175	23,405
95,250	95,300	24,398	21,172	26,113	22,491	98,250	98,300	25,328	22,012	27,193	23,421
95,300	95,350	24,413	21,186	26,131	22,506	98,300	98,350	25,343	22,026	27,211	23,436
95,350	95,400	24,429	21,200	26,149	22,522	98,350	98,400	25,359	22,040	27,229	23,452
95,400	95,450	24,444	21,214	26,167	22,537	98,400	98,450	25,374	22,054	27,247	23,467
95,450	95,500	24,460	21,228	26,185	22,553	98,450	98,500	25,390	22,068	27,265	23,483
95,500	95,550	24,475	21,242	26,203	22,568	98,500	98,550	25,405	22,082	27,283	23,498
95,550	95,600	24,491	21,256	26,221	22,584	98,550	98,600	25,421	22,096	27,301	23,514
95,600	95,650	24,506	21,270	26,239	22,599	98,600	98,650	25,436	22,110	27,319	23,529
95,650	95,700	24,522	21,284	26,257	22,615	98,650	98,700	25,452	22,124	27,337	23,545
95,700	95,750	24,537	21,298	26,275	22,630	98,700	98,750	25,467	22,138	27,355	23,560
95,750	95,800	24,553	21,312	26,293	22,646	98,750	98,800	25,483	22,152	27,373	23,576
95,800	95,850	24,568	21,326	26,311	22,661	98,800	98,850	25,498	22,166	27,391	23,591
95,850	95,900	24,584	21,340	26,329	22,677	98,850	98,900	25,514	22,180	27,409	23,607
95,900	95,950	24,599	21,354	26,347	22,692	98,900	98,950	25,529	22,194	27,427	23,622
95,950	96,000	24,615	21,368	26,365	22,708	98,950	99,000	25,545	22,208	27,445	23,638
96,000						**99,000**					
96,000	96,050	24,630	21,382	26,383	22,723	99,000	99,050	25,560	22,222	27,463	23,653
96,050	96,100	24,646	21,396	26,401	22,739	99,050	99,100	25,576	22,236	27,481	23,669
96,100	96,150	24,661	21,410	26,419	22,754	99,100	99,150	25,591	22,250	27,499	23,684
96,150	96,200	24,677	21,424	26,437	22,770	99,150	99,200	25,607	22,264	27,517	23,700
96,200	96,250	24,692	21,438	26,455	22,785	99,200	99,250	25,622	22,278	27,535	23,715
96,250	96,300	24,708	21,452	26,473	22,801	99,250	99,300	25,638	22,292	27,553	23,731
96,300	96,350	24,723	21,466	26,491	22,816	99,300	99,350	25,653	22,306	27,571	23,746
96,350	96,400	24,739	21,480	26,509	22,832	99,350	99,400	25,669	22,320	27,589	23,762
96,400	96,450	24,754	21,494	26,527	22,847	99,400	99,450	25,684	22,334	27,607	23,777
96,450	96,500	24,770	21,508	26,545	22,863	99,450	99,500	25,700	22,348	27,625	23,793
96,500	96,550	24,785	21,522	26,563	22,878	99,500	99,550	25,715	22,362	27,643	23,808
96,550	96,600	24,801	21,536	26,581	22,894	99,550	99,600	25,731	22,376	27,661	23,824
96,600	96,650	24,816	21,550	26,599	22,909	99,600	99,650	25,746	22,390	27,679	23,839
96,650	96,700	24,832	21,564	26,617	22,925	99,650	99,700	25,762	22,404	27,697	23,855
96,700	96,750	24,847	21,578	26,635	22,940	99,700	99,750	25,777	22,418	27,715	23,870
96,750	96,800	24,863	21,592	26,653	22,956	99,750	99,800	25,793	22,432	27,733	23,886
96,800	96,850	24,878	21,606	26,671	22,971	99,800	99,850	25,808	22,446	27,751	23,901
96,850	96,900	24,894	21,620	26,689	22,987	99,850	99,900	25,824	22,460	27,769	23,917
96,900	96,950	24,909	21,634	26,707	23,002	99,900	99,950	25,839	22,474	27,787	23,932
96,950	97,000	24,925	21,648	26,725	23,018	99,950	100,000	25,855	22,488	27,805	23,948
97,000											
97,000	97,050	24,940	21,662	26,743	23,033						
97,050	97,100	24,956	21,676	26,761	23,049						
97,100	97,150	24,971	21,690	26,779	23,064						
97,150	97,200	24,987	21,704	26,797	23,080						
97,200	97,250	25,002	21,718	26,815	23,095						
97,250	97,300	25,018	21,732	26,833	23,111						
97,300	97,350	25,033	21,746	26,851	23,126						
97,350	97,400	25,049	21,760	26,869	23,142						
97,400	97,450	25,064	21,774	26,887	23,157						
97,450	97,500	25,080	21,788	26,905	23,173						
97,500	97,550	25,095	21,802	26,923	23,188						
97,550	97,600	25,111	21,816	26,941	23,204						
97,600	97,650	25,126	21,830	26,959	23,219						
97,650	97,700	25,142	21,844	26,977	23,235						
97,700	97,750	25,157	21,858	26,995	23,250						
97,750	97,800	25,173	21,872	27,013	23,266						
97,800	97,850	25,188	21,886	27,031	23,281						
97,850	97,900	25,204	21,900	27,049	23,297						
97,900	97,950	25,219	21,914	27,067	23,312						
97,950	98,000	25,235	21,928	27,085	23,328						

$100,000 or over — use the Tax Rate Schedules x, y, or z.

* This column must also be used by a qualifying widow(er).

1998
Tax Rate
Schedules

Caution: *Use **only** if your taxable income (Form 1040, line 39) is $100,000 or more. If less, use the **Tax Table**. Even though you cannot use the Tax Rate Schedules below if your taxable income is less than $100,000, all levels of taxable income are shown so taxpayers can see the tax rate that applies to each level.*

Schedule X—Use if your filing status is **Single**

If the amount on Form 1040, line 39, is: Over—	But not over—	Enter on Form 1040, line 40		of the amount over—
$0	$25,350	.... 15%		$0
25,350	61,400	$3,802.50 +	28%	25,350
61,400	128,100	13,896.50 +	31%	61,400
128,100	278,450	34,573.50 +	36%	128,100
278,450	....	88,699.50 +	39.6%	278,450

Schedule Y-1—Use if your filing status is **Married filing jointly** or **Qualifying widow(er)**

If the amount on Form 1040, line 39, is: Over—	But not over—	Enter on Form 1040, line 40		of the amount over—
$0	$42,350	.... 15%		$0
42,350	102,300	$6,352.50 +	28%	42,350
102,300	155,950	23,138.50 +	31%	102,300
155,950	278,450	39,770.00 +	36%	155,950
278,450	....	83,870.00 +	39.6%	278,450

Schedule Y-2—Use if your filing status is **Married filing separately**

If the amount on Form 1040, line 39, is: Over—	But not over—	Enter on Form 1040, line 40		of the amount over—
$0	$21,175	.... 15%		$0
21,175	51,150	$3,176.25 +	28%	21,175
51,150	77,975	11,569.25 +	31%	51,150
77,975	139,225	19,885.00 +	36%	77,975
139,225	....	41,935.00 +	39.6%	139,225

Schedule Z—Use if your filing status is **Head of household**

If the amount on Form 1040, line 39, is: Over—	But not over—	Enter on Form 1040, line 40		of the amount over—
$0	$33,950	.... 15%		$0
33,950	87,700	$5,092.50 +	28%	33,950
87,700	142,000	20,142.50 +	31%	87,700
142,000	278,450	36,975.50 +	36%	142,000
278,450	....	86,097.50 +	39.6%	278,450

International Measurement and Monetary Exchange

METRIC UNITS OF MEASUREMENT

Throughout history systems have been devised to express units of measure. Most of the early systems were based on the human anatomy. For example, the inch as we know it today was originally the length between the knuckle and the tip of the thumb. A foot was the length of a person's foot, and a yard was the distance between the chin and the tip of the thumb when the arm was stretched outward from a person's body. The Romans later divided the foot measurement into twelve equal parts called *unciae*, and defined the mile as *milia passuum,* or a thousand paces. The pace was not accepted as a standard length, so the *milia passuum* was later defined as a furlong, which eventually led to the mile of 1,760 yards or 5,280 feet.

Many years later the English refined and expanded the Roman system. The *English system,* which includes such units of measure as anglicized ounces, pounds, pints, quarts, pecks, bushels, inches, feet, and miles, was eventually also adopted as the system of weights and measures of the United States. This system, also called the *U.S. Customary Measurements,* has been used by the U.S. business community to determine the quantity of sale and the unit price of sale for more than 200 years. However, due to the complexity of converting U.S. customary measurements (e.g., inches to feet, feet to yards, pints to quarts, quarts to gallons, etc.) the metric system is being adopted by most American industries as their official system of weights and measures.

Following the French Revolution in 1789, the French created a new unit of length based upon a natural measurement. This new unit was called the *metre,* which means measure. The *metric system* has become the official system of measurement of every major industrialized country in the world, except the United States. In 1975, the Metric Conversion Act was enacted by the U.S. government to encourage a complete conversion from the English system to the metric system, but that goal has not been realized and a dual system of measurement continues to exist in the United States. Many individual companies have converted to metric, especially those involved in international trade. In addition, the entire automobile, tool, beverage, and pharmaceutical industries have gone almost completely metric. It is not uncommon to see both metric and English units on food labels, or receive weather reports in both Celsius and Fahrenheit. While the conversion has not been as rapid as originally anticipated, the metric system will most likely become the official system of weights and measures of the United States in the future.

The metric system is a system of weights and measures based on decimals and powers of 10. Therefore, each unit of metric measure (e.g., meter,

Table E.1

Metric prefixes

Prefix	Power	Multiple	Symbols
kilo	10^3	1000.0	km, kl, kg
hecto	10^2	100.0	hm, hl, hg
deka	10^1	10.0	dkm, kal, dag
meter, liter, gram	10^0	0.0	m, l, g
deci	10^{-1}	.1	dm, dl, dg
centi	10^{-2}	.01	cm, cl, cg
milli	10^{-3}	.001	mm, ml, mg

liter, gram) can be multiplied or divided by factors of ten to get larger or smaller units. For example, in the English system short lengths are expressed in inches, or feet, and longer lengths are expressed in yards or miles. But in the metric system, short lengths are expressed in millimeters, decimeters, and centimeters, while longer lengths are expressed in meters, dekameters, hectometers, or kilometers. Notice that each of the metric measures of length contains a prefix that indicates its power of ten. The primary prefixes of the metric system are shown in Table E.1. These prefixes are used not only with *meter,* measuring length, but also with liter, the metric measure for volume, and *gram,* the metric measure for weight.

TEMPERATURE

In the metric system temperature is measured in degrees *Celsius.* On the English temperature scale water freezes at 32° and boils at 212° Fahrenheit. In comparison, on the metric scale water freezes at 0° and boils at 100° Celsius. To convert from degrees Celsius to degrees Fahrenheit we use the formula.

$$F° = \frac{9}{5} C° + 32°$$

For example, if the Celsius temperature is 20°, the equivalent Fahrenheit temperature can be calculated as follows:

$$F° = \frac{9}{5} C° + 32°$$

$$F° = \frac{9}{5} (20°) + 32°$$

$$= \frac{180°}{5} + 32°$$

$$= 36° + 32°$$

$$= 68°$$

Fahrenheit	Celsius	
212	100	water boils
194	90	
176	80	
158	70	
140	60	
122	50	
98.6	37	normal body temperature
86	30	
68	20	typical room temperature
50	10	
32	0	water freezes
14	−10	
0	−18	
−28	−30	
−40	−40	

Conversely, to convert from degrees Fahrenheit to degrees Celsius we would use the formula

$$C° = \frac{5}{9}(F° - 32°)$$

So, for a Fahrenheit temperature of 68°,

$$C° = \frac{5}{9}(68° - 32°)$$

$$C° = \frac{5}{9}(36°)$$

$$C° = \frac{180°}{9}$$

$$C° = 20°$$

METRIC CONVERSIONS

The metric system of measurement will some day be as familiar to U.S. consumers as is the English system. As the United States proceeds with the transition from the English system to the metric system, there is an ongoing need for conversion from one system to the other. To simplify this procedure, conversion tables such as Table E.2 provide approximate conversions between the two systems. For example, to convert from meters to feet the

Table E.2

Metric–English
conversion table

Metric to English			English to Metric		
Metric measure	**English** measure	**Conver-sion factor**	**English** measure	**Metric** measure	**Conver-sion factor**
Length			**Length**		
centimeter	inch	0.4	inch	centimeter	2.5
meter	foot	3.3	foot	meter	0.3
meter	yard	1.1	yard	meter	0.9
kilometer	mile	0.6	mile	kilometer	1.6
Weight			**Weight**		
gram	ounce	0.035	ounce	gram	2.8
gram	pound	0.002	pound	gram	454.0
kilogram	pound	2.2	pound	kilogram	0.454
Volume			**Volume**		
milliliter	ounce	0.03	ounce	milliliter	30.0
liter	pint	2.1	pint	liter	0.47
liter	quart	1.06	quart	liter	0.95
liter	gallon	0.26	gallon	liter	3.8

approximate conversion factor from Table E.2 is 3.3 In other words, there
are approximately 3.3 feet in 1 meter. So to convert 15 meters to approximate feet, we would calculate as follows:

$$15 \text{ meters} \times 3.3 \ \frac{\text{feet}}{\text{meter}} = 49.5 \text{ feet}$$

MONETARY EXCHANGE RATES

International trade involves the selling (exporting) and buying (importing) of
goods and services between businesses located in different countries. This
type of trading presents numerous economic considerations that affect both
the businesses and their nations. For example, if a nation's imports exceed
its exports, it has a negative *balance of trade* or a *trade deficit*. However, if a nation's exports exceed its imports, it has a positive balance of trade referred
to as a *trade surplus*. The money that is used to pay for imports or is received
for exports creates the flow of money into or out of a nation—or its *balance
of payments*. A country's balance of payments is also affected by buying and
selling currency in international money markets, by tourists, and funds allocated for foreign aid programs.

The balance of trade between two countries can be significantly affected by the exchange rates of their currency. An *exchange rate* is the rate at which the currency of one country can be exchanged for the currency of another country. Table E.3 lists the exchange rates for various countries on June 16, 1999, in terms of one U.S. dollar. For example, $1 on this date would purchase 6.37 francs in France. Thus, the exchange rate would be 6.37 to 1. The value of a country's currency is subject to change based on political and economic conditions to assure it is not *undervalued* or *overvalued*. When a nation's currency is undervalued or overvalued, its government either *revalues* or *devalues* the currency to correct its balance-of-payments problems. The objective of *revaluation* is to increase the exchange rate, making it more expensive for other countries to purchase goods and services. Whereas, the objective of *devaluation* is to decrease the exchange rate, which reduces the cost of goods and services purchased by other countries.

Payments for transactions between buyers and sellers in different nations are facilitated through the services of the *World Bank*, the *International Mone-*

Table E.3

Exchange rates

Country	Monetary unit	Value per U.S. dollar
Australia	dollar	1.51
Belgium	franc	.39
Bermuda	dollar	1.00
Canada	dollar	1.46
Denmark	kroner	7.22
England	pound	0.62
Europe	euro	1.0465
France	franc	6.37
Germany	mark	1.90
Greece	drachma	314.00
Ireland	pound	0.76
Italy	lira	1,881.00
Japan	yen	120.00
Mexico	peso	9.50
Netherlands	guilder	2.14
New Zealand	dollar	1.87
Norway	krone	7.95
Portugal	escudo	194.00
Saudi Arabia	riyal	3.75
Scotland	pound	0.62
Singapore	dollar	1.72
Spain	peseta	161.00
Sweden	kroner	8.63
Switzerland	franc	1.54

tary Fund, and local banks. For example, if a U.S. citizen were to plan a weekend trip to Canada, they could go to their local bank and exchange 500 U.S. dollars for 730 Canadian dollars based on the exchange rates in Table E.3 (500 × 1.46 = 730). The illustrations below demonstrate in more detail the payments for international transactions based on exchange rates listed in Table E.3.

Illustration 1

A U.S. importer in Boston purchased five cases of olive oil from an exporter in Barcelona, Spain, for 4,025 pesetas per case. Translate the price of the transaction into U.S. dollars.

Solution

Step 1: = 5 cases × 4,025 pesetas per case

= 20,125 pesetas

Step 2: = 20,125 pesetas ÷ 161 (161 pesetas per U.S. dollar)

= 125 U.S. dollars

Illustration 2

If a Big Mac value meal costs $3.99, how much would the meal cost in (a) Japanese yens, (b) French francs, and (c) English pounds?

Solution

a. = $3.99 × 6.37 francs

= 25.42 francs

b. = $3.99 × 120 yen

= 478.8 yen

c. = $3.99 × .62 pound

= 2.47 pounds

Basic Calculator Operations

This appendix acquaints you with different types of calculators and their various features. Each type has at least one storage register, which allows numbers to be saved or accumulated.

- *Algebraic calculators* perform mathematical calculations using the rules of algebra for basic operations ($+$, $-$, $\times$, $\div$) and operations with parentheses.

- *Scientific algebraic calculators* provide additional scientific operations, including the ability to raise numbers to a power (eg. $2^3 = 8$).

- *Scientific calculators with statistical functions* are calculators that provide the additional ability to determine the mean and standard deviation of a set of numbers easily.

- *Business analyst or financial calculators* are scientific calculators that allow certain common business functions to be calculated easily. They calculate percentage changes, markup and gross profit margin, compound interest, and future and present value annuity, for example.

You will want to read the owners' manual to determine how to derive the greatest benefits from your calculator. The following charts show some of the basic operations using three popular brands of algebraic calculators.

Some keys of the calculator can be used for more than one type of calculation. For example, the x^2 key can also be used to find the *square root* of a number. The second use, or second function, of the key is usually indicated by a symbol written on the body of the calculator either above or below the key. For example, the $\sqrt{}$ symbol usually appears above the $\boxed{x^2}$ key and is printed on the body of the calculator. To use the key for the second function, you must first depress another key on the calculator labeled, $\boxed{\text{2nd F}}$, or $\boxed{\text{shift}}$, or $\boxed{\text{inv}}$ (inverse), and then the desired key. For example, to find the square root of 49, execute the following sequence of steps: 49 $\boxed{\text{2nd F}}$ $\boxed{x^2}$. The display will show 7.

Although no specific calculator is favored by the authors of this text, the calculator examples in the text use $\boxed{\text{M in}}$ and $\boxed{\text{M out}}$ or $\boxed{\text{RCL}}$ memory keys.

Function	Casio		Sharp		Texas Instruments	
	Keyed Entry	**Display**	**Keyed Entry**	**Display**	**Keyed Entry**	**Display**
Addition: e.g., $3 + 5$	$3\boxed{+}5\boxed{=}$	8	$3\boxed{+}5\boxed{=}$	8	$3\boxed{+}5\boxed{=}$	8
Subtraction: e.g., $3 - 5$	$3\boxed{-}5\boxed{=}$	-2	$3\boxed{-}5\boxed{=}$	-2	$3\boxed{-}5\boxed{=}$	-2
Multiplication: e.g., 3×5	$3\boxed{\times}5\boxed{=}$	15	$3\boxed{\times}5\boxed{=}$	15	$3\boxed{\times}5\boxed{=}$	15
Division: e.g., $3 \div 5$	$3\boxed{\div}5\boxed{=}$	0.6	$3\boxed{\div}5\boxed{=}$	0.6	$3\boxed{\div}5\boxed{=}$	0.6
Change sign from $+$ to $-$, or from $-$ to $+$: e.g., change $+2$ to -2	$2\boxed{+/-}$	-2	$2\boxed{+/-}$	-2	$2\boxed{+/-}$	-2
Use parentheses: e.g., $(3 + 5) \times 4$	$\boxed{(}3\boxed{+}5\boxed{)}\boxed{\times}4\boxed{=}$	32	$\boxed{(}3\boxed{+}5\boxed{)}\boxed{\times}4\boxed{=}$	32	$\boxed{(}3\boxed{+}5\boxed{)}\boxed{\times}4\boxed{=}$	32
Square a number: e.g., 5^2	$5\boxed{x^2}$	25	$5\boxed{x^2}$	25	$5\boxed{x^2}$	25
Raise a number to a power: e.g., 5^3	$5\boxed{y^x}3\boxed{=}$	125	$5\boxed{y^x}3\boxed{=}$	125	$5\boxed{y^x}3\boxed{=}$	125
Obtain the square root of a number* e.g., $\sqrt{36}$	$36\boxed{\text{shift}}\overset{\sqrt{x}}{\boxed{x^2}}$	6	$36\boxed{\text{2ndF}}\overset{\sqrt{x}}{\boxed{x^2}}$	6	$36\boxed{\text{2ndF}}\overset{\sqrt{x}}{\boxed{x^2}}$	6

Memory	Casio		Sharp		Texas Instruments	
	Keyed Entry	Display	Keyed Entry	Display	Keyed Entry	Display
Store a number in memory and clear out previous numbers stored there: e.g., put 7 in memory.	7 [M in]	7	7 [x→M]	7	7 [Sto]	7
Look at what is stored in memory.	[M out]	7	[RM]	7	[RCL]	7
Add or accumulate numbers in memory: e.g., add 5, 7, and 20.	5 [M in] 7 [M+] 20 [M+]		5 [x→M] 7 [M+] 20 [M+]		5 [Sto] 7 [Sum] 20 [Sum]	
Look at the total of the accumulated sum in memory: e.g., 32 from above.	[M out]	32	[RM]	32	[RCL]	32

Appendix G
Answers to Odd-Numbered Exercises

CHAPTER 1

Section 1.1

1. −5 *3.* −36 *5.* 36 *7.* −4 *9.* 4 *11.* −8 *13.* −40 *15.* −2.5
17. −16.5 *19.* −2.1 *21.* 280 *23.* −135 *25.* −89 *27.* 8 *29.* −26
31. −1 *33.* −47 *35.* 18. *37.* 15x *39.* −11x + 6y *41.* 15x + 5y
43. 59x + 16y *45.* X^3 *47.* 2 − x + 5y *49.* 7x − 7/4 = 10

Section 1.2

1. x = 7 *3.* X = 8 *5.* X = 2 *7.* X = 9 *9.* X = 10 *11.* x = −51.5
13. X = −25 *15.* X = −38 *17.* X = 12 *19.* X = −62.4 *21.* R = G/H
23. G = N + D *25.* R = D/P *27.* I = M − P *29.* $P = A/(1 + i)^n$

Section 1.3

1. $100, $350 *3.* $480 *5.* $90, $270 *7.* $2,071.44 *9.* 3 hours 45 minutes
11. $94,000 *13.* 20 *15.* 18,000

Section 1.4

1. **a.** 21:42 or 21/42 = ½ = .5/1 or .5:1 **b.** 21:63 or .33:1 *3.* **a.** 150:100 or 1.5:1; **b.** 2:3;
c. 150:250 or .6:1 *5.* 175:800 or .22:1 *7.* 9 women *9.* 50,000:80,000:70,000 or 5:8:7
Sid earned $12,500
Ben earned $20,000
Hal earned $15,500
11. x = 18 *13.* P = 450 *15.* Y = 2 *17.* W = .5 *19.* **a.** b:20 = 4:5; b. b = 16
21. b:s = 4:9, b:$18,000 = 4:9 b = $8,000 *23.* b = $2,200 *25.* W = 105

CHAPTER 2

Section 2.1

1. .667 *3.* 148.141 *5.* .4 *7.* ½ *9.* ⁴/₅ *11.* ⅛ *13.* 30% *15.* 150%
17. 5% *19.* .06 *21.* .005 *23.* .20 *25.* 25% *27.* 68.8% *29.* 70% *31.* $6^{2/5}$
33. ¹/₂₀₀ *35.* .20, 20% *37.* .04, 4% *39.* 17.5, 1.750% *41.* .25 *43.* 65.769
45. 95.287 *47.* 119.19 *49.* ¹⁹/₂₅, ⅕, ¹/₂₅ *51.* ⁵/₂, 2 ½, ⁹²⁰/₃₆₈ *53.* $136.92
55. ⅓, $30,000

Section 2.2

1. 75 *3.* $6.93 *5.* 1,696.99 *7.* 16,000 *9.* 250 *11.* 30% *13.* 66.7%

15. 11%   *17.* 800   *19.* $340   *21.* 12.5%   *23.* 25%   *25.* 6.5%   *27.* 3,400

29. 800   *31.* $4,462.50   *33.* 1,600%   *35.* 7.5   *37.* 5,000   *39.* $1,000 (6% = $900)

41. 40   *43.* $8,880   *45.* 86.7%

Section 2.3

1. $.42 decrease, 31.8% decrease   *3.* 1,215, 18.8%   *5.* 3.45 decrease, 26.6% decrease

7. 65⅝, 3.4% increase   *9.* 16,250 decrease, 6.5% decrease   *11.* 100   *13.* 110.3

15. 141.2   *17.* $68.22   *19.* $1,000.00   *21.* 10%   *23.* 12.5%   *25.* 12.5% decrease

27. 50%   *29.* 140%   *31.* 157.6   *33.* $488.95   *35.* $33.55

CHAPTER 3

Section 3.1

1.

Class ages	f	Relative f	Percent %f
5–15	3	$\frac{3}{60}$ = .05	5%
15–25	6	$\frac{6}{60}$ = .10	10%
25–35	15	$\frac{15}{60}$ = .25	25%
35–45	20	$\frac{20}{60}$ = .33	33%
45–55	10	$\frac{10}{60} = \frac{1}{6}$ = .17	17%
55–65	3	$\frac{3}{60}$ = .05	5%
65–75	3	$\frac{3}{60}$ = .05	5%
	60	1.00	100%

3. **a.**

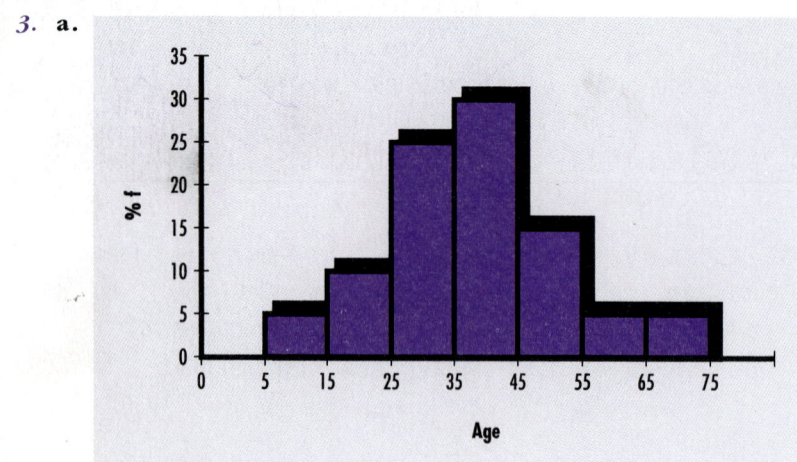

4. b.

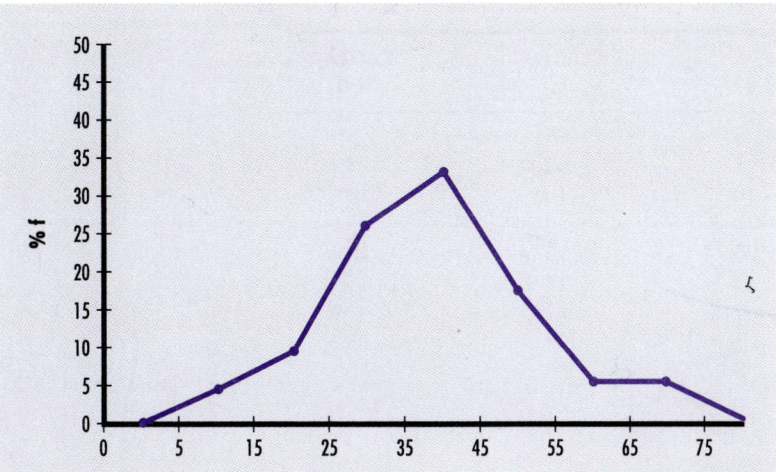

5.

Class	f
10–30	2
30–50	3
50–70	5
70–90	6
90–110	8
110–130	12
130–150	8
150–170	6
	50

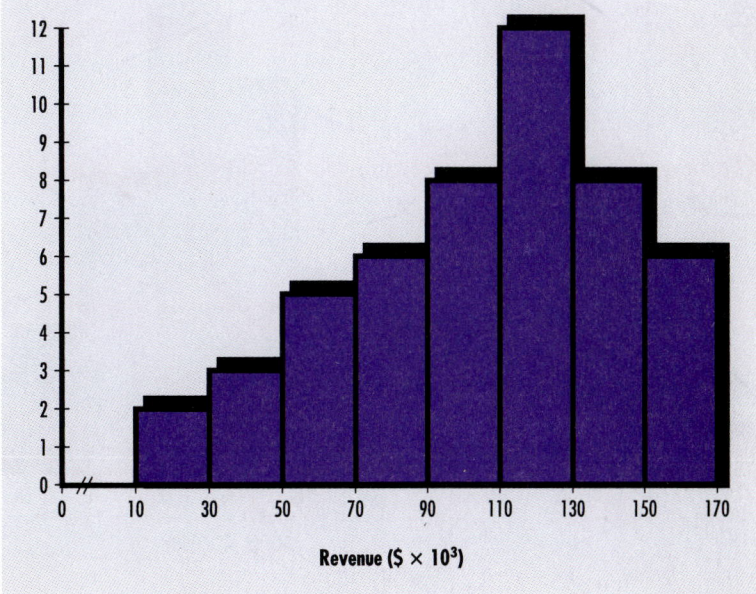

7.

Class	f	Relative f	Percent % f
10–20	5	.05	5%
20–30	5	.05	5%
30–40	10	.10	10%
40–50	20	.20	20%
50–60	25	.25	25%
60–70	15	.15	15%
70–80	7	.07	7%
80–90	5	.05	5%
90–100	8	.08	8%
	100	1.00	100%

9. a.

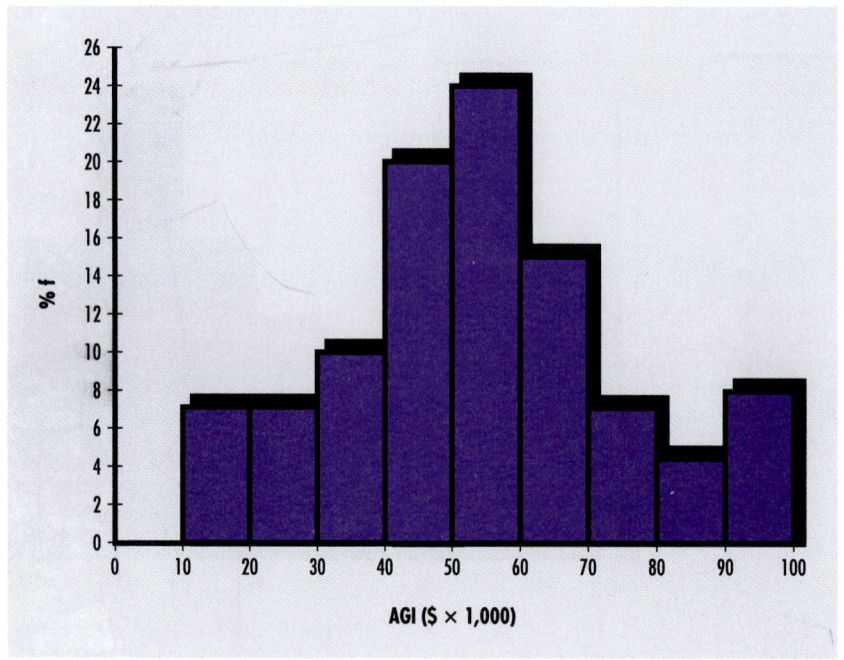

9. **b.**

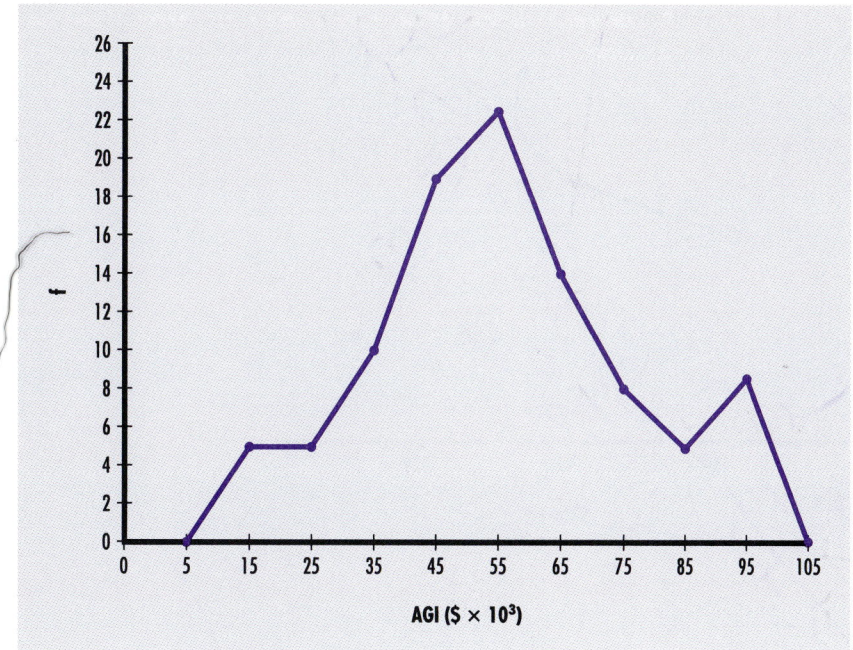

9. **c.**

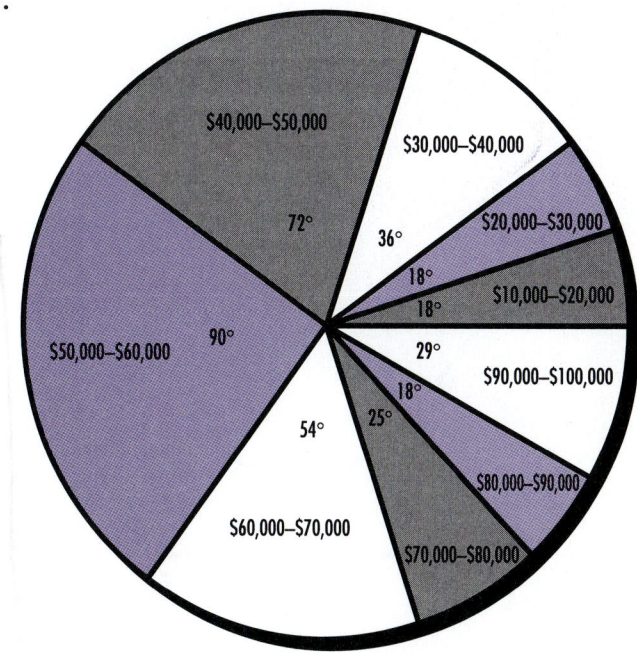

11.

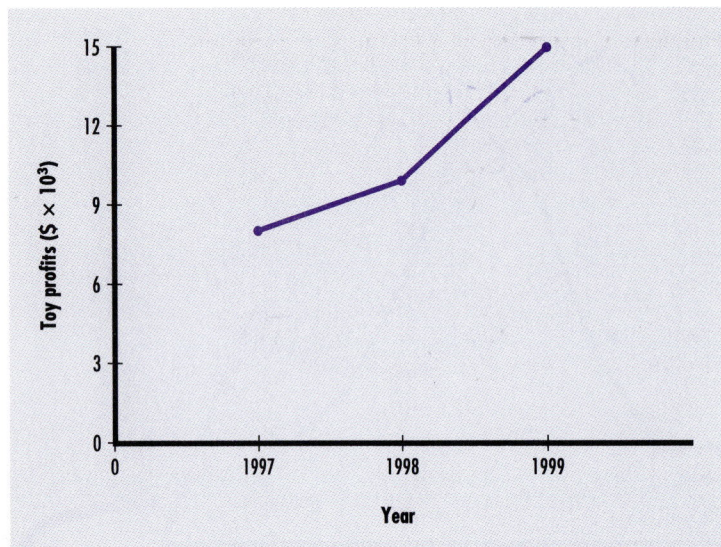

13.

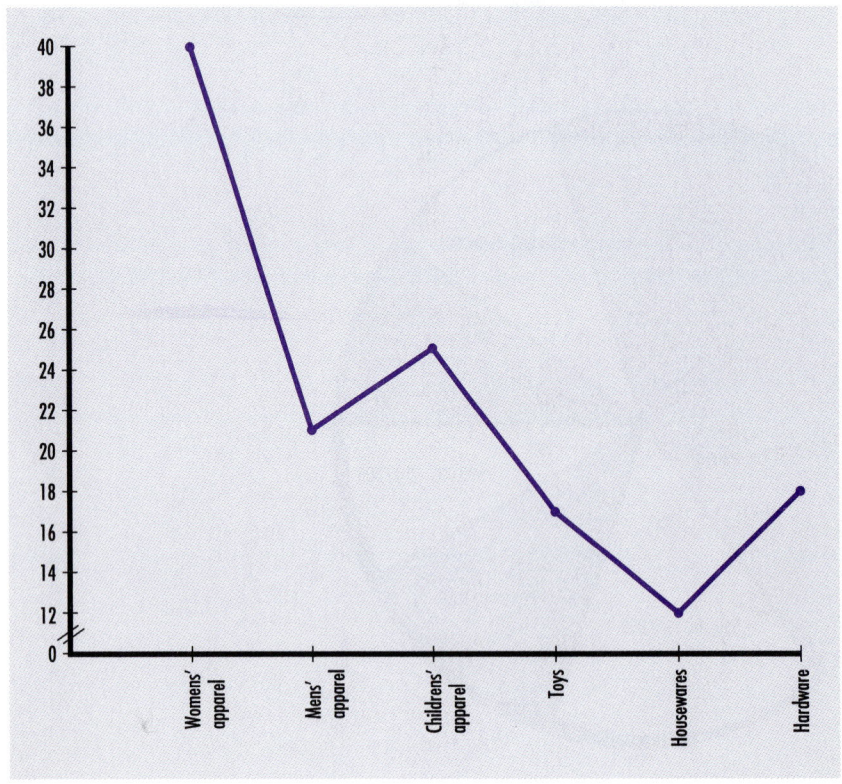

15.

1997 Department	Profit	% or Total	Degrees
Womens'	$30,000	.30	108°
Mens'	$15,000	.15	54°
Childrens'	$20,000	.20	72°
Toys	$10,000	.10	36°
Housewares	$10,000	.10	36°
Hardware	$15,000	.15	54°
	$100,000	1.00	360°

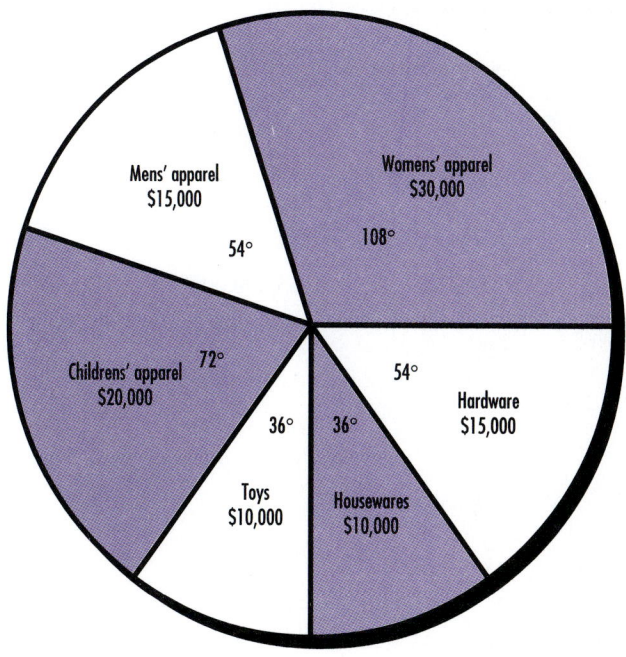

17.

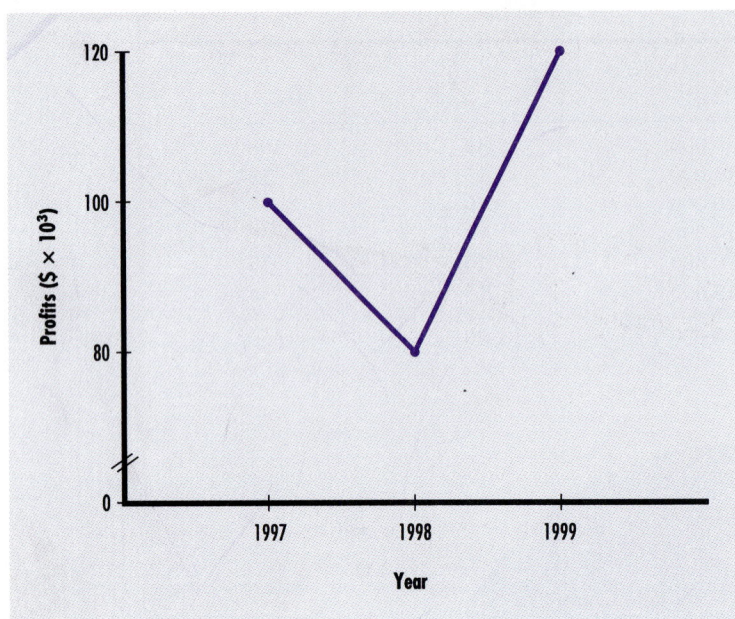

Section 3.2

1. **a.** 1.53; **b.** 1; **c.** 1   *3.* **a.** 150; **b.** 195; **c.** 200 and 220   *5.* **a.** $3,100,000; **b.** $800,000;
c. $800,000; **d.** mean; **e.** median or mode   *7.* 3.52, 3, 2   *9.* $63   *11.* 2.4   *13.* 2.06
15. 53.06

Section 3.3

1. 12, 17, 30, 5.48   *3.* 25, 10, 20.5, 4.53   *5.* 6, 1.5, 3.13   *7.* 29.6, 29   *9.* $V = 9, s = 3.58$
11. Yes   *13.* 76   *15.* 10.25   *17.* 7   *19.* $v = 4.386 \ s = 2.09$

Section 3.4

1. **a.** ¼; **b.** ¼; **c.** ½; **d.** ½; **e.** ¼   *3.* **a.** 13/30; **b.** 17/30   *5.* **a.** ¼; **b.** ¼; **c.** ¾; **d.** 0
7. **a.** 7/24; **b.** 5/24; **c.** 1/12; **d.** 0   *9.* .80   *11.* 0   *13.* .20   *15.* 90%   *17.* .52
19. .32   *21.* 0   *23.* 44%   *25.* 100%

CHAPTER 4

Section 4.1

1. $21,688   *3.* $31,795   *5.* $76,515   *7.* current liabilities   *9.* owner's equity
11. fixed asset   *13.* fixed asset   *15.* current asset   *17.* fixed asset   *19.* $39,570
21. $61,005   *23.* $25,800   *25.* $148.975

Section 4.2

1. $27,500, 9,150, 2,670   *3.* $4,625, 152,648, 72,142   *5.* $51,375, 42,625   *7.* $73,970, 36,520

9. $35,220   *11.* $170,130   *13.* $6,300   *15.* $4,265

Section 4.3

1. net purchases 2000, 44.7; 1999, 44.5

gross margin 2000, 55.9; 1999, 56.6

total operating expenses 2000, 25.6, 1999, 22.8

net income (2000), 15.1; (1999), 16.6

3. total current assets ($3,500), (4.6)%

total assets ($3,850), (1.5)%

total liabilities $2,100, 1.9%

total equity ($5,950), (4.1)%

Section 4.4

1. 2.75:1   *3.* .42:1   *5.* 84.9%   *7.* (2000) .91:1, (99) .88:1   *9.* (2000) .45:1, (99) .44:1

11. (99) 9.23 times, (2000) 8.29 times   *13.* (99) 23 days, (2000) 23.3 days   *15.* (99) 70.2%, (2000) 64.5%   *17.* $285,000   *19.* $108,857.14

CHAPTER 5

Section 5.1

1. $32.85   *3.* $32.50   *5.* $16.50   *7.* $170.85   *9.* $617.50   *11.* $1,337.50

13. $390.00   *15.* $137.50   *17.* $216.25   *19.* $1,952.00   *21.* $7.28   *23.* $43.50

25. $3.50   *27.* $2.58   *29.* $3.625, $.365   *31.* $5.00

Section 5.2

1. $225, $1,275   *3.* $286, $1,014   *5.* $1,406, $5,994   *7.* 33⅓%   *9.* 10%   *11.* 85%, .85   *13.* 78%, .78   *15.* 83%, .83   *17.* .80, $2,800   *19.* .70, $504   *21.* $300, $1,700, $170, $1,530, $470   *23.* $70, $280, $42, $238, $112   *25.* $165, $385, $19.25, $365.75, $184.25

27. 85/90, .765   *29.* 60/85/90, .459   *31.* .68, $2,040, $960   *33.* .57375, $516.38, $383.62

35. .7691, $6,992.08, $2,077.92   *37.* .72675, .27325   *39.* .66348, .33652   *41.* $168.75, $506.25

43. $39   *45.* $1,461.91   *47.* **a.** $2,275.88; **b.** $2,380   *49.* .2147, 21.5%

Section 5.3

1. June 10, June 30   *3.* April 10, April 30   *5.* Dec. 10, Dec. 30   *7.* May 15, June 29

9. Feb. 15, April 1   *11.* March 31, April 20   *13.* Oct. 20, Nov. 9   *15.* July 7, July 27

17. April 3, May 18   *19.* Aug. 30, Oct. 14   *21.* April 20, May 10   *23.* Oct. 25, Nov. 14

25. Dec. 25, Jan. 14   *27.* July 1, July 21   *29.* April 10, May 30   *31.* May 25, July 14

33. **a.** Oct. 12; **b.** Nov. 11; **c.** Dec. 11   *35.* **a.** $8,245; **b.** $8,500   *37.* $916.50

39. $765.33   *41.* $3,350   *43.* $2,516.64   *45.* $4,320

Section 5.4

1. $31.43   *3.* $215.08   *5.* $194.04   *7.* $845.94   *9.* **a.** $1,054; **b.** $9,946

11. $7,842.74   *13.* **a.** $550; **b.** $50   *15.* **a.** $83,475.18; **b.** $86,975.18; **c.** $4,375

CHAPTER 6

Section 6.1

1. $33.00, $5.00   *3.* $510.00, ($9.00)   *5.* ($300), ($625)   *7.* $12.50, 100%   *9.* $125.25,
35.8%   *11.* **a.** $600; **b.** $360; **c.** $240   *13.* $120.15   *15.* $4.76   *17.* $7.81

19. 53.1%   *21.* $4.52   *23.* $1.75   *25.* $239.15

Section 6.2

1. $24.00, $6.00   *3.* 16.74%, $164.50   *5.* $190.85, $284.85   *7.* $6.00, 50%, 33.3%

9. $174.23, $414.83, $72.4%   *11.* 6.7%   *13.* 33.4%   *15.* 67.6%   *17.* $36, $48

19. $5,000, $6,518.90   *21.* $650   *23.* $207.20, $77.70   *25.* $6.25   *27.* $650   *29.* 25%

31. 21.3%   *33.* $1,543.21

Section 6.3

1. $243.75, $304.69, 345   *3.* $8,160, $14,068.97, 7,705   *5.* $1,764.71   *7.* 115, $1,200

9. 76, $3.00   *11.* $8.71   *13.* $82.50, $206.25, $7.64   *15.* $1.70   *17.* $301.67

19. $1.78   *21.* $275.88

Section 6.4

1. $5.85, $13.65   *3.* 25%, $.20   *5.* $9.48, $6.32   *7.* 40%, $9.60   *9.* $15.00, $17.50, $2.50

11. $11.90, 0   *13.* $2,100   *15.* $267.75   *17.* 42.9%   *19.* $1,500, $7,125.00   *21.* $8.75

23. $.49, $.05   *25.* $296, 8%

CHAPTER 7

Section 7.1

1. $960   *3.* $382.50   *5.* $650   *7.* $920.31   *9.* $219.52   *11.* $12,960

13. $2,637.50   *15.* $11,625   *17.* **a.** $612.50; **b.** $8,112.50   *19.* $38,333.33

Section 7.2

1. $6,000   *3.* $2,250   *5.* $13,000   *7.* 7.5%   *9.* 10.5%   *11.* 2   *13.* 1   *15.* 1.5

17. $614.25   *19.* $201.67   *21.* 7.4%   *23.* $20,492.33   *25.* $2,666.67

Section 7.3

1. $427.40   *3.* $407.93   *5.* $98.19   *7.* $99.56   *9.* $233.33   *11.* 90   *13.* 184
15. 97   *17.* Dec. 23   *19.* Dec. 15   *21.* $292.60   *23.* $1,273.42   *25.* $1,017.12
27. **a.** August 30;   **b.** $3147.95

Section 7.4

1. $I = \$100$, $Fv = \$1,100$ *3.* $I = \$1,397.25$, $Fv = \$4,097.25$ *5.* $T = 4$ yrs. $I = \$2,880$
7. $R = 10\%$, $I = \$800$ *9.* $R = 8\%$, $I = b\$20$ *11.* $P = \$4,000$, $Fv = \$5,260$ *13.* $R = 12\%$,
$I = \$6,720$ *15.* $Fv = \$575$ *17.* $Pv = \$7,200$, $Fv = \$7,840$, $I = \$3,840$ *19.* $Pv = \$3,970$

CHAPTER 8

Section 8.1

1. $1,311.75, $9,688.25   *3.* $36, $864   *5.* $1,848, $10,152   *7.* $550, $1,650   *9.* $425.25,
$5,874.75   *11.* $11,428.57   *13.* $6,944.44   *15.* $8,750   *17.* $34,736.84   *19.* $20,800,
30.77%   *21.* $25,722.67, 13.28%   *23.* $20,152, 15.72%   *25.* $16,121.67, 16.34%
27. **a.** $2,291.25;   **b.** $21,208.75;   **c.** $23,500   *29.* 10%

Section 8.2

1. $16,500   *3.* $1,500   *5.* $R = 11.11\%$ *7.* $I = \$1,200$, $M = \$5,200$ *9.* 17.57%
11. $M = \$4,180$ *13.* 12.11%   *15.* 12.11%   *17.* $I = \$14,102.56$ *19.* $13,000

Section 8.3

1. $13,368.75   *3.* **a.** $21,137.12;   **b.** 13.31%   *5.* $41,221.47   *7.* $71,563.54
9. $76,427.08

Section 8.4

1. $7,800   *3.* $2,736   *5.* $6,862.50   *7.* $1,059.80   *9.* $6,918   *11.* $148.06
13. $1,314.08   *15.* $1,693.47   *17.* $207.58   *19.* $9,000   *21.* $15,300   *23.* $553.23
25. $2,381.75

CHAPTER 9

Section 9.1

1. .04; 18   *3.* .005; 120   *5.* .025; 60   *7.* $8,671.77; $1,671.77   *9.* $1,250.61; $250.61
11. $10,509.45, $509.45   *13.* $9,070.09   *15.* $8,954.24   *17.* $9,096.99   *19.* $9,030.56
21. $17,339.85   *23.* $15,643.10   *25.* $14,644.40   *27.* $1,179.22   *29.* $85,223.44
31. $11,716.59   *33.* $50,789.38

Section 9.2

1. $1,837.56   *3.* **a.** $1,346.86; **b.** $1,610.32; **c.** $1,814.02   *5.* **a.** $6,150; **b.** $14,076.53;
c. $39,629.73   *7.* $49,215.84   *9.* $148,450.56

Section 9.3

1. $804.96   *3.* $5,759.06   *5.* $69,907.41   *7.* $2,768.38   *9.* $5,254.52   *11.* $4,312.62
13. $14,851.49   *15.* $13,806.46   *17.* $4,610.50   *19.* $334.11   *21.* $2,280.57
23. $34,857.94   *25.* $5,836.53   *27.* $4,852.66   *29.* $14,948.19   *31.* $28.913.52

Section 9.4

1. 18.81%   *3.* 19.56%   *5.* 15.87%   *7.* 19.68%   *9.* 10.38%   *11.* 11.30%   *13.* 6.17%
15. 8.33%, $832.78

CHAPTER 10

Section 10.1

1. $3,038.59   *3.* $14,578.37   *5.* $5,900.42   *7.* $9,721.56   *9.* $228,102.00
11. **a.** $244,077.61; **b.** $104,077.61   *13.* **a.** $255,446.40; **b.** $55,446.40   *15.* $1,397,983.60
17. $37,278.46   *19.* $88,485.73   *21.* $155,001.52   *23.* **a.** $34,850.86; **b.** $149,872.22
25. $21,563.50

Section 10.2

1. $777.93   *3.* $11,348.33   *5.* $6,387.05   *7.* $101,147.17   *9.* $174,313.93
11. $44,378.12   *13.* $341,942.87   *15.* $45,409.71   *17.* $185,000   *19.* $7,594.79

Section 10.3

1. $113.06   *3.* $327.34   *5.* $153.82   *7.* $160.95   *9.* $1,131.74   *11.* $3,660.66
13. $7,761.96   *15.* $786.96

Section 10.4

1. $2,590.10   *3.* $5,753.60   *5.* $378.15   *7.* $1,000.98   *9.* $7,173.50   *11.* $360,820
13. $1,497.50   *15.* $223,400

CHAPTER 11

Section 11.1

1. $439.78   *3.* $2.32, $167.42   *5.* $1.21, $0.00   *7.* **a.** $201.70; **b.** $274.90; **c.** $364.50
9. $3.78   *11.* $464.64   *13.* **a.** $600.98; **b.** $712.63; **c.** $983.51   *15.* $12.75   *17.* $279.57

Section 11.2

1. $700   *3.* $120   *5.* $550   *7.* $4,680, $195   *9.* $6,080, $168.89   *11.* **a.** $891;
b. $89.10; **c.** $245.03   *13.* $108.10   *15.* $300   *17.* $864   *19.* $787.50

Section 11.3

1. $99.64   *3.* $44.65   *5.* $334.37   *7.* $642.73   *9.* **a.** $222.44; **b.** $3,346.40
11. **a.** $149.77; **b.** $594.48   *13.* **a.** $449.66; **b.** $86,877.60   *15.* **a.** $487.17; **b.** $48,920.80
17. **a.** $95,000; **b.** $977.18; **c.** $256,785.51   *19.* **a.** $110,000; **b.** $920.08; **c.** $220,820.18
21. $550.55   *23.* $514.30   *25.* $53,016   *27.* **a.** $372.54; **b.** $37,409.60
29. **a.** $22,827.60; **b.** $42,960; **c.** $20,132.40

Section 11.4

1. Interest rate per quarter $= i = \dfrac{.08}{4} = .02$

n	Principal	Interest	Payment	Principal payment	New principal	
1	$4,500.00	$90.00	$614.30	$524.30	$3,975.70	
2	3,975.70	79.51	614.30	534.79	3,440.91	
3	3,440.91	68.82	614.30	545.48	2,895.43	
4	2,895.43	57.91	614.30	556.39	2,339.04	
5	2,339.04	46.78	614.30	567.52	1,771.52	
6	1,771.52	35.43	614.30	578.87	1,192.65	
7	1,192.65	23.85	614.30	590.45	602.20	
8	602.20	12.04	614.30	602.26	−.06	(rounding error)

3. Payment $= \left(F, i\% = \dfrac{10\%}{2} = 5\%, n = 2 \times 3 = 6 \right) \times \$2,400 = .197018 \times \$2,400 = \$472.84$

n	Principal	Interest	Payment	Principal payment	New principal	
1	$2,400.00	$120.00	$472.84	$352.84	$2,047.16	
2	2,047.16	102.36	472.84	370.48	1,676.68	
3	1,676.68	83.83	472.84	389.01	1,287.67	
4	1,287.67	64.38	472.84	408.46	879.21	
5	879.21	43.96	472.84	428.88	450.33	
6	450.33	22.52	472.84	450.32	.01	(rounding error)

5. Payment $= \left(F, i\% = \dfrac{8\%}{4} = 2\%, n = 4 \times 2 = 8 \right) \times \$12{,}000 = .136510 \times \$12{,}000 = \$1{,}638.12$

n	Principal	Interest	Payment	Principal payment	New principal
1	$12,000.00	$240.00	$1,638.12	$1,398.12	$10,601.88
2	10,601.88	212.04	1,638.12	1,426.08	9,175.80
3	9,175.80	183.52	1,638.12	1,454.60	7,721.20
4	7,721.20	154.42	1,638.12	1,483.70	6,237.50
5	6,237.50	124.75	1,638.12	1,513.37	4,724.13
6	4,724.13	94.48	1,638.12	1,543.64	3,180.49
7	3,180.49	63.61	1,638.12	1,574.51	1,605.98
8	1,605.98	32.12	1,638.12	1,606.00	−.02 (rounding error)

7. $8,844.74, $6,712.91, $3,556.60, $2,417.93   **9.** $10,583.67, $5,607.39, $0   **11.** $953.64
13. $1,595.30   **15.** $8,827.20

Section 11.5

1. **a.** $1,396.52; **b.** $653.32; **c.** $176.55   **3.** **a.** $4,237.41; **b.** $2,490.41; **c.** $1,165.06;
d. $314.84   **5.** **a.** $232.40; **b.** $109.09; **c.** $30.24   **7.** **a.** $1,183.81; **b.** $560.01; **c.** $371.27;
d. $157.72   **9.** $43.32   **11.** $91.17   **13.** $59.44   **15.** **a.** $61.10; **b.** $35.64; **c.** $16.97;
d. $1.70   **17.** **a.** $190.93; **b.** $89.10; **c.** $25.46   **19.** **a.** $13,511.03; **b.** $4,488.97

Section 11.6

1. 26%   **3.** 19%   **5.** 10%   **7.** 11%   **9.** 27%   **11.** 22%   **13.** **a.** $2,513.10;
b. $413.10; **c.** 28%   **15.** **a.** $5,200; **b.** $2,600; **c.** 17.25%

CHAPTER 12

Section 12.2

1. $9.50   **3.** $16.40   **5.** $21.00   **7.** $42.61   **9.** $1,345.65   **11.** pay to the order of
Sutter's Hardware   **13.** for deposit only   **15.** $2,382.22   **17.** $77.28   **19.** $87.12 (Visa),
$64.04 (MasterCard)   **21.** $4.33 (Visa) + $15.56 (MasterCard) = $19.89   **23.** $9.40   **25.** $8.23
27. $27.64   **29.** $506.86

Section 12.3

1. $791.95   **3.** $25,309.23   **5.** $1,626.94   **7.** $2,923.75   **9.** $419.95   **11.** $1,478.70

CHAPTER 13

Section 13.1

1. $2.50; $62.50   *3.* $.40; $40   *5.* $4; $80   *7.* $10.50; $20; $280,300; $10.125, $1.375
9. $99.375; $9.25; $42,000; $98; −$1.375   *11.* $60.125; $1.20; $117,300; $55.75; $4.75
13. $73.25; $.04; $2,293,400; $72.625; $8   *15.* $17.25; $.32; $5,843,100; $16.50; $.50
17. $2,487.50; $74.625; $2,562.125   *19.* $36,550; $1,096.50; $39,646.50   *21.* $6,111; $6,300; $189
23. $21,370.31; $22,031.25; $660.94   *25.* $14,259; $14,700; $441   *27.* $109.50
29. $11,310 gain   *31.* −$2,230.50   *33.* −$1,067.48   *35.* −$1,555.20   *37.* **a.** $51;
b. $108,171;   **c.** $55.75;   **d.** $115,904.25;   **e.** $7,733.25   *39.* −$1,202

Section 13.2

1. **a.** 1999;   **b.** $93.75   *3.* **a.** 2009;   **b.** $93.75   *5.* **a.** 2005;   **b.** $72.50   *7.* **a.** 2011;
b. $57.5   *9.* **a.** 2015;   **b.** $121.25   *11.* $93.75; $1,030; 9.1%   *13.* $93.75; $1,050; 8.9%
15. $72.50; $625; 11.6%   *17.* $57.50; $850; 6.8%   *19.* $121.25; $1,010; 12%   *21.* $6.88;
$1,006.88; $966.88   *23.* $15.63; $5,278.15; $5,078.15   *25.* $38.33; $8,546.64; $8,226.64
27. $6.56, $12,918.72; $12,438.72   *29.* $34.90; $1,054.90; $1,014.90   *31.* $10,137.50; $9,937.50
33. $43,161.60; $42,121.60   *35.* $5,562; $5,262; $1,549.50

Section 13.3

1. $23.77   *3.* $0.26   *5.* B rating = second 20% group   *7.* $7.83   *9.* $.31, $6.56
11. 2.1%, $0.38   *13.* $62.35, .7%   *15.* $6.15   *17.* $14,735   *19.* No, $4.83
21. $26.40 loss

Section 13.4

1. $204,000; $258,000; $138,000   *3.* $12; $3,600; $11,400; $16.29   *5.* $625; $250; $375; $1,250
7. $2,400; $1,600; $800; $3,200   *9.* $4,545.45; $2,272.73; $1,363.64; $1,818.18   *11.* $68,832.60;
$103,248.89; $151,431.71; $105,933.37; $70,533.41   *13.* $9,263.16; $5,789.47; $6,947.37   *15.* $8.25;
$8,250; $71,750; $8.97   *17.* $335; $165   *19.* $148.30; $224.44; $296.59; $333.67   *21.* $59.27;
$7.73   *23.* $2,517.98; $7,553.96; $4,028.78; $6,043.16; $503.60; $352.52   *25.* $498.20; $195.68;
$124.40; $181.70; $237.61; $167.73; $321.47; $62.90; $237.61; $181.70

CHAPTER 14

Section 14.1

1. $3,000   *3.* $6,000   *5.* $19,170.83   *7.* $5,240; $10,480; $20,520   *9.* $5,240; $20,960;
$10,040   *11.* $1,500; $8,800   *13.* $11,000; $342,000   *15.* $1,080 annual depreciation expense

Section 14.2

1. $4,250   *3.* $4,196.88   *5.* $9,025   *7.* $5,880   *9.* $2,881.20   *11.* $4,800; $7,200
13. $71,000; $60,350   *15.* Book value yr 1, $7,250, year 3, $2,832.03

Section 14.3

1. 3/6   *3.* 5/55   *5.* 9/210   *7.* $2,560   *9.* $12,000   *11.* $8,000, $3,200, $3,200, $8,800
13. $8,000, $1,600, $7,200, $4,800   *15.* **a.** $5,400; **b.** $23,100   *17.* year 1 book value $2,400; year
3 book value $720; year 5 book value $0   *19.* **a.** $16,800; **b.** $62,400; **c.** $12,200

Section 14.4

1. $.089; $292.50   *3.* $.856; $5,521.20   *5.* $.288; $4,883.60   *7.* $16,461.55   *9.* $6,786.57
11. $945; $5,775   *13.* $1,620; $5,100   *15.* $2,787.20   *17.* $66,921.55   *19.* year 1 book
value $28,890; year 2 book value $12,255.20

Section 14.5

1. 5; $19.20   *3.* 3; $44.45   *5.* 5; $11.52   *7.* 20; $4.462   *9.* $18,500; 5; .20; $3,700; $14,800
11. $85,250; 20; .03750; $3,197; $82,053   *13.* $125,000; 15; 5; $6,250; $118,750   *15.* year 1 book
value $41,655; year 3 book value $21,253; year 5 book value $10,843; year 7 book value $2,168
17. $49,500 cost recovered; year 1 book value $39,600; year 3 book value $14,256; year 5 book value
$2,852   *19.* year 1 book value $108,000   *21.* year 1 book value $332,500   *23.* year 1 book
value $315,000; year 3 book value $201,600; year 5 book value $129,010; year 7 book value $80,290; year 9
book value $34,405   *25.* year 1 book value $842,129; year 3 book value $788,153; year 5 book value
$734,177

CHAPTER 15

Section 15.1

1. $56,925   *3.* $113,620   *5.* $478,000   *7.* $1,350   *9.* $800   *11.* $11,125
13. **a.** $165; **b.** $290; **c.** $383.33; **d.** $11,125   *15.* $1,325

Section 15.2

1. $25,550   *3.* $13,957   *5.* **a.** $800; **b.** $780   *7.* **a.** $12.25; **b.** $14.75; **c.** $10
9. **a.** $1,120,000; **b.** $1,425,000; **c.** $813,000   *11.* **a.** $3,300; **b.** $3,600   *13.* **a.** $4,300;
b. $3,800   *15.* **a.** $1,092; **b.** $1,105

Section 15.3

1. .725   *3.* .60   *5.* $43,500   *7.* $306,000   *9.* $91,000   *11.* $140,000   *13.* $9,000
15. $30,000

CHAPTER 16

Section 16.1

1. 37, $385.54   *3.* 35, $337.75   *5.* 24, $213.60   *7.* 40, 6.5, $12.15, $324, $78.98, $402.98
9. 40, 8, $10.20, $272, $81.60, $353.60   *11.* 40, 6.75, $11.40, $304, $76.95, $380.95   *13.* 4,686,
$9.22, $13.83, $405.81, $55.34, $461.15   *15.* 484, $10.55, $15.82, $421.84, 0, $421.84   *17.* 1,316,
$11.76, $17.63, $505.50, $52.90, $558.40   *19.* $325.80, $11.40, $314.40   *21.* $263.30, 0, $263.30
23. $442.05   *25.* $522.10   *27.* $468.20   *29.* $92.24

Section 16.2

1. $400, $433.33, $866.67, $10,400   *3.* $322.50, $698.75, $1,397.50, $16,770   *5.* $784.62,
$1,569.23, $1,700, ($40,800)   *7.* $51,225, $2,254.63   *9.* $42,050, $1,698.00   *11.* $22,005,
$20,805, $338.08, $550, $888.08   *13.* $39,270, $38,070, $618.64, $550, $1,168.64   *15.* $460.89,
$507.50   *17.* $2,925.25   *19.* $2,340.81

Section 16.3

1. $2.26, $9.68   *3.* $6.10, $26.09   *5.* $27.90, $6.53   *7.* $136.40, $45.60   *9.* $23.00
11. $64.00   *13.* $33.00   *15.* $28.35   *17.* $246.37   *19.* $118.91   *21.* $52, $14.65,
$62.65, $734.30   *23.* $171, $17.99, $76.93, $812.43   *25.* $17.68, $5.77, $24.65, $321.10
27. $6.23, $7.67, $32.80, $416.77   *29.* **a.** $94.29;   **b.** $22.05   *31.* $44   *33.* **a.** $490.41;
b. $201.50; **c.** $47.13   *35.* $301.87

Section 16.4

1. $414.62, $96.97, $53.50, $280.87   *3.* $239.64, $56.05, $30.92, $162.34   *5.* $325.33, $76.09,
$2.50, $13.13   *7.* $247.39, $57.86, $25.08, $131.66   *9.* **a.** $6,200;   **b.** $1,450;   **c.** $400;   **d.** $2,500
11. $88.90 ($62.00 + $14.50 + $12.40)   *13.* **a.** $2,212.50;   **b.** $32.08;   **c.** $137.18;   **d.** $17.70;
e. $95.14   *15.* **a.** $1,404.43;   **b.** $328.43;   **c.** $41.20;   **d.** $180.24;   **e.** $8,852.01

CHAPTER 17

Section 17.1

1. $30.60; $.92; $31.52   *3.* $7.35; $.51; $7.86   *5.* $97.44; $5.85; $103.29   *7.* $342.00; $345.42;
$352.33   *9.* $1,165.50; $1,182.98; $1,242.13   *11.* $248.40   *13.* $9.62   *15.* $10,710
17. $11,587.50   *19.* **a.** $1,380 Fed;   **b.** $937.50 state;   **c.** $7,125

Section 17.2

1. $3,000; $11,250; $60,000; $73,000

3.

Hundreds units	Thousands units
25	2.5
45	4.5
90	9
150	15
380	38
3640	364
1280	128
13000	1300

5. $4,250; $1,621.50; $1,725; $3,840; $4,550; $442.50   *7.*   **a.** $750,000;   **b.** 002;   **c.** $115,000;   **d.** $230
9.   **a.** $108,000,000;   **b.** .138888;   **c.** $175,000;   **d.** $2,430.56

Section 17.3

1. $10,397   *3.* $65,981   *5.* $79,916   *7.* $37,819; 5,674   *9.* $17,448; 2,614
11. $5,840; owe $566   *13.* $2,816; owe $324   *15.* $11,678; owe $868   *17.* $53,080
19. $580.90 tax owed   *21.* $882 refund due   *23.* $10,530   *25.* $4,324

CHAPTER 18

Section 18.1

1. $51.20; $964; $1,456; $2,182.40   *3.* $47.19; $827.97; $1,278.42; $2,089.23   *5.* $16,160
7. $130; $38,000   *9.* $115; $38,000

Section 18.2

1. $5,439; $7,938; $10,437   *3.* $971.25; $1,417.50; $1,863.75   *5.* $280,000; $160,714.29
7. $480,000; $400,000   *9.* $100,000; $40,000

Section 18.3

1. $13,000   *3.* $8,200

Glossary

Absolute Loss A financial condition that occurs when the reduced price is less than the actual cost of an item.

Accelerated Methods of Depreciation Depreciation methods that have a rate of depreciation greater than the straight line rate, such as declining balance, sum of years digits, and MACRS.

Account Form A format for preparing the bank reconciliation statement that lists adjustments to the bank balance and book balance separately.

Account Purchase A type of purchase arranged by a commission agent for the buyer.

Account Sale A type of sale arranged by a commission agent for the seller.

Accrued Interest The amount of interest the bond has earned on a given date but will not be paid until a future date.

Accumulated Depreciation The total depreciation of an asset as of a given date.

Acid Test Ratio A measure of the immediate debt-paying ability of a company by comparing quick assets to current liabilities.

Actuary A mathematician who computes insurance risk and premiums.

Adjusted Gross Income (AGI) The amount of income reported on a tax return that is the difference between the total income and total adjustments.

Aliquot Part Any number that will divide evenly into another number.

Amortization A present value annuity application that determines how large each equal payment must be to liquidate a lump sum of money in a given amount of time.

Amortization of a Loan The process of liquidating a debt with a series of regular payments or installments.

Annual Percentage Rate The rate of interest stated on a yearly basis and calculated on the unpaid principal balance of each period of loan payments.

Annuity Due An annuity that increases in value with regular periodic payments plus the interest on the principal of the annuity; the payments are made at the beginning of each period.

Annuity A series of periodic payments into a compound interest account or out of the account that add to or subtract from the principal of the account.

Annuity Life Insurance Life insurance that pays a regular periodic payment to the insured for the remainder of his/her life.

Assessment Rate The rate used to determine the assessment value of property for tax purposes.

Assessment Value The taxable value of real property that is found by multiplying the fair market value by the assessment rate.

Asset Turnover Ratio A financial ratio that identifies the number of times assets have turned over.

Assets Anything of value owned by a firm such as its cash, buildings, and land.

Associative Principle $A + (B + C) = (A + B) + C$ for addition, and $A(BC) = (AB)C$ for multiplication.

Automobile Insurance Insurance providing protection to offset financial losses due to liability and collision claims involving the insured's vehicle.

Automatic Teller Machine (ATM) A computerized machine that allows depositors to conveniently withdraw cash, make deposits, or transfer funds any time.

Average Cost Method A method used to calculate the value of ending inventory based on the average cost per unit of each item remaining in the inventory.

Average Daily Balance The average of the daily balances of a revolving charge account for the number of days in the billing period.

Back-end-load The commission is paid by the investor when the mutual fund shares are sold.

Balance of Payments The flow of money into or out of a country resulting from the buying and selling of goods, services, and currency.

Balance of Trade The total economic value of all products that a country imports, less the total economic value of all the products that it imports.

Balance Sheet   A report that identifies the financial position of a business at a specific point in time.

Bank Discount   The amount of interest charged by a bank on a note that is deducted from the maturity amount to determine the proceeds.

Bank Statement   A monthly record provided by banks to depositors that lists all transactions to the depositor's checking account.

Bar Graph   A graphical representation of data using bars to indicate the frequencies of each class.

Base   An amount that represents the total or whole and always equals 100%.

Beneficiary   The person designated to receive the insurance award (money).

Blank Endorsement   An endorsement that requires only the signature of the person or firm appearing on the check as the payee.

Bond   An interest-bearing loan that an investor makes to a company.

Book Value   The asset's net value after deducting the accumulated depreciation from the asset's total cost when placed in service.

Breakeven Point   The point where the reduced price is sufficient to cover the item's cost and operating expenses.

Business Insurance   Insurance to protect the assets and income of a business against the risks of accidental loss.

Cancelled Check   Check that has been paid by the bank and listed on the bank statement by date, number, and amount.

Capital   The owner's claim or investment in the business as reported in the balance sheet of a sole proprietorship or a partnership.

Capital Gain/Loss   The profit earned or the money lost from the sale of securities and real estate.

Cash Discount   A discount based on the net price of a sales invoice that encourages buyers to pay invoices before the net payment date.

Cash Surrender Value   The amount earned and paid to the insured upon surrendering a life insurance policy.

Certified Check   A person or business check that is certified by a bank that guarantees the payee payment.

Chain Discount   A multiple trade discount quoted as 5% less 10%, less 5%.

Chargebacks   An amount of money that is considered a penalty and, when applied, reduces an employee's earnings.

Check   A written order from an individual or a company to its bank to pay a designated party a specific amount of money kept on deposit.

Check Register   The individual checking account record used to record all information pertaining to the checking account.

Check Stub   The record attached to a check that is used to record the information written on the check.

Circle Graph   A graphical representation of data using sectors of the circle to indicate the proportion each class is of the whole.

Class Life   The classification of an asset and the time period over which the asset's total cost will be recovered under MACRS depreciation.

Closed-Ended Credit   Credit extended to a borrower that the borrower uses to purchase one specific item or service, such as an automobile, or auto insurance, etc., and pays off with a set number of installment payments.

Coinsurance Clause   A fire insurance clause that states the minimum amount a building must be insured for so that a claim will be fully paid.

Collateral   Something of value that is used as security for a loan and that will be surrendered to the lender if the borrower defaults on the loan.

Commission   A fee charged by a commission agent for the services required to arrange an account sale or an account purchase; an incentive wage based on the amount of net sales generated during a pay period.

Commission Agent   An individual or a business that is used by manufacturers and wholesalers to buy or sell their merchandise.

Common Stock   A stock that represents the largest portion of a corporation's ownership but does not share the privileges of preferred stock.

Commutative Principle   A mathematical principle which states that the order of addition or multiplication of two numbers gives the same results, $A + B = B + A$, and $AB = BA$.

Complement A number that represents the difference between a number and that number's power of 100 $(10 - 4 = 6)$

Compound Amount The principal plus interest, where the principal includes previous interest already earned.

Compound Interest Interest calculated on principal plus previous interest earned.

Consignee An individual or business (commission agent) that accepts merchandise on consignment for sale.

Consignor Manufacturers or wholesalers that sell their merchandise on consignment with the assistance of a consignee.

Constant A symbol that has a fixed value.

Consumer Price Index A price index compiled by the U.S. Bureau of Labor that includes a representative sample of goods purchased by households in the United States.

Cost of Goods Sold The section of the income statement that determines the cost of inventory that produced sales.

Current Market Value The value of an asset or stock in the marketplace as determined by the principles of supply and demand.

Current Ratio Identifies the amount of current assets available to pay off each $1 of current debt.

Current Value The value of an asset at the present when compared to the value of the asset at a previous point in time.

Current Yield The percent of dividends received based on the stocks current closing price.

Daily Balance The previous day's balance of a revolving charge account plus new purchases and cash advances, minus payments and credits credited to the account on that day.

Data Numerical information.

Decimal Equivalent The decimal value of a fraction a/b, after conversion.

Decimal Fraction A fraction that has a denominator of 1, 10, 100, 1,000, etc.

Decimal Number Any number written using the proper selection and combination of digits 0 through 9.

Decimal System A number system using the digits 0 through 9 also referred to as the Hindu-Arabic numeration system.

Declining-Balance Depreciation An accelerated method of depreciation using 200%, 175%, 150%, or 125% of the straight-line rate.

Decreasing Term Life Insurance Life insurance for which the premium remains the same throughout the life of the policy, but the face value decreases at the end of each specified term.

Deductions Allowable expenditures reported on the tax return made by a taxpayer during the tax period.

Deposit An amount of money (cash and checks) that increases the balance of a checking account.

Deposit Slip The official record of a deposit transaction listing currency, cash, and checks.

Depreciable Amount The amount of the asset to be depreciated (total cost-salvage value) over its useful life.

Depreciation The process of decreasing the cost of an asset over the useful life of the asset.

Depreciation Expense The amount of the assets cost that is charged to the revenues earned in a given period.

Depreciation Schedule A table showing the periodic calculations and amounts related to the depreciation of an asset over its useful life.

De-valuation Decreasing the exchange rate, which reduces the cost of goods and services purchased by other countries.

Difference An amount determined when one number is subtracted from another number.

Differential Piece-Rate A wage plan that incorporates a premium rate for units produced above established standards.

Direct Variation A mathematical equation that results in the dependent variable increasing when the independent variable increases, and decreasing when the independent variable decreases, eg. $Y = mx$.

Disability Income Insurance   Insurance to cover loss of earnings while a wage earner is unable to work due to illness or injury.

Discount Fee   A fee banks charge businesses to make funds immediately available for the net amount of credit card sales deposits.

Discounting   The process of reducing the price of an item on a note to receive a lesser amount earlier than the established time period.

Discount Price   The price of a bond when its market price is less than its face value.

Discount Rate   The interest rate that is used to determine the amount of discount to be deducted from the maturity value of a discounted loan.

Distributive Principle   A mathematical principle that states that when a number is multiplied by a sum it should be multiplied by each and then the products added. A(B + C) = AB + AC

Distributor   A wholesale business that stores and promotes the sale of goods to a retail business or directly to the consumer.

Dividends   A portion of a company's profits paid to stockholders in the form of cash or additional stock.

Double Time   A premium wage paid for holidays and weekend work that is twice the regular hourly rate.

Drawer   An individual or a company authorizing its bank to pay checks bearing their signatures.

Effective Rate   The simple interest rate that you would need to receive so that the simple interest earned in one year would equal the compound interest earned in the same time.

Elective Deductions   Amounts withheld from an employee's gross earnings that have been authorized by the employee.

Electronic Banking   A service available to depositors that uses computer technology to transfer funds into and out of checking accounts.

Employee Earnings Record   A payroll record that lists all payroll data for a single employee on a quarterly basis.

Employer's Annual Federal Unemployment Tax   A report (Form 940) that is sent to the IRS and includes the annual payment of unemployment taxes.

Employer's Quarterly Federal   A report (Form 941) that is sent to the IRS and includes quarterly payments of FICA.

Employer's Tax Guide   A booklet (circular E) provided by the IRS that contains tables employers use to determine FWT and FICA taxes.

End-of-Month Dating (EOM)   Cash discount terms that begin on the end of the month the invoice is dated. The terms "proximo" and "prox" are sometimes used instead of EOM.

Endorsement   A signature or instructions with a signature written on the back of a check legally transferring ownership.

Endowment Life Insurance   Life insurance that pays an amount of money to the insured at the end of a specified time if the insured is still living at that time.

Equation   A numeric sentence that has an equal sign.

Exact Interest   The simple interest determined by dividing the number of days by 365.

Exchange rate   The rate at which the currency of one country can be exchanged for the currency of another.

Excise tax   A type of tax levied on the sale or manufacture of specific goods and services generally by the federal government.

Exempt   A classification given to employees who are compensated for their work, not the number of hours worked.

Exemptions   The number of people in a household that can be claimed to reduce personal income taxes.

Exponents   Superscripts placed to the right and above a number to indicate a repeated multiplication of a number by itself.

Extension Amount   The quantity value of items sold under each description and unit price.

Extra Dating   A cash discount (2/10−30X) that allows the buyer an extra 30 days or a total of 40 days from the date of the invoice.

Face Value   The amount of money paid by a life insurance policy to a beneficiary upon the death of the insured person.

Face Value of Bond The principal value of the bond usually expressed in thousand dollar units.

Fair Labor Standards Act A federal law that defines work conditions, minimum wage, regular work hours, and overtime hours.

Fair Market Value The amount of money real property could be sold for on the real estate market.

Federal Insurance Contributions Act A federal act that requires employers to withhold monies from employee earnings for Social Security and Medicare benefits.

Federal Unemployment Tax Act The federal act (FUTA) that identifies all employer responsibilities concerning unemployment insurance.

Federal Withholding Tax (FWT) A payroll deduction required by the federal government that is applied to the employee's federal income tax.

FICA Tax A federal tax paid by both the employer and employee to finance the Social Security fund.

Filing Status The personal status of a taxpayer at the time they file their personal income tax report (single, married, etc.).

Finance Charge The amount of interest paid for a loan.

Fire Insurance Insurance coverage offsetting economic losses due to fire damage to property.

First-in First-out (FIFO) A method used to determine the value of ending inventory by costing out the units purchased most recently as the inventory.

FOB Destination Transportation terms that means the seller pays freight charges and retains title to the goods until it is delivered.

FOB Shipping Point Transportation terms that means the buyer pays the freight charges and ownership of the goods transfers to the buyer before shipment.

Form 1040 A federal form used to report personal income taxes when the taxpayer elects to itemize deductions and has taxable income of $50,000 or more.

Form 1040EZ A short-form federal report used by individuals to file their personal income taxes.

Form 1099 A federal tax report used by employers to report some types of wages interest, and dividends.

Formula An algebraic equation that represents a specific general principal.

Frequency Distribution A means of grouping data into classes and frequencies that indicate the number of pieces of data in that class.

Front-end Load The commission is paid by the investor when the mutual fund shares are purchased.

Future Value The value of an investment at the end of a given time frame.

Future Value Annuity A compound interest investment where regular deposits are made at the end of stipulated periods of time.

Future Value of a Compound Amount A time value expression for the compound amount of a compound interest investment at a specific time in the future.

Future Value of an Annuity Due A time value expression for the anticipated value of an annuity due at some time in the future.

Future Value of an Ordinary Annuity A time value expression for the anticipated value of an ordinary annuity at some time in the future.

Gross Cost The amount to be remitted to the commission agent by the consignor that includes the prime cost plus total charges.

Gross Earnings The total compensation earned by an employee before mandatory and elective deductions are withheld.

Gross income All income received in the form of money, goods, property, and services that is not exempt from income tax.

Gross Margin Method A method used to estimate the value of ending inventory. Gross margin percent is used to determine the value of cost of goods sold, providing the necessary information to calculate ending inventory.

Gross Profit (margin) The difference between the selling price of an item and the cost of the item.

Health Insurance Insurance that provides protection to offset economic losses due to injury or illness.

Horizontal Analysis An analysis that produces the percent increase or decrease of corresponding accounts reported in financial statements in two different periods.

Incentive Rate A wage based on job performance instead of time worked.

Income Statement A summary report that measures the progress of a firm by determining its profitability.

Inference A prediction or decision made from statistical data.

Installment loan Closed-ended credit generally used for short-term loans.

Insurance Financial coverage by an agreement binding one party to indemnify another against specific loss in return for premiums paid by the first party.

Intangible Property Property such as copyrights, franchise fees, designs, patents, and customer lists.

Interest Sum of money paid or charged for the use of money.

Interest Rate Percent of principal to be paid for the use of money.

Interest Rebate The amount of interest saved by paying off a loan prior to the full term of the loan.

International Monetary Fund (IMF) A U.N. agency that consists of nations who cooperate to promote stable exchange rates, short-term loans, and other financial services.

International Trade The exporting and importing of goods and services between businesses located in different countries.

Investment Objective The primary financial outcome (capital, income, growth) the mutual fund is designed to achieve.

Invoice The official record of a business transaction between the seller and the buyer.

Invoice Terms Statements listed on an invoice that identify the percent of the cash discount and the time period for which it applies.

Last-in First-out (LIFO) A method used to determine the value of ending inventory by costing the units purchased at the beginning of the period as the inventory.

Liabilities Amounts of money owed by a firm to its creditors.

Life Insurance Insurance offsetting the economic losses due to the death of a person.

Limited Pay Life Insurance Life Insurance protection for the insured's entire life, but that is paid up with a limited number of premium payments.

Line Graph A graphical representation of data using lines to indicate the frequencies of each class.

List Price The suggested selling price of goods offered for sale.

Load Fund A mutual fund that contains a load charge (commission) when the shares are either purchased or sold.

Long-Term Capital Gains The profit earned from the sale of assets, such as securities or real estate.

Lower of Cost or Market A method of determining the value of ending inventory by costing the units remaining in inventory at the lower of the original cost or of the current market cost of the units.

MACRS Depreciation Method The Modified Accelerated Cost Recovery System method is used to calculate cost recovery on all assets placed in service for tax purposed after 1986.

Mandatory Deductions Reductions in gross earnings required by federal, state, and local governments such Social Security taxes and withholding taxes.

Manufacturer A business that buys raw materials or parts and produces or assembles them into finished goods.

Markdown A reduction in the marked priced expressed as an amount or a percent.

Marked Price The price of an item before the markdown is applied.

Markup (margin) An amount added to an item's cost to cover operating expenses and profit.

Markup Conversion Equation A formula used to convert markup from cost to price or from price to cost.

Markup Equation A formula that summarizes the markup calculation (selling price = cost + markup).

Markup on Cost Markup that is based on the cost of an item.

Markup on Selling Price Markup that is based on the selling price of an item.

Markup Rate The markup expressed as a percent of cost or selling price.

Maturity Value The sum of the principal and interest; the proceeds plus the discount of a discounted loan.

Maximum Percent of Markdown The largest markdown percent that can be offered without incurring an operating loss.

Mean The sum of the values divided by the number of values.

Measures of Central Tendency Statistical measures used to locate the center of a set of data.

Measures of Dispersion Measures that indicate the spread of the data.

Median The middle value of a set of values that have been arranged in ascending or descending order.

Medicare Tax A tax held from an employee's earnings to fund health insurance for the elderly.

Merchants Deposit Summary The deposit slip used by businesses to deposit credit card transactions into their checking accounts.

Middleperson A business that is considered to be a wholesaler or a retailer and buys goods from the manufacturer.

Mills A property tax rate that is equal to 1/1000th of a dollar, or 1/10th of a cent.

Mode That value (or those values) in a data set that occur most frequently.

Modified Accelerated Cost Recovery System (MACRS) A tax-based method used to recover the cost of assets. This system is also referred to as the General Depreciation System (GDS).

Mortgage A pledge of property to a creditor to secure the repayment of a debt.

Mortgage Loan Closed-ended credit generally used for long-term loans, usually for the purchase of real property.

Mutual Fund An investment company that invests monies of individuals and organizations in a portfolio of stocks, bonds, and short-term financial securities.

Net Asset Value (NAV) The closing price quote per mutual fund share on a given date.

Net Cost The amount paid by the buyer after the trade discount has been deducted from the list price.

Net Income The difference between the inflow of net assets from the sale of goods and services and the outflow of net assets associated with activities of the period.

Net Payment Date The date noted on an invoice that the "amount due" must be remitted by the buyer.

Net Price The amount charged by the seller after the trade discount has been deducted from the list price.

No-Load Fund A mutual fund that does not charge investors a commission when they buy or sell fund shares.

Nominal Rate The compound interest rate for one year.

Nonexempt A classification given to employees who are compensated for the number of hours worked.

NOW Account An account (negotiable orders of withdrawal) that is a savings account with terms and conditions similar to an interesting bearing checking account.

Numerical Summary Measures Mathematical measures such as mean, median, mode, range, and standard deviation that summarize numerical date with a single number.

Offer Price The price (NAV plus commission) an investor will pay for a mutual fund share on a given date.

Open-Ended Credit Credit extended to a borrower that the borrower may use openly for purchases up to a specified credit limit.

Operating Loss A financial position that occurs when the reduced price is sufficient to cover cost but not all of the operating expenses.

Original Value The value of an asset at the time of purchase. This price is used to determine the percent change in the asset over a period of time.

Ordinary Annuity An annuity that increases in value with regular periodic payments plus the interest on the principal of the annuity; the payments are made at the end of each period.

Ordinary Interest The simple interest determined by dividing the number of days by 360.

Overdraft A financial position that occurs when deposits are insufficient to cover the amounts of all checks presented for payment.

Overdraft Protection The bank service of paying checks when the depositors account has insufficient funds.

Overtime The number of hours worked over 40 hours a week or eight hours a day.

Overtime Rate An amount equal to one and one-half times the regular pay rate.

Owners Equity The sum of investments and cumulative profits less withdrawals by the owner(s).

Part The portion of the base that is determined by the rate.

Partial Payment A payment made on account during the cash discount period that is only a part of the total amount due.

Partial Payment of Loan The process of paying off part of the principal of a loan prior to the maturity date of the loan.

Payee The party to whom a check is written (pay to the order of).

Payroll Register A payroll record indicating gross earnings, deductions, and net earnings for all employees during a specific time period.

Percent A numeral that represents a part of 100 and is written with a percent sign (%).

Percent Change The percent increase or decrease between the current value and the original value.

Percent Decrease The percentage change when the current value is less than the original value resulting in a negative difference.

Percent Increase The percent change when the current value is greater than the original value resulting in a positive difference.

Percent Equivalent A fraction converted to a percent equal to the value of the fraction.

Percent Formula A formula that expresses the relationship between the base, rate, and part $P = B \times R$.

Percent Frequency Distribution A frequency distribution in which the frequencies are percents of the total number of pieces of data.

Percentage Method A method of determining federal withholding tax that is based on taxable earnings and tax rates.

Periodic Interest Rate The annual interest rate divided by the number of compounding periods per year.

Periodic Inventory System An inventory system where the value of the inventory is determined periodically rather than continuously.

Perishables Merchandise purchased for resale that will spoil if not sold within a short period of time.

Personal Property Property such as equipment and machinery that is not considered real estate.

Piece Rate A wage based on the units produced.

Place Value A number's value is determined by its position relative to the decimal point.

Postdating Terms stated "as of" that extend the cash discount and net payment periods of an invoice by identifying the starting date of the discount period as the "as of" date.

Preferred Stock A stock that provides preference to dividends and early claim on assets if the company is dissolved.

Premium The amount of money paid to the insurance company to purchase insurance coverage.

Premium Price The price of a bond when its market value is higher than its face value.

Premium Rate A higher wage paid for additional hours worked or additional units produced.

Present Value The value of an investment at any specified time during the time frame of the investment.

Present Value Annuity An annuity established with a lump sum that periodically pays out of the account a specific number of equal payments which will liquidate the account.

Present Value of a Compound Amount A time value expression for the value of a compound interest investment right now.

Price Earnings Ratio A stock performance ratio that compares the stock's closing price with the company's earnings per share.

Price Relative The quotient of the current price and the price of a past year expressed as a decimal percent.

Prime Cost The cost of merchandise purchased by a commission agent as shown on an account purchase invoice.

Principal The initial amount of money invested or borrowed.

Probability A ratio of the number of outcomes in which an event can occur to the total number of outcomes of an experiment.

Proceeds The actual dollar amount of a simple discount loan that a borrower receives from a lender.

Promissory Note A written promise by a borrower (maker) to pay a sum of money (face value) to a designated person or bearer (payee) of the note, at a set time or on demand.

Property Assessor An individual who determines the value of a tax districts taxable land and buildings.

Property Insurance Insurance coverage offsetting economic losses due to damage to property.

Property Tax Rate A rate determined by dividing the amount of revenues needed by a tax district by the total assessed value of all taxable property.

Proportion Two ratios set equal to each other.

Prospectus A document that provides important information about a mutual funds composition, objectives, past performance and administration.

Range The difference between the largest and smallest values in a set of data.

Rate The percent, decimal, or fractional size of the part.

Ratio A comparison of one number to another.

Real Property Property that is land and generally anything that is erected on, growing on, or attached to land.

Receipt-of Goods (ROG) Cash discount terms where the discount period starts on the date the goods are received.

Reconciliation The process of analyzing the bank statement and the checkbook register and making the adjustments to bring the balances into agreement.

Recovery Period The number of years over which the cost of an asset is recovered under MACRS depreciation.

Reduced Net Profit A reduction of profit resulting from a markdown on an item's marked price.

Reduced Price The price of an item after the markdown has been deducted from the marked price.

Re-evaluation Increasing the exchange rate making it more expensive for other countries to purchase goods and services.

Refund The amount returned by the government to the taxpayer when the total payments exceeds the total tax.

Regular Rate The amount paid to an employee per hour for hours worked up to 40 hours per pay period.

Relative Frequency Distribution A frequency distribution in which the frequencies are decimal parts of the total number of pieces of data.

Residual Value The estimated value of an asset at the end of its useful life.

Restrictive Endorsement An endorsement that limits the ability to cash a check as it includes "for deposit only" and the depositor's signature.

Retailer A business that buys goods from manufacturers or wholesalers and resells them to consumers.

Retail Method A method used to estimate the value of ending inventory based on a cost ratio determined by dividing the cost of goods for sale at cost by the cost of goods available for sale at retail.

Return on Investment Ratio Identifies the amount of profit realized from each $1 invested by the owners.

Returned Check (NSF) A check returned to the depositor by its bank due to nonsufficient funds in the account of the drawer.

Returned Goods Goods that are returned to the seller for credit by the buyer that does not meet the buyers requirements.

Revolving Charge Account Open-ended credit in which the borrower is allowed to charge up to a certain limit and agrees to pay at least a percentage of the outstanding balance at the end of a set time period, e.g., monthly, quarterly, etc.

Risk The uncertainty of financial loss.

Risk Management The attempt to minimize financial loss by (a) controlling risk and (b) using insurance to compensate for any loss.

Risk Manager Person who determines the existence of risk and who recommends and implements plans of action to control risk.

Rule of 78's The sum of digits method for determining interest rebate on a loan of one year $(1 + 2 + 3 + \ldots + 12 = 78)$.

Salary A fixed amount of compensation expressed in terms of a week, month, or year.

Sales Tax A tax that is charged on the sale of particular goods and services sold by retail merchants.

Salvage Value The estimated value of an asset at the end of its useful life; sometimes referred to as residual value.

Secured Loan A loan that has collateral provided to the lender.

Securities A negotiable investment such as stocks or bonds issued by companies to acquire capital.

Service Charge A monthly fee charged by banks to maintain a customer's checking account.

Share Account A checking account offered by credit unions and savings banks where the customer is considered an owner of the institution.

Shift Differentials A premium wage paid to employees who work between the hours of 4 P.M. and 8 A.M.

Signature Card A form required by a bank that identifies the owner(s) of a checking account and the signatures used to sign checks.

Signed Numbers Positive and negative numbers in which a positive can be thought of as a gain and a negative as a loss.

Simple Discount The simple interest that is collected at the beginning of a discounted loan.

Simple Discount Note A promissory note in which the face value of the note and the maturity value are the same.

Simple Interest Interest that is computed on the original principal only for a set period of time.

Simple Interest Note A promissory note that includes simple interest added to the face value of the note; i.e., maturity value = face value plus interest.

Single-Equivalent Discount Rate A trade discount rate equal to a stated chain trade discount.

Sinking Fund A fund set up to save for an anticipated expense in the future.

Social Security Fund The fund established by the Federal Insurance Contributions Act (FICA) that collects payments from employees and employers to pay benefits to retired individuals.

Special Endorsement An endorsement that allows the payee to transfer the check to a third party.

Specific Identification Method A method used to determine the value of ending inventory based on the actual cost of the items remaining in inventory.

Standard Deduction A reduction in adjusted gross income that is established by the IRS and is based on the taxpayers filing.

Standard Deviation A measure of spread of the data that is the square root of the variance.

State Disability Insurance An insurance whose premiums are paid by employers based on payroll amounts that make payments to workers who have incurred job related disabilities.

State Unemployment Tax Act A state act (SUTA) that identifies all employer responsibilities concerning unemployment insurance.

State Withholding Tax (SWT) A payroll deduction required by state governments that is applied to an employee's state income tax.

Statistics That branch of mathematics which collects, organizes, analyzes, and interprets numerical data.

Stem and Leaf Display A graphical representation of data using the place values of the data for the stem and leaves.

Stock A share or portion of ownership of a particular company or corporation.

Stock Certificate A document issued by a company to the investor indicating stock ownership.

Stock Exchange/Market A place or institution where securities are bought and sold.

Stock Quotation Information about the value, performance, and cost of stocks available for purchase.

Stockbroker A person who buys and sells stocks and bonds for individuals and companies.

Stockholder An owner of a corporation determined by the number of shares purchased.

Stop Payment Order An order by the drawer requesting that the bank not pay a check that was previously written.

Straight Piecework A wage plan that pays an employee a fixed amount for each unit of production.

Straight-Line Method A depreciation method that allows a business to evenly distribute the cost of an asset over its useful life.

Sum of Digits Method A method for computing the amount of interest rebate a borrower is entitled to on an early pay loan, based on the sum of the digits 1 through n, which equals $\dfrac{[n(n + 1)]}{2}$.

Sum-of-the-year's Digits An accelerated depreciation method based on a fractional part of the asset's depreciable amount.

Take-Home Pay Net earnings received by an employee after deductions have been subtracted from gross earnings.

Tangible Property Property that can be touched, such as machinery, equipment, and vehicles.

Task and Bonus Plan A wage plan where the employee is paid a bonus when they perform equal to or greater than a set task standard.

Tax Due The amount of tax remaining unpaid after total payments have been deducted from the total tax.

Tax Return A report (form 1040, 1040A, or 1040EZ) used to identify the amount of income tax owed or refund due.

Taxable Income The amount of adjusted gross income that remains after deductions and total exemptions are subtracted.

Time and a Half Rate A premium wage of 1.5 times the regular rate of pay for any hours worked in excess of 40 hours per week or 8 hours per day.

Term Life Insurance Life insurance that pays the face value of the policy to the beneficiary if the insured dies within a specified time period.

Time Card A record of the number of hours and employee worked during a specific pay period.

Time Value of Money As time passes for a monetary investment, the value of the money changes.

Total Income The amount of all annual earnings received by a taxpayer from all sources that is considered taxable and is reported on forms 1040, 1040A, or 1040EZ.

Trade Deficit A condition that results when a country's imports exceed is exports.

Trade Discount A reduction in the list price of an item expressed as a percent or an amount.

Trade Surplus A condition that results when a country's exports exceed its imports.

Traditional Depreciation Methods Methods used to compute depreciation on assets that are required under generally accepted accounting principles guidelines.

Transaction Costs The cost of services required to buy and sell securities.

Trimmed Mean A mean calculated by eliminating the lowest and highest values of the data set.

True Rate of Interest The simple interest rate of a simple discounted loan using proceeds as the principal and the amount of the discount as the interest [true interest rate = (discount interest)/(proceeds times time)].

U.S. Rule A federal rule that states that a partial payment of a loan must be divided into two parts, accrued interest on the loan and a partial reduction of the principal.

Unadjusted Basis The amount of the asset's value to be depreciated under the MACRS depreciation method.

Unearned Income Personal income that comes from sources other than employment.

Unemployment Tax Return (FUTA) Federal tax form 940 used to report unemployment tax (FUTA).

Unit Price The selling price per measure of unit.

Units-of-production Depreciation A depreciation method based on an assets actual service rather than its years of service.

Universal Life Insurance A flexible form of insurance that incorporates protection, savings, and annuity benefits that can be adapted to the changing needs of the insured.

Unsecured Loan A loan that has no specific collateral provided to the lender.

Useful Life The number of years the asset is expected to be in service.

Variable A symbol that can have several different values.

Variance The average of the squared differences between each value and the mean of the set of data.

Vertical Analysis An analysis that identifies the percentage relationships of the component parts of financial statements to either net sales or total assets.

W-2 Form A tax form provided by employers used to report earned income and total federal and state taxes withheld.

Wage-Bracket Method A method of determining federal withholding tax based on tables identifying specific wage amounts.

Wages An employee's compensation determined by the hours worked and the rate paid per hour.

Weighted Mean A mean in which some values carry a greater weight than other values in determining the mean.

Whole (ordinary) Life Insurance Life insurance protection for the insured's entire life as long as the premiums are paid.

World Bank U.N. agency that provides financial services to fund national improvements in undeveloped countries.

Index

Summary of Formulas

Chapter Topic/Formula

1.1   **Commutative principle**

$a + b = b + a$

$a \times b = b \times a$

Associative principle

$a + (b + c) = (a + b) + c$

$a \times (b \times c) = (a \times b) \times c$

Distributive

$a \times (b + c) = (a \times b) + (a \times c)$

1.2   **Variation**

Direct $y = kx$

Inverse $y = k/x$

2.1   **Percent formula**

$\text{Part} = \text{Base} \times \text{Rate} \qquad P = BR$

2.2 $\text{Base} = \dfrac{\text{Part}}{\text{Rate}} \qquad B = \dfrac{P}{R}$

2.3 $\text{Rate} = \dfrac{\text{Part}}{\text{Base}} \qquad R = \dfrac{P}{B}$

2.4   **Rate of change**

$\text{Rate (increase or decrease)} = \dfrac{\text{Part (current value} - \text{original value)}}{\text{Base (original value)}}$

3.2   **Mean**

$\bar{x} = \dfrac{\Sigma x}{n}$

Trimmed mean

$\bar{x} = \dfrac{\Sigma x}{n - 2}$

Weighted mean

$\bar{x} = \dfrac{\Sigma(f \times M)}{n}$

3.3 **Range**

R = Largest value − Smallest value

Variance

$$s^2 = \frac{\Sigma(x - \overline{x})^2}{n - 1} = \frac{\Sigma x^2 - (\Sigma x)^2/n}{n - 1}$$

3.4 **Probability**

$$P(\text{event A will occur}) = \frac{\text{Number of ways A can occur}}{\text{Number of ways experiment can occur}}$$

4.1 **Balance sheet**

Assets = Liabilities + Owner's Equity   A = L + OE

4.2 **Income Statement**

Net income = Net sales − Cost of goods sold − Operating expenses

Net sales = Gross sales − Sales returns and allowances − Operating expenses

Cost of goods sold = Beginning inventory + Net purchases − Ending inventory

4.4 **Financial ratios**

$$\text{Current ratio} = \frac{\text{Current assets}}{\text{Current liabilities}}$$

$$\text{Acid test ratio} = \frac{\text{Liquid assets}}{\text{Current liabilities}}$$

$$\text{Debt to equity ratio} = \frac{\text{Total liabilities}}{\text{Total owner's equity}}$$

$$\text{Debt to assets ratio} = \frac{\text{Total liabilities}}{\text{Total assets}}$$

$$\text{Asset turnover ratio} = \frac{\text{Net sales}}{\text{Average total assets}}$$

$$\text{Inventory turnover ratio} = \frac{\text{Cost of goods sold}}{\text{Average inventory}}$$

$$\text{Accounts receivable turnover} = \frac{\text{Net credit sales}}{\text{Average net accounts receivable}}$$

$$\text{Average collection ratio} = \frac{\text{Days in the year}}{\text{Accounts receivable turnover ratio}}$$

$$\text{Return on net sales ratio} = \frac{\text{Net income after taxes}}{\text{Net sales}}$$

$$\text{Return on investment ratio} = \frac{\text{Net income after taxes}}{\text{Average owner's equity}}$$

5.2   Trade discounts

$$\text{Trade discount} = \text{List price} \times \text{Trade discount rate}$$

$$\text{Net price} = \text{List price} - \text{Trade discount}$$

$$\text{Single equivalent discount rate} = 1 - (\text{Net price equivalent rate})$$

5.3   Cash discounts

$$\text{Cash discount} = \text{Net price} \times \text{Cash discount rate}$$

Days in Month

	31	30	28
January	August	April	February
March	October	June	(29 leap year)
May	December	September	
July		November	

5.4   Invoice charges and adjustments

$$\text{Freight charge} = \text{Weight units (cwt)} \times \text{Unit price rate}$$

$$\text{Partial payment credit} = \frac{\text{Amount of partial payment}}{1 - \text{Cash discount rate}}$$

6.1   Markup

$$\text{Markup} = \text{Operating expenses} + \text{Net profit}$$

$$\text{Selling price} = \text{Cost} + \text{Markup} \ (S = C + M)$$

$$\text{Markup on cost} = \frac{\text{Price}}{1 + \text{Markup rate}}$$

6.2   Markup on selling price $= \dfrac{\text{Cost}}{1 - \text{Markup rate}}$

6.4   **Markdown**

Markdown = Marked price $\times$ Markdown rate

Reduced price = Marked price $-$ Markdown

Breakeven price = Cost + Operating expenses

Profit/loss = Reduced price $-$ Breakeven price

Maximum markdown = Marked price $-$ Breakeven price

7.1   **Simple interest**

$I = P \times R \times T$

Maturity value

$M = P + I = P + PRT = P(1 + RT)$

7.2   **Simple interest rate**

$R = I/PT$

Simple interest time

$T = I/PR$

Simple interest principal

$P = I/TR$

8.1   **Simple discount**

$D = M \times R \times T$

Proceeds

$P = M - D = M - MRT = M(1 - RT)$

True interest rate

$R = \dfrac{D}{P \times T}$

9.1   **Compound interest**

$A = P(1 + i)^n$

9.2   **Present value of compound amount**

$$P = A\frac{1}{(1 + i)^n}$$

Effective rate

$$R_{\text{effective}} = (1 + i)^m - 1$$

10.1   **Ordinary annuity**

$$S = p\frac{(1 + i)^n - 1}{i}$$

Annuity due

$$Sd = p\frac{(1 + i)^n - 1}{i} \times (1 + i)$$

10.2   **Present value annuity**

$$A = p\frac{1 - (1 + i)^{-n}}{i}$$

10.3   **Sinking fund annuity**

$$p = S\frac{i}{(1 + i)^n - 1}$$

10.4   **Amortization**

$$p = A\frac{1}{\dfrac{1 - (1 + i)^{-n}}{i}}$$

11.5   **Sum of digits 1 through n**

$$S = \frac{n(n + 1)}{2}$$

13.1   **Securities (stock, bonds, mutual funds)**

$$\text{Price earnings ratio} = \frac{\text{Price per share}}{\text{Annual earnings per share}}$$

$$\text{Capital gain/loss} = (\text{Total sales of shares} - \text{Commission}) - (\text{Total cost of shares} + \text{Commission})$$

13.2 $$\text{Current yield} = \frac{\text{Annual earnings per unit}}{\text{Current price}}$$